D1277994

Current Topics in Membranes, Volume 52

Peptide–Lipid Interactions

Current Topics in Membranes, Volume 52

Series Editors

Dale J. Benos
Department of Physiology and Biophysics
University of Alabama
Birmingham, Alabama

Sidney A. Simon
Department of Neurobiology
Duke University Medical Center
Durham, North Carolina

Current Topics in Membranes, Volume 52

Peptide–Lipid Interactions

Edited by

Sidney A. Simon
Department of Neurobiology
Duke University Medical Center
Durham, North Carolina

Thomas J. McIntosh
Department of Cell Biology
Duke University Medical Center
Durham, North Carolina

ACADEMIC PRESS

An Elsevier Science Imprint

San Diego   San Francisco   New York   Boston   London   Sydney   Tokyo

Academic Press
An Elsevier Science Imprint
525 B Street, Suite 1900, San Diego, California 92101-4495, USA
http://www.academicpress.com

Academic Press
32 Jamestown Road, London NW1 7BY, UK
http://www.academicpress.com

International Standard Book Number: 0-12-153352-2 (casebound)
International Standard Book Number: 0-12-643871-4 (paperback)

PRINTED IN THE UNITED STATES OF AMERICA
02 03 04 05 06 07 EB 9 8 7 6 5 4 3 2 1

Contents

Contributors   xiii
Preface   xvii
Previous Volumes in Series   xix

PART I Biophysical Techniques for Analyzing
Peptide-Lipid Interactions

CHAPTER 1 Peptide-Membrane Interactions Determined
Using Site-Directed Spin Labeling
David S. Cafiso

  I. Introduction   4
  II. Site-Directed Spin Labeling   5
  III. Membrane Interactions of Basic Peptides Derived from
   Cell-Signaling Motifs   12
  IV. Studies Directed at Channel-Forming and Other
   Membrane-Active Peptides   18
  V. Conclusions   24
   References   25

CHAPTER 2 Isothermal Titration Calorimetry for
Studying Interactions between Peptides
and Lipid Membranes
Torsten Wieprecht and Joachim Seelig

  I. Introduction to Isothermal Titration Calorimetry   32
  II. Measurement of the Reaction Enthalpy and the
   Binding Isotherm   33
  III. Binding Models for Analyzing Binding Isotherms   37
  IV. Complex Membrane-Binding/Membrane-Perturbation
   Equilibria   41
  V. Thermodynamics of the Membrane-Induced
   Coil–Helix Transition   48
  VI. Thermodynamics of Peptide Binding to Curved
   and Planar Membranes   52

VII. Concluding Remarks   54
     References   54

CHAPTER 3 Infrared Reflection–Absorption Spectroscopy
of Lipids, Peptides, and Proteins in
Aqueous Monolayers
Richard Mendelsohn and Carol R. Flach

I. Introduction   58
II. Experimental Approaches   61
III. Information from IRRAS Measurements   66
IV. IRRAS Applications to Lipid/Protein Langmuir Films   75
V. Current State and Future Prospects for IRRAS   83
References   84

CHAPTER 4 Measuring the Depth of Amino Acid Residues
in Membrane-Inserted Peptides by
Fluorescence Quenching
Erwin London and Alexey S. Ladokhin

I. Introduction   90
II. Determining Depth from Fluorescence Quenching by
Empirical Inspection of Quenching Level   90
III. Determining Depth by Parallax Analysis   91
IV. Determining Depth by Distribution Analysis   93
V. Comparing Parallax Analysis and Distribution
Analysis   97
VI. Investigating Membrane Protein Topography by Varying
Fluorophore Depth Instead of Quencher Depth   98
VII. A Cautionary Note Concerning Quenching Analysis   98
VIII. Problems That Can Affect the Accuracy of Quenching
Measurements   99
IX. Studies of Membrane-Inserted Peptides by Fluorescence
Quenching   100
X. Conclusions   111
References   111

CHAPTER 5 Surface-Sensitive X-Ray and Neutron Scattering
Characterization of Planar Lipid Model Membranes
and Lipid/Peptide Interactions
Mathias Lösche

I. Introduction   118
II. Data Inversion and Modeling in Reflectivity
Measurements   122

III. Characterization of Planar Lipid Model Systems   127
References   150

CHAPTER 6 Lipid-Peptide Interaction Investigated by NMR
Klaus Gawrisch and Bernd W. Koenig

I. Introduction   164
II. Peptide Structure in the Membrane-Bound State   165
III. Structure and Dynamics of the Lipid Matrix   176
IV. Future Directions   183
References   184

CHAPTER 7 Peptide-Lipid Interactions in Supported
Monolayers and Bilayers
Lukas K. Tamm

I. Peptide Insertion into Lipid Monolayers   191
II. Supported Bilayers   194
III. ATR-FTIR Spectroscopy of Peptides in Supported
Bilayers   195
IV. Analysis of Fusion Peptide–Lipid Interactions   198
References   201

PART II Theoretical and Computational Analyses of
Peptide-Lipid Interactions

CHAPTER 8 Free Energy Determinants of Peptide Association
with Lipid Bilayers
Amit Kessel and Nir Ben-Tal

I. Introduction   206
II. Theory   211
III. Mean-Field Studies of Peptide–Membrane Systems   223
IV. Open Questions   242
References   244

CHAPTER 9 Investigating Ion Channels Using
Computational Methods
Eric Jakobsson, R. Jay Mashl, and Tsai-Tien Tseng

I. Introduction   256
II. Understanding Channel Function from Structure   257
III. Understanding Channel Function from Structural
Homology   266

IV. Future Prospects   271
References   271

PART III Experimental Investigations of
Peptide-Lipid Interactions

CHAPTER 10 The Role of Electrostatic and Nonpolar Interactions
in the Association of Peripheral Proteins
with Membranes
*Diana Murray, Anna Arbuzova, Barry Honig,
and Stuart McLaughlin*

I. Introduction   278
II. Electrostatic and Hydrophobic Interactions Mediate the
Membrane Association of Important Biological
Proteins   279
III. The Membrane Interaction of Simple Basic Peptides:
Theory and Experiment   282
IV. The Combination of Experimental and Theoretical
Approaches Provides a Detailed Comprehensive
Picture of the Membrane Association of Myristoylated
Proteins   289
V. Future Directions   300
References   303

CHAPTER 11 The Energetics of Peptide–Lipid Interactions:
Modulation by Interfacial Dipoles and Cholesterol
Thomas J. McIntosh, Adriana Vidal, and Sidney A. Simon

I. Introduction   310
II. Results   318
III. Discussion   327
IV. Conclusions   332
References   333

CHAPTER 12 Transmembrane α Helices
Sanjay Mall, J. Malcolm East, and Anthony G. Lee

I. Intrinsic Membrane Proteins   339
II. Effects of Intrinsic Membrane Proteins on Lipid
Bilayers   345
III. Model Transmembrane α Helices   351
IV. Biological Consequences of Hydrophobic Mismatch   363
References   365

CHAPTER 13 Lipidated Peptides as Tools for Understanding the Membrane Interactions of Lipid-Modified Proteins
John R. Silvius

    I. Introduction 372
    II. Structures of Lipid Modifications of Proteins 372
    III. Chemistry of Peptide Lipidation 375
    IV. Thermodynamics of Association of Lipidated Peptides with Lipid Bilayers 377
    V. Kinetics of Dissociation of Lipidated Sequences from Bilayers: Protein Targeting by Kinetic Trapping 383
    VI. Partitioning of Lipidated Peptides into Liquid-Ordered Lipid Domains 387
    VII. Conclusion 389
    References 390

CHAPTER 14 Experimental and Computational Studies of the Interactions of Amphipathic Peptides with Lipid Surfaces
Jere P. Segrest, Martin K. Jones, Vinod K. Mishra, and G. M. Anantharamaiah

    I. Introduction 398
    II. Amphipathic α Helixes 398
    III. Amphipathic β Strands/Sheets 415
    References 430

CHAPTER 15 Interactions of pH-Sensitive Peptides and Polymers with Lipid Bilayers: Binding and Membrane Stability
Doncho V. Zhelev and David Needham

    I. Introduction 438
    II. Interaction of Viral Fusion Peptides with the Membrane 440
    III. Interaction of AcE4K with Lipid Bilayer Membranes (SOPC) 442
    IV. Effect of Peptide Binding on Membrane Stability 450
    V. Membrane Instability Induced by the pH-Sensitive Polymer Poly(2-Ethylacrylic Acid) (PEAA) 456
    VI. Conclusion 460
    References 460

CHAPTER 16 The Hydrophobicity Threshold for Peptide
Insertion into Membranes
*Charles M. Deber, Li-Ping Liu, Chen Wang, Natalie K. Goto,
and Reinhart A. F. Reithmeier*

    I. Introduction   466
    II. Results   467
    III. Discussion   473
    IV. Conclusions   476
       References   477

PART IV Specialized Topics of Biological Relevance

CHAPTER 17 Signal Sequence Function in the Mammalian
Endoplasmic Reticulum: A Biological Perspective
Christopher V. Nicchitta

    I. Signal Sequences: Passage and Coronation   484
    II. Signal Peptides: A Profile for Function   484
    III. Signal Peptide Recognition: Where Biophysics Meets
       Biology   487
    IV. Yes, But What Happens at the Membrane?   492
    V. Where to from Here?   494
    VI. Conclusion   496
       References   496

CHAPTER 18 The Process of Membrane Fusion: Nipples,
Hemifusion, Pores, and Pore Growth
Fredric S. Cohen, Ruben M. Markosyan, and Grigory B. Melikyan

    I. Introduction   502
    II. Structural Features and Conformational Changes of
      Fusion Proteins   503
    III. Hemifusion and Its Possible Role in the
       Fusion Process   506
    IV. The Observed Temporal Sequence of Lipid Dye Spread
      and Fusion Pore Formation Depends upon Target
      Membrane and the Chosen Lipid Dye   508
    V. The Dependence of Fusion on Spontaneous Curvature
      of Lipid Monolayers   510
    VI. An Ectodomain of HA Anchored to a Membrane Can Yield
      Fusion or End-State Hemifusion   514
    VII. Fusion Does Not Require a Precise Amino Acid Sequence
      of the TM Domain   515

VIII. Capturing Candidates of Transitional Hemifusion   515
  IX. The Role of Fusion Proteins in Hemifusion and Pore
       Formation   517
   X. The Effects of Lipid Composition on Pore Flickering
       and Growth   518
  XI. Energy and Shape Considerations in Pore Growth   522
 XII. Outlook   524
      References   524

CHAPTER 19 Prenylation of CaaX-Type Proteins: Basic Principles
through Clinical Applications
Herbert I. Hurwitz and Patrick J. Casey

   I. Introduction   532
  II. Enzymology of the CaaX Prenyltransferases: FTase and
       GGTase-I   534
 III. Development of FTase Inhibitors as Anticancer
       Agents   540
  IV. Clinical Results from Evaluation of FTIs as Antitumor
       Agents   544
   V. Summary and Conclusions   545
      References   546

CHAPTER 20 Peptides as Probes of Protein–Protein Interactions
Involved in Neurotransmitter Release
Thomas Kuner, Hiroshi Tokumaru, and George J. Augustine

   I. Introduction   552
  II. Peptide Design   554
 III. Characterization of Peptide Activity on Protein–Protein
       Interactions   562
  IV. Functional Analysis of Peptides in a Presynaptic
       Terminal   564
   V. Summary and Outlook   567
      References   568

Index   571

Contributors

Numbers in parentheses indicate the pages on which the authors' contributions begin.

G. M. Anantharamaiah (397), Department of Medicine and the Atherosclerosis Research Unit, and Department of Biochemistry and Molecular Genetics, UAB Medical Center, Birmingham, Alabama 35294

Anna Arbuzova (277), Department of Physiology and Biophysics, Health Sciences Center, SUNY, Stony Brook, New York 11794

George J. Augustine (551), Marine Biological Laboratory, Woods Hole, Massachusetts 02543; and Max-Planck-Institut für medizinische Forschung, 69120 Heidelberg, Germany

Nir Ben-Tal (205), Department of Biochemistry, George S. Wise Faculty of Life Sciences, Tel Aviv University, Ramat Aviv 69978, Israel

David S. Cafiso (3), Department of Chemistry, University of Virginia, Charlottesville, Virginia 22904

Patrick J. Casey (531), Department of Pharmacology and Cancer Biology and Department of Biochemistry, Duke University Medical Center, Durham, North Carolina 27710

Fredric S. Cohen (501), Department of Molecular Biophysics and Physiology, Rush Medical College, Chicago, Illinois 60612

Charles M. Deber (465), Structural Biology and Biochemistry, Research Institute, Hospital for Sick Children, Toronto M5G 1X8, Ontario, Canada; Department of Biochemistry, University of Toronto, Toronto M5S 1A8, Ontario, Canada

J. Malcolm East (339), Division of Biochemistry and Molecular Biology, School of Biological Sciences, University of Southampton, Southampton, SO16 7PX, United Kingdom

Carol R. Flach (57), Department of Chemistry, Newark College, Rutgers University, Newark, New Jersey 07102

Klaus Gawrisch (163), Laboratory of Membrane Biochemistry and Biophysics, NIAAA, National Institutes of Health, Rockville, Maryland 20852

Natalie K. Goto (465), Structural Biology and Biochemistry, Research Institute, Hospital for Sick Children, Toronto M5G 1X8, Ontario, Canada; Department of Biochemistry, University of Toronto, Toronto M5S 1A8, Ontario, Canada

Barry Honig (277), Department of Biochemistry and Molecular Biophysics, Columbia University, New York, New York 10032

Herbert I. Hurwitz (531), Department of Medicine, Duke University Medical Center, Durham, North Carolina 27710

Eric Jakobsson (255), Beckman Institute for Advanced Science and Technology, National Center for Supercomputing Applications, Department of Molecular and Integrative Physiology, Bioengineering Program, University of Illinois at Urbana-Champaign, Urbana, Illinois 61801

Martin K. Jones (397), Department of Medicine and the Atherosclerosis Research Unit, UAB Medical Center, Birmingham, Alabama 35294

Amit Kessel (205), Department of Biochemistry, George S. Wise Faculty of Life Sciences, Tel Aviv University, Ramat Aviv 69978, Israel

Bernd W. Koenig (163), IBI-2: Institute of Structural Biology, Research Center Jülich, D-52425 Jülich, Germany

Thomas Kuner (551), Department of Neurobiology, Duke University Medical Center, Durham, North Carolina 27710; Marine Biological Laboratory, Woods Hole, Massachusetts 02543; and Max-Planck-Institut für medizinische Forschung, 69120 Heidelberg, Germany

Alexey S. Ladokhin (89), Department of Physiology and Biophysics, University of California, Irvine, California 92697; on leave from Institute of Molecular Biology and Genetics, National Academy of Sciences of Ukraine, Kiev 252143, Ukraine

Anthony G. Lee (339), Division of Biochemistry and Molecular Biology, School of Biological Sciences, University of Southampton, Southampton, SO16 7PX, United Kingdom

Li-Ping Liu (465), Structural Biology and Biochemistry, Research Institute, Hospital for Sick Children, Toronto M5G 1X8, Ontario, Canada; Department of Biochemistry, University of Toronto, Toronto M5S 1A8, Ontario, Canada

Erwin London (89), Department of Biochemistry and Cell Biology and Department of Chemistry, State University of New York at Stony Brook, Stony Brook, New York 11794

Mathias Lösche (117), Institute of Experimental Physics I, Leipzig University, D-04103 Leipzig, Germany

Sanjay Mall (339), Division of Biochemistry and Molecular Biology, School of Biological Sciences, University of Southampton, Southampton, SO16 7PX, United Kingdom

Ruben M. Markosyan (501), Department of Molecular Biophysics and Physiology, Rush Medical College, Chicago, Illinois 60612

R. Jay Mashl (255), Beckman Institute for Advanced Science and Technology, National Center for Supercomputing Applications, University of Illinois at Urbana-Champaign, Urbana, Illinois 61801

Thomas J. McIntosh (309), Department of Cell Biology, Duke University Medical Center, Durham, North Carolina 27710

Stuart McLaughlin (277), Department of Physiology and Biophysics, Health Sciences Center, SUNY, Stony Brook, New York 11794

Grigory B. Melikyan (501), Department of Molecular Biophysics and Physiology, Rush Medical College, Chicago, Illinois 60612

Richard Mendelsohn (57), Department of Chemistry, Newark College, Rutgers University, Newark, New Jersey 07102

Vinod K. Mishra (397), Department of Medicine and the Atherosclerosis Research Unit, UAB Medical Center, Birmingham, Alabama 35294

Diana Murray (277), Department of Biochemistry and Molecular Biophysics, Columbia University, New York, New York 10032

David Needham (437), Department of Mechanical Engineering and Materials Science, Duke University, Durham, North Carolina 27708

Christopher V. Nicchitta (483), Department of Cell Biology, Duke University Medical Center, Durham, North Carolina 27710

Reinhart A. F. Reithmeier (465), Departments of Biochemistry and Medicine, University of Toronto, Toronto M5S 1A8, Ontario, Canada

Joachim Seelig (31), Department of Biophysical Chemistry, Biocenter of the University of Basel, CH-4056 Basel, Switzerland

Jere P. Segrest (397), Department of Medicine and the Atherosclerosis Research Unit, and Department of Biochemistry and Molecular Genetics, UAB Medical Center, Birmingham, Alabama 35294

John R. Silvius (371), Department of Biochemistry, McGill University, Montreal, Quebec, Canada H3G 1Y6

Sidney A. Simon (309), Department of Neurobiology, Duke University Medical Center, Durham, North Carolina 27710

Lukas K. Tamm (191), Department of Molecular Physiology and Biological Physics, University of Virginia, Charlottesville, Virginia 22908

Hiroshi Tokumaru (551), Marine Biological Laboratory, Woods Hole, Massachusetts 02543; and Max-Planck-Institut für medizinische Forschung, 69120 Heidelberg, Germany

Tsai-Tien Tseng (255), Center for Biophysics and Computational Biology Molecular Biophysics Training Program, University of Illinois at Urbana-Champaign, Urbana, Illinois 61801

Adriana Vidal (309), Department of Cell Biology, Duke University Medical Center, Durham, North Carolina 27710

Chen Wang (465), Structural Biology and Biochemistry, Research Institute, Hospital for Sick Children, Toronto M5G 1X8, Ontario, Canada; Department of Biochemistry, University of Toronto, Toronto M5S 1A8, Ontario, Canada

Torsten Wieprecht (31), Department of Biophysical Chemistry, Biocenter of the University of Basel, CH-4056 Basel, Switzerland

Doncho V. Zhelev (437), Department of Mechanical Engineering and Materials Science, Duke University, Durham, North Carolina 27708

Preface

In the past several years the field of peptide–lipid interactions has exploded as sophisticated experimental and theoretical approaches have provided fascinating new data and concepts. One cannot pick up an issue of the major biochemical or biophysical journals without finding an article dealing with some aspect of this topic. Two primary reasons for this interest are that peptide–lipid interactions involve fundamental physiochemical interactions occurring in a complex environment and that they are important in a variety of biological processes, including signal transduction, membrane raft formation, protein trafficking, cell adhesion, and cell fusion. We decided to edit this book because the topic of peptide–lipid interactions warrants an up-to-date reference encompassing many diverse areas of this fast-developing discipline.

The approaches used in this book include experimental methods, theoretical analyses, and computer simulations designed to investigate various aspects of peptide–lipid interactions. The book is divided into four sections, which analyze peptide–lipid interactions from somewhat different perspectives. The first section focuses on some of the most important methodologies currently used in studying peptide–lipid interactions and the diverse types of information that can be obtained from these techniques. Among these state-of-the-art methods are electron spin resonance, isothermal titration calorimetry, infrared spectroscopy, fluorescence quenching, X-ray and neutron scattering, and nuclear magnetic resonance. The second section involves theoretical and thermodynamic analyses and computer simulations of the numerous classes of interactions that occur when peptides come into contact with lipids. The third section includes chapters on experimental analyses of the interactions between specific classes of peptides and well-characterized lipid membranes. The last section contains several specialized topics of biological relevance that deal with peptide–lipid interactions taking place in a physiological environment.

We were very fortunate in getting acknowledged experts in the field to write chapters and are extremely grateful to these researchers for allocating their valuable time to this project and for educating us on a variety of topics critical to membrane biology. We also thank Mica Haley at Academic Press for her support and professionalism throughout the production of this book. Finally, we thank our wives, Virginia Knight and Debbie McIntosh, for their untiring support and encouragement.

Sidney A. Simon
Thomas J. McIntosh

Previous Volumes in Series

Current Topics in Membranes and Transport

Volume 23 Genes and Membranes: Transport Proteins and Receptors* (1985)
Edited by Edward A. Adelberg and Carolyn W. Slayman

Volume 24 Membrane Protein Biosynthesis and Turnover (1985)
Edited by Philip A. Knauf and John S. Cook

Volume 25 Regulation of Calcium Transport across Muscle Membranes
(1985)
Edited by Adil E. Shamoo

Volume 26 $Na^+ - H^+$ Exchange, Intracellular pH, and Cell Function* (1986)
Edited by Peter S. Aronson and Walter F. Boron

Volume 27 The Role of Membranes in Cell Growth and Differentiation
(1986)
Edited by Lazaro J. Mandel and Dale J. Benos

Volume 28 Potassium Transport: Physiology and Pathophysiology* (1987)
Edited by Gerhard Giebisch

Volume 29 Membrane Structure and Function (1987)
Edited by Richard D. Klausner, Christoph Kempf, and Jos van Renswoude

Volume 30 Cell Volume Control: Fundamental and Comparative Aspects in
Animal Cells (1987)
Edited by R. Gilles, Arnost Kleinzeller, and L. Bolis

Volume 31 Molecular Neurobiology: Endocrine Approaches (1987)
Edited by Jerome F. Strauss, III, and Donald W. Pfaff

Volume 32 Membrane Fusion in Fertilization, Cellular Transport, and Viral
Infection (1988)
Edited by Nejat Düzgünes and Felix Bronner

Volume 33 Molecular Biology of Ionic Channels* (1988)
Edited by William S. Agnew, Toni Claudio, and Frederick J. Sigworth

* Part of the series from the Yale Department of Cellular and Molecular Physiology

Volume 34   Cellular and Molecular Biology of Sodium Transport* (1989)
Edited by Stanley G. Schultz

Volume 35   Mechanisms of Leukocyte Activation (1990)
Edited by Sergio Grinstein and Ori D. Rotstein

Volume 36   Protein–Membrane Interactions* (1990)
Edited by Toni Claudio

Volume 37   Channels and Noise in Epithelial Tissues (1990)
Edited by Sandy I. Helman and Willy Van Driessche

Current Topics in Membranes

Volume 38   Ordering the Membrane Cytoskeleton Tri-layer* (1991)
Edited by Mark S. Mooseker and Jon S. Morrow

Volume 39   Developmental Biology of Membrane Transport Systems (1991)
Edited by Dale J. Benos

Volume 40   Cell Lipids (1994)
Edited by Dick Hoekstra

Volume 41   Cell Biology and Membrane Transport Processes* (1994)
Edited by Michael Caplan

Volume 42   Chloride Channels (1994)
Edited by William B. Guggino

Volume 43   Membrane Protein–Cytoskeleton Interactions (1996)
Edited by W. James Nelson

Volume 44   Lipid Polymorphism and Membrane Properties (1997)
Edited by Richard Epand

Volume 45   The Eye's Aqueous Humor: From Secretion to Glaucoma (1998)
Edited by Mortimer M. Civan

Volume 46   Potassium Ion Channels: Molecular Structure, Function, and
    Diseases (1999)
Edited by Yoshihisa Kurachi, Lily Yeh Jan, and Michel Lazdunski

Volume 47   Amiloride-Sensitive Sodium Channels: Physiology and
    Functional Diversity (1999)
Edited by Dale J. Benos

Volume 48   Membrane Permeability: 100 Years since Ernest Overton (1999)
Edited by David W. Deamer, Arnost Kleinzeller, and Douglas M. Fambrough

Volume 49   Gap Junctions: Molecular Basis of Cell Communication in Health
   and Disease
Edited by Camillo Peracchia

Volume 50   Gastrointestinal Transport: Molecular Physiology
Edited by Kim E. Barrett and Mark Donowitz

Volume 51   Aquaporins
Edited by Stefan Hohmann, Søren Nielsen and Peter Agre

PART I

Biophysical Techniques for Analyzing
Peptide-Lipid Interactions

CHAPTER 1

Peptide–Membrane Interactions Determined Using Site-Directed Spin Labeling

David S. Cafiso

Department of Chemistry, University of Virginia,
Charlottesville, Virginia 22904

I. Introduction
II. Site-Directed Spin Labeling
    A. Synthesis of Spin-Labeled Peptides
    B. Analysis of EPR Spectra
III. Membrane Interactions of Basic Peptides Derived from Cell-Signaling Motifs
    A. Membrane Interactions of Peptides Derived from MARCKS and Src
    B. Structure and Membrane Binding of a Cytochrome c Signal Sequence
IV. Studies Directed at Channel-Forming and Other Membrane-Active Peptides
    A. Alamethicin
    B. Cecropins
    C. Melittin
    D. Orientation and Position of a Lipopeptaibol
    E. Structure and Membrane Interactions of Fusion Peptides
    F. Other EPR Methods for Investigating Peptide–Membrane Interactions
V. Conclusions
    References

The study of peptide–membrane interactions is important for understanding a wide range of fundamental molecular mechanisms, including the action of antibiotic peptides, the association of proteins involved in cell signaling, and membrane fusion. This review discusses the use of site-directed spin labeling to investigate peptide–membrane interactions. Site-directed spin labeling involves the incorporation of a spin label in a site-specific fashion into a macromolecule, such as a protein or peptide, either through synthesis or site-directed mutagenesis. Electron paramagnetic resonance (EPR) spectroscopy is then used to determine molecular structure and dynamics. A number of examples discussed in this review demonstrate how this methodology can be used to determine the conformation of membrane-associated peptides, their position along the bilayer normal, and their state of aggregation. Recent studies have shown that this approach is not highly perturbing, and with new advances in EPR resonator design, it is a highly sensitive technique.

I. INTRODUCTION

The investigation of membrane–peptide interactions has addressed a wide range of fundamental questions. For example, classes of naturally occurring peptides, such as the magainins, cecropins, defensins, and peptiabols, have important antibiotic activities and are membrane-active (Bechinger, 1997). Studies of these peptides have focused on the mechanisms by which they destabilize biological membranes and alter ionic permeability. These peptides are of intrinsic interest because of their biological activities, but they also provide models for membrane ion channels and the mechanisms of voltage-dependent channel gating.

Peptides are also being used as models for protein–membrane interactions. For example, peptides derived from the influenza glycoprotein hemagglutinin are being used to investigate the mechanisms of viral fusion, and peptides derived from highly basic domains of certain soluble proteins are being used to investigate the mechanisms by which these proteins become attached to the membrane–solution interface during cell signaling. In general our knowledge of membrane protein structure is lacking, primarily because methods such as crystallography and nuclear magnetic resonance (NMR), which work well for many water-soluble proteins, fail when applied to membrane proteins. Because of the importance of characterizing and understanding membrane protein structure, peptide models are being used to understand the interactions and folding of proteins when they interact with bilayers. Peptides are being used to investigate the interactions of signal sequences with bilayers and to generate free energy estimates for the partitioning of amino acid side chains into bilayers.

The use of site-directed spin labeling (SDSL) has proven to be a particularly powerful approach to investigating the interactions of peptides with bilayers. This method yields information regarding peptide secondary structure, peptide orientation, position along the bilayer normal, and the aggregation state or tertiary structure of the membrane-associated peptide. There is no difficulty with light scattering and the method can be applied to highly turbid samples. This methodology works well for small as well as quite large macromolecular structures that are associated with membranes. In this chapter, we will discuss general aspects of the use of site-directed spin labeling, particularly as they apply to the study of membrane–peptide interactions. We will also provide a number of examples of the use of this methodology to characterize peptides that interact with membranes.

II. SITE-DIRECTED SPIN LABELING

Naturally occurring proteins and peptides do not usually exhibit EPR spectra, unless they contain a paramagnetic center such as those formed with certain metals. However, EPR spectra can be observed when nitroxide spin labels, which are synthetic stable organic radicals, are incorporated into a macromolecule. A wide range of reactive spin labels have been synthesized (Gaffney, 1976; Hideg and Hankovszky, 1989; Keana, 1979) and these stable radical species can be incorporated into a wide range of structures, including nucleic acids and proteins.

EPR spectroscopy has been in use for quite some time as a biophysical method. However, the general approach of site-directed spin labeling, which involves engineering the selective incorporation of spin labels into peptides or proteins, is more recent and was developed in the late 1980s. A number of excellent reviews on the use of this methodology, particularly as applied to protein dynamics, protein conformational changes, and protein structure, can be found elsewhere (Hubbell and Altenbach, 1994; Hubbell *et al.,* 1996, 1998, 2000; Hustedt and Beth, 1999). The remarkable success of this technique is primarily a result of two developments: the advent of loop-gap resonators, which offer improved sensitivity and high microwave power densities compared to conventional cavity resonators (Froncisz and Hyde, 1982; Hyde and Froncisz, 1986), and the ability to specifically incorporate spin labels into proteins or peptides through site-directed mutagenesis or chemical synthesis.

A. Synthesis of Spin-Labeled Peptides

In the peptide studies that will be discussed here, two approaches have generally been used to incorporate spin labels, as illustrated in Fig. 1. The first involves the synthesis of peptides containing a single cysteine residue, which is then derivatized with a sulfhydryl-specific spin label, such as the methanethiosulfonate spin label (MTSSL) (Berliner *et al.,* 1982), yielding the labeled side chain R1. The second approach involves the incorporation of a spin-labeled amino acid, such as 2,2,4,4-tetramethylpeperidine-*N*-oxyl-4-amino-4-carboxylic acid (TOAC) (Toniolo *et al.,* 1995).

Several other approaches have been used to attach spin labels in specific cases. For example, spin labels have been attached to sites on melittin and on a peptide derived from hemagglutinin by derivatizing its lysine residues with a succinimidyl nitroxide derivative (Altenbach and Hubbell, 1988; Luneberg *et al.,* 1995).

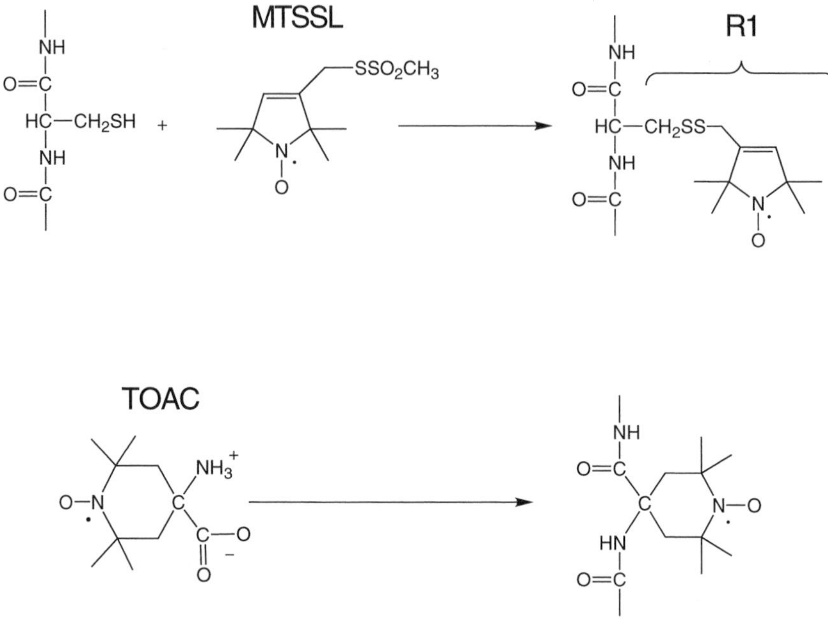

FIGURE 1   Spin labels have routinely been incorporated into peptides by the reaction of a cysteine residue with a sulfhydryl-specific methanethiosulfonate spin label (MTSSL) to produce the side chain R1. Alternatively, the use of a 9-fluorenylmethoxycarbonyl (FMOC)-protected spin-labeled amino acid during peptide synthesis has allowed the incorporation of the amino acid TOAC into a peptide.

Single-labeled peptides were then separated from the complex mixture by high-pressure liquid chromatography (HPLC). In the case of alamethicin, its C-terminal phenylalaninol could be derivatized by coupling it to a carboxylic acid proxyl nitroxide (Archer *et al.*, 1991; Wille *et al.*, 1989).

Among spin labels that have been used to derivatize cysteine, the MTSSL has been the most widely utilized. The probe shows good selectivity toward cysteine and it is often reactive to sites that are deeply buried with a protein structure. The EPR spectrum of R1 is highly sensitive to any tertiary contact that it makes, but it is also sensitive to the dynamics of the peptide backbone. This sensitivity to backbone motion is explained by recent crystallography on spin-labeled T4 lysozyme, which provides information on the likely conformations of R1 (Langen *et al.*, 2000). A sulfur on the R1 linkage is found to partially hydrogen bond to the Cα proton; thus, the atoms that make up the linkage between the peptide backbone and the spin label are not all freely rotatable, and the label motion is partially tied to backbone motion. It should be noted that a number of other spin-labeled side chains have also been incorporated into proteins, and several have been shown to be sensitive only to tertiary contact, having little or no sensitivity toward backbone dynamics

(Mchaourab *et al.*, 1996). Thus, it is possible to "tune" the nitroxide side chain to obtain sensitivity to the type of structural or dynamic information desired.

B. Analysis of EPR Spectra

The energy of the magnetic resonance transition in a nitroxide EPR spectrum is determined both by the *g* value, which defines the strength of the interaction between the electron and the applied static field, and by the hyperfine coupling constant or *A* value, which defines the interaction between the electron and the nearby nitrogen nucleus. The *g* and *A* values are dependent upon the orientation of the *N*-oxide bond with respect to the magnetic field, and they are described by second-rank tensors. If the rates of motion of the label are sufficiently rapid (less than 0.1 ns), the orientational dependence of these interactions is averaged and sharp resonance lines (three for each ^{14}N nuclear spin state) are observed. At slower rates of motion, the magnetic interactions defined by these tensors are not completely averaged and the EPR spectra broaden. Thus, the EPR spectra of nitroxides are dependent upon the rates of motion of the label. They are also dependent upon the polarity of the environment as well as proximity to other nitroxides and paramagnetic species.

1. Side-Chain Motion, Secondary Structure, and Tertiary Contact

Several pieces of information can be immediately derived from the EPR spectrum of a spin-labeled peptide. Shown in Fig. 2 are examples of EPR spectra of peptides spin-labeled with the MTSSL, free in solution, associated with membranes, and complexed with protein. The differences in nitroxide lineshape reflect differences in both the rate and the anisotropy of the nitroxide motion on the nanosecond time scale, and this motion is strongly affected by both the rate of peptide tumbling and the tertiary contact made by the nitroxide. The lineshapes of spin-labeled proteins are in fact exquisitely sensitive to structure, and several studies have demonstrated that the EPR lineshapes reflect the secondary structure and tertiary fold into which the labels are placed (Hubbell *et al.*, 1998; Mchaourab *et al.*, 1996).

Several different spectral parameters can be extracted from the ESR spectra that provide approximate measures of labeled side chain motion. For example, from the linewidth (ΔH or δ) of the central nitroxide resonance ($m_I = 0$), a parameter termed the scaled mobility M_S can be calculated. This parameter can be defined as $M_S = (\delta^{-1} - \delta_i^{-1})/(\delta_m^{-1} - \delta_i^{-1})$, where δ_i and δ_m represent, respectively, the linewidths of the side chain R1 at the most immobile and mobile sites found (Hubbell *et al.*, 2000). The parameter M_S provides a rough measure of the motion of the labeled side chain, where 0 corresponds to the most immobilized and 1 corresponds to the most mobile sites at which R1 is found. Sequence-dependent

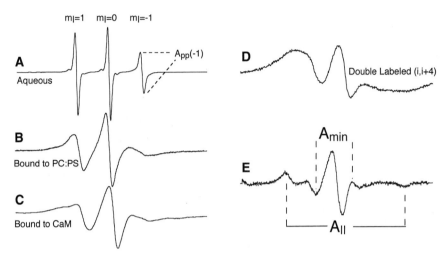

FIGURE 2 Examples of X-band EPR spectra of spin-labeled peptides. An EPR spectrum of MARCKS (151–175) labeled at position 23 (K23R1) in (A) aqueous solution and (B) bound to PC : PS (3 : 1). When a labeled peptide is in equilibrium between membrane and aqueous phases, the spectrum is a composite of the spectra shown in A and B. In this case a measure of the peak-to-peak high-field amplitude $A_{pp}(-1)$ can be used to determine its membrane partitioning as described in the text. (C) The EPR spectrum of the MARCKS peptide K23R1 bound to calmodulin. As indicated in the text, the peak-to-peak linewidth of the $m_I = 0$ resonance, δ, can be used to determine a scaled mobility parameter, M_S. (D) A spectrum of MARCKS (151–175) double-labeled at positions 12 and 16 in the presence of calmodulin. This peptide is in a helical configuration when bound to calmodulin and the spin labels in this case are separated by 6–8 Å (K. Zaiger and D. S. Cafiso, unpublished). (E) A spectrum of KFFxFFK, where x is the spin-labeled amino acid TOAC, bound to PC : PS (3 : 1) membranes. This spectrum is an example of a case where the nitroxide is partially oriented with respect to the bilayer normal.

differences in M_S reflect secondary structure and tertiary contact of the label. The second moment of the EPR spectrum provides similar information (Mchaourab *et al.*, 1996), but can be difficult to estimate for noisy spectra. Finally, the rotational correlation time and motion of the label can be estimated from spectral lineshapes either directly or by simulation (Schneider and Freed, 1989), although in some cases simulations of the nitroxide spectra may not be unique.

Another parameter that can provide useful information on peptide–membrane interactions is the magnitude of the hyperfine coupling constant. This constant reflects the unpaired electron density near nitrogen in the N—O bond. Because the N—O bond is polar, the electron distribution is dependent upon the polarity of the environment. For labels that are undergoing rapid isotropic motion, the average value of the hyperfine coupling constant a_0 can be read directly from the EPR spectrum and is determined by $a_0 = \frac{1}{3}(A_{\parallel} + 2A_{\perp})$, where the values $A_{\parallel}$ and $A_{\perp}$ are, respectively, the parallel and perpendicular components of the hyperfine coupling tensor. For labels that are not undergoing isotropic motion but are partially oriented with respect to the bilayer normal, spectral parameters corresponding to $A_{\parallel}$ and $A_{\min}$ can be extracted from the EPR spectrum, and the value of $A_{\perp}$ can then be determined as described previously, yielding a value for a_0 (Griffith and Jost, 1976) (see Fig. 2E).

2. Membrane–Aqueous Partitioning from EPR Spectra

An important feature of EPR is that the time scale of this spectroscopy is fast compared with the rates of exchange of small molecules on and off the membrane. As a result, a spin-labeled peptide that partitions between aqueous and membrane phases yields an EPR spectrum that is a composite of signals arising from both the bound and aqueous populations (Archer et al., 1991). It is therefore easy to accurately quantitate the peptide partitioning directly from the EPR spectrum. The faction of bound peptide in a sample f_b (or the ratio of bound to total spin N_b/N_{tot}) can be determined from

$$f_b = \frac{N_b}{N_{tot}} = \frac{A_{pp}(-1)_f - A_{pp}(-1)_{tot}}{A_{pp}(-1)_f - A_{pp}(-1)_b}, \tag{1}$$

where $A_{pp}(-1)_f$ is the peak-to-peak amplitude of the high-field ($m_I = -1$) resonance of the sample, and $A_{pp}(-1)_{tot}$ and $A_{pp}(-1)_b$ represent, respectively, the peak-to-peak amplitudes of the sample when the spin label is entirely aqueous or entirely membrane-bound (see Fig. 2A). If f_b is measured as a function of lipid concentration [lipid], the reciprocal molar binding constant K can then be determined by fitting the data to the following expression:

$$f_b = \frac{K_p[\text{Lipid}]}{1 + K_p[\text{Lipid}]}. \tag{2}$$

3. Interactions between Spin Labels

Nitroxides can interact through either a Heisenberg spin–spin exchange mechanism or a dipole–dipole mechanism, and under some conditions, both mechanisms come into play. In either case, these interactions produce line broadening in the EPR spectrum as shown in Fig. 2D. The Heisenberg exchange mechanism is a short-range interaction that occurs when nitroxides come into contact, and it is important in liquids. This spin exchange mechanism has been used to extract information regarding molecular interactions and diffusional rates within bilayers (Sachse et al., 1987). In the rigid lattice condition, where there is an absence of molecular motion, EPR spectra are broadened due to static dipolar interactions, and this interaction has an r^{-3} distance dependence. Under these conditions, dipolar interactions between spin labels have been used to accurately extract interspin distances in the range of 8–25 Å (Hustedt et al., 1997; Rabenstein and Shin, 1996; Steinhoff et al., 1997). In the limit where the rotational correlation time of the macromolecule is sufficiently rapid to average static dipolar interactions (molecular wegiht ≤15,000), a good theory exists to describe the dependence of line broadening upon distance. In this limit, the interaction depends upon r^{-6}, and accurate distance estimates can be obtained over a range comparable to that obtained in the rigid lattice condition (Mchaourab et al., 1997). A good theory does not exist for cases of intermediate motion, for example, a membrane-associated protein at room temperature. For this case, protein diffusion is usually slow on the EPR time scale ($t > 50$ ns), whereas the R1 side chains execute relatively rapid internal motion ($\tau \sim 1$ ns). Under these conditions, the rigid lattice condition may hold, because the interspin vector is in the slow motional limit. This appears to be the case, as this theory yields accurate interspin distances at least within experimental error (Hubbell et al., 2000).

These spin–spin interactions have been used to determine interspin distances in both water-soluble and membrane proteins, and more detailed discussions of the theory and its limitations can be found elsewhere (Hubbell et al., 1998; Hustedt and Beth, 1999). As we demonstrate below, these methods are not only useful for identifying the molecular structures of peptides, but they are very useful in identifying the molecular aggregation of peptides within the bilayer. This information can be particularly difficult to obtain and is not directly evident using other magnetic resonance approaches, such as solid-state NMR, or spectroscopic approaches, such as infrared or circular dichroism.

4. Nitroxide Accessibility and Bilayer Depth from EPR Power Saturation

A measurement of the electron spin–lattice relaxation rate T_{1e} can provide additional information regarding the position and structure of a membrane-associated spin label. T_{1e} is easily estimated in a conventional continuous-wave spectrometer by power saturating the EPR spectrum. This measurement is accomplished by measuring the peak-to-peak amplitude of the central nitroxide resonance as a function of microwave power P, and this behavior is then fit to the following function to yield $P_{1/2}$, which is related to $1/T_{1e}$:

$$A_{pp}(0) = I\sqrt{P}\left[1 + (2^{-\varepsilon} - 1)\frac{P}{P_{1/2}}\right]^{-\varepsilon}. \tag{3}$$

Here, I is a scaling factor, $P_{1/2}$ is the microwave power required to reduce the resonance amplitude to half its unsaturated value, and ε is a measure of the homogeneity of the saturation of the resonance (Altenbach et al., 1994). By allowing I, ε, and $P_{1/2}$ to be adjustable parameters in a fit of the data to Eq. (3), a characteristic $P_{1/2}$ is obtained. In the presence of a second paramagnetic species, such as molecular oxygen or paramagnetic metals, collisions between the second paramagnetic species and the spin label will shorten T_{1e} and increase $P_{1/2}$. The difference between $P_{1/2}$ in the presence and absence of a second paramagnetic species is

$\Delta P_{1/2}$. The parameter $P_{1/2}$ is proportional to $1/(T_{1e} T_{2e})$, but the effects of T_{2e} (the spin–spin relaxation rate) can be eliminated by dividing $P_{1/2}$ by the central EPR linewidth $\Delta H(0)$ (Farahbakhsh et al., 1992). In this way, a parameter $\Delta P'_{1/2}$ that is solely dependent upon the collision frequency between the label and the second paramagnetic species can be calculated using the equation

$$\Delta P'_{1/2} = \frac{\Delta P_{1/2}}{\Delta H(0)}. \tag{4}$$

Since different spectrometers and resonators can deliver different effective microwave powers, $\Delta P'_{1/2}$ will be unique to the instrument being used. However, a parameter that is independent of the instrument can easily be obtained by normalizing $\Delta P'_{1/2}$ against a standard sample, such as 2,2-diphenyl-1-picrylhydrazyl (DPPH), to yield the collision parameter Π:

$$\Pi = \frac{\Delta P'_{1/2}}{\Delta P'_{1/2}(\text{DPPH})}. \tag{5}$$

Here $\Delta P'_{1/2}$ (DPPH) is the $\Delta P'_{1/2}$ value for a sample of DPPH. The collision parameter Π provides information on accessibility of the spin-labeled side chain and on secondary structure.

The $\Delta P_{1/2}$ parameter can also be used to determine the depth or position of the peptide-associated nitroxide along the bilayer normal. Because oxygen concentrates within the bilayer interior and paramagnetic metal complexes such as Ni(II) ethylenediamine-N,N'-diacetic acid (EDDA) are primarily aqueous, labels that reside within the bilayer interior will experience high collision frequencies with O_2, whereas labels residing in an aqueous environment will have high collision frequencies with Ni(II)EDDA. If the standard-state chemical potentials of these paramagnetic reagents vary linearly with distance, the depth parameter Φ will have a linear dependence on distance along the bilayer normal:

$$\Phi = \ln \left[\frac{\Delta P_{1/2} (O_2)}{\Delta P_{1/2}(\text{NiEDDA})} \right]. \tag{6}$$

This is found to be the case for labels placed within the bilayer interior either on proteins of known structure or on spin labels attached to phospholipids (Altenbach et al., 1994). However, outside the membrane interface, the depth parameter appears to approach its bulk value quickly, and nitroxides placed 5–10 Å outside the membrane–solution interface give values of Φ near their bulk value (K. Victor and D. S. Cafiso, unpublished).

In proteins, substitution of the spin-labeled side chain R1 does not appear to be any more perturbing than any other single amino acid substitution (Mchaourab et al., 1996). Nonetheless, one must always be concerned that addition of the side chain R1 to a peptide may significantly alter its structure or position, particularly if R1 is significantly more hydrophobic or has a significantly larger volume that the side chain it is replacing. In addition to testing the labeled peptide function, the best test for ensuring that individual side-chain labels are not significantly altering the peptide is to make multiple measurements using several spin-labeled analogs and to check for consistency in the information provided by each label. A second concern regarding the use of the side chain R1 is that whereas information on position reflects the position of the nitroxide, it may not provide accurate information about the position of the peptide backbone. There are several ways to address this concern. First, as indicated above, high-resolution structures have recently been obtained for spin-labeled proteins, and these data indicate that a sulfur in the disulfide linkage of R1 can form a weak hydrogen bond with the $C\alpha$ proton. As a result, R1 is not highly flexible, but is restrained to a limited number of conformations (Langen *et al.*, 2000). Thus, the number of rotatable bonds is not as great as first appears. Second, if necessary, the spin-labeled amino acid TOAC (Fig. 1) can be incorporated into the sequence, and the position of the peptide backbone can be more precisely determined.

A number of the examples given below will discuss the application of site-directed spin labeling to investigate peptide–membrane interactions. The technique often provides information that is not available using other approaches, and it has a sensitivity that can come close to that of fluorescence methods. For other applications of site-directed spin labeling, the interested reader is referred to reviews that discuss the use of this technique as it is applied to membrane protein structure, protein–protein interactions, protein dynamics, and protein conformational changes (Hubbell *et al.*, 1998, 2000).

III. MEMBRANE INTERACTIONS OF BASIC PEPTIDES DERIVED FROM CELL-SIGNALING MOTIFS

The membrane attachment of water-soluble proteins to membrane interfaces is an important process in cell signaling. Attachment to the membrane interface facilitates appropriate protein–protein and protein–substrate interactions. One general mechanism for the attachment of proteins involves the electrostatic interaction of a basic protein domain with the membrane–solution interface often in combination with protein acylation (Kim *et al.*, 1991; Montich *et al.*, 1993; Mosior and McLaughlin, 1991, 1992). In the case of proteins such as the myristoylated alanine rich C-kinase substrate (MARCKS) and the Src tyrosine kinases, both protein acylation and electrostatic interactions are necessary for membrane attachment. Phosphorylation within these basic motifs eliminates the electrostatic interaction and can dissociate the protein from the interface. This type of mechanism functions as an electrostatic switch to control the association of proteins such as MARCKS (McLaughlin and Aderem, 1995).

A. Membrane Interactions of Peptides Derived from MARCKS and Src

Site-directed spin labeling has been used to study the membrane location and structure of peptides derived from protein domains that associate electrostatically with membranes containing acidic lipid. Shown in Fig. 3 are EPR spectra for the MTSSL spin label (Fig. 1) derivatized to single cysteine residues that have been incorporated into peptides derived from the membrane-binding domains of MARCKS and Src (Qin and Cafiso, 1996; Victor and Cafiso, 1998). These spectra span approximately an order of magnitude in rotational correlation time (0.4–4 ns) and are all characteristic of labels undergoing relatively fast motion. For the MARCKS peptide, the lineshapes are similar, except at the N-terminus, where they have shorter correlation times. There is no indication of secondary structure in these spectra, and interactions between two spin labels incorporated into these peptides indicate that the peptide is in an extended conformation when bound to the membrane. For the Src peptides, labels near the acylated N-terminus exhibit less motion than labels on the C-terminal end. These differences are also reflected in the scaled mobilities M_S for labels on these peptides as a function of sequence position (Fig. 3).

When membrane-bound, the MARCKS peptide penetrates the membrane–solution interface with its five phenylalanine residues buried within the interface. Power saturation of the EPR spectra indicates that at each position on the MARCKS peptide, the label is found to reside 5–10 Å within the membrane–solution interface, except at the N-terminus, where labels at positions 2 and 3 are located in the aqueous phase adjacent the lipid bilayer. The N-terminal end of MARCKS is the most highly charged portion of the peptide with the first six residues all being basic. At the N-terminal end of the Src peptide, the residues are located at the membrane interface, which is not surprising, given that this end of the peptide is acylated. However, residues on the C-terminal side of the peptide are located in the aqueous phase several angstroms away from the membrane–solution interface when the peptide is bound. Figure 4 shows the structures and positions of these two peptides based on mobility and depth measurements obtained from EPR.

The positions assumed by these peptides appear to be the result of an interplay among several forces. These peptides are attracted to the membrane interface as a result of a long-range Coulombic interaction between the positively charged peptide and the negatively charged lipid interface. However, as the peptide gets close to the bilayer interface, it experiences a repulsive dehydration force. This dehydration force may include a Born-image force that acts on the positively charged peptide as it approaches the low-dielectric interface, as well as an entropy loss that is suffered as the peptide becomes restricted at the interface (Ben-Tal *et al.*, 1996; Murray *et al.*, 1997, 1999; Qin and Cafiso, 1996; Victor *et al.*, 1999). The peptide can apparently overcome this dehydration force and attach to the membrane interface if it has sufficient hydrophobic character. Using this model, one

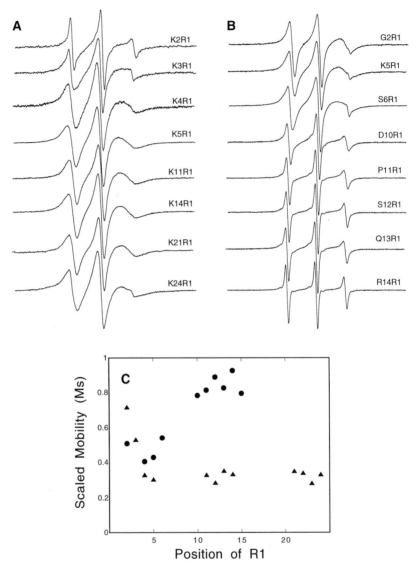

FIGURE 3   (A) EPR spectra of spin-labeled MARCKS (151–175) bound to membranes composed of PC : PS (3 : 1) (Qin and Cafiso, 1996). Labels at positions 2 and 3 have significantly shorter rotational correlation times than labels elsewhere on the peptide and are localized in the aqueous phase. (B) EPR spectra of spin-labeled Src (2–16) bound to membranes composed of PC : PS (3 : 1) (Victor and Cafiso, 1998). Labels near the N-terminal myristoylation have the slowest rotational correlation times and are located near the membrane interface. Labels near the C-terminal half of the peptide have short rotational correlation times and are located in the aqueous phase. (C) Scaled mobility M_S for spin-labeled MARCKS (151–175) (▲) and Src peptides (●) as a function of position.

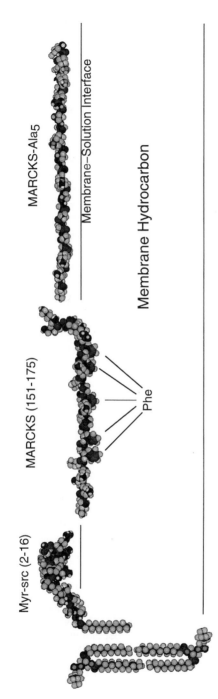

FIGURE 4 Membrane-associated structures of Src- and MARCKS-derived peptides determined by site-directed spin labeling. These structures are extended when bound to the membrane interface (Qin and Cafiso, 1996; Victor and Cafiso, 1998; Victor et al., 1999). Mry-Src (2–16) is positioned above the membrane interface, except at its N-terminal myristoylation site. MARCKS (151–175) is positioned at the interface with its highly charged N-terminus lying in the aqueous phase. Removal of the five phenylalanine residues from the native MARCKS sequence translocates this structure (MARCKS–Ala₅) from the interface to a position approximately 5 Å above the membrane–solution interface.

can account for the behavior of the Src and MARCKS peptides at the membrane–solution interface. Like the N-terminus of the MARCKS peptide, the Src peptide remains in the aqueous phase because it lacks sufficient hydrophobic character to attach it to the membrane interface. The central portion of MARCKS presumably attaches because its five phenylalanine residues overcome the dehydration force that is experienced near the interface.

To test this idea further, a variant of the MARCKS peptide lacking phenylalanine was synthesized and site-directed spin labeling was used to compare its position to that of the native peptide (Victor *et al.*, 1999). Unlike the native peptide, labeled side chains on the peptide lacking phenylalanine do not contact or penetrate the membrane interface, but reside approximately 5–10 Å on the aqueous side of the membrane–solution interface as illustrated in Fig. 4. EPR was also used to examine the binding of this peptide, which was found to have a binding constant only six fold less than that of the native MARCKS peptide. Addition of each phenylalanine appears to add 0.2 kcal/mol to the binding of this peptide, a value that is far less than 1.2 kcal/mol expected based on published interfacial hydrophobicity scales (Wimley and White, 1996). This somewhat surprising observation is consistent with differences in the binding position for these peptides. The decreased energy contribution for each phenylalanine can be understood if a large portion of the energy gained from the hydrophobic interaction of phenylalanine with the interface is utilized to overcome the dehydration force and attach the peptide to the membrane interface.

The results of this work indicate that highly charged, but polar peptides bind strongly to membranes, but do not physically contact the interface and are localized within the double layer. Peptides containing sufficient hydrophobicity attach to the membrane interface. To more accurately determine the position of the backbone of these highly basic polar peptides, a series of peptides containing lysine and phenylalanine was synthesized with the spin-labeled amino acid TOAC (Victor and Cafiso, 2000). Because this label is incorporated into the peptide backbone, it is highly sensitive to the attachment of the peptide to the membrane interface. Shown in Fig. 5 are EPR spectra of several model peptides completely bound to membranes composed of phosphatidylcholine (PC) and phosphatidylserine (PS). Peptides containing only lysine have strong binding constants, but exhibit dramatically different lineshapes than peptides containing lysine and phenylalanine. TOAC labels in lysine-containing peptides exhibit high degrees of mobility, suggesting that these peptides are freely rotating in their bound state. In contrast, peptides that incorporate two or more phenylalanine residues exhibit spectra that approach the rigid limit on the EPR time scale. Bilayer depth measurements determined by power saturating the EPR spectra show that the backbone of peptides containing only lysine resides approximately 5 Å above the membrane interface, whereas the backbone of peptides containing lysine and phenylalanine is localized within the interface. The binding of these charged peptides to the membrane interface likely

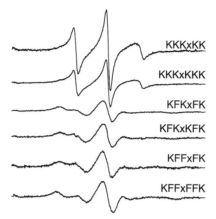

FIGURE 5 EPR spectra of model basic peptides containing lysine and phenylalanine completely bound to PC : PS (3 : 1) membranes. The spin-labeled amino acid TOAC (x) is incorporated into these peptides and their EPR spectra are highly sensitive to the attachment of the peptide to the membrane–solution interface. Peptides that contain only lysine (KKKxKK and KKKxKKK) are localized in the aqueous phase approximately 5 Å from the membrane interface. Other peptides containing two or more phenylalanines are found localized within the membrane–solution interface.

eliminates rapid tumbling of the peptide backbone about its long axis and accounts for the dramatic change in its EPR spectrum.

B. Structure and Membrane Binding of a Cytochrome c Signal Sequence

Signal sequences that are localized at the terminal end of mitochondrial membrane proteins direct the incorporation of these proteins into the bilayer. There has been considerable interest in how these peptides interact with membranes and are recognized by the protein synthesis machinery. Site-directed spin labeling has been used in several cases to gain insight into the interactions of peptide–signal sequences with membrane interfaces. For example, SDSL has been used to investigate the position and secondary structure of the yeast cytochrome c subunit IV (COX IV) signal sequence (Yu et al., 1994b). Depth parameters obtained from oxygen and NiEDDA collision frequencies were used to show that peptide was extended and positioned along the membrane interface with its hydrophobic side chains buried within the hydrocarbon and its charged side chains facing the membrane interface. Site-directed spin labeling was also used to study the role of electrostatic interactions in the association of this peptide with the membrane interface. At low surface potentials, the binding of this peptide was modeled accurately by the Gouy–Chapman Stern theory (Thorgeirsson et al., 1995), but the interaction was found to deviate from this ideal behavior at higher surface potentials.

These peptides also provided a tool for a study of the membrane binding free energies of individual amino acid side chains (Thorgeirsson *et al.,* 1996). Because these peptides are extended both in solution and on the membrane and have partition free energies that can easily be measured by binding experiments, they are ideally suited to measurements of amino acid affinity. Using EPR spectroscopy, the binding of a spin-labeled peptide based on the COX IV signal sequence was determined where a guest amino acid was incorporated at position 7. The data demonstrated a strong correlation with octanol–water partition coefficients, but the measured free energies did not correlate well with side-chain surface areas for some amino acids. In a similar study, these investigators designed a helical peptide that was oriented parallel to the membrane interface and investigated the contributions to membrane binding made by specific amino acids (Russell *et al.,* 1998, 1999). Guest amino acids could be placed at various depths within the lipid bilayer, and the data indicated that the free energy contributions were independent of depth. The free energy contributions to membrane binding were also found to be reduced by approximately two-thirds compared to the energies determined in extended peptide structures.

IV. STUDIES DIRECTED AT CHANNEL-FORMING AND OTHER MEMBRANE-ACTIVE PEPTIDES

Site-directed spin labeling has proven to be a valuable tool for the investigation of channel-forming peptides. This approach has provided information both on the orientation of channel-forming peptides within bilayers and on their position along the bilayer normal. As indicated above, the method is ideally suited to providing information regarding the state of aggregation of membrane-bound peptides. This feature has been particularly useful for the study of channel-forming molecules such as alamethicin and the cecropins, where the conductive state is believed to occur through the association of peptide monomers.

A. Alamethicin

Alamethicin is a voltage-gated channel-forming peptide that is a natural product of the fungus *Trichoderma viride*. This peptide belongs to a family of peptides called peptaibols, which have a reduced C-terminus and contain aminoisobutryic acid (Aib), an amino acid having a tetrasubstituted α carbon (Bechinger, 1997; Cafiso, 1994; Sansom, 1993). When membrane-bound, the peptide is largely helical and amphipathic, and monomers of the peptide are thought to form channels when four or more monomers aggregate with their nonpolar surfaces facing the membrane hydrocarbon.

One fascinating observation regarding alamethicin is that it exhibits cooperativity in its membrane association (Stankowski and Schwarz, 1989), and one explanation for this behavior is that membrane-bound monomers undergo a concentration-dependent aggregation within the bilayer. EPR spectroscopy has been used to investigate the aggregation state of alamethicin within bilayers, by the use of a derivative where the C-terminus of the peptide is selectively derivatized with a proxyl spin label. This spin label should exhibit strong collision exchange and dipolar broadening when it is present in an aggregated form. Like the native peptide, this spin-labeled derivative exhibits voltage-dependent gating and exhibits cooperativity in its membrane binding (Archer *et al.*, 1991). However, a careful examination of the EPR spectra of this derviative shows no indication of aggregation. In fact, alamethicin can be titrated into bilayers of PC at concentration ratios ranging from 1 : 400 to 1 : 7 lipids : peptide, with no indication of aggregation (Barranger-Mathys and Cafiso, 1994). As shown in Fig. 6A, the EPR linewidths increase linearly as a function of the concentration of peptide within the bilayer. This increase is a result of spin exchange that arises from collisions between labeled peptides (Barranger-Mathys and Cafiso, 1994). Furthermore, this linewidth increase can be used to estimate a frequency of collisions between labeled peptides, and this frequency is consistent with the diffusion coefficient expected for an alamethicin monomer. Thus, in spite of the fact that alamethicin conducts in an aggregated form, the great majority of peptide exists as a monomer in the membrane, at least in the absence of a membrane potential.

Site-directed spin labeling has been used to investigate the position, orientation, and structure of alamethicin in lipid bilayers (Barranger-Mathys and Cafiso, 1996). This was accomplished by synthesizing derivatives of alamethicin where cysteine residues were incorporated into the alamethicin sequence. In a fashion identical to that described above, the MTSSL reagent was used to place the R1 side chain at various positions along the peptide sequence. From the positions of labels determined by power saturation, the peptide was shown to be oriented across the bilayer in an α-helical configuration with its C-terminus located several angstroms above the aqueous interface as shown in Fig. 6B. This structure was also found to be consistent with solid-state NMR data (North *et al.*, 1995) and with the results of a recent computational analysis (Kessel *et al.*, 2000).

Partition measurements using EPR have recently been carried out in membranes composed of mixtures of PC and phosphatidylethanolamine (PE) (Lewis and Cafiso, 1999). Addition of PE to these lipid mixtures both increases the bilayer thickness and changes a membrane property termed the spontaneous curvature. The spontaneous curvature is the curvature that one monolayer in the bilayer would assume if it were not restrained in a bilayer (Gruner, 1989). As a result, spontaneous curvature provides an indication of the curvature stress that is present within the bilayer. Binding data show that the free energy of alamethicin is dependent upon the

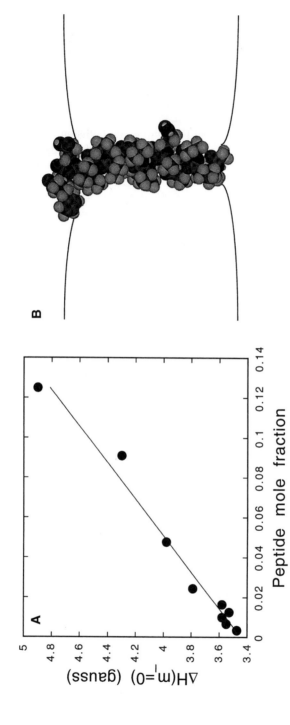

FIGURE 6 (A) Central EPR linewidth ΔH for a spin-labeled alamethicin analog as a function of peptide concentration within a PC bilayer (Barranger-Mathys and Cafiso, 1994). The linewidth increases in a linear fashion as expected for a peptide that is freely diffusing as a monomer within the bilayer. The change in EPR linewidth as a function of peptide concentration is consistent with the diffusion constant expected for an alamethicin monomer. (B) Position and structure of alamethicin within bilayers based in part upon the results of site-directed spin labeling (Barranger-Mathys and Cafiso, 1996). There is a hydrophobic mismatch between peptide and membrane that could account for the sensitivity of the peptide binding and channel conductance to membrane spontaneous curvature (Lewis and Cafiso, 1999).

spontaneous curvature, such that the free energy increases linearly as the curvature strain is increased. Surprisingly, channels formed by this peptide become more stable as the curvature strain is increased (Keller *et al.,* 1993). A simple explanation for these results is that the peptide inserts across the bilayer, but experiences a mismatch between its hydrophobic length and the thickness of the bilayer as shown in Fig. 6B. A hydrophobic mismatch may explain the cooperativity that is seen in the membrane binding and would explain why the conductive (aggregated) form of the peptide is favored as the spontaneous curvature is increased.

The mechanisms that lead to the voltage-dependent gating of this peptide are not understood. In principle, EPR methods are ideally suited to investigations of voltage-dependent changes in peptide conformation or aggregation state. However, one difficulty with these experiments is that potentials are difficult to set up in model membrane systems in the presence of alamethicin. Early work carried out in chloroplasts made use of these systems to create membrane potentials and investigated the effect of energizing the membranes on the EPR spectra of a C-terminal spin-labeled analog of alamethicin (Wille *et al.,* 1989). These data suggested that alamethicin became more deeply buried within the bilayer as potentials were applied.

B. Cecropins

The cecropins play a major role in the immune system of a number of insect species, and they appear to act at the level of the lipid bilayer by producing channels that are voltage-dependent. EPR spectroscopy has been used to investigate the binding and state of aggregation of cecropin AD within the bilayer (Mchaourab *et al.,* 1994). Cecropin AD was synthesized with a cysteine reside in its C-terminal domain at position 33 and labeled with the MTSSL. Using an approach identical to that described above, the partitioning of this peptide was then directly measured by EPR spectroscopy. The data indicated that binding was driven by interactions between the positively charged N-terminus and the acidic lipid surface and that binding was highly dependent upon the membrane thickness. Further, the binding displayed a cooperativity in membranes composed of palmitoyloleoyl PC (POPC)/palmitoyloleoyl phosphatidylglycerol (POPG), but did not in thinner bilayers composed of dilauroyl PC (DLPC)/dilauroyl PG (DLPG). This cooperativity, which could be a result of aggregation (Stankowski and Schwarz, 1989), was found to depend upon the ionization state of the peptide. Direct EPR measurements on the membrane-bound peptide demonstrated that the peptide aggregated in longer, but not in shorter membranes, indicating that the cooperativity was likely the result of the aggregation of the peptide. The EPR spectra and membrane orientation of the peptide suggest a model where the peptide forms dimers through the formation of an intermolecular salt bridge.

C. Melittin

Melittin is the main polypeptide component of the venom of the European honeybee and it exhibits a number of activities on biological membranes, including cell lysis, membrane fusion, and the activation of phospholipase A2 (Bechinger, 1997; Dempsey, 1990). Site-directed spin labeling was used to investigate both the orientation and the aggregation state of this peptide in lipid bilayers (Altenbach and Hubbell, 1988). In this case, lysine residues on native melittin were reacted with succinimidyl-2,2,5,5-tetramethyl-3-pyrroline-1-oxyl-3-carboxylate under conditions that produced on average one labeled site per peptide. Derivatives were isolated by HPLC and identified, yielding single-labeled peptides at the N-terminus, lysine-7, lysine-21, and lysine-23. From the EPR spectra of these peptides, conversion of the peptide from a random coil to an α-helical tetramer was observed to occur in solution in an ionic-strength-dependent fashion. However, when the peptide was bound to the lipid bilayer, the absence of interactions between spin-labeled sites indicated that the peptide was monomeric under a wide range of conditions. Measurements of the collision frequency between these spin labels with aqueous chromium oxalate (a charged paramagnetic reagent) indicated that the peptide was α-helical and oriented with its axis parallel to the membrane surface.

D. Orientation and Position of a Lipopeptaibol

Incorporation of the spin-labeled amino acid TOAC (see Fig. 1) was used to characterize the orientation and membrane depth of trichogin GA IV, a lipopeptaibol (Monaco et al., 1999). This peptide is one of the shortest members of the peptaibol family, consisting of 10 residues and an N-terminal acylation. Surprisingly, this peptide shows membrane activity similar to that of the longer members of this family, such as alamethicin. Incorporation of TOAC into the peptide sequence did not significantly alter either the structure or the activity of the peptide, a result that may in part be due to the conformational similarity between TOAC and Aib. Following an approach used previously (Luneberg et al., 1995), these authors made use of the dependence of the hyperfine coupling constant on the polarity of the environment to position the peptide within the bilayer interior. They generated a calibration curve using steric acid spin labels, where spin labels were placed at different depths within the bilayer. To correct for the difference between the doxyl spin labels used on the steric acids and the TOAC spin label, they calculated Δa_0, the change in the hyperfine coupling constant between the membrane and aqueous environments. Using this approach, they found that trichogin, which is helical on the membrane interface, lies parallel to the surface of the bilayer with its glycine-rich face oriented toward the aqueous phase and its more hydrophobic face positioned toward the bilayer interior.

E. Structure and Membrane Interactions of Fusion Peptides

The membrane interactions of hemagglutinin (HA), the major glycoprotein of influenza, mediate the fusion of the virus with the target cell membrane. HA consists of two subunits, HA1 and HA2, and the N-terminal end of HA2 forms a domain that is highly conserved and appears to be critical for the fusion event. This domain is buried within the HA homotrimer structure at neutral pH, but lowering the pH is thought to expose this domain to the aqueous phase, making it accessible to the target membrane. Site-directed spin labeling has been used in a number of studies to investigate the interactions of this N-terminal fusion peptide with membranes. For example, spin labels were incorporated into a peptide consisting of the N-terminal 40 residues of HA2 (Yu *et al.,* 1994a). From the EPR lineshapes and collision frequencies with chromium oxalate, both the C- and N-terminal segments of this peptide were shown to be inserted into the bilayer, but unlike the aqueous form of this peptide, there was no evidence that it forms trimers. This work also demonstrated that the pH-dependent insertion of this peptide correlated with the pH dependence of the fusion activity of HA.

In another study directed at a 20-amino-acid fusion peptide derived from the N-terminal fragment of HA2, spin labels were incorporated by derivatizing lysines in the peptide with a succinimidyl nitroxide (Luneberg *et al.,* 1995), following a procedure that was used for labeling melittin (Altenbach *et al.,* 1989). From the EPR lineshapes and from the hyperfine coupling constant a_0, the EPR data demonstrated that the peptide inserted into bilayers with its N-terminus buried within the membrane hydrocarbon core. Subsequently, a more detailed EPR study directed at a 127-amino-acid fragment of HA2 placed eight spin labels within its N-terminal fusion domain by the specific labeling of eight cysteine mutants with the MTSSL (Macosko *et al.,* 1997). These investigators found that this segment was a flexible monomer at both neutral and fusogenic pH. EPR power saturation methods were used to measure the depths of the nitroxides at each site and the data were consistent with the N-terminus being structured as an α helix that was tilted approximately 25° from the plane of membrane interface so that the maximum penetration of the N-terminal segment was approximately 15 Å from the phosphate group. The tilt and orientation of this helical region were found to be consistent with the amphiphilicity of this sequence.

The envelope glycoprotein gp41 of the human immunodeficiency virus (HIV) is homologous to the HA2 glycoprotein of influenza, and the membrane interactions of a 38-residue, leucine zipper-like heptad repeat region from gp41 were investigated by site-directed spin labeling (Rabenstein and Shin, 1995). This peptide was labeled at three positions by incorporation of cysteine mutants, and EPR was used to investigate the membrane binding and structure of the peptide. The hyperfine coupling constants indicated that the peptides were bound parallel to the membrane interface. In contrast to the solution structure, where the peptide is assembled as a tetrameric coiled coil, the membrane-bound peptide was found to be monomeric, in a manner similar to that found for the HA-derived peptides. These investigators also demonstrated that mutations that destabilized the coiled-coil structure of gp41 in solution also reduced its membrane affinity.

F. Other EPR Methods for Investigating Peptide-Membrane Interactions

The discussion in this chapter has focused on site-directed spin-labeling studies where nitroxide labels are incorporated into specific sites in peptides for the purpose of investigating peptide structure and placement within membranes. It should be noted that a wide range of studies not specifically discussed here have been carried out to investigate peptide–membrane interactions by examining the EPR spectra of spin-labeled lipids. The EPR spectra of doxyl-labeled lipids reveal information on lipid segmental order and are therefore sensitive to the molecular dynamics and packing of the lipid acyl chains (Griffith and Jost, 1976). Changes in lipid packing and acyl chain dynamics accompany the incorporation of a peptide, and the EPR spectra of these spin-labeled lipids can therefore be used to provide information on the incorporation of peptides. Using this approach, a number of important studies have been carried out on a wide range of peptides, including melittin (Kleinschmidt et al., 1997; Watala and Gwozdzinski, 1992) and cecropin (Hung et al., 1999), a peptide derived from the N-terminal end of the HIV-1 gp41 glycoprotein (Gordon et al., 1992, 1993), gramicidin A (Earle et al., 1994), signal peptides (Sankaram and Jones, 1994; Sankaram et al., 1994; Snel et al., 1995), peptides derived from small K^+ channels (Aggeli et al., 1996; Horvath et al., 1995, 1997), melanotropic peptides (Biaggi et al., 1996), basic lysine-containing peptides (Kleinschmidt and Marsh, 1997), and a model transmembrane peptide (Subczynski et al., 1998). Studies of this type have revealed the depth of interaction of peptides into lipid bilayers, the selective interactions of peptides with specific lipid headgroups, the effect of peptides on membrane domain structure, the stoichiometry of lipid–peptide interactions, and the ordering of lipids associated with membrane peptides.

V. CONCLUSIONS

Site-directed spin labeling is a particularly powerful approach to investigating the conformation, position, and tertiary organization of peptides that interact with membranes. Although the method requires the incorporation of a nitroxide spin label, numerous studies have shown that these labels are not highly perturbing and that peptide function is usually maintained. EPR lineshapes and spin–lattice relaxation rates provide information on peptide conformation and position within bilayers, information which can be difficult to obtain using other methods. Spin labels that are in close proximity to each other exhibit dramatic line broadening effects, and these can be used to accurately determine intermolecular distances. These interactions also make this method uniquely suited to investigations of peptide–peptide interactions and peptide aggregation within the lipid bilayer.

Acknowledgments

The author thanks Drs. Stuart McLaughlin, Wayne Hubbell, and Nir Ben-Tal for helpful discussions. Work from the author's laboratory was supported in part by grants from the NIH (GM35215) and NSF (MCB 9728083).

References

Aggeli, A., Boden, N., Cheng, Y. L., Findlay, J. B., Knowles, P. F., Kovatchev, P., and Turnbull, P. J. (1996). Peptides modeled on the transmembrane region of the slow voltage-gated IsK potassium channel: Structural characterization of peptide assemblies in the beta-strand conformation. *Biochemistry* **35**, 16213–16221.

Altenbach, C., and Hubbell, W. L. (1988). The aggregation state of spin-labeled melittin in solution and bound to phospholipid membranes: Evidence that membrane-bound melittin is monomeric. *Proteins* **3**, 230–242.

Altenbach, C., Froncisz, W., Hyde, J. S., and Hubbell, W. L. (1989). Conformation of spin-labeled melittin at membrane surfaces investigated by pulse saturation recovery and continuous wave power saturation electron paramagnetic resonance. *Biophys. J.* **56**, 1183–1191.

Altenbach, C., Greenhalgh, D. A., Khorana, H. G., and Hubbell, W. L. (1994). A collision gradient-method to determine the immersion depth of nitroxides in lipid bilayers: Application to spin-labeled mutants of bacteriorhodopsin. *Proc. Natl. Acad. Sci. USA* **91**, 1667–1671.

Archer, S. J., Ellena, J. F., and Cafiso, D. S. (1991). Dynamics and aggregation of the peptide ion channel alamethicin. *Biophys. J.* **60**, 389–398.

Barranger-Mathys, M., and Cafiso, D. S. (1994). Collisions between helical peptides in membranes monitored using electron paramagnetic resonance: Evidence that alamethicin is monomeric in the absence of a membrane potential. *Biophys. J.* **67**, 172–176.

Barranger-Mathys, M., and Cafiso, D. S. (1996). Membrane structure of voltage-gated channel forming peptides revealed by site-directed spin labeling. *Biochemistry* **35**, 498–505.

Bechinger, B. (1997). Structure and functions of channel-forming peptides: Magainins, cecropins, melittin and alamethicin. *J. Membr. Biol.* **156**, 197–211.

Ben-Tal, N., Honig, B., Peitzsch, R. M., Denisov, G., and McLaughlin, S. (1996). Binding of small basic peptides to membranes containing acidic lipids: Theoretical models and experimental results. *Biophys. J.* **71**, 561–575.

Berliner, L. J., Grinwald, J., Hankovszky, H. O., and Hideg, K. (1982). A novel reversible thiol-specific spin label: Papain active-site labeling and inhibition. *Analyt. Biochem.* **119**, 450–455.

Biaggi, M. H., Pinheiro, T. J., Watts, A., and Lamy-Freund, M. T. (1996). Spin label and ^{2}H-NMR studies on the interaction of melanotropic peptides with lipid bilayers. *Eur. Biophys. J.* **24**, 251–259.

Cafiso, D. S. (1994). Alamethicin: A peptide model for voltage-gating and protein membrane electrostatic interactions. *Annu. Rev. Biophys. Biomol. Struct.* **23**, 141–165.

Dempsey, C. E. (1990). The actions of melittin on membranes. *Biochim. Biophys. Acta* **1031**, 143–161.

Earle, K. A., Moscicki, J. K., Ge, M., Budil, D. E., and Freed, J. H. (1994). 250-GHz electron spin resonance studies of polarity gradients along the aliphatic chains in phospholipid membranes. *Biophys. J.* **66**, 1213–1221.

Farahbakhsh, Z. T., Altenbach, C., and Hubbell, W. L. (1992). Spin-labeled cysteines as sensors for protein–lipid interaction and conformation in rhodopsin. *Photochem. Photobiol.* **56**, 1019–1033.

Froncisz, W., and Hyde, J. S. (1982). The loop-gap resonator: A new microwave lumped circuit ESR sample structure. *J. Magn. Reson.* **47**, 515–521.

Gaffney, B. J. (1976). The chemistry of spin labels. *In* "Spin Labeling, Theory and Applications" (L. Berliner, ed.), pp. 339–372. Academic Press, New York.

Gordon, L. M., Curtain, C. C., Zhong, Y. C., Kirkpatrick, A., Mobley, P. W., and Waring, A. J. (1992). The amino-terminal peptide of HIV-1 glycoprotein 41 interacts with human erythrocyte membranes: Peptide conformation, orientation and aggregation. *Biochim. Biophys. Acta* **1139**, 257–274.

Gordon, L. M., Curtain, C. C., McCloyn, V., Kirkpatrick, A., Mobley, P. W., and Waring, A. J. (1993). The amino-terminal peptide of HIV-1 gp41 interacts with human serum albumin. *AIDS Res. Hum. Retrovir.* **9**, 1145–1156.

Griffith, O. H., and Jost, P. C. (1976). Lipid spin labels in biological membranes. *In* "Spin Labeling, Theory and Applications" (L. Berliner, ed.), pp. 339–372. Academic Press, New York.

Gruner, S. M. (1989). Stability of lyotropic phases with curved interfaces. *J. Phys. Chem.* **93**, 7562–7570.

Hideg, K., and Hankovszky, O. H. (1989). Chemistry of spin labeled amino acids and peptides. *In* "Biological Magnetic Resonance: Spin Labeling Theory and Applications" (L. J. Berliner and J. Reuben, eds.), pp. 427–488. Plenum Press, New York.

Horvath, L. I., Heimburg, T., Kovachev, P., Findlay, J. B., Hideg, K., and Marsh, D. (1995). Integration of a K^+ channel-associated peptide in a lipid bilayer: Conformation, lipid–protein interactions, and rotational diffusion. *Biochemistry* **34**, 3893–3898.

Horvath, L. I., Knowles, P. F., Kovachev, P., Findlay, J. B., and Marsh, D. (1997). A single-residue deletion alters the lipid selectivity of a K^+ channel-associated peptide in the beta-conformation: Spin label electron spin resonance studies. *Biophys. J.* **73**, 2588–2594.

Hubbell, W., and Altenbach, C. (1994). Investigation of structure and dynamics in membrane proteins using site-directed spin labeling. *Curr. Opin. Struct. Biol.* **4**, 566–578.

Hubbell, W. L., McHaourab, H. S., Altenbach, C., and Lietzow, M. A. (1996). Watching proteins move using site-directed spin labeling. *Structure* **4**, 779–783.

Hubbell, W. L., Gross, A., Langen, R., and Lietzow, M. A. (1998). Recent advances in site-directed spin labeling of proteins. *Curr. Opin. Struct. Biol.* **8**, 649–656.

Hubbell, W. L., Cafiso, D. S., and Altenbach, C. (2000). Identifying conformational changes with site-directed spin labeling. *Nature Struct. Biol.* **7**, 735–739.

Hung, S. C., Wang, W., Chan, S. I., and Chen, H. M. (1999). Membrane lysis by the antibacterial peptides cecropins B1 and B3: A spin-label electron spin resonance study on phospholipid bilayers. *Biophys. J.* **77**, 3120–3133.

Hustedt, E. J., and Beth, A. H. (1999). Nitroxide spin–spin interactions: Applications to protein structure and dynamics. *Annu. Rev. Biophys. Biomol. Struct.* **28**, 139–153.

Hustedt, E., Smirnov, A. I., Laub, C. F., Cobb, C. E., and Beth, A. H. (1997). Molecular distances from dipolar coupled spin-labels: The global analysis of multifrequency continuous wave electron paramagnetic resonance data. *Biophys. J.* **74**, 1861–1877.

Hyde, J. S., and Froncisz, W. (1986). Loop-gap resonators. *Electron Spin Reson.* **10A**, 175–184.

Keana, J. F. W. (1979). New aspects of nitroxide chemistry. *In* "Spin Labeling. II. Theory and Applications" (L. J. Berliner, eds.), pp. 115–172. Academic Press, New York.

Keller, S. L., Bezrukov, S. M., Gruner, S. M., Tate, M. W., Vodyanoy, I., and Parsegian, V. A. (1993). Probablilty of alamethicin conductance states varies with nonlamellar tendency of bilayer phospholipids. *Biophys. J.* **65**, 23–27.

Kessel, A., Cafiso, D. S., and Ben-Tal, N. (2000). Continuum solvent model calculations of alamethicin–membrane interactions: Thermodynamic aspects. *Biophys. J.* **78**, 571–583.

Kim, J., Mosior, M., Chung, L. A., Wu, H., and McLaughlin, S. A. (1991). Binding of peptides with basic residues to membranes containing acidic phospholipids. *Biophys. J.* **60,** 135–148.

Kleinschmidt, J. H., and Marsh, D. (1997). Spin-label electron spin resonance studies on the interactions of lysine peptides with phospholipid membranes. *Biophys. J.* **73,** 2546–2555.

Kleinschmidt, J. H., Mahaney, J. E., Thomas, D. D., and Marsh, D. (1997). Interaction of bee venom melittin with zwitterionic and negatively charged phospholipid bilayers: A spin-label electron spin resonance study. *Biophys. J.* **72,** 767–778.

Langen, R., Oh, K. J., Cascio, D., and Hubbell, W. L. (2000). Crystal structures of spin labeled T4 lysozyme mutants: Implications for the interpretation of EPR spectra in terms of structure. *Biochemistry* **39,** 8396–8405.

Lewis, J. R., and Cafiso, D. S. (1999). Membrane spontaneous curvature modulates the binding energy of a channel forming voltage-gated peptide. *Biochemistry* **38,** 5932–5938.

Luneberg, J., Martin, I., Nubler, F., Ruysschaert, J.-M., and Herrmann, A. (1995). Structure and topology of the influenza virus fusion peptide in lipid bilayers. *J. Biol. Chem.* **17,** 27606–27614.

Macosko, J. C., Kim, C. H., and Shin, Y. K. (1997). The membrane topology of the fusion peptide region of influenza hemagglutinin determined by spin-labeling EPR. *J. Mol. Biol.* **267,** 1139–1148.

Mchaourab, H. S., Hyde, J. S., and Feix, J. B. (1994). Binding and state of aggregation of spin-labeled cecropin AD in phospholipid bilayers: Effects of surface charge and fatty acyl chain length. *Biochemistry* **33,** 6691–6699.

Mchaourab, H., Lietzow, M., Hideg, K., and Hubbell, W. (1996). Motion of spin-labeled side-chains in T4 lysozyme. I. Correlation with protein structure and dynamics. *Biochemistry* **35,** 7692–7704.

Mchaourab, H. S., Oh, K. J., Fang, C. J., and Hubbell, W. L. (1997). Conformation of T4 lysozyme in solution. Hinge-bending motion and the substrate-induced conformational transition studied by site-directed spin labeling. *Biochemistry* **36,** 307–316.

McLaughlin, S., and Aderem, A. (1995). The myristoyl-electrostatic switch: A modulator of reversible protein–membrane interactions. *TIBS* **20,** 272–276.

Monaco, V., Formaggio, F., Crisma, M., Toniolo, C., Hanson, P., and Millhauser, G. L. (1999). Orientation and immersion depth of a helical lipopeptaibol in membranes using TOAC as an ESR probe. *Biopolymers* **50,** 239–253.

Montich, G., Scarlata, S., and McLaughlin, S. (1993). Thermodynamic characterization of the association of small basic peptides with membranes containing acidic lipids. *Biochim. Biophys. Acta* **1146,** 17–24.

Mosior, M., and McLaughlin, S. A. (1991). Peptides that mimic the pseudosubstrate region of protein kinase C bind to acidic lipids in membranes. *Biophys. J.* **60,** 149–159.

Mosior, M., and McLaughlin, S. A. (1992). Electrostatics and dimensionality can produce apparent cooperativity when protein kinase C and its substrates bind to acidic lipids in membranes. *In* "Protein Kinase C: Current Topics and Future Perspectives" (R. Epand and D. Lester, eds.), pp. 154–180. Ellis Horwood, Chester, England.

Murray, D., Ben-Tal, N., Honig, B., and McLaughlin, S. (1997). Electrostatic interaction of myristoylated proteins with membranes: Simple physics, complicated biology. *Structure* **5,** 985–989.

Murray, D., Arbuzova, A., Hangyas-Mihalyne, G., Gambhir, A., Ben-Tal, N., Honig, B., and McLaughlin, S. (1999). Electrostatic properties of membranes containing acidic lipids and adsorbed basic peptides: Theory and experiment. *Biophys. J.* **77,** 3176–3188.

North, C. L., Barranger-Mathys, M., and Cafiso, D. S. (1995). Membrane orientation of the N-terminal segment of alamethicin determined by solid-state ^{15}N NMR. *Biophys. J.* **69,** 2392–2397.

Qin, Z., and Cafiso, D. S. (1996). Membrane structure of protein kinase C and calmodulin binding domain of myristoylated alanine rich C kinase substrate determined by site-directed spin labeling. *Biochemistry* **35,** 2917–2925.

Rabenstein, M., and Shin, Y. K. (1995). A peptide from the heptad repeat of human immunodeficiency virus gp41 shows both membrane binding and coiled-coil formation. *Biochemistry* **34,** 13390–13397.

Rabenstein, M. D., and Shin, Y.-K. (1996). Determination of the distance between two spin labels attached to a macromolecule. *Proc. Natl. Acad. Sci. USA* **92,** 8239–8243.

Russell, C. J., King, D. S., Thorgeirsson, T. E., and Shin, Y. K. (1998). *De novo* design of a peptide which partitions between water and phospholipid bilayers as a monomeric alpha-helix. *Protein Eng.* **11,** 539–547.

Russell, C. J., Thorgeirsson, T. E., and Shin, Y. K. (1999). The membrane affinities of the aliphatic amino acid side chains in an alpha-helical context are independent of membrane immersion depth. *Biochemistry* **38,** 337–346.

Sachse, J.-H., King, M. D., and Marsh, D. (1987). ESR determination of lipid translational diffusion coefficients at low spin-label concentrations in biological membranes, using exchange broadening, exchange narrowing, and dipole–dipole interactions. *J. Magn. Reson.* **71,** 385–404.

Sankaram, M. B., and Jones, J. D. (1994). Mode of membrane interaction of wild-type and mutant signal peptides of the *Escherichia coli* outer membrane protein A. *J. Biol. Chem.* **269,** 23477–23483.

Sankaram, M. B., Marsh, D., Gierasch, L. M., and Thompson, T. E. (1994). Reorganization of lipid domain structure in membranes by a transmembrane peptide: An ESR spin label study on the effect of the *Escherichia coli* outer membrane protein A signal peptide on the fluid lipid domain connectivity in binary mixtures of dimyristoyl phosphatidylcholine and distearoyl phosphatidylcholine. *Biophys. J.* **66,** 1959–1968.

Sansom, M. S. P. (1993). Alamethicin and related peptaibols—Model ion channels. *Eur. Biophys. J.* **22,** 105–124.

Schneider, D. J., and Freed, J. H. (1989). Calculating slow motional magnetic resonance spectra. *In* "Biological Magnetic Resonance: Spin Labeling Theory and Applications" (L. J. Berliner and R. J., eds.), pp. 1–76. Plenum Press, New York.

Snel, M. M., de Kroon, A. I., and Marsh, D. (1995). Mitochondrial presequence inserts differently into membranes containing cardiolipin and phosphatidylglycerol. *Biochemistry* **34,** 3605–3613.

Stankowski, S., and Schwarz, G. (1989). Lipid dependence of peptide–membrane interactions. Bilayer affinity and aggregation of the peptide alamethicin. *FEBS Lett.* **250,** 556–560.

Steinhoff, H. J., Radzwill, N., Thevis, W., Lenz, B., Brandenburg, D., Antson, A., Dodson, G., and Wollmer, A. (1997). Determination of interspin distances between spin labels attached to insulin: Comparison of the electron paramagnetic resonance data with X-ray structure. *Biophys. J.* **73,** 3287–3298.

Subczynski, W. K., Lewis, R. N., McElhaney, R. N., Hodges, R. S., Hyde, J. S., and Kusumi, A. (1998). Molecular organization and dynamics of 1-palmitoyl-2-oleoylphosphatidylcholine bilayers containing a transmembrane alpha-helical peptide. *Biochemistry* **37,** 3156–3164.

Thorgeirsson, T. E., Yu, Y. G., and Shin, Y. K. (1995). A limiting law for the electrostatics of the binding of polypeptides to phospholipid bilayers. *Biochemistry* **34,** 5518–5522.

Thorgeirsson, T. E., Russell, C. J., King, D. S., and Shin, Y. K. (1996). Direct determination of the membrane affinities of individual amino acids. *Biochemistry* **35,** 1803–1809.

Toniolo, C., Valente, E., Formaggio, F., Crisma, M., Pilloni, G., Corvaja, C., Toffoletti, A., Martinez, G. V., Hanson, M. P., and Millhauser, G. L. (1995). Synthesis and conformational studies of peptides containing TOAC, a spin-labelled C alpha, alpha-disubstituted glycine. *J. Peptide Sci.* **1,** 45–57.

Victor, K., and Cafiso, D. S. (1998). Structure and position of the N-terminal binding domain of pp60src at the membrane interface. *Biochemistry* **37,** 3402–3410.

Victor, K., and Cafiso, D. S. (2000). An EPR investigation of the electrochemical interactions between TOAC model peptides and acidic phospholipid vesicles. *Biophys. J.* **78,** 413A.

Victor, K., Jacob, J., and Cafiso, D. S. (1999). Interactions controlling the membrane binding of basic protein domains. *Biochemistry* **38,** 12527–12536.

Watala, C., and Gwozdzinski, K. (1992). Melittin-induced alterations in dynamic properties of human red blood cell membranes. *Chem. Biol. Interact.* **82,** 135–149.

Wille, B., Franz, B., and Jung, G. (1989). Location and dynamics of alamethicin in unilamellar vesicles and thylakoids as model systems: A spin label study. *Biochim. Biophys. Acta* **986,** 47–60.

Wimley, W. C., and White, S. H. (1996). Experimentally determined hydrophobicity scale for proteins at membrane interfaces. *Nature Struct. Biol.* **3,** 842–848.

Yu, Y. G., King, D. S., and Shin, Y. K. (1994a). Insertion of a coiled-coil peptide from influenza virus hemagglutinin into membranes. *Science* **266,** 274–276.

Yu, Y. G., Thorgeirsson, T. E., and Shin, Y. K. (1994b). Topology of an amphiphilic mitochondrial signal sequence in the membrane- inserted state: A spin labeling study. *Biochemistry* **33,** 14221–14226.

CHAPTER 2

Isothermal Titration Calorimetry for Studying Interactions between Peptides and Lipid Membranes

Torsten Wieprecht and Joachim Seelig

Department of Biophysical Chemistry, Biocenter of the University of Basel,
CH-4056 Basel, Switzerland

I. Introduction to Isothermal Titration Calorimetry
II. Measurement of the Reaction Enthalpy and the Binding Isotherm
III. Binding Models for Analyzing Binding Isotherms
    A. Partition Equilibrium versus Langmuir Adsorption Equilibrium
    B. Surface Partition Equilibrium with Electrostatic Correction
IV. Complex Membrane-Binding/Membrane-Perturbation Equilibria
    A. Membrane Permeabilization
    B. Peptide/Protein-Induced Membrane Phase Transitions
    C. Membrane-Induced Peptide Aggregation
    D. Protonation Reaction at the Membrane Surface
V. Thermodynamics of the Membrane-Induced Coil–Helix Transition
VI. Thermodynamics of Peptide Binding to Curved and Planar Membranes
VII. Concluding Remarks
    References

The availability of new, high-sensitivity titration calorimeters has made isothermal titration calorimetry increasingly popular for the study of peptide–membrane interactions. Unlike most other methods, titration calorimetry not only allows a determination of the binding/partitioning isotherm, but provides a complete thermodynamic analysis of the binding reaction, including the free energy of binding ΔG^0, the enthalpy of binding ΔH^0, the entropy of binding ΔS^0, and the heat capacity change ΔC_p. Here, we summarize the experimental approaches used in titration calorimetry for deriving the binding isotherm and ΔH^0. We discuss frequently employed binding models and show that electrostatic interactions between charged peptides and membranes can be distinguished from other interactions by combining a partition equilibrium with the Gouy–Chapman theory. Titration calorimetry can also be used to investigate secondary processes accompanying peptide/protein–membrane binding, such as membrane permeabilization, peptide-induced lipid phase transitions, peptide aggregation at the membrane surface, protonation/deprotonation reactions, and peptide conformational changes.

I. INTRODUCTION TO ISOTHERMAL TITRATION CALORIMETRY

Over the last decade isothermal titration calorimetry (ITC) has become increasingly popular for the study of all kinds of "binding" reactions including protein–ligand interactions and peptide–membrane interactions. Mainly three reasons account for this trend: (i) The development of high-sensitivity titration calorimeters allows the measurement of binding reactions with solutions in the micromolar concentration range (Wiseman *et al.,* 1989). (ii) Almost all chemical and physical processes are accompanied by the absorption or release of heat. Binding can hence be measured without the need for a specific spectroscopic or isotopic labeling. (iii) In addition to providing the binding enthalpy ΔH^0, ITC measurement can provide the binding isotherm. From the latter, the binding constant and the binding mechanism can be deduced and a full thermodynamic description of the system (ΔH^0, ΔG^0, ΔS^0) is possible.

Modern ITC instruments consist of a measuring cell and a reference cell, both with a cell volume of about 1 ml. The reactant is contained in the measuring cell and the substrate is injected via a Hamilton syringe (3–30 μl per injection), which, at the same time, acts as a stirrer providing a fast and efficient mixing of reactant and substrate. The heat absorbed or released upon interaction is measured using an electronic feedback system, which keeps the reference cell and the measuring cell at the same temperature. While the reference cell is heated with a very small constant power, the heating power of the measuring cell is adjusted to minimize the temperature difference to the reference cell. An exothermic reaction leads hence to a reduced heating power (= measured signal) of the measuring cell compared to the reference cell, whereas an endothermic reaction requires an enhanced heating power. The instrument is highly sensitive and measures heats of reaction of ~ 1 μcal, which corresponds to the detection of a temperature change of 10^{-6} K (1-ml cell) (Wiseman *et al.,* 1989). The equilibration time between two consecutive injections is normally in the range 4–10 min. A typical binding experiment consisting of about 10–20 injections can be finished within 40–200 min.

Isothermal titration calorimetry has proven to be extremely useful for the study of peptide/protein–membrane interactions. The typical application is the study of thermodynamic parameters of a binding reaction. A complete thermodynamic picture of the binding reaction, usually given by the free energy ΔG^0, the enthalpy ΔH^0, the entropy of binding ΔS^0, and the heat capacity change ΔC_p, is obtained and provides information about the driving forces of the reaction. Moreover, ITC has been used to study secondary processes accompanying peptide–membrane binding, that is, peptide-induced membrane permeabilizations, lipid phase changes, membrane-induced peptide–peptide associations, protonation reactions at the membrane surface, and peptide conformational changes. In this review, we will describe in detail ITC experiments used to study peptide–membrane interactions. We will discuss binding models for the quantitative description of peptide–membrane binding and summarize studies aimed at elucidating secondary reactions accompanying the binding reaction.

II. MEASUREMENT OF THE REACTION ENTHALPY AND THE BINDING ISOTHERM

In general, two different ITC experiments are used to derive the reaction enthalpy and the binding isotherm (Beschiaschvili and Seelig, 1992; J. Seelig, 1997). In order to determine the reaction enthalpy ΔH in a single injection experiment, the calorimeter cell contains a suspension of lipid vesicles, typically at a concentration between 10 and 20 mM. Small aliquots (5–10 μl) of a peptide solution (between 100–200 μM) are injected into the calorimeter cell. The peptide is hence diluted by a factor of 100–200 and the lipid/peptide ratio in the cell is then of the order of 10,000. If the lipid affinity of the peptide is large enough, the injected peptide can be assumed to completely bind to the vesicles. A typical example of such an experiment is given in Fig. 1A, which shows the result of the injection of 5 μl of a 200 μM solution of the antibacterial peptide magainin 2 amide (M2a) into 1-palmitoyl-2-oleoyl-sn-glyero-3-phosphocholine (POPC) small unilamellar vesicles (SUVs) (Wieprecht et $al.$, 1999b). Figure 1B shows the heat of injection as evaluated from the integration of the calorimeter traces. After subtraction of the heat of dilution, observed in a separate control experiment by injecting peptide into buffer only, the binding enthalpy ΔH^0 can be calculated by dividing the heat of injection by the molar amount of injected peptide. For the interaction of M2a with POPC SUVs at 15°C, binding was accompanied by an exothermic enthalpy change ΔH^0 of -18.4 kcal/mol.

The binding isotherm can be determined in a second type of experiment, if the two solutions are exchanged. The calorimeter cell contains the peptide solution (typically 6–40 μM), and aliquots of a concentrated lipid suspension (10–40 mM) are injected via the Hamilton syringe. After each injection, peptide is removed from the bulk solution and binds to the injected vesicles. Therefore, with increasing

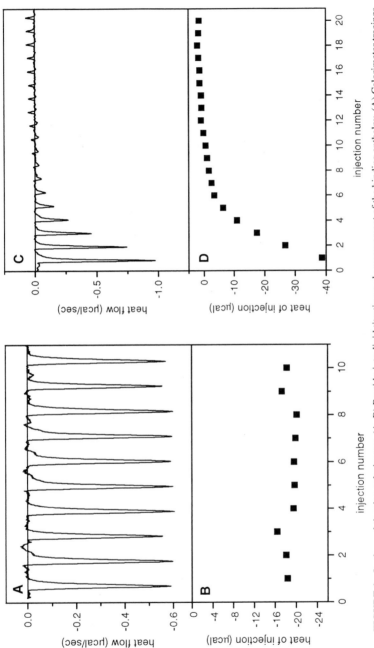

FIGURE 1 Isothermal titration calorimetry. (A, B) Peptide-into-lipid calorimetry. (A) Calorimeter tracings for the injection of 5 μl of a 200 μM magainin 2 amide solution into a calorimeter cell containing a 20 mM POPC SUV suspension at 15°C. (B) Heat of reaction per injection as evaluated by integration of the calorimeter tracings given in A. The heat of dilution (~2 μcal) was measured in a separate control experiment and the data in B were correspondingly corrected (Wieprecht et al., 1999b). (C, D) Lipid-into-peptide titration and measurement of the binding isotherm. (C) Calorimeter tracings for the injection of 5-μl aliquots of a 38 mM POPC SUV suspension into a reaction cell containing a 6 μM solution of the magainin analogue I[6]V[9]W[12]T[15]I[17] M2a at 15°C. (D) Heat of reaction as a function of injection number, obtained by integration of the calorimeter tracings.

injection number, less and less peptide is free in solution and the heat of injection is expected to decrease with increasing injection number. A typical measurement is given in Fig. 1C. The figure shows the result of the injection of 5 μl of a 38 mM POPC SUV suspension into 6 μM of the magainin analogue $I^6V^9W^{12}T^{15}I^{17}$ M2a (Wieprecht et al., 1999b). The exothermic heat of injection, corrected for the heat of vesicle dilution, approaches zero after about 15 injections (Fig. 1D), because virtually all peptide is now bound to the lipid vesicles. It is thus possible to calculate the enthalpy of binding ΔH^0 according to

$$\Delta H^0 = \sum_i \delta h_i \Big/ \left(c_{pep}^0 V_{cell}\right), \tag{1}$$

where the δh_i are the heats of injection (heats of dilution subtracted), c_{pep}^0 is the total peptide concentration within the calorimeter cell, and V_{cell} is the cell volume. The ΔH^0 values determined by lipid-into-peptide-titrations should be in agreement with those obtained in the first type of experiment. The two experiments are performed under different conditions with respect to the bound peptide-to-lipid ratio X_b. Whereas X_b for lipid-into-peptide titrations is typically in the range 0.04–0.005, it is only $\sim$0.0001 in the reverse titration. The agreement between both ΔH^0 values reveals that the binding enthalpy is largely independent of X_b. This is a prerequisite for the derivation of the binding isotherm.

The binding isotherm can be derived from lipid-into-peptide titrations according to the following procedure (J. Seelig, 1997). After i injections, the fraction of peptide $X_P^{(i)}$ bound to lipid vesicles is given by

$$X_P^{(i)} = \frac{n_{P,b}^{(i)}}{n_{pep}^0} = \frac{\sum_{k=1}^i \delta h_k}{\Delta H^0 \cdot V_{cell} c_{pep}^0}, \tag{2}$$

where $n_{P,b}^{(i)}$ is the molar amount of bound peptide after i injections, n_{pep}^0 and c_{pep}^0 are the total amount and concentration, respectively, of the peptide in the calorimeter cell, and $\sum_{k=1}^i \delta h_k$ is the sum of the first i reaction heats. The concentration of peptide remaining free in solution, $c_f^{(i)}$, is given by

$$c_f^{(i)} = f_{dil}^{(i)} c_{pep}^{(0)} \left(1 - X_P^{(i)}\right). \tag{3}$$

The dilution factor takes into account the increase in volume due to vesicle injection and is defined as

$$f_{dil}^{(i)} = \left(\frac{V_{cell}}{V_{cell} + V_{inj}}\right)^i \approx \left(\frac{V_{cell}}{V_{cell} + i V_{inj}}\right), \tag{4}$$

where V_{inj} is the injected volume per injection step ($V_{inj}/V_{cell} \ll 1$). The degree of peptide binding $X_b^{(i)}$ is defined as the mole ratio of bound peptide per lipid. Because the lipid content increases with the number of injections, $X_b^{(i)}$ is given by

$$X_b^{(i)} = \frac{n_{pep,b}^{(i)}}{n_L^{(i)}} = X_P^{(i)} \frac{c_{pep}^0 V_{cell}}{i V_{inj} c_L^0}, \tag{5}$$

where c_L^0 is the concentration of the lipid stock solution in the injection syringe. A plot of $X_b^{(i)}$ versus $c_f^{(i)}$ finally yields the desired binding isotherm, that is, $X_b = f(c_f)$. It should be noted that no assumptions on the molecular nature of the binding/adsorption process need be made in order to derive the binding/adsorption isotherm from ITC measurements. Figure 2 shows, as an example, the binding isotherm for binding of the magainin analogue $I^6 V^9 W^{12} T^{15} I^{17}$ M2a to POPC SUVs as derived from the titration experiment in Figs. 1C and 1D (Wieprecht et al., 1999b).

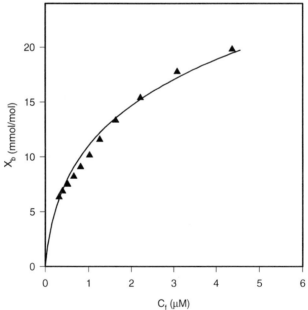

FIGURE 2 Binding isotherm of the magainin analogue $I^6 V^9 W^{12} T^{15} I^{17}$ M2a for POPC SUVs at 15°C (Wieprecht et al., 1999b). The extent of binding X_b is the molar ratio of bound peptide per lipid. c_f is the free peptide concentration in solution. The binding isotherm was derived from the experiment shown in Fig. 1, using the procedure described in the text. The solid line corresponds to a calculated binding isotherm based on a model which combines a surface partition equilibrium with the Gouy–Chapman theory. The fit parameters are $K = 45000 \, M^{-1}$ and $z = 3.3$.

III. BINDING MODELS FOR ANALYZING BINDING ISOTHERMS

The thermodynamic analysis of the binding reaction requires the assumption of a specific binding model. If the binding model is consistent with the experimental results, the binding constant K, standard free energy ΔG^0, and entropy of binding ΔS^0 can be derived. In the following paragraphs we will distinguish two basic types of binding models: (i) binding models that do not consider electrostatic effects between peptides and the membrane surface and (ii) models that account for the interaction of charged peptides with a charged membrane surface.

A. Partition Equilibrium versus Langmuir Adsorption Equilibrium

The simplest "binding" model is a partitioning of a peptide between the aqueous phase and the lipid membrane. This model assumes a linear relationship between the molar ratio of membrane-bound peptide per lipid X_b and the free peptide concentration in solution C_f, according to

$$X_b = K_p C_f. \tag{6}$$

The free energy of the binding reaction can then be calculated from the partition coefficient K_p using the equation

$$\Delta G^0 = -RT \ln 55.5 K_p, \tag{7}$$

where R is the universal gas constant and T is the temperature. The factor 55.5 is the molar concentration of water; including it is equivalent to taking the concentration of peptide in aqueous solution in the form of its mole fraction (cratic contribution to the binding; cf. Cantor and Schimmel, 1980). Finally, the binding entropy can be calculated from

$$\Delta S^0 = (\Delta H^0 - \Delta G^0)/T. \tag{8}$$

A linear relationship between X_b and C_f is anticipated for the interaction of non-charged peptides with lipid vesicles at low X_b values. An example is the binding of the immunosuppressive peptide cyclosporin A to POPC SUVs (J. Seelig, 1997). Similarly, the binding of model peptides derived from the amphipathic α helixes of apolipoproteins to POPC vesicles could be described well by a simple partition equilibrium (Gazzara et al., 1997b). Even though the latter peptides contain several charged residues, the overall charge is zero and electrostatic attraction/repulsion effects at the membrane surface can be neglected.

As an alternative model, binding may follow a Langmuir adsorption isotherm rather than a partition equilibrium. An adsorption model assumes that the number of binding sites in the membrane is limited. If a binding site is made up of n

energetically equivalent and noninteracting lipid molecules, the Langmuir binding isotherm can be written as

$$\frac{\theta}{1-\theta} = K c_f, \tag{9}$$

where $\theta = nC_{P,b}/C_L$ is the mole fraction of occupied binding sites. It should be mentioned that this model is mathematically equivalent to the "chemical" equilibrium frequently used in the literature. The latter assumes that one peptide "reacts" with a cluster of n lipids to form a PL_n complex. The adsorption model is useful for describing the binding of large proteins to membranes, where only a limited number of binding sites is available per lipid vesicle for steric reasons. The model was employed to describe binding of the apolipoprotein Apo A1 to large unilamellar vesicles composed of dimyrostoylphosphatidylglycerol. Apo A1 was found to bind to a cluster of 21 lipid molecules with a binding constant of 17,700 M^{-1} at 30°C (Epand et al., 1990). The adsorption model also could be used to fit the titration curves for the interaction of a mitochondrial leader sequence with negatively charged membranes (Myers et al., 1987). A binding model that assumes binding of a peptide to a specific number of lipids was also used to fit titration calorimetry data for the interaction of the fusion-inhibiting peptide Z-D-Phe–L-Phe–Gly with DOPC unilamellar vesicles (Turner et al., 1995). Z-D-Phe-L-Phe–Gly interacts with a cluster of 5.6 lipid molecules with an association constant of only 290 M^{-1}. In this study, the ITC experiments were done under conditions such that always less than 50% of the binding sites were occupied (nonsaturation conditions).

B. Surface Partition Equilibrium with Electrostatic Correction

Membrane-binding peptides often carry a net positive charge due to the presence of lysine and arginine residues, which, in turn, influences the interaction with biological membranes. Many biological membranes contain negatively charged lipids and are hence characterized by a negative surface potential. The surface potential leads to an attraction of positively charged peptide to the membrane surface, and thus the peptide concentration near the membrane surface, denoted c_M in the following, is enhanced compared to the bulk concentration, c_f. For a partition equilibrium, one would hence expect that the ratio of bound peptide per lipid X_b is no longer proportional to c_f, but is proportional to c_M (J. Seelig et al., 1993):

$$X_b = K_p c_M. \tag{10}$$

With increasing amount of bound peptide, the membrane surface potential and the electrostatic attraction decrease. The characteristic property of the electrostatic interactions is thus a downward-bending of the binding isotherm. A similar effect can be seen for the interaction of charged peptides with electrically neutral membranes. Without peptide, the surface potential of a neutral membrane is zero. However, after binding of the first positively charged peptide, the membrane becomes positively charged, resulting in a repulsion of further peptide from the membrane surface and hence $c_M < c_f$. The repulsion becomes stronger as more positively charged peptide is bound to the membrane, leading again to a downward bending of the binding isotherm, provided that binding follows a simple partitioning.

Electrostatic effects can be incorporated into the binding model as follows. The surface concentration c_M depends on (i) the free peptide concentration in bulk solution, (ii) the peptide charge, and (iii) the membrane surface potential. C_M can be calculated using the Boltzmann equation,

$$c_M = c_f \exp(-z F_0 \psi_0 / RT), \tag{11}$$

where z is the peptide charge, F_0 is the Faraday constant, ψ_0 is the surface potential, and RT is the thermal energy. Unfortunately, the membrane surface potential ψ_0 cannot be measured directly. However, ψ_0 is connected to the membrane surface charge density σ, which, in turn, is linearly related to the extent of binding X_b. The quantitative relation between σ and ψ_0 is provided by the Gouy–Chapman theory (for reviews see McLaughlin, 1977, 1989; Aveyard and Haydon, 1973),

$$\sigma^2 = 2000\varepsilon_0\varepsilon_r RT \sum_i c_i \left(e^{-z_i F_0 \psi_0 / RT} - 1\right), \tag{12}$$

where c_i is the concentration of the ith electrolyte in the bulk aqueous phase, z_i is the signed valency of the ith species, ε_0 is the electric permittivity of free space, and ε_r is the dielectric constant of water. Combining Eqs. (10)–(12) and calculating σ from the experimentally accessible X_b, we can evaluate ψ_0, c_M, and, finally, K_p. It should be emphasized that the partition coefficient $K_p = X_b/c_M$ describes transfer of the peptide from the membrane *surface* into the membrane. Binding obeys a *surface* partitioning equilibrium, if K_p is independent of the peptide concentration, and the model can then be used to fit the experimental binding isotherm. The free energy and entropy of binding can finally be derived from K_p using Eqs. (7) and (8).

In recent years, this model has been employed to describe binding of a multitude of structurally very different charged molecules to model membranes. Among them are cyclic somatostatin analogues antibacterial peptides, β-amyloid analogues and substance P agonists (A. Seelig *et al.,* 1996; J. Seelig *et al.,* 1993; Terzi *et al.,* 1994, 1995). Figure 3 shows, as an example, the binding isotherms for the binding of the antibacterial peptide magainin 2 amide (M2a) to POPC SUVs and to SUVs containing 15% and 25% of the negatively charged lipid

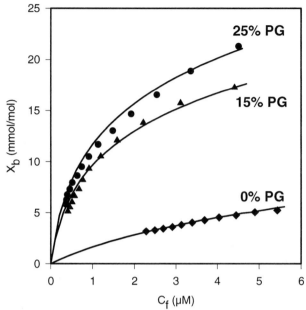

FIGURE 3 Binding isotherms of magainin 2 amide to electrically neutral POPC SUVs and to POPC SUVs containing 15% and 25% of the negatively charged lipid POPG. The solid lines correspond to the best theoretical fits to the experimental data. The binding model combines a surface partition equilibrium with the Gouy–Chapman theory (electrostatic attraction). The binding constants corrected for electrostatic effects are 2000 M^{-1} for POPC, 280 M^{-1} for POPC/POPG (85:15), and 50 M^{-1} for POPC/POPG (75:25) (Wenk and Seelig, 1998; Wieprecht *et al.*, 1999b).

1-palmitoyl-2-oleoyl-*sn*-glycero-3-phosphoglycerol (POPG), respectively (Wenk and Seelig, 1998; Wieprecht *et al.*, 1999b). Binding of the positively charged M2a increases with increasing content of negatively charged phospholipids. The solid lines correspond to the best fits of the surface partition model to the binding isotherms and the fit parameters are summarized in the legend to Fig. 3. For these simulations it was assumed that the peptides are bound to the outer leaflet of the bilayer only. Interestingly, whereas binding increases with increasing POPG content, the binding constant, which is corrected for electrostatic interactions, decreases: $K_p(0\%\ \text{PG}) = 2000\ M^{-1}$, $K_p(15\%\ \text{PG}) = 280\ M^{-1}$, and $K_p(25\%\ \text{PG}) = 50\ M^{-1}$ (Wenk and Seelig, 1998; Wieprecht *et al.*, 1999b). The hydrophobic contribution to the binding constant decreases with increasing PG content as the peptide maximizes its nonspecific electrostatic interaction (Wieprecht *et al.*, 1999b). The enhanced overall binding in the presence of POPG can be attributed to the electrostatic attraction of the positively charged peptide to the membrane surface.

IV. COMPLEX MEMBRANE-BINDING/MEMBRANE-PERTURBATION EQUILIBRIA

A. Membrane Permeabilization

The molar heat of peptide binding should be independent of the peptide concentration. Only under this condition is a smooth decrease in the magnitude of the heat of reaction expected with increasing injection number (cf. Figs. 1C and 1D). However, studies with membrane-permeabilizing peptides have revealed the existence of rather complex titration curves. Figure 4A shows the calorimeter traces and Fig. 4B the heats of reaction for the injection of 5 μl of a 12.14 mM POPC/POPG (3:1) SUV suspension into a 12.86 μM M2a solution (Wenk and Seelig, 1998). The heat of injection does not decrease smoothly with each injection, but shows an inflection point after about 20 injections. Figure 4B shows a comparison between the experimental δh_i and the simulated titration curve based on a surface partition equilibrium with a correction of electrostatic effects by means of the Gouy–Chapman theory. The binding parameters used for this calculation ($K = 60\ M^{-1}$, $\Delta H = -17$ kcal/mol, and $z = 3.7$–3.8) were obtained from measurements with M2a concentrations below 7 μM. At peptide concentrations below 7 μM, a continuous decrease of the heat of injection were observed, permitting the derivation of the binding isotherm (Wenk and Seelig, 1998). A comparison between simulated and measured titration curves revealed that for the first injections, the measured heat was considerably less exothermic than expected. However, during the second part of the experiment, the experimental heats were more exothermic than predicted. The overall heat released during the experiment corresponds to the heat of binding. The biphasic behavior of the titration curve was explained by a reversible pore formation/pore disintegration process of M2a (Wenk and Seelig, 1998). At high ratios of bound peptide per lipid (first injections), the exothermic peptide binding is accompanied by an endothermic pore formation process, leading to a reduced exothermic interaction enthalpy. With increasing injection number, the ratio of bound peptide per lipid decreases and at a critical X_b, the pores start to disintegrate. Pore disintegration returns the heat of pore formation consumed in the first part of the experiment, resulting in an enhanced exothermic heat of reaction.

A similar behavior was observed for the antibacterial membrane-permeabilizing peptide PGLa (Figs. 4C and 4D) (Wieprecht et al., 2000a). Again, the heats of injection observed for the first injections were considerably less exothermic than expected for the binding process proper (solid lines), but were more exothermic for higher injection numbers. In fact, a second, endothermic process, which is somewhat slower than the exothermic binding, is even resolved for the first few injections. The endothermic heat of pore formation is somewhat larger for PGLa (9.7 kcal/mol) than for M2a (6.2 kcal/mol).

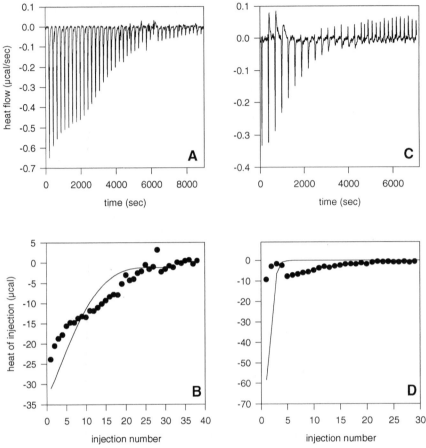

FIGURE 4 Pore formation and/or lipid perturbation. Injection of POPC/POPG (3:1) SUVs into solutions of the antibacterial peptide magainin 2 amide (A, B) and PGLa (C, D) (Wenk and Seelig, 1998; Wieprecht *et al.,* 2000a). (A) Calorimeter traces for the injection of 5-μl aliquots of a 12.14 mM SUV suspension into 12.86 μM magainin 2 amide at 30°C. (B) Heat of injection of the experiment shown in (A). The heat of dilution was subtracted. The solid line corresponds to the calculated heat of binding based on the surface partition equilibrium with $K = 60 \, M^{-1}$, $\Delta H = -16$ kcal/mol, and $z = 3.6$. (C) Calorimeter traces for the injection of 5-μl aliquots of a 24.53 mM SUV suspension into 6 μM PGLa at 30°C. (D) Heat of injection of the experiment shown in C after subtraction of the heat of dilution. The solid line corresponds to the calculated heat of binding based on the surface partition equilibrium with $K = 1500 \, M^{-1}$, $\Delta H = -12.3$ kcal/mol, and $z = 4.6$. The deviation of the experimental data from the calculated binding isotherms can be explained by a pore formation/pore disintegration process (see text).

For PGLa, the ITC results were compared to a dye release assay used to assess the membrane permeability. The critical limit of bound peptide per lipid where the pores start to disintegrate was found to be $X_b \approx 18$ mmol/mol by means of titration calorimetry, suggesting a markedly enhanced membrane permeability for $X_b \geq 18$ mmol/mol. This is in excellent agreement with the dye release experiments. A 50% dye release was observed for $X_b \approx 17$ mmol/mol.

It should be noted that the mechanism of pore formation is still a matter of debate. It was shown for M2a (Matsuzaki *et al.*, 1996) that peptide-induced membrane permeabilization is coupled to a transfer of peptide and lipid from the outer leaflet to the inner leaflet of the membrane. To account for these findings, the toroidal pore model has been suggested (Matsuzaki *et al.*, 1996). However, direct evidence for a pore of well-defined structure does not exist. Hence, the endothermic heats measured in the ITC experiments cannot be attributed to a specific molecular structure and the term "pore formation" is only used to indicate the state of enhanced membrane permeability. Most likely, the enthalpy of "pore formation" contains contributions from changes in both the structure of the peptides and the lipid matrix.

B. Peptide/Protein-Induced Membrane Phase Transitions

Most studies discussed so far were performed between 25°C and 35°C with fluid lipid membranes which have their main transition temperature well below 0°C (e.g., POPC at −5°C). A peptide-induced lipid phase transition is hence unlikely to occur in these studies. This is different in studies of peptide/protein–membrane binding with lipids carrying two saturated acyl chains. With batch calorimetry, the interaction of the peptide hormone glucagon with dimyristoylphosphatidyl-choline (DMPC) suspensions was studied below and above the main transition temperature T_M of DMPC ($T_M = 24$°C) (Epand and Sturtevant, 1981). The standard enthalpy of interaction was found to be +80 kcal/mol below T_M at 23°C, but −150 kcal/mol above T_M at 25°C. Differential scanning calorimetry revealed a pronounced broadening of the phase transition in presence of the peptide. Hence, the endothermic reaction enthalpy at 23°C was mainly attributed to a partial protein-induced melting of the lipid, whereas the highly exothermic interaction enthalpy at 25°C was suggested to reflect a peptide-induced transfer of lipid into the gel phase. At these temperatures, a significant contribution from the actual binding reaction, that is, the noncovalent bonding between peptide and lipids, or from peptide conformational changes was excluded. Similar results were obtained for the interaction of the plasma apolipoprotein A2 with DMPC (Massey *et al.*, 1981). A large endothermic heat of reaction (+90 kcal/mol) was observed just below T_M and was explained by the transition of gel-phase lipids into more fluid-like lipids. The highly exothermic heat of reaction of −260 kcal/mol at 24.5°C was suggested to arise mainly from an isothermal acyl chain crystallization. A further temperature increase to 30°C decreased the exothermic interaction enthalpy to -62 kcal/mol. At this temperature, lipid phase transitions are no longer involved, and the enthalpy change was attributed exclusively to α-helix formation (cf. Section V).

A more complicated situation was reported for the interaction of myelin-basic protein with sonicated vesicles made from brain phosphatidylserine (PS) or dimyristoylphosphatidylserine (DMPS) (Ramsay *et al.*, 1986). Whereas the interaction of myelin basic protein with fluid PS SUVs was exothermic over the whole concentration range, the interaction with gel-phase DMPS was exothermic at low peptide concentrations, but became endothermic at larger protein-to-lipid ratios. This process was correlated with the formation of large aggregates in the vesicle suspension due to protein-mediated vesicle–vesicle association. Essentially the same titration pattern was found for the titration of apolipoprotein Apo A1 with LUVs made from the negatively charged lipid dimyristoylphosphatidylglycerol (DMPG) above T_M (Epand *et al.*, 1990). For the first injections (high protein/lipid ratio), the heat of injection was endothermic, but it became exothermic for higher injection numbers. The authors explained this behavior by the existence of two different molecular processes. The first, endothermic process reflects the transfer of vesicular DMPG to the high-affinity binding sites of Apo A1, whereas the exothermic heat of reaction is attributed to the transfer of the lipid from the vesicle into the lipid matrix of discoidal lipoprotein micelles. The lack of curvature in the latter may allow the lipid to pack in a more gel-like arrangement, giving rise to the release of heat.

Different types of secondary interactions have been identified for the interaction of cytochrome c with negatively charged vesicles (Heimburg and Biltonen, 1994; Zhang and Rowe, 1994). The enthalpy of binding of cytochrome c to fluid dioleoylphosphatidylglycerol (DOPG) LUVs was found to be $+15$ kcal/mol at 25°C. Since differential scanning experiments indicated a drastically reduced thermal stability of membrane-bound cytochrome c, the enthalpy change was suggested to be due to a partial "unfolding" or loosening of the protein structure upon membrane binding (Zhang and Rowe, 1994). However, at temperatures below the main transition of DMPG, the large endothermic heats of binding (50–70 kcal/mol) were explained by changes in the lipid structural state (Heimburg and Biltonen, 1994).

In a recent study, Boots *et al.* (1999) investigated the interaction of the protein β-lactoglobulin with vesicles composed of monostearoylglycerol and dicetylphosphate by means of ITC. The authors found significant differences in the binding thermodynamics derived with titration calorimetry and from a classical binding assay. The fit of the ITC curves yielded a stoichiometry of 900 lipid molecules per protein, whereas the conventional binding assay suggested 90 lipid molecules per protein. These differences were explained by the existence of two different binding modes. In the first mode, binding is accompanied by a strong exothermic heat of reaction, but involves only 10% of the bound β-lactoglobulin. The second binding mode, involving 90% of the bound protein, is described as nonspecific electrostatic attraction between β-lactoglobulin and the negatively charged vesicles. The second mode did not give rise to a detectable heat effect. A second remarkable result was the large negative interaction enthalpy of about -760 kcal/mol of bound protein. This unusual exothermic enthalpy change was traced back to a protein-induced transition of the lipid from the L_β (gel) phase into a coagel phase.

These examples illustrate the rather complex heat responses observed if the binding step is accompanied by changes in the structure and phase behavior of the lipids. The determination of the binding isotherm may be difficult or even impossible under these conditions, but ITC has the advantage of detecting even very small and subtle changes in the lipid structure. However, additional information from other techniques (e.g., spectroscopy) is needed to get a detailed molecular picture of the induced lipid structural changes.

C. Membrane-Induced Peptide Aggregation

Binding of peptides to membranes may be accompanied by a peptide–peptide aggregation driven, for example, by an enhanced local peptide concentration at the membrane, by the hydrophobic membrane environment favoring peptide–peptide aggregation via electrostatic interactions, or by a favorable preorientation of peptide at the lipid–water interface. A membrane-induced peptide–peptide aggregation was observed for some Alzheimer β-amyloid peptides (Terzi *et al.*, 1994). The 11-amino-acid fragment βAP(25–35)OH, which has similar neurotoxic and neurotrophic activities to those of the parent peptide (Yankner *et al.*, 1990), was shown to bind to negatively charged POPC/POPG (3:1) SUVs with an enthalpy change of about -2 kcal/mol. The intrinsic binding constant calculated after correction of electrostatic effects (peptide charge $z = +1.5$) by means of the Gouy–Chapman theory was about $10 \, M^{-1}$ (Terzi *et al.*, 1994). Circular dichroism (CD) experiments revealed that binding to the membrane was accompanied by a random coil $\rightarrow \beta$ sheet transition (Fig. 5). The CD spectra were similar to those measured at high peptide concentrations in the absence of lipid. ITC experiments where a high peptide concentration was injected into buffer revealed reaction enthalpies of between 1 and 3 kcal/mol, which were attributed to the β sheet $\rightarrow$ random coil transition. These data are in remarkable accordance with the lipid titration experiments, suggesting that the ΔH observed in the latter experiments mainly arises from the β-sheet formation. The enthalpy change of the actual binding reaction (without β-sheet formation) is hence close to zero. This is rather unusual, because binding of peptides and organic molecules to SUVs is normally accompanied by a distinctly exothermic heat of reaction (Beschiaschvili and Seelig, 1992; J. Seelig

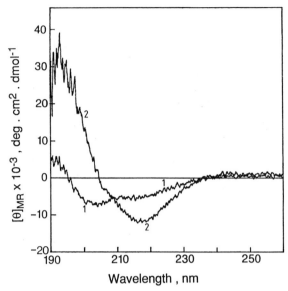

FIGURE 5   Lipid-induced random coil → β sheet transition of the Alzheimer peptide βAP(25–35)OH as detected with circular dichroism spectroscopy. (1) CD spectrum of the peptide (50 μM) in buffer (5 mM sodium acetate, pH 5.0), (2) CD spectrum of the peptide (50 μM) in the presence of 1.38 mM POPC/POPG (3:1) SUVs (buffer: 10 mM sodium acetate, pH 5.0). Taken from Terzi et al. (1994).

and Ganz, 1991), and argues against a lipid penetration of βAP(25–35)OH. Taken together, the result suggest that the membrane acts as a catalyst of the peptide–peptide association by (i) enhancing the local peptide concentration in the membrane vicinity due to an electrostatic attraction of the positively charged peptide to the negatively charged membrane and (ii) promoting a parallel alignment of the lipid molecules (Terzi et al., 1994). A membrane-induced β-sheet formation was also found for the 40-amino-acid β-amyloid analogue βAP(1–40) (Terzi et al., 1995).

D. Protonation Reaction at the Membrane Surface

Basic and acid groups in a given peptide may experience protonation/deprotonation reactions as the peptide binds to a membrane (Cevc and Marsh, 1987; Fernandez and Fromherz, 1977; Fromherz, 1989). Two different mechanisms may contribute to these reactions: (i) A translocation of peptide from water into a medium of lower dielectric constant will destabilize charged residues (e.g., $-NH_3^+$, $-COO^-$) and favor the corresponding uncharged groups. Consequently, the pK_a

value of the group will be shifted. (ii) Many biological membranes are negatively charged and hence attract protons to the membrane surface. The pH in the vicinity of a negatively charged membrane will be lower than in bulk solution. This mechanism also leads to a protonation of carboxylate groups, but will stabilize NH_3^+ groups. The two mechanisms are synergetic for COOH groups, but counteract each other for amino groups.

Titration calorimetry is a direct and elegant method for studying membrane-induced protonation or deprotonation reactions (Flogel and Biltonen, 1975; Morin and Freire, 1991; J. Seelig et al., 1993). Let us assume that binding of a peptide to a membrane is accompanied by a protonation step. Then, the measured reaction enthalpy ΔH is composed of the intrinsic binding enthalpy ΔH_1, the enthalpy change attributed to the peptide protonation ΔH_2, and the buffer dissociation enthalpy ΔH_{diss} (see Beschiaschvili and Seelig, 1992), according to the equation

$$\Delta H = \Delta H_1 + n(\Delta H_{\text{diss}} + \Delta H_2), \tag{13}$$

where n is the molar amount of protons accepted per mole of bound peptide. n can be determined if the ITC experiments are performed in a series of buffers with different dissociation enthalpies. According to Eq. (13), ΔH will vary with the buffer dissociation enthalpy, and a plot of ΔH versus ΔH_{diss} will give a straight line with the slope n.

This approach has been used to study protonation reactions coupled to membrane binding of a variety of peptides (Beschiaschvili and Seelig, 1992; J. Seelig et al., 1993; Terzi et al., 1994, 1995). As an example, Fig. 6 shows the variation of the binding enthalpy of the somatostatin analogue peptide SDZ 206–276 for POPC/POPG (3:1) SUVs with the buffer dissociation enthalpy at pH 6.25 and 7.4 (J. Seelig et al., 1993). The buffers used were Tris ($\Delta H_{\text{diss}} = 11.51$ kcal/mol), 3-(N-morpholino)propanesulfonic acid (MOPS; $\Delta H_{\text{diss}} = 5.29$ kcal/mol), and phosphate buffer ($\Delta H_{\text{diss}} = 1.22$ kcal/mol) (Morin and Freire, 1991). The peptide SDZ 206–276 has a free N-terminal amino group with a pK_a of about 7.2. At pH 6.25, the measured reaction enthalpy is independent of the type of buffer, providing evidence for a fully charged terminal amino group at this pH. In contrast, at pH 7.4 a linear variation of ΔH with the buffer dissociation enthalpy was observed with

$$\Delta H = -6.9 + 0.23 \Delta H_{\text{diss}} \quad \text{(kcal/mol)}. \tag{14}$$

The slope of $n = 0.23$ reveals that 0.23 proton is accepted per binding of one peptide molecule to the membrane (J. Seelig et al., 1993). This protonation reaction can well be explained by the reduced pH at the negatively charged membrane surface.

It should be noted that protonation/deprotonation of the N-terminus ($pK_a \approx 7.2$) is the most likely reaction accompanying binding of peptides to membranes at

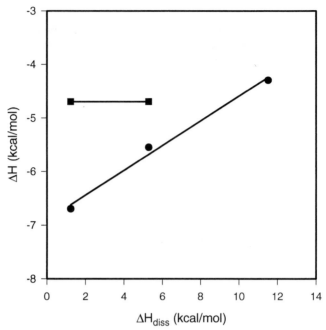

FIGURE 6 Variation of the binding enthalpy of the somatostatin analogue peptide SDZ 206–276 to POPC/POPG (3:1) SUVs with the buffer dissociation enthalpy. Measurements were made in Tris ($\Delta H_{\mathrm{diss}} = 1.22$ kcal/mol), MOPS ($\Delta H_{\mathrm{diss}} = 5.29$ kcal/mol), and phosphate buffer ($\Delta H_{\mathrm{diss}} = 11.51$ kcal/mol). ($\blacksquare$) Measurements were performed at pH 6.25. The peptide is fully protonated at this pH and the state of protonation is not changed upon membrane binding. ($\bullet$) Measurements were made at pH 7.4. The binding enthalpy is described as $\Delta H = 0.23\Delta H_{\mathrm{diss}} - 6.9$ kcal/mol, revealing that 0.23 proton is absorbed per peptide upon membrane binding. Data were taken from J. Seelig *et al.* (1993).

pH $\approx$ 7. Dissociation enthalpies of N-terminal amino groups are of the order of 11 ± 2 kcal/mol (Martin, 1964). In order to measure the intrinsic binding enthalpy ΔH_1 (without contributions from protonation/deprotonation), one should perform the measurements preferentially in Tris buffer. The ΔH_{diss} of Tris (11.51 kcal/mol) is very similar to that of N-terminal amino groups, resulting in $\Delta H_{\mathrm{diss}} + \Delta H_2 \approx 0$ and $\Delta H \approx \Delta H_1$ [cf. Eq. (13)].

V. THERMODYNAMICS OF THE MEMBRANE-INDUCED COIL-HELIX TRANSITION

Binding of peptides and proteins to biological membranes is often accompanied by a random coil–α helix transition. The main reason for a membrane-induced helix formation is the fact that the helix-forming tendency of peptides is generally low in water, but considerably larger in the hydrophobic environment of a biological membrane. Examples of peptides/proteins undergoing helix formation upon membrane binding are plasma apolipoproteins, mitochondrial presequences, virus fusion peptides, and antibacterial peptides (Anantharamaiah *et al.*, 1985; Hammen *et al.*, 1996; Keller *et al.*, 1992; Martin *et al.*, 1994; Segrest *et al.*, 1974; Wieprecht *et al.*, 1996). For these peptides, the energetics of membrane binding are superimposed on the energetics of the coil → α helix transition.

In an early study, Massey *et al.* (1979) correlated the enthalpies of interaction (determined by batch calorimetry) between different apolipoproteins and various micelle-forming lipids with the helicity change upon lipid binding. They observed a linear correlation between enthalpy and helicity change and calculated an enthalpy of helix formation of -1.3 kcal/mol per residue. Using a similar approach, they later estimated the contribution of helix formation to the enthalpy of interaction between apolipoprotein ApoA-II and phospholipids to be -2 kcal/mol per residue (Massey *et al.*, 1981). The helicity of ApoA-II was modulated in buffer by the addition of different concentrations of guanidine hydrochloride. However, the complexes formed between lipids and apolipoproteins were mostly non-bilayer structures.

Recently, we have proposed a new method for analyzing the thermodynamics of the membrane-induced α-helix formation and have applied it to the coil–helix transition of the antibacterial, amphipathic peptide magainin 2 amide (M2a) (Wieprecht *et al.*, 1999a). We have systematically varied the helix content of M2a by synthesizing analogues where two adjacent amino acid residues were substituted by their corresponding D-enantiomers. Double-D-substitution led to a local disturbance of the helical conformation, but did not modify the overall hydrophobicity and side-chain functionality (Rothemund *et al.*, 1995; Wieprecht *et al.*, 1996). In the ITC study, the parent compound M2a was compared with its analogues d4,5 M2a, d11,12 M2a, and d16,17 M2a (Wieprecht *et al.*, 1999a). The binding parameters ΔH^0, ΔG^0, and ΔS^0 of these peptides for POPC/POPG (3:1) small unilamellar vesicles (SUVs) were linearly related to the extent of helix formation (Fig. 7). From the slopes of the regression plots, an enthalpy of helix formation $\Delta H_{\text{helix}} = -0.7$ kcal/mol per residue, an entropy change $\Delta S_{\text{helix}} = -1.9$ cal/mol K per residue, and a free energy change $\Delta G_{\text{helix}} = -0.14$ kcal/mol per residue were calculated. Hence, helix formation in a membrane environment is driven by enthalpy, but opposed by entropy. From the intercepts, the thermodynamic parameters of binding of a hypothetical nonhelical magainin peptide were estimated to be $\Delta H_0 = -4.7$ kcal/mol, $\Delta G_0 = -2.4$ kcal/mol, and $\Delta S_0 = -7.4$ cal/mol K. It is important to note that this method allows an uncoupling of the conformational transition from other contributions to the binding thermodynamics. This is illustrated in Fig. 8 for the free energy of binding and helix formation of M2a. The free

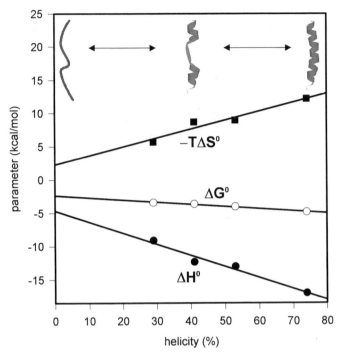

FIGURE 7 Thermodynamics of the coil → α helix transition in a membrane environment. Dependence of the enthalpy of binding ΔH^0, the free energy of binding ΔG^0, and the entropy of binding $-T\Delta S^0$ on the helicity of magainin analogue peptides in the lipid-bound state at 30°C (Wieprecht et al., 1999a).

energy of binding of M2a was found to be −4.8 kcal/mol. From the slope of the regression plot (Fig. 7), a contribution of helix formation of −2.4 kcal/mol to the free energy of binding was calculated and, as already mentioned, the free energy of binding of a hypothetically nonhelical M2a was estimated to be −2.4 kcal/mol. Therefore, the membrane-induced helix formation accounts for ∼50% of the free energy of binding and is hence a major driving force of the binding reaction.

We have repeated these investigations with large unilamellar vesicles (LUVs) with a diameter of 100 nm, which are more closely related to planar membranes (T. Wieprecht et al., in press). Despite differences in the absolute values of the thermodynamic parameters of LUVs compared to SUVs (discussed below), the slopes remained unchanged and essentially the same parameters of helix formation were found.

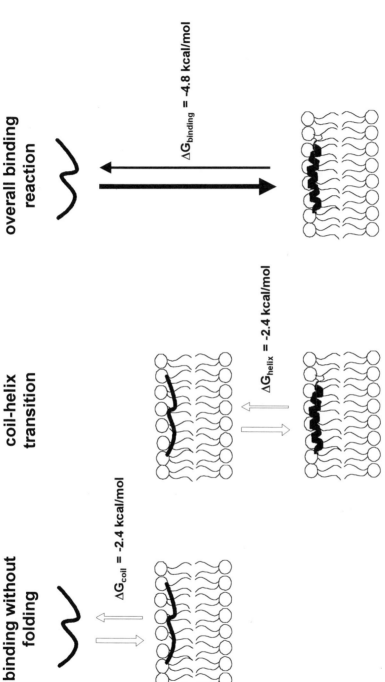

FIGURE 8 Schematic diagram of the binding of magainin 2 amide (M2a) to a lipid membrane. The figure is based on the data shown in Figs. 4 and 8. It illustrates that helix formation accounts for about 50% (−2.4 kcal/mol) of the total free energy of binding of M2a to negatively charged POPC/POPG (3:1) SUVs (−4.8 kcal/mol).

VI. THERMODYNAMICS OF PEPTIDE BINDING TO CURVED
AND PLANAR MEMBRANES

Binding of the potential-sensitive dye 2-p-toluidinylnaphthalene-6-sulfonate (TNS) to sonicated SUVs composed of phosphatidylcholine was found to be accompanied by a strong exothermic reaction enthalpy of about −9 kcal/mol as determined by gel filtration and, later, by ITC (Huang and Charlton, 1972; J. Seelig and Ganz, 1991). Because the free energy of binding was only −6.2 kcal/mol, binding was opposed by an entropy change of −11.1 cal/mol K (Huang and Charlton, 1972). This finding was in contrast to the understanding of the hydrophobic effect as the major driving force of membrane binding. A nonpolar molecule is thought to partition from water into a nonpolar phase, because its hydration shell is released upon transition (Tanford, 1980). Partition should hence be accompanied by a large positive entropy change and by only a very small enthalpy effect at room temperature. Because of the discrepancies observed for TNS–POPC binding, a "nonclassical" hydrophobic-type reaction was suggested (Huang and Charlton, 1972).

These studies were extended to a variety of chemically quite different membrane-binding molecules, including small organic molecules and peptides. For most of these substances, binding to sonicated unilamellar vesicles of ∼30 nm diameter was driven by a large exothermic heat of reaction (J. Seelig, 1997). The binding entropy was close to zero or even negative, in the latter case even opposing membrane binding. Examples are the binding of the drug amlodipine to POPC SUVs ($\Delta H = -9.2$ kcal/mol; Bauerle and Seelig, 1991), of a somatostatin peptide to POPC/POPG (3:1) SUVs ($\Delta H = -7.2$ kcal/mol; Beschiaschvili and Seelig, 1992), and of the substance P analogues (Arg^9)SP to POPC/POPG (3:1) SUVs ($\Delta H = -7.2$ kcal/mol; A. Seelig et al., 1996). The nonclassical hydrophobic effect observed with SUVs was explained by increased van der Waals interactions between the hydrophobic residue of the solute and the inner core of the lipid bilayer (J. Seelig, 1997; J. Seelig and Ganz, 1991).

Several studies have compared the thermodynamics of binding of molecules to SUVs and to LUVs (Beschiaschvili and Seelig, 1992; Keller et al., 1992; Wieprecht et al., 2000b). LUVs prepared by the extrusion technique typically have a mean diameter of ∼100 nm. They are less curved than SUVs (reduced curvature stress) and the lipid packing density is higher. LUVs are more similar to planar membranes and are hence more closely related to biological membranes. The enthalpy of binding of different substances to LUVs was found to be less exothermic than that to SUVs. For some compounds ΔH was <0 for SUVs, but >0 for LUVs. The ΔH of the binding of the somatostatin analogue SMS 201–995 was found to be −7.2 kcal/mol for SUVs, but only −1.4 kcal/mol for 100-nm LUVs. In contrast, the Gibbs free energy of binding was almost constant for both membrane systems (Beschiaschvili and Seelig, 1992). The difference in the two binding enthalpies increases with the size of the molecules investigated. One explanation of this enthalpy–entropy compensation mechanism is given by the thermoelastic properties of a bilayer. Using standard thermodynamic relations, it can be shown that the inner energy and the entropy of a bilayer vary considerably with the lipid area A_L, whereas the free energy is only slightly dependence on A_L (Beschiaschvili and Seelig, 1992; Bloom *et al.*, 1991; Davies and Jones, 1992). The change in the inner energy U accompanying (reversible) expansion of a bilayer at constant temperature is given by

$$(\delta U/\delta A_L)_T = -\pi + (\delta Q/\delta A_L)_T = -\pi + T\alpha/\chi = \pi_i, \qquad (15)$$

where π is the surface pressure, δQ is the heat of expansion, α is the area expansivity at constant membrane tension, χ is the isothermal area compressibility, and π_i is the internal tension. Two assumptions were made (Beschiaschvili and Seelig, 1992): (i) Binding of SMS 201–995 to bilayers leads to a small disordering of the acyl chains, thereby slightly increasing the area per lipid A_L. (ii) The internal tension π_i, which is a measure of the cohesive forces in a sample, is about 20% larger in well-packed LUVs than in highly curved SUVs. For typical values of π_i, this corresponds to a difference in the internal tension between LUVs and SUVs of about $\Delta\pi_i \approx -70\ \text{mJ/m}^2$. It should be mentioned that $\Delta\delta U \approx \Delta\delta Q = \Delta H_{\text{LUVs}} - \Delta H_{\text{SUVs}}$, because the difference between the membrane tension of LUVs and SUVs, $\Delta\pi$, is much smaller than the difference in the internal tension, $\Delta\pi_i$ [cf. Eq. (13)]. Assuming δA_L to be approximately the same in LUVs and SUVs, one can explain the differences in the binding enthalpies between LUVs and SUVs by differences in the internal tension of both model membranes (Beschiaschvili and Seelig, 1992). However, it should be mentioned that this model assumes an *endothermic* contribution to ΔH for both SUVs and LUVs arising from the lipid area increase. Since the overall ΔH values are negative, other significant contributions must exist, which are responsible for the negative overall binding enthalpies. Similar models have been used to account for the enthalpy–entropy compensation mechanisms found for binding of apolipoprotein model peptides (Gazzara *et al.*, 1997a) to POPC vesicles and of M2a to POPC/POPG (3:1) vesicles at 45°C (Wieprecht *et al.*, 2000b).

It should be noted that the thermodynamic binding parameters of the latter peptides are superimposed with a coil–helix transition accompanying membrane binding. In Section V we summarized investigations which allowed an uncoupling of the thermodynamic parameters of the coil–helix transition from the actual binding reaction. The enthalpy of binding of a hypothetical nonhelical M2a (binding without folding) to SUVs was estimated to be $\Delta H_0 = -4.7$ kcal/mol (Wieprecht *et al.*, 1999a), whereas the corresponding value for LUVs was $+10.5$ kcal/mol (T. Wieprecht *et al.*, in press). Again, the free energy of binding was very similar for SUVs and LUVs (-2.4 and -2.0 kcal/mol, respectively). Therefore, while the binding of the nonhelical peptide to SUVs is driven by enthalpy, but opposed by entropy, the opposite holds true for LUVs. A qualitative explanation for these differences in the binding thermodynamics can again be sought in the different cohesive forces in SUVs and LUVs, as discussed above.

VII. CONCLUDING REMARKS

Isothermal titration calorimetry has become a powerful tool for the study of binding reactions between peptides/proteins and lipid bilayers. ITC allows a simultaneous determination of the enthalpy of binding and the binding isotherm. Provided that a proper binding model can be found, all relevant thermodynamic parameters of the binding reaction are easily accessible. In addition to measuring the actual binding process, ITC is also valuable for the study of secondary effects accompanying membrane binding. ITC has proven to be extremely sensitive in detecting changes and differences in bilayer structure, such as differences between curved and planar membranes, peptide-induced lipid-phase transitions, or membrane permeabilization. Titration calorimetry can also detect protonation and deprotonation of peptides upon lipid binding. Furthermore, the method can be used to study the thermodynamics of membrane-induced peptide and protein folding reactions.

References

Anantharamaiah, G. M., Jones, J. L., Brouillette, C. G., Schmidt, C. F., Chung, B. H., Hughes, T. A., Bhown, A. S., and Segrest, J. P. (1985). Studies of synthetic peptide analogs of the amphipathic helix: Structure of complexes with dimyristoyl phosphatidylcholine. *J. Biol. Chem.* **260**, 10248–10255.

Aveyard, R., and Haydon, D. A. (1973). "An Introduction to the Principles of Surface Chemistry." Cambridge University Press, Cambridge.

Bauerle, H. D., and Seelig, J. (1991). Interaction of charged and uncharged calcium channel antagonists with phospholipid membranes: Binding equilibrium, binding enthalpy, and membrane location. *Biochemistry* **30**, 7203–7211.

Beschiaschvili, G., and Seelig, J. (1992). Peptide binding to lipid bilayers Nonclassical hydrophobic effect and membrane-induced pK shifts. *Biochemistry* **31**, 10044–10053.

Bloom, M., Evans, E., and Mouritsen, O. G. (1991). Physical properties of the fluid lipid-bilayer component of cell membranes: A perspective. *Q. Rev. Biophys.* **24**, 293–397.

Boots, J. W., Chupin, V., Killian, J. A., Demel, R. A., and de Kruijff, B. (1999). Interaction mode specific reorganization of gel phase monoglyceride bilayers by beta-lactoglobulin. *Biochim. Biophys. Acta* **1420**, 241–251.

Cantor, C. R., and Schimmel, P. R. (1980). "Biophysical Chemistry." Freeman, San Francisco.

Cevc, G., and Marsh, D. (1987). "Phospholipid Bilayers. Physical Principles and Models." Wiley-Interscience, New York.

Davies, R. J., and Jones, M. N. (1992). The thermal behaviour of phosphatidylcholine–glycophorin monolayers in relation to monolayer and bilayer internal pressure. *Biochim. Biophys. Acta* **1103**, 8–12.

Epand, R. M., and Sturtevant, J. M. (1981). A calorimetric study of peptide–phospholipid interactions: The glucagon–dimyristoylphosphatidylcholine complex. *Biochemistry* **20,** 4603–4606.

Epand, R. M., Segrest, J. P., and Anantharamaiah, G. M. (1990). Thermodynamics of the binding of human apolipoprotein A-I to dimyristoylphosphatidylglycerol. *J. Biol. Chem.* **265,** 20829–20832.

Fernandez, M. S., and Fromherz, P. (1977). *J. Phys. Chem.* **81,** 1755–1761.

Flogel, M., and Biltonen, R. L. (1975). The pH dependence of the thermodynamics of the interaction of 3'-cytidine monophosphate with ribonuclease A. *Biochemistry* **14,** 2610–2615.

Fromherz, P. (1989). Lipid coumarin dye as a probe of interfacial electrical potential in biomembranes. *Meth. Enzymol.* **171,** 376–387.

Gazzara, J. A., Phillips, M. C., Lund-Katz, S., Palgunachari, M. N., Segrest, J. P., Anantharamaiah, G. M., Rodrigueza, W. V., and Snow, J. W. (1997a). Effect of vesicle size on their interaction with class A amphipathic helical peptides. *J. Lipid. Res.* **38,** 2147–2154.

Gazzara, J. A., Phillips, M. C., Lund-Katz, S., Palgunachari, M. N., Segrest, J. P., Anantharamaiah, G. M., and Snow, J. W. (1997b). Interaction of class A amphipathic helical peptides with phospholipid unilamellar vesicles. *J. Lipid. Res.* **38,** 2134–2146.

Hammen, P. K., Gorenstein, D. G., and Weiner, H. (1996). Amphiphilicity determines binding properties of three mitochondrial presequences to lipid surfaces. *Biochemistry* **35,** 3772–3781.

Heimburg, T., and Biltonen, R. L. (1994). Thermotropic behavior of dimyristoylphosphatidylglycerol and its interaction with cytochrome *c*. *Biochemistry* **33,** 9477–9488.

Huang, C. H., and Charlton, J. P. (1972). Interactions of phosphatidylcholine vesicles with 2-*p*-toluidinylnaphthalene-6-sulfonate. *Biochemistry* **11,** 735–740.

Keller, R. C., Killian, J. A., and de Kruijff, B. (1992). Anionic phospholipids are essential for α-helix formation of the signal peptide of prePhoE upon interaction with phospholipid vesicles. *Biochemistry* **31,** 1672–1677.

Martin, B. R. (1964). "Introduction to Biophysical Chemistry." McGraw-Hill, New York.

Martin, I., Dubois, M. C., Defrise-Quertain, F., Saermark, T., Burny, A., Brasseur, R., and Ruysschaert, J. M. (1994). Correlation between fusogenicity of synthetic modified peptides corresponding to the NH_2-terminal extremity of simian immunodeficiency virus gp32 and their mode of insertion into the lipid bilayer: An infrared spectroscopy study. *J. Virol.* **68,** 1139–1148.

Massey, J. B., Gotto, A. M., Jr., and Pownall, H. J. (1979). Contribution of α helix formation in human plasma apolipoproteins to their enthalpy of association with phospholipids. *J. Biol. Chem.* **254,** 9559–9561.

Massey, J. B., Gotto, A. M., Jr., and Pownall, H. J. (1981). Thermodynamics of lipid–protein interactions: Interaction of apolipoprotein A-II from human plasma high-density lipoproteins with dimyristoylphosphatidylcholine. *Biochemistry* **20,** 1575–1584.

Matsuzaki, K., Murase, O., Fujii, N., and Miyajima, K. (1996). An antimicrobial peptide, magainin 2, induced rapid flip-flop of phospholipids coupled with pore formation and peptide translocation. *Biochemistry* **35,** 11361–11368.

McLaughlin, S. (1977). Electrostatic potentials at membrane–solution interfaces. *Curr. Top. Membr. Transp.* **9,** 71–144.

McLaughlin, S. (1989). The electrostatic properties of membranes. *Annu. Rev. Biophys. Biophys. Chem.* **18,** 113–136.

Morin, P. E., and Freire, E. (1991). Direct calorimetric analysis of the enzymatic activity of yeast cytochrome *c* oxidase. *Biochemistry* **30,** 8494–8500.

Myers, M., Mayorga, O. L., Emtage, J., and Freire, E. (1987). Thermodynamic characterization of interactions between ornithine transcarbamylase leader peptide and phospholipid bilayer membranes. *Biochemistry* **26,** 4309–4315.

Ramsay, G., Prabhu, R., and Freire, E. (1986). Direct measurement of the energetics of association between myelin basic protein and phosphatidylserine vesicles. *Biochemistry* **25,** 2265–2670.

Rothemund, S., Beyermann, M., Krause, E., Krause, G., Bienert, M., Hodges, R. S., Sykes, B. D., and Sonnichsen, F. D. (1995). Structure effects of double D-amino acid replacements: A nuclear magnetic resonance and circular dichroism study using amphipathic model helices. *Biochemistry* **34**, 12954–12962.

Seelig, A., Alt, T., Lotz, S., and Holzemann, G. (1996). Binding of substance P agonists to lipid membranes and to the neurokinin-1 receptor. *Biochemistry* **35**, 4365–4374.

Seelig, J. (1997). Titration calorimetry of lipid–peptide interactions. *Biochim. Biophys. Acta* **1331**, 103–116.

Seelig, J., and Ganz, P. (1991). Nonclassical hydrophobic effect in membrane binding equilibria. *Biochemistry* **30**, 9354–9359.

Seelig, J., Nebel, S., Ganz, P., and Bruns, C. (1993). Electrostatic and nonpolar peptide–membrane interactions: Lipid binding and functional properties of somatostatin analogues of charge $z = +1$ to $z = +3$. *Biochemistry* **32**, 9714–9721.

Segrest, J. P., Jackson, R. L., Morrisett, J. D., and Gotto, A. M., Jr. (1974). A molecular theory of lipid–protein interactions in the plasma lipoproteins. *FEBS Lett.* **38**, 247–258.

Tanford, F. (1980). "The Hydrophobic Effect: Formation of Micelles and Biological Membranes." Wiley, New York.

Terzi, E., Holzemann, G., and Seelig, J. (1994). Alzheimer beta-amyloid peptide 25–35: Electrostatic interactions with phospholipid membranes. *Biochemistry* **33**, 7434–7441.

Terzi, E., Holzemann, G., and Seelig, J. (1995). Self-association of beta-amyloid peptide (1–40) in solution and binding to lipid membranes. *J. Mol. Biol.* **252**, 633–642.

Turner, D. C., Straume, M., Kasimova, M. R., and Gaber, B. P. (1995). Thermodynamics of interaction of the fusion-inhibiting peptide Z-D-Phe–L-Phe–Gly with dioleoylphosphatidylcholine vesicles: Direct calorimetric determination. *Biochemistry* **34**, 9517–9525.

Wenk, M. R., and Seelig, J. (1998). Magainin 2 amide interaction with lipid membranes: Calorimetric detection of peptide binding and pore formation. *Biochemistry* **37**, 3909–3916.

Wieprecht, T., Dathe, M., Schumann, M., Krause, E., Beyermann, M., and Bienert, M. (1996). Conformational and functional study of magainin 2 in model membrane environments using the new approach of systematic double-D-amino acid replacement. *Biochemistry* **35**, 10844–10853.

Wieprecht, T., Apostolov, O., Beyermann, M., and Seelig, J. (1999a). Thermodynamics of the alpha-helix–coil transition of amphipathic peptides in a membrane environment: Implications for the peptide-membrane binding equilibrium. *J. Mol. Biol.* **294**, 785–794.

Wieprecht, T., Beyermann, M., and Seelig, J. (1999b). Binding of antibacterial magainin peptides to electrically neutral membranes: Thermodynamics and structure. *Biochemistry* **38**, 10377–10387.

Wieprecht, T., Apostolov, O., Beyermann, M., and Seelig, J. (2000a). Membrane binding and pore formation of the antibacterial peptide PGLa: Thermodynamic and mechanistic aspects. *Biochemistry* **39**, 442–452.

Wieprecht, T., Apostolov, O., and Seelig, J. (2000b). Binding of the antibacterial peptide magainin 2 amide to small and large unilamellar vesicles. *Biophys. Chem.* **85**, 187–198.

Wieprecht, T., Beyermann, M., and Seelig, J. (2002). Thermodynamics of the coil-α-helix transition of amphipathic peptides in a membrane environment: the role of vesicle curvature. *Biophys. Chem.* In press.

Wiseman, T., Williston, S., Brandts, J. F., and Lin, L. N. (1989). Rapid measurement of binding constants and heats of binding using a new titration calorimeter. *Analyt. Biochem.* **179**, 131–137.

Yankner, B. A., Duffy, L. K., and Kirschner, D. A. (1990). Neurotrophic and neurotoxic effects of amyloid beta protein: Reversal by tachykinin neuropeptides. *Science* **250**, 279–282.

Zhang, F., and Rowe, E. S. (1994). Calorimetric studies of the interactions of cytochrome *c* with dioleoylphosphatidylglycerol extruded vesicles: Ionic strength effects. *Biochim. Biophys. Acta* **1193**, 219–225.

CHAPTER 3

Infrared Reflection–Absorption Spectroscopy of Lipids, Peptides, and Proteins in Aqueous Monolayers

Richard Mendelsohn and Carol R. Flach

Department of Chemistry, Newark College, Rutgers University, Newark, New Jersey 07102

I. Introduction
II. Experimental Approaches
    A. Basic Concepts
    B. Instrumentation
    C. Reduction of Spectral Interference from Water Vapor
    D. Langmuir Films
III. Information from IRRAS Measurements
    A. Frequencies
    B. Intensities
    C. IRRAS Studies of Molecular Conformation and Interactions in Monolayer Films
IV. IRRAS Applications to Lipid/Protein Langmuir Films
    A. The Pulmonary Surfactant System
    B. Further IRRAS Applications to Peptide and Proteins in Monolayer Films
V. Current State and Future Prospects for IRRAS
    References

Infrared reflection–absorption spectroscopy (IRRAS) provides a unique means of monitoring the chain conformation and orientation of lipids as well as the secondary structure and orientation of peptides and proteins *in situ* in monolayers at the air/water interface. This chapter describes the experimental approaches and instrumentation used to acquire IRRAS spectra. Spectra–structure correlations are reviewed. Several applications of the technique are discussed, including determinations of lipid chain conformation, lipid headgroup interactions, and protein and peptide secondary structure.

Whereas the measured IRRAS frequencies may be correlated with molecular structure information in the usual way, determination of molecular orientation requires a detailed analysis of the reflected light intensities. Simulations included in this chapter show that the infrared (IR) bands may be distorted from their transmission spectroscopy bandshapes, especially when measurements are made at angles of incidence near the Brewster angle. The measured intensities have been quantitatively used to monitor the tilt angle of the hydrophobic pulmonary surfactant protein SP-C in lipid monolayers. Additional applications to peptide orientation are reviewed.

I. INTRODUCTION

Monolayers at the air/water interface are widely employed as convenient experimental paradigms that mimic many vital biological processes. These include the adsorption of pulmonary surfactant to the air/alveolar lining, the interaction of peripheral membrane proteins with phospholipids, and the mechanism of action of enzyme classes such as the phospholipases. The advantages realized when using monolayers as models for biological interfaces arise primarily from the ease with which experimental variables may be manipulated. These include parameters that are not readily controlled in bulk phases or in films prepared on solid substrates, such as lateral pressure, surface area, and domain size and shape.

Since the days of Langmuir, a variety of methods have been employed to evaluate the physical state of monolayer films. These are summarized in Table I (references are included for each method) and tend to fall into three major classes:

1. Techniques that measure surface film thermodynamics, thickness, and phase behavior: surface pressure–molecular area (π–A) isotherms, ellipsometry, surface potential, and X-ray reflectivity
2. Techniques that monitor domain structure, chain tilt angles, and phase structure: fluorescence and Brewster angle microscopies and X-ray and neutron reflectivities
3. Techniques based on vibrational spectroscopy, which provide information about the molecular conformation of film constituents: infrared reflection–absorption spectroscopy (IRRAS) and the nonlinear approaches of sum frequency and second harmonic generation

Whereas the determination of film thermodynamics and domain structure (classes 1 and 2 above) are necessary for understanding the physics of monolayer films and for the construction of phase diagrams, it seems fair to note that the continuing biochemical utility of lipid and protein films as experimental monolayer models requires application and refinement of the vibrational optical spectroscopic

TABLE I
Techniques for the Physical Characterization of Monolayer Films

Physical methods	Structural information	Reference
Pressure–area isotherms	Surface thermodynamics, molecular areas, phase transitions, compressibilities	Phillips and Chapman, 1968; Kaganer et al., 1999
Surface potential	Film homogeneity plus limited information about molecular orientation	Oliviera and Bonardi, 1997
Surface viscosity	Viscoelasticity of films	Gaines, 1966
Ellipsometry	Film optical constants, phase transitions	Azzam and Bashara, 1987
X-ray reflectivity	In-plane molecular order, molecular orientation, and subcell structure	Als-Nielsen et al., 1994; Kaganer et al., 1999; Möhwald, 1990
Neutron reflectivity	Surface densities, domain formation	Thomas and Penfold, 1996
Brewster angle microscopy	Domain structure, size distribution	Möbius, 1998; Lautz et al., 1998
Fluorescence microscopy	Domain organization	McConnell, 1991
Sum frequency generation and second harmonic generation	Chain conformational order, liquid–liquid interfaces	Richmond, 1997; Shen, 1997; Bain, 1998
IRRAS	Chain conformation and orientation, protein/peptide secondary structure, functional group orientation	See text

techniques. The capability of these methods to directly monitor molecular structure elements, such as lipid chain conformation, orientation, and headgroup interactions, along with protein/peptide secondary structure and orientation, makes their contribution unique.

Until the mid 1980s, techniques for the determination of film molecular properties were unavailable. At that time, Dluhy and his co-workers (Dluhy, 1986; Dluhy and Cornell, 1985; Dluhy et al., 1988; Mitchell and Dluhy, 1988) demonstrated the feasibility of obtaining IR spectra from Langmuir films of fatty acids and phospholipids in situ at the air/water interface. Spectra of these classes of molecules could then be acquired as a function of physiologically relevant variables. IR data interpretation for hydrocarbon chain-bearing amphiphiles is facilitated by the availability of extensive correlations between the chain structure and the IR spectra of the molecules (Mendelsohn and Mantsch, 1986; Mendelsohn et al., 1995a). In addition to the structural information available from the hydrophobic regions of these molecules, several groups demonstrated the sensitivity of the headgroup modes to ionic interactions (Flach et al., 1993; Gericke and Hühnerfuss, 1994, 1995; Hunt et al., 1989). The next generation of IRRAS applications involved determination of peptide or protein secondary structure in these films (Cornut et al.,

TABLE II

IR Modes Observable in IRRAS Spectra

Mode	Wavenumber (cm^{-1})	Sensitivity/structural information
Lipids		
Chain modes		
CH_2 symmetric stretch	2848–2854	The frequencies are qualitative markers of conformational
CH_2 asymmetric stretch	2916–2924	disorder; in addition, solid-solid transitions due to
CD_2 symmetric stretch	2090–2100	packing changes may be inferred if the final frequency
CD_2 asymmetric stretch	2195–2200	of the CH_2 symmetric stretch is $<2850\ cm^{-1}$
CH_2 scissoring	1462, 1473	Orthorhombic phase doublet (perpendicular subcell packing)
	1468	Hexagonal or triclinic phase
CD_2 scissoring	1086, 1094	Orthorhombic phase doublet (perpendicular subcell packing)
	1089	Hexagonal or triclinic phases
Polar region modes		
PO_2^- asymmetric stretch	1220–1250	The frequency is sensitive to ion binding/hydration
PO_2^- symmetric stretch	~1090	—
C=O stretch (fatty acids)	1690–1740	The frequency is sensitive to protonation state
C=O stretch of esters	1710–1740	The frequency is sensitive to hydration and hydrogen bonding
Peptide bond modes		
Amide I (mostly C=O stretch)	1610–1690	Frequency reflects secondary structure, as follows: (a) α helix (unexchanged, unhydrated) ~1655 cm^{-1}, hydration can lower the frequency by up to ~20 cm^{-1}; (b) antiparallel β sheet, doublet (1620–1630 cm^{-1}, strong; 1680–1690 cm^{-1}, weak); (c) random coil (1640–1650 cm^{-1}); hydration, H→D exchange lower the frequencies
Amide II (N—H in-plane bend + C—N stretch)	1520–1560	Primarily used to monitor H→D exchange
Amide A (N—H stretch)	3200–3400	Limited sensitivity to secondary structure

1996; Flach *et al.,* 1994; Pastrana-Rios *et al.,* 1995). Acquisition of data from the conformation-sensitive regions of protein spectra required the development of methods for eliminating interference from water vapor absorption (Buffeteau *et al.,* 1991; Flach *et al.,* 1994). In addition, theoretical approaches for the analysis of IRRAS intensities were adapted for the determination of molecular and functional group orientation. A summary of the spectra–structure correlations for those vibrational modes of lipids and proteins currently accessible by IRRAS is given in Table II.

This chapter describes the experimental approaches and instrumentation used to acquire IRRAS spectra, and their application to the study of lipid/peptide and lipid/protein films in monolayers. We thus confine the discussion to Langmuir (aqueous) films. Applications of IR spectroscopy to Langmuir–Blodgett or multi-lamellar films on solid surfaces using experimental approaches such as attenuated total reflectance (ATR) infrared spectroscopy have been reviewed elsewhere (Axelsen and Citra, 1997; Tamm and Tatulian, 1997). Detailed listings of published articles have appeared in recent reviews (Blaudez *et al.*, 1999; Dluhy *et al.*, 1995; Mendelsohn and Flach, in press; Mendelsohn *et al.*, 1995a; Ohe *et al.*, 1999). This chapter emphasizes applications to the determination of lipid, peptide, and protein structure and orientation.

II. EXPERIMENTAL APPROACHES

A. Basic Concepts

The basic idea underlying an IRRAS experiment is straightforward. When IR radiation illuminates an aqueous film, a small fraction of the incident light is reflected. The reflected intensity depends on the experimental geometry (angle of incidence and state of polarization of the radiation), the optical constants (real and imaginary parts of the refractive indices) of the film and subphase, and the orientation of the transition moments. The sample vibrational modes (i.e., the IR spectrum of the film constituents) appear in the spectrum of reflected light. IRRAS spectra are presented as plots of reflectance–absorbance (RA) versus wavenumber. Reflectance–absorbance is defined as $-\log_{10}(R/R_o)$, where R is the reflectivity of the film-covered surface and R_o is the reflectivity of the water. In normal transmission spectroscopy, absorbance is a positive number. In contrast, RA values in IRRAS measurements may take on positive or negative values, depending on the optical parameters of the experiment, as noted above.

B. Instrumentation

A schematic of the IRRAS instrument currently in use at Rutgers is shown in Fig. 1. Whereas details of design vary among the dozen or so groups around the world utilizing this technique, a minimal set of component requirements is represented in the figure. The Langmuir trough must be of dimensions sufficiently large to reduce surface curvature by elimination of the meniscus. For Teflon troughs, this linear dimension is $\sim$6 cm. In addition, the trough should incorporate one or more of the following features for particular experimental applications: an adjustable barrier for control of surface area, a Wilhelmy plate for surface pressure measurements, temperature control of the subphase, humidity control of the environment, and an injection port for protein adsorption

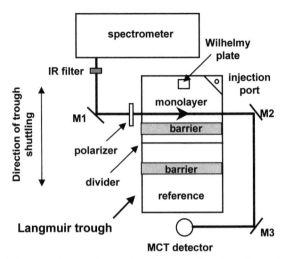

FIGURE 1 Schematic of a typical IRRAS optical setup. For adsorption studies, proteins may be introduced into the subphase through the injection port of the Langmuir trough. For reduction of spectral interference from water vapor, the trough is shuttled between two fixed positions along the direction indicated. Spectra are co-added in alternating fashion between the film-containing and reference surfaces. In practice, the entire apparatus is under dry air or N_2 purge. Ambient and subphase temperature and relative humidity may be controlled as required.

experiments. The trough and entire optical setup should be enclosed and purged with either nitrogen gas or humidity-controlled air. This is required as part of the protocol for studies of proteins and peptides in monolayers to reduce interference in the 1500- to 1800-cm^{-1} region from the rotation–vibration spectrum of water vapor.

The optical components should be mounted so that the angle of incidence may be varied between 20° and 70°, and a position in the optical train should be available for insertion of a polarizer. In addition, the nature of the reflectivity from an aqueous subphase for the two components of plane-polarized light is such that the polarizer must be of extremely high quality. The following numerical example emphasizes this point. The standard nomenclature used is as follows: p-polarized (parallel) radiation has its electric vector oscillating in the plane of incidence, whereas s-polarized (perpendicular) radiation has its electric vector perpendicular to the plane of incidence. The Brewster angle for external reflection at an air/water interface is the arctangent of the real part of the refractive index of the subphase and defines the angle of incidence resulting in minimum reflectivity for the p-polarized component of the incident radiation. The intensity of the reflected p-polarized light close to the Brewster angle may be 100 times weaker than the s-polarized reflected intensity. If the polarizer permits 1% of s-polarized light to be transmitted when the polarizer is oriented to transmit "only" p-polarized radiation, then the putatively pure *p*-polarized component will in fact contain an equal amount of the unwanted *s*-polarized component. The situation evidently worsens for polarizers of poorer quality. It is important to measure the polarizer efficiency across the spectral range of interest. A final requirement for polarization measurements is that the beam divergence needs to be controlled to $\leq 1°$ with an iris diaphragm.

C. Reduction of Spectral Interference from Water Vapor

The acquisition of protein IRRAS spectra presents a major technical challenge. The problem arises because the absorption spectrum of the rotation–vibration bands of water vapor overlap the spectral regions (1500–1700 cm^{-1}) where the conformation-sensitive amide I and II modes of the peptide bond appear. Substitution of an H$_2$O subphase with D$_2$O lessens, but does not eliminate, the interference. Although the rotation–vibration bands of D$_2$O are indeed shifted to lower frequency compared with those of H$_2$O, the inevitable presence of HDO (due to isotope exchange processes) and residual atmospheric water vapor results in overlapped bands. Consequently, a great deal of attention has been given to elimination of the water vapor signal. Two approaches to the problem have been successful (Blaudez *et al.*, 1994; Buffeteau *et al.*, 1991; Flach *et al.*, 1994).

The Rutgers group (Flach *et al.*, 1994) constructed a sample shuttle (see Fig. 1) in which two Langmuir troughs are used in tandem, or equivalently, a single trough is divided by a barrier separating a film-covered region from a film-free surface. The sample channel holds the film-covered subphase and the reference channel contains only the subphase. A computer-controlled direct drive (servo) motor is used to acquire IRRAS spectra from the two channels. Signals from the light reflected from each channel are co-added separately, then ratioed to provide the RA values for IRRAS spectra. The advantage of the shuttle arrangement device is shown in Fig. 2. In the figure, IRRAS spectra of the amide I region from a monolayer of the pulmonary surfactant protein SP-A is shown with two different shuttling protocols. The peptide amide I mode (~1650 cm^{-1}) is in general overlapped by many sharp features arising from the rotation–vibration spectrum of water vapor, which are quite visible in the bottom trace. These features are reduced by at least an order of magnitude as a result of the dual-channel shuttle operation and by selection of the appropriate cycle of co-additions. For the example shown in Fig. 2 (top), excellent water vapor compensation was achieved with eight cycles of 128 scans each from the reference and sample compartments of the trough. With this approach, protein amide I features as weak as 0.3 mRA may be detected.

The second approach, developed by the Bordeaux group (Blaudez *et al.*, 1994; Buffeteau *et al.*, 1991), uses a polarization modulation (PM) technique and is termed PM-IRRAS. The polarization of the incident electric field is rapidly

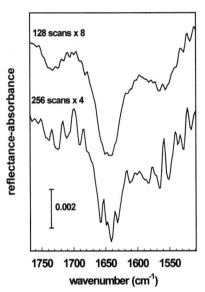

FIGURE 2 Compensation of water vapor. The bottom trace shows the IRRAS spectrum of pulmonary surfactant SP-A in which water vapor has been poorly compensated by co-addition of four cycles of 256 scans from the sample and reference portions of the Langmuir trough. The sharp peaks in this spectrum are the rotation–vibration bands of vapor-phase water in the optical path. The top trace shows the improvement gained in this instance from co-addition of eight cycles of 128 scans. The vertical line is 2 mRA units.

modulated between s- and p-polarized radiation, and the field is reflected from the water surface. The reflected beam signal is electronically filtered and demodulated with a lock-in amplifier. Following Fourier transformation, a differential reflectivity spectrum is computed as $\Delta R/R = (R_p - R_s)/(R_p + R_s)$. A detailed discussion of the theory and practical details of the approach is presented in the references noted above.

Each method of water vapor suppression has its advantages and disadvantages. The incorporation of a shuttle system is technically straightforward, is easily automated, and requires minimal modifications of the optical path of the spectrometer. The dual-trough arrangement tends to enlarge the volume that has to be purged and to increase data collection times. In addition, the shuttle must be operated with a direct drive motor (stepping motors tend to disturb the film) at sufficiently slow speeds to preclude disturbance of the film.

PM-IRRAS requires the incorporation of additional, fairly expensive elements in the optical system. Polarization modulation efficiency is perfect only at a single frequency, so that perfect compensation of water vapor cannot be achieved in practice, although the suppression is generally very good. Because PM-IRRAS suppresses signals from randomly oriented molecules in the beam, it also tends to suppress vibrations from randomly oriented species in the monolayer film.

D. Langmuir Films

The physical characteristics of monolayers may be altered by the means used for monolayer formation and compression. When a sufficient amount of a surface-active material is placed at the air/water interface, it spreads until the surface pressure reaches a constant value, the equilibrium spreading pressure. When isotherms are to be acquired for insoluble monolayers, the molecules should be spread over a large enough surface area that the surface pressure is much less than the equilibrium spreading pressure. In reality, this is not always possible, due to the nature of the film constituents and experimental constraints, such as the geometry of the trough. For example, when phosphatidylcholines (PC) and phosphatidylethanolamines (PE) are spread from hexane–ethanol mixtures, if the hydrocarbon chains are sufficiently long, condensed monolayers are formed, whereas shorter chains tend to produce expanded films. Transitions between expanded and condensed states may be induced. For example, upon compression of 1,2-dipalmitoylphosphatidylcholine (DPPC) films at 22°C, the monolayers undergo a transition from an expanded to a condensed film. The thermodynamics of these transitions has been extensively studied. An excellent recent review has been presented by Kaganer *et al.* (1999).

The method by which Langmuir films are formed can substantially affect their molecular properties. IRRAS has been used to investigate differences in molecular structure resulting from the application of two widely used approaches for spreading and/or compression (Gericke *et al.,* 1993a,b). In the single-shot method, a known amount of sample is spread to a fixed surface area. New monolayers are spread to cover a range of molecular areas. In the discontinuous compression method, a monolayer is spread over relatively large molecular areas and the barrier is moved to a particular location, where measurements are begun after the pressure has become constant. Intermittent film compression continues as different positions along the π–A isotherm are sampled. Polarized IRRAS measurements were used to determine chain tilt angles for amphiphilic monolayers relative to the surface normal. For monolayers of 1-hexadecanol prepared by the single-shot method, smaller tilt angles were found than for discontinuous compression, even after allowing relatively long times (80 min) for film relaxation. In contrast, the conformational order of the acyl chains determined from methylene stretching frequencies (see following section) showed a high degree of order in both cases. Similar experiments with DPPC show different behavior than with the alcohol. DPPC monolayers spread in the liquid condensed (LC)/liquid expanded (LE) coexistence region using the single-shot method relax to a more ordered state in terms of tilt angle and acyl chain conformation, without, however, reaching the same degree of conformational order as when discontinuous compression is applied. The method of film preparation is probably best viewed as an experimental variable and must be carefully considered in any IRRAS investigation of monolayer structure.

III. INFORMATION FROM IRRAS MEASUREMENTS

IRRAS measurements provide the two basic types of information traditionally acquired from molecular infrared absorption spectroscopy: the frequencies of molecular vibrations and the band intensities (transmittance or absorbance), which in IRRAS are calculated as reflectance–absorbance intensities. These two experimental observables are discussed in turn.

A. Frequencies

As is well known, IR radiation is absorbed by molecular vibrations, provided that the oscillating nuclei produce a dipole moment change during the normal mode. Due primarily to anomalous dispersion effects in the real part of the refractive index, the absorption frequencies measured from IRRAS measurements at angles of incidence far from the Brewster angle may differ slightly from those observed in transmission studies of the same material. These effects are usually small (<2–3 cm^{-1}). In contrast, for angles of incidence near the Brewster angle, the bandshapes may be dramatically distorted by the appearance of derivative-like features. These effects are illustrated in spectral simulations of lipid CH$_2$ symmetric stretching and protein amide I modes in Fig. 3, as discussed below. For lipids and proteins, the observed vibrational frequencies may be quite sensitive to molecular conformation (see Table II). This sensitivity provides the basic reason for the utility of IR spectroscopy in structural studies of biological materials.

FIGURE 3 Simulations of IRRAS spectral bands, as follows: (A) CH$_2$ symmetric stretch (*s*-polarization) at five angles of incidence as shown. The initial frequency was assumed to be 2850 cm^{-1}, and the chains were assumed to be 30 Å long and tilted at 28° to the surface normal. The real parts of the film refractive index were assumed to be 1.41. The subphase (H$_2$O) refractive index was calculated from Bertie *et al.* (1989). The transition moment is perpendicular to the chain direction. A Lorentzian lineshape (10 cm^{-1} halfwidth) was assumed. (B) CH$_2$ symmetric stretch (*p*-polarization) at five angles of incidence, as shown. Parameters are as in A. (C) Amide I band (*p*-polarization) of an α helix on an H$_2$O subphase at five angles of incidence, with the helix axis oriented perpendicular to the interface. The amide I transition moment is assumed to lie at 28° from the C=O bond direction in the peptide group plane. Other parameters as in A. Initial frequency was set at 1650 cm^{-1}. (D) Amide I band (*p*-polarization) of an α-helix on a D$_2$O subphase at five angles of incidence, with the helix axis oriented perpendicular to the interface. Other parameters as in A and C.

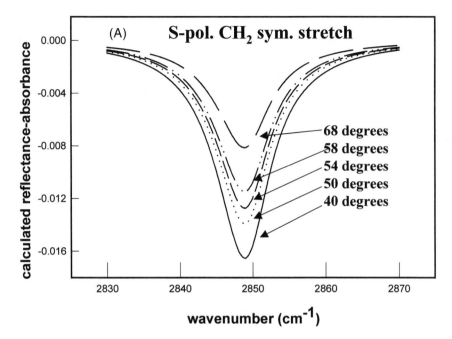

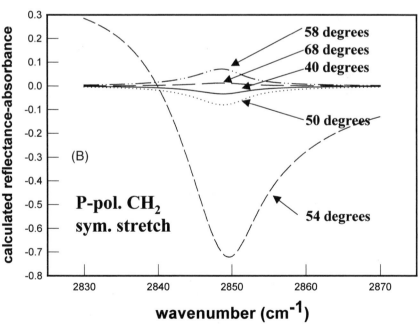

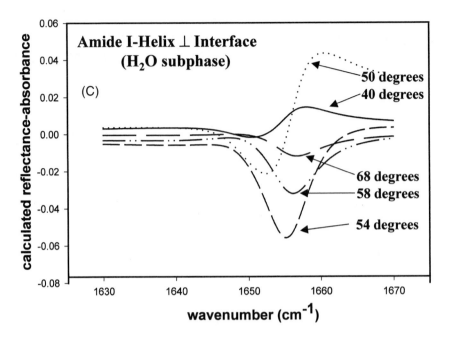

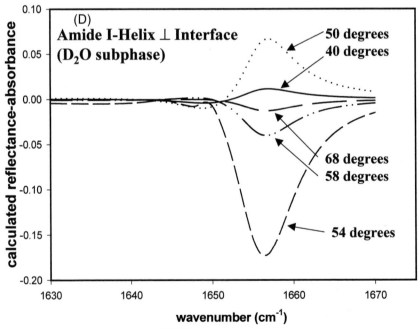

FIGURE 3 (*continued*)

There are several levels of sophistication at which vibrational spectra may be interpreted. For small molecules, vibrational spectral analysis is based on an assumed molecular geometry. Spectra are interpreted as completely as possible using symmetry considerations, spectroscopic selection rules, and additional theoretical arguments. Each observed molecular vibration is a normal mode, that is, a linear combination of the internal motions of small groups of atoms (internal coordinates). Calculation of the normal modes of vibration requires solving Newton's equations of motion for all atoms in the molecule. After this exercise is carried out for many small molecules, a large basis set of information is available for understanding spectra of large molecules. Some vibrations, known as group frequencies, involve only a few closely connected atoms, and always occur at approximately the same band position. For example, the symmetric and asymmetric methylene stretching vibrations from lipid or fatty acid acyl chains may be confidently assumed to occur at about 2850 and 2920 cm^{-1}, respectively. For lipids, peptides, and proteins, the "normal coordinate problem" is substantially underdetermined, even with extensive isotopic labeling. That is, there are many fewer observed vibrational frequencies than there are force constants to describe them. Less rigorous, more empirical levels of analysis are then often applied, based on the assumption that spectra–structure correlations established for small molecules may be directly transferred to lipid assemblies or proteins. The added notion that must be considered in extrapolating spectra–structure correlations from small to large molecules is the occurrence of vibrational coupling between (nearly) identical chemical units, as discussed by Zbinden (1964). For example, the peptide bond amide I (mostly C=O stretch) vibrations of an antiparallel β-sheet structure are coupled due to through-space and through-bond interactions (Brauner *et al.*, 2000; Miyazawa and Blout, 1961), so that an initially isolated single amide I mode (near 1650 cm^{-1}) splits and results in an IR spectrum with a strong peak at $\sim$1625 cm^{-1} and a weak peak at $\sim$1690 cm^{-1}. This pattern is diagnostic for antiparallel β-sheet secondary structure.

B. Intensities

Whereas frequency information from IRRAS experiments may be interpreted in terms of molecular structure using the well-developed approaches outlined above, the extraction of molecular orientational information from IRRAS intensities requires a detailed consideration of the reflection properties of the interface. The quantitative details of this process are described in several publications from this laboratory and others (for a review, see Mendelsohn *et al.*, 1995a) and will not be repeated here. A general outline is provided below to facilitate the reader's understanding of our interpretations of the experimental results.

For understanding reflection properties, insoluble monolayer films are usually treated as the central layer of a three layer (air–film–water) interface. Reflection coefficients (the electric field amplitude ratios of the reflected ray to the incident ray) for *s*- and *p*-polarized radiation can be directly calculated from one of several existing modifications of Fresnel's classical equations. The initial theoretical approaches outlined in Born and Wolf (1980) were modified for anisotropic films by Schopper (1952), Kuzmin and Michailov (1981), Kuzmin *et al.* (1992), and Ishida and associates (Yamamoto and Ishida, 1994). Although the equations presented by Schopper and Kuzmin *et al.* are somewhat different in their mathematical forms, comparative simulations at angles of incidence far from the Brewster angle for the methylene asymmetric stretching mode using the same set of optical constants and film thickness lead to predicted RA values that agree to better than 0.25%. However, discrepancies near the Brewster angle are substantial, and experimental approaches (not yet undertaken) are required to decide which formalism best describes IRRAS. We have used Kuzmin's equations as described previously (Mendelsohn *et al.*, 1995a) for the simulations presented below.

To illustrate the effects of angle of incidence φ and state of incident light polarization on IR spectral features, simulations of the symmetric CH_2 stretching mode at 2850 cm^{-1} are shown as a function of φ for *s*-polarized (perpendicular to the plane of incidence) and *p*-polarized (parallel to the plane of incidence) radiation in Figs. 3A and 3B, respectively. For the simulation, the chains were assumed to be tilted at 28° to the surface normal, whereas the transition moment direction for this mode is perpendicular to the chain direction. The input absorption frequency of 2850 cm^{-1} is shifted down by about 2 cm^{-1} in the simulated IRRAS spectra for *s*-polarized incident radiation. In addition, the peak intensity decreases monotonically as φ increases from 40° to 68°. In particular, close to the Brewster angle ($\sim$54°), nothing unusual happens, that is, the bandshape is not altered. In contrast, major changes occur in the *p*-polarized spectra as φ is altered (Fig. 3B). At $\varphi = 40°$, the RA is negative, and increases in magnitude when φ is increased to 50°. Close to the Brewster angle, the bandshape is completely distorted from the initial symmetric Lorentzian function, and the RA is increased (in a negative sense). It is noted that although the RA is greater for the *p* component, the actual reflectivity from the monolayer is much greater for the *s* component, especially close to the Brewster angle. Finally, at angles of incidence greater than the Brewster angle, the RA for *p*-polarized radiation becomes positive and diminishes in intensity with increasing angle of incidence.

Similar effects are noted for simulations of the IRRAS spectrum for *p*-polarized radiation of the amide I mode on both H_2O and D_2O subphases in Figs. 3C and 3D, respectively. The direction of the helix amide I transition moment is assumed to lie at 28° from the C=O bond, whereas the helix axis is assumed to be strictly perpendicular to the monolayer surface. An initial amide I frequency of 1650 cm^{-1} was assumed, so that significant alterations in the peak position (at least 5 cm^{-1}) are observed in the IRRAS spectrum. When the simulation (Fig. 3C) was carried out for an H_2O subphase at $\varphi = 40°$, a negative-going feature was observed on the low-frequency side of the peak, and the band maximum (positive RA) was shifted to higher frequencies than its initial value. The derivative-like features were enhanced at an angle of incidence of 50°. The RA contour shows a negative RA with a progressively diminishing (negative) peak intensity as φ is increased from 54° to 68°.

The origin of these effects may be traced in part to the refractive index of the H_2O subphase. As shown in Fig. 3D, when a D_2O subphase is assumed, the overall intensities at each angle of incidence are significantly higher than on H_2O, whereas the derivative-like features are reduced in magnitude. Peak shifts are still evident. The increased intensity of the amide I mode for a protein monolayer on D_2O is due to a greater mismatch between the film and subphase refractive indices than for a film on an H_2O subphase. This feature is helpful in detection of the amide I modes on a D_2O subphase.

The above simulations show that the shape and intensity of IRRAS bands depend markedly on several parameters, including the tilt angle of the absorbing functional group with respect to the normal to the monolayer. For chain molecules under certain conditions (e.g., for the CH_2 stretching modes of an ordered, all-*trans* chain), the tilt angle of the absorbing dipole reflects the tilt angle of the chain.

The operational procedure for extracting tilt angles from measured IRRAS intensities is as follows:

1. IRRAS measurements (both polarizations) are performed for a series of angles of incidence of the incoming IR radiation. These data constitute the primary experimental observations.

2. As shown in the caption to Fig. 3, several parameters are required for simulations using the aforementioned theoretical descriptions of the modified Fresnel equations. The following parameters are required to calculate a single RA value: the angle of incidence, the mean tilt angle of the molecular axis relative to the surface normal, the angle that the transition dipole makes with the molecular axis, the vacuum wavelength of the light, the film thickness, the indices of refraction and extinction coefficients of the film and subphases, and the directional refractive indices and extinction coefficients of the film. The refractive indices of the aqueous subphases are available from the measurements of Bertie *et al.* (1989).

3. Comparison of the experimental data with the simulations is performed. In general the only unknowns in simulations are the tilt angles of the functional group and the film extinction coefficients. Because there are more experimental data points (two per angle of incidence) than unknowns in the optical system, a least squares approach for the comparison can be undertaken and the best value of tilt angle thereby extracted from the experimental intensity measurements.

C. IRRAS Studies of Molecular Conformation and Interactions in Monolayer Films

IRRAS frequencies are often used to evaluate molecular conformation. As an example, we refer to the CH_2 stretching modes for phospholipid acyl chains, which monitor conformational order and chain packing (Mendelsohn and Mantsch, 1986). In bulk phases, the symmetric stretching mode provides a convenient means to detect the occurrence of acyl chain conformational disordering during thermotropic gel→liquid crystal transitions. This transition results in an increase in the symmetric stretching mode frequency, from ~2850 to ~2852.6 cm^{-1}, and is illustrated for 1,2-dipalmitoylphosphatidylglycerol (DPPG) in Fig. 4A. In Langmuir monolayers, compression of a DPPG monolayer film results in a frequency decrease from ~2854 to 2850 cm^{-1} as the surface pressure is increased from ~3 to 10 mN/m (see Fig. 4B). Comparison of the bulk-phase data (frequency increase upon chain disordering) with the IRRAS measurements (frequency decrease upon compression) thus provides direct evidence for a surface pressure-induced conformational ordering of the acyl chains in this molecule.

In bulk phases, the methylene frequencies occasionally detect structural alterations at temperatures well below that of the gel→liquid crystal phase transition.

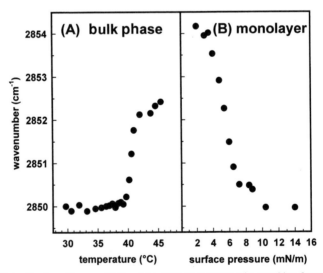

FIGURE 4 Conformation sensitivity of the methylene symmetric stretching frequency of 1,2-dipalmitoylphosphatidylglycerol (DPPG). (A) Variation in band position during the thermotropic gel–liquid crystal phase transition. The frequency increase monitors the formation of *gauche* rotations in the acyl chains. (B) Variation in band position, revealing chain ordering during isothermal compression of the DPPG monolayer.

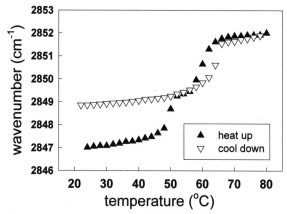

FIGURE 5 The thermotropic transition of the ceramide 5 component in a 1 : 1 mixture of ceramide 5 with perdeuterated palmitic acid. The initial heating cycle reveals the main order–disorder transition at ~60°, whereas the lower temperature transition at 50° is accompanied by a change in methylene frequencies typical of a solid–solid phase transition in the chains. The solid–solid transition is not observed during the cooling process.

An example is shown for the ceramide 5 component of an equimolar ceramide 5/deuterated palmitic acid mixture in Fig. 5. Two transitions are observed characterized by a midpoints of about 50°C and 60°C, respectively. The higher temperature transition is accompanied by a frequency increase from 2849.5 to 2852 cm^{-1} and arises from a (nearly) thermally reversible order–disorder transition. The lower temperature transition is accompanied by a frequency increase from ~2847.8 to 2849 cm^{-1}. Although this frequency shift may seem small, band positions may be easily measured with a precision of 0.05 cm^{-1} in transmission and 0.1 cm^{-1} in IRRAS. Thus, the order–disorder transition is easily detected in either case. In general, transitions detected from cooperative changes in methylene stretching frequencies are assumed to arise from packing (solid–solid transitions) alterations if the final frequency is less than 2850 cm^{-1}, and from the introduction of conformational disorder into the acyl chains if the final frequency is greater than this value. The lower temperature transition in this instance is not readily reversible when the sample is cooled.

Although the CH$_2$ stretching modes provide a convenient means of detecting structural and packing alterations in chains, the information from this measurement is essentially qualitative. For example, the increase in frequency at the gel→liquid crystal transition has proven difficult to correlate quantitatively with the number of *gauche* rotations incorporated into the acyl chains. Similarly, the observation by IR spectroscopy of a solid–solid phase transition (resulting in altered chain packing) with a final frequency of <2850 cm^{-1} cannot be used to identify the particular phases that occur.

Of significant advantage for IRRAS and bulk phase IR studies of chain conformation is the possibility of incorporating deuterium into the chains to distinguish conformational differences between particular regions of a molecule or between two different molecular species in a mixture. Deuteration shifts the methylene symmetric and asymmetric stretching vibrations to $\sim$2090 and $\sim$2200 cm^{-1}, respectively, causing the bands to appear in a region of the spectrum free of other potentially interfering vibrations. These modes display a sensitivity to structural and packing changes similar to that of the CH_2 stretching modes. Two IRRAS reports (Gericke and Mendelsohn, 1996; Gericke *et al.*, 1996) have examined differences in conformational order in different regions of phospholipid chains. Tail-end perdeuterated DPPC (last seven CD_2 groups) was examined by IRRAS. Examination of the CH_2 and CD_2 frequencies and comparisons with appropriate reference compounds revealed that the chains in the Langmuir films showed more conformational order adjacent to the interface. Studies were also undertaken of the acyl chain conformations of the positional isomers 1-palmitoyl-d_{31}-2-oleoylphosphatidylcholine (P-d_{31}OPC) and 1-oleoyl-2-palmitoyl-d_{31}-phosphatidylcholine (OP-d_{31}PC). The oleoyl chains of the former were slightly more ordered at a surface pressure of >25 mN/m, whereas the palmitoyl chains possessed a similar level of conformational order in the two derivatives.

The reason for the widespread use of the methylene frequencies to detect conformational transitions in monolayers is simply that these modes are the most intense in the IRRAS spectra of phospholipids and fatty acids. In situations where greater signal/noise ratios may be achieved, other chain vibrations provide more quantitative structural information. These include the methylene scissoring modes, which occur at 1462–1474 cm^{-1} for proteated chains and 1085–1092 cm^{-1} for perdeuterated chains. These modes have been found to split into doublets if the acyl chains are packed in orthorhombic perpendicular subcells. Snyder and his co-workers (Mendelsohn and Snyder, 1996; Mendelsohn *et al.*, 1995b; Snyder *et al.*, 1992, 1994, 1995, 1996) have investigated this phenomenon in detail and determined that the effect is of short range and perpendicular to the chain direction. If deuterated and proteated chains are mixed in the same sample, vibrational interaction does not occur between their respective scissoring vibrations, due to mismatch in oscillator frequencies. In addition, the magnitude of the splitting reflects the number of interacting chains of a particular isotope, and hence measures the size of the domain in which the particular chain is located. Although the scissoring bands are three times weaker than the methylene stretching vibrations, splitting has been detected (Flach *et al.*, 1997) in monolayers of behenic acid methyl ester, providing direct evidence for orthorhombic perpendicular-type chain packing. These modes are shown in Fig. 6.

In addition to the chain modes, vibrations arising from the phosphate and carbonyl moieties of phospholipids are suitable for investigations of hydration and ion binding in this region of the molecule. The effect of ionic interactions with

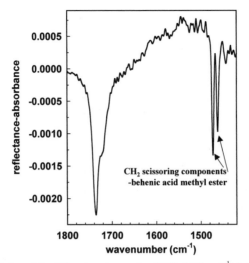

FIGURE 6 Splitting of the CH_2 scissoring modes (1462, 1474 cm^{-1}) in behenic acid methyl ester. Such splitting provides strong evidence for chains packed in (two-dimensional) orthorhombic perpendicular subcells.

phospholipid headgroups was first demonstrated by Dluhy and co-workers (Hunt *et al.*, 1989). The pattern of asymmetric phosphate stretching vibrations at 1220–1260 cm^{-1} responds to changes in the hydration state and/or cation interaction with the lipid headgroup. The carboxylate frequencies in fatty acids have also been shown to be sensitive to the protonation state of the molecule. In particular, the effects of divalent cations on fatty acid monolayers may be elucidated through studies of the asymmetric and symmetric COO^- modes (Gericke and Hühnerfuss, 1994; Simon-Kutscher *et al.*, 1996).

IV. IRRAS APPLICATIONS TO LIPID/PROTEIN LANGMUIR FILMS

A. The Pulmonary Surfactant System

An original motivation for the development of IRRAS was its relevance for studies of monolayers related to the pulmonary surfactant system. Some background relating to this system is therefore introduced. A recent comprehensive review was given by Johansson and Curstedt (1997).

Pulmonary surfactant is a lipid/protein complex assembled from components (lamellar bodies, tubular myelin, and vesicles) located within the alveolar subphase. As the surface-active constituents of surfactant organize at the air/alveolar interface, a monolayer is suggested to form and to provide the functional form of surfactant. The monolayer allows the integrity of the alveolar surface to be maintained by withstanding high surface pressures upon expiration and by facilitating the spreading of constituents across the interface during inhalation.

Pulmonary surfactant is ~90% lipid and ~10% protein by weight. The main lipid components are DPPC, unsaturated PC, and phosphatidylglycerol (PG). Four surfactant-specific proteins have been isolated. Surfactant proteins SP-B and SP-C are hydrophobic, low molecular weight substances known to facilitate adsorption and spreading of phospholipids such as DPPC at the air/water interface. SP-A and SP-D are multimeric, hydrophilic, collagenous proteins with the ability to bind lipids and are thought to play a role in host-defense activities. SP-A and SP-B have also been suggested to be necessary for the formation of tubular myelin, a complex structure from which the surface-active monolayer is presumed to form.

Animal-derived surfactant preparations are currently used therapeutically for treatment of respiratory distress syndrome, a pathological condition caused by a deficiency in surfactant levels in premature infants. Supplementation of exogenous surfactant with specific lipids and/or SP-B and SP-C is under evaluation. However, a coherent molecular-level understanding of protein function and lipid/protein interaction in lung surfactant does not exist. Thus, a wide variety of biophysical techniques have been used in an attempt to elucidate structure–function relationships of the surfactant components. The major advantage of IRRAS for these investigations is that the approach permits the evaluation of the secondary structure of proteins (and peptides derived from them) in environments (aqueous monolayers) that mimic those suggested to be important *in vivo*.

Two sets of experiments from our laboratory serve to illustrate the nature of the information provided by IRRAS measurements from peptides and proteins related to he surfactant system.

1. Secondary Structure of a Peptide Derived from Pulmonary Surfactant SP-B

SP-B is isolated as a positively charged, disulfide-linked homodimer with each 79-amino-acid monomer containing three intramolecular disulfide bonds. Circular dichroism (CD) and Fourier transform infrared (FTIR) spectroscopic studies of SP-B in organic solutions, lipid micelles, and bilayers show that the protein possesses 40–60% α-helical content with varying amounts of random coil, β-sheet, and β-turn structure. Peptide analogues of SP-B are currently being evaluated for both therapeutic use and in attempts to elucidate structure–function relationships in the native protein.

The choice of specific peptide sequences for IRRAS study is based on model amphipathic helical motifs and/or the native amino acid sequence. However, the secondary structure determined in bulk-phase environments may differ from that found in monolayers, thereby possibly invalidating any correlations drawn between bulk-phase structure and surface activity.

PRIMARY SEQUENCES OF SP-B$_{9-36}$ AND SP-C FROM PULMONARY SURFACTANT

SP-B $_{9-36}$: WLARA^{5}LIKRI^{10}QAMIP^{15}KGALA^{20}VAVAQ^{25}VAR

SP-C : LRIPC^{5}CPVNL^{10}KRLLV^{15}VVVVV^{20}VLVVV^{25}VIVGA^{30} LLMGL^{35}

FIGURE 7 Primary sequences of a synthetic fragment of SP-B$_{9-36}$ and of SP-C from pulmonary surfactant.

This laboratory has undertaken studies of the conformation of several synthetic peptide fragments, including SP-B$_{9-36}$, whose primary sequence is shown in Fig. 7. Secondary structures in Langmuir films were determined in the presence and the absence of lipids under conditions of varying surface pressure. The most striking behavior was a surface pressure-induced conformational change in the peptide monolayer. IRRAS spectra are shown in Fig. 8 for a pure peptide monolayer at a

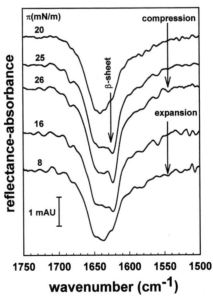

FIGURE 8 IRRAS spectra (unpolarized radiation, angle of incidence 30°) of a synthetic fragment from pulmonary surfactant, SP-B$_{9-36}$. The peptide exhibits a mixed (helix + random) secondary structure at a surface pressure of 20 mN/m. As the surface pressure is increased, an increase in the proportion of β strands or sheets is observed, as evidenced by the appearance and increasing intensity of a sharp feature at 1624 cm^{-1}. As the surface pressure is reduced, this feature diminishes in intensity, showing the reversibility of this secondary structure alteration.

series of surface pressures during compression and subsequent expansion. At initial pressures of 20 mN/m, a broad mixed helix/random contour is evident (broad spectral feature centered at 1640 cm^{-1}). As the pressure is increased to 25 mN/m, a new feature attributable to β strands or sheets appears at 1624 cm^{-1}. This feature disappears when the film is reexpanded. The cycle of reversible conformational change during compression and expansion can be repeated. Similar patterns are noted for the peptide in a mixed film with DPPC. Although the biological significance of this surface pressure-induced reversible secondary structure change remains obscure, it is evident that IRRAS provided a unique means of studying this phenomenon. Furthermore, the bulk-phase conformation of the protein differs from that in the monolayer, thereby revealing the importance of utilizing a technique that provides direct characterization of secondary structure at the air/water interface.

2. Orientation of SP-C in Lipid Environments

SP-C is proteolytically generated *in vivo* from a 21-kDa precursor and has a molecular weight of ~3.5 kDa. The protein is S-palmitoylated and assumes a primarily α-helical secondary structure in bulk phases. The helix orientation in multilayers of DPPC has been determined by attenuated total reflectance (ATR) infrared spectroscopy (Pastrana *et al.,* 1991) and is found to be tilted by about 24° from the normal to the bilayer plane, that is, it is essentially parallel to the DPPC acyl chains, because DPPC is known to be tilted by about 28° from the bilayer normal. The relatively polar N-terminal, possessing two prolines and three positively charged residues, is assumed to be unordered and located closer to the membrane surface. The sequence of human SP-C is shown in Fig. 7.

As with all the surfactant proteins, the *in vivo* function of SP-C is not known with certainty; inferences are drawn from *in vitro* experiments. SP-C has been shown to promote spreading of surface-active phospholipids. The molecular basis for this process is not well understood, but the problem is important, in that it may aid design of therapeutic agents for respiratory distress syndrome. Thus, this laboratory (Flach *et al.,* 1999; Gericke *et al.,* 1997) undertook quantitative IRRAS measurements to determine the orientation of SP-C in phospholipid monolayer films.

IRRAS data in the 1600- to 1800-cm^{-1} region for SP-C/DPPC monolayers at an air/D$_2$O interface after compression to a surface pressure of 28 mN/m are shown for *s*- and *p*-polarization an angle of incidence of 40° in Fig. 9 (top). Quantitative analysis of the amide I mode intensities was used to determine the tilt of the SP-C helix. The simulated and measured RA values for this band at various incident angles are shown in Fig. 9 (bottom). Details of the calculation are given in the cited work. The best fit for the data was obtained for a helix tilt angle of 70°.

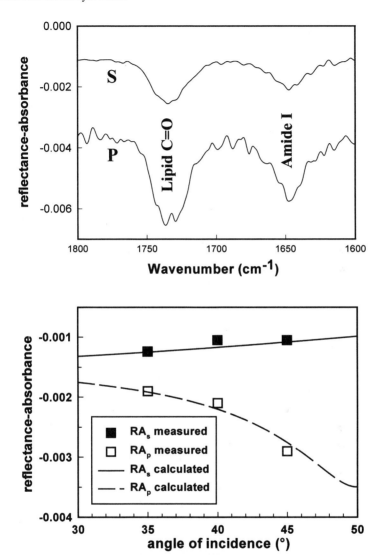

FIGURE 9 Calculation of the tilt angle of the SP-C helix. Top: IRRAS data (polarization as indicated) for SP-C/DPPC (1 : 20 mole ratio) mixed monolayer. The angle of incidence was 40°. The vibrational modes are marked. The spectra are offset for clarity. Bottom: Calculated intensities for s-polarized (*solid line*) and p-polarized (*dashed line*) radiation for an assumed helix tilt angle of 70° from the normal to the monolayer. The experimental points are shown for s-polarized (■) and p-polarized (□) light at various angles of incidence. The match between the calculated and experimental points represents the best fit to the data.

Orientation measurements were also carried out on the acyl chains of DPPC in monolayers in the presence and absence of SP-C. The molecular chain tilt angle decreased from $\sim26°$ in pure lipid monolayers (comparable to bilayers) to $\sim10°$ in the mixed monolayer films.

Overall, the combination of polarized ATR and IRRAS data reveal a remarkable change in the tilt angle of the protein, from $\sim24°$ (i.e., parallel to the acyl chains) to the interface normal in lipid bilayers to $\sim70°$ in the mixed monolayer films. This major change in helix orientation provides clues to the mechanism by which SP-C facilitates the spreading of DPPC at the air/water interface. The observed helix tilt angles for both the monolayer and bilayer preparations allow for optimal hydrophobic matching of the lipid acyl chains with the protein hydrophobic helical regions. As the protein tilts upon insertion into the monolayer, it occupies a greater surface area than at low tilt angles and thereby facilitates the spreading of lipids across the interface. Thus the model for the interaction is one of a hydrophobic lever. Simultaneously, the DPPC acyl chains are forced into a more vertical position to maximize the interaction, permitting a high density of lipid molecules, which is necessary to achieve the low surface tensions found in the lung. A recent extension of this work (Flach *et al.,* 1999) has shown that the absence of S-palmitoylation does not alter helix orientation, although it produces large changes in the pressure–area isotherms of SP-C monolayers.

B. Further IRRAS Applications to Peptide and Proteins in Monolayer Films

In addition to the pulmonary surfactant system, a variety of other applications employing IRRAS and PM-IRRAS to study peptide and protein conformation and orientation have appeared.Ulrich and Vogel (1999) reported PM-IRRAS studies of the interaction of gramicidin A with 1,2-dimyristoylphosphatidylcholine (DMPC). Intensity modeling of the amide I frequencies led to the conclusion that the peptide adopted a $\beta^{6.3}$ helical conformation. A dependence of the peptide orientation on the surface pressure in DMPC monolayers was suggested. At low pressure, the helix lay flat on the surface, while at higher pressures, the helix was oriented more parallel to the surface normal.

The Bordeaux group (Castano *et al.,* 1999; Cornut *et al.,* 1996) have used PM-IRRAS to examine the behavior of the lytic, amphipathic peptide $L_{15}K_7$ and the related series L_iK_j ($i = 2j$). Peptides longer than 12 residues adopted helical conformations, whereas shorter peptides formed intermolecular antiparallel β sheets. The observed amide I/amide II mode intensity ratios were used to determine that the preferential orientation adopted by these peptides was parallel to the interface. Similar IRRAS studies from the Rutgers group (Dieudonné *et al.,* 1998) probed the propensity for helix formation in monolayers of the hydrophobic peptide series $K_2(LA)_x$ ($x = 6, 8, 10, 12$). The secondary structure adopted by the two longest peptides depended remarkably on the initial spreading pressures; at high pressure, the molecules were α-helical, whereas at low spreading pressures, an extended sheet structure was detected. This observation provides an example in which the spreading conditions dramatically alter the observed secondary structure of the peptide film.

In addition to studies of peptide films, the secondary structures of (lipid-free) proteins at the air/water interface have been reported. Schladitz *et al.* (1999) recently examined the secondary structure of the first 40 residues of the amyloid (prion) β-sheet protein. In bulk phase, the protein was shown, through CD measurements, to be rich in coil, turn, and α helix. However, when spread at the air/water interface, monolayers with high proportions of β-sheet secondary structure resulted, due to surface denaturation. The study again illustrates that conformations adopted at the interface may differ dramatically from those in the bulk phase, and in this instance, the differences may have relevance to physiological events.

In a study designed to evaluate protein secondary structure and stability in response to film spreading conditions, Gallant *et al.* (1998) examined the photosystem II core complex (PS II CC) under two conditions of spreading. When a complex was compressed immediately after spreading at 5.7 mN/m, the IR-RAS data showed that the native (predominantly α-helical) secondary structure persisted upon compression. In contrast, when the monolayer was incubated for 30 min at an initial pressure of 0.6 mN/m, denatured protein, as revealed by large amounts of β sheet, were evident. In a follow-up study (Lavoie *et al.,* 1999), the propensity for surface denaturation of PS II CC was compared to that of rhodopsin and bacteriorhodopsin. The last of these was found to retain its native secondary structure under conditions that denatured the other proteins. The authors suggest that this behavior is related to the two-dimensional crystalline structure of bacteriorhodopsin.

The majority of IRRAS studies reported involve hydrophobic proteins either in single-component monolayers or in mixed monolayers with lipids. A different class of experiments currently feasible involves studies of soluble proteins that nevertheless carry out important biological functions via lipid interactions. An example is the Ca^{2+}-mediated interaction of annexin V(A×V) with acidic phospholipids (Wu *et al.,* 1998, 1999). A×V possesses a secondary structure made up of four-helix bundles in four core domains. The Ca^{2+}-dependent binding of the protein to phospholipids is suggested to play an important role in many of the putative cellular functions of the annexins. IRRAS was applied to monitor the conformation and stability of the protein upon interaction with dimyristoylphosphatidic acid (DMPA). A×V was injected into the D_2O subphase beneath either a film-free surface or a DMPA monolayer; adsorption of protein to the surface was characterized by an increase in surface pressure with time and (in the presence of lipid) by the observation of domains with Brewster angle microscopy (BAM). The secondary structure adopted by the protein as detected by IRRAS

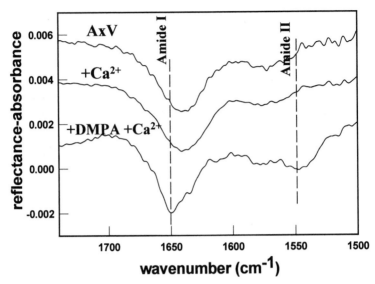

FIGURE 10 IRRAS of annexin V in three monolayer states, as indicated. The shift to lower frequencies of the amide I mode in the lipid-free films reflects surface unfolding. The residual amide II intensity near 1550 cm^{-1} in the ternary AxV/DMPA/Ca^{2+} system provides evidence that the protein under these conditions is partially protected from H$\rightarrow$D exchange, because these experiments were carried out on a D$_2$O subphase.

(see Fig. 10) depended significantly on the presence of the phospholipid mono-layer. AxV films in the presence and absence of Ca^{2+} without the phospholipid monolayer exhibit a loss of α-helical secondary structure and an increase of random coil forms, as indicated by a frequency shift to lower wavenumbers in the amide I vibration. The presence of DMPA monolayers limited the extent of this denaturation event and thus stabilized AxV structure in the monolayer. The combined presence of the DMPA monolayer and Ca^{2+} also enhanced the sta-bility of the native protein, as monitored by a reduction in peptide bond H$\rightarrow$D exchange. The latter was measured through the residual intensity of the amide II mode ($\sim$1550 cm^{-1}). The nature of the amide II vibration (mixed N—H bending and C—N stretch) is such that H$\rightarrow$D exchange shifts the mode to much lower frequen-cies. This effect permits the amide II intensity to be a useful marker for exchange processes.

Recent reports have begun to specify shifts in amide I band positions due to interactions between neighboring groups within a peptide and the accessibility of peptide-group hydrogen bonding sites to the surrounding medium (Brauner *et al.*, 2000; Williams *et al.*, 1996). The bulk-phase IR amide I frequency dependence of the central β-sheet region in the antimicrobial peptide gramicidin S on the hydrogen bonding potential of different environments was corroborated by IRRAS results for monolayers of varying composition (Lewis *et al.,* 1999). It was shown that the characteristic secondary structure of the peptide was maintained in a variety of environments. A shift in the amide I band to lower frequency was ascribed to an increase in the hydrogen bonding potential of the environment. In particular, the position of the amide I band, observed at approximately 1643 cm^{-1} in binary DPPG/gramicidin S(GS) monolayers, shifted to 1630 cm^{-1} for pure GS monolayers due to the increase in polarity of the environment. Thus, the amide I frequency was shown to be sensitive to hydration factors as well as to secondary structure.

V. CURRENT STATE AND FUTURE PROSPECTS FOR IRRAS

IRRAS is at a state of development where spectra of lipid films may be acquired in a straightforward manner. Adequate knowledge of both the spectra–structure relationships and the direction of transition moments is available for evaluation of chain structure and orientation. When reasonable measures are applied to eliminate interference from water vapor, spectra can also be acquired from pure protein or mixed lipid/protein films. The theoretical models for reflected intensities are adequate for determining the molecular orientation of α-helical structural elements. The situation for β sheets is a little more difficult, due to loss of uniaxial symmetry. For a general orientation of the transition moment of the amide I mode, two angular variables must be determined. The appropriate geometric equations have been described by Marsh (1997). An initial attempt at a qualitative evaluation of antiparallel β-sheet orientation using PM-IRRAS has been reported for a series of synthetic, amphipathic peptides, (KL)$_m$K in lipid monolayers on an H$_2$O subphase (Castano *et al.,* 2000). The β-sheet was characterized as having a slight tilt or lying flat in the plane of the interface. To completely characterize the orientation of antiparallel β sheets, both the amide I and amide II bands, or the two split components of amide I corresponding to the $v_\perp(\pi, 0)$ and $v_\parallel(0, \pi)$ absorption bands (Miyazawa and Blout, 1961), must be examined.

In addition to the determination of secondary structure and orientation, sufficient sensitivity exists for the determination of the conformation of particular regions within peptides. The approach requires synthesis of peptides with peptide bond ^{13}C labels at particular sites. The effects of such labels are quite different for α helices and β sheets. For helices, a new band of the appropriate (weak) intensity appears shifted to lower frequency by an amount ($\sim$40 cm^{-1}) anticipated from a simple calculation that treats the C=O bond as a diatomic molecule. In contrast, the effects of isotopic substitution on β sheets are much more dramatic. The entire pattern of inter- and intrastrand vibrational coupling is altered, so that the entire problem must be treated as a combined through-space and through-bond coupled oscillator problem (Brauner *et al.,* 2000).

A difficulty with current tilt angle measurements is the necessity to collect IRRAS data at various angles of incidence. Until recently, this has required breaking down the monolayer experiment for rearrangement of the optical system at each new angle. Bruker Instruments is marketing a device which will permit varying the angle of incidence under computer control without the necessity of breaking down the monolayer. The availability of this unit will speed up IRRAS data collection at varying angles of incidence by an order of magnitude.

In summary, IRRAS offers the unique opportunity to directly examine peptide or protein conformation and orientation in monolayer films *in situ* at the air/water interface. Although originally developed for examination of the interaction between components of the pulmonary surfactant system, IRRAS offers more general utility, for example, in studies of peripheral membrane proteins interacting with lipids. As the sensitivity of the instrumentation improves, the use of isotopically labeled peptides may permit studies of conformational changes in each component during peptide–protein interaction in monolayer environments.

Acknowledgment

The experiments and apparatus at Rutgers University described in this chapter were funded by a grant to R.M. (GM 29864) from the U.S. Public Health Service.

References

Als-Neilsen, J., Jacquemain, D., Kjaer, K., Leveiller, F., Lahav, M., and Leiserowitz, L. (1994). Principles and applications of grazing incidence X-ray and neutron scattering from ordered molecular monolayers at the air–water interface. *Phys. Lett.* **246,** 252–313.

Axelsen, P. H., and Citra, M. J. (1997). Orientational order determination by internal reflection infrared spectroscopy. *Prog. Biophys. Mol. Biol.* **66,** 227–253.

Azzam, R. M. A., and Bashara, N. M. (1977). "Ellipsometry and Polarized Light." North-Holland, Amsterdam.

Bain, C. D. (1998). Studies of adsorption at interfaces by optical techniques-ellipsometry, second harmonic generation, and sum–Frequency generation. *Curr. Opin. Colloid Interface Sci.* **3,** 287–292.

Bertie, J. E., Ahmed, M. K., and Eysel, H. H. (1989). Infrared intensities of liquids. 5. Optical and dielectric constants, integrated intensities and dipole moment derivatives of H_2O and D_2O. *J. Phys. Chem.* **93,** 2210–2218.

Blaudez, D., Buffeteau, T., Cornut, J. C., Desbat, B., Escafre, N., Pezolet, M., and Turlet, J. M. (1994). Polarization modulation FTIR spectroscopy at the air/water interface. *Thin Solid Films* **242,** 146–150.

Blaudez, D., Buffeteau, T., Desbat, B., and Turlet, J. M. (1999). Infrared and Raman spectroscopies of monolayers at the air–water interface. *Curr. Opin. Colloid Interface Sci.* **4,** 265–272.

Born, M., and Wolf, E. (1980). "Principles of Optics," 6th ed. Pergamon Press, Oxford.

Brauner, J. W., Dugan, C., and Mendelsohn, R. (2000). [13]C isotope labeling of hydrophobic peptides: Origin of the anomalous intensity distribution in the infrared amide I spectral region of β-sheet structures. *J. Am. Chem. Soc.* **122,** 677–683.

Buffeteau, T., Desbat, B., and Turlet, J.-M. (1991). Polarization modulation FT-IR spectroscopy of surfaces and ultra-thin films: Experimental procedure and quantitative analysis. *Appl. Spectrosc.* **45,** 380–389.

Castano, S., Desbat, B., Laguerre, M., and Dufourcq, J. (1999). Structure, orientation, and affinity for interfaces and lipids of ideally amphipathic lytic L_iK_j ($i = 2j$) peptides. *Biochim. Biophys. Acta* **1416,** 176–194.

Castano, S., Desbat, B., and Dufourcq, J. (2000). Ideally amphipathic beta-sheeted peptides at interfaces: Structure, orientation, affinities for lipids and hemolytic activity of $(KL)_mK$ peptides. *Biochim. Biophys. Acta* **1463,** 65–80.

Cornut, I., Desbat, B., Turlet, J.-M., and Dufourcq, J. (1996). *In situ* study by polarization modulated Fourier transform infrared spectroscopy of the structure and orientation of lipids and amphipathic peptides at the air–water interface. *Biophys. J.* **70,** 305–312.

Dieudonné, D., Gericke, A., Flach, C. R., Jiang, X., Farid, R. S., and Mendelsohn, R. (1998). Propensity for helix formation in the hydrophobic peptides $K_2(LA)_x$ ($x = 6, 8, 10, 12$) in monolayer, bulk and lipid-containing phases: Infrared and circular dichroism studies. *J. Am. Chem. Soc.* **120,** 792–799.

Dluhy, R. A. (1986). Quantitative external reflection infrared spectroscopic analysis of insoluble monolayers spread at the air/water interface. *J. Phys. Chem.* **90,** 1373–1379.

Dluhy, R. A., and Cornell, D. G. (1985). *In situ* measurement of insoluble monolayers at the air/water interface. *J. Phys. Chem.* **89,** 3195–3197.

Dluhy, R. A., Wright, N. A., and Griffiths, P. R. (1988). *In situ* measurement of the FT-IR spectra of phospholipid monolayers at the air/water interface. *Appl. Spectrosc.* **42,** 138–141.

Dluhy, R. A., Stephens, S. M., Widayati, S., and Williams, A. D. (1995). Vibrational spectroscopy of biophysical monolayers: Applications of IR and Raman spectroscopy to biomembrane model systems at interfaces. *Spectrochim. Acta (A)* **51,** 1413–1447.

Flach, C. R., Brauner, J. W., and Mendelsohn, R. (1993). Calcium ion interactions with insoluble phospholipid monolayer films at the A/W interface: External reflection–absorption IR studies. *Biophys. J.* **65,** 1994–2001.

Flach, C. R., Brauner, J. W., Taylor, J. W., Baldwin, R. C., and Mendelsohn, R. (1994). External reflection FT-IR of peptide monolayers *in situ* at the A/W interface, experimental design, spectra–structure correlations, and effects of hydrogen–deuterium exchange. *Biophys. J.* **67,** 402–410.

Flach, C. R., Gericke, A., and Mendelsohn, R. (1997). Quantitative determination of chain molecular tilt angles in monolayer films at the air/water interface: IRRAS of behenic acid methyl ester. *J. Phys. Chem. B* **101,** 58–65.

Flach, C. R., Gericke, A., Keough, K. M. W., and Mendelsohn, R. (1999). Palmitoylation of lung surfactant protein SP-C alters surface thermodynamics but not protein secondary structure or orientation in 1,2-dipalmitoylphosphatidylcholine langmuir films. *Biochim. Biophys. Acta* **1416,** 11–20.

Gaines, G. L., Jr. (1966). "Insoluble Monolayers at Liquid-Gas Interfaces." Interscience, New York.

Gallant, J., Desbat, B., Vaknin, D., and Salesse, C. (1998). Polarization-modulated infrared spectroscopy and X-ray reflectivity of photosystem II core complex at the gas–water interface. *Biophys. J.* **75,** 2888–2899.

Gericke, A., and Hühnerfuss, H. (1994). The effect of cations on the order of saturated fatty acid monolayers at the air/water interface as determined by infrared reflection–absorption spectroscopy. *Thin Solid Films* **245,** 74–82.

Gericke, A., and Hühnerfuss, H. (1995). The conformational order and headgroup structure of long-chain alkanoic ester monolayers at the air water interface. *Ber. Bunsen Ges. Physikal. Chem.* **99,** 641–650.

Gericke, A., and Mendelsohn, R. (1996). Partial chain deuteration as an IRRAS probe of conformational order of different regions in hexadecanoic acid monolayers at the air/water interface. *Langmuir* **12,** 758–762.

Gericke, A., Simon-Kutscher, J., and Hühnerfuss, H. (1993a). Influence of the spreading solvent on the properties of monolayers at the air/water interface. *Langmuir* **9,** 2119–2127.

Gericke, A., Simon-Kutsher, J., and Hühnerfuss, H. (1993b). Comparison of different spreading techniques for monolayers at the air water interface by external infrared reflection absorption spectroscopy. *Langmuir* **9**, 3115–3121.

Gericke, A., Moore, D. J., Erukulla, R., Bittman, R., and Mendelsohn, R. (1996). Partially deuterated phospholipids as IR structure probes of conformational order in bulk and monolayer phases. *J. Mol. Struct.* **379**, 227–239.

Gericke, A., Flach, C. R., and Mendelsohn, R. (1997). Structure and orientation of lung surfactant SP-C and L-α-dipalmitoylphosphatidylcholine in aqueous monolayers. *Biophys. J.* **73**, 492–499.

Hunt, R. A., Mitchell, M. L., and Dluhy, R. A. (1989). The interfacial structure of phospholipid monolayer films: An infrared reflectance study. *J. Mol. Struct.* **214**, 93–109.

Johansson, J., and Curstedt, T. (1997). Molecular structures and interactions of pulmonary surfactant components. *Eur. J. Biochem.* **244**, 675–693.

Kaganer, V. M., Möhwald, H., and Dutta, P. (1999). Structure and phase transitions in langmuir monolayers. *Rev. Mod. Phys.* **71**, 779–819.

Kuzmin, V. L., and Michailov, A. V. (1981). Molecular theory of light reflection and applicability limits of the macroscopic approach. *Opt. Spectrosc. (USSR)* **51**, 383–385.

Kuzmin, V. L., Romanov, V. P., and Michailov, A. V. (1992). Reflection of light at the boundary of liquid systems and structure of the surface layer: A review. *Opt. Spectrosc.* **73**, 1–26 [Translated from *Opt. Spektrosk.* (1992) **73**, 3–47].

Lautz, C., Fisher, T. M., Weygand, M., Losche, M., Howes, P. B., and Kjaer, K. (1998). Determination of alkyl chain tilt angles in langmuir monolayers: A comparison of Brewster angle autocorrelation spectroscopy and X-ray diffraction. *J. Chem. Phys.* **108**, 4640–4646.

Lavoie, H., Gallant, J., Grandbois, M., Blaudez, D., Desbat, B., Boucher, F., and Salesse, C. (1999). The behavior of membrane proteins in monolayers at the gas–water interface: Comparison between photosystem II, rhodopsin and bacteriorhodopsin. *Mater. Sci. Eng.* **10**, 147–154.

Lewis, R. N. A. H., Prenner, E. J., Kondejewski, L. H., Flach, C. R., Mendelsohn, R., Hodges, R. S., and McElhaney, R. N. (1999). Fourier transform infrared spectroscopic studies of the interaction of the antimicrobial peptide gramicidin S with lipid micelles and with lipid monolayer and bilayer membranes. *Biochemistry* **38**, 15193–15203.

Marsh, D. (1997). Dichroic ratios in polarized Fourier transform infrared for nonaxial symmetry of β-sheet structures. *Biophys. J.* **72**, 2710–2718.

McConnell, H. (1991). Structures and transitions in lipid monolayers at the air–water interface. *Annu. Rev. Phys. Chem.* **42**, 171–195.

Mendelsohn, R., and Flach, C. (in press). Infrared reflection–absorption spectroscopy of monolayer films at the air–water interface. *In* "Handbook on Vibrational Spectroscopy" (P. R. Griffiths and J. M. Chalmers, eds.). Wiley, Chichester, England.

Mendelsohn, R., and Mantsch, H. H. (1986). Fourier transform infrared studies of lipid–protein interaction. *In* "Progress in Protein–Lipid Interactions 2" (A. Watts and J. J. H. H. M. dePont eds.), pp. 103–146. Elsevier, Amsterdam.

Mendelsohn, R., and Snyder, R. G., Jr. (1996). Infrared spectroscopic determination of conformational disorder and lateral phase separation in phospholipid acyl chains. *In* "Biological Membranes: A Molecular Perspective from Computation and Experiments" (K. M. Merz, Jr., and B. Roux, eds.), pp. 145–174. Birkhauser, Boston.

Mendelsohn, R., Brauner, J. W., and Gericke, A. (1995a). External infrared reflection absorption spectrometry of monolayer films at the air–water interface. *Annu. Rev. Phys. Chem.* **46**, 305–334.

Mendelsohn, R., Liang, L., Strauss, H. L., and Snyder, R. G., Jr. (1995b). IR spectroscopic determination of gel state miscibilities in mixtures of disaturated long–chain phosphatidylcholines. *Biophys. J.* **69**, 1987–1998.

Mitchell, M. L., and Dluhy, R. A. (1988). *In situ* FT-IR investigation of phospholipid phase transitions at the air/water interface. *J. Am. Chem. Soc.* **110**, 712–718.

Miyazawa, T., and Blout, E. R. (1961). The infrared spectra of polypeptides in various conformations: Amide I and II bands. *J. Am. Chem. Soc.* **29**, 611–616.

Möbius, D. (1998). Morphology and structural characterization of organized monolayers by Brewster angle microscopy. *Curr. Opin. Colloid Interface Sci.* **3**, 137–142.

Möhwald, H. (1990). Phospholipid and phospholipid–protein monolayers at the air–water interface. *Annu. Rev. Phys. Chem.* **41**, 441–476.

Ohe, C., Ando, H., Sato, N., Urai, Y., Yamammoto, M., and Itoh, K. (1999). Corboxylate counter-ion interactions and changes in these interactions during photopolymerization of a long-chain diacetylene monocarboxylic acid at air–water interfaces: External infrared reflection–absorption spectroscopic study. *J. Phys. Chem. (B)* **103**, 435–444.

Oliveira, O. N., and Bonardi, C. (1997). The surface potential of langmuir monolayers revisited. *Langmuir* **13**, 5920–5924.

Pastrana, B., Mautone, A. J., and Mendelsohn, R. (1991). Fourier transform infrared studies of secondary structure and orientation of pulmonary surfactant SP-C and its effect on the dynamic surface properties of phospholipids. *Biochemistry* **30**, 10058–10064.

Pastrana-Rios, B., Taneva, S., Keough, K. M. W., Mautone, A. J., and Mendelsohn, R. (1995). External reflection absorption infrared spectroscopy study of lung surfactant proteins SP-B and SP-C in phospholipid monolayers at the A/W interface. *Biophys. J.* **69**, 2531–2540.

Phillips, M. C., and Chapman, D. (1968). Monolayer characteristics of saturated 1,2 diacyl phosphatidylcholines (lecithins) and phosphatidylehanolamines at the air–water interface. *Biochim. Biophys. Acta* **163**, 301–313.

Richmond, G. L. (1997). Vibrational spectroscopy of molecules at liquid/liquid interfaces. *Anal. Chem.* **69**, A536–A543.

Schladitz, C., Vieria, E. P., Hermel, H., and Möhwald, H. (1999). Amyloid beta-sheet formation at the air–water interface. *Biophys. J.* **77**, 3305–3310.

Schopper, H. (1952). Zur Optik Dünner Doppelbrechender und Dichroitischer Schichten. *Z. Phys.* **132**, 146–170.

Shen, Y. R. (1997). Wave mixing spectroscopy for surface studies. *Solid State Commun.* **102**, 221–229.

Simon-Kutscher, J., Gericke, A., and Hühnerfuss, H. (1996). Effect of bivalent Ba, Cu, Ni and Zn cations on the structure of octadecanoic acid monolayers at the air–water interface as determined by external infrared reflection-absorption spectroscopy. *Langmuir* **12**, 1027–1034.

Snyder, R. G., Goh, M. C., Srivatsavoy, V. J. P., Strauss, H. L., and Dorset, D. L. (1992). Measurement of the growth kinetics of microdomains in binary *n*-alkane solid solutions by infrared spectroscopy. *J. Phys. Chem.* **96**, 10008–10018.

Snyder, R. G., Srivatsavoy, V. J. P., Cates, D. A., Strauss, H. L., White, J. W., and Dorset, D. L. (1994). Hydrogen/deuterium isotope effects on microphase separation in unstable crystalline mixtures of binary *n*-alkanes. *J. Phys. Chem.* **98**, 674–684.

Snyder, R. G., Strauss, H. L., and Cates, D. A. (1995). Detection and measurement of microaggregation in binary mixtures of esters and phospholipid dispersions. *J. Phys. Chem.* **99**, 8432–8439.

Snyder, R. G., Liang, L., Strauss, H. L., and Mendelsohn, R. (1996). IR spectroscopic study of the structure and phase behavior of long-chain diacylphosphatidylcholines in the gel phase. *Biophys. J.* **71**, 3186–3196.

Tamm, L. K., and Tatulian, S. A. (1997). Infrared spectroscopy of proteins and peptides in lipid bilayers. *Q. Rev. Biophys.* **30**, 365–429.

Thomas, R. K., and Penfold, J. (1996). Neutron and X-ray reflectometry of interfacial systems in colloid and polymer chemistry. *Curr. Opin. Colloid Interface Sci.* **1**, 23–33.

Ulrich, W.-P., and Vogel, H. (1999). Polarization-modulated FTIR spectroscopy of lipid/gramicidin monolayers at the air/water interface. *Biophys. J.* **76**, 1639–1647.

Williams, S., Causgrove, T. P., Gilmanshin, R., Fang, K. S., Callender, R. H., Woodruff, W. H., and Dyer, R. B. (1996). Fast events in protein folding: Helix melting and formation in a small peptide. *Biochemistry* **35,** 691–697.

Wu, F. J., Gericke, A., Flach, C. R., Mearly, T. R., Seaton, B. A., and Mendelsohn, R. (1998). Domain structure and molecular conformation in aqueous monolayers of annexin V/ dimyristoylphosphatidic acid/Ca^{2+} complexes: A Brewster angle microscopy/IR reflection–absorption spectroscopy study. *Biophys. J.* **74,** 3273–3281.

Wu, F., Flach, C. R., Seaton, B. A., Mealy, T. R., and Mendelsohn, R. (1999). Stability of annexin V in binary and ternary complexes with Ca^{2+} and anionic phospholipids: IR studies of monolayer and bulk phases. *Biochemistry* **38,** 792–799.

Yamamoto, K., and Ishida, H. (1994). Optical theory applied to infrared spectroscopy. *Vibrational Spectrosc.* **8,** 1–36.

Zbinden, R. (1964). "Infrared Spectroscopy of High Polymers." Academic Press, New York.

CHAPTER 4

Measuring the Depth of Amino Acid Residues in Membrane-Inserted Peptides by Fluorescence Quenching

Erwin London* and Alexey S. Ladokhin†
*Department of Biochemistry and Cell Biology and Department of Chemistry, State University of New York at Stony Brook, Stony Brook, New York 11794-5215; †Department of Physiology and Biophysics, University of California, Irvine, California 92697; on leave from Institute of Molecular Biology and Genetics, National Academy of Sciences of Ukraine, Kiev 252143, Ukraine

I. Introduction
II. Determining Depth from Fluorescence Quenching by Empirical Inspection of Quenching Level
III. Determining Depth by Parallax Analysis
IV. Determining Depth by Distribution Analysis
   A. Physical Principles: Depth Distributions in Membranes
   B. Theory
   C. Definition of Quenching Profile: $\ln(F_0/F)$ or $F_0/F - 1$
V. Comparing Parallax Analysis and Distribution Analysis
VI. Investigating Membrane Protein Topography by Varying Fluorophore Depth Instead of Quencher Depth
VII. A Cautionary Note Concerning Quenching Analysis
VIII. Problems That Can Affect the Accuracy of Quenching Measurements
IX. Studies of Membrane-Inserted Peptides by Fluorescence Quenching
   A. Introduction
   B. Amphipathic Peptides
   C. Bioactive Peptides
   D. Model Hydrophobic/Transmembrane Helices
   E. Membrane Penetration Depth of Very Small Synthetic Peptides
   F. Signal Sequence Peptides
   G. Studies on Peptide Sequences Found in Viral Membrane Proteins
   H. Other Peptide Sequences Derived from Membrane-Associating Proteins
   I. Peptide Mimics
X. Conclusions
   References

Current Topics in Membranes, Volume 52

I. INTRODUCTION

Fluorescence quenching is the process whereby a quenching molecule prevents fluorescence emission from an otherwise fluorescent molecule. Quenching is a powerful method for determining the depth of a fluorescent molecule, or that of a fluorescent group attached to a larger molecule, within a lipid bilayer. The depth of a group within a bilayer can illuminate several aspects of membrane structure, including the details of the topography, orientation, and folding of membrane-inserted proteins and peptides. In this chapter, we first discuss the principles and limitations of the analysis of fluorescence quenching in membranes. The second part describes applications of quenching to the structure of model membrane-bound peptides.

In a typical quenching experiment using model membranes, a series of quencher-carrying molecules with quenchers that occupy different depths in the bilayer are incorporated into an artificial lipid bilayer that also contains the fluorescent molecule of interest. Often the quenchers are derivatives of fatty acids or phospholipids, with the quencher attached either to the polar headgroup or to a specific fatty acyl carbon atom. This attachment gives the quencher a relatively "fixed" depth. Fatty acids have the advantage that they can be added to preformed bilayers. On the other hand, they have the disadvantages that their binding to the bilayer must be quantitated in order to obtain accurate quenching data (Blatt and Sawyer, 1985) and that high fatty acid levels can perturb membrane structure (London, 1982). The quencher groups most commonly used are dibromo or nitroxide derivatives. These groups quench only nearby fluorophores, although contact between quencher and fluorophore is not necessarily required (Green *et al.*, 1990).

The amount of quenching is determined from the ratio of fluorescence in a sample containing the quencher (defined as intensity F) to that in a similar sample in which the quencher is omitted (defined as F_0). The pattern of F/F_0 versus depth of the quenching group is used to calculate the depth of the fluorescent group, as described below.

II. DETERMINING DEPTH FROM FLUORESCENCE QUENCHING BY EMPIRICAL INSPECTION OF QUENCHING LEVEL

The first fluorescence quenching method developed to measure depth compares quenching by a series of fatty acid-attached quencher groups located at different depths, in order to identify the quencher that gives the most quenching, usually as judged from a Stern–Volmer analysis (for reviews see Blatt and Sawyer, 1985; London, 1982). It should be noted that, despite the use of the Stern-Volmer analysis, quenching by nitroxide fatty acids is unlikely to involve a true collisional process (Chattopadhyay and London, 1987; London, 1982). In a later variation, a method was developed involving identification of the strongest quencher among a series of brominated lipids (Markello *et al.*, 1985). In either method, the depth of the most strongly quenching quencher defines the approximate depth of the fluorophore. A limitation of these methods is that if the fluorescent molecule is either shallower or deeper than all of the quenchers used, only a lower or upper limit to fluorophore depth may be obtained. A second limitation arises from the fact that deeply located fluorescent molecules can be significantly quenched by quenchers in both halves (i.e., both leaflets or monolayers) of the bilayer. This can result in strongest quenching by the deepest quencher in cases in which a fluorophore is actually closer in depth to a shallower quencher. The final limitation is that a series of quenchers located at as many depths as possible must be used in order to pinpoint depth to high resolution.

III. DETERMINING DEPTH BY PARALLAX ANALYSIS

Parallax analysis (PA) was developed in order to calculate a more precise average depth using a minimal number of quenchers. It applies most precisely when quenching has a simple radial dependence on fluorophore–quencher distance, when the lateral distribution of quenchers is random, and when the fluorophore–quencher distance does not appreciably change during the time the fluorophore is in the excited state [and thus it is unsuitable for analysis of quenching of long-lifetime pyrene derivatives (Sassaroli *et al.*, 1985)]. The appropriate analytical expression for the distance of a fluorophore from the center of the bilayer (Z_{cf}) is (Chattopadhyay and London, 1987)

$$Z_{cf} = [-\ln(F_1/F_2)/\pi C - L_{21}]/2L_{21} + L_{c1},$$

where F_1 is the fluorescence intensity in the presence of a shallower quencher, F_2 is the fluorescence intensity in the presence of a deeper quencher, C is the concentration of the quencher in two dimensions, L_{21} is the difference in the depths of the deep and shallow quenchers, and L_{c1} is the distance of the shallow quencher from the bilayer center.

In principle, PA only requires two fluorescence measurements, one with each of two quenchers. However, in practice, three quenchers, well spread out in terms of depth, are usually necessary. This is because the accuracy of the analysis is compromised if a quencher is used that is very far from the fluorophore in terms of depth. The nitroxide quenchers most commonly used for parallax analysis are derivatives of phosphatidylcholine (PC). Some examples are TempoPC, which has a tempocholine headgroup (located about 20 Å from the bilayer center), 5SLPC, which has a doxyl group on the 5-carbon of the 2-position fatty acyl chain (located about 12 Å from the bilayer center), and 12SLPC, which has a doxyl group on the 12-carbon of the 2-position fatty acyl chain (located about 6 Å from the bilayer center). Depth is calculated by applying the parallax equation to the fluorescence intensity data from the samples incorporating the two quenchers closest to the depth of the fluorophore. The appropriate quenchers are easily identified, because quenchers giving the strongest quenching are almost always the closest in depth to the fluorophore. In other words, for a deep fluorophore, depth is determined using data from one sample containing 5SLPC and a second containing 12SLPC, whereas for a shallow fluorophore a sample containing TempoPC and a second containing 5SLPC are used. PA can also be applied to brominated lipid quenching, but the shorter range of quenching means that the analysis is inaccurate if the two quenchers used to calculate depth are not within a few angstroms of the fluorophore depth (Abrams and London, 1992; Ladokhin and Holloway, 1995b).

One case that must be treated specially is when a fluorophore is close to the 5SLPC in depth and also almost equidistant from the TempoPC and 12SLPC quenchers, so the latter two molecules quench to a similar degree. In these cases, the average depth calculated from both of the quencher pairs noted above is more accurate than that from either of the individual pairs. Somewhat arbitrarily, averaging is applied when the ratio of fluorescence intensity in the presence of TempoPC to that in the presence of 12SLPC falls within the range 0.95–1.05 (Abrams and London, 1993).

A second case that must be treated specially is that of a very deeply located fluorescent group. This is the case in which a fluorophore can be quenched both by those quenchers in the half of the bilayer containing the fluorophore (the *cis* monolayer) and those quenchers in the half not containing the fluorophore (the *trans* monolayer). Depending on the space taken up by the molecule carrying the fluorophore, different equations must be used for the determination of depth when *trans* quenching is significant. One expression applies if a molecule is restricted to one monolayer (e.g., a fluorescent lipid) (Chattopadhyay and London, 1987) and a different one when it is transmembranous (e.g., a transmembrane helix) (Ren *et al.,* 1997). In either case, when a Z_{cf} of less than about 5 Å is obtained using the simple parallax analysis equation shown above, then a more accurate Z_{cf} must be calculated using the modified parallax analysis equations.

The accuracy of PA is affected by a number of factors. For example, both the distribution of fluorophores over depth and fast anisotropic motions can affect quenching, and their effects have been analyzed theoretically (Abrams and London, 1992). Restrictions on the lateral approach of fluorophore and quencher must also be considered (see below). Many of these variables average out, and for small molecules and peptides, it appears that accurate depths can usually be calculated. This is confirmed to some degree by experimental agreement with depths calculated by other methods, including other quenching analyses (e.g., Abrams and London, 1992; Kaiser and London, 1998a). Studies with a series of transmembrane peptides in which the Trp position is varied have given the best data in this regard, because values for the depth of Trp residues can be estimated from helix dimensions (Ren *et al.*, 1997). Under favorable conditions it appears that average depth can be determined to an accuracy of 1–2 Å.

IV. DETERMINING DEPTH BY DISTRIBUTION ANALYSIS

A. Physical Principles: Depth Distributions in Membranes

The modern view of membrane structure has come a long way in the last decade, revealing the great complexity of transverse organization seen both experimentally (White and Wiener, 1995) and computationally (Pastor *et al.*, 1991). We no longer consider the lipid bilayer to be a two-dimensional slab with well-defined boundaries between hydrocarbon, polar headgroups and aqueous phase. The joint refinement method of Wiener and White (1991, 1992) for X-ray and neutron data, collected on oriented hydrated lipid multilayers, allows determination of fully resolved bilayer "structures." Because of the great thermal disorder associated with fluid bilayers, such "structures" are defined operationally as the time-averaged spatial distributions of the principal structural groups projected onto an axis normal to the bilayer plane, represented as Gaussian functions. Consonant with this "distributional" view of the bilayer, the distribution analysis for membrane quenching (Ladokhin, 1993, 1997, 1999a,b; Ladokhin and Holloway, 1995b; Ladokhin *et al.*, 1993a) uses the Gaussian function to fit the fluorescence quenching profile.

An example of a recently resolved structure of a bilayer formed of brominated lipids and containing an amphipathic helical peptide 18A (Hristova *et al.*, 1999) is presented in Fig. 1 (lower panel). A peptide characterized by well-demarcated hydrophobic and hydrophilic helical sides was found to be located on the level of the glycerol group of the lipid. The quenching profiles resulting from application of distribution analysis (DA) (Ladokhin, 1997) to data reported by Mishra and Palgunachari (1996) for three peptides of the same family are presented in the upper panel. The insert in the middle is an in-scale molecular model of helical peptide (shown as a backbone ribbon with a single tryptophan side chain on the hydrophobic face) oriented with helical axes perpendicular to the plane of the figure. When the helix is placed at the depth determined by X-ray refinement, the position of Trp corresponds nicely to the peaks of quenching profiles. This comparison not only validates the lower resolution fluorescence method, but, more importantly, it provides an illustration of what kind of answer

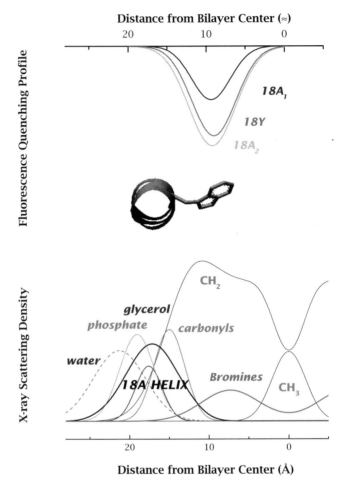

FIGURE 1 Comparison of X-ray diffraction and fluorescence quenching for determining the membrane penetration of an amphipathic helical peptide. The transverse distribution of 18A helix and lipid-attached bromines, as well as those for the other membrane components, as determined by X-ray diffraction on oriented multilayers (Hristova *et al.,* 1999), are shown in the lower panel. The upper panel contains tryptophan quenching profiles generated by the distribution analysis technique (Ladokhin, 1997) from measurements of three analogue peptides, $18A_1$, $18A_2$, and 18Y, bound to vesicles in solution (Mishra and Palgunachari, 1996). An in-scale molecular model of 18A-like peptide (*insert*) represents an axial view of an α helix with a single tryptophan side chain, located on the hydrophobic face. Placing the helix at a depth determined by X-ray diffraction naturally positions the tryptophan side chains at a depth determined in a fluorescence quenching experiment.

one should expect from depth-dependent quenching. The resulting profile is expected to be a fairly broad distribution, covering a substantial range of bilayer depth.

B. Theory

In a depth-dependent fluorescence quenching experiment one determines the fluorescence intensity F of a probe as a function of the known depth of the quencher h. The intensity in the absence of quenching is F_0. In its more general form, DA fits the quenching profile with the sum of two mirror-symmetrical Gaussian distributions G (with a total of three independent parameters),

$$\ln \frac{F_0}{F(h)} = \text{quenching profile} = G(h) + G(-h), \tag{1}$$

where

$$G(h) = \frac{S}{\sigma\sqrt{2\pi}} \exp\left[-\frac{(h - h_m)^2}{2\sigma^2} \right]. \tag{2}$$

The requirement of two symmetrical Gaussian components in Eq. (1) arises from the need to account for the transleaflet quenching. Regardless of whether the peptide is distributed in one leaflet or two, there is always going to be a quencher in the opposite leaflet. Due to the inherent thermal disorder of the bilayer (Pastor *et al.*, 1991; White and Wiener, 1995), this quenching should not be neglected (Ladokhin, 1997, 1999a).

The three fitting parameters of DA have the following meaning:

1. The mean depth h_m corresponds to the most probable transverse position of the fluorescent probe.

2. The dispersion σ arises from several broadening terms, such as the finite size of the probe and the quencher and the widths of their thermal envelopes. A substantial inherent width of the thermal envelope is a universal feature of the transverse distribution of membrane moieties (see Fig. 1). Additional heterogeneity could arise from multiple conformations, as has been suggested for interfacially bound and inserted melittin (Ladokhin and Holloway, 1995a; Ladokhin *et al.*, 1993a).

3. The area under the curve S is proportional to the total quenching. It can be represented as a product of the inherent quenching constant γ (determined by the nature of the quenching mechanism), the excited-state lifetime in the absence of quenching τ, the degree to which the probe is exposed to the lipid phase ω, and the concentration of the quencher C:

$$S = \gamma \cdot \omega \cdot \tau \cdot C. \tag{3}$$

The S parameter can be used to quantitate the exposure of a fluorophore to the lipid phase and determine the shielding of a tryptophan side chain by a protein moiety (Ladokhin, 1999b).

C. Definition of Quenching Profile: $\ln(F_0/F)$ or $F_0/F - 1$

The DA method traditionally defines the quenching profile as the logarithm of the ratio of intensities with and without quenching [Eq. (1)] and fits it with the Gaussian function. The closely related method of Sassaroli et al. (1995) fits a Gaussian to a depth profile of Stern–Volmer constants ($F_0/F - 1$). The original rationale for using the logarithm in distribution analysis was based on a modified formalism of "volume of action" (Ladokhin, 1997), which assumes that a restricted transverse diffusion is the principal event in dynamic quenching. This and other theoretical quenching equations (Chattopadhyay and London, 1987) are supported by the empirical observation that the *logarithm of intensity* is proportional to the surface concentration of quenchers (Chung et al., 1992). This was demonstrated most clearly for tryptophan fluorescence, which has a relatively short lifetime. However, for pyrene, a fluorophore with a long lifetime, the ratio of intensities was found to show linear change with quencher concentration (Sassaroli et al., 1995). Perhaps the difference lies in various relative contributions of lateral diffusion, which is expected to contribute substantially to the quenching of pyrene, but not that of tryptophan.

Let us consider what kind of variation in DA parameters results from these two definitions of the quenching profile, using the data for bromine quenching of 18A-like peptides (Mishra and Palgunachari, 1996) discussed above (see Fig. 1). The parameters of the Gaussian fits for the $\ln(F_0/F)$ or for $F_0/F - 1$ are presented in Table I. The results indicate that the choice has absolutely no effect on the

TABLE I

Effect of the Choice of the Quenching Profile (QP) Definition on the Resulting DA Parameters[a]

Peptide	QP $= \ln(F_0/F)$			QP $= F_0/F - 1$		
	h_m (Å)	σ (Å)	S/S_{18A1}	h_m (Å)	σ (Å)	S/S_{18A1}
18A$_1$	9.4	2.5	1.0	9.4	2.3	1.0
18A$_2$	9.2	3.0	2.3	9.2	2.6	2.5
18Y	9.0	2.8	1.9	9.0	2.4	2.1

[a] A set of quenching data for a family of closely related amphipathic helical peptides from Mishra and Palgunachari (1996) was analyzed using distribution analysis [Eqs. (1) and (2)] assuming QP $= \ln(F_0/F)$ or QP $= F_0/F - 1$. In both cases, the resulting parameters are very close to each other, if not identical.

determination of the mean depth, and affects the other parameters slightly. Apparently either definition can be used to describe a quenching experiment.

V. COMPARING PARALLAX ANALYSIS AND DISTRIBUTION ANALYSIS

In many cases, parallax analysis and distribution analysis give the same answer (Breukink *et al.,* 1998; Kaiser and London, 1998a). However, in other cases, one method clearly performs better than the other. In several situations, DA has an advantage. For example, when there is incomplete binding of a fluorophore, PA requires a correction for the fluorescence of unbound molecules, but DA does not. In another example, brominated lipids quench over such a short range that PA can only be applied to bromine quenching from quenchers very close to the fluorophore in depth (Abrams and London, 1992; Ladokhin and Holloway, 1995b). DA does not suffer from this limitation (Ladokhin and Holloway, 1995b). DA can also more easily handle cases in which the lateral approach of a quencher to the fluorophore is (uniformly) limited by steric effects, as when a fluorophore is attached to a protein (Ladokhin, 1999a). A different issue involving accessibility to quencher occurs when a fluorophore-bearing molecule binds the quencher-bearing lipid more strongly than unlabeled lipid, or vice versa (see below). Again, this poses a difficulty for PA, but not for DA. It should also be noted that DA is more useful for deriving quenching parameters other than average fluorophore depth (Ladokhin, 1999a,b; Ladokhin and Holloway, 1995b).

PA has advantages when data are only available from a small set of quenchers. When the quenchers used are either all deeper than the fluorophore or all shallower than the fluorophore, DA can be too sensitive to both the shape of the function describing the amount of quenching versus depth and any experimental error in the quenching measurements to allow accurate measurement of depth (Kaiser and London, 1998a). In addition, although DA can be superior at defining depths when multiple populations of fluorophore exist at different depths (Ladokhin, 1999a), sometimes only PA can be used to obtain (crude) distributional information (Asuncion-Punzalan *et al.,* 1998; Kaiser and London, 1999b), because quenchers are available at too few depths and/or are spaced too closely to one another in depth to use DA.

We recommend trying to apply both methods to quenching data whenever possible. Agreement between the results greatly reduces the possibility of misinterpretation of the data due to type of analysis used. When the results disagree, a decision can be made about which method is more accurate for the particular experimental conditions.

VI. INVESTIGATING MEMBRANE PROTEIN TOPOGRAPHY BY VARYING FLUOROPHORE DEPTH INSTEAD OF QUENCHER DEPTH

Another approach to the analysis of the quenching of peptides or proteins uses a single quencher with a series of fluorophores placed at different positions along the amino acid sequence. The closer a fluorophore is to a quencher in depth, the greater should be the quenching. This type of experiment requires the assumption that the depth of a fluorescent group does not affect its intrinsic sensitivity to quenching. This assumption can fail when sensitivity to quenching is affected by the properties and environment of the fluorophore. For example, sensitivity to quenching increases when fluorescence lifetime increases. Deeply buried fluorophores could have very long lifetimes, due to the decreased polarity in the bilayer interior, and thus be extra sensitive to quenching.

One way to circumvent complications due to increasing sensitivity of fluorophore to quenching as a function of depth is to use a very deep quencher (McKnight *et al.,* 1991a; Shatursky *et al.,* 1999). In such a case, as the fluorophore is positioned more and more deeply quenching increases both due to increasing proximity to the quencher and increased vulnerability to quenching. This utility of this approach is demonstrated by the elegant study of perfringolysin-O structure by Shatursky *et al.* (1999). They were able to define the depth of each residue in transmembrane β-sheet hairpins by the identification of the residues that 12SLPC quenched maximally (i.e., those at the bilayer center) and minimally (i.e., those at the membrane surface).

This method is attractive for a number of reasons. First, it only requires one quencher. Second, the exact concentration of quencher is not important, only that it be the same in each sample. Third, no theoretical assumptions are needed. On the other hand, this method requires preparing a whole series of fluorescently labeled proteins. Furthermore, exposure to lipid must be equivalent for each residue examined. These requirements can be met with transmembrane β sheets and in simple transmembrane helices (Ren *et al.,* 1997), but perhaps not for complex multihelix proteins, in which a section of a transmembrane helix may be partly shielded from lipid by contact with other helices. Despite these limitations, this method should often be attractive for analysis of membrane protein topography.

VII. A CAUTIONARY NOTE CONCERNING QUENCHING ANALYSIS

It should be pointed out that the first and most important type of analysis is visual examination of the raw quenching data. This is especially important for depth-dependent fluorescence quenching in membranes, because of the limited number of quenchers available. Any mathematical treatment of a set of data from two to five quenchers inevitably relies on certain assumptions. In our experience, visual examination of the quenching profile can be instrumental in avoiding application of an inappropriate mathematical analysis. There are cases in which the raw data clearly invalidate an indiscriminately applied quenching analysis. For example, quenching profiles that are flat (quencher depth-independent) or concave (stronger quenching by shallow and deep quenchers than for those at intermediate depths) are usually indicative of multiple fluorophore populations, and calculation of a single average depth from pairwise analysis and subsequent averaging will most often result in misleading conclusions.

VIII. PROBLEMS THAT CAN AFFECT THE ACCURACY OF QUENCHING MEASUREMENTS

There are a number of problems that can plague quenching experiments no matter what type of analysis is used. For example, it is not necessarily true that each lipid in a preparation of quencher carries the quencher group. Therefore, it is important to calibrate the concentration of the quencher (Abrams and London, 1992, 1993). The fewer the number of quenchers used, the more critical the calibration. Another concern is the presence of unwanted chemical reactions that either can occur between the quencher and fluorophore or be affected by the presence of the quencher (Kachel et al., 1998). Appropriate controls and/or conditions that avoid such reactions must be chosen (Kachel et al., 1998).

Several other problems can arise because a quencher lipid may not behave identically to natural lipids. This is generally a more serious problem for nitroxide-labeled lipids than for brominated lipids. In one case, it was found that peptide binding to model membranes was sufficiently perturbed by nitroxide lipids (but not brominated lipids) to prevent accurate measurement of depth (Volgino et al., 1999). The possibility of quencher lipids perturbing the conformation of the molecule to which the fluorophore is attached is also of concern. These studies might make it seem that brominated lipids should be preferred for quenching studies in general. This is not the case, because brominated lipids can only quench Trp and related fluorophores, whereas nitroxide lipids can quench almost any type of fluorophore.

Another difficulty arising from the difference between quencher and natural lipids is the assumption that the labeled and natural lipids are mixed randomly within the bilayer. For example, specific binding of an anionic lipid to a cationic peptide would result in a nonrandom distribution if lipids with different polar headgroups are present in a sample. More exaggerated nonrandom distributions occur in cases in which membranes contain lipids in an ordered state, such as the gel state or liquid ordered state. Quenching groups generally impart a low melting temperature T_m to the lipids to which they are attached, resulting in their tendency to form the ordinary fluid liquid crystalline state at room temperature. Thus, they can form separate highly fluid domains that are laterally separated from those enriched in high-T_m unlabeled lipids (Ahmed et al., 1997).

The effect of protein on the lipid bilayer must also be considered. For example, lipids in contact with protein may not have the same conformation they have in the rest of the bilayer (Mangavel *et al.*, 1998; Tsukihara *et al.*, 1996) and this would affect quencher depth. This will be especially serious in cases in which polar headgroups are dragged into the core of the bilayer. A related problem is that a protein may have a shape resulting in a depth-dependent steric hindrance of close approach between quencher and protein-linked fluorophore. In this case, quenchers at different depths cannot approach the fluorophore laterally to the same degree. For example, a protein that only partly penetrates a bilayer could selectively block a shallow quencher from approaching a fluorophore, and as a result, the fluorescent group would appear to be deeper than its true position. This is particularly serious, because it is not possible to correct for this effect. As a result, the apparent depths obtained when depth is measured for fluorescent groups attached to large, complex proteins may be not be as accurate as those obtained for small fluorescent molecules. However, it is not clear that this type of perturbation is serious for peptides.

Generally one would expect a small perturbation of the position of quencher by proteins that have uniform cross section along the transverse axis (α-helical bundle, β barrel) relative to interfacial α helices. Surprisingly, even in the latter case, the perturbation could be quite modest. For example, the diffraction study of Hristova *et al.* (1999) shows a <1-Å shift in position of bromines upon incorporation of a large amount of an 18A-type peptide.

Limitations of fluorescent labeling itself must not be overlooked. Even if there is no perturbation of protein or peptide structure, a fluorescent probe attached to a protein by a flexible linker may occupy a depth different than that of the residue to which it is attached.

Finally, accurate measurements of intensity can be a problem in membranes. Many membrane quenching studies involve steady-state fluorescence measurements in the presence of a significant level of light scattering. One often overlooked effect of scattering is the wavelength-dependent loss of emitted light, which hinders measurements of both spectral position and fluorescence intensity. Experimental schemes that minimize and/or correct for scattering artifacts through manipulation of optical path, polarizer orientation, or the use of a reference fluorophore can be necessary (Ladokhin *et al.*, 2000).

IX. STUDIES OF MEMBRANE-INSERTED PEPTIDES BY FLUORESCENCE QUENCHING

A. Introduction

Membrane-interactive peptides are an attractive target for quenching studies. The relatively small size of peptides alleviates some of the difficulties that plague quenching studies of proteins. In addition, the ability to control peptide sequence provides an excellent opportunity to answer systematic questions about the sequence/behavior relationship. Control over sequence also makes it quite practical to generate a series of appropriate analogues with single Trp or single fluorescently labeled Cys (or sometimes Lys) residues.

Furthermore, studies on peptides can yield much information. There are numerous natural bioactive peptides that interact with membranes, and how they interact with membranes is critical for understanding their function. Studies on synthetic membrane-inserting peptides are just as important, because they have been an important tool for understanding membrane protein structure and function.

B. Amphipathic Peptides

Amphipathic helices tend to locate close to the membrane surface. Quenching has been used to determine how deeply such peptides are buried in the membrane. For example, Johnson and Cornell (1994) investigated an amphipathic helix-forming peptide derived from the sequence of CTP:phosphocholine cytidylyltransferase. In the native protein, this sequence has been proposed to form the membrane binding domain. The depth of Trp on the hydrophobic face of the helix was assessed using a brominated PC (BrPC) mixed in a 1:1 ratio with phosphatidylglycerol (PG). Quenching by bromines was strongest when they were attached to both the 9 and 10 fatty acyl carbons (9,10-bromo). The weaker quenching by bromines on the 6- and 7-carbons (6,7-bromo) was similar to that of bromines at the 11- and 12-carbons (11,12-bromo). Together these results suggest a Trp depth closest to the 9-carbon of the fatty acyl chains.

Several groups have studied the behavior of synthetic amphipathic peptides. In one of the more comprehensive and careful studies, Chung et al. (1992) examined an amphipathic pore-forming peptide with alternating blocks of Leu and Ser. Analogues with single Trp substitutions at a series of consecutive positions were synthesized to allow analysis of peptide structure within the membrane. These peptides had a helical structure, and conventional fluorescence techniques readily demonstrated that the Ser residues were exposed to water, whereas the Leu residues were exposed to lipid. However, these methods were unable to distinguish between a transmembrane structure with an aqueous pore and one in which helices were oriented parallel to the membrane surface. Analysis of quenching by nitroxide-labeled phospholipids demonstrated that the peptide was oriented parallel to the membrane surface, rather than in the transmembrane form.

Parente et al. (1990) studied a 30-residue artificial amphipathic sequence that had the ability to trigger membrane fusion at low pH. To examine membrane penetration by this peptide, the quenching of a Trp residue on the N-terminus by a cholesterol dibrominated on the 6- and 7-carbons was compared to the quenching by 9,10-BrPC. On a molar basis, the cholesterol was a stronger quencher than the PC, suggesting a relatively shallow location of the Trp residue. As the authors realized, the quenching of these two molecules is difficult to compare, given the differences in their structure, and more precise conclusions would require a more completely analogous and complete quencher series.

Mishra and Palunachari (1996) studied the behavior of amphipathic peptides meant to mimic the amphipathic helices of apolipoproteins. Three peptides, representing the A1, A2, or Y helix classes, were examined. Analysis of quenching by brominated lipids showed relatively small differences in the depth of a Trp residue place in position 6 on each helix, which always fell 8–10 Å from the bilayer center.

Mangavel *et al.* (1998) examined the behavior of an amphipathic peptide composed of alternating Lys and Leu blocks. A single Trp was introduced at four different positions along the helix. A helical wheel diagram predicted that the Trp should have given a series of different (although similar) depths 1–5 Å deeper than the depth of the helix center axis if the peptide orients parallel to the bilayer surface. In each case, stronger quenching was observed with a 6,7- and 9,10-BrPC than with an 11,12-BrPC. Close inspection of the quenching data suggests that the depths of the different Trp residues did vary to some degree. However, the efficiency of quenching by brominated lipids did not exactly parallel predicted values, leading the investigators to speculate that specific electrostatic interactions between the lipid phosphate and lysine were important in determining lipid (and therefore bromine) position relative to the peptide, and thus that these interactions, rather than peptide depth, controlled the level of quenching observed. It is also possible that the flexible positioning of the Trp side chains relative to the bilayer and its natural preference for the membrane interface (Kachel *et al.*, 1995; Yau *et al.*, 1998) also contributed to a difference in Trp depth from predicted values.

C. Bioactive Peptides

Because of their importance in drug development, the membrane interactions of bioactive peptides have been extensively studied. Often, they are hydrophobic molecules that bind strongly to lipid bilayers and form pores. However, proof that pore formation is their mechanism of action *in vivo* is often difficult. Furthermore, these peptides often appear to be amphipathic molecules, and may predominantly localize near the membrane surface. Many studies are still at the stage of characterizing what structure predominates in membranes. Even when this is known, it should be kept in mind that determining whether the predominant and the pore-forming structures are equivalent is a difficult problem, and one that is often left unsolved.

In what may have been the first example of combining a systematic series of Trp substitutions with lipid quenchers, Voges *et al.* (1987) used quenching by nitroxide fatty acids to investigate the orientation of alamethicin analogues in which the Trp position was varied. They found a pattern of quenching dependent upon Trp position, which (after correction for fatty-acid-binding differences) was consistent with a transmembrane structure, with Trps close to the center of the sequence being more strongly quenched by deeply locating nitroxides, and Trps at the ends of the peptide being more strongly quenched by shallow nitroxide groups. There was some variability in the exact quenching profiles, and this limited the resolution of the data in terms of pinpointing Trp depth.

Nitroxide quenching has also been used to study nisin, a 34-residue peptide with activity against Gram-positive bacteria. Nisin is a hydrophobic, cationic peptide with several complex post-translational modifications and no Trp. Recently, it has been shown that the lipid A of the bacterial outer membrane is a functionally important binding site for nisin (Breukink *et al.,* 1999).

Studies of single-Trp analogues have been undertaken to examine the orientation of membrane-inserted nisin. In one study a Trp was placed at residue 30, near the C-terminus (Martin *et al.,* 1996). A deep location was found for this Trp, as ascertained by parallax analysis of nitroxide phospholipid quenching. This led to a proposal of deep membrane penetration by the C terminus of nisin. Deeper penetration was obtained in PC/cardiolipin mixtures than in PC/phosphatidylethanolamine (PE) mixtures.

In a second study, Trp was placed near the N-terminus (residue 1), C-terminus (residue 32), or peptide center (residue 17). Quenching by nitroxide-labeled PCs was measured in two different PG/PC mixtures, and depth calculated both by parallax analysis and distribution analysis (Breukink *et al.,* 1998). The two analyses gave very similar results. It was found that the Trp residues are close to the polar region of the bilayer in 1:1 PC/PG mixtures, but all become more deeply buried in the bilayer as PG content is increased, in agreement with acrylamide quenching experiments. Judged from the three Trp analogues, the N-terminal of the peptides seemed to insert more deeply than the remainder of the peptide.

It should be noted that the quencher concentrations were not calibrated in this study, so the absolute values of depth have some uncertainty. Thus, it is unclear whether there is a real difference in depth between residue 30 in this study and residue 32 in the previous study. On the other hand, the relative differences observed in depths by Breukink *et al.* (1998) should not be significantly affected by calibration. Futhermore, both studies suggest deeper penetration of the bilayer when anionic lipid concentration is high.

Another study measured the location of a Tyr residue in the center of the antimicrobial peptide epilancin K7 (Driessen *et al.,* 1995). From inspection of quenching levels by nitroxide-labeled PC, Tyr was found to locate deeply in PC vesicles (near the depth of the 10-carbon of an acyl chain), but more shallowly in PG-containing vesicles (closer to the 5-carbon of an acyl chain).

Depth studies have also been carried out on magainin 2 (Matsuzuki *et al.,* 1994). This antimicrobial peptide from *Xenopus* forms an amphipathic helix in bilayers. Several Trp substitutions were made on the putative lipid-facing side of the helix. Each gave a moderate depth for what is likely to be the most deeply embedded portion of the peptide, and all were consistent with a non-transmembrane surface location of the helix. However, it must be pointed out this report provided calculated Z_{cf} values without providing the raw quenching data. This makes it difficult to assess whether there are multiple peptide conformations (see Section VII), a possibility suggested in other studies (Hirsh *et al.,* 1996; Ludtke *et al.,* 1996). Raw data should be included when reporting quenching results.

The existence of multiple conformations for the bee venom peptide melittin when membrane-bound has been suggested from a combination of iodide and BrPC quenching studies of its single tryptophan (Ladokhin and Holloway, 1995a; Ladokhin *et al.,* 1993a). Application of distribution analysis to a collection of quenching profiles measured with a set of four bromolipids at different incubation times allowed the monitoring of the dynamics of the conversion from the interfacial to transmembrane conformation. A transmembrane form has been implicated in pore formation by melittin and many other host-defense peptides (Huang, 2000; Ladokhin *et al.,* 1997; Vogel and Jähnig, 1986).

Mastroparans are peptide components of wasp venom. Fujita *et al.* (1994) found that the native Trp of mastoparan X, which is close to the N-terminus of this 14-residue peptide, was quenched much more strongly by a nitroxide attached to the 5-carbon of a free fatty acid than by a nitroxide attached to the fatty acid 12- or 16-carbon, consistent with a very shallow Trp location. (However, no corrections for differential fatty acid binding to membranes were made.) An anthracene group attached at the C-terminus was also preferentially quenched by a shallow nitroxide. However, the difference between quenching by the shallow and deeper nitroxides for the anthracene was much smaller than that for the Trp.

In another example, a 24-residue fragment of the peptide hormone adrenocorticotropin (ACTH) was studied (Moreno and Prieto, 1993). A Trp at position 9 was substantially more quenched by a nitroxide attached to the 5-carbon of a fatty acid chain than by one attached to the 16-carbon. This indicated a shallow location, and was in agreement with the fairly red emission of the Trp. Quenching values were corrected for both fatty acid unbound to membranes (using literature values for binding) and the fluorescence of unbound ACTH. However, the physical state of the membrane, which was composed primarily of the high-T_m lipid dipalmitoyl PC, could have affected the lateral distribution of quenchers (see above) and thus the quenching, which was apparently studied at room temperature.

D. Model Hydrophobic/Transmembrane Helices

Another type of membrane-inserted peptide that has become important in studies of membrane protein folding is the synthetic transmembrane helix. Simple model peptides containing a long poly(Leu) sequence flanked by Lys residues have become useful model systems for examining transmembrane helices. By inspection of relative quenching levels, Bolen and Holloway (1990) showed that a Trp placed in the center of the Leu sequence was located at the bilayer center as assessed by BrPC quenching. This meant the single Trp did not significantly disrupt the transmembrane structure of the peptide, although a minor population of peptide at the membrane surface was also detected. The distance dependence of quenching was interpreted in terms of an energy transfer-like process solely to explain the broad quenching profile. However, the quenching profiles have a substantial inherent width due to distributional factors, and so the same data can be explained assuming collisional or other near-contact-dependent mechanisms and accounting for additional broadening caused by the transleaflet quenching (Ladokhin, 1999a).

An additional study of the depth of a poly(Leu) peptide with 17 hydrophobic residues and a central Trp residue was performed using nitroxide-labeled fatty acids by Grenier *et al.* (1998b). As in the study of Bolen and Holloway, populations with a Trp location at the bilayer center and a surface bound state were proposed.

Ren *et al.* (1997) looked at a series of poly(Leu) peptides with Trp placed in different positions. These peptides had a slightly longer poly(Leu) sequence than that used by Bolen and Holloway (19 vs. 17 hydrophobic residues). The depths calculated by parallax analysis of quenching by nitroxide-labeled phospholipids were in reasonable agreement with the expected transmembrane structure. This allowed the authors to determine the correlation between the λ_{max} of a lipid-exposed Trp and its depth in the bilayer. A second study by Ren *et al.* (1999) showed that the Trp λ_{max} of poly(Leu) peptides could be affected by oligomerization under conditions where the Trp remained at the bilayer center. Presumably, this results from the difference in local environment of a Trp in contact with a neighboring polypeptide chain and one in contact with lipid. Whether or not Trp–Trp contacts contribute to the effect of oligomerization on λ_{max} was not determined. It was also found in this study that a very short poly(Leu) sequence (11 hydrophobic residues) resulted in a peptide location close to the bilayer surface, as judged by parallax analysis of quenching induced by nitroxide-labeled phospholipids.

Quenching has also been applied to poly(Leu) peptides in which different polar or ionizable residues are introduced into the hydrophobic core of the peptide (Lew and London, 2000). At neutral pH, these peptides all maintained a transmembrane state based on the observation that parallax analysis of quenching by nitroxide-labeled phospholipids of a Trp residue at the center of the peptide located it at the bilayer center. Quenching also indicated that the transmembrane conformation was maintained after deprotonation of the helix-flanking Lys residues at high pH, except in the case of a peptide with an Asp in the hydrophobic core.

Substitution of guest residues into a poly(Ala) helix was studied by Liu and Deber (1997). Increasing the polarity of a polar substituent increased the exposure of a Trp in the hydrophobic core to aqueous quenching agents. Quenching by nitroxide-labeled phospholipids showed quenching becoming progressively stronger as the nitroxide location was moved toward the end of an acyl chain, consistent with a deep Trp location. Surprisingly, there was no significant difference in the depth dependence of nitroxide quenching with substituents of different polarity. This may partly be a result of the Trp being offset three residues from the center of the hydrophobic sequence. This would give a Trp location at approximately 4–5 Å from the bilayer center in the transmembrane state, and this would make it relatively difficult to distinguish it from a non-transmembrane state in which the Trp was still buried in the bilayer. The authors concluded that these peptides might form a mixture of transmembrane and surface-bound orientations.

E. Membrane Penetration Depth of Very Small Synthetic Peptides

Small hydrophobic peptides have been valuable for studies of the influence of hydrophobic interactions on the stability of peptide association with the membrane surface. In two studies, De Kroon et al. (1990, 1991) investigated the behavior of hydrophobic penta- and hexapeptides with a Trp one residue away from the C-terminal. In all cases, a shallow depth was found for the Trp using brominated lipids. Strongest quenching was observed with 6,7-bromoPC and a PC carrying one bromine attached adjacent to the carboxyl group, that is, to the 2-carbon. (This molecule was not fully brominated at the 2-carbon.) Taking into account the intrinsically weaker quenching of a single bromine relative to a dibrominated grouping (Bolen and Holloway, 1990) made it seem likely that these peptides locate very shallowly at the membrane/solution interface. When hydrophobicity was increased by blocking the N- and C-termini, quenching indicated a somewhat deeper peptide location.

F. Signal Sequence Peptides

Signal (leader) sequences are generally 20- to 25-residue N-terminal extensions of proteins that play an important role in targeting proteins for translocation across the membrane bilayer. Although it is now clear that their interactions with proteins are critical to their function, they have a highly hydrophobic core and can interact with lipids. Understanding this interaction could help define how lipids influence signal sequence behavior and function in vivo. Signal sequences are also intriguing, because their hydrophobic cores tend to be shorter than is typical for transmembrane proteins, and it is not immediately obvious what structure they would take upon interaction with a lipid bilayer.

Killian *et al.* (1990) studied the behavior of a PhoE signal sequence peptide. A Trp residue was introduced toward the C-terminal of the hydrophobic core as a probe. This substitution did not alter signal peptide function. Using the same series of brominated molecules as De Kroon *et al.* (see above), they found that the Trp occupied a shallow location (perhaps near the lipid/water interface) in cardiolipin/PC vesicles and an even more shallow location in pure PC vesicles. (Quenching was not extremely depth-dependent, which may mean that there was also a population much more deeply inserted than the predominant conformation.) It should be noted that these conclusions are somewhat dependent on the ability to interpret the quenching by the bromine on the 2-carbon.

Wang *et al.* (1997) studied the signal sequence of *E. coli* glucitol permease, using an analogue with a Trp in the center of the peptide. Their vesicles contained 25% phosphatidylserine (PS)/75% PC. They found strong and equal quenching by nitroxide labels on the 5- and 10-carbons of PC, suggesting a significant degree of insertion, with a Trp depth halfway between that of the two quenchers.

Gierasch and colleagues examined the behavior of the LamB signal sequence peptide in a series of studies. McKnight *et al.* (1991a) studied analogues with a single Trp residue placed near the N-terminal, C-terminal, and center of the peptide. Circular dichroism indicated that these analogues adopted a conformation similar to that of the wild-type peptide in terms of helix content, and the strength of binding to lipid vesicles was similar for all three Trp analogues. Quenching by aqueous iodide was consistent with the center Trp (18W) being most deeply buried within the bilayer and the C-terminal Trp (24W) being the shallowest. Quenching by nitroxide-labeled lipids led to similar conclusions. Quenching by PCs with nitroxides attached to either the 12- or 16-carbon was in the order 18W > 5W > 24W. As noted above, it is possible to order residue depth by a comparison of quenching by a single deeply buried nitroxide in this way.

An implied assumption of this approach is that the intrinsic sensitivity of Trp to quenching does not decrease with increasing depth. This assumption is of concern, because comparison of the data in McKnight *et al.* (1991a) for quenching by nitroxides attached to the 5- and 12-carbons of the acyl chain (i.e., 5SLPC and 12SLPC) suggests a slightly different order of depths. The ratio of quenching by the shallower nitroxide to that of the deeper nitroxide a parameter closely related to depth (see above) was lowest for the 18W, consistent with its deep location, but was lower for the 24W than the 5W Trp. By this criterion, the 24W residue was located more deeply than the 5W.

On the other hand, in a second study by McKnight *et al.* (1991b), the order of quenching of 18W by nitroxides attached to the 5- and 12-carbons was reversed. As the authors noted, this suggested that calibration of quencher concentration was an issue (and, as noted earlier, the calibration is critical in this type of analysis), and so they correctly did not compare 5SLPC and 12SLPC quenching values. In this study, the effect of attachment of the signal peptide to the mature protein on the membrane insertion of the signal peptide was studied. Based on a combination of fluorescence and quenching data, the authors concluded the mature protein had little effect on signal peptide interaction with lipid.

In a more detailed study, Jones and Gierasch (1994) revisited the issues of determining LamB signal peptide conformation by fluorescence and fluorescence quenching. They calibrated the quenchers, which again were PCs carrying nitroxides attached to the 5- or 12-carbon, by examining the quenching of a hydrophobic peptide likely to form a transmembrane structure. Varients with a Trp near the N-terminus, C-terminus, and peptide center were used. As expected, shallower depths were measured for the N- and C-terminal regions. They then examined signal peptides with a Trp in the 18 position, and variants also containing one acidic or one basic residue. Strongest quenching was obtained with the nitroxide attached to the 12-carbon in the absence of ionizable residues, suggesting deep insertion. Shallower depths were observed in the presence of the ionizable residues.

One intriguing result in this study was that an analogue with an Asp residue seemed to result in deep insertion, but without the expected blue shift in Trp fluorescence. The authors suggested that hydration around the Trp due to the presence of the COOH group could explain the anomalous red shift. Other possible explanations noted by the authors included direct interaction between the Asp COOH and Trp and decreased association with membranes. The fact that a strong blue shift was observed at low pH, which would protonate the Asp COOH and increase hydrophobicity, is consistent with any of these explanations. The possibility of a red shift due to a low degree of insertion is supported by the weak level of quenching achieved with nitroxides, but, as the authors noted, this could reflect a high degree of oligomerization, which would shield the Trp from lipid. It should be noted that weak quenching due to partial insertion would distort results obtained by parallax analysis unless correction is made for the fluorescence of unbound fluorophore (see above).

Voglino *et al.* (1999) reinvestigated LamB signal peptide orientation. They found that a PC carrying a nitroxide group attached to the carbon-7 of the fatty acyl inhibited the binding of a LamB peptide containing the Trp at position 18 to small unilamellar vesicles. Even 10% of this nitroxide lipid had an appreciable effect on binding to PC vesicles. The effect was alleviated when vesicles contained 35% PG, presumably due to electrostatic attraction of the peptide to the anionic PG molecules. An effect of quencher on peptide binding to PC vesicles was not observed with brominated lipids. Using brominated lipids, it was found that 6,7-BrPC gave more quenching than 9,10-BrPC. This suggested a shallow insertion, consistent with a "hammock" model in which the peptide would have both ends exposed to the aqueous solution on the same side of the membrane and a peptide center at the polar/hydrocarbon interface.

G. Studies on Peptide Sequences Found in Viral Membrane Proteins

Clague *et al.* (1991) studied the fusion peptide from the influenza hemagglutinin protein. This peptide is believed to trigger fusion between the viral and target membranes. They estimated that the Trp near the center of the peptide penetrated to a depth corresponding to the middle of a fatty acyl chain at both neutral and low pH. It should be noted that these investigators evaluated quenching via the effect of nitroxide groups attached to free fatty acids on the fluorescence lifetime. Several studies have reported a reduction in excited-state lifetime upon addition of lipid-attached quenchers (Clague *et al.,* 1991; Ladokhin and Holloway, 1995a,b; Ladokhin *et al.,* 1992; Sassaroli *et al.,* 1995). Under certain conditions, such as when quenching is fully collisional or when quencher/fluorophore distance is fixed at a single value, the fractional decrease in lifetime parallels the fractional decrease in intensity, and is subject to fewer of the artifacts that can invalidate intensity measurements. For example, lifetime measurements circumvent errors in determination of fluorescent probe concentration. They could also allow an easy correction for the contribution of unbound probe (provided the lifetime changes sufficiently upon membrane partitioning). These theoretical considerations become practical advantages mainly when the fluorescence decay is monoexponential in the absence of quenching. However, decay can be multiexponential, as predicted, for example, when there is a random lateral distribution of quenchers that exhibit a gradual distance-dependent quenching, or when a fluorophore (e.g., Trp) exhibits intrinsically multiexponential decay behavior. In such cases, much care must be taken in interpreting the lifetime data (Ladokhin, 1999a, 2000; Ladokhin and Holloway, 1995a).

Ruiz-Arguello *et al.* (1998) demonstrated that a putative fusion peptide from Ebola virus penetrated model membranes deeply, by comparing quenching by two brominated phospholipids. They found strong, almost equal quenching by 5,6- and 11,12-BrPC.

Kliger and Shai (1997) examined the membrane association behavior of a peptide corresponding to a sequence in the cytoplasmic tail of the human immunodeficiency virus (HIV) envelope protein. This 27-residue peptide had a leucine zipper-like sequence and four Trps in the N-terminal and central portions of the sequence. Quenching with a series of brominated lipids showed strongest quenching by the shallowest quencher groups (on 6,7-BrPC), but even the deepest quenchers gave significant quenching. Combining these results with other experimental data, they interpreted the results in terms of a non-transmembrane structure with shallow insertion by the N-terminal sequence and deeper insertion by the center of the peptide.

H. Other Peptide Sequences Derived from Membrane-Associating Proteins

Hanyu *et al.* (1998) studied the isolated S4 segment of a sodium channel. This segment is transmembranous in the whole protein, and is supposed to undergo a structural change in response to membrane potential. The investigators evaluated the fluorescence of a rhodamine derivative attached to the N-terminus of the peptide. Their study was particularly interesting because it involved simultaneous fluorescence and current measurements on peptides incorporated into planar bilayers. They used this system to assess the dependence of penetration depth on membrane potential. By comparing the quenching of iodide to that of 5SLPC, they concluded that there was a positive potential-induced deeper insertion, whereas quenching with a negative potential was no different from that without a membrane potential.

Ben-Efraim and Shai (1996) examined the MO portion of the ROMK 1 K channel. This segment is N-terminal to a transmembrane sequence. A Trp in the middle of the MO segment appeared to be located close to the membrane surface, as shown by the fact that quenching was strongest for 6,7-BrPC and negligible for 11,12-BrPC.

Various aspects of membrane penetration of cytochrome $b5$ and its membrane-binding nonpolar peptide (NPP) have been subjects of multiple quenching studies (Ladokhin 1999b; Ladokhin *et al.*, 1991, 1993b; Markello *et al.*, 1985; Tretyachenko-Ladokhina *et al.*, 1993). Tryptophans in various mutants were shown to be located 9–11 Å from the bilayer center. Distribution analysis revealed that the quenching profile of a single tryptophan in the NPP mutant has the same mean depth as in the entire protein. However, the dispersion is higher for the peptide, suggesting a higher degree of conformational flexibility. Heating results in additional disorder and an increase of the lipid exposure of the tryptophan side chain (Ladokhin, 1999b). It has been suggested that this conformational change might be related to temperature-induced conversion from exchangeable to nonexchangeable forms of cytochrome $b5$ (Ladokhin *et al.*, 1993b).

I. Peptide Mimics

It should also be noted that investigators have studied molecules that can mimic the properties of membrane-inserted peptides. Abel *et al.* (1999) used quenching to investigate the structure of channel-forming crown ether compounds in which a dansyl group was attached to the two macrocycle units at each end of a Tris macrocycle compound. The dansyl groups were estimated to be 14 Å from the bilayer center, based on analysis of quenching by nitroxide-labeled lipids, suggesting the potential for the Tris macrocycle to form a transmembrane structure.

Baumeister *et al.* (2000) synthesized a hybrid molecule with a peptide linked to a polyphenyl backbone. This molecule forms channels within lipid bilayers. Quenching of the octaphenylene core by nitroxides gave strong quenching that was almost equal for 5SLPC and 12SLPC. Combining these results with circular dichroism data indicating formation of a β-strand-dominated structure, they interpreted the quenching result as supportive of a β-barrel structure in the bilayer with a membrane-spanning octaphenylene orientation. Such an orientation would place the octaphenylene group in more or less equal proximity to quenchers at any depth in the bilayer. An alternative is that the octaphenylene group locates parallel to the membrane surface at a depth halfway between the nitroxide groups attached to the 5- and 12-carbons. However, this model is difficult to reconcile with the other structural data.

X. CONCLUSIONS

Membrane-inserted peptides are good targets for quenching studies, and the difficulties encountered when using large proteins may be less serious when small peptides are used. Nevertheless, an overview of the actual applications to peptide structure shows that a complete picture of the predominant conformation of a peptide in the bilayer has rarely been obtained. Trying to extrapolate the depth of the entire molecule from the depth of one or a very few specific positions is clearly one difficulty. Use of a systematic series of fluorescently labeled analogues, while difficult, may be necessary in many cases. A more careful application of quenching methods is also required. We hope a wider appreciation of the issues described in this review will lead to a more precise application of quenching methods to membrane structure.

Acknowledgments

A.S.L. is grateful to Drs. Kalina Hristova and Stephen H. White for providing the original data used in Fig. 1. This work was supported by NIH grants GM 48596 (to E.L.) and GM 46823 (A.S.L., grant to S. H. White).

References

Abel, E., Maguire, G. E. M., Murillo, O., Suzuki, I., De Wall, S. L., and Gokel, G. W. (1999). Hydraphile channels: Structural and fluorescent probes of position and function in a phospholipid bilayer. *J. Am. Chem. Soc.* **121,** 9043–9052.

Abrams, F. S., and London, E. (1992). Calibration of the parallax fluorescence quenching method for determination of membrane penetration depth: Refinement and comparison of quenching by spin-labeled and brominated phospholipids. *Biochemistry* **31,** 5312–5322.

Abrams, F. S., and London, E. (1993). Extension of the parallax analysis of membrane penetration depth to the polar region of membranes: Use of fluorescence quenching by a spin-label attached to the phospholipid polar headgroup. *Biochemistry* **32,** 10826–10831.

Abrams, F. S., Chattopadhyay, A., and London, E. (1992). Determination of the location of fluorescent probes attached to fatty acids using the parallax analysis of fluorescence quenching: Effect of carboxyl ionization and environment on depth. *Biochemistry* **31,** 5322–5327.

Ahmed, S. N., Brown, D. A., and London, E. (1997). On the origin of sphingolipid/cholesterol rich detergent-insoluble domains in cell membranes: physiological concentrations of cholesterol and sphingolipid induce formation of a detergent-insoluble liquid ordered phase in model membranes. *Biochemistry* **36,** 10944–10953.

Asuncion-Punzalan, E., and London, E. (1995). Control of the depth of molecules within membranes by polar groups: Determination of the location of anthracene labeled probes in model membranes by parallax analysis of nitroxide-labeled phospholipid induced fluorescence quenching. *Biochemistry* **34,** 11460–11466.

Asuncion-Punzalan, E., Kachel, K., and London, E. (1998). Polar molecules can locate at both shallow and deep locations in membranes: The behavior of dansyl and related probes. *Biochemistry* **37,** 4603–4611.

Baumeister, B., Sakai, N., and Matile, S. (2000). Giant artificial ion channels formed by self-assembled, cationic rigid-rod beta barrels. *Angew. Chem. Int. Ed.* **39,** 1955–1958.

Ben-Efraim, I., and Shai, Y. (1996). Secondary structure, membrane location, and coassembly within phospholipid membranes of synthetic segments derived from the N- and C-termini regions of the ROMK1 K$^+$ Channel. *Protein Sci.* **5,** 2287–2297.

Blatt, E., and Sawyer, W. H. (1985). Depth-dependent fluorescence quenching in micelles and membranes. *Biochim. Biophys. Acta* **822,** 43–62.

Bolen, E. J., and Holloway, P. W. (1990). Quenching of tryptophan fluorescence by brominated phospholipid. *Biochemistry* **29,** 9638–9643.

Breukink, E., van Kranij, C., van Dalen, A., Demel, R. A., Siezen, R. J., de Kruijff, B., and Kuipers, O. P. (1998). The orientation of nisin in membranes. *Biochemistry* **37,** 8153–8162.

Breukink, E., Wiedemann, I., van Kraaij, C., Kuipers, O. P., Sahl, H., and de Kruijff, B. (1999). Use of the cell wall precursor lipid II by a pore-forming peptide antibiotic. *Science* **286,** 2361–2364.

Chattopadhyay, A., and London, E. (1987). Parallax method for direct measurement of membrane penetration depth utilizing fluorescence quenching by spin-labeled phospholipids. *Biochemistry* **26,** 39–45.

Chung, L. A., Lear, J. D., and DeGrado, W. F. (1992). Fluorescence studies of the secondary structure and orientation of a model ion channel peptide in phospholipid vesicles. *Biochemistry* **31,** 6608–6616.

Clague, M. J., Knutson, J. R., Blumenthal, R., and Hermann, A. (1991). Interaction of influenza hemagglutinin amino-terminal peptide with phospholipid vesicles: A fluorescence study. *Biochemistry* **30,** 5491–5497.

De Kroon, A. I. P. M., Soekarjo, M. W., De Gier, J., and De Kruijff, B. (1990). The role of charge and hydrophobicity in peptide–lipid interaction: A comparative study based on tryptophan fluorescence quenching measurements combined with the use of aqueous and hydrophobic quenchers. *Biochemistry* **29,** 8228–8240.

De Kroon, A. I. P. M., Vogt, B., van't Hof, R., De Kruijff, B., and De Gier, J. (1991). Ion gradient-induced membrane translocation of model peptides. *Biophys. J.* **60,** 525–537.

Driessen, A. J. M., van den Hooven, H. W., Kuiper, W., van de Kamp, M., Saho, H.-G., Konings, R. N. H., and Konings, W. N. (1995). Mechanistic studies of lantibiotic-induced permeabilization of phospholipid vesicles. *Biochemistry* **34,** 1606–1614.

Fujita, K., Kimura, S., and Imanishi, Y. (1994). Self-assembly of mastoparan X derivative having fluorescence probe in lipid bilayer membrane. *Biochim. Biophys. Acta* **1195,** 157–163.

Green, S. A., Simpson, D. J., Zhou, G., Ho, P. S., and Blough, N. V. (1990). Intramolecular quenching of excited singlet states by stable nitroxyl radicals. *J. Am. Chem. Soc.* **112,** 7337–7346.

Grenier, S., Desmeules, P., Dutta, A. K., Yamazaki, A., and Salesse, C. (1998a). Determination of the depth of penetration of the alpha subunit of retinal G protein in membranes: A spectroscopic study. *Biochim. Biophys. Acta* **1370**, 199–206.

Grenier, S., Dutta, A. K., and Salesse, C. (1998b). Evaluation of membrane penetration depth utilizing fluorescence quenching by doxylated fatty acids. *Langmuir* **14**, 4643–4649.

Hanyu, Y., Yamada, T., and Matsumoto, G. (1998). Simultaneous measurement of spectroscopic and physiological signals from a planar bilayer system: Detecting voltage-dependent movement a membrane-incorporated peptide. *Biochemistry* **37**, 15376–15382.

Hirsh, D. J., Hammer, J., Maloy, W. L., Blazyk, J., and Schaefer, J. (1996). Secondary structure and location of a magainin analogue in synthetic phospholipid bilayers. *Biochemistry* **35**, 12733–12741.

Hristova, K., Wimley, W. C., Mishra, V. K., Anantharamaiah, G. M., Segrest, J. P., and White, S. H. (1999). An amphipathic alpha-helix at a membrane interface: A structural study using a novel X-ray diffraction method. *J. Mol. Biol.* **290**, 99–117.

Huang, H. W. (2000). Action of antimicrobial peptides: Two-state model. *Biochemistry* **39**, 8347–8352.

Johnson, J. E., and Cornell, R. B. (1994). Membrane binding amphipathic alpha-helical peptide derived from CTP:phosphocholine cytidylyltransferase. *Biochemistry* **33**, 4327–4335.

Jones, J. D., and Gierasch, L. M. (1994). Effect of charged residue substitutions on the membrane-interactive properties of signal sequences of the *Escherichia coli* LamB protein. *Biophys. J.* **67**, 1534–1545.

Kachel, K., Asuncion-Punzalan, E., and London, E. (1995). Anchoring of Trp and Tyr analogs at the hydrocarbon-polar boundary in membranes: Parallax analysis of fluorescence quenching induced by nitroxide-labeled phospholipids. *Biochemistry* **34**, 15475–15479.

Kachel, K., Asuncion-Punzalan, E., and London, E. (1998). The location of molecules with charged groups in membranes. *Biochim. Biophys. Acta* **1374**, 63–76.

Kaiser, R. D., and London, E. (1998a). Location of diphenylhexatriene (DPH) derivatives within membranes: Comparison of different fluorescence quenching analyses of membrane depth. *Biochemistry* **37**, 8180–8190.

Kaiser, R. D., and London, E. (1998b). Determination of the depth of BODIPY probes in model membranes by parallax analysis of fluorescence quenching. *Biochim. Biophys. Acta* **1375**, 13–22.

Killian, J. A., Keller, R. C. A., Struyve, M., de Kroon, A. I. P. M., Tommassen, J., and de Kruijff, B. (1990). Tryptophan fluoresecence study on the interaction of the signal peptide of the *Escherichia coli* outer membrane protein PhoE with model membranes. *Biochemistry* **29**, 8131–8137.

Kleinschmidt, J. H., and Tamm, L. K. (1999). Time-resolved distance determination by tryptophan fluorescence quenching: Probing intermediates in membrane protein folding. *Biochemistry* **38**, 4996–5005.

Kleinschmidt, J. H., den Blaauwen, T., Driessen, A. J. M., and Tamm, L. K. (1999). Outer membrane protein A of *Escherichia coli* inserts and folds into lipid bilayers by a concerted mechanism. *Biochemistry* **38**, 5006–5016.

Kliger, Y., and Shai, Y. (1997). A leucine zipper-like sequence for the cytoplasmic tail of the HIV-1 envelope glycoprotein binds and perturbs lipid bilayers. *Biochemistry* **36**, 5157–5169.

Ladokhin, A. S. (1993). Distribution analysis of membrane penetration by depth dependent fluorescence quenching. *Biophys. J.* **64**, A290.

Ladokhin, A. S. (1997). Distribution analysis of depth-dependent fluorescence quenching in membranes: A practical guide. *Meth. Enzymol.* **278**, 462–473.

Ladokhin, A. S. (1999a). Analysis of protein and peptide penetration into membranes by depth-dependent fluorescence quenching: Theoretical considerations. *Biophys. J.* **76**, 946–955.

Ladokhin, A. S. (1999b). Evaluation of lipid exposure of tryptophan residues in membrane peptides and proteins. *Anal. Biochem.* **276**, 65–71.

Ladokhin, A. S. (1999c). Red-edge excitation study of nonexponential fluorescence decay of indole in solution and in a protein. *J. Fluorescence* **9**, 1–9.

Ladokhin, A. S. (2000). Fluorescence spectroscopy in peptide and protein analysis. *In* "Encyclopedia of Analytical Chemistry" (R. A. Meyers, ed.), pp. 5762–5779. Wiley, New York.

Ladokhin, A. S., and Holloway, P. W. (1995a). Fluorescence quenching study of melittin-membrane interactions. *Ukrainian Biochem. J.* **67**, 34–40.

Ladokhin, A. S., and Holloway, P. W. (1995b). Fluorescence of membrane-bound tryptophan octyl ester. A model for studying intrinsic fluorescence of protein-membrane interactions. *Biophys. J.* **69**, 506–517.

Ladokhin, A. S., Wang, L., Steggles, A. W., and Holloway, P. W. (1991). Fluorescence study of a mutant cytochrome b_5 with a single tryptophan in the membrane-binding domain. *Biochemistry* **30**, 10200–10206.

Ladokhin, A. S., Malak, H., Johnson, M. L., Lakowicz, J. R., Wang, L., Steggles, A. W., and Holloway, P. W. (1992). Frequency-domain fluorescence of mutant cytochrome b_5. *In* "Time-Resolved Laser Spectroscopy in Biochemistry III" (J. R. Lakowicz, ed.), pp. 562–569. SPIE, Bellingham, WA.

Ladokhin, A. S., Holloway, P. W., and Kostrzhevska, E. G. (1993a). Distribution analysis of membrane penetration of proteins by depth-dependent fluorescence quenching. *J. Fluorescence* **3**, 195–197.

Ladokhin, A. S., Wang, L., Steggles, A. W., Malak, H., and Holloway, P. W. (1993b). Fluorescence study of a temperature-induced conversion from the "loose" to the "tight" binding form of membrane-bound cytochrome b_5. *Biochemistry* **32**, 6951–6956.

Ladokhin, A. S., Selsted, M. E., and White, S. H. (1997). Sizing membrane pores in lipid vesicles by leakage of co-encapsulated markers: Pore formation by melittin. *Biophys. J.* **72**, 1762–1766.

Ladokhin, A. S., Jayasinghe, S., and White, S. H. (2000). How to measure and analyze tryptophan fluorescence in membranes properly, and why bother? *Anal. Biochem.* **285**, 235–245.

Lew, S., and London, E. (2000). The effect of polar/ionizable residues within the core of hydrophobic helices on their behavior within lipid bilayers. *Biochemistry* **39**, 9632–9640.

Liu, L.-P., and Deber, C. M. (1997). Anionic phospholipids modulate peptide insertion into membranes. *Biochemistry* **36**, 5476–5482.

London, E. (1982). Investigation of membrane structure using fluorescence quenching by spin-labels: A review of recent studies. *Cell. Mol. Biochem.* **45**, 181–188.

London, E., and Feigenson, G. W. (1981). An analysis of the local phospholipid environment of diphenylhexatriene and gramicidin A. *Biochim. Biophys. Acta* **649**, 89–97.

Ludtke, S. J., He, K., Heller, W. T., Harroun, T. A., Yang, L., and Huang, H. W. (1996). Membrane pores induced by magainin. *Biochemistry* **35**, 13723–13728.

Mangavel, C., Maget-Dana, R., Tauc, P., Brochon, J.-C., Sy, D., and Reynaud, J. A. (1998). Structural investigations of basic amphipathic model peptides in the presence of lipid vesicles studied by circular dichroism, fluorescence, monolayer and modeling. *Biochim. Biophys. Acta* **1371**, 265–283.

Markello, T., Zlotnick, A., Everett, J., Tennyson, J., and Holloway, P. W. (1985). Determination of the topography of cytochrome *b*5 in lipid vesicles by fluorescence quenching. *Biochemistry* **24**, 2895–2901.

Martin, I., Ruysschaert, J.-M., Sanders, D., and Giffard, C. J. (1996). Interaction of the lantibiotic nisin with membranes revealed by fluorescence quenching of an introduced tryptophan. *Eur. J. Biochem.* **239**, 156–164.

Matsuzuki, K., Murase, O., Tokuda, H., Funakoshi, S., Fujii, N., and Miyajima, K. (1994). Orientational and aggregational states of magainin 2 in phopsholipid bilayers. *Biochemistry* **33**, 3342–3349.

McKnight, C. J., Rafalski, M., and Gierasch, L. M. (1991a). Fluorescence analysis of tryptophan-containing variants of the LamB signal sequence upon insertion into a lipid bilayer. *Biochemistry* **30**, 6241–6246.

McKnight, C. J., Stradley, S. J., Jones, J. D., and Gierasch, L. M. (1991b). Conformational and membrane-binding properties of a signal sequence are largely unaltered by its adjacent mature region. *Proc. Natl. Acad. Sci. USA* **88,** 5799–5803.

Mishra, V. K., and Palgunachari, M. N. (1996). Interaction of model class A1, class A2, and class Y amphipathic helical peptides with membranes. *Biochemistry* **35,** 11210–11220.

Moreno, M. J., and Prieto, M. (1993). Interaction of the peptide hormone adrenocorticotropin ACTH(1–24), with a model system: A fluorescence study. *Photochem. Photobiol.* **57,** 431–437.

Parente, R. A., Nadasdi, L., Subbarao, N. K., and Szoka, Jr., F. C. (1990). Association of a pH-sensitive peptide with membrane vesicles: Role of amino acid sequence. *Biochemistry* **29,** 8713–8719.

Pastor, R. W., Venable, R. M., and Karplus., M. (1991). Model for the structure of the lipid bilayer. *Proc. Natl. Acad. Sci. USA* **88,** 892–896.

Ren, J., Lew, S., Wang, Z., and London, E. (1997). Transmembrane orientation of hydrophobic alpha-helices is regulated by the relationship of helix length to bilayer thickness and by cholesterol concentration. *Biochemistry* **36,** 10213–10220.

Ren, J., Lew, S., Wang, J., and London, E. (1999). Control of transmembrane orientation and interhelical interactions within membranes by hydrophobic helix length. *Biochemistry* **38,** 5905–5912.

Ruiz-Arguello, M. B., Goni, F. M., Pereira, F. B., and Nieva, J. L. (1998). Phosphatidylinositol-dependent membrane fusion induced by a putative fusogenic sequence of ebola virus. *J. Virology* **72,** 1775–1781.

Sassaroli, M., Ruonala, M., Virtanen, J., Vauhkonen, M., and Somerharju, P. (1995). Transversal distribution of acyl-linked pyrene moieties in liquid-crystalline phosphatidylcholine bilayers. A fluorescence quenching study. *Biochemistry* **34,** 8843–8851.

Shatursky, O., Heuck, A. P., Shepard, L. A., Rossjohn, J., Parker, M. W., Johnson, A. E., and Tweten, R. K. (1999). The mechanism of membrane insertion for a cholesterol-dependent cytolysin: A novel paradigm for pore-forming toxins. *Cell* **99,** 293–299.

Tretyachenko-Ladokhina, V. G., Ladokhin, A. S., Wang, L., Steggles, A. W., and Holloway, P. W. (1993). Amino acid substitutions in the membrane-binding domain of cytochrome b_5 alter its membrane-binding properties *Biochim. Biophys. Acta* **1153,** 163–169.

Tsukihara, T., Aoyama, H., Yamashita, E., Tomizaki, T., Yamaguchi, H., Shinzawa-Itoh, K., Nakashima, R., Yaono, R., and Yoshikawa, S. (1996). The whole structure of the 13-subunit oxidized cytochrome *c* oxidase at 2.8 Å. *Science* **272,** 1136–1144.

Vogel, H., and Jähnig, F. (1986). The Structure of melittin in membranes. *Biophys. J.* **50,** 573–582.

Voges, K.-P., Jung, G., and Sawyer, W. H. (1987). Depth-dependent fluorescent quenching of a tryptophan residue located at defined positions on a rigid 21-peptide helix in liposomes. *Biochim. Biophys. Acta* **896,** 64–76.

Volgino, L., Simon, S. A., and McIntosh, T. J. (1999). Orientation of LamB signal peptide in bilayers: Influence on lipid probes on peptide binding and interpretation of fluorescence quenching data. *Biochemistry* **38,** 7509–7516.

Wang, Q., Cui, D., and Lin, Q. (1997). Fluorescence studies on the interaction of a synthetic signal peptide and its analog with liposomes. *Biochim. Biophys. Acta* **1324,** 69–75.

White, S. H., and Wiener, M. C. (1995). Determination of the structure of fluid lipid bilayer membranes. *In* "Permeability and Stability of Lipid Bilayers" (E. A. Disalvo and S. A. Simon, eds.), pp. 1–19. CRC Press, Boca Raton, FL.

Wiener, M. C., and White, S. H. (1991). Fluid bilayer structure determination by the combined use of X-ray and neutron diffraction. I. Fluid bilayer models and the limits of resolution. *Biophys. J.* **59,** 162–173.

Wiener, M. C., and White, S. H. (1992). "Structure of a fluid dioleoylphosphatidylcholine bilayer determined by joint refinement of X-ray and neutron diffraction data. III. Complete structure. *Biophys. J.* **61,** 434–447.

Yau, W.-M., Wimley, W. C., Gawrisch, K., and White, S. H. (1998). The preference of tryptophan for membrane interfaces. *Biochemistry* **37,** 14713–14718.

CHAPTER 5

Surface-Sensitive X-Ray and Neutron Scattering Characterization of Planar Lipid Model Membranes and Lipid/Peptide Interactions

Mathias Lösche

Institute of Experimental Physics I, Leipzig University, D-04103 Leipzig, Germany

I. Introduction
II. Data Inversion and Modeling in Reflectivity Measurements
III. Characterization of Planar Lipid Model Systems
    A. Surface Monolayer Systems
    B. Lipid Bilayers and Multi-Bilayer Stacks
    C. SAMs, HBMs, and Grafted Tethered Lipid Bilayers
    References

Recent progress in submolecular-level structural investigations of planar lipid model membranes using scattering techniques is reviewed. Particular emphasis is placed on the quantification of peptide/lipid and protein/lipid interactions. Floating phospholipid monolayers on aqueous subphases ("Langmuir monolayers") enable full physicochemical control of a membrane mimic, thus providing unique opportunities for investigations of the interaction of peptides with biomembrane surfaces. Particular progress has been recently made with surface-sensitive diffraction methods for characterizing molecularly thin protein crystal sheets. We describe (multi-) bilayer systems for which the distribution of submolecular fragments has been determined, providing a fully resolved picture of their thermally disordered structure using the "liquid-crystallography" approach of White and co-workers. This approach has recently been extended to tackle basic problems of membrane protein folding. Applied to solid-state interfaces with a fluid or gas

phase, scattering methods provide invaluable tools for the Ångstrom-scale characterization of systems as diverse as uniaxially oriented proteins in molecular
layers attached to self-assembled monolayers (SAMs), hybrid bilayer membranes
(HBMs), and polymer-cushioned, tethered bilayer membranes.

I. INTRODUCTION

By providing detailed three-dimensional structures of proteins and nucleic acids,
as well as protein/nucleic acid complexes, on the scale of individual atoms, molecular biology and high-resolution X-ray crystallography have revolutionized our
views of biology, medicine, disease, and morphogenesis in terms of biological
supramolecular self-assembly. Molecular machines that mend the games of life
have thus become comprehensible to a level of detail that had not been dreamt
of a mere few decades ago. Yet, the availability of such structures, beautiful as
they are, solves only one facet of the problem that involves the question, how
does life work? Another key ingredient for a deeper understanding is the spatial
organization of these machines within the cell, mediated largely by membrane
structures, composed of molecular bilayers only 5 nm thick, which are laterally
disordered and thus resist analysis with the methods that have been so successfully
employed to explore the internal structure of proteins. As if devised to complicate
their molecular-scale analysis, such lipid membranes are highly complex blends
of several classes of amphiphilic building blocks that undergo continuous changes
in their compositions, topology, and, thus, functionality. In order to at least unveil
general principles, if not the full details, of their functioning in terms of their
physical and physicochemical properties, model systems have been successfully
employed to start to unravel this constituent of the secret of life.

Planar lipid model systems, such as lipid Langmuir monolayers (Möhwald,
1990; Möhwald et al., 1995), substrate-supported lipid bilayers or multi-bilayer
stacks (Nagle and Tristram-Nagle, 2000; Safinya, 1997), bilayer lipid membranes
(BLMs; Bezrukov and Winterhalter, 2000; Diederich et al., 1998; Hanyu et al.,
1998; Krylov et al., 2000), and their inside-out counterparts, Newton black films
(NBFs; Bélorgey and Benattar, 1991; Benattar et al., 1999; Cuvilliers et al., 2000a),
have been frequently used as models, of different degrees of significance, for biological membranes and particularly employed to study lipid–lipid interactions
and the self-organization of lipids into biomimetic supramolecular aggregates.
More recently, complex supramolecular architectures, such as polymer-cushioned
or tethered lipid bilayers (Knoll et al., 2000; Sackmann and Tanaka, 2000), self-
assembled monolayers (SAMs) covalently linked to planar solid surfaces (Ulman,
1991), or hybrid bilayer membranes (HBMs) composed of SAMs and transferred

bilayers (Meuse *et al.,* 1998), have been employed to create stable and versatile membrane mimics. All of these systems, with the exception of BLMs, which are too small in lateral extension, are very amenable to structural characterization by scattering methods. Thus, surface-sensitive X-ray and neutron scattering methods (Als-Nielsen and Möhwald, 1991; Als-Nielsen *et al.,* 1994; Blasie and Timmins, 1999; Tolan, 1999) have made important inroads into various disciplines for the characterization of surfaces and interfaces in molecular and submolecular detail. Lipid/peptide interactions determine many structural and functional aspects in modern life science. This chapter summarizes recent developments in surface-sensitive scattering, with an emphasis on reflectivity measurements, for the structural and functional characterization of planar lipid systems, in particular in the context of investigations of lipid/peptide interactions.

Langmuir monolayers of amphiphilic compounds, such as fatty acids and esters or phospholipids, floating at the air/water interface are frequent objects of research both in materials and life science as well as in the physics of low-dimensional systems (Knobler and Desai, 1992; McConnell, 1991; Möhwald, 1995). Among the most powerful methods for the investigation of their structures on a submolecular level are surface-sensitive scattering techniques (Als-Nielsen and Möhwald, 1991; Als-Nielsen *et al.,* 1994; Weissbuch *et al.,* 1997): the diffraction of synchrotron X-ray radiation upon grazing incidence (GIXD), specular reflectivity measurements using X-rays or cold neutron beams, and diffuse (near-specular) scattering (see Fig. 1).

GIXD has attracted extensive attention recently for its unique capabilities in determining the spatial organization of linear alkane compounds in partially ordered—hexatic or crystalline—surface phases (Dutta *et al.,* 1987; Kaganer *et al.,* 1998, 1999; Kjaer *et al.,* 1987). It has been considerably boosted by the advent of third generation synchrotron radiation sources (Frahm *et al.,* 1995), which enable high-quality measurements within short illumination times and also permit diffraction measurements from more demanding systems, such as molecular protein crystal sheets (Haas *et al.,* 1995; Lenne *et al.,* 2000; Verclas *et al.,* 1999; Weygand *et al.,* 1999). Generally, domain sizes are in the micrometer range, and because beam footprints on planar samples are of the order of square centimeters in size, this implies that diffraction occurs from a randomly oriented in-plane crystal powder. For two-dimensional (2D) lattices of simple molecules, such as linear alkanes, the structure factors

$$F_{hk}(Q_z) = \sum_j f_j \exp[i(Q_{hk}R + Q_z z_j)] \tag{1}$$

can be worked out in detail (Als-Nielsen and Kjaer, 1989). As the scattering centers (characterized by the atom form factors f_j) are essentially confined to a plane, extending along the $\hat{x}$ and $\hat{y}$ axes at $z = 0$ and thus sampled by $R = (x, y)$, the

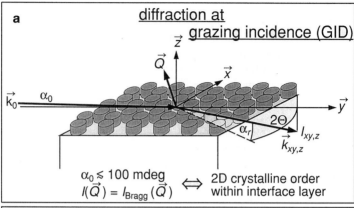

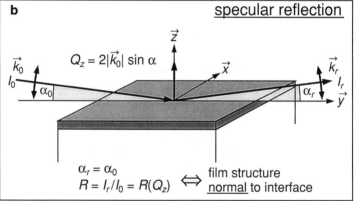

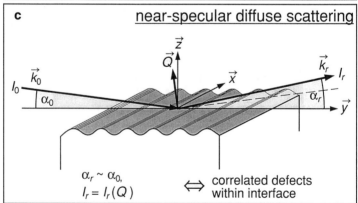

FIGURE 1 Surface-sensitive X-ray or neutron scattering geometries for various types of experiments.

diffracted intensity $I_{hk}(Q_{hk})$ is spread out in 1D Bragg rods along Q_z, where $\vec{Q}$ is the momentum transfer and h and k are the Miller indices. Thus,

$$I_{hk}(Q_{hk}) = |V(Q_z)F_{hk}(Q_z)|^2 \exp\left(-Q_z^2\sigma^2\right), \tag{2}$$

where V accounts for the interference between waves propagating upward and downward within the surface film upon multiple scattering events (Vineyard, 1982) and σ is a measure of the surface roughness (see below). The coherence of the lattice, characterized by the lattice constants $d_{hk} = 2\pi/Q_{hk}$, is given via the Scherrer equation (Guinier, 1968) by

$$L \approx 0.9 \frac{\lambda}{\Delta \cos\theta_{hk}}, \tag{3}$$

where $2\theta_{hk}$ is the horizontal scattering angle and Δ is the resolution-corrected full-width at half-maximum (FWHM) linewidth of the Bragg rod.

Specular reflectivity measurements in turn give information on the electron density distribution along the surface normal. Moreover, utilizing knowledge of chemical structure of film-forming compounds, one can extract information on the three-dimensional organization in a quite general approach, composition-space refinement (Vaknin et al., 1991b; Wiener and White, 1991b). An overview of a basic quantitative assessment is given below. In contrast to diffraction, reflectivity measurements are not limited to highly ordered systems, but also provide information on molecular subfragments that do not participate in the ordering within a hexatic or crystalline phase, such as the lipid headgroups. Thus, reflection measurements have great potential in particular for probing the static structure of the headgroups in lipid surface monolayers on the submolecular level. This is particularly valuable for investigations of lipid surfaces in a biophysical context: Biomembranes are disordered systems and many physiologically important interactions take place at the interface between the membrane and the aqueous compartment, that is, at the lipid headgroups. With the development of synchrotron insertion devices, boosting the available beam intensities by orders of magnitude, the quality of X-ray reflectivity measurements has also been dramatically improved by almost doubling the accessible Q_z range. Similarly, the available Q_z range in surface-sensitive neutron scattering has been continuously extended, owing to a careful sample cell design (Krueger et al., 1995; Meuse et al., 1998). By boosting the available neutron flux, the commissioning of next-generation neutron sources will drive this development even further (Blasie and Timmins, 1999).

These developments have resulted in a considerable gain in resolution of the underlying structures and required a reexamination of the models used in the past for the evaluation of reflectivity data. Although the so-called two-box model, which has been extensively used to describe the submolecular organization of (phospho-) lipid surface monolayers (Daillant et al., 1991; Helm et al., 1987, 1991;

Möhwald *et al.*, 1990; Rieu *et al.*, 1995; Vaknin *et al.*, 1991b) has been very capable of describing X-ray datasets that extend up to $Q_z^{\max} \sim 0.5\,\text{Å}^{-1}$, such models fail for the description of the high-resolution data now available (Weygand *et al.*, 1999). In response, novel data modeling strategies have been developed (Schalke *et al.*, 2000; Wiener and White, 1991b) that use distribution functions to describe the organization of submolecular fragments, particularly of phospholipid headgroup substructures, normal to the interface. Such a development has been greatly facilitated by the recent availability of both model-free data inversion techniques (Berk and Majkrzak, 1995; Chou *et al.*, 1997; Hamley and Skov Pedersen, 1994; Skov Pedersen and Hamley, 1994a,b; Zhou and Chen, 1993)—enabling an analysis of the inherent weaknesses of the two-box approaches—and volumetric information on lipid substructures from molecular modeling (Armen *et al.*, 1998; Petrache *et al.*, 1997)—used to reduce the uncertainties in the complex distribution-function model.

Lateral correlations of in-plane structural properties are tested by nonspecular scattering of X-rays. Within the distorted-wave Born approximation (DWBA), the observed intensity I is related to the structure factor of a rough surface,

$$S(Q) = \frac{\exp(-\,{}^1\!/_2|Q_z|^2\sigma^2)}{|Q_z|^2}$$
$$\times \iint \{\exp[|Q_z|^2 C(R) - 1]\} \exp[i(Q_x x + Q_y y)]\, dx\, dy, \qquad (4)$$

which is characterized by height–height correlations $C(R) = \langle z(0)z(R)\rangle$, by (Sinha *et al.*, 1988)

$$I \propto |T_{\mathrm{i}}(Q)|^2 |T_{\mathrm{f}}(Q)|^2 S(Q), \qquad (5)$$

where $T_{\mathrm{i/f}}$ are the Fresnel transmission functions.

II. DATA INVERSION AND MODELING IN REFLECTIVITY MEASUREMENTS

Biological membranes are generally fluid systems with a high degree of in-plane disorder. Thus, in-plane diffraction methods are often not suited to characterize the physiologically relevant states of such systems. Reflectivity measurements, on the other hand, are sensitive to the electron density profile across an interface and are thus perfectly suited to determine the structural organization of laterally disordered interface films. The optics of X-rays and neutron beams at surfaces has been extensively dealt with in the literature (Als-Nielsen and Kjaer, 1989; Als-Nielsen *et al.*, 1994; Penfold and Thomas, 1990). Briefly, the real part of the refractive index n is slightly different from (and usually lower than) unity for the relevant frequencies, and the imaginary part of n, that is, the absorption coefficient β, is negligibly small,

$$n = 1 - \delta + i\beta \qquad \text{(with } \beta \approx 0\text{)}. \tag{6}$$

This implies *external* total reflection of a plane wave impinging on a planar, ideally sharp interface between two media with refractive indices of 1 and n, respectively, as long as the incident angle α is below the critical angle

$$\alpha_c = \sqrt{2\delta} = \frac{\sqrt{4\pi\rho_{av}}}{k} = \begin{cases} \dfrac{\sqrt{4\pi\rho_{el}r_0}}{k} & \text{for X-rays} \\[2ex] \dfrac{\sqrt{4\pi\rho_n}}{k} & \text{for neutrons.} \end{cases} \tag{7}$$

Here, $k = 2\pi/\lambda$ is the magnitude of the wavevector, ρ_{av} is the average scattering length density (SLD) of a molecular-size volume V at the interface, ρ_{el} is the electron density, $r_0 = e^2/m_ec^2 \approx 2.82$ fm is the classical Thomson electron radius, and $\rho_n = (1/V)\sum_i \nu_i b_i$ is the neutron SLD, as computed from the atomic content of V (ν atoms of the species i that is characterized by the neutron scattering length b are contained in V). While for X-rays, α_c is always real, it can be imaginary in neutron experiments, for example, if the beam is reflected from the surface of H_2O (for which $\rho_n < 0$). In this case, there is no Q_z regime of total external reflection. Because δ is of the order of 10^{-5}, α_c is typically in the millidegree range.

If one is interested in specular reflectivity, the momentum transfer $\vec{Q} = \vec{k}_{out} - \vec{k}_{in}$ is strictly normal to the interface,

$$Q_z = 2k\sin\alpha = \frac{4\pi}{\lambda}\sin\alpha, \tag{8}$$

where in a local coordinate system $\hat{x}$ and $\hat{y}$ define the interface and $\hat{z}$ points in the perpendicular direction. Hence,

$$Q_c = \sqrt{4\pi\rho_{av}} \tag{9}$$

is a quantity characteristic of the medium, since it is independent of the wavelength λ (for example, $Q_c^{\text{X-rays}} \approx 0.0217$ Å^{-1} for water). From Fresnel's law, neglecting differences between different polarization directions and higher orders in δ, which is justified, as $\delta \approx 10^{-5}$, one finds the reflection amplitude $r_{1,2}$ between media with the indices n_1 and n_2,

$$r_{1,2} \approx \frac{\alpha - \sqrt{\alpha^2 + 2(\delta_1 - \delta_2)}}{\alpha + \sqrt{\alpha^2 + 2(\delta_1 - \delta_2)}} \approx \frac{Q_{c,1}^2 - Q_{c,2}^2}{4Q_z^2} = \frac{4\pi}{Q_z^2}(\rho_{av,1} - \rho_{av,2}), \tag{10}$$

and the reflectivity

$$R = |r|^2. \tag{11}$$

The Fresnel reflectivity R_F of an ideal surface in vacuum (or air) is derived from Eqs. (10) and (11) as

$$R_F \approx \left(\frac{Q_c}{2Q_z}\right)^4 \tag{11a}$$

as long as Q_z is sufficiently large (e.g., $>5Q_c$).

In reality, interfaces between two media are not mathematically sharp, but are graded on an Ångstrom-length scale, due to atomic roughness in the case of planar solid-state surfaces or thermally excited capillary waves in the case of fluid surfaces (Meunier, 1987).[1] This is phenomenologically taken into account by convolution of the step function $\Theta(z - z_0)$, which describes the ideal interface, with a Gaussian yielding the error function as the relevant profile describing the interface (Als-Nielsen and Möhwald, 1991),

$$\Theta(z - z_0) \rightarrow \frac{1}{2} \operatorname{erf}\left(\frac{z - z_0}{\sqrt{2}\sigma}\right) + \frac{1}{2}, \tag{12}$$

where

$$\operatorname{erf}(z) = \frac{2}{\sqrt{\pi}} \int_0^z e^{-t^2} dt$$

and σ is a parameter related to the amplitude of the roughness. In comparison with the reflectivity from an ideal interface, the surface roughness (s.r.) leads to a Debye–Waller-like damping of the reflection amplitude (Névot and Croce, 1980),

$$r(\text{w/ s.r.}) = r(\text{w/o s.r.}) \exp\left(-\frac{Q_z^2 \sigma^2}{2}\right). \tag{13}$$

If a molecularly thin homogeneous film (index n_2) is located on a semiinfinite substrate (index n_3), reflection according to Eq. (10) occurs at both the front and back faces, giving rise to interference with an intensity pattern in the far-field characteristic of the index and the thickness d_2 of the film. The interference originates from a phase factor that takes into account the propagation of the wave in the medium,

$$r_{1,2} = \frac{r_2 + r_3 \exp(2ik_{z,n_2}d_2)}{1 + r_2 r_3 \exp(2ik_{z,n_2}d_2)}, \tag{10a}$$

and the absolute square yields the reflectivity,

$$R = \frac{r_2^2 + r_3^2 + 2r_2 r_3 \cos(2k_{z,n_2}d_2)}{1 + r_2^2 r_3^2 + 2r_2 r_3 \cos(2k_{z,n_2}d_2)}. \tag{10b}$$

---

[1] A rigorous theoretical assessment of the liquid–vapor interface of simple fluids has been recently developed by Mecke and Dietrich (1999).

Stratified surface films can obviously be accounted for by recursive application of Eqs. (10), taking into account a global surface roughness by means of Eq. (13), as first suggested by Parratt (1954). Arbitrary SLD profiles at the interface may be treated either by slicing the profiles into a sequence of thin layers and determining the reflectivity by using the Parratt recursion algorithm or by using the kinematic approximation (Als-Nielsen and Kjaer, 1989; Als-Nielsen and Möhwald, 1991; Als-Nielsen *et al.*, 1994),

$$\frac{R(Q_z)}{R_F(Q_z)} \approx \frac{1}{\rho_{\text{substrate}}^2} \left| \int \frac{d\rho(z)}{dz} e^{i Q_z z} \, dz \right|^2, \tag{14}$$

which is only valid for $Q_z > 5Q_c$, but has the advantages of allowing for any analytical form of $\rho(z)$ and being at the same time more intuitive than the recursion formalism.

As in any scattering experiment, the experimental data cannot be directly translated into the underlying structure, because of the "phase problem," that is, the loss of phase information upon obtaining the scattered intensities from the amplitudes by using Eqs. (10a), (10b), or (14). Data quality and resolution as well as recent developments in the modeling of membrane surfaces have been recently reviewed (Schalke and Lösche, 2000). For modeling of both X-ray and neutron reflectivity data, "box" or slab models (Als-Nielsen and Kjaer, 1989; Als-Nielsen and Möhwald, 1991; Als-Nielsen *et al.*, 1994; Möhwald, 1990), which describe a lipid monolayer as two contiguous slabs, one hydrophobic and one hydrophilic, have frequently been used to describe aqueous surface monolayers (Bayerl *et al.*, 1990; Brumm *et al.*, 1994; Grundy *et al.*, 1988; Helm *et al.*, 1987, 1991; Kjaer *et al.*, 1989; Vaknin *et al.*, 1991b) or water/oil interfaces (Thoma *et al.*, 1996) as well as protein or polymer interactions with lipid surface layers (Fukuto *et al.*, 1997; Gallant *et al.*, 1998; Gidalevitz *et al.*, 1999a,b; Johnson *et al.*, 1991b; Kuhl *et al.*, 1999; Lösche *et al.*, 1993; Naumann *et al.*, 1996; Weygand *et al.*, 1999; Vaknin *et al.*, 1991a). Analogous models for data inversion have also been used to describe molecular layer systems at solid/fluid interfaces (Diederich and Lösche, 1997; Fragneto *et al.*, 1995; Johnson *et al.*, 1991a; Krueger *et al.*, 1995; Kuhl *et al.*, 1998; Malik *et al.*, 1997; Schmidt *et al.*, 1992).

More recently, it has been reported that the simple box approach is rather inadequate for the description of surface monolayers from data collected at third generation synchrotron sources (Frahm *et al.*, 1995) at high momentum transfer (Krüger *et al.*, 2001; Schalke and Lösche, 2000; Schalke *et al.*, 2000), and a more sophisticated model that describes the interface film in terms of volume-restricted distribution functions (VRDF) across the interface has been developed (Schalke *et al.*, 2000). The novel data refinement technique is inspired by ideas of Wiener and White (1991a,b, 1992a,b; Wiener *et al.*, 1991) for the evaluation of X-ray and neutron small-angle scattering from planar multi-bilayer systems. In addition, it takes into account recent developments in molecular dynamics simulations of bilayers (Armen *et al.*, 1998; Petrache *et al.*, 1997), from which volumetric information has been derived and utilized to interpret the small-angle scattering data with a higher confidence level. In its implementation for floating surface monolayers, the VRDF model treats the aliphatic chains just as they are accounted for in the box-model approach; the phospholipid backbone and lower headgroup, however, are parsed into fragments that are placed into distribution functions along the interface normal (Schalke *et al.*, 2000). In contrast to the box model, which satisfies space-filling automatically within the chemical approach to its interpretation, the filling of space has to be explicitly taken into account in the VRDF approach, because the distribution functions may partially or entirely interpenetrate each other, and even if they are separated along z, space-filling is usually not fulfilled at their perimeters. In addition to a satisfactory description of the experimental data, this requires

$$\sum_{\xi} n_{\xi}(z) \cdot V_{\xi} = 1, \tag{15}$$

where $n_{\xi}(z) = N_{\xi}(z)/(A_{\text{lipid}}\, dz)$ is the number density of the fragment ξ within the plane located at a distance z from the interface and V_{ξ} is its partial volume. Handling of this additional constraint obviously requires *a priori* knowledge of V_{ξ} or its determination from the fit.

Thermal broadening of the interface structure in the distribution function model derives from two contributions that are well distinguished: a broadening by capillary waves σ_{cw} and an intrinsic broadening of fragment positions σ_{int}, which is the value one would expect to observe *without* capillary waves, that is, within an interface film at an ideally flat surface. If one assumes that typical wavelengths for the two processes are well-separated in real space, the upper and lower interfaces of the alkane slab, considered to be atomically flat, at least for the ordered phases LC and SC,[2] are only affected by σ_{cw}, whereas the distributions of the headgroup fragments are affected by both contributions, which are summed geometrically:

$$\sigma_{\text{total},\xi} = \sqrt{\sigma_{\text{cw}}^2 + \sigma_{\text{int},\xi}^2}. \tag{16}$$

This separation of the contributions to interfacial broadening permits one both to determine the evolution of intrinsic broadening along the isotherm and to check whether the capillary wave's amplitudes depend on π and T as predicted by theory (Pershan, 1990, 2000; Plech *et al.*, 2000):

$$C(R) = \frac{k_{\text{B}} T}{2\pi \gamma} K_0(R\sqrt{\delta \rho g \gamma}). \tag{17}$$

Here, k_B, γ, and $\delta\rho$ denote the Boltzmann constant, surface tension, and mass density difference between fluid subphase and its vapor phase, respectively. K_0 is the zeroth-order modified Bessel function of the second kind.

---

[2]Phospholipid monolayer phases are denoted using the nomenclature introduced by Cadenhead *et al.* (1980): G, gaseous; LE, liquid expanded; LC, liquid condensed; SC, solid condensed.

III. CHARACTERIZATION OF PLANAR LIPID MODEL SYSTEMS

A. Surface Monolayer Systems

1. Pure Lipid Monolayers

Monomolecular (phospho-) lipid surface layers (Knobler and Desai, 1992; McConnell, 1991; Möhwald, 1990, 1995) have been intensively studied for their unique properties as quasi-2D molecular systems and for the ease by which their molecular properties may be controlled. A rich phase diagram has been established (Kaganer et al., 1999), primarily by characterizing the order of the amphiphiles' aliphatic chain using GIXD (Kaganer et al., 1999) and refining this assessment with optical microscopy techniques (Fischer et al., 1995; Marshall et al., 2000; Teer et al., 1997), and it is fairly well understood in terms of Landau theory with a quite limited number of translational and orientational order parameters (Kaganer and Loginov, 1995; Kaganer and Osipov, 1998). It has thus been recently established in great detail that molecular order within Langmuir monolayers is primarily driven by van der Waals interactions between the linear alkane moieties and is controlled by a subtle balance between the packing properties of the chains and the headgroups (Weidemann et al., 1998, 1999). Since lipid bilayer membranes may be regarded as a set of two weakly coupled monolayers facing each other, such studies are of potential interest for the comprehension of biological systems. On the other hand, because biological membranes possess a large degree of in-plane disorder, the crystalline and semicrystalline phases observed in the monolayers are of somewhat limited interest.

Because diffraction from the hydrophilic lipid headgroups within monolayers has not yet been observed, it is generally assumed that they do not participate in lipid ordering. Nevertheless, since monolayers are regarded as relevant model systems particularly for the interface of biomembranes with their aqueous environment, the structural and dynamic properties of the headgroups are of particular interest. Thus, numerous reflectivity studies have been undertaken to elucidate the structure of this interface on a submolecular level. For their interpretation, box models (Als-Nielsen and Kjaer, 1989) provide a relatively simple, rather intuitive approach to data interpretation in submolecular terms with the additional benefit of easy implementation (Parratt, 1954). This is why they have been extensively used for the inversion of X-ray and neutron reflectivity data in studies of lipid monolayers at aqueous surfaces (Bayerl et al., 1990; Brumm et al., 1994; Grundy et al., 1988; Helm et al., 1987, 1991; Kjaer et al., 1989; Vaknin and Kelley, 2000; Vaknin et al., 1991b) or water/oil interfaces (Thoma et al., 1996), as well as peptide or polymer interactions with lipid surface layers (Fukuto et al., 1997; Gallant et al., 1998; Gidalevitz et al., 1999a,b; Johnson et al., 1991b; Kuhl et al., 1999; Lösche et al., 1993; Naumann et al., 1996; Vaknin et al., 1991a; Weygand et al., 1999). Analogous models for data inversion have also been used to describe the structure of (inside-out oriented) opposing monolayer pairs within NBFs (Cuvilliers *et al.,* 2000a).

One potential problem in the data inversion with box models is a breakdown of the chemical interpretation if the local SLD in the real structure shows large variations along the $\hat{z}$ direction within one box. Before the arrival of third generation synchrotron sources, such variations could not be experimentally determined, due to the lack of resolution, and may thus have resulted in misinterpretations of the atomic contents of the boxes, because the interface between them is not naturally determined by chemical composition, but rather by the gradients of the SLD [see Eq. (14)]. The steepest gradient within a phospholipid film in which the headgroup extends substantially into the subphase need not be located at the hydrophilic/hydrophobic interface, but may occur closer to the phosphate moiety, in which case the chemical interpretation of the headgroup slab as one homogeneous structure may be dangerously misleading. In this context, one may challenge earlier results on the structural reorganization of dimyristoylphosphatidylethanolamine (DMPE) and dilauroylphosphatidylethanolamine (DLPE) monolayers upon monolayer compression (i.e., reduction of the area A_{lipid} per lipid molecule in the surface film) (Helm *et al.,* 1991; Möhwald, 1990): It has been reported that for DMPE the hydration number n_{w} changes from $\sim$24 at $A_{\text{lipid}} \approx 80 \text{ Å}^2$ to $\sim$1.5 for at $A_{\text{lipid}} \approx 40 \text{ Å}^2$ while undergoing the main monolayer phase transition, LE $\rightarrow$ LC. Similar dramatic changes in phospholipid headgroup hydration at the LE/LC phase transition were reported for dipalmitoylphosphatidylcholine (DPPC) as observed by analyzing box models describing the neutron reflectivity of DPPC monolayers (Naumann *et al.,* 1995). The results from these box model studies have in common that the apparent thickness of the headgroup layer, d_{hphil} is as large as, or even larger than (>11 Å), the extended length of the headgroup fragments, particularly in the disordered phases, and that d_{hphil} *decreases* upon reduction of A_{lipid}. This would imply a tilting movement of the headgroup dipoles *away* from the surface normal, whereas the average chain directions tilt toward the surface normal upon reduction of the available area per molecule. Moreover, the apparent value of d_{hphil} was invariably close to, or even larger than, the extended length of the PE headgroup. Together with the notion that only 1.5 water molecules should hydrate the phosphate in a compressed monolayer *on top of an infinite reservoir of water,* these results raise severe doubts about the validity of the model. A careful examination of this counterintuitive situation by the authors of the first study concluded with the statement that *within the two-box model approach,* results were robust against variations of details of the model (Helm *et al.,* 1991).

With the development of the VRDF approach to reflectivity data evaluation, more intuitive and more consistent results on the submolecular structure of phospholipid monolayers began to appear (Krüger *et al.,* 2001; Schalke, 2000; Schalke and

Lösche, 2000; Schalke *et al.*, 2000). For a number of distinct phospholipids it has been reported that the phosphate hydration is *constant* along the isotherm—the number of water molecules associated with the phosphate is $n_w^p \approx 5-6$ (Krüger *et al.*, 2001; Schalke *et al.*, 2000)—whereas a decrease of the average chain tilt angle from the surface normal, β, is accompanied by a corresponding decrease of the average headgroup tilt α (see Fig. 2). For the two anionic phospholipid headgroups, phosphatidic acid (PA) and the larger phosphatidylglycerol (PG), it has been reported that an unexpectedly large amount of ions is bound on concentrated salt solutions (10 mM BaCl$_2$), which are continuously squeezed out of the headgroup region upon compression. Only at high surface pressures π has the expected stochiometric ratio of $2:1$ (PA$^-$:Ba^{2+}) been observed. In both cases it was observed that α stays essentially constant over a large range of π and the decrease of A_{lipid} is

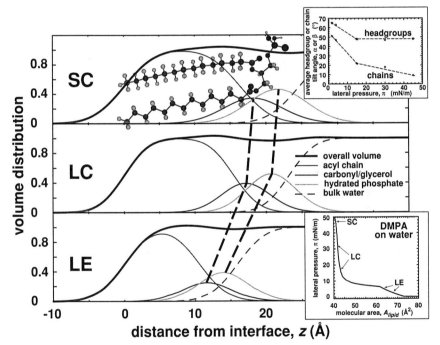

FIGURE 2 Volume-restricted distribution function approach for the modeling of planar lipid layer systems as exemplified for DMPA on water (data from Schalke *et al.*, 2000). Dashed thick lines in the main panel indicate the development of the center locations of the fragment distributions upon film compression, from which the average headgroup tilt angle α is deduced. Insets indicate the surface isotherm (*bottom*) and the concurrent tilt of the lipid headgroups (α) and chains (β) toward the surface normal upon compression. The LE $\rightarrow$ LC phase transition shows clearly in the development of the tilt angles.

compensated for by the exclusion of ions (Krüger *et al.,* 2001; Schalke and Lösche, 2000). Consistent with intuitive expectation, the value of α is greater for the small PA headgroup ($\alpha \approx 35°$ in the condensed phase on 10 mM BaCl$_2$; Schalke and Lösche, 2000) than for the PG headgroup ($\alpha \approx 25°$ under the same nominal conditions; Krüger *et al.,* 2001). In contrast, values of $\alpha \approx 65°$ and $45°$ are observed for PA on pure water subphase in the LE and LC phases, respectively (Schalke *et al.,* 2000). In all of these cases, hydration of the carbonyl moieties is small and was thus neglected in the models. The distribution width of the phosphate position as projected on the surface normal was found to be significant in all these structures (FWHM width $\sigma_{int} \approx 2$ Å) and was generally observed to increase upon compression. It appears, thus, that a fair amount of headgroup disorder characterizes the structure of such monolayers.

One important caveat that applies to all of the models for phospholipids with headgroups more complex than the simplest (PA) is that the terminal moiety, such as the secondary glycerol in PG or the choline in PC, is quite similar in electron density to the surrounding water. This implies that these models are relatively vulnerable, because they are based entirely on X-ray reflectivity results at this point, and a more definite structural assessment will have to await comprehensive neutron reflectivity measurements on headgroup-deuterated lipids.

Contrast variation using X-ray synchrotron radiation, on the other hand, is feasible, as recently demonstrated at the 6-ID beamline of the Advanced Photon Source (APS) at Argonne National Laboratory: In an exploratory experiment, the anomalous scattering of Ba^{2+} ions due to their L_{I-III} absorption lines has been utilized to characterize the binding of such ions to anionic dimyristoylphosphatidic acid (DMPA) headgroups in great detail and without any assumptions (Schalke and Lösche, 2000) on the charge distribution at the interface (P. Krüger, D. Vaknin, and M. Lösche, unpublished results). Such experiments should pave the way, for example, for detailed studies of ion binding to lipid membranes or the molecular mechanisms of biomineralization at molecularly well-defined model interfaces (Lochhead *et al.,* 1997).

2. Composite Monolayer Systems

Composite surface monolayer systems have been frequently studied to better comprehend the various interactions of peptides (and other macromolecular components of biological relevance) with membranes (Möhwald, 1990; Möhwald *et al.,* 1995). Of particular significance are systems that form monolayers in their physiological environment, such as pulmonary surfactant models (Galla *et al.,* 1998; Krüger *et al.,* 1999; Lee *et al.,* 1997), or that mimic the association of peripheral ligands, such as components of the cellular exoskeleton (Sackmann, 1995) or membrane fusion mediators (Arnold, 1995; Gerke and Moss, 1997). Other studies concern the action of phospholipases at membrane surfaces and address particularly the interdependence of lipid order and enzymatic activity.

Finally, a number of studies are dedicated to the elucidation of the surface dependence of peptide folding (White and Wimley, 1998) and of surface-induced peptide or protein crystallization processes (Brisson *et al.,* 1999; Kühlbrandt, 1992).

Pulmonary surfactants offer one of the rare cases where a monomolecular lipid layer is the equivalent of a physiological system. However, during the respiration cycle, this monolayer undergoes a reversible, peptide-mediated transformation into a state of (what is believed to be) higher organizational order. Particularly the surfactant peptides SP-B and SP-C are implicated in a refinement of the lipid mixture that involves a high percentage of PCs and PGs and has an unusually large proportion of saturated chains. In fact, this is the only known example where DPPC seems to be a major component of a physiological system. Recently, an explicit model of the topological changes and a mechanism that involves the hydrophobic SP fractions have been proposed (Galla *et al.,* 1998). A particular model system, phosphatidic acid (PA) and a truncated form of synthetic SP-B (SP-B$_{1-25}$), which has been extensively studied (Lipp *et al.,* 1996) as a replacement formulation of natural pulmonary surfactant for the treatment of acute respiratory distress syndrome (ARDS), has recently been characterized using X-ray scattering (Lee *et al.,* 2000, 2001). In connection with the molecular structure of the peptide, as it appears from Fourier transform infrared (FTIR) spectroscopy and molecular dynamics simulations (Gordon *et al.,* 2000)—a rather hydrophobic α helix adjacent to a short (seven amino acids) β structure at the N-terminus that is believed to promote the incorporation of the peptide into the monolayer (Hawgood *et al.,* 1998)—it was deduced from the electron density profile using a box model that the peptide, which was included at 20 wt%, well above the physiological concentration of SP-B, intercalates the headgroup and is oriented under an oblique angle, as visualized in Fig. 3. By and large, however, the results point to differences of the mesoscopic organization of the surfactant film upon interaction of the peptide rather than changes at the molecular level.

Another field of intense research concerns the molecular-level structure and molecular interactions of components of the cytoskeleton with the lipid membrane. Neutron reflection measurements have been conducted to reveal particularly the structural features of spectrin (Johnson *et al.,* 1991b), hisactophilin (Naumann *et al.,* 1996), and actin filament (Demé *et al.,* 2000) binding to floating monolayers. At the current stage, as long as box modeling is the main tool for data inversion, these studies are essentially limited to determining whether or not the peptides bind in homogeneous layers at the interface, quantifying apparent layer thickness values (and comparing those with dimensions of the respective protein molecules) as well as equivalent volume occupation of the peptide in the surface-adsorbed layer, and optimizing experimental conditions for peptide binding. In particular, a recent paper (Demé *et al.,* 2000) reports a comprehensive study on the ionic strength dependence of the electrostatic interaction of a peptide (F actin) with charged interfaces.

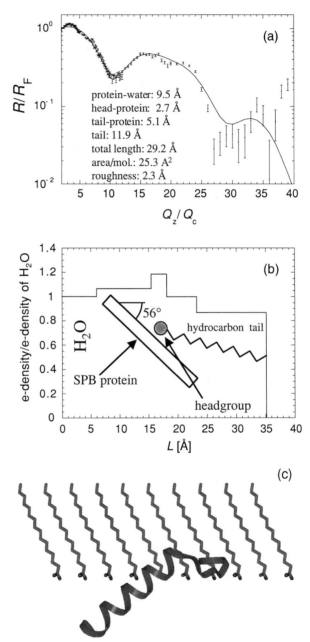

FIGURE 3 Association of the recombinant, truncated surfactant peptide B (SP-B$_{1-25}$) with palmitic acid monolayers according to Lee *et al.* (2001). (a) X-ray reflectometry data and (b) deduced SLD profile from a box modeling approach. (c) Molecular model of the complex surface monolayer. Based on figures in Lee *et al.* (2001), with permission.

Annexins are a class of protein known to promote vesicle aggregation and fusion *in vitro* in a Ca^{2+}-dependent manner and to be implicated in membrane trafficking (Gerke and Moss, 1997). Although they have been extensively studied in recent years, little is known about the molecular details of their membrane interaction. A recent study addresses such questions by characterizing the membrane binding of annexin I in X-ray reflectivity measurements (Bitto *et al.*, 2000). Because annexin I has two inequivalent membrane-binding sites, a number of different scenarios for the mediation of close contact between two fusing membranes by the protein are viable. Of these, experimental results in which membrane fusion is promoted by the formation of a laterally aggregated protein monolayer on one membrane surface favor the mechanism by which annexin I is uniaxially attached via its primary binding site, and the subsequent binding of the juxtaposed membrane occurs via interactions with the secondary binding site of the proteins (Bitto *et al.*, 2000).

The interaction of phospholipase A_2 (PLA_2) with PC monolayers has been studied by Brezesinski and co-workers with a host of surface-sensitive characterization techniques, including GIXD (Dahmen-Levison *et al.*, 1998; Peters *et al.*, 2000).

Protein unfolding or refolding at aqueous surfaces, neat or covered with lipid films, has been studied by various groups (Gidalevitz *et al.*, 1999a,b; Holt *et al.*, 2000; Strzalka *et al.*, 2000). Gidalevitz and co-workers studied the denaturation of glucose oxidase (GOx), alcohol dehydrogenase, and urease at bare buffer surfaces and reported that the deposition of a Langmuir monolayer does not prevent such denaturation (Gidalevitz *et al.*, 1999a), whereas crosslinking of GOx with glutaraldehyde does, as assessed by the observation of a protein layer that resembles closely the molecular dimensions of the respective species; denaturation is conjectured from the observation of peptide monolayers whose thickness resembles the approximate diameter of an α helix and is thus far below the protein dimensions. Qualitatively similar behavior has been reported for myoglobin at aqueous surfaces (Holt *et al.*, 2000). Strzalka and co-workers studied the association of synthetic α-helical peptides, so-called maquettes (Robertson *et al.*, 1994), at clean and DLPE-covered buffer surfaces. Although the peptides are found to adsorb with their helices coplanar and form coherent peptide layers at low surface pressures, they are desorbed into the subphase upon subsequent compression. A variant of the parent maquette that bears hydrophobic alkyl anchors, however, is stabilized at the interface by incorporation of the anchors into a DLPE surface film and is reported to undergo a transition upon compression in which the coplanar orientation of the helices switches to preferentially normal. No in-plane correlation of the proteins, however, has been reported in any of these cases.

The crystallization of proteins underneath Langmuir monolayers (Darst *et al.*, 1991; Vénien-Bryan *et al.*, 1998) and subsequent characterization at interfaces using GIXD is emerging as an effective tool for structural studies in the work of a growing number of groups. The Möhwald group first reported X-ray diffraction from a monomolecular crystal sheet of streptavidin formed underneath a surface monolayer of a biotinylated lipid (Haas *et al.*, 1995). The monodomain size of such crystal sheets is of the same order as the domain sizes of ordered lipid domains (Krüger and Lösche, 2000), $\phi \approx 10~\mu$m (Lösche *et al.*, 1993), and the beam footprint in a typical synchrotron X-ray experiment is of the order of 1 cm^2, implying that 2D powder diffraction is observed in these experiments. Working with the well-characterized bacteriorhodopsin (bR) system (Grigorieff *et al.*, 1996; Pebay-Peyroula *et al.*, 1997), Verclas and co-workers have considerably pushed the technology (Verclas *et al.*, 1999). For purple membrane (PM) deposited on a clean buffer surface, they reported diffraction out to the prominent h, $k = 4$, 3 peak at $Q_{xy} \approx 0.72$ Å^{-1} (overlaid with the h, $k = 6, 0$ diffraction peak of the crystal powder) of the hexagonal lattice of the protein/lipid cocrystal (see Fig. 4a). Although the same lattice symmetry was observed as in earlier electron microscopy work (Henderson *et al.*, 1990), the lattice constant was reported to be a few percent smaller (61.3 Å vs. 62.5 Å) in PM patches deposited at the air/water interface. Generally, this work demonstrates that a resolution is attainable in such studies that approaches 8 Å, such that structural elements such as α helices can be located within molecular films of crystalline protein (Fig. 4b). Berge and co-workers at the European Synchrotron Radiation Facility (Grenoble, France) have studied GIXD from molecular protein sheet crystals of streptavidin (for which they were also able to attain a maximum in-plane momentum transfer corresponding to a resolution of $\sim$10 Å), annexin V, and the RNA transcription factor HupR ligated via an attached multihistidin tag nickel-chelated to a functionalized lipid headgroup in the surface monolayer (Lenne *et al.*, 2000). For the streptavidin diffraction pattern, Bragg rods have been extracted from the spectra. Similarly, Weygand and co-workers have studied the binding of bacterial S-layer (Sleytr *et al.*, 1999) protomers to phospholipid surface monolayers and their subsequent recrystallization into coherent S-layer lattices (Diederich *et al.*, 1996; Wetzer *et al.*, 1998; Weygand *et al.*, 1999, 2000) at the HASYLAB BW1 beamline (Frahm *et al.*, 1995) of the Deutsches Elektronensynchrotron (DESY; Hamburg, Germany). For S-layers of the Gram-positive

---

FIGURE 4 (a) The Q_z-averaged, background-corrected small-angle GIXD intensity measured for scattering from a PM film at $\pi = 10$ mN/m on an aqueous surface (250 mM KCl). The Bragg peaks have been indexed under the assumption of a hexagonal lattice with a lattice constant $a = 62.5$ Å. The linearity of the Q_{xy} values of the peak positions versus $\sqrt{h^2 + hk + k^2}$, shown in the inset, verifies the assumption of a hexagonal lattice and quantifies the deviation of the lattice constant from that measured in electron microscopy. Reproduced from Verclas *et al.* (1999) with permission. (b) Projected electron density map using the intensities shown in (a) and phases derived from electron microscopy (Henderson *et al.*, 1986). The annular structure of the trimeric seven helix bundles within the unit cell is clearly identified. (See color plate.)

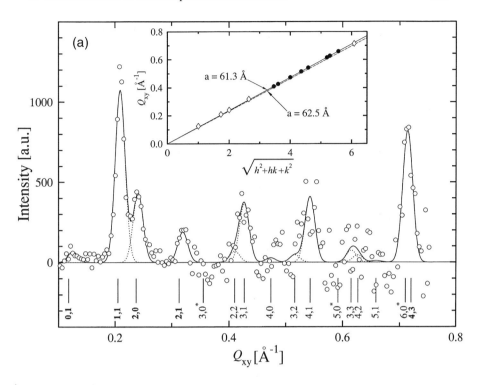

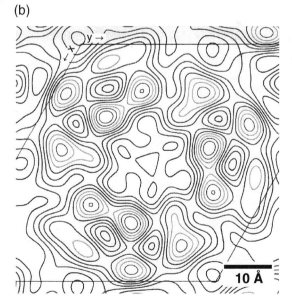

eubacterium *Bacillus sphaericus* CCM2177, which is the best-studied example to date, the coupling of the protein to the lipid monolayer[3] and the recrystallized S-layer structure has been studied in detail. Background-corrected Bragg rods determined from the small-angle GIXD of such S-layers recrystallized under a DPPE monolayer are shown in Fig. 5a (Weygand, 2000). In FTIR and X-ray scattering studies it has been observed that the peptide side chains on top of the attached S-layer interact with the phospholipid headgroups while leaving the acyl chains essentially unaffected (a minor increase in chain order has been reported; Diederich *et al.*, 1996; Weygand *et al.*, 1999). The protein, whose atomic 3D structure is unknown due its apparent incapability to crystallize in three dimensions, forms a porous layer at the lipid that controls transport to the membrane surface. The protein's volume density distribution normal to the interface has been determined (Weygand *et al.*, 1999) by using the model-independent data inversion technique of Skov Pedersen and Hamley (1994b). It shows high protein density both close to the lipid membrane surface and at the far face of the porous layer as well as water-filled cavities near the center of the recrystallized S-layer. From complementing neutron reflectometry measurements on hetero- (lipid/S-layer protein) bilayer systems at various contrasts, the interaction of the peptide with the lipid headgroups emerges in great detail, as shown in Fig. 5b (Weygand, 2000; Weygand *et al.*, 2002). This is consistent with the observation that particularly the asymmetric and symmetric P—O stretch vibrations of the lipid headgroup phosphate, as observed in surface-sensitive FTIR measurements (Fourier transform infrared reflection–absorption spectroscopy, FT-IRRAS; Mendelsohn *et al.*, 1995), are sensitive to the S-layer recrystallization at the interface (Weygand *et al.*, 2000).

Other recent investigations of protein/lipid systems at the air/water interface include an FTIR and X-ray reflectivity characterization of photosystem II core complex (Gallant *et al.*, 1998), the binding of cytochrome *c* to mixed, phase-separated phospholipid monolayers using neutron reflectivity and FTIR measurements (Maierhofer *et al.*, 2000), and the membrane association of the human immunodeficiency virus-1 (HIV-1)-related accessory protein virus protein U (Vpu; Marassi *et al.*, 1999), which helps facilitate virus release from infected cells (Zheng *et al.*, 2001). For the latter system it has recently been established that the two amphiphatic helices of Vpu located at the C-terminal extend into the subphase and its N-terminal hydrophobic helix intercalates the hydrophobic chain slab. The average tilt angle of this helix was observed to depend on surface pressure, and a continuous decrease of the tilt from the surface normal with increasing pressure

---

[3]Note that the S-protein/phospholipid interaction is a *nonnatural* coupling, because the S-layer resides at the outer face of a *peptidoglycan* cell wall in Gram-positive bacteria (Sleytr *et al.*, 1999). The motivation for studying S-layers in such nonnatural supramolecular complexes is their potential for the stabilization of lipid bilayer membranes (Mader *et al.*, 1999; Schuster *et al.*, 1998a).

(a)

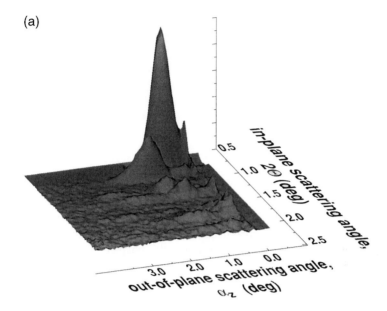

(b)

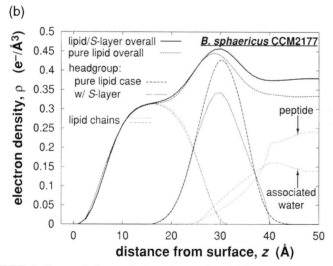

FIGURE 5 Structural characterization of reconstituted S-layers (*B. sphaericus* CCM2177) underneath DPPE monolayers on an aqueous subphase containing 10 m*M* CaCl$_2$. (a) Small-angle GIXD showing the background-corrected Bragg rods deriving from scattering from the monomolecular protein crystal sheet. (b) Peptide insertion into the lipid layer as revealed from a joint model refinement using X-ray and neutron reflectivity data. Both adapted from Weygand (2000).

has been reported. Concomitantly the cytoplasmatic helices were found to flatten progressively toward the membrane surface, attributed to a comparably restrictive hinge sequence between the hydrophilic and the first amphiphatic helix. Only at the highest surface pressure studied, $\pi = 55$ mN/m, did the electron density profile indicate the displacement of the C-terminal helix from the monolayer, presumably owing to steric hindrance among the helices at high peptide concentrations at the membrane surface.

B. Lipid Bilayers and Multi-Bilayer Stacks

1. Pure Lipid Systems

Phospholipid bilayers prepared at planar (or macroscopically curved) solid supports, and particularly multi-bilayer stacks, which may add up to a few micrometers in total thickness, lend themselves to a much more facile, and thus more complete, characterization with the methods covered in this chapter. Exploiting X-ray and neutron diffraction from multi-bilayer stacks on an absolute scale (Franks *et al.,* 1978), Wiener and White perfected the physical characterization of fluid bilayers in a fully resolved and complete structural model by determining the transmembrane distribution of individual submolecular fragments (Wiener and White, 1992b, and references therein). Such fluid bilayers resemble closely the physical state of actual biomembranes, and in particular the transmembrane distribution of double bonds of the unsaturated chains contained in physiological lipids is a sensitive indicator of the partitioning of peptides into the membrane (Jacobs and White, 1989). These distributions, which represent the time-averaged projection of the molecular fragments on the bilayer normal, may be determined by neutron diffraction from samples containing specifically deuterated fragments (Wiener *et al.,* 1991) or by X-ray diffraction from samples containing specifically brominated lipids that have been shown to be isomorphous for the case of dioleoylphosphatidylcholine (DOPC) (Wiener and White, 1991c). From this work, a general picture has appeared that divides the bilayer structure into a hydrocarbon core region $\sim$30 Å thick and an interface region sandwiching the core on both sides with $\sim$15-Å-thick slabs that include both the hydrophilic headgroups and membrane-bound water. These two distinct regions may in some sense be treated as two distinct "phases."

A limitation of Wiener–White approach to fluid bilayer structure elucidation is that only samples at reduced hydration, controlled via the relative humidity (RH) of the sample environment, diffract sufficiently well for the determination of a fully resolved picture[4] and that the close apposition of neighboring headgroup layers within the multilayer stacks—typically separated by $\sim$10 Å of headgroup-bound water—limits the accessibility of the membrane surface to external solutes.

---

[4] At higher hydration, thermal fluctuations degrade the intensities of higher diffraction orders (Zhang *et al.,* 1994).

A fully resolved model of such bilayer membranes (Fig. 6a) has thus far only been achieved at RH values that correspond to $\sim$5 water molecules per lipid (Wiener and White, 1992a,b; Wiener *et al.,* 1991). This picture has been considerably refined with the help of molecular dynamics simulations (Armen *et al.,* 1998; Petrache *et al.,* 1997). Underresolved bilayer models at higher hydration have recently been determined up to 93% RH, corresponding to 9.4 water molecules per lipid, using X-ray diffraction from oriented samples and extended into the regime of even higher hydration by X-ray diffraction from unoriented membranes in the form of lipid suspensions (McIntosh *et al.,* 1987; 1989) in poly(vinylpyrrolidone) (PVP) solutions (Hristova and White, 1998). The dependence of the Bragg spacing on the sample water content (Fig. 6b) suggests that 12 water molecules complete the hydration shell of the PC headgroups. In connection with the finding from monolayer studies that $\sim$6 water molecules hydrate the phosphate within phospholipid headgroups independent of the molecular area within the film (Schalke and Lösche, 2000), this suggests that an extra $\sim$6 water molecules are tightly bound at the headgroup adjacent to an (infinite) fluid compartment.

The crossover from partially to fully hydrated membrane stacks has been recently studied in detail using nonspecular neutron and X-ray scattering measurements, which enable the quantification of lateral height–height correlation functions, as well as specular reflectivity (Münster *et al.,* 1999; Salditt *et al.,* 1999, 2002). For DMPC multi-bilayer membranes at a RH slightly below 100% it has been observed that thermal fluctuations are greatly suppressed, resulting in an absence of the Landau–Peierls effect that is typical of liquid crystalline order (Als-Nielsen *et al.,* 1980). Thus, significant deviations from the standard Caillé theory (Caillé, 1972) have been observed (Salditt *et al.,* 1999). Rather, it is concluded that local static defects dominate the lineshape of the diffuse intensity in both neutron (Münster *et al.,* 1999) and X-ray scattering (Salditt *et al.,* 1999). In the course of that work, the unbinding of the multilamellar stacks as a function of temperature has been observed with such reflectivity measurements (Vogel *et al.,* 2000). On the other hand, inelastic X-ray scattering has been utilized to study the collective thermal motion of bilayer stacks (Chen *et al.,* 2001), and a dispersion relation for the acoustic modes has been experimentally derived. In this dispersion relation, a softening of the excitation around $k = 1.4 \text{ Å}^{-1}$ has been observed that has been implied to play a role in the transport of small molecules, such as H_2O, across the membrane in the liquid crystalline (L_α) phase. Quasielastic neutron scattering spectra of two-component (DPPC/cholesterol) membrane stacks have been interpreted as indicative of an anisotropic out-of-plane motion of the sterol molecules (Gliss *et al.,* 1999), conceivably similar to a "wobbling" motion of cholesterol implied from biochemical investigations of sterol transfer between membranes (Steck *et al.,* 1988).

Due to the greatly reduced volume density of scattering centers, solid-supported lipid bilayers are even more difficult to characterize than multi-bilayer stacks. The internal (solid support/lipid membrane) is inaccessible to X-rays in the mid-energy

(a)

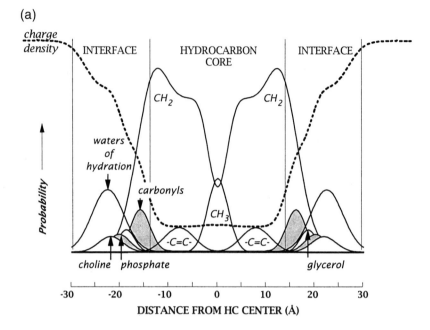

(b)

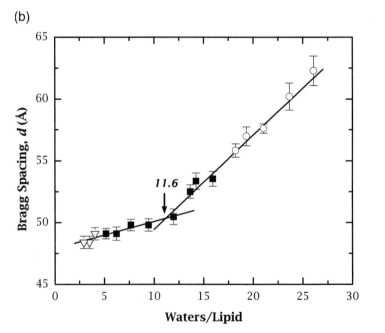

range ($E_{\text{X-ray}} \approx 10\,\text{keV}$, corresponding roughly to a wavelength of $\lambda_{\text{X-ray}} \approx 1.2\,\text{Å}$) that is most easily available and has generally been used for the measurements reported thus far. Neutron scattering provides easy access to buried interfaces; however, due to the relatively low brilliance of even the most advanced neutron sources, the information content of reflectivity spectra is severely limited (Schalke and Lösche, 2000). On the other hand, the great flexibility of contrast enhancement via specific deuteration makes neutron scattering measurements well worth the effort. Recent years have seen a continuous increase in the performance of neutron scattering experiments (Koenig *et al.*, 1996; Krueger *et al.*, 1995)—more through a careful engineering of the sample environment primarily in order to reduce incoherent sample background than through an increase of source performance— as well as data evaluation procedures (Koenig *et al.*, 1996; Lösche *et al.*, 1998; Schalke *et al.*, 2000; Vaknin *et al.*, 1991b; Wiener and White, 1991b) that are generally applicable in neutron or X-ray scattering. Thus, the relatively coarse picture that has been developed for PC bilayers formed at a silicon wafer surface (Johnson *et al.*, 1991a; Naumann *et al.*, 1995) has been considerably refined in recent work (Koenig *et al.*, 1996), taking account of lateral inhomogeneities of the membrane as assessed by independent atomic force microscopy (AFM) character-ization. Weakly bound, freely floating bilayers in close vicinity to bilayer-covered silicon surfaces have been reported (Charitat *et al.*, 1999) and studied with respect to their thermal responses (Fragneto *et al.*, 2001). Such systems are thought to provide an alternate opportunity for the investigation of peptide interactions with bilayer membranes (Fragneto *et al.*, 2000) (see below).

2. Peptide/Lipid Bilayer Systems

The structural and thermodynamic principles of membrane protein folding and insertion as well as the basis of their stability within membranes make up an area that contains some of the most pressing open questions in structural biology. The problems involved have been recently spelled out in detail in insightful papers by White and Wimley (1998, 1999). In brief, while the thermodynamic stability of soluble proteins is rather straightforward to determine from bulk-phase partitioning measurements, the situation is much more complex for membrane proteins (see Fig. 7), because lipid membranes are characterized by thermodynamic properties

---

FIGURE 6   (a) Distribution of submolecular lipid fragments across a DOPC bilayer as determined from the joint refinement of X-ray and neutron diffraction. Overlaid as a dashed line is the derived polarity ("charge density") profile for the bilayer structure. Membrane-adsorbed helical peptides, as characterized in X-ray diffraction from bilayer-peptide systems (Hristova *et al.*, 1999), are located in the headgroup region ("INTERFACE"), where the steepest gradient of the polarity profile occurs. Based on Fig. 6 of White and Wimley (1999). (b) Crossover from partial dehydration to full hydration of phospholipid (DOPC) membrane surfaces as observed in X-ray membrane diffraction. Reprinted from Hristova and White (1998); copyright 1998, with permission from Elsevier Science.

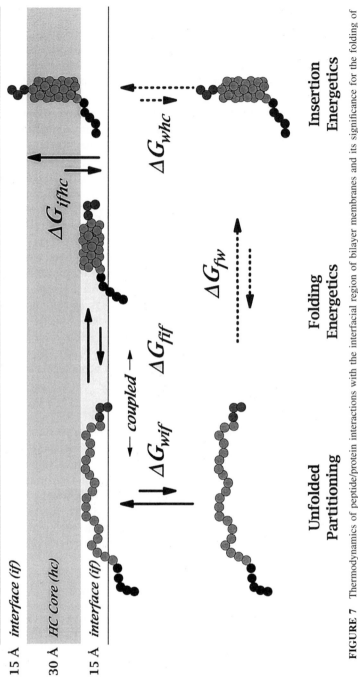

FIGURE 7 Thermodynamics of peptide/protein interactions with the interfacial region of bilayer membranes and its significance for the folding of membrane proteins. Reproduced from White and Wimley (1998) with permission.

of their own: They exist in a free energy minimum that derives from small differences between large enthalpic and entropic terms. As a consequence, free energies of partitioning of membrane solutes, such as peptides or proteins, include changes in bilayer free energy deriving from perturbations by the solute. These contributions have collectively been referred to as the "bilayer effect" (Wimley and White, 1993) and have to be distinguished from the hydrophobic effect that arises from the water interactions with the bilayer constituents. The thermodynamic pathway connecting an entirely unfolded polypeptide chain in aqueous solvent and its correctly folded intramembrane protein form involves thus a transient state in which the unfolded peptide is transferred into a prefolded state, mediated by the presence of the membrane interface, rather than consisting of a simple two-step transformation (folding in solution and membrane insertion; see Fig. 7) (White and Wimley, 1998). Hence, the whole process may be divided into three major intermediate steps, which depend implicitly on the presence of the membrane interface: (i) binding of the unfolded peptide to the interface, associated with the energy term ΔG_{wif}, (ii) folding of the peptide that resides at the interface into a compact structure (ΔG_{fif}), and (iii) membrane insertion of the folded, interface-bound protein structure (ΔG_{ifhc}). Along this line, studies of the hydrophobic binding of small peptides at bilayer interfaces have revealed an experimental hydrophobicity scale (Wimley and White, 1996), serving as a reference from which ΔG_{wif} may be estimated. Structural issues enter as far as the association of larger peptides with membrane interfaces and their aggregation into complexes is concerned (Chung and Thompson, 1996). The bilayer interface regime, as distinguished from the hydrophobic core in the assessment of the bilayer structure (Wiener and White, 1992b), has been shown to account for roughly 50% of the total thickness of the thermal bilayer structure (see Fig. 6a) and is very capable of accommodating a folded peptide helix. Of particular importance for the partitioning of peptide into this region is the fact that it consists of a complex mixture of glyceryl, carbonyl, phosphate, choline, and water; even some of the methylenes penetrate this region, as revealed by the time-averaged distribution of molecular fragments shown in Fig. 6a. The region is thus rich in optional noncovalent interaction sites for peptides attracted to the membrane surface. Thus, even peptides attracted to the bilayer membrane primarily by the hydrophobic effect are found located within its interfacial region, not the hydrophobic core, as determined from neutron diffraction (Jacobs and White, 1989). This may be attributed, at least for small peptides, to their inaptitude to form stable *intra*molecular hydrogen bonds.

Another important property of the bilayer's interfacial region that bears important implications for the trapping of structured (i.e., larger) peptide molecules is the fact that this region experiences the steepest part of the charge density gradient, and thus the steepest gradient in polarity, that occurs across the bilayer (see Fig. 6a). Consequently, X-ray diffraction measurements on DOPC multilayers at 66% RH containing the ideally amphiphatic synthetic peptide poly(A) (5 mol%

with respect to total lipid) showed that the α helix is well aligned with the membrane surface (inclination angle from the surface $\gamma < 15°$), where it is perfectly trapped within this polarity gradient: The helix center was found located between the center positions of the carbonyl and phosphate moieties and largely overlapping with the glycerol backbone of the PC (Hristova *et al.,* 1999). A very similar result, even on a quantitative scale, was found for mellitin (1 mol%) association with DOPC bilayers (Hristova *et al.,* 2001), whereas an artificial mellitin dimer derived from the Q25C mutant of the native peptide, although similarly aligned with the membrane, caused such strong perturbations of the bilayer that no satisfactory model for the overall structure was identified. For the native mellitin, on the other hand, model refinement showed that the mellitin helix is in a conformation consistent with the nuclear magnetic resonance (NMR) structure (Okada *et al.,* 1994) rather than the X-ray crystal structure (Terwilliger and Eisenberg, 1982a,b).

While, as summarized above, Steve White's laboratory has systematically developed the issue of peptide/lipid bilayer interactions in general terms, Huey Huang's lab at Rice University has specifically investigated the mode of action of endogenous gene-encoded antimicrobial peptides, which are key components of the mammalian immune system (Huang, 2000). X-ray and neutron scattering measurements have been extensively used to clarify the molecular organization of model peptides, such as mellitin, alamethicin, gramicidin, magainin, and protegrin, at or within planar bilayer membranes. The most interesting question in this area is the origin of the specificity with which lysis of target bacterial cells is achieved, given the relatively unspecific nature of peptide/lipid interactions. It has been conjectured that such peptide may occur in two states at the membrane: (1) a dormant state in which the molecules are trapped at the membrane surface, conceivably in an arrangement similar to that described above for poly(A) (Hristova *et al.,* 1999) and mellitin (Hristova *et al.,* 2001), and (2) an active, lethal state in which the peptides, whose lengths match that of the bilayer width in most cases, are inserted across the membrane and are thought to form pores that mediate cell lysis (Boman, 1994). Thus, using a geometry that is essentially orthogonal to that in GIXD, He *et al.* have studied the in-plane X-ray scattering of gramicidin inserted into DLPC bilayers, organized in multi-bilayer stacks (He *et al.,* 1993a,b). Using the power of neutron scattering in the same geometry, they were able to directly demonstrate pore formation of alamethicin (He *et al.,* 1995, 1996) in membranes by utilizing the high contrast between the water (D_2O)-filled channels and their membrane surrounding. It was thus determined that eight monomers of the peptide are most likely involved in channel formation (He *et al.,* 1996). Using these tools, experimental evidence for the "mattress effect" (Mouritsen and Bloom, 1984) of peptide influence on the lipid bilayer has been reported and in-plane correlation between the channels has been detected, which originates presumably from the strain field (Aranda-Espinoza *et al.,* 1996) created by the gramicidin pores within the bilayer (Harroun *et al.,* 1999). Interbilayer correlation, finally, has been detected using neutron scattering (Yang *et al.*, 1999) upon moderate dehydration of the multi-bilayer samples, indicating the formation of 3D ordered peptide/lipid samples (Yang *et al.*, 2000), and interpreted to originate due to hydration forces (Yang *et al.*, 1999). A progressively more detailed comprehension of the action of these antimicrobial peptides on membranes as well as a molecular-scale structure of peptide/membrane assemblies are thus emerging.

GIXD has also been applied to study the order of aligned DMPC multi-bilayer membranes and the effect magainin 2 exerts on such membrane models (Münster *et al.*, 2000). However, in this case only the decay of correlations between the acyl chains upon peptide association with the membrane has been reported. More interesting are reports on the diffuse and specular reflectivity (Koltover *et al.*, 1998) as well as GIXD (Koltover *et al.*, 1999) from highly aligned bR multilayer stacks, in which the thermal disintegration of the PM lattice has been studied and a phase diagram indicating the stability of the protein/lipid cocrystal has been derived showing that the thermal denaturation of the protein depends critically on inter-bilayer distance (Koltover *et al.*, 1999). The thermal denaturation of bR within PM multi-bilayer crystals has been successively studied in detail (Müller *et al.*, 2000), and it has been revealed that although long-range correlation is lost, the protein disintegrates only partially and that only some sections of a partially disintegrated secondary structure are exposed to the aqueous phase, which explains the increased sensitivity of the denatured protein to proteolysis upon denaturation. On the other hand, even in the denatured state at temperatures as high as $T \approx$ 100°C peptide/lipid interactions play a major role in preventing the system from undergoing total loss of supramolecular organization (Müller *et al.*, 2000).

Other recent studies of interest include neutron scattering studies of feline leukemia virus fusion peptide (Davies *et al.*, 1998) and of neurokinin A (Darkes *et al.*, 2000) bound to aligned planar DOPG and DOPC multilayers. Fragneto and co-workers have characterized the structure of single bilayers of DPPC or DPPC/dipalmitoylphosphatidylserine (DPPS) (9:1) on flat Si wafers and have investigated the impact of a 16-amino-acid peptide, helix III of the antennapedia homeodomain, which has been used as a transmembrane vector (Fragneto *et al.*, 2000). A water layer of 5 Å thickness was observed between the substrate and the bilayer, independent of headgroup composition. After correction for incomplete coverage (Charitat *et al.*, 1999), the revealed bilayer structure was found to be in line with that observed in earlier work on such systems (Koenig *et al.*, 1996), with an average tilt of the acyl chains around 40° from the surface normal. Upon introduction of the peptide, changes were most pronounced in the headgroup region, and a significant increase of the apparent roughness of the bilayer structure was reported. Due to the macroscopic heterogeneity of the bilayer, conclusions on the peptide's impact remain somewhat vague, however.

A rather quaint system in the context of lipid/peptide interactions, in which the hydrophilic/hydrophobic bilayer sandwich structure manifests itself in an inside-out fashion as a free-standing film in air, has been investigated by Benattar and co-workers in their structural studies of NBFs using X-ray reflectivity (Bélorgey and Benattar, 1991). Although the structural assessment of the pure lipid systems (Cuvilliers *et al.*, 2000a) remains on the level of box models, resulting in a similarly unrealistic picture of the headgroup as given by box models of Langmuir monolayers (Schalke and Lösche, 2000), protein (bovine serum albumin) insertion in such systems has been reported (Benattar *et al.*, 1999) and has been observed to lead to the formation of coherent and macroscopically homogeneous peptide layer within the inside-out structure (Cuvilliers *et al.*, 2000b). Whether this system is capable of promoting the 2D crystallization of proteins or leads to new insights in protein/lipid interactions remains to be seen.

C. SAMs, HBMs, and Grafted Tethered Lipid Bilayers

Soft matter functionalized surfaces (Diederich and Lösche, 1997) are generally considered better systems than bare solid-state (e.g., Si, Au, or glass) surfaces for the preparation of substrate-supported functional, biomimetic peptide/lipid model membranes (Knoll *et al.*, 2000; Sackmann and Tanaka, 2000). Various approaches lend themselves to that end. A general tradeoff exists between the complexity of surface functionalization, resulting in a progressively increasing flexibility for the incorporation of functions into the interface structures, and the level of effort in terms of surface chemistry on the one hand and in terms of surface characterization on the other by which the resultant structures may be characterized on the molecular level. Naturally, an aim is thus to achieve advanced and stable functional surface architectures while using minimum effort for the modification.

1. SAMs Utilized for the Implementation of Vectorially Oriented Protein Arrays

Organic monolayers, self-assembled and tethered via thiol chemistry to Au or Si surfaces, provide a minimalistic approach for the "softening" of "hard" (i.e., solid-state) surfaces for stable functionalization with proteins (Amador *et al.*, 1993). Although not at the center of interest of this review, because no peptide/lipid interactions are involved, the opportunities to prepare organized protein layer structures and to approach their molecular-scale structural characterization are well worth mentioning. Thus, the unspecific binding of human serum albumin (HSA) to hydrophobically terminated SAMs has been studied using X-ray reflectometry, and significant differences in the tenacity of adsorbed layers as a function of the SAM packing density has been reported (Petrash *et al.*, 1997). If one single amino acid residue on a protein's surface is utilized for the ligation to the surface, uniaxial orientation may be achieved (Stayton *et al.*, 1992). Reflectivity and interferometry, in which the phasing is achieved through interference of the beam reflected from the surface structure under investigation with a beam reflected at a high-contrast feature buried at a specific position underneath the surface or (neutrons only) at a reference structure prepared with different spin orientations (de Haan *et al.,* 1995; Majkrzak and Berk, 1995; Majkrzak *et al.,* 1998), have been exploited to characterize the association of cytochrome oxidase with SAM surfaces (Edwards *et al.,* 1997, 1998; Kneller *et al.,* 2001) and the subsequent formation of a bimolecular complex of cytochrome oxidase with cytochrome *c* (Edwards *et al.,* 1998). Yeast cytochrome *c* by itself has been immobilized on sulfhydryl-terminated SAMs by its reaction via a naturally occurring, unique surface cysteine residue and its heme site investigated with polarized X-ray absorption fine structure (XAFS) in reflection mode (Edwards *et al.,* 2000). Neutron reflectometry revealed the hydration in terms of water distribution profiles within the protein surface layers (Kneller *et al.,* 2001). SAMs have also been used for interferometry studies of Ca^{2+}-ATPase (Prokop *et al.,* 1996) as well as maquette peptides (Robertson *et al.,* 1994), the latter complementing the characterization at the air/water interface (Strzalka *et al.,* 2000) after transfer to Si wafers incorporating molecular beam epitaxy (MBE)-grown Ge–Si superstructures. From the deduced SLD profiles it was concluded that the helical bundles incorporated within the maquettes extend normal to the solid-state surface (Strzalka *et al.,* 2001), even in cases when they were oriented parallel to the aqueous surface prior to the transfer (Strzalka *et al.,* 2000). This is taken as evidence that substantial molecular rearrangement may occur upon transfer of even carefully engineered surface structures.

2. HBMs as Synthetic Membrane Model Mimics

A hybrid approach to mimicking lipid membrane surfaces by a solid supported system has been taken in Anne Plant's lab at the National Institute of Standards and Technology (Gaithersburg, MD) by developing hybrid bilayer membranes (HBMs). In this approach, an alkanethiol SAM is prepared on a flat Au substrate, followed by the transfer of a (phospho-) lipid monolayer from the air/water interface (Plant, 1993). After drainage of surplus lipid material with filter paper, the resulting dual-layer structure is stable even in an ambient environment where the hydrophilic lipid headgroup is exposed to the gas phase. The molecular-scale structure of such systems has been investigated by a host of characterization methods (Meuse *et al.,* 1998), including neutron reflectometry (Majkrzak *et al.,* 2000). In very recent work it has been demonstrated by studying the association of mellitin with HBMs that such systems provide a promising and versatile tool for the investigation of peptide/lipid interactions (Krueger *et al.,* 2001). At the same time, this work provides a reference for both the state of the art of technical performance in neutron reflectometry measurements and sensible and realistic data evaluation. An example of the structure of the parent HBM system, as it emerges in terms of a neutron SLD distribution in an isotopically well-designed system (hydrogenated

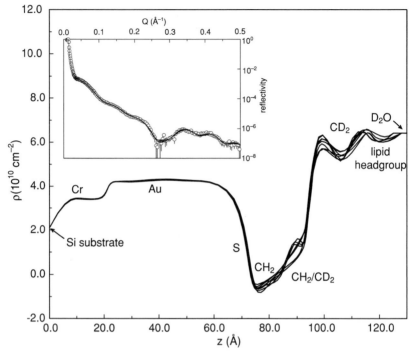

FIGURE 8 Structure of an octadecanethiol/DMPC HBM on a planar Au substrate as derived from neutron reflectometry. Displayed in the main figure is a family of neutron SLD profiles describing the structure of an HBM, composed of octadecanethiol and chain-perdeuterated DMPC, in contact with D_2O. Each SLD profile is consistent with the experimental data shown in the inset. From Krueger *et al.* (2001) with permission.

SAM, chain-perdeuterated lipid in the transferred monolayer adjacent to D_2O) is shown in Fig. 8.

3. Tethered Functional Lipid Bilayers

Continuous lipid bilayers that are solid-supported but simultaneously spatially decoupled from the solid substrate (Sackmann, 1996) are a biophysicist's dream of a perfect mimic of membrane functions and at the same time a synthetic chemist's and structural experimentalist's nightmare to synthesize and characterize in molecular terms. Whereas Langmuir monolayers, SAMs, and HBMs are all quite obviously inadequate systems for the incorporation of large membrane proteins, even bilayers immediately adsorbed to solid substrates, such as the ones referred to above (Charitat *et al.*, 1999; Johnson *et al.*, 1991a; Koenig *et al.*, 1996), are clearly inadequate for the implementation of membrane functions, because the

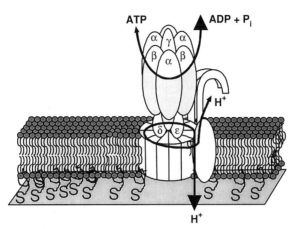

FIGURE 9 Schematic representation of an ATPase-functionalized, polymer-tethered lipid bilayer. Reprinted from Knoll *et al.* (2000), copyright 2000, with permission from Elsevier Science.

molecularly thin water layer that persists between the proximal bilayer leaflet and the solid substrate (5–10 Å in thickness) is clearly insufficient to allow the high lateral mobility of the lipid components within the structure required to retain most membrane functions and even more so to accommodate extramembrane sections that most membrane proteins incorporate. Thus, various concepts to incorporate tethering layers between the solid substrate and the membrane have been put forward, for example, in the form of soft (polymer) cushions required to attach the membrane to the substrate by establishing covalent links between the solid and some of the lipid molecules while creating a stable and well-defined hydration layer of thickness 10–100 Å between substrate and membrane (Knoll *et al.*, 2000) (see Fig. 9). Successful attempts at such schemes have attracted considerable attention (Cornell *et al.*, 1997; Raguse *et al.*, 1998).

Surface-sensitive scattering methods, and specifically neutron reflectometry, are particularly well suited to study the nanometer-scale structure of tethered lipid bilayer systems and their precursors. The adsorption of polyelectrolytes, such as poly(styrenesulfonate), at the water surface (Saville *et al.*, 1994; Yim *et al.*, 2000), the molecular-scale conformation of hydrophilic polymer brushes at the air/water interface (Kuhl *et al.*, 1999; Majewski *et al.*, 1997; Politsch *et al.*, 2001; Wurlitzer *et al.*, 2001), and structured polyelectrolyte films deposited at wafer surfaces (Lösche *et al.*, 1998; Schmitt *et al.*, 1993; Steitz *et al.*, 2000) have been extensively studied. Polymer-supported monolayers in air and bilayers under water have been shown to form on adsorbed polyelectrolyte cushions (Wong *et al.*, 1999a) and have been characterized with respect to their molecular interactions with juxtaposed bilayers (Wong *et al.*, 1999b), where membrane fusion has been shown to occur. Synthetic routes for the formation of crosslinked polymer cushions have been described (Seitz *et al.*, 2000). Alternate promising strategies for the formation of stabilized (Küpcü *et al.*, 1995; Schuster *et al.*, 1998b) bilayer membranes involve S-layers as carrier materials (Wetzer *et al.*, 1997). The incorporation of membrane proteins in such systems has already been demonstrated (Schuster *et al.*, 1998a) and the lateral mobility of the constituent phospholipids has been characterized via fluorescence recovery after photobleaching (FRAP) measurements (Györvary *et al.*, 1999; Starr and Thompson, 2000). For a deeper understanding of the self-assembly process that leads to the formation of S-layer-supported membranes a molecular-scale comprehension of the lipid/peptide interface, as derived from surface-sensitive scattering methods (Weygand *et al.*, 1999, 2000), has proven invaluable. The future holds great promises and great challenges in this rapidly developing field. Surface-sensitive X-ray and neutron scattering methods will be among the primary tools that drive this development.

Acknowledgments

A warm thank you to J. Kent Blasie, Giovanna Fragneto-Cusani, Ka-Yee C. Lee, Tim Salditt, Stephan Verclas, Markus Weygand, and Steve H. White for fruitful discussions, for making available preprints, and for the communication of unpublished work. I thank Kristian Kjaer devising the liquid surface experimental station at BW1 and for valuable advice on the experimental work, as well as Markus Weygand and Peter Krüger for measurements. I am very grateful to U. B. Sleytr, B. Wetzer, and D. Pum for a longstanding and fruitful collaboration in the field of S-layer characterization. Beam time at HASYLAB (DESY, Hamburg, Germany) under project no. II-99-078 is gratefully acknowledged. The work has been financially supported by the DFG through the SFB 294 (TP F3) and by the Fonds der Chemischen Industrie (Frankfurt, Germany).

References

Als-Nielsen, J., and Kjaer, K. (1989). X-ray reflectivity and diffraction studies of liquid surfaces and surfactant monolayers. *In* "Phase Transitions in Soft Condensed Matter" (T. Riste and D. Sherrington eds.), pp. 113–138. Plenum Press, New York.

Als-Nielsen, J., and Möhwald, H. (1991). Synchrotron X-ray scattering studies of Langmuir-films. *In* "Handbook on Synchrotron Radiation" (S. Ebashi, M. Koch, and E. Rubinstein, eds.), Vol. 4, pp. 1–53. Elsevier North-Holland, Amsterdam.

Als-Nielsen, J., Litster, J. D., Birgeneau, R. J., Kaplan, M., Safinya, C. R., Lindegaard-Andersen, A., and Mathiesen, S. (1980). Observation of algebraic decay of positional order in a smectic liquid crystal. *Phys. Rev. B* **22,** 312–320.

Als-Nielsen, J., Jacquemain, D., Kjaer, K., Lahav, M., Leveiller, F., and Leiserowitz, L. (1994). Principles and applications of grazing incidence X-ray and neutron scattering from ordered molecular monolayers at the air–water interface. *Phys. Rep.* **246,** 251–313.

Amador, S. M., Pachence, J. M., Fischetti, R. F., McCauley, J. P., Smith III, A. B., and Blasie, J. K. (1993). The use of self-assembled monolayers to covalently tether protein monolayers to the surface of solid substrates. *Langmuir* **9,** 812–817.

Aranda-Espinoza, H., Berman, A., Dan, N., Pincus, P., and Safran, S. (1996). Interaction between inclusions embedded in membranes. *Biophys. J.* **71,** 648–656.

Armen, R. S., Uitto, O. D., and Feller, S. E. (1998). Phospholipid component volumes: Determination and application to bilayer structure calculations. *Biophys. J.* **75,** 734–744.

Arnold, K. (1995). Cation-induced vesicle fusion modulated by polymers and proteins. *In* "Structure and Dynamics of Membranes: Generic and Specific Interactions" (R. Lipowsky and E. Sackmann, eds.), Vol. 1B, pp. 903–957. North Holland, Amsterdam.

Bayerl, T. M., Thomas, R. K., Penfold, J., Rennie, A., and Sackmann, E. (1990). Specular reflection of neutrons at phospholipid monolayers: Changes of monolayer structure and headgroup hydration at the phase transition from the expanded to the condensed phase state. *Biophys. J.* **57,** 1095–1098.

Bélorgey, O., and Benattar, J.-J. (1991). Structural properties of soap black films investigated by X-ray reflectivity. *Phys. Rev. Lett.* **66,** 313–316.

Benattar, J.-J., Nedyalkov, M., Prost, J., Tiss, A., Verger, R., and Guilbert, C. (1999). Insertion process of a protein single layer within a Newton black film. *Phys. Rev. Lett.* **82,** 5297–5300.

Berk, N. F., and Majkrzak, C. F. (1995). Using parametric B splines to fit specular reflectivities. *Phys. Rev. B* **51,** 11296–11309.

Bezrukov, S. M., and Winterhalter, M. (2000). Examining noise sources at the single-molecule level: $1/f$ noise of an open maltoporin channel. *Phys. Rev. Lett.* **85,** 202–205.

Bitto, E., Li, M., Tikhonov, M., Schlossman, M. L., and Cho, W. (2000). Mechanism of annexin I-mediated membrane aggregation. *Biochemistry* **39,** 13469–13477.

Blasie, J. K., and Timmins, P. (1999). Neutron scattering in structural biology and biomolecular materials. *MRS Bull.* **24,** 40–47.

Boman, H. G. (1994). "Antimicrobial Peptides." Wiley, Chichester, England.

Brisson, A., Bergsma-Schutter, W., Oling, F., Lambert, O., and Reviakine, I. (1999). Two-dimensional crystallization of proteins on lipid monolayers at the air–water interface and transfer to an electron microscopy grid. *J. Cryst. Growth* **196,** 456–470.

Brumm, T., Naumann, C., Rennie, A. R., Thomas, R. K., Kanellas, D., Penfold, J., and Bayerl, T. M. (1994). Conformational changes of the lecithin headgroup in monolayers at the air/water interface. A neutron reflection study. *Eur. Biophys. J.* **23,** 289–295.

Cadenhead, D. A., Müller-Landau, F., and Kellner, B. M. J. (1980). Phase transitions in insoluble one and two-component films at the air/water interface. *In* "Ordering in Two Dimensions" (S. K. Sinha, ed.), pp. 73–81. Elsevier North Holland, Amsterdam.

Caillé, A. (1972). Physique cristalline—Remarques sur la diffusion des rayons X dans les smectiques A. *Compt. Rend. Acad. Sci. Paris B* **274,** 891–893.

Charitat, T., Bellet-Amalric, E., Fragneto, G., and Graner, F. (1999). Adsorbed and free lipid bilayers at the solid–liquid interface. *Eur. Phys. J. B* **8,** 583–593.

Chen, S. H., Liao, C. Y., Huang, H. W., Weiss, T. M., Bellisent-Funel, M. C., and Sette, F. (2001). Collective dynamics in fully hydrated phospholipid bilayers studied by inelastic X-ray scattering. *Phys. Rev. Lett.* **86,** 740–743.

Chou, C.-H., Regan, M. J., Pershan, P. S., and Zhou, X.-L. (1997). Model-independent reconstruction of smooth electron density profiles from reflectivity data of liquid surfaces. *Phys. Rev. E* **55,** 7212–7216.

Chung, L. A., and Thompson, T. E. (1996). Design of membrane-inserting peptides: Spectroscopic characterization with and without lipid bilayers. *Biochemistry* **35,** 11343–11354.

Cornell, B. A., Braach-Maksvytis, V. L. B., King, L. B., Osman, P. D. J., Raguse, B., Wieczorek, L., and Pace, R. J. (1997). A biosensor that uses ion-channel switches. *Nature* **387,** 580–583.

Cuvilliers, N., Millet, F., Petkova, V., Nedyalkov, M., and Benattar, J.-J. (2000a). Structure of free-standing phospholipidic bilayer films. *Langmuir* **16,** 5029–5035.

Cuvilliers, N., Petkova, V., Nedyalkov, M., Millet, F., and Benattar, J.-J. (2000b). Protein insertion within a biological freestanding film. *Physica B* **283,** 1–5.

Dahmen-Levison, U., Brezesinski, G., and Möhwald, H. (1998). Specific adsorption of PLA$_2$ at monolayers. *Thin Solid Films* **327–329,** 616–620.

Daillant, J., Bosio, L., Harzallah, B., and Benattar, J.-J. (1991). Structural properties and elasticity of amphiphilics on water. *J. Phys. II France* **1,** 149–170.

Darkes, M. J. M., Hauss, T., Dante, S., and Bradshaw, J. P. (2000). Revealing the membrane-bound structure of neurokinin A using neutron diffraction. *Physica B* **278–278,** 505–507.

Darst, S. A., Ahlers, M., Meller, P. H., Kubalek, E. W., Blankenburg, R., Ribi, H. O., Ringsdorf, H., and Kornberg, R. D. (1991). Two-dimensional crystals of streptavidin on biotinylated lipid layers and their interaction with biotinylated macromolecules. *Biophys. J.* **59,** 387–396.

Davies, S. M. A., Darkes, M. J. M., and Bradshaw, J. P. (1998). A neutron study of the feline leukemia virus fusion peptide: Implications for biological fusion? *Physica B* **241–243,** 1148–1151.

de Haan, V. O., van Well, A. A., Adenwall, S., and Felcher, G. P. (1995). Retrieval of phase information in neutron reflectometry. *Phys. Rev. B* **52,** 10831–10833.

Demé, B., Hess, D., Tristl, M., Lee, L.-T., and Sackmann, E. (2000). Binding of actin filaments to charged lipid monolayers: Film balance experiments combined with neutron reflectivity. *Eur. Phys. J. E* **2,** 125–136.

Diederich, A., and Lösche, M. (1997). Novel biosensoric devices based on molecular protein hetero-multilayer films. *In* "Protein Array: An Alternate Biomolecular System" (K. Nagayama, ed.), Vol. 34, pp. 205–230. Japan Scientific Societies Press/Elsevier, Tokyo/Limerick.

Diederich, A., Sponer, C., Pum, D., Sleytr, U. B., and Lösche, M. (1996). Reciprocal influence between the protein and lipid components of a lipid–protein membrane model. *Colloids Surf. B Biointerfaces* **6,** 335–346.

Diederich, A., Bähr, G., and Winterhalter, M. (1998). Influence of surface charges on the rupture of black lipid membranes. *Phys. Rev. E* **58,** 4883–4889.

Dutta, P., Peng, J. B., Lin, B., Ketterson, J. B., Prakash, M., Georgopoulos, P., and Ehrlich, S. (1987). X-ray diffraction studies of organic monolayers on the surface of water. *Phys. Rev. Lett.* **58,** 2228–2231.

Edwards, A. M., Chupa, J. A., Strongin, R. M., Smith, A. B., Blasie, J. K., and Bean, J. C. (1997). Vectorially-oriented monolayers of cytochrome oxidase: Fabrication and profile structures. *Langmuir* **13,** 1634–1643.

Edwards, A. M., Blasie, J. K., and Bean, J. C. (1998). Vectorially oriented monolayers of the cytochrome *c*/cytochrome oxidase biomolecular complex. *Biophys. J.* **74,** 1346–1357.

Edwards, A. M., Zhang, K., Nordgren, C. E., and Blasie, J. K. (2000). Heme structure and orientation in single monolayers of cytochrome *c* on polar and nonpolar soft surfaces. *Biophys. J.* **79,** 3105–3117.

Fischer, B., Teer, E., and Knobler, C. M. (1995). Optical measurements of the phase diagram of Langmuir monolayers of fatty acid–alcohol mixtures. *J. Chem. Phys.* **103,** 2365–2368.

Fragneto, G., Thomas, R. K., Rennie, A. R., and Penfold, J. (1995). Neutron reflectivity study of bovine β-casein adsorbed on OTS self-assembled monolayers. *Science* **267,** 657–660.

Fragneto, G., Graner, F., Charitat, T., Dubos, P., and Bellet-Amalric, E. (2000). Interaction of the third helix of antennapedia homeodomain with a deposited phospholipid bilayer: A neutron reflectivity structural study. *Langmuir* **16,** 4581–4588.

Fragneto, G., Charitat, T., Graner, F., Mecke, K., Perino-Galice, L., and Bellet-Amalric, E. (2001). A floating lipid bilayer. *Europhys. Lett.* **53,** 100–106.

Frahm, R., Weigelt, J., Meyer, G., and Materlik, G. (1995). X-ray undulator beamline BW1 at DORIS III. *Rev. Sci. Instrum.* **66,** 1677–1680.

Franks, N. P., Arunachalam, T., and Caspi, E. (1978). A direct method for determination of electron density profiles on an absolute scale. *Nature* **276,** 530–532.

Fukuto, M., Penanen, K., Heilmann, R. K., Pershan, P. S., and Vaknin, D. (1997). C_{60}-propylamine adduct monolayers at the gas/water interface: A Brewster angle microscopy and X-ray scattering study. *J. Chem. Phys.* **107,** 5531–5546.

Galla, H.-J., Bourdos, N., von Nahmen, A., Amrein, M., and Sieber, M. (1998). The role of pulmonary surfactant protein C during the breathing cycle. *Thin Solid Films* **327–329,** 632–635.

Gallant, J., Desbat, B., Vaknin, D., and Salesse, C. (1998). Polarization-modulated infrared spectroscopy and X-ray reflectivity of photosystem II core complex at the gas–water interface. *Biophys. J.* **75,** 2888–2899.

Gerke, V., and Moss, S. E. (1997). Annexins and membrane dynamics. *Biochim. Biophys. Acta* **1357**, 129–154.

Gidalevitz, D., Huang, Z., and Rice, S. A. (1999a). Protein folding at the air–water interface studied with X-ray reflectivity. *Proc. Natl. Acad. Sci. USA* **96**, 2608–2611.

Gidalevitz, D., Huang, Z., and Rice, S. A. (1999b). Urease and hexadecylamine–urease films at the air–water interface: An X-ray reflection and grazing incidence X-ray diffraction study. *Biophys. J.* **76**, 2797–2802.

Gliss, C., Randel, O., Casalta, H., Sackmann, E., Zorn, R., and Bayerl, T. M. (1999). Anisotropic motion of cholesterol in oriented DPPC bilayers studied by quasielastic neutron scattering: The liquid-ordered phase. *Biophys. J.* **77**, 331–340.

Gordon, L. M., Lee, K. Y. C., Lipp, M. M., Zasadzinski, J. A., Walther, F. J., Sherman, M. A., and Waring, A. J. (2000). Conformational mapping of the N-terminal segment of surfactant protein B in lipid using ^{13}C-enhanced Fourier transform infrared spectroscopy. *J. Peptide Res.* **55**, 330–347.

Grigorieff, N., Ceska, T. A., Downing, K. H., Baldwin, J. M., and Henderson, R. (1996). Electron-crystallographic refinement of the structure of bacteriorhodopsin. *J. Mol. Biol.* **259**, 393–421.

Grundy, M. J., Richardson, R. M., Roser, S. J., Penfold, J., and Ward, R. C. (1988). X-ray and neutron reflectivity from spread monolayers. *Thin Solid Films* **159**, 43–52.

Guinier, A. (1968). "X-Ray Diffraction." Freeman, San Francisco.

Györvary, E., Wetzer, B., Sleytr, U. B., Sinner, A., Offenhäuser, A., and Knoll, W. (1999). Lateral diffusion of lipids in silane-, dextran-, and S-layer-supported mono- and bilayers. *Langmuir* **15**, 1337–1347.

Haas, H., Brezesinski, G., and Möhwald, H. (1995). X-ray diffraction of a protein crystal anchored at the air/water interface. *Biophys. J.* **68**, 312–314.

Hamley, I. W., and Skov Pedersen, J. (1994). Analysis of neutron and X-ray reflectivity data. I. Theory. *J. Appl. Crystallogr.* **27**, 29–35.

Hanyu, Y., Yamada, T., and Matsumoto, G. (1998). Simultaneous measurement of spectroscopic and physiological signals from a planar bilayer system: Detecting voltage-dependent movement of a membrane-incorporated peptide. *Biochemistry* **37**, 15376–15382.

Harroun, T. A., Heller, W. T., Weiss, T. M., Yang, L., and Huang, H. W. (1999). Experimental evidence for hydrophobic matching and membrane-mediated interaction in lipid bilayers containing gramicidin. *Biophys. J.* **76**, 937–945.

Hawgood, S., Derrick, M., and Poulain, F. (1998). Structure and properties of surfactant protein B. *Biochim. Biophys. Acta* **1408**, 150–160.

He, K., Ludtke, S. J., Wu, Y., and Huang, H. W. (1993a). X-ray scattering in the plane of membrane. *J. Phys. (France) IV* **3**, 265–270.

He, K., Ludtke, S. J., Wu, Y., and Huang, H. W. (1993b). X-ray scattering with momentum transfer in the plane of membrane: Application to gramicidin organization. *Biophys. J.* **64**, 157–162.

He, K., Ludtke, S. J., Worcester, D. L., and Huang, H. W. (1995). Antimicrobial peptide pores in membranes detected by neutron in-plane scattering. *Biochemistry* **34**, 15614–15618.

He, K., Ludtke, S. J., Worcester, D. L., and Huang, H. W. (1996). Neutron scattering in the plane of membranes: Structure of alamethicin pores. *Biophys. J.* **70**, 2659–2666.

Helm, C. A., Möhwald, H., Kjaer, K., and Als-Nielsen, J. (1987). *Phospholipid monolayer density distribution perpendicular to the water surface:* A synchrotron X-ray reflectivity study. *Europhys. Lett.* **4**, 697–703.

Helm, C. A., Tippmann-Krayer, P., Möhwald, H., Als-Nielsen, J., and Kjaer, K. (1991). Phases of phosphatidyl ethanolamine monolayers studied by synchrotron X-ray scattering. *Biophys. J.* **60**, 1457–1476.

Henderson, R., Baldwin, J. M., Downing, K. H., Lepault, J., and Zemlin, F. (1986). Structure of the purple membrane from *Halobacterium halobium:* Recording, measurement and evaluation of electron micrographs at 3.5 Å resolution. *Ultramicroscopy* **19**, 147–178.

Henderson, R., Baldwin, J. M., Ceska, T. A., Zemlin, F., Beckmann, E., and Downing, K. H. (1990). Model for the structure of bacteriorhodopsin based on high-resolution cryo-microscopy. *J. Mol. Biol.* **213**, 899–929.

Holt, S. A., McGillivray, D. J., Poon, S., and White, J. W. (2000). Protein deformation and surfactancy at an interface. *J. Phys. Chem. B* **104**, 7431–7438.

Hristova, K., and White, S. H. (1998). Determination of the hydrocarbon core structure of fluid dioleoylphosphocholine (DOPC) bilayers by X-ray diffraction using specific bromination of the double-bonds: Effect of hydration. *Biophys. J.* **74**, 2419–2433.

Hristova, K., Wimley, W. C., Mishra, V. K., Anantharamiah, G. M., Segrest, J. P., and White, S. H. (1999). An amphipathic α-helix at a membrane interface: A structural study using a novel X-ray diffraction method. *J. Mol. Biol.* **290**, 99–117.

Hristova, K., Dempsey, C. E., and White, S. H. (2001). Structure, location, and lipid perturbations of melittin at the membrane interface. *Biophys. J.* **80**, 801–811.

Huang, H. W. (2000). Action of antimicrobial peptides: Two-state models. *Biochemistry* **39**, 8347–8352.

Jacobs, R. E., and White, S. E. (1989). The nature of the hydrophobic binding of small peptides at the bilayer interface: Implications for the insertion of transbilayer helices. *Biochemistry* **28**, 3421–3437.

Johnson, S. J., Bayerl, T. M., McDermott, D. C., Adam, G. W., Rennie, A. R., Thomas, R. K., and Sackmann, E. (1991a). Structure of an adsorbed dimyristoylphosphatidylcholine bilayer measured with specular reflection of neutrons. *Biophys. J.* **59**, 289–294.

Johnson, S. J., Bayerl, T. M., Weihan, W., Noack, H., Penfold, J., Thomas, R. K., Kanellas, D., Rennie, A. R., and Sackmann, E. (1991b). Coupling of spectrin and polylysine to phospholipid monolayers studied by specular reflection of neutrons. *Biophys. J.* **60**, 1017–1025.

Kaganer, V. M., and Loginov, E. B. (1995). Symmetry and phase transitions in Langmuir monolayers: The Landau theory. *Phys. Rev. E* **51**, 2237–2249.

Kaganer, V. M., and Osipov, M. A. (1998). Molecular model for the simultaneous orientational and translational ordering in a two-dimensional liquid. *J. Chem. Phys.* **109**, 2600–2603.

Kaganer, V. M., Brezesinski, G., Möhwald, H., Howes, P. B., and Kjaer, K. (1998). Positional order in Langmuir monolayers. *Phys. Rev. Lett.* **81**, 5864–5867.

Kaganer, V. M., Möhwald, H., and Dutta, P. (1999). Structure and phase transitions in Langmuir monolayers. *Rev. Mod. Phys.* **71**, 779–819.

Kjaer, K., Als-Nielsen, J., Helm, C. A., Laxhuber, L. A., and Möhwald, H. (1987). Ordering in lipid monolayers studied by synchrotron X-ray diffraction and fluorescence microscopy. *Phys. Rev. Lett.* **58**, 2224–2228.

Kjaer, K., Als-Nielsen, J., Helm, C. A., Tippmann-Krayer, P., and Möhwald, H. (1989). Synchrotron X-ray diffraction and reflection studies of arachidic acid monolayers at the air–water interface. *J. Phys. Chem.* **93**, 3200–3206.

Kneller, L. R., Edwards, A. M., Nordgren, C. E., Blasie, J. K., Berk, N. F., Krueger, S., and Majkrzak, C. F. (2001). Hydration state of single cytochrome c monolayers on soft interfaces via neutron interferometry. *Biophys. J.* **80**, 2248–2261.

Knobler, C. M., and Desai, R. C. (1992). Phase transitions in monolayers. *Annu. Rev. Phys. Chem.* **43**, 207–236.

Knoll, W., Frank, C. W., Heibel, C., Naumann, R., Offenhäuser, A., Rühe, J., Schmidt, E. K., Shen, W. W., and Sinner, A. (2000). Functional tethered lipid bilayers. *Rev. Mol. Biotechnol.* **74**, 137–158.

Koenig, B. W., Krueger, S., Orts, W. J., Majkrzak, C. F., Berk, N. F., Silverton, J. V., and Gawrisch, K. (1996). Neutron reflectivity and atomic force microscopy studies of a lipid bilayer in water adsorbed to the surface of a silicon single crystal. *Langmuir* **12**, 1343–1350.

Koltover, I., Salditt, T., Rigaud, J.-L., and Safinya, C. R. (1998). Stacked 2D crystalline sheets of the membrane-protein bacteriorhodopsin: A specular and diffuse reflectivity study. *Phys. Rev. Lett.* **81**, 2494–2497.

Koltover, I., Raedler, J. O., Salditt, T., Rothschild, K. J., and Safinya, C. R. (1999). Phase behavior and interactions of the membrane-protein bacteriorhodopsin. *Phys. Rev. Lett.* **82**, 3184–3187.

Krueger, S., Ankner, J. F., Satija, S. K., Majkrzak, C. F., Gurley, D., and Colombini, M. (1995). Extending the angular range of neutron reflectivity measurements from planar lipid bilayers: Application to a model biological membrane. *Langmuir* **11**, 3218–3222.

Krueger, S., Meuse, C. W., Majkrzak, C. F., Dura, J. A., Berk, N. F., Tarek, M., and Plant, A. L. (2001). Investigation of hybrid bilayer membranes with neutron reflectometry: Probing the interactions of melittin. *Langmuir* **17**, 511–521.

Krüger, P., and Lösche, M. (2000). Molecular chirality and the domain shapes in lipid monolayers on aqueous surfaces. *Phys. Rev. E* **62**, 7031–7043.

Krüger, P., Schalke, M., Wang, Z., Notter, R. H., Dluhy, R. A., and Lösche, M. (1999). Effect of hydrophobic surfactant proteins SP-B and SP-C on binary phospholipid monolayers. I. Fluorescence and dark-field microscopy. *Biophys. J.* **77**, 903–914.

Krüger, P., Schalke, M., Linderholm, J., and Lösche, M. (2001). Multipurpose X-ray reflectometer optimized for the characterization of organic surface films on aqueous subphases. *Rev. Sci. Instrum.* **72**, 184–192.

Krylov, A. V., Kotova, E. A., Yaroslavov, A. A., and Antonenko, Y. N. (2000). Stabilization of O-pyromellitylgramicidin channels in bilayer lipid membranes through electrostatic interaction with polylysines of different chain lengths. *Biochim. Biophys. Acta* **1509**, 373–384.

Kuhl, T. L., Majewski, J., Wong, J. Y., Steinberg, S., Leckband, D. E., Israelachvili, J. N., and Smith, G. S. (1998). A neutron reflectivity study of polymer–modified phospholipid monolayers at the solid–solution interface: Polyethylene glycol–lipids on silane-modified substrates. *Biophys. J.* **75**, 2352–2362.

Kuhl, T. L., Majewski, J., Howes, P. B., Kjaer, K., von Nahmen, A., Lee, K. Y. C., Ocko, B., Israelachvili, J. N., and Smith, G. S. (1999). Packing stress relaxation in polymer–lipid monolayers at the air–water interface: An X-ray grazing-incidence diffraction and reflectivity study. *J. Am. Chem. Soc.* **121**, 7682–7688.

Kühlbrandt, W. (1992). Two-dimensional crystallization of membrane proteins. *Q. Rev. Biophys.* **25**, 1–49.

Küpcü, S., Sára, M., and Sleytr, U. B. (1995). Liposomes coated with crystalline bacterial cell surface protein (S-layer) as immobilization structures for macromolecules. *Biochim. Biophys. Acta* **1235**, 263–269.

Lee, K. Y. C., Lipp, M. M., Zasadzinski, J. A., and Waring, A. J. (1997). Effects of lung surfactant specific protein SP-B and model SP-B peptide on lipid monolayers at the air–water interface. *Colloids Surf. A: Physicochem. Eng. Aspects* **128**, 225–242.

Lee, K. Y. C., Majewski, J., Kuhl, T. L., Howes, P. B., Kjaer, K., Lipp, M. M., Waring, A. J., Zasadzinski, J. A., and Smith, G. S. (2000). The incorporation of lung surfactant specific protein SP-B into lipid monolayers at the air–fluid interface: A grazing incidence X-ray diffraction study. *Mat. Res. Soc. Symp. Proc.* **590**, 177–182.

Lee, K. Y. C., Majewski, J., Kuhl, T. L., Howes, P. B., Kjaer, K., Lipp, M. M., Waring, A. J., Zasadzinski, J. A., and Smith, G. S. (2001). A synchrotron X-ray study of lung surfactant specific protein SP-B in lipid monolayers. *Biophys. J.* **81**, 572–585.

Lenne, P.-F., Berge, B., Renault, A., Zakri, C., Vénien-Bryan, C., Courty, S., Balavoine, F., Bergsma-Schutter, W., Brisson, A., Grübel, G., Boudet, N., Konovalov, O., and Legrand, J.-F. (2000). Synchrotron radiation diffraction from two-dimensional protein crystals at the air/water interface. *Biophys. J.* **79**, 496–500.

Lipp, M. M., Lee, K. Y. C., Zasadzinski, J. A., and Waring, A. J. (1996). Phase and morphology changes in lipid monolayers induced by SP-B protein and its amino-terminal peptide. *Science* **273**, 1196–1199.

Lochhead, M. J., Letellier, S. R., and Vogel, V. (1997). Assessing the role of interfacial electrostatics in oriented mineral nucleation at charged organic monolayers. *J. Phys. Chem. B* **101**, 10821–10827.

Lösche, M., Piepenstock, M., Diederich, A., Grünewald, T., Kjaer, K., and Vaknin, D. (1993). Influence of surface chemistry on the structural organization of monomolecular protein layers adsorbed to functionalized aqueous interfaces. *Biophys. J.* **65**, 2160–2177.

Lösche, M., Schmitt, J., Decher, G., Bouwman, W. G., and Kjaer, K. (1998). Detailed structure of molecularly thin polyelectrolyte multilayer films on solid substrates as revealed by neutron reflectometry. *Macromolecules* **31**, 8893–8906.

Mader, C., Küpcü, S., Sára, M., and Sleytr, U. B. (1999). Stabilizing effect of an S-layer on liposomes towards thermal or mechanical stress. *Biochim. Biophys. Acta* **1418**, 106–116.

Maierhofer, A. P., Bucknall, D. G., and Bayerl, T. M. (2000). Modulation of cytochrome c coupling to anionic lipid monolayers by a change of the phase state: A combined neutron and infrared reflection study. *Biophys. J.* **79**, 1428–1437.

Majewski, J., Kuhl, T. L., Gerstenberg, M. C., Israelachvili, J. N., and Smith, G. S. (1997). Structure of phospholipid monolayers containing poly(ethylene glycol) lipids at the air–water interface. *J. Phys. Chem. B* **101**, 3122–3129.

Majkrzak, C. F., and Berk, N. F. (1995). Exact determination of the phase in neutron reflectometry. *Phys. Rev. B* **52**, 10827–10830.

Majkrzak, C. F., Berk, N. F., Dura, J. A., Satija, S. K., Karim, A., Pedulla, J., and Deslattes, R. D. (1998). Phase determination and inversion in specular neutron reflectometry. *Physica B* **246**, 338–342.

Majkrzak, C. F., Berk, N. F., Krueger, S., Dura, J. A., Tarek, M., Tobias, D., Silin, V., Meuse, C. W., Woodward, J., and Plant, A. L. (2000). First-principles determination of hybrid bilayer membrane structure by phase-sensitive neutron reflectometry. *Biophys. J.* **79**, 3330–3340.

Malik, A., Lin, W., Durbin, M. K., Marks, T. J., and Dutta, P. (1997). Specular X-ray reflectivity studies of microstructure and ordering in self-assembled multilayers. *J. Chem. Phys.* **107**, 645–652.

Marassi, F. M., Ma, C., and Gratkowski, H. (1999). Correlation of the structural and functional domains in the membrane protein Vpu from HIV-1. *Proc. Natl. Acad. Sci. USA* **96**, 14336–14341.

Marshall, G., Teer, E., Knobler, C. M., Schalke, M., and Lösche, M. (2000). Phase diagrams for monolayers of long-chain thioacetates. *Colloids Surf. A Physicochem. Eng. Aspects* **171**, 41–48.

McConnell, H. M. (1991). Structures and transitions in lipid monolayers at the air–water interface. *Annu. Rev. Phys. Chem.* **42**, 171–195.

McIntosh, T. J., Magid, A. D., and Simon, S. A. (1987). Steric repulsion between phosphatidylcholine bilayers. *Biochemistry* **26**, 7325–7332.

McIntosh, T. J., Magid, A. D., and Simon, S. A. (1989). Repulsive interaction between uncharged bilayers: Hydration and fluctuation pressures for monoglycerides. *Biophys. J.* **55**, 897–904.

Mecke, K. R., and Dietrich, S. (1999). Effective Hamiltonian for liquid–vapor interfaces. *Phys. Rev. E* **59**, 6766–6784.

Mendelsohn, R., Brauner, J. W., and Gericke, A. (1995). External infrared reflection absorption spectrometry of monolayer films at the air–water interface. *Annu. Rev. Phys. Chem.* **46**, 305–334.

Meunier, J. (1987). Liquid interfaces: Role of the fluctuations and analysis of ellipsometry and reflectivity measurements. *J. Phys. France* **48**, 1819–1831.

Meuse, C. W., Krueger, S., Majkrzak, C. F., Dura, J. A., Fu, J., Connor, J. T., and Plant, A. L. (1998). Hybrid bilayer membranes in air and water: Infrared spectroscopy and neutron reflectivity studies. *Biophys. J.* **74**, 1388–1398.

Möhwald, H. (1990). Phospholipid and phospholipid–protein monolayers at the air/water interface. *Annu. Rev. Phys. Chem.* **41**, 441–476.

Möhwald, H. (1995). Phospholipid monolayers. *In* "Structure and Dynamics of Membranes: From Cells to Vesicles" (R. Lipowsky and E. Sackmann, eds.), Vol. 1A, pp. 161–211. North Holland, Amsterdam.

Möhwald, H., Kenn, R. M., Degenhardt, D., Kjaer, K., and Als-Nielsen, J. (1990). Partial order in phospholipid monolayers. *Physica A* **168**, 127–139.

Möhwald, H., Baltes, H., Schwendler, M., Helm, C. A., Brezesinski, G., and Haas, H. (1995). Phospholipid and protein monolayers. *Jpn. J. Appl. Phys.* **34**, 3906–3913.

Mouritsen, O. G., and Bloom, M. (1984). Mattress model of lipid–protein interactions in membranes. *Biophys. J.* **46**, 141–153.

Müller, J., Münster, C., and Salditt, T. (2000). Thermal denaturing of bacteriorhodopsin by X-ray scattering from oriented purple membranes. *Biophys. J.* **78**, 3208–3217.

Münster, C., Salditt, T., Vogel, M., Siebrecht, R., and Peisl, J. (1999). Nonspecular neutron scattering from highly aligned phospholipid membranes. *Europhys. Lett.* **46**, 486–492.

Münster, C., Lu, J., Schinzel, S., Bechinger, B., and Salditt, T. (2000). Grazing incidence X-ray diffraction of highly aligned phospholipid membranes containing the antimicrobial peptide magainin 2. *Eur. Biophys. J.* **28**, 683–688.

Nagle, J. F., and Tristram-Nagle, S. (2000). Structure of lipid bilayers. *Biochim. Biophys. Acta* **1469**, 159–195.

Naumann, C., Brumm, T., Rennie, A. R., Penfold, J., and Bayerl, T. M. (1995). Hydration of DPPC monolayers at the air/water interface and its modulation by the nonionic surfactant $C_{12}E_4$: A neutron reflectivity study. *Langmuir* **11**, 3948–3952.

Naumann, C., Dietrich, C., Behrisch, A., Bayerl, T. M., Schleicher, M., Bucknall, D., and Sackmann, E. (1996). Hisactophilin-mediated binding of actin to lipid lamellae: A neutron reflectivity study of protein membrane coupling. *Biophys. J.* **71**, 811–823.

Névot, L., and Croce, P. (1980). Caractérisation des surfaces par réflexion rasante de rayons X. Application à l'étude du polissage de quelques verres silicates. *Rev. Phys. Appl.* **15**, 761–779.

Okada, A., Wakamatsu, K., Miyazawa, T., and Higashijima, T. (1994). Vesicle-bound conformation of mellitin: Transferred nuclear Overhauser enhancement analysis in the presence of perdeuterated phosphatidylcholine vesicles. *Biochemistry* **33**, 9438–9446.

Parratt, L. G. (1954). Surface studies of solids by total reflection of X-rays. *Phys. Rev.* **95**, 359–369.

Pebay-Peyroula, E., Rummel, G., Rosenbusch, J. P., and Landau, E. M. (1997). X-ray structure of bacteriorhodopsin at 2.5 Å resolution from microcrystals grown in lipidic cubic phases. *Science* **277**, 1676–1681.

Penfold, J., and Thomas, R. K. (1990). The application of the specular reflection of neutrons to the study of interfaces. *J. Phys. Condens. Matter* **2**, 1369–1412.

Pershan, P. S. (1990). Structure of surfaces and interfaces as studied using synchrotron radiation. *Faraday Disc. Chem. Soc.* **89**, 231–245.

Pershan, P. S. (2000). Effects of thermal roughness on X-ray studies of liquid surfaces. *Colloids Surf. A Physicochem. Eng. Aspects* **171**, 149–157.

Peters, G. H., Dahmen-Levison, U., de Meijere, K., Brezesinski, G., Toxvaerd, S., Möhwald, H., Svendsen, A., and Kinnunen, P. K. J. (2000). Influence of surface properties of mixed monolayers on lipolytic hydrolysis. *Langmuir* **16**, 2779–2788.

Petrache, H. I., Feller, S. E., and Nagle, J. F. (1997). Determination of component volumes of lipid bilayers from simulations. *Biophys. J.* **70**, 2237–2242.

Petrash, S., Sheller, N. B., Dando, W., and Foster, M. D. (1997). Variation in tenacity of protein adsorption on self-assembled monolayers with monolayer order as observed by X-ray reflectivity. *Langmuir* **13**, 1881–1883.

Plant, A. L. (1993). Self-assembled phospholipid/alkanethiol biomimetic bilayers on gold. *Langmuir* **9**, 2764–2767.

Plech, A., Salditt, T., Münster, C., and Peisl, J. (2000). Investigation of structure and growth of self-assembled polyelectrolyte layers by X-ray and neutron scattering under grazing angles. *J. Colloid Interface Sci.* **223**, 74–82.

Politsch, E., Cevc, G., Wurlitzer, A., and Lösche, M. (2001). Conformation of polymer brushes at aqueous surfaces determined with X-ray and neutron reflectometry. I. Novel data evaluation procedure for polymers at interfaces. *Macromolecules* **34**, 1328–1333.

Prokop, L. A., Strongin, R. M., Smith, A. B., Blasie, J. K., Peticolas, L. J., and Bean, J. C. (1996). Vectorially oriented monolayers of detergent-solubilized Ca^{2+}-ATPase from sarcoplasmic reticulum. *Biophys. J.* **70**, 2131–2143.

Raguse, B., Braach-Maksvytis, V. L. B., Cornell, B. A., King, L. B., Osman, P. D. J., Pace, R. J., and Wieczorek, L. (1998). Tethered lipid bilayer membranes: Formation and ionic reservoir characterization. *Langmuir* **14**, 648–659.

Rieu, J. P., Legrand, J. F., Renault, A., Berge, B., Ocko, B. M., Wu, X. Z., and Deutsch, M. (1995). Melting of 1-alcohol monolayers at the air–water interface. I. X-ray reflectivity investigations. *J. Phys. II France* **5**, 607–615.

Robertson, D. E., Farid, R. S., Moser, C. C., Urbauer, J. L., Mulholland, S. E., Pidikiti, R., Lear, J. D., Wand, A. J., DeGrado, W. D., and Dutton, P. L. (1994). Design and synthesis of multi-haem proteins. *Nature* **368**, 425–432.

Sackmann, E. (1995). Biological membranes. Architecture and function. *In* "Structure and Dynamics of Membranes: From Cells to Vesicles" (R. Lipowsky and E. Sackmann, eds.), Vol. 1A, pp. 1–63. North Holland, Amsterdam.

Sackmann, E. (1996). Supported membranes: Scientific and practical applications. *Science* **271**, 43–48.

Sackmann, E., and Tanaka, M. (2000). Supported membranes on soft polymer cushions: Fabrication, characterization and applications. *Trends Biotechnol.* **18**, 58–64.

Safinya, C. R. (1997). Biomolecular materials: Structure, interactions and higher order self-assembly. *Colloids Surf. A Physicochem. Eng. Aspects* **128**, 183–195.

Salditt, T., Münster, C., Lu, J., Vogel, M., Frenzl, W., and Souvorov, A. (1999). Specular and diffuse scattering of highly aligned phospholipid membranes. *Phys. Rev. E* **60**, 7285–7289.

Salditt, T., Li, C., Spaar, A., and Mennicke, U. (2002). X-ray reflectivity of solid-supported, multilamellar membranes. *Eur. Phys. J. E.* Submitted.

Saville, P. M., Gentle, I. R., White, J. W., Penfold, J., and Webster, J. R. P. (1994). Specular and off-specular neutron reflectivity of a low molecular weight polystyrene surfactant at the air–water interface. *J. Phys. Chem.* **98**, 5935–5942.

Schalke, M. (2000). Konformation und Hydratation von Phospholipiden in Oberflächenmonoschichten: Röntgenreflexion und IR-Spektroskopie. Ph.D. thesis, Leipzig University, Leipzig.

Schalke, M., and Lösche, M. (2000). Structural models of lipid surface monolayers from X-ray and neutron reflectivity measurements. *Adv. Colloid Interface Sci.* **88**, 243–274.

Schalke, M., Krüger, P., Weygand, M., and Lösche, M. (2000). Submolecular organization of DMPA in surface monolayers: Beyond the two-layer model. *Biochim. Biophys. Acta* **1464**, 113–126.

Schmidt, A., Spinke, J., Bayerl, T., Sackmann, E., and Knoll, W. (1992). Streptavidin binding to biotinylated lipid layers on solid supports. A neutron and surface plasmon optical study. *Biophys. J.* **63**, 1185–1192.

Schmitt, J., Grünewald, T., Decher, G., Pershan, P. S., Kjaer, K., and Lösche, M. (1993). Internal structure of layer-by-layer adsorbed polyelectrolyte films: A neutron and X-ray reflectivity study. *Macromolecules* **26**, 7058–7063.

Schuster, B., Pum, D., Braha, O., Bayley, H., and Sleytr, U. B. (1998a). Self-assembled α-hemolysin pores in an S-layer supported lipid bilayer. *Biochim. Biophys. Acta* **1370**, 280–288.

Schuster, B., Pum, D., and Sleytr, U. B. (1998b). Voltage-clamp studies on S-layer supported tetraether lipid membranes. *Biochim. Biophys. Acta* **1369**, 51–60.

Seitz, M., Ter-Ovanesyan, E., Hausch, M., Park, C. K., Zasadzinski, J. A., Zentel, R., and Israelachvili, J. N. (2000). Formation of tethered supported bilayers by vesicle fusion onto lipopolymer monolayers promoted by osmotic stress. *Langmuir* **16**, 6067–6070.

Sinha, S. K., Sirota, E. B., Garoff, S., and Stanley, H. B. (1988). X-ray and neutron scattering from rough surfaces. *Phys. Rev. B* **38**, 2297–2311.

Skov Pedersen, J., and Hamley, I. W. (1994a). Analysis of neutron and X-ray reflectivity data by constrained least-squares methods. *Physica B* **198**, 16–23.

Skov Pedersen, J., and Hamley, I. W. (1994b). Analysis of neutron and X-ray reflectivity data. II. Constrained least-square methods. *J. Appl. Crystallogr.* **27**, 36–49.

Sleytr, U. B., Messner, P., Pum, D., and Sára, M. (1999). Crystalline bacterial cell surface layers (S-layers): From supramolecular cell structure to biomimetics and nanotechnology. *Angew. Chem. Int. Ed. Engl.* **38**, 1034–1054.

Starr, T. E., and Thompson, N. L. (2000). Formation and characterization of planar phospholipid bilayers supported on TiO_2 and $SrTiO_3$ single crystals. *Langmuir* **16**, 10301–10308.

Stayton, P. S., Ohliger, J. M., Jiang, M., Bohn, P. W., and Sligar, S. G. (1992). Genetic engineering of surface attachment site yields oriented protein monolayers. *J. Am. Chem. Soc.* **114**, 9298–9299.

Steck, T. L., Kezdy, F. J., and Lange, Y. (1988). An activation–collision mechanism for cholesterol transfer between membranes. *J. Biol. Chem.* **263**, 13023–13031.

Steitz, R., Leiner, V., Siebrecht, R., and Klitzing, R. v. (2000). Influence of the ionic strength on the structure of polyelectrolyte films at the solid/liquid interface. *Colloids Surf. A Physicochem. Eng. Aspects* **163**, 63–70.

Strzalka, J., Chen, X., Moser, C. C., Dutton, P. L., Ocko, B. M., and Blasie, J. K. (2000). X-ray scattering studies of maquette peptide monolayers. 1. Reflectivity and grazing incidence diffraction at the air/water interface. *Langmuir* **16**, 10404–10418.

Strzalka, J., Chen, X., Moser, C. C., Dutton, P. L., Bean, J. C., and Blasie, J. K. (2001). X-ray scattering studies of maquette peptide monolayers. 2. Interferometry at the vapor/solid interface. *Langmuir* **17**, 1193–1199.

Teer, E., Knobler, C. M., Lautz, C., Wurlitzer, S., Kildae, J., and Fischer, T. M. (1997). Optical measurements of the phase diagram of Langmuir monolayers of fatty acid, ester, and alcohol mixtures by Brewster-angle microscopy. *J. Chem. Phys.* **106**, 1913–1920.

Terwilliger, T. C., and Eisenberg, D. (1982a). The structure of mellitin. I. Structure determination and partial refinement. *J. Biol. Chem.* **257**, 6010–6015.

Terwilliger, T. C., and Eisenberg, D. (1982b). The structure of mellitin. II. Interpretation of the structure. *J. Biol. Chem.* **257**, 6016–6022.

Thoma, M., Schwendler, M., Baltes, H., Helm, C. A., Pfohl, T., Riegler, H., and Möhwald, H. (1996). Ellipsometry and X-ray reflectivity studies on monolayers of phosphatidylethanolamine and phosphatidylcholine in contact with n-dodecane, n-hexadecane, and bicyclohexyl. *Langmuir* **12**, 1722–1728.

Tolan, M. (1999). "X-Ray Scattering from Soft-Matter Thin Films." Springer, Berlin.

Ulman, A. (1991). "Ultrathin Organic Films." Academic Press, San Diego, CA.

Vaknin, D., and Kelley, M. S. (2000). The structure of D-erythro-C_{18} ceramide at the air–water interface. *Biophys. J.* **79**, 2616–2623.

Vaknin, D., Als-Nielsen, J., Piepenstock, M., and Lösche, M. (1991a). Recognition processes at a functionalized lipid surface observed with molecular resolution. *Biophys. J.* **60**, 1545–1552.

Vaknin, D., Kjaer, K., Als-Nielsen, J., and Lösche, M. (1991b). Structural properties of phosphatidylcholine in a monolayer at the air/water interface. Neutron reflection study and reexamination of X-ray reflection experiments. *Biophys. J.* **59**, 1325–1332.

Vénien-Bryan, C., Lenne, P.-F., Zakri, C., Renault, A., Brisson, A., Legrand, J.-F., and Berge, B. (1998). Characterization of the growth of 2D protein crystals on a lipid monolayer by ellipsometry and rigidity measurements coupled to electron microscopy. *Biophys. J.* **74**, 2649–2657.

Verclas, S. A. W., Howes, P. B., Kjaer, K., Wurlitzer, A., Weygand, M., Büldt, G., Dencher, N. A., and Lösche, M. (1999). Grazing incidence X-ray diffraction from a monolayer of purple membrane at the air/water interface. *J. Mol. Biol.* **287**, 837–843.

Vineyard, G. H. (1982). Grazing-incidence diffraction and the distorted-wave approximation for the study of surfaces. *Phys. Rev. B* **26**, 4146–4159.

Vogel, M., Münster, C., Frenzl, W., and Salditt, T. (2000). Thermal unbinding of highly oriented phospholipid membranes. *Phys. Rev. Lett.* **84**, 390–393.

Weidemann, G., Brezesinski, G., Vollhardt, D., and Möhwald, H. (1998). Disorder in Langmuir mono-layers. 1. Disordered packing of alkyl chains. *Langmuir* **14**, 6485–6492.

Weidemann, G., Brezesinski, G., Vollhardt, D., DeWolf, C., and Möhwald, H. (1999). Disorder in Langmuir monolayers. 2. Relation between disordered alkyl chain packing and the loss of long-range tilt orientational order. *Langmuir* **15**, 2901–2910.

Weissbuch, I., Popovitz-Biro, R., Lahav, M., Leiserowitz, L., Kjaer, K., and Als-Nielsen, J. (1997). Molecular self assembly into crystals at air–liquid interfaces. *In* "Advances in Chemical Physics" (I. Prigogine and S. A. Rice, eds.), Vol. 102, pp. 39–120. Wiley, New York.

Wetzer, B., Pum, D., and Sleytr, U. B. (1997). S-layer stabilized solid supported lipid bilayers. *J. Struct. Biol.* **119**, 123–128.

Wetzer, B., Pfandler, A., Györvary, E., Pum, D., Lösche, M., and Sleytr, U. B. (1998). S-layer recon-stitution at phospholipid monolayers. *Langmuir* **14**, 6899–6906.

Weygand, M. (2000). Struktur und mikroskopische Charakterisierung der Wechselwirkung von S-Schicht-Proteinen mit Phospholipiden: Röntgen- und Neutronenstreuuntersuchungen. Ph.D. thesis, Leipzig University, Leipzig.

Weygand, M., Wetzer, B., Pum, D., Sleytr, U. B., Cuvillier, N., Kjaer, K., Howes, P. B., and Lösche, M. (1999). Bacterial S-layer protein coupling to lipids: X-ray reflectivity and grazing incidence diffraction studies. *Biophys. J.* **76**, 458–468.

Weygand, M., Schalke, M., Howes, P. B., Kjaer, K., Friedmann, J., Wetzer, B., Pum, D., Sleytr, U. B., and Lösche, M. (2000). Coupling of protein sheet crystals (S-layers) to phospholipid monolayers. *J. Mater. Chem.* **10**, 141–148.

Weygand, M., Kjaer, K., Howes, P. B., Wetzer, B., Pum, D., Sleytr, U. B., and Lösche, M. (2002). Structural reorganization of phospholipid headgroups upon recrystallization of an S-layer lattice. *J. Phys. Chem. B.* Submitted.

White, S. H., and Wimley, W. C. (1998). Hydrophobicinteractions of peptides with membrane inter-faces. *Biochim. Biophys. Acta* **1376**, 339–352.

White, S. H., and Wimley, W. C. (1999). Membrane protein folding and stability: Physical principles. *Annu. Rev. Biomol. Struct.* **28**, 319–365.

Wiener, M. C., and White, S. H. (1991a). Fluid bilayer structure determination by the combined use of X-ray and neutron diffraction. I. Fluid bilayer models and the limits of resolution. *Biophys. J.* **59**, 162–173.

Wiener, M. C., and White, S. H. (1991b). Fluid bilayer structure determination by the combined use of X-ray and neutron diffraction. II. "Composition-space" refinement method. *Biophys. J.* **59**, 174–185.

Wiener, M. C., and White, S. H. (1991c). Transbilayer distribution of bromine in fluid bilayers contain-ing a specifically brominated analog of dioleoylphosphatidylcholine. *Biochemistry* **30**, 6997–7008.

Wiener, M. C., and White, S. H. (1992a). Structure of a fluid dioleoylphosphatidylcholine bilayer determined by joint refinement of X-ray and neutron diffraction data. II. Distribution and packing of terminal methyl groups. *Biophys. J.* **61**, 428–433.

Wiener, M. C., and White, S. H. (1992b). Structure of a fluid dioleoylphosphatidylcholine bilayer determined by joint refinement of X-ray and neutron diffraction data. III. Complete structure. *Biophys. J.* **61**, 434–447.

Wiener, M. C., King, G. I., and White, S. H. (1991). Structure of a fluid dioleoylphosphatidylcholine bilayer determined by joint refinement of X-ray and neutron diffraction data. I. Scaling of neutron data and the distribution of double bonds and water. *Biophys. J.* **60**, 568–576.

Wimley, W. C., and White, S. H. (1993). Membrane partitioning: Distinguishing bilayer effects from the hydrophobic effect. *Biochemistry* **32**, 6307–6312.

Wimley, W. C., and White, S. H. (1996). Experimentally determined hydrophobicity scale for proteins at membrane interfaces. *Nat. Struct. Biol.* **3**, 842–848.

Wong, J. Y., Majewski, J., Seitz, M., Park, C. K., Israelachvili, J. N., and Smith, G. S. (1999a). Polymer-cushioned bilayers. I. A structural study of various preparation methods using neutron reflectometry. *Biophys. J.* **77,** 1445–1457.
Wong, J. Y., Park, C. K., Seitz, M., and Israelachvili, J. N. (1999b). Polymer-cushioned bilayers. II. An investigation of interaction forces using the surface forces apparatus. *Biophys. J.* **77,** 1458–1468.
Wurlitzer, A., Politsch, E., Hübner, S., Krüger, P., Weygand, M., Kjaer, K., Hommes, P., Nuyken, O., Cevc, G., and Lösche, M. (2001). Conformation of polymer brushes at aqueous surfaces determined with X-ray and neutron reflectometry. II. High density phase transition of lipopoly-oxazolines. *Macromolecules* **34,** 1334–1342.
Yang, L., Weiss, T. M., Harroun, T. A., Heller, W. T., and Huang, H. W. (1999). Supramolecular structures of peptide assemblies in membranes by neutron off-plane scattering: Method of analysis. *Biophys. J.* **77,** 2648–2656.
Yang, L., Weiss, T. M., Lehrer, R. I., and Huang, H. W. (2000). Crystallization of antimicrobial pores in membranes: Magainin and protegrin. *Biophys. J.* **79,** 2002–2009.
Yim, H., Matheson, A., Ivkov, R., Satija, S., Majewski, J., and Smith, G. S. (2000). Adsorption of poly(styrenesulfonate) to the air surface of water by neutron reflectivity. *Macromolecules* **33,** 6126–6133.
Zhang, R., Suter, R. M., and Nagle, J. F. (1994). Theory of the structure factor of lipid bilayers. *Phys. Rev. E* **50,** 5047–5060.
Zheng, S., Strzalka, J., Ma, C., Opella, S. J., Ocko, B. M., and Blasie, J. K. (2001). Structural studies of the HIV-1 accessory protein Vpu in Langmuir monolayers: Synchrotron X-ray reflectivity. *Biophys. J.* **80,** 1837–1850.
Zhou, X.-L., and Chen, S.-H. (1993). Model-independent method for reconstruction of scattering-length-density profiles using neutron and X-ray reflectivity data. *Phys. Rev. E* **47,** 3174–3190.

CHAPTER 6

Lipid–Peptide Interaction Investigated by NMR

Klaus Gawrisch* and Bernd W. Koenig†

*Laboratory of Membrane Biochemistry and Biophysics, NIAAA, National Institutes of Health, Rockville, Maryland 20852; †IBI-2: Institute of Structural Biology, Research Center Jülich, D-52425 Jülich, Germany

I. Introduction
II. Peptide Structure in the Membrane-Bound State
    A. Peptide Structure in Membrane-Mimetic Environments Studied by High-Resolution NMR
    B. Membrane-Bound Structure of Weakly Interacting Peptides
    C. Solid-State NMR Approaches to Studying Peptide Structure in the Membrane-Bound State
III. Structure and Dynamics of the Lipid Matrix
    A. ^{31}P Anisotropy of Chemical Shift
    B. ^{2}H NMR
    C. Magic Angle Spinning NMR
    D. Influence of Peptide Binding on Lipid Structure
IV. Future Directions
    References

Nuclear magnetic resonance (NMR) enables the investigation of peptides as well as the lipid matrix to which the peptides bind. Peptide structure is studied by both high-resolution and solid-state NMR methods. Application of high-resolution NMR requires rapid isotropic peptide motions, achieved by conducting experiments in membrane-mimetic environments, that is, organic solvents or detergent solutions. Solid-state NMR investigations are directly conducted on peptides in the membrane-bound state, but experiments often require elimination of peptide motions by freezing or dehydrating samples. Novel NMR techniques utilizing rapid magic angle spinning or oriented peptide/lipid samples have dramatically

Current Topics in Membranes, Volume 52

improved resolution of solid-state NMR experiments in recent years. Peptide binding to lipids changes the membrane hydrophobic thickness and area per molecule, both reflected in lipid order parameter changes. Nuclear Overhauser enhancement spectroscopy (NOESY), in combination with magic angle spinning, is a novel approach to studying structure and dynamics of the lipid matrix. The experiments provide deeper insight into the tremendous conformational and spatial disorder of lipids in biomembranes.

I. INTRODUCTION

High-resolution NMR is a reliable tool for structural investigations of peptides and small proteins in solution. The molecular weight limit for complete resolution of structure is near 30,000 daltons. Under favorable circumstances, this range can be extended to values near 100,000 daltons using transverse relaxation-optimized spectroscopy (TROSY) sequences (Pervushin *et al.*, 1997) in combination with high magnetic field strength and extensive isotopic labeling of the protein. The complexes of peptides with micelles fall within the permissible weight range and can be studied by proven high-resolution NMR approaches. Furthermore, peptide structures similar to those of the membrane-bound state can be investigated on dissolved peptides under solvent conditions that simulate a membrane environment.

The aggregates of peptides with lipid bilayers have molecular weights that far exceed the limits of high-resolution NMR. Even the complexes between peptides and small sonicated liposomes, with diameters of less than 1000 Å, tumble too slowly for high-resolution NMR applications. Solid-state NMR is used to study the structure of peptides that are bound to membranes. The method takes advantage of the spatially anisotropic interactions that shift, split, and broaden NMR resonance lines. Conformation and dynamics of lipids and changes in lipid packing as a result of peptide binding can be studied as well.

The term "structure" in the context of peptide–lipid interaction must be used with great caution. Models that show lipids and peptides in static arrangements have shaped our perception. In reality, the lipids in the matrix of biomembranes are liquid-crystalline and exchange rapidly among an infinite number of conformations. Peptides may have rigid backbones, but can also retain significant degrees of flexibility. Lifetimes of noncovalent interactions between peptides and lipids range from pico- to milliseconds. The lipid matrix easily adjusts to the needs of peptide–lipid interaction. Average lipid conformation and lipid lateral organization may change as a result of interactions with peptides. Although NMR competes with other methods for the structural investigation of peptides, it is the method of choice for studying the dynamics of lipid–peptide complexes. Motional correlation times that cover 12 orders of magnitude, from picoseconds to seconds, are accessible by NMR.

The goal of this chapter is to provide general knowledge about opportunities and limitations of NMR in the study of peptide–lipid interactions. Technical details that would appeal mostly to NMR spectroscopists have been avoided, but can be found in the cited literature. Because of the authors' involvement in studies on the interaction of amphipathic peptides with lipid/water interfaces, most of the experimental examples are taken from this field of research.

II. PEPTIDE STRUCTURE IN THE MEMBRANE-BOUND STATE

A. Peptide Structure in Membrane-Mimetic Environments Studied by High-Resolution NMR

The majority of short peptides in aqueous solution are very flexible. Peptide binding to a lipid membrane often results in the formation of secondary structure or even of a unique peptide conformation. The structural transition reflects the change in the microenvironment of the peptide upon membrane binding, that is, the transfer down a steep polarity gradient from the very polar aqueous phase to the complex environment of a hydrophobic/hydrophilic interface.

Organic solvents of low polarity or detergent micelles in water may promote the formation of peptide structures similar to what is found in peptide–membrane complexes. Studies that mimic membrane conditions are frequently referred to as studies in membrane-mimetic environments. For example, methanol or trifluoroethanol/water mixtures facilitate formation of α helices, provided the peptide has helical propensity. The environment of detergent micelles is a better choice for structural studies on membrane-spanning segments. In addition, micelles are also an appropriate medium for amphipathic peptides that require a polar and/or a charged interface region for proper folding. The NH—NH region of a two-dimensional (2D) ^{1}H NOESY spectrum of a 21-residue amphipathic peptide from the envelope glycoprotein gp41 in sodium dodecylsulfate (SDS) solution is presented in Fig. 1.

The choice of experimental conditions is of fundamental importance. The peptide structure may depend on solvent properties (Gesell *et al.,* 1997) as well as pH, ionic strength, and temperature. When detergent micelles are used, detergent concentration is particularly important to ensure both a unique peptide structure as well as a proper aggregate size to enable detection of well-resolved NMR spectra (McDonnell and Opella, 1993). Sodium dodecylsulfate and dodecylphosphocholine (DPC) micelles are most frequently used to mimic negatively charged and zwitterionic membranes, respectively. Experiments may benefit from the use

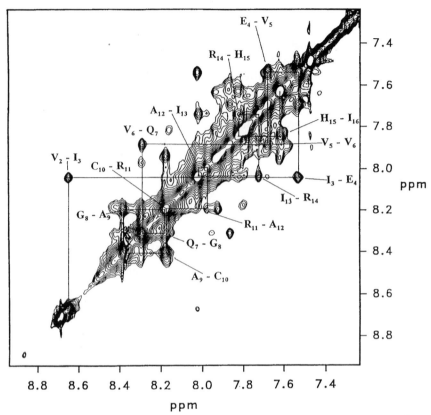

FIGURE 1   High-resolution NMR study of a cytolytic peptide fragment, 828–848 (P828), from the carboxy terminus of the envelope glycoprotein gp41 of HIV-1 in SDS micellar solution. Shown is the amide region of the ^{1}H 2D NOESY NMR spectrum of 5.3 mM P828 in 277 mM 1-stearoyl-d$_{35}$-2-oleolyl-*sn*-glycero-3-phosphocholine solution at $T = 60°$C. Sequential NH-to-NH crosspeaks are labeled. Figure reproduced from Koenig *et al.* (1995).

of perdeuterated detergent molecules to prevent signal superposition with the much weaker peptide signals. Equivalence of peptide structure in the chosen membrane-mimetic environment with the membrane-bound structure must be demonstrated. This is conveniently done by methods, like circular dichroism (CD), which can be applied to solutions of peptide interacting with small unilamellar liposomes as well as to peptides in organic solvent or in micelle solution. Matching of the CD spectra of a peptide in a membrane-mimetic environment and in a real membrane strongly suggests similar peptide conformations (Gawrisch *et al.*, 1993; Koenig *et al.*, 1995).

The rationale behind the use of membrane-mimetic environments is their compatibility with high-resolution NMR experiments for peptide structural investigations. Peptides in organic solvents or micelles perform rapid, isotropic rotational diffusion. The rotational correlation time of a micelle-bound peptide depends on the size of aggregates and the viscosity of the medium. Micelle size is critical: A small peptide with a weight of 2000 daltons bound to a micelle may behave spectroscopically like a protein in water with a weight of 20,000–30,000 daltons. At such molecular weights, NMR spectra may suffer from line broadening and short spin–spin relaxation times, and NOESY cross-relaxation may be under the influence of spin diffusion.

Structural determination of peptides adopting a unique conformation in membrane-mimetic environment follows the same strategy as NMR structure determination on soluble proteins (Wüthrich, 1986). Two-dimensional ^1H–^1H experiments [total correlation spectroscopy (TOCSY) or homonuclear Hartmann–Hahn (HOHAHA) spectroscopy (Bax, 1989), correlation spectroscopy (COSY; Wüthrich, 1986), NOESY (Kumar et al., 1980), and rotating-frame nuclear Overhauser enhancement spectroscopy (ROESY; Bothner-By et al., 1984)] on unlabeled peptide will usually provide sufficient resolution for complete resonance assignment of all peptide protons using the sequential resonance assignment technique (Wüthrich, 1986). In the case of severe proton signal overlap, the larger dispersion in the ^{15}N and ^{13}C dimensions is exploited to assign the protons that are directly bound to ^{15}N and ^{13}C nuclei, for example, by two-dimensional experiments with inverse detection of the insensitive nuclei via the resonance of protons, like heteronuclear single quantum coherence (HSQC) and heteronuclear multiple quantum coherence (HMQC) (Bax et al., 1989), or by three- and four-dimensional, heteronuclear-edited, proton-detected experiments (Clore and Gronenborn, 1991). At peptide concentrations in the millimolar range, the ^{13}C signals can be detected at natural abundance (Koenig et al., 1999a). However, inverse experiments with ^{15}N spins require isotope labeling.

The chemical shift values of ^1H (Wishart et al., 1991) and ^{13}C spins (Spera and Bax, 1991; Wishart and Sykes, 1994) are site-specific indicators of the presence of secondary structure (see Fig. 2). Similarly, the ^1H–^1H nuclear Overhauser effect (NOE) connectivity pattern, which reflects spatial proximity of peptide protons, provides information on the local secondary structure (Wüthrich, 1986). The combination of strong NH/NH(i, $i + 1$) NOE crosspeaks with weak αH/NH(i, $i + 1$) crosspeaks indicates α-helical structures (see Figs. 1 and 3), whereas the combination of strong αH/NH and weak NH/NH NOE crosspeaks is expected for β-sheet structures. In addition to approximate interproton distance constraints from NOESY or ROESY spectra, torsion angles (see Fig. 4) in rigid peptides can be restrained. The $^3J_{NH-H\alpha}$ and $^3J_{H\alpha-H\beta}$ scalar coupling constants are related to protein φ and χ_1 torsion angles via empirical Karplus relations

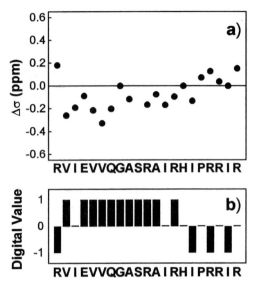

FIGURE 2 The peptide P828 is not structured in water, but converts to a partially helical conformation in SDS micelles. The structural transition is reflected in characteristic changes in backbone chemical shifts. (a) Conformation-induced H_α secondary shift. Shown is the difference between the chemical shift values observed in water and in SDS solution. (b) Digital values representing the $^{13}C_\alpha$ secondary shifts of P828S in SDS micelles. Depending on the deviation of the chemical shift from average values, a digital value of 1, 0, or −1 is assigned to every C_α carbon. The digital value +1 for amino acids from valine V_2 to arginine R_{14} confirms the existence of a helical conformation in this region of the molecule. Figure reproduced from Koenig *et al.* (1999a).

(Pardi *et al.*, 1984). They are obtained from $^1H-^{15}N$ HMQC (Forman-Kay *et al.*, 1990; Kay and Bax, 1990) and homonuclear primitive exclusive COSY (PE.COSY; Müller, 1987) correlation spectra, respectively. Torsion angles ψ have been determined from cross-correlated relaxation rates (Reif *et al.*, 1997; Sprangers *et al.*, 2000).

The NMR-derived geometric constraints are inserted as interaction potentials into molecular dynamics calculations (Nilges, 1996), which provide sets of low-energy peptide structures. Alternatively, peptide conformation can be evaluated by distance geometry calculations based on NMR distance constraints (Okada *et al.*, 1994).

In a membrane-mimetic environment, and also when bound to membranes, small peptides often retain a significant degree of flexibility rather than form a rigid structure. Fluctuations in the micelle–water interface are significantly larger than structural fluctuations in the lipid/water interface. Therefore it is likely that peptides in micelles show increased conformational flexibility. When peptides are flexible, spectroscopic parameters reflect the averaging over the entire set of

FIGURE 3 Summary of ^{1}H–^{1}H NOE connectivities of peptide P828S, which enable identification of secondary structure. The index i, is the number of the amino acid in the peptide sequence. For the crosspeaks between neighboring amino acids, NH(i)/NH(i + 1), and for H(i)/NH(i + 1), the height of the boxes is proportional to the measured NOE intensities. Question marks indicate crosspeaks that cannot be assessed due to signal overlap. Medium-range NOE interactions over three and four amino acids that are characteristic of helices are summarized in the four lower rows. Solid bars indicate unique crosspeaks, and dotted lines are used for connectivities whose presence or absence cannot be assessed, due to spectral overlap. Observation of medium-range crosspeaks of the peptide region near the C-terminus was not possible due to the proximity of the corresponding H resonances to water. According to induced changes in chemical shift, this region is unstructured. Figure reproduced from Koenig *et al.* (1999a).

peptide structures. Quantitative analysis of spin–lattice and spin–spin relaxation data provides a more detailed description of the dynamics of conformational transition (Ishima and Torchia, 2000). The amplitude of the heteronuclear ^{15}N–^{1}H nuclear Overhauser enhancement is very sensitive to motions with frequencies near 10^9 Hz, which allows a qualitative distinction between mobile and rigid sites in a peptide (Bogusky *et al.*, 1987, 1988). The relative intensities of sequential NH/NH and αH/NH crosspeaks in homonuclear NOESY spectra reveal conformational averaging (Dyson and Wright, 1991; Koenig *et al.*, 1999a). The temperature dependence of the chemical shift of peptide amide protons is sensitive to the formation of hydrogen bonds. A linear dependence with small temperature coefficients between 0 and -3 ppb/K indicates regular secondary structure elements stabilized by intramolecular hydrogen bonds, whereas values between -6 and -10 ppb/K are typical of hydrogen bonding between peptide and solvent (Deslauriers and Smith, 1980). Intermediate values may reflect conformational flexibility (Koenig *et al.*, 1999a).

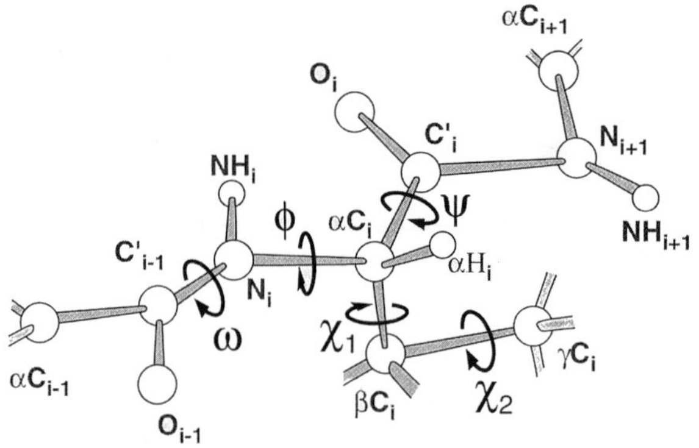

FIGURE 4 Standard nomenclature for atoms and torsion angles in a peptide chain. A torsion angle defines the amount of twist about a bond axis and is defined, for a bonded series of four atoms (A—B—C—D), as the angle of rotation about bond B—C required to make the projection of bond axis B—A coincide with the projection of bond axis C—D, when viewed along the B—C direction. The angle is positive for clockwise rotation. Backbone torsion angles ϕ, that is, $(C'_{i-1}—N_i—\alpha C_i—C'_i)$, and ψ, that is, $(N_i—\alpha C_i—C'_i—N_{i+1})$, and side-chain torsion angle χ_1, that is, $(N_i—\alpha C_i—\beta C_i—\gamma C_i)$, are particularly useful.

B. Membrane-Bound Structure of Weakly Interacting Peptides

Resonance signals of peptides that interact weakly with the surface of liposomes, but spend most of their time in solution, remain highly resolved. However, the membrane–peptide interactions cause significant NMR relaxation enhancement. The NOE cross-relaxation rates depend on the motional correlation time τ_c of the internuclear vector pairs of interacting protons. The nuclear Overhauser effect is positive for short τ_c and negative for long τ_c, because of differences in the correlation time dependence of relaxation pathways (Neuhaus and Williamson, 1989). At the field strength of modern high-field NMR spectrometers, the crossover in the sign of NOE, where crosspeaks are very weak or undetectable, occurs at τ_c between 10^{-9} and 10^{-10} s. These are typical correlation times of short peptides with about 10 amino acids in aqueous solution.

Cross-relaxation of peptides that briefly bind to membranes is dominated by the much longer correlation times of slowly tumbling peptide/lipid complexes in the membrane-bound state. The proton distance information of the membrane-bound state is "transferred" to the solution-state NMR signals. Therefore, such experiments are referred to as transferred NOE (TrNOE) measurements (Clore and Gronenborn, 1982, 1983; Ni, 1994). Observation of TrNOEs requires that the peptide off-rate is fast relative to both the longitudinal relaxation rate of the bound peptide and the inverse of the NOE mixing time. It is often necessary to lower the natural affinity of the peptide for membrane interaction to meet the stringent off-rate requirements. This may be achieved by modifying the lipid composition (Z. Wang *et al.,* 1993), varying the pH or ionic strength of the solution (Anglister and Zilber, 1990; Koenig *et al.,* 1999b), or inducing conservative mutations in the peptide sequence (Anglister and Zilber, 1990; Koenig *et al.,* 2000). Higher temperature also enhances the off-rate (Koenig *et al.,* 2000). It may be advantageous to record NOESY spectra of the peptide both in the presence and the absence of membranes and to subtract the NOE of peptide acquired without the presence of membranes to remove NOE contributions from the free peptide.

The TrNOE method has been successfully used to study the membrane-bound conformation of peptides such as mastoparan-X (Wakamatsu *et al.,* 1992), melittin (Okada *et al.,* 1994), neuropeptides (Bersch *et al.,* 1993), and signal peptides (Z. Wang *et al.,* 1993), as well as the structure of peptides bound to antibody fragments (Anglister and Zilber, 1990). It was observed that very low tumbling rates of peptides in the membrane-bound state make TrNOE experiments prone to spin diffusion, potentially resulting in crosspeaks between protons farther apart in space than the longest measurable NOE distances of 5 Å. Great care must be taken to properly design and interpret TrNOE experiments in order to avoid or at least to recognize such artifacts (Arepalli *et al.,* 1995; Ni, 1994).

C. Solid-State NMR Approaches to Studying Peptide Structure in the Membrane-Bound State

Permanent attachment of peptides to membranes or incorporation into membranes results in an anisotropy of peptide orientation that is not averaged out by fast motions. Solid-state NMR is the method of choice for investigating long-lived peptide–lipid complexes. It allows characterizing the structure of peptides that are immobilized on the relevant NMR time scale. There is no need to use membrane-mimetic environments. Many solid-state NMR methods work best with frozen or dehydrated samples, where most peptide motions are suppressed. Unfortunately, freezing and dehydration can alter the structure and dynamics of peptide–membrane complexes.

Anisotropic interactions between magnetic nuclei and their environment dominate the appearance of solid-state NMR spectra. They result in orientation-dependent shifts and splittings of peptide resonances. The magnitude of these effects is very sensitive to the orientation of bond vectors relative to the external magnetic field and to the distance between interacting nuclei. NMR parameters of interest include the ^{15}N-chemical shift anisotropy (CSA) of peptide amides, the ^{13}C-CSA of peptide carbonyl and C_α carbons, dipolar couplings between ^{1}H, ^{15}N, and ^{13}C spins of the peptide backbone, and quadrupolar splittings of ^{2}H nuclei attached to amide nitrogens, C_α carbons, or side-chain carbons.

There is a tradeoff between the ability to study peptides in the membrane-bound state and spectral resolution. Anisotropic interactions in solid-state NMR severely broaden resonance lines, resulting in low sensitivity and signal overlap. In recent years these limitations have been partially overcome by new approaches, for example, by using uniaxially oriented membranes or by applying magic angle spinning (MAS), which eliminates excessive signal broadening from anisotropic interactions. Although structural studies on large membrane proteins are still challenging, investigation of membrane-bound peptides with these techniques is very rewarding.

1. Angular Dependence of Interactions

a. Chemical Shift. The chemical shift interaction of a nuclear spin is a tensor quantity. The experimentally observed chemical shift value is a function of both the magnitude of principal tensor components and their orientation relative to the direction of the external magnetic field B_0. Orientation of tensor axes relative to the molecular frame of peptides is determined experimentally or by quantum chemical calculations. Amide ^{15}N and carbonyl ^{13}C chemical shifts contain valuable information on the orientation of the peptide plane relative to the B_0 field. In general, the amide ^{15}N chemical shift tensor ($\sigma_{33} \geq \sigma_{22} \geq \sigma_{11}$) in peptide bonds is almost axially symmetric ($\sigma_{22} \approx \sigma_{11}$), with an asymmetry $\Delta\sigma = \sigma_{33} - 1/2(\sigma_{22} + \sigma_{11}) \approx 150$ ppm (Harbison *et al.*, 1984; Oas *et al.*, 1987a; Wu *et al.*, 1995). The principal axis, σ_{33}, is oriented close to the peptide plane and forms an angle of about 15–20° with the N–H bond axis (Brender *et al.*, 2001). The dispersion of the isotropic amide ^{15}N chemical shift in proteins is small compared to $\Delta\sigma$ and amounts to ~20 ppm (Brender *et al.*, 2001). Therefore, the orientation of an immobilized N–H bond vector relative to the external magnetic field can be estimated from a single amide ^{15}N resonance frequency measurement. For example, amide ^{15}N frequencies of peptides immobilized in uniaxially aligned membranes have been used to distinguish transmembrane and membrane-parallel orientations of helical segments (Bechinger *et al.*, 1991, 1996). Membrane alignment with low mosaic spread is critical in such experiments and may be accomplished by spreading the membranes between stacked glass slides (McDonnell *et al.*, 1993). Samples are oriented with the membrane normal parallel to the B_0 field of the NMR magnet, using flat-coil probes (Bechinger and Opella, 1991). The N–H bond in α helices is, in first approximation, oriented parallel to the helix axis (Shon *et al.*, 1991). Amide ^{15}N resonance frequencies close to 220 and 80 ppm (relative to ^{15}NH$_3$) are expected for the transmembrane and surface-parallel orientations of α helices, respectively. Alternatively, alignment can be achieved by using magnetically oriented lipid bilayer fragments in solution (bicelles), allowing either parallel or perpendicular orientation of the membrane normal relative to B_0 (Howard and Opella, 1996; Prosser *et al.*, 1998; Sanders and Schwonek, 1992).

Bicelle samples may allow for a higher filling factor of NMR coils, resulting in higher sensitivity.

The ^{13}C-CSA of the carbonyl carbon in peptide bonds can also be used to estimate the orientation of the peptide plane relative to the B_0 field. Data show only moderate variation of the carbonyl ^{13}C chemical shift interaction tensors in peptide bonds (Oas *et al.*, 1987b; Separovic *et al.*, 1990; Stark *et al.*, 1983). The σ_{11} and σ_{22} components are located in the peptide plane with σ_{22} almost parallel to the C=O bond. The principal value and the orientation of the σ_{22} component are affected by hydrogen-bonding interactions (Asakawa *et al.*, 1992). Analysis of the carbonyl ^{13}C chemical shift has been used to study the orientation of gramicidin and melittin in uniaxially oriented membranes (R. Smith *et al.*, 1989, 1994).

b. [2]*H NMR Quadrupolar Interactions.* Orientation of [2]H—N and [2]H—C bonds with respect to the external magnetic field is derived from the strength of the quadrupolar interaction, which splits ^{2}H NMR signals into a doublet. Typical static quadrupolar coupling constants are 150 kHz for [2]H—N bonds (LoGrasso *et al.*, 1988) and 168 kHz for [2]H—C bonds (Burnett and Muller, 1971). In first approximation, the electric field gradients in [2]H—N or [2]H—C bonds, which act on the quadrupole moment of ^{2}H nuclei, are axially symmetric around the bond axis. The angular dependence of quadrupolar interactions on the angle θ between the direction of B_0 and the chemical bond axis is given by a second-order Legendre polynomial function, $P_2 = [3\cos^2(\theta) - 1]/2$. This function changes sign at the magic angle of 54.7°. Measurement of the sign of quadrupolar splittings is difficult and often omitted. This results in an additional degeneracy of bond orientation. Bond vector motions reduce quadrupolar splittings; the measured values reflect the dynamic average of bond vector orientation.

Labile amide protons of peptides are easily deuterated by ^{2}H exchange labeling (Datema *et al.*, 1986). Signal superposition with other deuterated residues, such as hydroxyl groups, amino groups, and [2]H$_2$O, as well as rapid back-exchange in H$_2$O may complicate the NMR study. Peptide amide [2]H—N bond orientation relative to the membrane normal is reflected in ^{2}H NMR spectra of [2]H-exchange-labeled peptides immobilized in uniaxially aligned membranes. The technique was used to discriminate among several structural models of gramicidin in membranes (Prosser *et al.*, 1994). Specific deuterium labeling of nonlabile peptide hydrogens in combination with ^{2}H NMR on gramicidin immobilized in oriented membranes was used to study the average orientation of [2]H—C$_\alpha$ bonds (Hing *et al.*, 1990), amino acid side chains (Killian *et al.*, 1992; Lee *et al.*, 1995), and tryptophane indole rings (Koeppe *et al.*, 1994). Deuterium spectra of [2]H-labeled peptides in non-oriented membranes report on the dynamics and on the local membrane environment of the labeled site (Dempsey *et al.*, 1987; Koenig *et al.*, 1999a).

Recently, an approach that greatly enhances sensitivity and selectivity of detection of quadrupolar splittings was developed. The ^{2}H NMR spectrum of a nonspinning solid was recorded indirectly through its modulation of the much more intense signal of a nearby ^{1}H nucleus. A 15-fold enhancement of sensitivity was reported (S. Liu and K. Schmidt-Rohr, personal communication, 2001). Furthermore, this technique allows suppression of signals of highly mobile nuclei (e.g., the intense ^{2}H$_2$O signal).

   c. Dipolar Interactions.   The strong distance dependence of the dipolar interaction can be used to determine distances within peptides. Nuclei with nonzero spin possess a magnetic moment, causing a dipole field, which modifies the effective field at the site of nearby nuclei. The strength of the dipole field H_{ij} depends on the inverse third power of the internuclear distance r_{ij}^{-3} and on the angle θ between the internuclear vector and external B_0 field: $H_{ij} = \gamma_I[3\cos^2(\theta) - 1]/r_{ij}^3$. The upper limit of accessible distances is 13 Å or less, depending on the gyromagnetic ratio γ_I of the interacting nuclei. Although measurement of dipolar interactions between nuclei in oriented samples is feasible, experiments are almost exclusively done under MAS conditions, which greatly improve the resolution and the intensity of resonance lines. Samples are packed into rotors that spin at rates in the kilohertz range with the rotor axis oriented under the magic angle, $\theta = 54.7°$, relative to the external magnetic field. Fast rotation reduces the resonance linewidth by averaging anisotropic interactions. The strength of dipolar interactions can be measured with the following two techniques.

2. Rotational Resonance (RR) and Rotational-Echo Double Resonance (REDOR)

   In the RR experiment, homonuclear couplings (e.g., the coupling between pairs of ^{13}C nuclei) are measured by matching the rotational frequency with the chemical shift difference of the interacting nuclei (Creuzet *et al.*, 1991; Levitt *et al.*, 1990; Spencer *et al.*, 1994). This produces a transfer of magnetization between the interacting nuclei that can be measured with great precision, equivalent to distance measurement with subangstrom resolution. Measurement of ^{13}C–^{13}C distances is limited to 6 Å or less. Recently, this technique was improved to gauge the longer intermolecular distances reliably (Balazs and Thompson, 1999).

   The REDOR experiments measures heteronuclear dipolar couplings, for example, the couplings between ^{15}N and ^{13}C nuclei in the peptide backbone. The method relies on the dephasing of magnetization of the observed spin through dipolar coupling to a second spin (Gullion and Schaefer, 1989). Accurate, high-sensitivity measurement of distances in peptides has been achieved (Hing and Schaefer, 1993; Marshall *et al.*, 1990). The method was used successfully to study the secondary structure of magainins (Hirsh *et al.*, 1996). Measurement of ^{15}N–^{13}C distances is limited to 5 Å or less. A complication of the original experiment is its sensitivity to the natural background signals of nuclei that are not participating in the dipolar interaction. Modifications allow the selection of dipole-coupled spins against a background of uncoupled spins (Holl et al., 1990).

3. Correlation of Anisotropy of Chemical Shift and Dipolar Interactions

The purpose of the polarization inversion with spin exchange at the magic angle (PISEMA) experiment is to improve spectral resolution by correlating two orientation-dependent interactions, the ^{15}N-chemical shift frequency and the strength of the ^{1}H–^{15}N dipolar coupling. PISEMA requires preparation of well-oriented samples, which are investigated without magic angle spinning. The experiment is capable of resolving a large number of backbone ^{15}N resonances, in particular for α helices that are oriented parallel to the magnetic field (Wu et al., 1994). To improve resolution, the experiment was extended to a third dimension by coupling ^{15}N chemical shift, ^{1}H–^{15}N dipolar interaction, and ^{1}H or ^{13}C chemical shift (Gu and Opella, 1999; Ramamoorthy et al., 1995). Transmembrane helices that are oriented parallel to the magnetic field form regular patterns in PISEMA spectra called polarity index slant angle (PISA) wheels (Wang et al., 2000). Location of resonances in the two-dimensional contour plots allows precise determination of helical tilt in the membrane (Marassi and Opella, 2000). After assignment of resonance signals, even the angle of rotational orientation of the helix can be determined (Wang et al., 2000). The method has great potential for the study of helix orientation in bundles of helices.

4. Measurement of ϕ and ψ Torsion Angles of the Peptide Backbone

Peptide secondary structure is linked to combinations of the sterically permitted peptide backbone torsion angles ϕ and ψ (Ramachandran et al., 1963) (see Fig. 4). The torsion angles can be measured by solid-state NMR experiments that correlate orientation-dependent interactions, such as the CSA of amide ^{15}N, carbonyl ^{13}C, and α-carbon ^{13}C, or the dipolar coupling between pairs of nuclei, for exmple, ^{1}H–^{15}N or ^{1}H–^{13}C$_\alpha$ (Feng et al., 1997; Reif et al., 2000; Weliky and Tycko, 1996). The ^{13}C$_\alpha$-CSA for α-helical residues is smaller than for β-sheet structures. This feature was used in a two-dimensional ^{15}N–^{13}C correlation experiment which selects the signals of helical residues (Hong, 2000). Correlation of relative orientations between ^{1}H–^{15}N and ^{1}H$_\alpha$–^{13}C$_\alpha$ bonds allows identification of β-sheet residues (Huster et al., 2000). The torsion angle ψ in ^{13}C$_\alpha$ and carbonyl ^{13}C doubly labeled amino acids was measured in experiments that correlate orientation of ^{1}H$_\alpha$–^{13}C$_\alpha$ bonds with the orientation of the carbonyl ^{13}C-CSA tensor (Ishii et al., 1996; Schmidt-Rohr, 1996). Only one ^{13}C-labeled amino acid per ψ angle is needed for this experiment. In contrast, distance measurements for determining peptide backbone structure by RR and REDOR experiments require labeling of two sequential residues. The experiment with double-quantum selection (Schmidt-Rohr, 1996) also suppresses ^{13}C background signals and improves resolution by partially removing the inhomogenous spectral broadening of resonance lines.

III. STRUCTURE AND DYNAMICS OF THE LIPID MATRIX

Peptide binding to membranes depends on the properties of the lipid matrix, and peptide binding alters lipid bilayer structure. In particular, the binding of amphipathic α helices to the lipid/water interface has a profound influence on lipid organization. The lipid matrix of biomembranes belongs to the class of liquid-crystalline phases, which are very well suited for NMR structural studies. Lipid molecules have almost liquid-like degrees of freedom: vibrational changes of bond length and bond orientation with correlation times of a few picoseconds, *gauche/trans* isomerization of carbon–carbon bonds with correlation times on the order of 100 ps, molecular rotation and wobble with correlation times on the order of nanoseconds, and lateral diffusion of molecules within the plane of membranes at rates of $D \approx 1 \times 10^{-8}$ cm^2 s^{-1} (Pastor and Feller, 1996) (see Fig. 5). However, all of these motions have a certain degree of spatial anisotropy. They reduce the magnitude of anisotropic interactions, but do not average them out entirely, as observed for molecules that perform isotropic tumbling motions. Therefore, dipole–dipole interactions, quadrupole interactions, and the anisotropies of

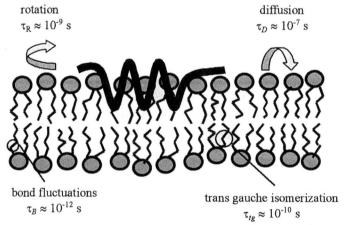

rotation
$\tau_R \approx 10^{-9}$ s

diffusion
$\tau_D \approx 10^{-7}$ s

bond fluctuations
$\tau_B \approx 10^{-12}$ s

trans gauche isomerization
$\tau_{tg} \approx 10^{-10}$ s

FIGURE 5 A biological membrane with phospholipid molecules and an attached amphiphatic peptide. Lipids in the membrane matrix exist in a multitude of conformations with rapid transition among them. They rotate, tumble, and perform rapid lateral diffusion. Correlation times that are of relevance for NOESY NMR are given. Motions of protein segments and of entire proteins are more restricted and much slower.

chemical shift still broaden, split, and shift NMR resonance signals of lipids, but with reduced magnitude. The orientation of partially averaged interaction tensors is determined by the orientation of the symmetry axis of motions, for example, the normal to the lipid bilayer. The state of the lipid matrix is judged by the degree of averaging of anisotropic interactions.

A. ^{31}P Anisotropy of Chemical Shift

The ^{31}P-anisotropy of the chemical shift (^{31}P-CSA) of lipid phosphate groups provides information on the lipid phase state and on peptide interaction with the lipid/water interface. Typical rigid-lattice ^{31}P-CSA values of phosphatidylcholines are $\sigma_{11} = -81$ ppm, $\sigma_{22} = -25$ ppm, and $\sigma_{33} = 108$ ppm (convention: chemical shift values increase with increasing field strength) (Kohler and Klein, 1976). Values for phosphatidylethanolamines and phosphatidylserines are similar. Fast rotational diffusion of lipids about the bilayer normal and headgroup wobble reduce these values to a tensor with apparent axial symmetry, $\Delta\sigma = \sigma_{\parallel} - \sigma_{\perp} \approx -45$ ppm, where $\sigma_{\parallel}$ is the tensor component oriented parallel to the bilayer normal and $\sigma_{\perp}$ is the tensor component perpendicular to it (J. Seelig, 1978). Lipid motions in lamellar, inverse hexagonal, cubic, and micellar phases have different symmetry, resulting in easily distinguishable degrees of averaging of ^{31}P-CSA. Therefore ^{31}P NMR is a convenient tool for studying the phase state of lipids (Cullis and de Kruijff, 1978). Furthermore, ^{31}P spectra enable the determination of mosaic spread in bilayer orientation for oriented membrane samples (Arnold *et al.*, 1979).

Interaction of peptides with the lipid/water interface alters the motionally averaged ^{31}P-CSA, because of changes in headgroup orientation and changes in lipid phase state, and also by inducing membrane curvature in the lamellar phase. The fast lateral diffusion of lipids over lamellar surfaces with small radius of curvature is equivalent to lipid reorientation in the outer magnetic field. Effective anisotropies show first signs of a reduction for radii of curvature that are smaller than 5000 Å. Anisotropies disappear completely for radii smaller than 1000 Å, that is, for small unilamellar liposomes (Burnell *et al.*, 1980; Gawrisch *et al.*, 1986).

The ^{13}C atoms of lipid carbonyl groups have characteristic anisotropies of chemical shift as well. Effective values of carbonyl anisotropies of chemical shift are linked to the conformation and the mobility of the glycerol backbone in lipids (S. O. Smith *et al.*, 1992).

B. ^{2}H NMR

The prevalent approach to studying the conformation and dynamics of lipid molecules in membranes is specific deuteration of lipids combined with the measurement of ^{2}H NMR quadrupole splittings. Very well resolved quadrupole splittings of deuterated lipid headgroups (Gally *et al.*, 1975), hydrocarbon chains (A. Seelig and Seelig, 1975), and the glycerol backbone (J. Seelig, 1977) are detected. "High- fidelity" NMR spectra without baseline distortions and intensity and phase errors are obtained by the quadrupolar echo sequence (Davis *et al.*, 1976). Experiments are mostly conducted on nonoriented lipid samples, but resolution of superimposed splittings and signal intensity improve considerably with the use of oriented membranes. The ^{2}H NMR spectra of randomly oriented bilayers can be mathematically transformed into virtual spectra of oriented bilayers by a procedure called dePakeing (Sternin *et al.*, 1983). More recently dePakeing was improved to analyze quadrupolar splittings and the distribution function of bilayer normals for samples that orient spontaneously in strong magnetic fields (Schäfer *et al.*, 1998) and for oriented bilayers at the surface of solid substrates and in bicelles (Sternin *et al.*, 2001).

The value of the quadrupole splitting $\Delta \nu$ is conveniently expressed as an order parameter S_{CD} that is linked to the orientation and motions of a C—D bond, $\Delta \nu = \frac{3}{4}(e^2 Q/h)S_{CD}$ and $S_{CD} = \langle \frac{1}{2}(3 \cos^2 \theta - 1) \rangle$. The term $e^2 Q/h$ is the quadrupole coupling constant, equal to 168 kHz for a typical C—D bond (Burnett and Muller, 1971), and θ is the angle between the orientation of the C—D bond and the bilayer normal. The bracket indicates time averaging. The order parameter is $+1$ for bonds that are oriented parallel to the external magnetic field, vanishes to zero at a bond orientation of 54.7° (the magic angle), and is -0.5 for bonds oriented perpendicular to the magnetic field. Note that a reduced order parameter may indicate a particular orientation, but also motional averaging. For example, the decrease in chain order of lipid hydrocarbon chains toward the terminal methyl group has been linked to a probability of *gauche–trans* isomerization in chains that is higher at the terminal methyl end. Order in the upper half of saturated chains, that is, near the lipid/water interface, is high and almost constant (order parameter plateau), whereas the order in the second half of the chain decreases with a steep gradient toward the terminal methyl group (A. Seelig and Seelig, 1975).

Chain order is very sensitive to changes in area per molecule, for example, as a result of changes in temperature or hydration (Koenig *et al.*, 1997), or as the result of interaction with amphipathic peptides (Koenig *et al.*, 1999a). Changes in lipid headgroup orientation have been followed via measurement of quadrupole splittings of choline α- and β-methylene groups (Gally *et al.*, 1975).

C. Magic Angle Spinning NMR

A significant increase in the number of resolved lipid NMR resonances has been achieved by MAS. MAS NMR reduces the linewidth of lipid resonances to about

10 Hz (Holte and Gawrisch, 1997), which is close to high-resolution NMR conditions. Resolution of lipid resonances in MAS experiments is better than resolution of spectra from very small unlilamellar liposomes, which tumble rapidly enough to eliminate anisotropic interactions. Magic angle spinning frequencies up to 10 kHz have negligible influence on lipid packing in bilayers. Experiments can be conducted at any level of hydration, provided that the lipids are in the biologically relevant liquid crystalline phase.

1. Dipolar Recoupling On-Axis with Scaling and Shape Preservation (DROSS)

The two-dimensional DROSS experiment determines $^1H-^{13}C$ dipolar interactions in lipid molecules (Gross et al., 1997; Holte et al., 1998). The strength of dipolar interaction is a measure of the H—C bond order parameter, equivalent to order parameters determined by ^2H NMR. Unlike ^2H NMR, the DROSS experiment enables assignment of order parameters to lipid segments without specific labeling. The experiment takes advantage of the much greater chemical shift dispersion of ^{13}C compared to ^2H nuclei. Recoupling of dipolar interactions is achieved by application of radiofrequency pulses that are synchronized to the phase of the spinning rotor. For technical reasons, the precision of order parameters determined by DROSS is somewhat lower than the precision of order parameters determined by ^2H NMR.

2. Nuclear Overhauser Enhancement Spectroscopy (NOESY)

The excellent resolution of lipid ^1H NMR resonances (see Fig. 6) permits the application of techniques that probe magnetization transfer between protons. One such technique that is well known for its application to high-resolution NMR structural studies of soluble proteins is nuclear Overhauser enhancement spectroscopy (NOESY) (see Fig. 7). We developed MAS NOESY into a tool for membrane structural studies (Feller et al., 1999; Huster and Gawrisch, 1999; Huster et al., 1999; Yau and Gawrisch, 2000).

Magnetization transfer is the result of interactions between the magnetic dipoles of the protons in lipids. Rates of transfer become observable when the protons approach each other to distances of 5 Å or less. The surprising observation has been that magnetization is transferred between all lipid resonances, but at different rates. Even the most distant protons, such as methyl groups of the choline headgroup and methyl groups at the end of lipid hydrocarbon chains, exchange magnetization. Previously, this observation was explained by a process called spin diffusion, in which magnetization between protons is transferred indirectly as a multiple-step process. However, our experiments on protonated lipid in a deuterated matrix (Huster et al., 1999) and on binary mixtures of specifically deuterated lipids (Huster and Gawrisch, 1999) demonstrated unambiguously that spin diffusion is

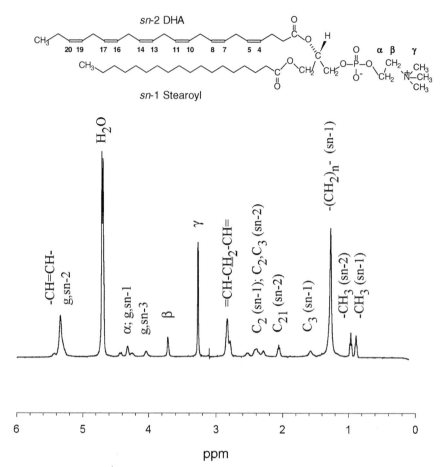

FIGURE 6 The ^{1}H MAS NMR spectrum of 1-stearoyl-2-docosahexaenoyl-*sn*-glycero-3-phosphocholine in 50 wt% D_2O recorded at a resonance frequency of 500.13 MHz, a temperature of 25 °C, and an MAS spinning speed of 10 kHz. The resolution of ^{1}H MAS spectra is better than the resolution of proton spectra from small unilamellar liposomes.

insignificant under most experimental conditions. The direct intermolecular transfer of magnetization between lipid protons is more efficient.

Cross-relaxation rates between lipid segments measured by ^{1}H NMR NOESY report the statistics of contacts between neighboring lipid molecules. The only conceivable model that would allow the bewildering multitude of cross-relaxations to happen is one of a lipid bilayer with a very disordered arrangement of lipids (Holte and Gawrisch, 1997, 1999; Huster *et al.*, 1999). Even protons that are separated by 20 Å or more, according to bilayer structures in the crystalline phase, can approach each other to less than 5 Å in the liquid crystalline phase, albeit with

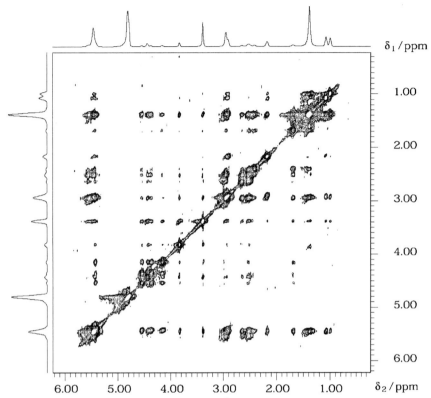

FIGURE 7 Two-dimensional NOESY MAS ^{1}H NMR spectrum of 1-stearoyl-2-docosahex-aenoyl-*sn*-glycero-3-phosphocholine bilayers recorded at a mixing time of 300 ms. The rate of magnetization transfer is reflected in the intensity of the off-diagonal crosspeaks. For cross-relaxation rate analysis, the peak volumes are measured as a function of mixing time. High cross-relaxation rates reflect frequent close contact between interacting protons. The high degree of lipid motional disorder in biomembranes results in measurable cross-relaxation among all lipid resonances. However, lipid segments that are, on average, further apart exchange magnetization at a lower rate.

low probability. Precise quantitative interpretation of these headgroup-to-chain contacts is somewhat influenced by the secondary dependence of cross-relaxation on differences in correlation times and differences in proton–proton distances of closest approach.

Measurement of ^{1}H MAS NOESY cross-relaxation rates permits the study of lipid structure and dynamics in bilayers as well as lateral lipid organization in complex lipid mixtures (Huster *et al.,* 1998). Furthermore, we successfully used this technique to locate small molecules, such as ethanol, and indole analogues in membranes with atomic resolution (Holte and Gawrisch, 1997; Yau *et al.,* 1998).

D. Influence of Peptide Binding on Lipid Structure

Interaction of peptides with the lipid matrix is reflected in changes of lipid order parameters. Amphipathic helical peptides have a preference for binding to the lipid/water interface region and tend to intercalate their hydrophobic side chains between lipid hydrocarbon chains to hide them from water. The partial penetration of the peptide into the lipid matrix increases area per lipid molecule at the lipid/water interface. The amino acid side chains are shorter than most lipid hydrocarbon chains and therefore do not reach to the bilayer center. This extra volume is filled by more-disordered hydrocarbon chain segments near the terminal methyl groups. It is reflected in a decrease of chain order parameters in ^{2}H NMR experiments on perdeuterated hydrocarbon chains (see Fig. 8) (Koenig *et al.*, 1999a). Depending on peptide penetration depth, such partial penetration creates positive membrane curvature stress, which has been linked to pore formation and membrane lysis.

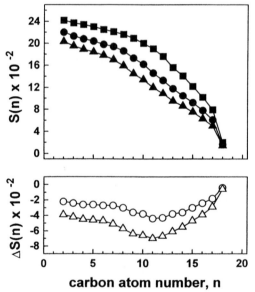

FIGURE 8   Influence of peptide P828S on the *sn*-1 hydrocarbon chain order of 1-stearoyl-d$_{35}$-2-oleoly-*sn*-glycero-3-phosphocholine (SOPC-d$_{35}$) investigated at 32°C. The ^{2}H NMR order parameter profiles of SOPS-d$_{35}$ in the absence of P828S (■) and at molar lipid : peptide ratios of 20 : 1 (●), and 10 : 1 (▲), respectively, are shown in the upper panel. The peptide-induced difference in order parameters along the chain at molar lipid : peptide ratios of 20 : 1 (○), and 10 : 1 (△) is shown in the lower panel. Peptide-induced order changes are largest in the bilayer center, suggesting that the peptide acts as a spacer that is located in the membrane's interface region. Figure reproduced from Koenig *et al.* (1999a).

Charged amphipathic peptides have a tendency to change the orientation and flexibility of lipid headgroups, which is easily observed by ^{2}H NMR on headgroup-labeled phospholipids and ^{31}P NMR of lipid phosphate groups (Wieprecht et al., 2000). Lipid/water interface properties can be also conveniently studied via the ^{2}H NMR quadrupole splittings of deuterated water (Gawrisch et al., 1992).

Helical peptides that span the bilayer have less influence on the packing of lipid hydrocarbon chains. Membranes have a tendency to adjust their hydrophobic thickness to match the hydrophobic length of the peptide. Long peptides increase bilayer thickness and reduce area per lipid molecule, as reflected in an increase of chain order parameters. Peptides that are shorter have the opposite effect (de Planque et al., 1998; Nezil and Bloom, 1992). Furthermore, peptides that are significantly shorter than the hydrophobic thickness of lamellar bilayers promote the formation of inverse-hexagonal and cubic lipid phases, which have been detected by ^{31}P NMR measurements (Prenner et al., 1999; van der Wel et al., 2000).

IV. FUTURE DIRECTIONS

The improvements in resolution and sensitivity of solid-state NMR experiments over the past 10 years are remarkable. However, it still takes considerable effort to synthesize the specifically labeled peptides required to obtain structural information from a sufficient number of peptide segments. High-resolution NMR on rapidly tumbling peptides provides a larger number of constraints in a single experiment. At the same time, the precision of distances, bond orientations, and dihedral bond angles measured by solid-state NMR is superior over equivalent values from high-resolution NMR measurements, enabling peptide structural determination by solid-state NMR with fewer constraints. Both techniques require special sample preparation procedures that may interfere with interpretation. High-resolution NMR relies on sufficiently fast isotropic motions of the peptide, whereas solid-state NMR requires peptide immobilization.

The introduction of multidimensional experiments in combination with magic angle spinning, such as MAS NOESY, has increased resolution of lipid structural investigations. New insights into the conformational freedom of lipids, lateral lipid organization, and location of small membrane-bound molecules have been obtained.

Future development of high-resolution approaches to peptide structural studies may benefit from utilizing bicelles that spontaneously orient in the strong magnetic fields applied to NMR samples (Sanders and Landis, 1995; Tjandra and Bax, 1997). High-resolution experiments on peptides that weakly interact with bicelles benefit from the small anisotropic NMR interactions, which can be measured without losing spectral resolution. Solid-state NMR experiments on peptides that are strongly bound to bicelles or incorporated into the bilayer region of bicelles benefit from the very high degree of bilayer orientation of bicelles in the magnetic field. Bicelles have the additional advantage that membrane surfaces are freely accessible from the water phase on both sides of the bilayer, enabling functional studies under almost physiological conditions.

Acknowledgment
K.G. thanks Dr. Daniel Huster for a valuable discussion on solid-state NMR approaches to studying peptide structure.

References
Anglister, J., and Zilber, B. (1990). Antibodies against a peptide of cholera toxin differing in cross-reactivity with the toxin differ in their specific interactions with the peptide as observed by ^{1}H NMR spectroscopy. *Biochemistry* **29**, 921–928.

Arepalli, S. R., Glaudemans, C. P. J., Daves, G. D., Kovac, P., and Bax, A. (1995). Identification of protein-mediated indirect NOE effects in a disaccharide–Fab' complex by transferred ROESY. *J. Magn. Reson. B* **106**, 195–198.

Arnold, K., Gawrisch, K., and Volke, F. (1979). ^{31}P NMR investigations of phospholipids. I. Dipolar interactions and the ^{31}P NMR lineshape of oriented phospholipid/water dispersions. *Studia Biophys.* **75**, 189–197.

Asakawa, N., Kuroki, S., Kurosu, H., Ando, I., Shoji, A., and Ozaki, T. (1992). Hydrogen-bonding effect on ^{13}C NMR chemical-shifts of L-alanine residue carbonyl carbons of peptides in the solid-state. *J. Am. Chem. Soc.* **114**, 3261–3265.

Balazs, Y. S., and Thompson, L. K. (1999). Practical methods for solid-state NMR distance measurements on large biomolecules: Constant-time rotational resonance. *J. Magn. Reson.* **139**, 371–376.

Bax, A. (1989). Homonuclear Hartmann–Hahn experiments. *Meth. Enzymol.* **176**, 151–168.

Bax, A., Sparks, S. W., and Torchia, D. A. (1989). Detection of insensitive nuclei. *Meth. Enzymol.* **176**, 134–150.

Bechinger, B., and Opella, S. J. (1991). Flat-coil probe for NMR spectroscopy of oriented membrane samples. *J. Magn. Reson.* **95**, 585–588.

Bechinger, B., Kim, Y., Chirlian, L. E., Gesell, J., Neumann, J. M., Montal, M., Tomich, J., Zasloff, M., and Opella, S. J. (1991). Orientations of amphipathic helical peptides in membrane bilayers determined by solid-state NMR spectroscopy. *J. Biomol. NMR* **1**, 167–173.

Bechinger, B., Gierasch, L. M., Montal, M., Zasloff, M., and Opella, S. J. (1996). Orientations of helical peptides in membrane bilayers by solid state NMR spectroscopy. *Solid State Nucl. Magn. Reson.* **7**, 185–191.

Bersch, B., Koehl, P., Nakatani, Y., Ourisson, G., and Milon, A. (1993). ^{1}H nuclear magnetic resonance determination of the membrane-bound conformation of senktide, a highly selective neurokinin-B agonist. *J. Biomol. NMR* **3**, 443–461.

Bogusky, M. J., Schiksnis, R. A., Leo, G. C., and Opella, S. J. (1987). Protein backbone dynamics by solid-state and solution ^{15}N NMR spectroscopy. *J. Magn. Reson.* **72**, 186–190.

Bogusky, M. J., Leo, G. C., and Opella, S. J. (1988). Comparison of the dynamics of the membrane-bound form of fd coat protein in micelles and in bilayers by solution and solid-state nitrogen-15 nuclear magnetic resonance spectroscopy. *Proteins* **4**, 123–130.

Bothner-By, A. A., Stephens, R. L., Lee, J., Warren, Ch. D., and Jeanloz, R. W. (1984). Structure determination of a tetrasaccharide: Transient nuclear Overhauser effects in the rotating frame. *J. Am. Chem. Soc.* **106**, 811–813.

Brender, J. R., Taylor, D. M., and Ramamoorthy, A. (2001). Orientation of amide-nitrogen-15 chemical shift tensors in peptides: A quantum chemical study. *J. Am. Chem. Soc.* **123**, 914–922.

Burnett, L. J., and Muller, B. H. (1971). Deuteron quadrupole coupling constants in three solid deuterated paraffin hydrocarbons: C_2D_6, C_4D_{10}, C_6D_{14}. *J. Chem. Phys.* **55**, 5829–5831.

Burnell, E. E., Cullis, P. R., and de Kruijff, B. (1980). Effects of tumbling and lateral diffusion on phosphatidylcholine model membranes ^{31}P NMR lineshapes. *Biochim. Biophys. Acta* **603**, 63–69.

Clore, G. M., and Gronenborn, A. M. (1982). Theory and application of the transferred nuclear Overhauser effect to the study of the conformations of small ligands bound to proteins. *J. Magn. Reson.* **48**, 402–417.

Clore, G. M., and Gronenborn, A. M. (1983). Theory of the time dependent transferred nuclear Overhauser effect: Applications to structural analysis of ligand–protein complexes in solution. *J. Magn. Reson.* **53**, 423–442.

Clore, G. M., and Gronenborn, A. M. (1991). Applications of three- and four-dimensional heteronuclear NMR spectroscopy to protein structure determination. *Prog. NMR Spectrosc.* **23**, 43–92.

Creuzet, F., McDermott, A., Gebhard, R., van der Hoef, K., Spijker-Assink, M. B., Herzfeld, J., Lugtenburg, J., Levitt, M. H., and Griffin, R. G. (1991). Determination of membrane-protein structure by rotational resonance NMR. Bacteriorhodopsin. *Science* **251**, 783–786.

Cullis, P. R., and de Kruijff, B. (1978). The polymorphic phase behaviour of phosphatidylethanolamines of natural and synthetic origin. A ^{31}P NMR study. *Biochim. Biophys. Acta* **513**, 31–42.

Datema, K. P., Pauls, K. P., and Bloom, M. (1986). Deuterium nuclear magnetic resonance investigation of the exchangeable sites on gramicidin A and gramicidin S in multilamellar vesicles of dipalmitoylphosphatidylcholine. *Biochemistry* **25**, 3796–3803.

Davis, J. H., Jeffrey, K. R., Bloom, M., Valic, M. I., and Higgs, T. P. (1976). Quadrupolar echo deuteron magnetic resonance spectroscopy in ordered hydrocarbon chains. *Chem. Phys. Lett.* **42**, 390–394.

Dempsey, C. E., Cryer, G. D., and Watts, A. (1987). The interaction of amino-deuteromethylated melittin with phospholipid membranes studied by deuterium NMR. *FEBS Lett.* **218**, 173–177.

de Planque, M. R. R., Greathouse, D. V., Koeppe, R. E., Schäfer, H., Marsh, D., and Killian, J. A. (1998). Influence of lipid/peptide hydrophobic mismatch on the thickness of diacylphosphatidylcholine bilayers: A ^2H NMR and ESR study using designed transmembrane alpha-helical peptides and gramicidin A. *Biochemistry* **37**, 9333–9345.

Deslauriers, R., and Smith, I. C. P. (1980). The multinuclear NMR approach to peptides. Structures, conformations, and dynamics. *In* "Biological Magnetic Resonance" (L. J. Berliner and J. Reuben, eds.), Vol. 2, pp. 243–344. Plenum Press, New York.

Dyson, H. J., and Wright, P. E. (1991). Defining solution conformations of small linear peptides. *Annu. Rev. Biophys. Biophys. Chem.* **20**, 519–538.

Feller, S. E., Huster, D., and Gawrisch, K. (1999). Interpretation of NOESY cross-relaxation rates from molecular dynamics simulation of a lipid bilayer. *J. Am. Chem. Soc.* **121**, 8963–8964.

Feng, X., Eden, M., Brinkmann, A., Luthman, H., Eriksson, L., Graslund, A., Antzutkin, O. N., and Levitt, M. H. (1997). Direct determination of a peptide torsional angle psi by double-quantum solid-state NMR. *J. Am. Chem. Soc.* **119**, 12006–12007.

Forman-Kay, J. D., Gronenborn, A. M., Kay, L. E., Wingfield, P. T., and Clore, G. M. (1990). Studies on the solution conformation of human thioredoxin using heteronuclear ^{15}N–^1H nuclear magnetic resonance spectroscopy. *Biochemistry* **29**, 1566–1572.

Gally, H. U., Niederberger, W., and Seelig, J. (1975). Conformation and motion of the choline head group in bilayers of dipalmitoyl-3-*sn*-phosphatidylcholine. *Biochemistry* **14**, 3647–3652.

Gawrisch, K., Stibenz, D., Möps, A., Arnold, K., Linss, W., and Halbhuber, K. J. (1986). The rate of lateral diffusion of phospholipids in erythrocyte microvesicles. *Biochim. Biophys. Acta* **856**, 443–447.

Gawrisch, K., Ruston, D., Zimmerberg, J., Parsegian, V. A., Rand, R. P., and Fuller, N. (1992). Membrane dipole potentials, hydration forces, and the ordering of water at membrane surfaces. *Biophys. J.* **61**, 1213–1223.

Gawrisch, K., Han, K.-H., Yang, J.-S., Bergelson, L. D., and Ferretti, J. A. (1993). Interaction of peptide fragment 828–848 of the envelope glycoprotein of human immunodeficiency virus type I with lipid bilayers. *Biochemistry* **32**, 3112–3118.

Gesell, J., Zasloff, M., and Opella, S. J. (1997). Two-dimensional ^{1}H NMR experiments show that the 23-residue magainin antibiotic peptide is an alpha-helix in dodecylphosphocholine micelles, sodium dodecylsulfate micelles, and trifluoroethanol/water solution. *J. Biomol. NMR* **9**, 127–135.

Gross, J. D., Warschawski, D. E., and Griffin, R. G. (1997). Dipolar recoupling in MAS NMR: A probe for segmental order in lipid bilayers. *J. Am. Chem. Soc.* **119**, 796–802.

Gu, Z., and Opella, S. J. (1999). Three-dimensional ^{13}C shift/^{1}H–^{15}N coupling/^{15}N shift solid-state NMR correlation spectroscopy. *J. Magn. Reson.* **138**, 193–198.

Gullion, T., and Schaefer, J. (1989). Rotational-echo double-resonance NMR. *J. Magn. Reson.* **81**, 196–200.

Harbison, G. S., Jelinski, L. W., Stark, R. E., Torchia, D. A., Herzfeld, J., and Griffin, R. G. (1984). ^{15}N chemical-shift and ^{15}N–^{13}C dipolar tensors for the peptide-bond in [1-^{13}C]glycyl[^{15}N]glycine hydrochloride monohydrate. *J. Magn. Reson.* **60**, 79–82.

Hing, A. W., and Schaefer, J. (1993). 2-Dimensional rotational-echo double-resonance of val-1 [1-^{13}C]gly2-[^{15}N]ala3-gramicidin-A in multilamellar dimyristoylphosphatidylcholine dispersions. *Biochemistry* **32**, 7593–7604.

Hing, A. W., Adams, S. P., Silbert, D. F., and Norberg, R. E. (1990). Deuterium NMR of Val1... (2-^{2}H)Ala3...gramicidin A in oriented DMPC bilayers. *Biochemistry* **29**, 4144–4156.

Hirsh, D. J., Hammer, J., Maloy, W. L., Blazyk, J., and Schaefer, J. (1996). Secondary structure and location of a magainin analogue in synthetic phospholipid bilayers. *Biochemistry* **35**, 12733–12741.

Holl, S. M., McKay, R. A., Gullion, T., and Schaefer, J. (1990). Rotational-echo triple-resonance NMR. *J. Magn. Reson.* **89**, 620–626.

Holte, L. L., and Gawrisch, K. (1997). Determining ethanol distribution in phospholipid multilayers with MAS-NOESY spectra. *Biochemistry* **36**, 4669–4674.

Holte, L. L., Koenig, B. W., Strey, H. H., and Gawrisch, K. (1998). Structure and dynamics of the docosahexaenoic acid chain in bilayers studied by NMR and X-ray diffraction. *Biophys. J.* **74**, A371.

Hong, M. (2000). Solid-state NMR determination of ^{13}C alpha chemical shift anisotropies for the identification of protein secondary structure. *J. Am. Chem. Soc.* **122**, 3762–3770.

Howard, K. P., and Opella, S. J. (1996). High-resolution solid-state NMR spectra of integral membrane proteins reconstituted into magnetically oriented phospholipid bilayers. *J. Magn. Reson. B* **112**, 91–94.

Huster, D., and Gawrisch, K. (1999). NOESY NMR crosspeaks between lipid headgroups and hydrocarbon chains: Spin diffusion or molecular disorder? *J. Am. Chem. Soc.* **121**, 1992–1993.

Huster, D., Arnold, K., and Gawrisch, K. (1998). Influence of docosahexaenoic acid and cholesterol on lateral lipid organization in phospholipid mixtures. *Biochemistry* **37**, 17299–17308.

Huster, D., Arnold, K., and Gawrisch, K. (1999). Investigation of lipid organization in biological membranes by two-dimensional nuclear Overhauser enhancement spectroscopy. *J. Phys. Chem. B* **103**, 243–251.

Huster, D., Yamaguchi, S., and Hong, M. (2000). Efficient beta-sheet identification in proteins by solid-state NMR spectroscopy. *J. Am. Chem. Soc.* **122**, 11320–11327.

Ishii, Y., Terao, T., and Kainosho, M. (1996). Relayed anisotropy correlation NMR: Determination of dihedral angles in solids. *Chem. Phys. Lett.* **256**, 133–140.

Ishima, R., and Torchia, D. A. (2000). Protein dynamics from NMR. *Nat. Struct. Biol.* **7**, 740–743.

Kay, L. E., and Bax, A. (1990). New methods for the measurement of NH—C$_\alpha$H J couplings in ^{15}N labeled proteins. *J. Magn. Reson.* **86**, 110–126.

Killian, J. A., Taylor, M. J., and Koeppe, R. E. (1992). Orientation of the valine-1 side chain of the gramicidin transmembrane channel and implications for channel functioning. A ^{2}H NMR study. *Biochemistry* **31,** 11283–11290.

Koenig, B. W., Bergelson, L. D., Gawrisch, K., Ward, J., and Ferretti, J. A. (1995). Effect of the conformation of a peptide from gp41 on binding and domain formation in model membranes. *Mol. Membrane Biol.* **12,** 77–82.

Koenig, B. W., Strey, H. H., and Gawrisch, K. (1997). Membrane lateral compressibility determined by NMR and X-ray diffraction: Effect of acyl chain polyunsaturation. *Biophys. J.* **73,** 1954–1966.

Koenig, B. W., Ferretti, J. A., and Gawrisch, K. (1999a). Site-specific deuterium order parameters and membrane-bound behavior of a peptide fragment from the intracellular domain of HIV-1 gp41. *Biochemistry* **38,** 6327–6334.

Koenig, B. W., Hu, J. S., Ottiger, M., Bose, S., Hendler, R. W., and Bax, A. (1999b). NMR measurement of dipolar couplings in proteins aligned by transient binding to purple membrane fragments. *J. Am. Chem. Soc.* **121,** 1385–1386.

Koenig, B. W., Mitchell, D. C., König, S., Grzesiek, S., Litman, B. J., and Bax, A. (2000). Measurement of dipolar couplings in a transducin peptide fragment weakly bound to oriented photo-activated rhodopsin. *J. Biomol. NMR* **16,** 121–125.

Koeppe, R. E., Killian, J. A., and Greathouse, D. V. (1994). Orientations of the tryptophan 9 and 11 side chains of the gramicidin channel based on deuterium nuclear magnetic resonance spectroscopy. *Biophys. J.* **66,** 14–24.

Kohler, S. J., and Klein, M. P. (1976). ^{31}P nuclear magnetic resonance chemical shielding tensors of phosphorylethanolamine, lecithin, and related compounds: Applications to headgroup motion in model membranes. *Biochemistry* **15,** 967–974.

Kumar, A., Ernst, R. R., and Wüthrich, K. (1980). A two-dimensional nuclear Overhauser enhancement (2D NOE) experiment for the elucidation of complete proton–proton cross-relaxation networks in biological macromolecules. *Biochem. Biophys. Res. Commun.* **95,** 1–6.

Lee, K. C., Huo, S., and Cross, T. A. (1995). Lipid–peptide interface: Valine conformation and dynamics in the gramicidin channel. *Biochemistry* **34,** 857–867.

Levitt, M. H., Raleigh, D. P., Creuzet, F., and Griffin, R. G. (1990). Theory and simulations of homonuclear spin pair systems in rotating solids. *J. Chem. Phys.* **92,** 6347–6364.

LoGrasso, P. V., Moll, F., and Cross, T. A. (1988). Solvent history dependence of gramicidin A conformations in hydrated lipid bilayers. *Biophys. J.* **54,** 259–267.

Marassi, F. M., and Opella, S. J. (2000). A solid-state NMR index of helical membrane protein structure and topology. *J. Magn. Reson.* **144,** 150–155.

Marshall, G. R., Beusen, D. D., Kociolek, K., Redlinski, A. S., Leplawy, M. T., Pan, Y., and Schaefer, J. (1990). Determination of a precise interatomic distance in a helical peptide by REDOR NMR. *J. Am. Chem. Soc.* **112,** 963–966.

McDonnell, P. A., and Opella, S. J. (1993). Effect of detergent concentration on multidimensional solution NMR spectra of membrane proteins in micelles. *J. Magn. Reson. B* **102,** 120–125.

McDonnell, P. A., Shon, K., Kim, Y., and Opella, S. J. (1993). fd coat protein structure in membrane environments. *J. Mol. Biol.* **233,** 447–463.

Müller, L. (1987). PE.COSY: A simple alternative to E.COSY. *J. Magn. Reson.* **72,** 191–196.

Neuhaus, D., and Williamson, M. P. (1989). "The Nuclear Overhauser Effect in Structural and Conformational Analysis." VHC, New York.

Nezil, F. A., and Bloom, M. (1992). Combined influence of cholesterol and synthetic amphiphillic peptides upon bilayer thickness in model membranes. *Biophys. J.* **61,** 1176–1183.

Ni, F. (1994). Recent developments in transferred NOE methods. *Prog. NMR Spectrosc.* **26,** 517–606.

Nilges, M. (1996). Structure calculation from NMR data. *Curr. Opin. Struct. Biol.* **6,** 617–623.

Oas, T. G., Hartzell, C. J., Dahlquist, F. W., and Drobny, G. P. (1987a). The amide ^{15}N chemical-shift tensors of 4 peptides determined from ^{13}C dipole-coupled chemical-shift powder patterns. *J. Am. Chem. Soc.* **109**, 5962–5966.

Oas, T. G., Hartzell, C. J., McMahon, T. J., Drobny, G. P., and Dahlquist, F. W. (1987b). The carbonyl ^{13}C chemical-shift tensor of 5 peptides determined from ^{15}N dipole-coupled chemical-shift powder patterns. *J. Am. Chem. Soc.* **109**, 5956–5962.

Okada, A., Wakamatsu, K., Miyazawa, T., and Higashijima, T. (1994). Vesicle-bound conformation of melittin: Transferred nuclear Overhauser enhancement analysis in the presence of perdeuterated phosphatidylcholine vesicles. *Biochemistry* **33**, 9438–9446.

Pardi, A., Billetter, and Wüthrich, K. (1984). Calibration of the angular dependence of the amide proton–C_α proton coupling constants, $^3J_{HN\alpha}$, in a globular protein. *J. Mol. Biol.* **180**, 741–751.

Pastor, R. W., and Feller, S. E. (1996). Time scales of lipid dynamics and molecular dynamics. *In* "Biological Membranes: A Molecular Perspective from Computation and Experiment" (K. M. Merz and B. Roux, eds.), pp. 3–30. Birkhauser, Boston.

Pervushin, K., Riek, R., Wider, G., and Wüthrich, K. (1997). Attenuated T_2 relaxation by mutual cancellation of dipole–dipole coupling and chemical shift anisotropy indicates an avenue to NMR structures of very large biological macromolecules in solution. *Proc. Natl. Acad. Sci. USA* **94**, 12366–12371.

Prenner, E. J., Lewis, R. N. A. H., and McElhaney, R. N. (1999). The interaction of the antimicrobial peptide gramicidin S with lipid bilayer model and biological membranes. *Biochim. Biophys. Acta* **1462**, 201–221.

Prosser, R. S., Daleman, S. I., and Davis, J. H. (1994). The structure of an integral membrane peptide: A deuterium NMR study of gramicidin. *Biophys. J.* **66**, 1415–1428.

Prosser, R. S., Hwang, J. S., and Vold, R. R. (1998). Magnetically aligned phospholipid bilayers with positive ordering: A new model membrane system. *Biophys. J.* **74**, 2405–2418.

Ramachandran, G. N., Ramakrishnan, C., and Sasisekharan, V. (1963). Stereochemistry of polypeptide chain configurations. *J. Mol. Biol.* **7**, 95–99.

Ramamoorthy, A., Marassi, F. M., Zasloff, M., and Opella, S. J. (1995). Three-dimensional solid-state NMR-spectroscopy of a peptide oriented in membrane bilayers. *J. Biomol. NMR* **6**, 329–334.

Reif, B., Hennig, M., and Griesinger, C. (1997). Direct measurement of angles between bond vectors in high-resolution NMR. *Science* **276**, 1230–1233.

Reif, B., Hohwy, M., Jaroniec, C. P., Rienstra, C. M., and Griffin, R. G. (2000). NH–NH vector correlation in peptides by solid-state NMR. *J. Magn Reson.* **145**, 132–141.

Sanders, C. R., and Landis, G. C. (1995). Reconstitution of membrane-proteins into lipid-rich bilayered mixed micelles for NMR-studies. *Biochemistry* **34**, 4030–4040.

Sanders, C. R., and Schwonek, J. P. (1992). Characterization of magnetically orientable bilayers in mixtures of dihexanoylphosphatidylcholine and dimyristoylphosphatidylcholine by solid-state NMR. *Biochemistry* **31**, 8898–8905.

Schäfer, H., Madler, B., and Sternin, E. (1998). Determination of orientational order parameters from ^{2}H NMR spectra of magnetically partially oriented lipid bilayers. *Biophys. J.* **74**, 1007–1014.

Schmidt-Rohr, K. (1996). Torsion angle determination in solid ^{13}C-labeled amino acids and peptides by separated-local-field double-quantum NMR. *J. Am. Chem. Soc.* **118**, 7601–7603.

Seelig, A., and Seelig, J. (1975). Bilayers of dipalmitoyl-3-*sn*-phosphatidylcholine. Conformational differences between the fatty acyl chains. *Biochim. Biophys. Acta* **406**, 1–5.

Seelig, J. (1977). Deuterium magnetic resonance: Theory and application to lipid membranes. *Q. Rev. Biophys.* **10**, 353–418.

Seelig, J. (1978). ^{31}P nuclear magnetic resonance and the head group structure of phospholipids in membranes. *Biochim. Biophys. Acta* **515**, 105–140.

Separovic, F., Smith, R., Yannoni, C. S., and Cornell, B. A. (1990). Molecular sequence effect on the ^{13}C carbonyl chemical shift shielding tensor. *J. Am. Chem. Soc.* **112**, 8324–8328.

Shon, K. J., Kim, Y. G., Colnago, L. A., and Opella, S. J. (1991). NMR studies of the structure and dynamics of membrane-bound bacteriophage Pf1 coat protein. *Science* **252**, 1303–1304.

Smith, R., Thomas, D. E., Separovic, F., Atkins, A. R., and Cornell, B. A. (1989). Determination of the structure of a membrane-incorporated ion channel. Solid-state nuclear magnetic resonance studies of gramicidin A. *Biophys. J.* **56**, 307–314.

Smith, R., Separovic, F., Milne, T. J., Whittaker, A., Bennett, F. M., Cornell, B. A., and Makriyannis, A. (1994). Structure and orientation of the pore-forming peptide, melittin, in lipid bilayers. *J. Mol. Biol.* **241**, 456–466.

Smith, S. O., Kustanovich, I., Bhamidipati, S., Salmon, A., and Hamilton, J. A. (1992). Interfacial conformation of dipalmitoylglycerol and dipalmitoylphosphatidylcholine in phospholipid bilayers. *Biochemistry* **31**, 11660–11664.

Spencer, R. G. S., Fishbein, K. W., Levitt, M. H., and Griffin, R. G. (1994). Rotational resonance with multiplepulse scaling in solid-state nuclear magnetic resonance. *J. Chem. Phys.* **100**, 5533–5545.

Spera, S., and Bax, A. (1991). Empirical correlation between protein backbone conformation and C_α and C_β ^{13}C nuclear magnetic resonance chemical shifts. *J. Am. Chem. Soc.* **113**, 5490–5492.

Sprangers, R., Bottomley, M. J., Linge, J. P., Schultz, J., Nilges, M., and Sattler, M. (2000). Refinement of the protein backbone angle psi in NMR structure calculations. *J. Biomol. NMR* **16**, 47–58.

Stark, R. E., Jelinski, L. W., Ruben, D. J., Torchia, D. A., and Griffin, R. G. (1983). ^{13}C chemical shift and $^{13}C–^{15}N$ dipolar tensors for the peptide bond: [1-^{13}C]glycyl[^{15}N]glycine·HCl·H$_2$O. *J. Magn. Reson.* **55**, 266–273.

Sternin, E., Bloom, M., and MacKay, A. L. (1983). De-Pake-ing of NMR spectra. *J. Magn. Reson.* **55**, 274–282.

Sternin, E., Schäfer, H., Polozov, I. V., and Gawrisch, K. (2001). Simultaneous determination of orientational and order parameter distributions from NMR spectra of partially oriented model membranes. *J. Magn. Reson.* **149**, 110–113.

Tjandra, N., and Bax, A. (1997). Direct measurement of distances and angles in biomolecules by NMR in a dilute liquid crystalline medium. *Science* **278**, 1111–1114.

van der Wel, P. C. A., Pott, T., Morein, S., Greathouse, D. V., Koeppe, R. E., and Killian, J. A. (2000). Tryptophan-anchored transmembrane peptides promote formation of nonlamellar phases in phosphatidylethanolamine model membranes in a mismatch-dependent manner. *Biochemistry* **39**, 3124–3133.

Wakamatsu, K., Okada, A., Miyazawa, T., Ohya, M., and Higashijima, T. (1992). Membrane-bound conformation of mastoparan-X, a G-protein-activating peptide. *Biochemistry* **31**, 5654–5660.

Wang, J., Denny, J., Tian, C., Kim, S., Mo, Y., Kovacs, F., Song, Z., Nishimura, K., Gan, Z., Fu, R., Quine, J. R., and Cross, T. A. (2000). Imaging membrane protein helical wheels. *J. Magn. Reson.* **144**, 162–167.

Wang, Z., Jones, J. D., Rizo, J., and Gierasch, L. M. (1993). Membrane-bound conformation of a signal peptide: A transferred nuclear Overhauser effect analysis. *Biochemistry* **32**, 13991–13999.

Weliky, D. P., and Tycko, R. (1996). Determination of peptide conformations by two-dimensional magic angle spinning NMR exchange spectroscopy with rotor synchronization. *J. Am. Chem. Soc.* **118**, 8487–8488.

Wieprecht, T., Apostolov, O., Beyermann, M., and Seelig, J. (2000). Membrane binding and pore formation of the antibacterial peptide PGLa: Thermodynamic and mechanistic aspects. *Biochemistry* **39**, 442–452.

Wishart, D. S., and Sykes, B. D. (1994). The ^{13}C chemical-shift index: A simple method for the identification of protein secondary structure using ^{13}C chemical-shift data. *J. Biomol. NMR* **4**, 171–180.

Wishart, D. S., Sykes, B. D., and Richards, F. M. (1991). Relationship between nuclear magnetic resonance chemical shift and protein secondary structure. *J. Mol. Biol.* **222,** 311–333.

Wu, C. H., Ramamoorthy, A., and Opella, S. J. (1994). High-resolution heteronuclear dipolar solid-state NMR-spectroscopy. *J. Magn. Reson. A* **109,** 270–272.

Wu, C. H., Ramamoorthy, A., Gierasch, L. M., and Opella, S. J. (1995). Simultaneous characterization of the amide ^{1}H chemical shift, ^{1}H–^{15}N dipolar, and ^{15}N chemical-shift interaction tensors in a peptide-bond by 3-dimensional solid-state NMR-spectroscopy. *J. Am. Chem. Soc.* **117,** 6148–6149.

Wüthrich, K. (1986). "NMR of Proteins and Nucleic Acids." Wiley, New York.

Yau, W. M., and Gawrisch, K. (2000). Lateral lipid diffusion dominates NOESY cross-relaxation in membranes. *J. Am. Chem. Soc.* **122,** 3971–3972.

Yau, W. M., Wimley, W. C., Gawrisch, K., and White, S. H. (1998). The preference of tryptophan for membrane interfaces. *Biochemistry* **37,** 14713–14718.

CHAPTER 7

Peptide-Lipid Interactions in Supported Monolayers and Bilayers

Lukas K. Tamm

Department of Molecular Physiology and Biological Physics, University of Virginia, Charlottesville, Virginia 22908

  I. Peptide Insertion into Lipid Monolayers
  II. Supported Bilayers
  III. ATR-FTIR Spectroscopy of Peptides in Supported Bilayers
  IV. Analysis of Fusion Peptide—Lipid Interactions
     References

This chapter describes monolayer and bilayer techniques for measuring the insertion and structure of peptides in lipid membranes. Measurements of lipid monolayer expansion together with molecular area measurements of the peptide yield binding isotherms, thermodynamic partition coefficients, and free energies of peptide insertion. Supported bilayers on solid substrates provide a relatively new membrane model system that has become particularly useful for measuring secondary structures and orientations of peptides in lipid bilayers by attenuated total reflection Fourier transform infrared (ATR-FTIR) spectroscopy in physiological environments. As an illustrative example, I discuss the application of these and other techniques to determine the structure and interactions of the fusion peptide of the influenza hemagglutinin in lipid bilayers.

I. PEPTIDE INSERTION INTO LIPID MONOLAYERS

Lipid monolayers are useful systems for studying thermodynamic and structural aspects of peptide–lipid interactions. Because monolayers are models of only half

Current Topics in Membranes, Volume 52

a bilayer, they serve best for studying interactions of amphipathic peptides that reside in only one leaflet of the bilayer and do not penetrate across the entire bilayer. Monolayers at the air–water interface of a Langmuir trough are characterized by the area/molecule, A and the surface pressure $\pi = \gamma_0 - \gamma$, where γ_0 is the surface tension of pure water (or buffer) and γ is the surface tension with the spread lipid monolayer. Each lipid/subphase system has its characteristic $\pi–A$ isotherm. In most of the older and some of the current literature, peptide insertion into monolayers has been measured by recording the pressure increase $\Delta\pi$ as a function of the initial pressure π_i at constant area and temperature. Straight lines that intercept the abscissa at a characteristic cutoff pressure are usually obtained from these measurements (see Fig. 1 for an example). This cutoff pressure can be taken as a qualitative measure of the penetration power of a peptide into a lipid monolayer of a particular lipid composition. For example, the bee venom peptide melittin, which has served as the "prototype" amphipathic peptide in many studies, penetrates negatively charged lipid monolayers up to 43 mN/m at a 0.1 μM subphase concentration (Bougis *et al.,* 1981).

Surface pressures are not defined in lipid bilayers. The surface pressure of a lipid monolayer that produces the same lipid packing density and therefore most closely reflects the pressure in lipid bilayers is between 32 and 36 mN/m. Apart from direct lipid area comparisons, one arrives at this conclusion by comparing the activity of phospholipases (Demel *et al.,* 1975) or the binding of local anesthetics (Seelig, 1987) in the two systems. Therefore, binding of peptides to lipid monolayers is best measured at a constant surface pressure of 32–36 mN/m, if one intends to

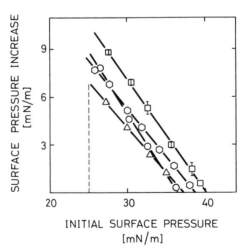

FIGURE 1 Insertion of the glucitol permease signal peptide into monolayers of different lipid composition. The lipids were (□) POPC, (○) POPE : POPG : cardiolipin (80 : 15 : 5), (○) POPG, (△) POPE. The concentration of the peptide in the subphase was 0.55 μM. From Tamm *et al.* (1989).

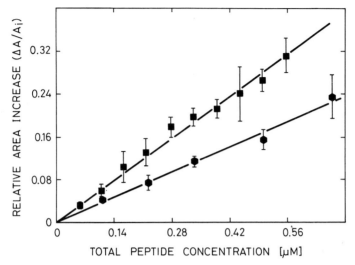

FIGURE 2 Binding isotherm of the insertion of the glucitol permease signal peptide into lipid monolayers at constant surface pressure (28.4 mN/m). The partition constants derived from Eq. (1) were 88,000 M^{-1} for partitioning into monolayers of POPC (■) and 52,000 M^{-1} for partitioning into monolayers of POPE : POPG : cardiolipin (80 : 15 : 5) (●). From Tamm *et al.* (1989).

extrapolate to how these peptides may bind to cell membranes. This is achieved on a Langmuir trough by keeping the surface pressure constant by expanding the surface area with an electronic feedback circuit as the peptide inserts (Tamm, 1986). Thermodynamically meaningful binding isotherms can thus be easily measured in this system. Figure 2 shows an example of an amphipathic signal peptide inserting into monolayers of different lipid composition at constant surface pressure. These isotherms can be analyzed by a simple peptide partition equilibrium between the aqueous phase and the lipid monolayer (Tamm, 1986),

$$\Delta A/A_i = (a_P/a_L)K C_P^0/(1 + K C_L^0), \qquad (1)$$

where $\Delta A/A_i$ is the relative area increase of the monolayer, K is the partition constant (in units of M^{-1}), a_P and a_L are the molecular areas in the monolayer, respectively, and C_P^0 and C_L^0 are the total concentrations of the peptide and lipid, respectively. The molecular area of the peptide can be determined from surface activity measurements with radioactively labeled peptides (Tamm, 1986) or by measuring the work of insertion πa_P, which can be determined from the slope of a plot of ln $\Delta A/A_i$ versus π, because $K = K_0 \exp(-\pi a_P/kT)$ (Boguslavsky *et al.*, 1994). Fits of the measured isotherms of Fig. 2 to Eq. (1) yield partition constants of 88,000 and 52,000 M^{-1} for partitioning of the negatively charged peptide into monolayers of the zwitterionic lipid 1-palmitoyl-2-oleoyl-3-*sn*-phosphatidylcholine (POPC) and

the negatively charged mixture of 1-palmitoyl-2-oleoyl-3-*sn*-phosphatidyletha-
nolamine (POPE):1-palmitoyl-2-oleoyl-3-*sn*-phosphatidylglycerol (POPG):cardi-
olipin, respectively (Tamm *et al.,* 1989). Obviously, electrostatic repulsion decreases the partitioning of the peptide into the lipid mixture. Electrostatic effects
of surface binding are generally well treated in quantitative terms by applying
the Gouy–Chapman theory of the diffuse ionic double layer at charged interfaces
(McLaughlin, 1989). A recent example of a monolayer study of charged peptide–
lipid interactions that incorporates the Gouy–Chapman theory is given by Seelig
et al. (2000).

II. SUPPORTED BILAYERS

Supported lipid bilayers have been developed as an alternative membrane model
system to complement studies that can be performed in lipid monolayers or lipid
vesicles (Tamm and McConnell, 1985). Supported bilayers are single unilamellar
lipid bilayer membranes supported on a hydrophilic solid substrate, such as quartz,
glass, or oxidized silicon chips. They share the planarity of lipid monolayers and
the bilayer structure of lipid vesicles. Because supported bilayers extend over large
areas (square centimeters), it is easy to observe their microscopic structure with a
light microscope. This is usually done by epifluorescence microscopy after labeling
with a suitable fluorescent lipid analogue. Epifluorescence microscopy is useful for
checking the quality of the supported bilayer and, in conjunction with fluorescence
recovery after photobleaching, can be used to monitor the lateral diffusion of lipid
and protein components in these systems. It has been determined by these methods
that the lipids in both leaflets of the bilayer are laterally mobile and exhibit lateral
diffusion coefficients that are very similar to those in free (unsupported) bilayers
(Tamm, 1988). Therefore, a thin lubricating layer of water exists between the
bilayer and solid support.

Supported bilayers may be prepared by three different methods (Tamm and
Tatulian, 1997). In the Langmuir–Blodgett/Schaefer method (Tamm and
McConnell, 1985), a monolayer is first transferred from the air–water interface
of a Langmuir trough to the hydrophilic support at a constant surface pressure
of 32–36 mN/m, that is, the bilayer equivalence pressure. A second monolayer is
deposited onto this surface by horizontal apposition of the substrate to a monolayer
at the same surface pressure. A second method of preparing supported bilayers
is to spread vesicles on a hydrophilic substrate (Brian and McConnell, 1984). A
dispersion of small unilamellar lipid vesicles spontaneously spreads and forms
a continuous planar bilayer on the surface when brought into contact with a hydrophilic solid substrate. Excess vesicles are easily removed by flushing the system
with buffer. The third method is a combination of the first two and involves the
spreading of vesicles on a preexisting supported monolayer (Kalb *et al.,* 1992).
This method is the preferred method for incorporating hydrophobic peptides into

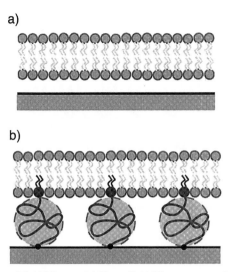

FIGURE 3   Supported lipid bilayers. (a) Planar lipid bilayer supported on a hydrophilic solid substrate, such as glass, quartz, or germanium oxide. An ~1-nm water-filled gap exists between the solid surface and the bilayer. (b) Tethered polymer-supported lipid bilayer. A 3400-molecular-weight polyethyleneglycol is covalently (silane) linked to the glass and to one in 33 lipids in the supported bilayer. The Flory radius of the polymer is 4.8 nm. The soft polymer cushion is supposed to increase the distance between the bilayer and support and to provide for a more natural environment for protein incorporation into supported bilayers. Adapted from Wagner and Tamm (2000).

supported lipid bilayers. Water-soluble membrane-interactive peptides are best bound to supported bilayers from solution after they have been be prepared by any of the three methods. The most recent methodological development in supported bilayer technology has been the advent of polymer-supported bilayers (Wagner and Tamm, 2000). In order to increase the spacing between the hydrophilic support and the supported bilayer, a random polymer (polyethyleneglycol) is introduced in this space in a controlled manner (Fig. 3). This is supposed to increase the dynamic flexibility of the supported bilayer and allow for the incorporation of larger membrane proteins. This new concept has been used to incorporate several integral membrane proteins (Wagner and Tamm, 2000, 2001), but has not yet been tested in peptide–lipid interaction studies.

III. ATR-FTIR SPECTROSCOPY OF PEPTIDES IN SUPPORTED BILAYERS

Supported bilayers are highly adequate model systems for investigating the secondary structure and orientation of peptides in membranes by Fourier transform infrared (FTIR) spectroscopy. A specially cut germanium plate is usually chosen

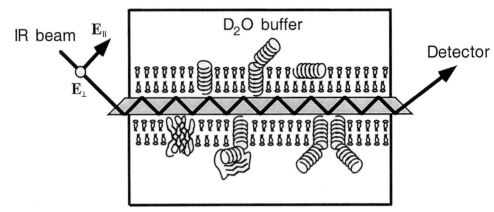

FIGURE 4 Configuration of a polarized ATR-FTIR experiment on supported lipid bilayers with included peptides and proteins. The plane-polarized ($E_\parallel$, $E_\perp$) infrared beam propagates via multiple internal reflections within the trapezoidal ATR germanium plate. The plate is assembled in a perfusable liquid holding cell.

as a substrate (Fig. 4). The thin natural oxide layer on germanium has similar surface properties to silicon oxide, that is, quartz or glass. The infrared beam is guided through beveled edges into the plate and propagates by multiple total internal reflections until it exits to the detector. An evanescent wave that penetrates a short distance ($\sim$400 nm) into the adjacent membrane and solution probes matter that is deposited at the surface. The IR signal is attenuated at all frequencies where surface-bound materials absorb. Spectra obtained in this configuration are therefore called attenuated total reflection (ATR)-FTIR spectra. Although there are some peculiarities associated with this mode of detection, ATR spectra are generally very similar to regular transmission spectra. Information on the secondary structure of peptides is usually obtained from the amide I band, a mode that is dominated by the peptide carbonyl stretching vibration. Since the peptide carbonyls are involved in hydrogen bonding in regular secondary structures, the amide I frequency reports on the secondary structure of the peptide. Typical absorption frequencies of β sheets (1625–1640 cm^{-1}) are generally well separated from those of α helices (1650–1660 cm^{-1}) and unordered structures (1640–1648 cm^{-1} in D$_2$O). More detailed tables of the relevant vibrational frequencies of various chemical groupings of peptides and lipids can be found in a comprehensive review on this topic (Tamm and Tatulian, 1997).

A second parameter obtained from polarized ATR-FTIR spectra of peptides in supported bilayers is the order parameter, which reflects the orientational distribution of peptides in the bilayer. The order parameter S is defined as

$$S = \tfrac{1}{2} \langle 3 \cos^2 \theta - 1 \rangle, \tag{2}$$

where θ is the angle between the peptide principal axis and the bilayer normal, and the angular brackets denote the ensemble average in the sample. An order parameter of 1.0 reflects a perfect perpendicular alignment and an order parameter of -0.5 reflects a perfect parallel alignment of the peptide with the membrane plane. Order parameters are experimentally obtained from the ratio of the absorbance measured with parallel-polarized light to the absorbance measured with perpendicular-polarized light, the so-called dichroic ratio (see Tamm and Tatulian, 1997, for equations and pertinent optical parameters).

Figure 5 shows the amide I band of the transmembrane domain of influenza hemagglutinin (Tatulian and Tamm, 2000). The peak frequency at 1655 cm^{-1} indicates that the peptide is mostly α-helical and the high dichroic ratio indicates an order parameter of 0.95, which translates into an "average" angle of $\sim$10° of the helix from the membrane normal. Melittin bound to supported lipid bilayers was also mostly α-helical, but had a helical order parameter of -0.45, indicating an "average" angle of $\sim$80° of the helix from the membrane normal (Frey and Tamm, 1991). Interestingly, melittin was found to be oriented more perpendicular to the plane of the membrane when measured in dehydrated membrane preparations. This result indicates that in some systems the state of hydration of the membrane is a very critical determinant of peptide structure and orientation. This has often been neglected in the pertinent literature.

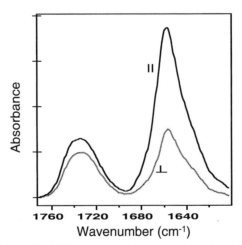

FIGURE 5 Polarized ATR-FTIR spectra of the transmembrane domain of influenza hemagglutinin in a supported lipid bilayer of 1,2-myristoyl-3-*sn*-phosphatidylcholine:1,2-myristoyl-3-*sn*-phosphatidylglycerol (4 : 1). The peak frequency at 1655 cm^{-1} indicates a predominantly α-helical conformation and the large dichroic ratio indicates a nearly perpendicular ($\sim$10° from bilayer normal) orientation of the peptide in the lipid bilayer. Adapted from Tatulian and Tamm (2000).

Changes of the lipid order in response to the incorporation of peptides into lipid bilayers are best determined by measuring the dichroic ratios of the two methylene stretching vibrations at $\sim$2850 and $\sim$2920 cm^{-1}. Melittin did not significantly affect this parameter (Frey and Tamm, 1991), but the transmembrane domain of influenza hemagglutinin increased the lipid chain order parameter from $\sim$0.35 to $\sim$0.5 (Tatulian and Tamm, 2000). Influenza hemagglutinin is one of the best-studied of integral membrane proteins that segregate into lipid rafts, which are thought to consist of more ordered lipids in cell membranes (Simons and Ikonen, 1997). Therefore, the increase of the lipid order parameter in the presence of the hemagglutinin transmembrane domain may be related to its localization in ordered lipid domains in cell membranes. Another prominent lipid band that is sometimes affected by incorporated peptides is the ester carbonyl stretching vibration. This band is sensitive to changes in hydrogen bonding (e.g., to water) in the membrane interface. Some viral fusion peptides cause a splitting of this band, indicating two different environments with different hydrogen-bonding patterns of the lipid ester carbonyl groups (Gray *et al.*, 1996)

IV. ANALYSIS OF FUSION PEPTIDE–LIPID INTERACTIONS

Fusion peptides are an interesting class of peptides, which upon activation become exposed from the ectodomains of fusion proteins and subsequently insert into the bilayer of target membranes in order to fuse target and host membranes. The sequences of fusion peptides are generally highly conserved within a given class of fusion proteins. A widely studied prototype fusion peptide is that of influenza hemagglutinin, which fuses membranes at pH 5, but not at pH 7 (see Chapter 18). Its sequence is

GlyLeuPheGlyAlaIleAlaGlyPheIleGluAsnGlyTrpGluGlyMetIleAspGly.

Most fusion peptides are more hydrophobic than melittin, but less hydrophobic than typical transmembrane domains, such as the previously discussed transmembrane domain of influenza hemagglutinin. This intermediate characteristic is likely important for their function, to fuse membranes, but makes them difficult to study experimentally. Most fusion peptides are not soluble in water and they also tend to aggregate in organic solvents. These problems can be circumvented by designing host–guest fusion peptides, which are soluble and still show very similar biological activities to their parent peptides (Han and Tamm, 2000a). In our design, we have appended the host sequence

H7 :   GlyCysGlyLysLysLysLys

to the C-terminus of the fusion (guest) peptides. This appendix is polar enough to solubilize a whole range of fusion peptide sequences without significantly affecting their biological activity. Since these peptides are highly charged, their solubility can actually be manipulated by changing the ionic strength of the buffer. The influenza host–guest fusion peptide, which we call P20H7, is completely soluble and assumes a random coil structure in low-ionic-strength buffers, but aggregates into β sheets in high-ionic-strength buffers (Han and Tamm, 2000b). Binding of the monomeric form to lipid bilayers is best studied by a fluorescence assay (Frey and Tamm, 1990) using a 7-nitrobenz-2-oxa-1,3-diazole (NBD)-labeled (at its single Cys) version of P20H7. After correcting by Gouy–Chapman theory for electrostatic effects and subtraction of the contribution of the host segment, one finds that the free energy of binding of the fusion peptide is -7.6 kcal/mol at pH 5 (Han and Tamm, 2000a). Isothermal titration calorimetry studies show that -3.0 kcal/mol is the enthalpic and -4.6 kcal/mol is the entropic contribution to the total binding energy (Y.-J. Chen and L. K. Tamm, unpublished results).

The peptide undergoes a major structural transition and becomes mostly α-helical upon binding and insertion into the membrane. Recent high-resolution nuclear magnetic resonance (NMR) studies at pH 7 and 5 revealed that the peptide adopts in dodecylphosphocholine micelles an angled helical structure at both pHs (Han *et al.*, 2001). A kink is located at Asn-12 in both structures, as shown in Fig. 6. The left arm of the structure is α-helical and almost unchanged

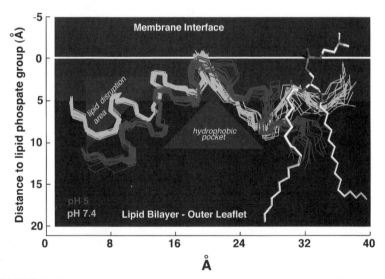

FIGURE 6 Structure and conformational change of the fusion peptide of influenza hemagglutinin in lipid bilayers at pH 7.4 (*yellow*) and pH 5 (*red*). The structures of the peptides in dodecylphosphocholine micelles were determined by ^{1}H-NMR spectroscopy. Site-directed spin-labeling at 18 positions was then used to determine the structures in lipid bilayers by electron paramagnetic resonance spectroscopy. Both structures are angled, but the fusion-triggering pH 5 structure has a sharper bend and inserts more deeply into the hydrophobic core of the bilayer than the inactive pH 7.4 structure. A phospholipid molecule in a fluid conformation is shown for reference. Adapted from Han and Tamm (2001). (See color plate.)

between the pH 7 and 5 structures, but a second, single-turn helix is induced in the right arm at pH 5, but not at pH 7. This conformational change repositions the charged residues in the right arm of the structure and creates a hydrophobic pocket at pH 5, which in turn allows both arms to penetrate into the hydrophobic core of the bilayer more deeply at pH 5 than at pH 7. The precise positioning of the influenza fusion peptides has been measured by site-directed spin labeling (Han *et al.*, 2001). A scale and a phospholipid molecule in a fluid conformation are provided in Fig. 6 as references to indicate the results of these measurements. The more angled shape and the more deeply inserted arms of the fusion peptide at pH 5 must somehow provide a mechanism for fusing membranes at low, but not at neutral pH.

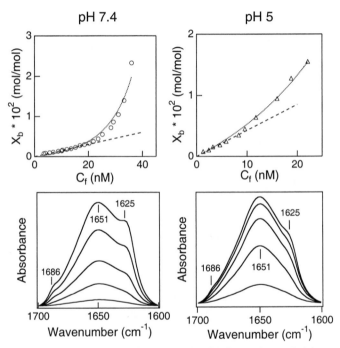

FIGURE 7 Self-association of the fusion peptide of influenza hemagglutinin at membrane surfaces at pH 7.4 and 5. Top: Binding curves, measured by NBD fluorescence, show deviations from straight lines, indicating self-association of the peptides at the membrane surface. The data were fitted with a cooperative self-association model. X_b and C_f designate bound and free peptide concentrations, respectively. Bottom: ATR-FTIR spectra of the fusion peptide in lipid bilayers at increasing peptide concentrations. At low concentrations the peptides are in a predominantly α-helical conformation (bands at 1651 cm^{-1}), but at higher concentrations antiparallel β structures develop (bands at 1625 and 1686 cm^{-1}). Adapted from Han and Tamm (2000b).

Fusion peptides are conformationally quite flexible. At higher concentrations, they aggregate at the membrane surface into β sheets (Han and Tamm, 2000b). This can be demonstrated by a quantitative analysis of the upward-bent binding isotherms (showing cooperativity of binding) and the development of an additional band at $\sim 1625\ cm^{-1}$ in ATR-FTIR spectra of peptides that are bound to the membrane at increasing concentrations (Fig. 7). It may well be that this conformational plasticity of fusion peptides is an important ingredient of the fusion mechanism, because several fusion proteins, which each bear three fusion peptides, must assemble in a single fusion site in cellular membrane fusion (Danieli *et al.*, 1996; Melikyan *et al.*, 1995). This will increase the local concentration of fusion peptides into a range where we observe some self-assembly into β sheets at the membrane surface. However, the precise role of each of these forms and the mechanism by which the conformational equilibria of fusion peptides couple into the lipid bilayer structure remain to be elucidated.

Acknowledgments

I thank the members of this laboratory, past and present, for their contributions and many stimulating discussions. This work was supported by NIH grants AI30557 and GM51329.

References

Boguslavsky, V., Rebecchi, M., Morris, A. J., Jhon, D.-Y., Rhee, S. G., and McLaughlin, S. H. (1994). Effect of monolayer surface pressure on the activities of phosphoinositide-specific phospholipase C-β_1, -γ_1, and -δ_1. *Biochemistry* **33**, 3032–3037.

Bougis, P., Rochat, H., Pieroni, G., and Verger, R. (1981). Penetration of phospholipid monolayers by cardiotoxins. *Biochemistry* **20**, 4915–4920.

Brian, A., and McConnell, H. M. (1984). Allogenic stimulation of cytotoxic T cells by supported planar membranes. *Proc. Natl. Acad. Sci. USA* **81**, 6159–6163.

Danieli, T., Pelletier, S. L., Henis, Y. I., and White, J. M. (1996). Membrane fusion mediated by the influenza virus hemagglutinin requires the concerted action of at least three hemagglutinin trimers. *J. Cell Biol.* **133**, 559–569.

Demel, R. A., Geurts Van Kessel, W. S. M., Zwaal, R. F. A., Roelogsen, B., and Van Deenen, L. L. M. (1975). Relation between various phospholipase actions on human red cell membranes and the interfacial phospholipid pressure in monolayers. *Biochim. Biophys. Acta* **406**, 97–107.

Frey, S., and Tamm, L. K. (1990). Membrane insertion and lateral diffusion of fluorescence-labeled cytochrome *c* oxidase subunit IV signal peptide in charged and uncharged phospholipid bilayers. *Biochem. J.* **272**, 713–719.

Frey, S., and Tamm, L. K. (1991). Orientation of melittin in phospholipid bilayers: A polarized attenuated total reflection infrared study. *Biophys. J.* **60**, 922–930.

Gray, C., Tatulian, S. A., Wharton, S. A., and Tamm, L. K. (1996). Effect of the N-terminal glycine on the secondary structure, orientation, and interaction of the influenza hemagglutinin fusion peptide with lipid bilayers. *Biophys. J.* **70**, 2275–2286.

Han, X., and Tamm, L. K. (2000a). A host-guest system to study structure–function relationships of membrane fusion peptides. *Proc. Natl. Acad. Sci. USA* **97**, 13097–13102.

Han, X., and Tamm, L. K. (2000b). pH-dependent self-association of influenza hemagglutinin fusion peptides in lipid bilayers. *J. Mol. Biol.* **304**, 953–965.

Han, X., Bushweller, J. H., Cafiso, D. S., and Tamm, L. K. (2001). Membrane structure and fusion-triggering conformational change of the fusion domain from influenza hemaagglutinin. *Nature Struct. Biol.* **8,** 715–720.

Kalb, E., Frey, S., and Tamm, L. K. (1992). Formation of supported planar bilayers by fusion of vesicles to supported phospholipid monolayers. *Biochim. Biophys. Acta* **1103,** 307–317.

McLaughlin, S. H. (1989). The electrostatic properties of membranes. *Annu. Rev. Biophys. Biophys. Chem.* **18,** 113–136.

Melikyan, G. B., Niles, W. D., and Cohen, F. S. (1995). The fusion kinetics of influenza hemagglutinin expression cells to planar bilayer membranes is affected by HA density and host cell surfaces. *J. Gen. Physiol.* **106,** 783–802.

Seelig, A. (1987). Local anesthetics and pressure: A comparison of dibucaine binding to lipid mono-layers and bilayers. *Biochim. Biophys. Acta* **899,** 196–204.

Seelig, A., Blatter, X. K., Frentzel, A., and Isenberg, G. (2000). Phospholipid binding of synthetic talin peptides provides evidence for an intrinsic membrane anchor of talin. *J. Biol. Chem.* **275,** 17954–17961.

Simons, K., and Ikonen, E. (1997). Functional rafts in cell membranes. *Nature* **387,** 569–572.

Tamm, L. K. (1986). Incorporation of a synthetic mitochondrial signal peptide into charged and uncharged phospholipid monolayers. *Biochemistry* **25,** 7470–7476.

Tamm, L. K. (1988). Lateral diffusion and fluorescence microscope studies on a monoclonal antibody specifically bound to supported phospholipid bilayers. *Biochemistry* **27,** 1450–1457.

Tamm, L. K., and McConnell, H. M. (1985). Supported phospholipid bilayers. *Biophys. J.* **47,** 105–113.

Tamm, L. K., and Tatulian, S. A. (1997). Infrared spectroscopy of proteins and peptides in lipid bilayers. *Q. Rev. Biophys.* **30,** 365–429.

Tamm, L. K., Tomich, J. M., and Saier, Jr., M. H. (1989). Membrane incorporation and induction of secondary structure of synthetic peptides corresponding to the N-terminal signal sequences of the glucitol and mannitol permeases of *Escherichia coli. J. Biol. Chem.* **264,** 2587–2592.

Tatulian, S. A., and Tamm, L. K. (2000). Secondary structure, orientation, oligomerization, and lipid interactions of the transmembrane domain of influenza hemagglutinin. *Biochemistry* **39,** 496–507.

Wagner, M. L., and Tamm, L. K. (2000). Tethered polymer-supported planar lipid bilayers for reconstitution of integral membrane proteins: Silane–polyethyleneglycol–lipid as a cushion and covalent linker. *Biophys. J.* **79,** 1400–1414.

Wagner, M. L., and Tamm, L. K. (2001). Reconstituted syntaxin1A/SNAP25 interacts with negatively charged lipids as measured by lateral diffusion in planar supported bilayers. *Biophys. J.* **81,** 266–275.

PART II

Theoretical and Computational
Analyses of Peptide-Lipid Interactions

CHAPTER 8

Free Energy Determinants of Peptide Association with Lipid Bilayers

Amit Kessel and Nir Ben-Tal

Department of Biochemistry, George S. Wise Faculty of Life Sciences, Tel Aviv University, Ramat Aviv 69978, Israel

I. Introduction
   A. Category A: Peptides That Interact Predominantly with the Hydrocarbon Region of the Bilayer
   B. Category B: Peptides That Interact Predominantly with the Polar Headgroup Region of the Bilayer
   C. Category C: Peptides That Interact with Both the Hydrocarbon and Polar Headgroup Regions of the Bilayer
   D. Theoretical and Computational Studies
II. Theory
   A. Effects Due to pK_a changes: ΔG_{pK_a}
   B. Peptide Conformation Effects: ΔG_{con}
   C. Peptide Immobilization Effects: ΔG_{imm}
   D. Hydrophobicity, Hydropathy Plots, and the Solvation Free Energy: ΔG_{sol}
   E. Lipid Perturbation Effects: ΔG_{lip}
   F. Membrane Deformation: ΔG_{def}
   G. Polar Headgroup Effects: ΔG_{hg}
   H. Specific Peptide–Lipid Interactions: ΔG_{sp}
   I. Peptide–Peptide Interactions and Lipid Demixing
III. Mean-Field Studies of Peptide–Membrane Systems
   A. Category A: Peptides That Interact Predominantly with the Hydrocarbon Region of the Bilayer
   B. Category B: Peptides That Interact Predominantly with the Polar Headgroup Region of the Bilayer
   C. Category C: Peptides That Interact with Both the Hydrocarbon and Polar Headgroup Regions of the Bilayer
IV. Open Questions
   References

In this chapter we describe different contributions to the free energy of peptide–membrane association and review computational methods for evaluating their magnitude at the mean-field level. The nature of interaction of different peptides with the bilayer, as reflected by the relative importance of each of the free energy contributions to peptide–membrane association, correlates strongly with the location of the peptide in the bilayer. Thus, we classify peptides into three different categories depending on their location in the membrane. The first category includes peptides that interact mainly with the hydrocarbon region of the membrane, the second includes peptides that interact predominantly with the polar headgroup region, and the third includes peptides that interact with both. We provide a review of recent studies of the energetics of peptide–membrane systems of each of the three classes at the mean-field level, emphasizing the relative importance of the free energy terms in each case, and highlighting pros and cons of the approach used. The emerging conclusion from this overview is that theoretical–computational treatment of peptides of the first two categories, that is, those that interact essentially either with the hydrocarbon region or with the polar headgroups only, is reasonably accurate. However, the theoretical treatment of peptides that interact with the whole bilayer structure (the hydrocarbon and the polar headgroup regions together) is substantially more difficult and current methodology is usually successful only if crucial, experimentally derived assumptions are made regarding the specific system at hand.

I. INTRODUCTION

A central problem in membrane biophysics is the fundamental understanding of the energetics of peptide–membrane systems. This has a bearing on understanding the interactions of signal peptides and signal anchors with lipid bilayers, the effects of antimicrobial peptides on cell membranes, the mechanism of voltage gating in ion channels, the nature of the immune response, and signal transduction, as well as on the prediction of structure and function in membrane proteins. A variety of powerful biochemical and biophysical (mainly spectroscopic) tools have been used for gathering information on peptide–membrane systems (reviewed by Cafiso, 1999; de Kroon et al., 1993; White and Wimley, 1999). The ample experimental data thus collected, together with data from computer simulations and theoretical studies of peptide–membrane systems (reviewed by Forrest and Sansom, 2000; Sansom, 1998; White and Wimley, 1999), provide at least partial understanding of the energetics of such systems. In this chapter we discuss the various free energy contributions to peptide–membrane association and provide examples to illustrate their relative importance in different cases.

The lipid bilayer has a complex structure, including a central hydrocarbon region, which is hydrophobic in nature, and two flanking polar headgroup regions

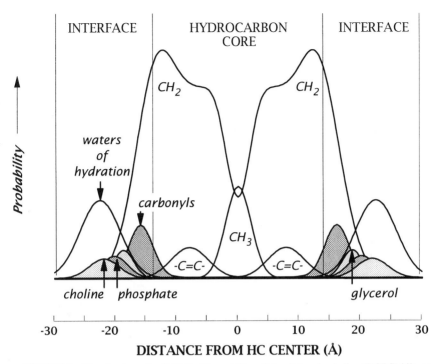

FIGURE 1 The structure of a fluid liquid crystalline dioleoylphosphocholine (DOPC) bilayer determined by the joint refinement of X-ray and neutron diffraction data (Wiener and White, 1992) (adapted from White and Wimley, 1999). The "structure" consists of the time-averaged distributions of the principal (quasi-molecular) structural groups of the lipid, projected onto an axis normal to the bilayer plane. The areas of the Gaussian distributions equal the number of structural groups represented by the Gaussians (one phosphate, two carbonyls, etc.); the distributions therefore represent the probability of finding a chemical group at a particular location. The interfaces of the bilayer, referred to in the text as the polar headgroup regions, are defined as the regions occupied by the headgroup's water of hydration. Although this structural image was obtained at low hydration (5.4 waters per lipid), more recent work demonstrates that the overall structure changes in relatively minor ways as the water content is increased (Hristova and White, 1998). Reprinted, with permission, from White and Wimley (1999). Copyright 1999 by Annual Reviews.

(Fig. 1). In most cases there is a strong correlation between the peptide's location in the membrane, as detected experimentally, and the dominant contributions to the free energy of its membrane association. For example, calculations suggest that the dominant contribution to the membrane association of alamethicin, a short transmembrane hydrophobic peptide (Barranger-Mathys and Cafiso, 1996; Huang and Wu, 1991; North *et al.*, 1995), comes from its interactions with the hydrocarbon region of the bilayer (Kessel *et al.*, 2000a). In this respect, it is useful to classify peptides according to their location in the membrane. However, it should be noted that the location of the peptide in the bilayer and the dominant contribution to its free energy of membrane association may not always be correlated. For example, the interactions of a peptide with the polar headgroups of the bilayer may be negligible even in cases in which it is in direct contact with both the polar headgroup and the hydrocarbon regions simultaneously.

For our purposes, it is convenient to distinguish among three hypothetical categories of peptide: first, peptides that are inserted into the hydrocarbon region of the bilayer and interact predominantly with this region; second, peptides that are adsorbed onto the bilayer and interact predominantly with the polar headgroups; and third, peptides that interact with both regions. Obviously, the theoretical/computational treatment of the first and second categories is less complicated than that of the third.

A. Category A: Peptides That Interact Predominantly with the Hydrocarbon Region of the Bilayer

In the first category, the free energy of the system is dominated by the interactions between the peptide and the hydrophobic core of the lipid bilayer. Consequently, it is hardly affected by the chemical nature of the polar headgroup region of the bilayer. The association of peptides in this category with membranes is driven by the nonpolar contributions to the association free energy, commonly referred to as "hydrophobic interactions." In most cases these contributions are partially balanced by the electrostatic free energy penalty resulting from the transfer of polar groups, such as those in the peptide backbone, from the aqueous phase into the hydrocarbon region of the bilayer.

Biological systems that belong to this category include transmembrane pore-forming peptides, such as gramicidin (Wallace, 1990), alamethicin (Cafiso, 1994), and phospholamban (Arkin *et al.,* 1997; Herzyk and Hubbard, 1998). Certain membrane proteins (e.g., ion channels and receptors) are composed of α-helical transmembrane segments, which are synthesized separately and assemble inside the lipid bilayer to form the intact protein. The general behavior of these transmembrane segments and their interactions with the lipid bilayer are similar to those of membrane-associated peptides, and therefore can be included in this category. A well-known example is the acetylcholine receptor (AChR), which is composed of five subunits that assemble inside the lipid bilayer to form an ion-conducting pore (Hucho *et al.,* 1996). Glycophorin provides another example; it forms homodimers of known three-dimensional structure (MacKenzie *et al.,* 1997).

The transmembrane domain of many membrane proteins is composed of a bundle of hydrophobic α helices, which originate from the same polypeptide chain. Studies of these proteins, such as bacteriorhodopsin, show that at least some of the transmembrane α helices can separately refold into lipid bilayers when they have been disconnected from each other by the removal of the connecting loops.

Moreover, when these fragments are subsequently brought together they spontaneously yield a functional protein (Popot, 1993; Popot and Engelman, 1990; Popot *et al.*, 1987). Such transmembrane helices may also be included in category A. Studies suggest that it should be possible to predict the locations of at least some of the transmembrane helices in the sequence of membrane proteins based on energetic considerations (see Section III.A.4 below).

B. Category B: Peptides That Interact Predominantly with the Polar Headgroup Region of the Bilayer

In the second category, the free energy of the system is dominated by the interactions between the peptide and the polar headgroup of the lipid bilayer. In such cases, the peptides are generally adsorbed onto the surface of the lipid bilayer, without significant penetration into the polar headgroup region. Consequently, they hardly interact with the hydrocarbon region of the bilayer. For example, studies indicate that the neuronal anchoring protein, AKAP79, associates mainly electrostatically with lipid bilayers; it contains three regions enriched with basic residues that interact with the acidic lipid phosphatidylinositol 4,5-bisphosphate (PIP$_2$) (Dell'Acqua *et al.*, 1998). In addition, many important peripheral proteins are membrane-anchored by a combination of a hydrocarbon chain, which interacts with the hydrophobic core of the bilayer, and a region of positively charged residues that interact (mainly) electrostatically with negatively charged lipids in the membrane (reviewed by McLaughlin and Aderem, 1995). Examples include cytochrome *c* (Heimburg and Marsh, 1995, 1996; Pinheiro, 1994; Pinheiro and Watts, 1994a,b), myelin basic protein (MacNaughtan *et al.*, 1985), phospholipases (Roberts, 1996), Src (Buser *et al.*, 1994; Resh, 1993, 1994; Sigal *et al.*, 1994), myristoylated alanine-rich C-kinase substrate (MARCKS) (Aderem, 1992; Blackshear, 1993; McLaughlin and Aderem, 1995), human immunodeficiency virus (HIV) matrix protein (C. P. Hill *et al.*, 1996; Massiah *et al.*, 1994; Zhou *et al.*, 1994), K-Ras (Cadwallader *et al.*, 1994; Hancock *et al.*, 1990), and human carbonic anhydrase IV (Stams *et al.*, 1996). In some of these cases it is possible to consider the contribution of the region of basic residues to the association free energy separately, and there is ample experimental and computational data suggesting that the region of basic residues interacts predominantly with the polar headgroup of the bilayer (Murray *et al.*, 1997, 1998).

C. Category C: Peptides That Interact with Both the Hydrocarbon and Polar Headgroup Regions of the Bilayer

In the third category, the peptide partially penetrates the lipid bilayer and interacts both with the polar headgroups and the hydrocarbon region of the bilayer.

Antibacterial peptides, such as magainin (Bechinger *et al.*, 1998), dermaseptin, and mammalian cecropin, belong to this category (Shai, 1999). These basic peptides are synthesized by a wide range of organisms as part of their defense system against invading bacteria (Bechinger, 1997; Saberwal and Nagaraj, 1994). The antibacterial action of such peptides is believed to be the result of their ability to induce structural disorder in the packing of lipid chains in the bacterial plasma membrane. According to the "carpet-like mechanism" suggested by Shai and co-workers (Pouny *et al.*, 1992; Shai, 1995), the antibacterial peptides are adsorbed, via their basic domain, to the negatively charged bacterial plasma membrane, each of them partially penetrating the hydrophobic core of the bilayer. The combined action of these peptides on the lipid bilayer eventually leads to its lysis and subsequently to the death of the bacterium.

Certain protein segments behave in a similar fashion. For example, the fusion peptide of influenza hemagglutinin mediates the fusion of internalized virions with the mature endosomal membrane (Carr and Kim, 1993, 1994), probably through structural rearrangements of the lipid bilayer. Several studies indicate that the fusion peptide penetrates the endosomal membrane, interacting with both the polar and nonpolar regions of the lipid bilayer (Durell *et al.*, 1997; Han *et al.*, 1999, 2001; Luneberg *et al.*, 1995; Macosko *et al.*, 1997).

D. Theoretical and Computational Studies

The structural complexity of peptide–membrane systems makes them a challenge for theoretical treatments. Thus, state-of-the-art computational methods usually involve *ad hoc* assumptions derived from experimental data on the particular system at hand. For example, in all-atom molecular dynamics (MD) simulations, the method of choice in theoretical studies of peptide–membrane systems, the choice of the initial peptide–membrane configuration is often based on available experimental data. Some MD simulations even involve the addition of new energy terms tailored for the specific system involved.

The theoretical methods that are currently used for the study of peptide–membrane systems differ from each other with respect to the level of detail used for representing the peptide and the bilayer. Straightforward MD simulations are based on atomic-detail models of both the peptide and the hydrated lipid bilayer (reviewed by Forrest and Sansom, 2000; Roux and Woolf, 1996; Sansom, 1998). MD simulations may provide insight into some aspects of peptide–membrane interactions (e.g., hydrogen bonding), but they do not usually provide energetic guidelines for processes such as peptide adsorption onto, and insertion into, lipid bilayers, primarily due to limitations on simulation time and size.

Continuum solvent models (e.g., Ben-Tal *et al.*, 1996a,b, 2000a; Berneche *et al.*, 1998; La Rocca *et al.*, 1999; May and Ben-Shaul, 1999; see also Section III) are at the opposite extreme in terms of the level of detail used for describing the system, relying on implicit description of the aqueous phase and, in some cases, of the bilayer. Thus, simulations that are based on these models are more feasible for the study of biological events that occur on long time scales and for thermodynamic analysis.

Monte Carlo (MC) simulations usually involve an intermediate level of representation. Preliminary tests have demonstrated their ability to deal with peptide–membrane systems (reviewed by Skolnick and Milik, 1996). However, their accuracy depends strongly on the quality of the energy function, which is, as yet, of questionable quality (Pastor, 1994), although efforts are being made to generate new and better energy functions (e.g., Schlenkrich et al., 1996).

Biggin and Sansom (1999) have recently reviewed various theoretical treatments of peptide–membrane and protein–membrane systems. They concentrated mainly on MD simulations and generally avoided thermodynamic aspects. This chapter is complementary to their review, in that it focuses mainly on the thermodynamic (and long-term kinetic) aspects and reviews recent attempts to estimate the major free energy components of peptide–membrane interactions at the mean-field level. We review recent continuum solvent models and Monte Carlo studies to provide a general framework for mean-field treatment of peptide–membrane interactions. This framework should hold for peptide–membrane systems of each of the three categories mentioned above, and it may be useful in the critical assessment of many simple model studies of peptide–membrane systems.

II. THEORY

The total free energy difference between a peptide in contact with the membrane and one in the aqueous phase (ΔG_{tot}) can be decomposed into components in several ways (Ben-Tal et al., 1996a; Cafiso, 1999; Engelman and Steitz, 1981; Fattal and Ben-Shaul, 1993; Honig and Hubbell, 1984; Jacobs and White, 1989; Jähnig, 1983; Milik and Skolnick, 1993; White and Wimley, 1999). Figure 2 presents an option that is compatible with the mean-field approach used here. ΔG_{tot} is a sum of differences of (1) effects due to changes in the pK_a of titratable residues (ΔG_{pKa}), (2) peptide conformation (ΔG_{con}) and immobilization (ΔG_{imm}) effects, (3) electrostatic and nonpolar contributions to the solvation free energy (ΔG_{sol}), (4) lipid perturbation effects (ΔG_{lip}), (5) membrane deformation effects (ΔG_{def}), (6) effects resulting from interactions between the peptide and lipid headgroups (ΔG_{hg}), and (7) effects resulting from specific peptide–lipid interactions (ΔG_{sp}). Thus,

$$\Delta G_{tot} = \Delta G_{pKa} + \Delta G_{con} + \Delta G_{imm} + \Delta G_{sol} + \Delta G_{lip} + \Delta G_{def}$$
$$+ \Delta G_{hg} + \Delta G_{sp}. \tag{1}$$

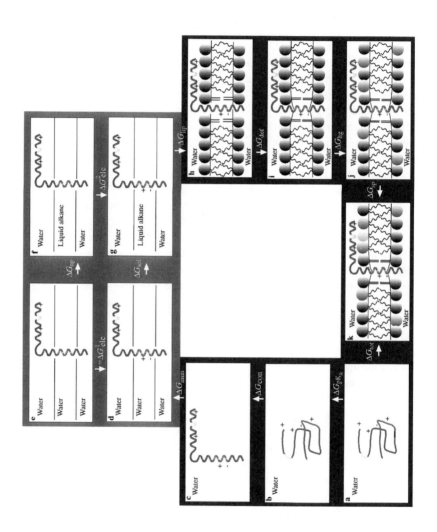

FIGURE 2 A schematic diagram describing the various contributions to the total free energy of association of a peptide with the lipid bilayer. The association process is depicted as a sequence of different thermodynamic substates, designated a–k. The various contributions to the total free energy of peptide–membrane association characterize the passages between the different substates. The peptide is schematically depicted as a green curve (in its unfolded states a and b) or as two joined green helices (in its folded states c–k), and ΔG_{con} is the free energy due to conformational change. The black (in state a) and pink (in b–k) patches on the peptide represent two different protonation states of a titratable residue, and ΔG_{pKa} is the free energy associated with the change of protonation state. The central hydrophobic region of the peptide is colored blue, and the two horizontal lines represent the boundaries of the hydrocarbon region of the membrane. The free energy penalty of peptide immobilization (c–d) is denoted ΔG_{imm}. The solvation free energy (d–g) is denoted ΔG_{sol}. Here d–g, surrounded by light blue shading, describe a thermodynamic cycle based on the separation of ΔG_{sol} into electrostatic and nonpolar contributions. The hydrophobic core of the peptide is discharged in water (d–e; $-\Delta G_{elc}^{1}$). It is then transferred into a liquid alkane (e–f; ΔG_{np}) and recharged again in the liquid alkane (f–g; ΔG_{elc}^{2}). The unperturbed lipid chains are represented by the black wavy lines. Peptide-induced lipid perturbation effects (g–h) are denoted ΔG_{lip}; state h shows lipid chains in the vicinity of the peptide, which become more rigid due to their interactions with the peptide core; the ΔG_{lip} contributions reflect the associated entropic penalty. Membrane deformation effects (h–i) are denoted ΔG_{def}. The neutral and acidic headgroups are represented by gray and red spheres, respectively. Effects resulting from the interactions between the peptide and lipid headgroups, for example, positively charged residues of the peptide with negatively charged lipids (i–j), are denoted ΔG_{hg}. The yellow sphere represents a polar headgroup that interacts specifically with the peptide (the peptide segment that is involved in this interaction is colored yellow). Effects resulting from specific peptide–lipid interactions (j–k) are denoted ΔG_{sp}. (See color plate.)

A. Effects Due to pK_a Changes: ΔG_{pKa}

The transfer of a charged residue from water to a less polar medium, such as the lipid bilayer, involves a large electrostatic desolvation free energy penalty. This free energy penalty may be lowered significantly by protonation or deprotonation of the residue, which leads to its neutralization (e.g., Honig and Hubbell, 1984). Thus, peptide–membrane association may induce changes in the protonation state of titratable residues in the peptide (Fig. 2a to Fig. 2b). These changes may be a direct consequence of the peptide interactions with either the polar headgroup region (Esmann and Marsh, 1985; Horvath et al., 1988) or the hydrocarbon region of the membrane (Honig and Hubbell, 1984). Alternatively, they may be associated with conformational changes that the lipid bilayer induces in the peptide (White and Wimley, 1999). Regardless of the cause of the change in the protonation state of an amino acid, the free energy change associated with it, ΔG_{pKa}, may be calculated as

$$\Delta G_{pKa} = -2.3kT(pK_a - pH), \qquad (2)$$

where k is the Boltzmann constant, T is the absolute temperature, and K_a is the ionization equilibrium constant of the residue. For example, ΔG_{pKa} for the neutralization of aspartic or glutamic acid ($pK_a \approx 4$) at neutral pH is ~ 4 kcal/mol.

B. Peptide Conformation Effects: ΔG_{con}

The transfer of a peptide from a polar environment, such as water, to a less polar medium, such as the membrane, may involve conformational changes in the peptide, with the resulting free energy contribution ΔG_{con}. Indeed, lipid bilayers are known to induce formation of secondary structure elements (e.g., Deber and Goto, 1996; Engelman and Steitz, 1981; Popot, 1993; White and Wimley, 1999). The conformational changes usually take place to lower the free energy of transfer of unsatisfied hydrogen bonds from the aqueous phase into the bilayer. Experimental data (reviewed by Engelman et al., 1986; Popot, 1993; Popot and Engelman, 1990; White and Wimley, 1999) and simulations (Milik and Skolnick, 1993, 1995; see also review by Biggin and Sansom, 1999) suggest that the sequence of events is probably such that the conformational changes in the peptide follow its association with the membrane. Nevertheless, in many cases it is computationally more convenient to account for ΔG_{con} in the aqueous phase (Fig. 2b to Fig. 2c). In some cases the structural changes are limited to a short amino acid stretch going, say, from random coil to α-helix conformation upon membrane insertion. Both experiments (Yang and Honig, 1995) and theory (Wojcik et al., 1990) suggest that for stretches of up to about 15 residues, ΔG_{con} in the aqueous phase is small and sometimes even negligible. Thus, in such cases it is reasonable to assume that $\Delta G_{con} \approx 0$.

It may be difficult to get an estimate of ΔG_{con} associated with the folding of a membrane protein in the aqueous phase. In this respect, the approach of White and Wimley (1999), who experimentally derived an estimate for the reduction in free energy due to secondary structure formation in bilayers, may be a more feasible option for treating proteins.

C. Peptide Immobilization Effects: ΔG_{imm}

Any association process—the binding of a ligand to a protein, the dimerization of two proteins, the insertion of a peptide into a lipid bilayer (Fig. 2c to Fig. 2d), or the adsorption of a peptide onto a bilayer—involves an entropy loss, reflecting the conversion of free translational and rotational, degrees of freedom into bound motions. For example, in bulk solution, a ligand has three translational degrees of freedom of its center of mass and three rotational degrees of freedom. Thus, depending on its concentration (i.e., on the choice of the standard state), a certain volume V and all of the "orientational space," that is, $8\pi^2$, will be accessible to the ligand molecule when it is free in solution. Upon binding to a receptor, some, usually all, of these degrees of freedom become confined. Thus, only a portion V_b of the volume and a portion γ of orientational space are accessible for the ligand molecule. The immobilization free energy penalty ΔG_{imm} is a sum of contributions from the restrictions on translational, ΔG_{tr}, and rotational, ΔG_{rot}, degrees of freedom imposed on the ligand upon binding (Brady and Sharp, 1997; Erickson, 1989; Gilson et al., 1997; T. L. Hill, 1985; Janin, 1996):

$$\Delta G_{imm} = \Delta G_{tr} + \Delta G_{rot} = -kT \, \ln(V_b/V) - kT \, \ln(\gamma/8\pi^2). \tag{3}$$

It is rather difficult to measure ΔG_{imm} directly, because it is usually not the only entropy-based contribution to association; for example, ΔG_{sol} and ΔG_{con} usually involve significant entropic components. Finkelstein and Janin (1989) derived a theoretical estimate of $\Delta G_{imm} \approx 15$ kcal/mol for protein dimerization by calculating V_b and γ from the root mean square fluctuations (B-factors) in protein crystals. Their estimate seems reasonable for tight binding and when all six external translational and rotational degrees of freedom are confined due to binding (Holtzer, 1994; Horton and Lewis, 1992; Murphy et al., 1994; Novotny et al., 1989; Page and Jencks, 1971; Searle and Williams, 1992; Searle et al., 1992; Tidor and Karplus, 1994; Vajda et al., 1994, Weng et al., 1996).

The planar pseudo-symmetry of bilayers and the nature of peptide–membrane interactions mean that peptide–membrane associations differ from, for example, binding of a ligand to a receptor. Of the three center-of-mass translational degrees of freedom of a peptide in solution, two, associated with translation in the membrane plane (x–y), are retained. Similarly, one of the three rotational degrees of freedom of a peptide in solution—the one associated with peptide rotation around the

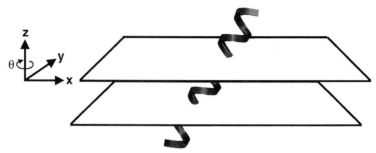

FIGURE 3 Peptide immobilization effects. Upon membrane association, the peptide may still freely translate as a rigid body in the membrane plane (x–y), but its motion along the membrane normal (z) becomes limited. Similarly, the peptide can freely rotate around the normal (θ) even when membrane-associated, but its two other rigid-body rotational motions become restricted.

membrane normal (θ)—does not markedly change its character upon association (Fig. 3). Therefore, because of the pseudo-symmetry of the membrane, usually only three degrees of freedom—one translation (in the z axis) and two rotations—are confined upon association.

The upper bound of the value of ΔG_{imm} for the insertion of a 25-mer polyalanine α helix into a lipid bilayer has been estimated to be ~3.7 kcal/mol (Ben-Shaul *et al.*, 1996; Ben-Tal *et al.*, 1996a). This estimate is much smaller than that obtained for binding processes, because to the fluidity of the bilayer enables it to tolerate many peptide configurations prohibited by a rigid receptor. Recently, an even smaller value of $\Delta G_{imm} \approx 1.3$ kcal/mol was estimated for the electrostatically driven adsorption of the positively charged peptide pentalysine onto membranes containing negatively charged lipids (Ben-Tal *et al.*, 2000b). Membrane fluidity was not taken into account in this study. The peptide resides about 3 Å from the van der Waals surface of the bilayer, and the small value of ΔG_{imm} in this case was attributed to the screening effect of water, which significantly reduces the Coulomb attraction between the peptide and the membrane.

The $\Delta G_{imm} \approx 1.3$ kcal/mol estimate for pentalysine is comparable to the data obtained by Peitzsch and McLaughlin's (1993) estimate based on the measured partitioning of fatty acids and acylated peptides between the aqueous phase and phospholipid vesicles; in both studies the same number of degrees of freedom is restricted. They reported that their measured association free energy is consistent with an empirical rule derived by Tanford (1991) from the partitioning of fatty acids into *n*-heptane, in which association entropy plays no part. They therefore concluded that ΔG_{imm} should be small, ~2 kcal/mol. A similar estimate can be derived from measurements of the association of model acylated proteins with lipid bilayers (Pool and Thompson, 1998).

D. Hydrophobicity, Hydropathy Plots, and the Solvation Free Energy: ΔG_{sol}

The water environment has significant effects on soluble proteins, and solvation free energy ΔG_{sol} has primary importance in stability and binding in globular proteins (Eisenberg and McLachlan, 1986; Honig and Nicholls, 1995; Honig et al., 1993; Kellis et al., 1988; Kyte and Doolittle, 1982; McCammon, 1998; Sheinerman et al., 2000; Warshel, 1991; Yue and Dill, 1996). An estimate of ΔG_{sol} may be obtained from hydrophobicity scales; indeed, hydropathy plots have been proved valuable in the discrimination between residues that are solvent-exposed and residues that are buried in the core of soluble proteins (e.g., Eisenberg and McLachlan, 1986; Kyte and Doolittle, 1982).

Hydrophobicity scales are also commonly used in structure predictions of membrane proteins. An implicit assumption involved in the use of these scales is that a particular solvent mimics the membrane environment. For example, the widely used Kyte and Doolittle scale (Kyte and Doolittle, 1982) uses vacuum-to-water transfer free energies to assign a hydrophobicity value to each amino acid. A moving window is then applied to the sequence to determine which segments are most likely to appear within the bilayer. A different approach, taken by von Heijne (1981a,b) and Engelman et al. (1986), is based on defining free energy contributions of individual chemical groups to partitioning. Another approach is based on the Eisenberg and McLachlan (1986) and Efremov et al. (1999c) schemes, which assume that transfer free energies are proportional to a sum of atomic properties that are derived from a least squares fit to experimental transfer free energies.

All of these schemes share a common feature: Free energies of transfer are assumed to be proportional to some inherent property, whether that of an individual atom, a chemical group, or an entire amino acid side chain. Although they work fairly well in predicting the helical spanning regions of membrane proteins (Fasman and Gilbert, 1990; Jähnig, 1990; White and Wimley, 1999), they fail in complex cases, where residues are involved in interactions and cannot be accurately represented by simple numbers or categories (e.g., Rose, 1987). Clearly, a prediction based on the transfer free energies of individual groups into a bulk phase cannot account for specific interactions between groups that characterize heterogeneous systems (see discussions by White and his co-workers: Jacobs and White, 1989; White and Wimley, 1999). Thus, transfer free energies of individual amino or carbonyl groups cannot be used to predict the free energy of transfer of a molecule in which these groups are hydrogen-bonded to one another. Similarly, a group that partitions near an interface and may form specific interactions with lipid headgroups cannot be treated as if it were completely immersed in a bilayer. Moreover, groups that are buried near an interface can still interact quite strongly with water (Gilson et al., 1985), but this interaction clearly cannot be accounted for with numbers based on partitioning into a bulk phase.

Hydrophobicity scales are commonly used to approximate ΔG_{sol} (Fig. 2d to Fig. 2g) in simulations of peptide–membrane systems (Biggin *et al.*, 1997; Edholm and Jähnig, 1988; Efremov *et al.*, 1999a–c; Gersappe *et al.*, 1993; Milik and Skolnick, 1993, 1995; Seagraves and Reinhardt, 1995; Skolnick and Milik, 1996). Such simulations are based on a simplified description of the aqueous phase and the lipid bilayer as two structureless media connected to each other by a smooth boundary (e.g., La Rocca *et al.*, 1999; Skolnick and Milik, 1996). Numbers associated with partitioning between polar and nonpolar media are assigned to atoms, chemical groups, or whole residues (depending on the resolution of the simulation) as they cross the bilayer–water boundary. The outcome of this approach is encouraging, in that in all of the reported simulations, the calculations are consistent with available experimental data (e.g., regarding the most likely conformation and orientation of the peptide in the bilayer or its free energy of association with the bilayer). However, in view of the limited number of peptide–membrane systems studied using this approach, and particularly the small number of cases tested using a single variant of this approach, it is premature to evaluate its usefulness at this point.

ΔG_{sol} of Fig. 2d to Fig. 2g is the free energy of transfer of the hydrophobic core of the peptide from water into a bulk hydrocarbon phase. It accounts both for electrostatic contributions (ΔG_{elc}; Fig. 2d to Fig. 2e, and Fig. 2f to Fig. 2g) resulting from changes in the solvent dielectric constant and for van der Waals and solvent structure effects, which are grouped in the nonpolar term (ΔG_{np}; Fig. 2e to Fig. 2f); together they define the classic hydrophobic effect:

$$\Delta G_{sol} = \Delta G_{elc} + \Delta G_{np} = \Delta G_{elc}^{1} - \Delta G_{elc}^{2} + \Delta G_{np}. \tag{4}$$

The main contributions to ΔG_{sol} usually come from the interactions of the peptide with the hydrocarbon region of the bilayer (e.g., Fig. 2; the blue segment of the peptide). ΔG_{sol} usually involves an electrostatic penalty, such as that due to the partitioning of backbone hydrogen bonds into the hydrocarbon region of the membrane. However, for hydrophobic peptides this penalty is overcompensated for by the nonpolar contributions, resulting in net attraction between the peptide and the membrane.

Mean-field calculations show that for hydrophobic peptides ΔG_{sol} is usually the dominant contribution to ΔG_{tot} and its value for membrane-associated peptides of about 20 residues is typically -10 to -20 kcal/mol (Ben-Tal *et al.*, 1996a; Kessel *et al.*, 2000a; see also below).

Since it is known that peptide adsorbing to the polar headgroup region often induces secondary structure formation (White and Wimley, 1999, and references therein), it is obvious that interactions of peptides with the polar headgroups involve

ΔG_{sol} components. Headgroup contributions to ΔG_{sol} may be due to the removal of water molecules from a peptide segment, the headgroups, or both (Ben-Tal *et al.*, 1996b, 1997a). Because of the polar nature of the headgroups, it is anticipated that both the electrostatic desolvation penalty and the nonpolar contributions associated with partitioning into the headgroup region will be smaller in magnitude than those associated with partitioning into the hydrocarbon region. The headgroups' contributions to the solvation free energy may be included in the ΔG_{sol} or in the ΔG_{hg} term of Eq. (1).

1. Electrostatic Contributions to Solvation: ΔG_{elc}

The electrostatic contribution to the solvation free energy (ΔG_{elc}; Fig. 2d to Fig. 2e, and Fig. 2f to Fig. 2g) can be obtained from finite-difference (FD) solutions to the Poisson–Boltzmann (PB) equation (the FDPB method) (Honig and Nicholls, 1995; Honig *et al.*, 1993):

$$\nabla \cdot [\varepsilon(r) \nabla \phi(r)] - \varepsilon_r \kappa(r)^2 \sinh[\phi(r)] + e^2 \rho^f(r)/(\varepsilon_0 kT) = 0. \qquad (5)$$

Here $\varepsilon(r)$ is the spatially dependent dielectric constant, $\kappa(r)$ is the Debye–Hückel parameter, $\rho^f(r)$ is the distribution of fixed charges, $\phi(r)$ is the electrostatic potential, k is the Boltzmann constant, e is the electron charge, ε_r is the dielectric constant of the aqueous phase, ε_0 is the permittivity of free space, and T is the absolute temperature. The spatial distributions $\varepsilon(r)$, $\kappa(r)$, and $\rho^f(r)$ are obtained from the structure of the peptide and membrane and the equation is solved for $\phi(r)$. The electrostatic free energy, ΔG_{elc} is obtained, by integration, from the spatial distributions of $\rho^f(r)$ and $\phi(r)$. In the absence of salt ions [or equivalently when $\phi(r)$ is small and the Poisson–Boltzmann equation can be linearized], ΔG_{elc} is given by

$$\Delta G_{elc} = 1/2 \int dr\, \rho^f(r) \cdot \phi(r). \qquad (6)$$

2. Nonpolar Contributions to Solvation: ΔG_{np}

By definition, the ΔG_{np} term in Eq. (4) includes all the contributions to the solvation free energy that are missing in the ΔG_{elc} term. Thus, ΔG_{np} (Fig. 2e to Fig. 2f) includes contributions of electrostatic origin, such as changes in the van der Waals (dispersion) force in the transfer from the aqueous phase to the membrane, as well as contributions from solvent structure effects, collectively referred to as the classic hydrophobic effect. ΔG_{np} can be derived from the partitioning of nonpolar molecules, such as alkanes, between water and nonpolar solvents. Normally there is a linear relation between the measured log partition coefficient (or equivalently the transfer free energy) and the alkane size. A common method to account for the ΔG_{np} contribution is to relate it to the water accessible surface area A of the alkanes,

$$\Delta G_{np} = \gamma A + b, \tag{7}$$

where γ is a surface tension coefficient and b is an intercept.

A choice of alkane partitioning between water and liquid alkanes, giving $\gamma = 0.0278$ kcal/(mol $\mathring{A}^2$) and $b = -1.71$ kcal/mol (Sitkoff *et al.*, 1996), is particularly suitable for calculations with peptides and small molecules that interact predominantly with the hydrocarbon region of the bilayer (Ben-Tal *et al.*, 1996a, 2000a; Kessel *et al.*, 2000a,b). Experimental data suggest that the partitioning of amino acids between the aqueous phase and the polar headgroup region of palmitoyloleoylphosphatidylcholine (POPC) bilayers can be approximated using $\gamma = 0.012$ kcal/(mol $\mathring{A}^2$) (Wimley and White, 1996).

E. Lipid Perturbation Effects: ΔG_{lip}

On average, the hydrophobic tails of the lipid molecules in a membrane are oriented along the membrane normal. The extent of chain stretching is often measured in terms of the average orientational order parameter of the tail S, which lies between 0 and 1. In the solid (or "gel") phase of the membrane, the lipid chains are nearly all in an "all-*trans*" conformation, lying fully stretched along the membrane normal, with order parameter $S = 0.8$ (Jähnig, 1983). In the fluid (or "liquid crystalline") phase of the membrane, which is of greater biological interest, the order parameter is considerably smaller, typically $S = 0.2$ (Jähnig, 1983). This lower value indicates that the partially stretched lipid chains possess considerable conformational freedom. Their entropy is thus higher than in the solid phase. The insertion of a rigid inclusion, such as a hydrophobic peptide, into a fluid membrane may stiffen the lipid chains in its immediate vicinity, thus lowering the conformational entropy of the system (Fig. 2g to Fig. 2h). This peptide–lipid interaction results from the lower elasticity (chain conformational freedom) of the membrane, which gives rise to a positive contribution to ΔG_{tot}.

ΔG_{lip} is obviously entropic in origin and its magnitude is proportional to the contact area between the peptide and the lipid tails, just as ΔG_{np} is proportional to the water-accessible surface area. Thus, one should expect a negligible value of ΔG_{lip} for peptides that are adsorbed at the membrane–water interface and do not interact with the lipid chains directly (Ben-Tal *et al.*, 1996a). For the transmembrane insertion of a 25-mer polyalanine α helix into a model bilayer, ΔG_{lip} has been estimated to be 2.3 kcal/mol (Ben-Shaul *et al.*, 1996). Since the effective circumference of all transmembrane helices is roughly the same and since there is very little variation in the width of the hydrocarbon region of biological membranes, this estimate may be viewed as typical for ΔG_{lip} associated with helix insertion into biological membranes.

Such treatment does not take into account the extra work that must be done against surface pressure when inserting a protein or a peptide into only one side of a bilayer. It has been suggested that this is the reson for the 100-fold change in the activity of phospholipase C upon changing the pressure of a monolayer (Boguslavsky *et al.,* 1994).

F. Membrane Deformation: ΔG_{def}

When the length of the hydrophobic core of the peptide exactly matches the hydrophobic thickness of the membrane, the free energy contribution associated with the effect of the peptide on the organization of the lipid chains is generally, although not always (Fattal and Ben-Shaul, 1993), minimal, and includes only ΔG_{lip}. Negative hydrophobic mismatch, which occurs when the width of the hydrocarbon region of the bilayer is larger than the peptide's hydrophobic core (Fig. 2h), results in an additional free energy penalty (ΔG_{def}). This is associated with excess compression of the lipid molecules around the protein to achieve perfect hydrophobic matching (Fig. 2i), following the "mattress model" (Mouritsen and Bloom, 1984). This perfect matching facilitates the removal of polar regions in the peptide from the hydrocarbon region of the membrane into the aqueous phase, which would otherwise involve a high free energy penalty contribution from ΔG_{el}. A molecular model of the lipid chains can be used to estimate the dependence of ΔG_{def} on the mismatch between a lipid bilayer and a flat hydrophobic inclusion (Fattal and Ben-Shaul, 1993) (Table I). ΔG_{def} is usually small; even a severe helix-induced membrane deformation of 5 Å results in a free energy penalty of less than 1 kcal/mol.

The inverse situation, positive mismatch, in which the width of the hydrocarbon region of the bilayer is smaller than the peptide's hydrophobic core, is also energetically unfavorable: It involves the exposure of hydrophobic regions in the peptide core to the polar aqueous phase or water–membrane interface. The lipid tails may be expected to compensate by expanding in response to positive mismatch and it is possible to derive the corresponding ΔG_{def} contributions from Fatal and Ben-Shaul's (1993) calculations (Table I). Alternatively, the peptide may adopt a tilted orientation in the membrane (Killian, 1998). In fact, most transmembrane helices in membrane proteins are somewhat tilted (Bowie, 1999; Killian, 1998; White and Wimley, 1999). Similarly, peptides like alamethicin and isolated protein segments like the M2 segment of the nicotinic acetylcholine receptor have been shown to adopt a tilted orientation inside lipid bilayers (Kessel *et al.,* 2000a; Opella *et al.,* 1999).

TABLE I

Dependence of the Membrane Deformation Free Energy ΔG_{def} on the
Hydrophobic Mismatch between the Lipid Bilayer and a Transmembrane
Peptide/Protein with a 30-Å-Long Hydrophobic Core[a]

Membrane width[b] (Å)	$\Delta G_{def}/l^c$ (kcal/mol Å)	$\Delta G_{def}(5\ \text{Å})^d$ (kcal/mol)	$\Delta G_{def}(17\ \text{Å})^e$ (kcal/mol)
40	0.33	3.57	28.76
38	0.22	2.34	18.80
36	0.13	1.36	10.96
34	0.06	0.65	5.22
32	0.02	0.20	1.58
30	0.00	0.00	0.00
28	0.01	0.08	0.64
26	0.04	0.41	3.33
24	0.09	1.01	8.13
22	0.18	1.87	15.03
20	0.28	2.99	24.04

[a] The dependence is based on the calculations of Fattal and Ben-Shaul (1993), in which the transmembrane protein/peptide was treated as a flat inclusion of infinite radius. In later studies, we applied curvature corrections of the form $2\pi R_P(1 - R_L/R_P)$, where $R_L = 3.3$ Å and R_P are the radii of the lipid chain and the helical peptide, respectively.
[b] The width of the hydrocarbon region of the lipid bilayer.
[c] The results of Fattal and Ben-Shaul (1993) for membrane deformation induced by a hydrophobic inclusion of infinite radius. l is the inclusion's circumference.
[d] ΔG_{def} associated with a hydrophobic inclusion of radius 5 Å, corresponding to a monomeric transmembrane helix.
[e] ΔG_{def} due to a hydrophobic inclusion of radius 17 Å, corresponding to a transmembrane hexameric helical bundle.

G. Polar Headgroup Effects: ΔG_{hg}

Peptide–membrane association may involve interactions with the lipid head-groups (Fig. 2i to Fig. 2j). The free energy contributions associated with these interactions are collectively referred to as ΔG_{hg}. The desolvation contributions to these interactions (e.g., Ben-Tal *et al.,* 1997a) may be considered as part of ΔG_{sol}. However, the polar headgroups may add unique contributions, such as the nonspecific attraction between acidic lipid heads and basic residues in the peptide (McLaughlin and Aderem, 1995; Murray *et al.,* 1997). Many peptides contain basic residues that interact electrostatically with acidic headgroups in lipid bilayers. Examples include signal sequences (von Heijne, 1998; Zheng and Gierasch, 1996) and antimicrobial peptides (Shai, 1999; White *et al.,* 1995).

Chapter 3 in this book focuses on the headgroups' contributions to peptide and protein association with lipid bilayers. Here, we briefly present a few examples, focusing on the contributions of basic residues to the free energy of membrane association. Under physiological conditions (membranes containing 25–30 mol% of monovalent acidic lipids at a concentration of 150 mM monovalent salt ions), each positively charged residue (lysine or arginine) in the vicinity of the bilayer plane contributes about 1 kcal/mol to ΔG_{tot} (e.g., Ben-Tal *et al.*, 1996b; see Section III.B.3 below).

H. Specific Peptide-Lipid Interactions: ΔG_{sp}

Certain peptides or protein segments may interact specifically with membrane lipids (Fig. 2j to Fig. 2k). For example, the PH domain of PLC-δ_1 has been demonstrated to interact with high affinity and stereospecificity with both membrane phosphatidylinositol 4,5-bisphosphate and D-*myo*-inositol 1,4,5-trisphosphate (Lemmon *et al.*, 1995). Furthermore, the specific interaction between a protein and a certain lipid molecule may affect the activity of the protein, as in, for example, protein kinase C (Newton, 1993) and β-hydroxybutyrate (Cortese *et al.*, 1989). Specific peptide–lipid interactions, which may involve different parts of the lipid molecule, may be electrostatic or nonpolar in nature. However, for computational convenience they may be considered as a distinct free energy contribution, denoted ΔG_{sp}.

I. Peptide-Peptide Interactions and Lipid Demixing

Throughout this chapter we consider the interactions between a single peptide molecule and the lipid bilayer and do not take into account direct or membrane-mediated interactions between peptide molecules. Similarly, we do not take into account the ability of the peptide to attract certain kinds of lipids (e.g., acidic) into its vicinity. These contributions have been considered in previous studies (e.g., Denisov *et al.*, 1998; Heimburg and Marsh, 1995; Heimburg *et al.*, 1999; May *et al.*, 2000; Murray *et al.*, 2001).

III. MEAN-FIELD STUDIES OF PEPTIDE-MEMBRANE SYSTEMS

We present here a summary of recent studies of peptide–membrane systems of the three categories mentioned in Section I, that is, systems in which the peptide interacts with the hydrocarbon region only, the headgroup region only, or both. We selected simple model studies of several peptides that were carried out using different methodologies. We describe each of the methodologies in terms of the general framework of Fig. 2 and evaluate the relative importance of the free energy terms of Eq. (1) in each case.

A. Category A: Peptides That Interact Predominantly with the Hydrocarbon Region of the Bilayer

For peptides (and protein fragments) that interact predominantly with the hydrocarbon region of the bilayer, the free energy of the peptide–membrane system should be dominated by the solvation free energy term ΔG_{sol} and should hardly be affected by the chemical nature of the polar headgroups of the bilayer. We review three sets of studies: (1) Monte Carlo (MC) simulations, incorporating a hydrophobicity scale, of two bacteriophage coat proteins, *Pf1* and *fd*, each of which contains a short transmembrane segment of about 20 residues, which dominates the energetics of the system (Milik and Skolnick, 1993, 1995); they therefore fall into category A, even though they may appear to interact with the entire lipid bilayer; (2) MD simulations, using a "hydrophobicity potential" to represent the membrane, on the pore-forming antimicrobial peptide alamethicin, which is typical for this category; (3) continuum solvent model studies, on a model α-helical peptide, $(Ala)_{25}$, and on alamethicin.

1. Pf1 and fd Coat Proteins

Pf1 and *fd* are coat proteins which belong to two classes of filamentous bacteriophages (Clark and Gray, 1989; Nakashima *et al.*, 1975). These proteins are relatively short and contain two linked α-helical fragments. One of the segments is hydrophobic and the other is amphipathic. The sequence of *Pf1* is

GVI*DT*[SAV*E*SAIT]*DGQGD*[M**K**AIGGYIVGALVILAVAGLIYSM]L**RKA**,

where strictly hydrophobic (A, I, L, P, and V) and aromatic (F, Y, and W) residues are in bold black, positively charged residues (K and R) are underlined, and negatively charged residues (E and D) are in italic. The brackets indicate sequences that assume helical conformation (Roux and Woolf, 1996). The sequence of *fd* is

A*EGDD*P[A**K**AAF*D*SLQASAT*E*][YIGYAWAMVVVIVGATIGI]

KLF**KK**FTS**K**AS,

where the left and right brackets indicate sequences that may form amphipathic and hydrophobic α helices, respectively (McDonnell *et al.*, 1993). When membrane-associated, the hydrophobic helix of each protein assumes a transmembrane orientation, whereas the amphipathic helix of each becomes surface-oriented (Shon *et al.*, 1991).

Milik and Skolnick used MC simulations to study the membrane association of these proteins (Milik and Skolnick, 1993, 1995). They used a simplified model of the proteins and the lipid bilayer and an approximated form of some of the free energy terms in Eq. (1) and Fig. 2. The proteins were described as chains of spheres with centers at the C_α positions. The ΔG_{sol} contributions of the side chains were derived from Roseman's octanol/water hydrophobicity scale (Roseman, 1988a). Different ΔG_{sol} contributions of the backbone were used depending on the secondary structure, that is, α-helix versus extended conformations. The ΔG_{hg}, ΔG_{lip}, and ΔG_{imm} contributions were estimated from experimental data on the partitioning of very short peptides into dioleylphosphatidylcholine (DOPC) lipid vesicles (Jacobs and White, 1989). The ΔG_{con} component was taken into account by the addition of an empirical energy term that was set to maintain the helical structure of the proteins. These free energy terms were incorporated into MC simulations to search the energetically favored peptide conformations and protein–membrane configurations.

The simulations indicated that the transfer of the C-terminal transmembrane fragments of *Pf1* and *fd* from the aqueous solution into the lipid bilayer proceeds through the following steps. First, the sequence is adsorbed onto the surface of the lipid bilayer in a random-coil conformation. Then, the fragment assumes a distorted-coil conformation. Finally, it is inserted into the bilayer while assuming an approximate α-helical conformation.

Milik and Skolnick's (1993, 1995) simulations were heavily based on water–octanol partitioning data, assuming that the amphiphilic medium provided by octanol is a good approximation of the hydrocarbon region of the lipid bilayer. In fact, octanol mimics the water–bilayer interface well, but it is significantly more polar than the hydrocarbon region of the bilayer (Thorgeirsson *et al.,* 1996; White and Wimley, 1999; Wimley and White, 1996; Yu *et al.,* 1994). The results of the simulations were in accord with the available experimental data, in spite of the crude representation of the membrane (and the peptide). However, this may be attributed in part to the nature of the peptides chosen for the study: Because they are predominantly helical, they fit well with the enhancement of α-helix conformations in the model. In addition, they consist of a highly hydrophobic (transmembrane) segment flanked by regions of polar and charged residues and are therefore bound to assume transmembrane orientations.

A more accurate treatment of protein–membrane interactions would probably involve the combination of two different hydrophobicity scales, one for residue partitioning into the interface and the other for residue partitioning into the hydrocarbon core (Efremov *et al.,* 1999a; White and Wimley, 1999). The incorporation of both scales (which have only recently become available) into Milik and Skolnick's energy function might have provided more realistic simulations.

2. Polyalanine α Helices

As a first stage in determining the energetics of membrane protein stability, Ben-Tal *et al.* (1996a) and Ben-Shaul *et al.* (1996) studied polyalanine α helices to determine factors that are common to the insertion of all peptides and proteins, such as those due to the peptide backbone. They used continuum solvent model calculations and the thermodynamic cycle of Fig. 2 to estimate ΔG_{tot} of Eq. (1). ΔG_{elc} was obtained from finite-difference solutions to the Poisson–Boltzmann equation (the FDPB method) (Honig and Nicholls, 1995; Honig et al., 1993). The (Ala)$_{25}$ α helix was represented in atomic detail, with atomic radii and partial charges defined at the coordinates of each nucleus. The charges and radii were taken from PARSE, a parameter set that was derived to reproduce vacuum-to-water (Sitkoff et al., 1994) and liquid alkane-to-water (Sitkoff et al., 1996) solvation free energies of small organic molecules. The lipid bilayer was described implicitly, as a low-dielectric slab. ΔG_{np} was calculated from the change in the water-accessible surface area of (Ala)$_{25}$ upon membrane association, using Eq. (7). The peptide free energy contributions due to peptide immobilization and lipid perturbation were calculated from a statistical thermodynamic model of the lipid tails (Ben-Shaul et al., 1996; Fattal and Ben-Shaul, 1993).

Two configurations of a membrane-associated 25-mer polyalanine helix were found to be lower in free energy than the isolated helix in the aqueous phase. The first corresponded to a transmembrane orientation, in which the helix termini protruded from either side of the bilayer. The second minimum was for a surface orientation, in which the helix was adsorbed onto the surface of the bilayer with its principal axis parallel to the membrane plane. This orientation is similar to that found in the crystal structures of the amphipathic α helices in the membrane-exposed faces of the peripheral membrane proteins Cox1 and Cox2 (Kurumbail et al., 1996; Picot et al., 1994) and in the interfacial helix in the structure of rhodopsin (Palczewski et al., 2000).

Previous attempts to estimate the free energy of association of polyalanine α helices with bilayers (e.g., Engelman et al., 1986) produced values of ~ -30 kcal/ mol, which are much more negative than the values of ~ -5 kcal/mol obtained from measurements on related systems (e.g., Moll and Thompson, 1994; White and Wimley, 1999, and references therein). The main reason for the discrepancy between the calculated and measured values was the neglect of the electrostatic penalty associated with the transfer of the helix backbone from the aqueous phase into the bilayer. This penalty was taken into account by Ben-Tal et al. (1996a) and the calculated values of the free energy in both minima, -4 kcal/mol for the transmembrane orientation (Ben-Tal et al., 1996a) and -3 kcal/mol for the horizontal orientation (N. Ben-Tal, unpublished results), corresponded well with the measured values.

3. The Contribution of the Backbone Hydrogen Bond

As mentioned in Section II.D, many attempts have been made to develop polarity (hydrophobicity) scales that indicate the relative tendencies of the amino acids to be inserted into lipid bilayers. The scales are based on the implicit assumption that each side chain has a fixed free energy contribution regardless of its neighbors and that, by adding the contribution of the backbone, it is possible to estimate the total free energy of transfer of the amino acid. However, this assumption is misleading, because transfer free energies are context-dependent. For example, the free energies of transfer of the hydrogen-bonded (N—H···O=C) and nonbonded (N—H, O=C) groups between two solvents depends quite sensitively on their environment. The water-to-liquid alkane free energies of transfer of these groups were calculated as 3.8 and 7.8 kcal/mol, respectively, when they were part of either an N-methylacetamide (NMA) dimer or monomer (Ben-Tal *et al.,* 1997b). Much smaller values, 2.1 and 6.4 kcal/mol, respectively, were calculated when these groups were part of a polyalanine α helix (Ben-Tal *et al.,* 1996a). The reason for the difference is that NMA and the helix form differently shaped cavities in the solvent. Specifically, the amide groups in an α helix are surrounded by a low-dielectric region even in the aqueous phase, such that their free energies of transfer into a liquid alkane are smaller than those of NMA. The same logic dictates that the transfer free energies of amino acid chains that are part of an α helix would be different from those of the isolated amino acid. The situation is compounded by the possibility of side chain–side chain and side chain–backbone hydrogen-bonding interactions. It may be possible, however, to take these effects into account in devising context-dependent hydrophobicity scales, such as the one presented in Section III.A.4 below.

4. Computationally Derived Hydrophobicity Scale

Using the same methodology as presented in Sections III.A.2 and III.A.3, Ben-Tal (1997, and unpublished data) developed a hydrophobicity scale based on the free energies of transfer of the amino acids from aqueous phase into lipid bilayers. Unlike other hydrophobicity scales (e.g., Kyte and Doolittle, 1982), which assume that the free energies of transfer are proportional to some inherent property (of an individual atom, a chemical group, or an amino acid), this scale was derived using the theoretical model presented above (Section III.A.2), and it accounts for the amino acids being located at the center of an α helix. The scale is given in Table II and is compared to three other scales in Fig. 4. The scales are generally similar, all having the hydrophobic and charged amino acids at the two extremes. Ben-Tal's scale (filled circles in Fig. 4) is the least hydrophobic of the four. This is mainly because, unlike other scales, it includes the free energy penalty of inserting the helix backbone into the bilayer.

The most significant difference between Ben-Tal's scale and the others is the energy penalty of inserting an arginine (R) residue into the bilayer; it is much larger in Ben-Tal's scale than in the others. It should also be noted that the energy penalty of inserting arginine into the bilayer significantly exceeds that of lysine (K) in Ben-Tal's scale, both residues being taken in their deprotonated (neutral) form to avoid the penalty of inserting a charge into the hydrocarbon region of the

TABLE II

Hydrophobicity Scale Representing Free Energies of
Transfer of Each of the 20 Amino Acids from Water
into the Center of the Hydrocarbon Region of a Model
Lipid Bilayer[a]

Amino acid	ΔG_{tot} (kcal/mol)
I	−2.6
L	−2.6
F	−1.5
V	−1.2
A	−0.2
G	0.0
C	+0.4
S	+0.8
T	+1.1
M	+1.3
W	+1.3
P	+2.8
Y	+4.3
Q	+5.4
H	+6.8
K	+7.4
N	+7.7
E	+9.5
D	+11.5
R	+19.8

[a]The scale was computationally derived, as described in
Section III.A.4. The amino acid residues are presented using a
single-letter code.

bilayer. The source of the difference between the values of arginine and lysine
in Ben-Tal's scale is the different charge distributions between these two residues,
guanidinium versus amine. However, it may well be that the difference was
overexpressed in the PARSE set of partial charges used in the calculations (Sitkoff
et al., 1996). Further theoretical and experimental work is needed to clarify this and
other issues regarding the relative hydrophobicity of the amino acids in the scale.
Nevertheless, preliminary tests have shown that the scale is significantly more
potent in detecting transmembrane helices in the sequence of membrane than
other hydrophobicity scales (H. Ben-Ami, T. Seifer, B. Honig and N. Ben-Tal,
manuscript in preparation).

5. Alamethicin

Alamethicin is a 20-amino-acid antibiotic peptide produced by the fungus
Trichoderma viride, which self-assembles to form ion channels in lipid bilayers.

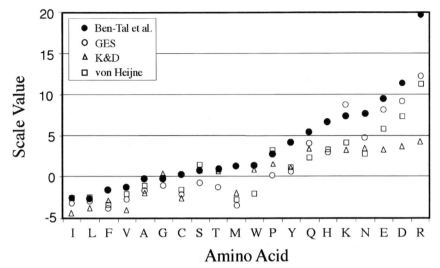

FIGURE 4   Comparison of four different hydrophobicity scales of the amino acid residues. Values for the scales, on the vertical axis, are as follows: free energies in kcal/mol for Ben-Tal *et al.*'s (unpublished data) scale (filled circles), for the Goldman–Engelman–Steitz (GES; Engelman *et al.*, 1986) scale (open circles), and for the von Heijne (1981a,b) scale (open squares), and scale index for the Kyte–Doolittle (Kyte and Doolittle, 1982) scale (open triangles).

The sequence of alamethicin is Ac-**UPUAUAQUVUGLUPVUUQQF**-OH, where Ac is acetyl, U is α-amino isobutyric acid, and F-OH is phenylalaninol, and, again, the hydrophobic and aromatic residues are in bold. The peptide is predominantly α-helical in solution (Banerjee and Chan, 1983; Esposito *et al.*, 1987; Fox and Richards, 1982; Yee and O'Neil, 1992), and the helical structure of its N-terminal segment is maintained in dimyristoylphosphatidylcholine (DMPC) (North *et al.*, 1995). The small size of alamethicin makes it an attractive model for the investigation of peptide–membrane interactions, and indeed, it has been studied intensively both theoretically and experimentally (see reviews by Cafiso, 1994, Sansom, 1998).

Sansom and his co-workers (Biggin *et al.*, 1997) studied alamethicin using a simplified representation of the membrane as a "hydrophobic potential," adapted from Milik and Skolnick's (1993, 1995) Monte Carlo simulations of peptide–membrane systems, described above (Section III.A.1). In this study, alamethicin structures were generated by simulated annealing, and a restrained molecular dynamics simulation, in the presence of a transbilayer voltage, was carried out. The interactions between alamethicin and the lipid bilayer were represented by a hydrophobicity index assigned to the side chains of the amino acid residues, and a constant value of 4.1 kcal/mol was assigned to each potential hydrogen-bonding group exposed to the bilayer. The authors stated that the main limitation of this

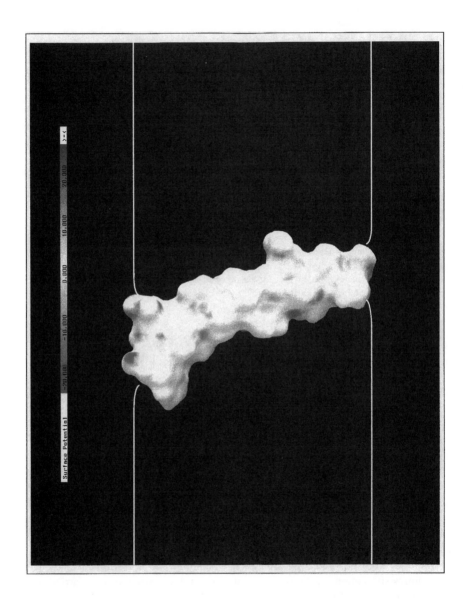

study was the coarse-grained treatment of the interactions between alamethicin and the lipid bilayer, namely, the use of a residue-based hydrophobicity index, which obscures the atomic detail of these interactions.

Free energy calculations have very recently been carried out within the framework of the mean-field approach of Eq. (1) and Fig. 2 to study alamethicin–membrane interactions in the limit of low peptide concentration (Kessel *et al.*, 2000a,b). The main results are described below.

a. ΔG_{tot} and the Most Likely Alamethicin–Bilayer Configuration. For the alamethicin–membrane system, the total free energy of Eq. (1) can be approximated as $\Delta G_{\text{tot}} \approx \Delta G_{\text{sol}} + \Delta G_{\text{imm}} + \Delta G_{\text{lip}} + \Delta G_{\text{def}}$ (Kessel *et al.*, 2000a). (Note that $\Delta G_{pKa} = 0$, because the alamethicin isoform used here does not contain titratable residues). Recent studies suggest that alamethicin interacts mainly with the hydrocarbon region of the bilayer and that its conformation does not change significantly upon association with the membrane (Barranger-Mathys and Cafiso, 1996; Jayasinghe *et al.*, 1998; Kessel *et al.*, 2000a,b; J. R. Lewis and Cafiso, 1999). Thus, the ΔG_{con}, ΔG_{hg}, and ΔG_{sp} terms are approximately zero (discussed in Kessel *et al.*, 2000a). The ΔG_{sol} term of Eq. (4) was calculated with the same method used for the polyalanine α-helix calculations described in Section III.A.2. The contributions of the ΔG_{imm}, ΔG_{lip}, and ΔG_{def} terms were estimated based on a statistical thermodynamic model of the lipid bilayer (Ben-Shaul *et al.*, 1996; Fattal and Ben-Shaul, 1993). The values were given in Sections II.C, II.E, and II.F.

Again, two configurations of a membrane-associated alamethicin were found to be lower in free energy than isolated alamethicin in the aqueous phase: transmembrane and surface configurations. The most stable of these, associated with the most negative ΔG_{tot} value of Kessel *et al.*'s (2000a) calculations, was the transmembrane orientation (Fig. 5), in agreement with experimental measurements (Barranger-Mathys and Cafiso, 1996; Huang and Wu, 1991; North *et al.*,

---

FIGURE 5 Schematic representation of the orientation of alamethicin in the 2-Å deformed bilayer. Alamethicin is depicted with the electrostatic potential ϕ color-coded on its molecular surface. This was calculated using DelPhi (Nicholls and Honig, 1991) and is displayed on the molecular surface using GRASP (Nicholls *et al.*, 1991). Negative potentials ($0 > \phi > -20$ kT/e) are red, positive potentials ($0 < \phi < 20$ kT/e) are blue, and neutral potentials are white. Calculations have been carried out for the conformation found in the crystal structure (Protein Data Bank, accession number 1AMT). The peptide is shown with its N-terminus pointing up and its more polar C-terminus pointing down. The two white lines represent the boundaries of the hydrocarbon region of the lipid bilayer. The polar termini of alamethicin are outside the hydrocarbon region of the deformed lipid bilayer, and the hydrophobic core of the peptide is immersed inside this region. (See color plate.)

1995). The ΔG_{tot} value associated with this configuration was -5.5 kcal/mol, again in agreement with the value derived from nuclear magnetic resonance (NMR) measurements (J. R. Lewis and Cafiso, 1999).

The hydrophobic region of alamethicin is about 3 Å shorter than the native width of the hydrocarbon region of the lipid bilayer (Fig. 5). One of the interesting results of Kessel et al.'s (2000a) calculations is that the transmembrane insertion of alamethicin into a lipid bilayer of native hydrocarbon region of 30 Å is likely to involve an approximately 2-Å deformation of the bilayer, to match the hydrophobic length of the peptide (Fig. 5). This is in agreement with experimental data (He et al., 1996; J. R. Lewis and Cafiso, 1999; Wu et al., 1995). A 2-Å deformation of the lipid bilayer may be insignificant, considering the dynamic nature of the bilayer. However, recent calculations with other hydrophobic peptides, such as the M2δ segment of the acetylcholine receptor (A. Kessel and N. Ben-Tal, unpublished data) and gramicidin (S. Bransburg-Zabary, A. Kessel, M. Gutman and N. Ben-Tal, unpublished data), indicate that the transmembrane insertion of such peptides may lead to membrane deformation of as much as 10 Å. These preliminary results suggest that peptide-induced membrane deformation is a general characteristic of such systems.

The hydrophobic mismatch between alamethicin and the lipid bilayer, which is responsible for the membrane deformation upon alamethicin insertion, may explain the assembly of the intact channel. The mismatch results mainly from the limitation of the length of the peptide's central hydrophobic region by the C-terminal polar Gln18 side chain. The conservation of this polar residue throughout evolution suggests an evolutionary advantage for this mismatch. The formation of the ion channel results from aggregation of the alamethicin monomers, which reduces the extent of peptide–bilayer interactions. Therefore the deformation of the membrane should also be decreased. Thus, a peptide such as alamethicin is likely to oligomerize and form ion channels to reduce its unfavorable interactions with the lipid bilayer. This hypothesis is supported by experimental studies of alamethicin (Keller et al., 1993; J. R. Lewis and Cafiso, 1999). The general role of the hydrophobic mismatch in protein oligomerization is also demonstrated by a study of bacteriorhodopsin (B. A. Lewis and Engelman, 1983a).

b. Flip-Flop across the Lipid Bilayer.   Alamethicin can reverse its orientation in the bilayer, that is, undergo a "flip-flop" motion. The flip-flop of alamethicin in POPC vesicles has been studied by NMR spectroscopy (Jayasinghe et al., 1998), and the results suggest that it involves an energy barrier, presumably due to the free energy of transfer of the peptide termini across the bilayer. Recent mean-field calculations, following the framework of Eq. (1) and Fig. 2, have suggested a path for the flip-flop motion (Kessel et al., 2000b).

The obvious path, which involves the rotation of the peptide around its center of mass, is characterized by a very high free energy barrier ($\sim$30 kcal/mol), resulting from burying the two termini in the bilayer simultaneously (Kessel *et al.*, 2000a). A more plausible option is a sequential rotation of the peptide around one terminus at a time (Fig. 6); This flip-flop path involves two free energy barriers, each associated with inserting one of the peptide termini from the aqueous phase into the bilayer. The flip-flop rate was estimated by calculating the ΔG_{tot} values associated with the configurations above and below each of these two free energy barriers (Kessel *et al.*, 2000b). The ΔG_{tot} values and the barrier heights are given in Fig. 6. The flip-flop rate derived from these calculations, $\sim$$10^{-7}$/s, is close to the measured value of 1.7×10^{-6}/s (Jayasinghe *et al.*, 1998). Selective charge neutralization of polar groups in alamethicin confirms that the main free energy barrier of the flip-flop process results from the water-to-lipid bilayer transfer of the C-terminus.

6. Conclusions

The studies presented in Sections III.A.1–III.A.5 used three different mean-field approaches to treat membrane systems in which the energetics is dominated by the interactions of a transmembrane hydrophobic span with the hydrocarbon region of the lipid bilayer.

The common advantage of these approaches is that their simplicity makes it possible to use them to describe both thermodynamic and long-term kinetic aspects of peptide–membrane systems. The common disadvantage of these approaches is the lack of atomic description of the lipid bilayer. This is especially important in the description of the interactions between the peptide and the chemically diverse polar headgroup region of the bilayer (Fig. 1). Although these interactions are insignificant in the systems presented above, this may not always be the case (e.g., see peptides of category C, Section III.C below).

Although the three approaches have many qualities in common, they do differ in some respects. The continuum solvent model of Section III.A.2–III.A.4 is based on atomic description of the peptide and a slab representation of the lipid bilayer. Its limitations have been discussed at length (Ben-Tal *et al.*, 1996a, 2000a). The major shortcomings are the model's complete neglect of the polar headgroup region and the fact that, at least in its current implementation, the model requires a known high-resolution structure of the peptide. The model of Milik and Skolnick (1993, 1995) is based on an approximate description of both the protein and the bilayer, and the model used by Sansom and his co-workers (Biggin *et al.*, 1997) relies on a residue-based hydrophobicity index, which obscures much of the atomic detail of the interactions between the peptide and the bilayer. However, although these two models do not provide an accurate estimate of ΔG_{sol}, both provide a means with which to consider conformational changes of

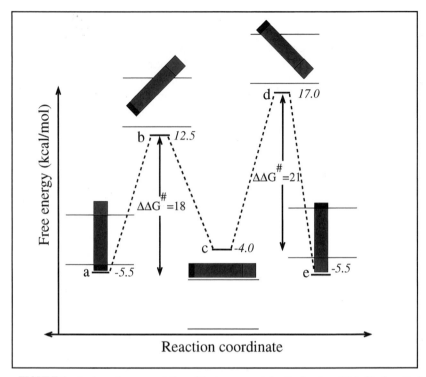

FIGURE 6 A schematic representation of the two suggested paths for alamethicin flip-flop. Each of the two paths involves the following configurations, each of which is obtained from the previous one by rotation: (a) The initial (or final) transmembrane orientation of the peptide in the bilayer, with the C-terminus (medium gray) facing upward and the N-terminus (black) downward. (b) Insertion of the N-terminus of the peptide into the lipid bilayer, while the C-terminus remains at the water–bilayer interface. The long axis of the peptide is tilted approximately 45° with respect to the bilayer normal. The path from configuration a to b involves rotation around the C-terminus residue. (c) Peptide adsorption on the surface of the lipid bilayer. The path from configuration b to c involves further rotation around the C-terminus residue. (d) Insertion of the C-terminus of the peptide into the lipid bilayer, while the N-terminus remains at the water–bilayer interface. This orientation is the reciprocal of the orientation described in b. The path from configuration c to d involves rotation around the N-terminus residue. (e) The final (or initial) transmembrane orientation of the peptide in the bilayer, with the C-terminus facing downward and the N-terminus upward. This transmembrane orientation is the reciprocal of orientation a. The path from configuration d to e involves further rotation around the N-terminus residue. The two paths differ from one another in the order of occurrence of the configurations a → e versus e → a. Alamethicin is schematically depicted as a rectangle. The central hydrophobic region of the peptide is dark gray, the C-terminus of the peptide is medium gray, and the N-terminus is black. The borders of the hydrophobic core of the lipid bilayer are marked by the two horizontal lines. The calculated free energy values of the alamethicin–membrane system in each configuration are given in italic. The values of the free energy barriers of the most probable path, from the left-hand side to the right-hand side, are given in roman.

the peptide upon membrane insertion. The approach of Sansom and co-workers also considers transbilayer voltage effects, which are important in many biological systems.

B. Category B: Peptides That Interact Predominantly with the Polar Headgroup Region of the Bilayer

For peptides (and protein fragments) that interact mainly with the polar headgroups of bilayer lipids, the free energy of the peptide–membrane system should be dominated, in essence, by the ΔG_{hg} and ΔG_{sp} contributions and should hardly be affected by the hydrocarbon region of the bilayer.

Short basic peptides, such as pentalysine, which interact electrostatically with acidic lipids in the membrane, are typical examples for this category. Because the exact chemical nature of the lipid does not matter, provided that it is negatively charged (as opposed to zwitterionic) and that it can electrostatically attract the positively charged peptides, the ΔG_{sp} contributions should be negligible. For example, pentalysine associates with a negatively charged membrane made up of phosphatidylserine (PS) or phosphatidylglycerol (PG) with essentially the same ΔG_{tot} (Kim *et al.,* 1991; Mosior and McLaughlin, 1992). A brief summary of some studies on the interaction of such peptides with the bilayer follows. A more detailed description can be found in Chapter 3.

1. Short Basic Peptides

Many important peripheral proteins contain clusters of basic residues that interact electrostatically with acidic lipids in the membrane. Tri-, penta-, and heptalysine have been studied to determine factors that are common to the membrane association of all of these peripheral proteins (Ben-Tal *et al.,* 1996b; Murray *et al.,* 1999). The electrostatic free energy of interaction of the basic peptides with phospholipid bilayers, that is, the electrostatic contribution to ΔG_{hg}, was calculated by applying the (nonlinear) Poisson–Boltzmann equation [Eq. (5)] to atomic models of the peptides and bilayers in aqueous solution. The electrostatic free energy of interaction, which arises from both the long-range Coulombic attraction between the positively charged peptide and the negatively charged lipid bilayer and the short-range Born repulsion, had a minimum at a distance of $\sim$3 Å (i.e., one layer of water) between the van der Waals surfaces of the peptide and the lipid bilayer. The molar association constant K, calculated as the Boltzmann average over many peptide–membrane configurations, was in accord with the measured values: K was typically about 10-fold smaller than the experimental value (i.e., a difference of about 1.5 kcal/mol in the free energy of association). The predicted dependence of K (or the association free energies) on the ionic strength of the solution, on the mole percent of acidic lipids in the membrane, and on the number of basic residues in the peptide agreed very well with experimental measurements.

Furthermore, the same model has been applied to calculate the electrostatic free energy of membrane association of the positively charged toxin charybdotoxin (CTX) and analogues of known NMR structure (Ben-Tal et al., 1997a) and of a basic peptide corresponding to residues 2–19 of Src (Ben-Tal et al., 1997a; Murray et al., 1998). Again, the calculated and measured binding free energies agreed and the calculations provided a molecular interpretation of the experimental data.

The limitations of the model have been discussed in detail elsewhere (Ben-Tal et al., 1996b, 1997a). The main weakness of the model is apparently the static representation used for the polar headgroups. As a result, any peptide-induced structural modifications, including small adjustments in the location of the polar headgroups that are in direct contact with the peptide, are not taken into account.

2. Neuromodulin

Neuromodulin, a calmodulin substrate, is thought to play an important role in the calcium/phospholipid second-messenger signaling system (Liu and Storm, 1990). It is believed to be attached to the cell membrane by the two palmitoyl groups that are covalently attached to cysteine residues near its N-terminus (Houbre et al., 1991; Skene and Virag, 1989). The calmodulin-binding domain of human neuromodulin (residues 39–55) contains a cluster of basic residues; when neuromodulin is free, rather than bound, these residues may interact with acidic lipids in the bilayer.

The peptide KIQASFRGHITRKKLKG, which corresponds to residues 37–53 of bovine neuromodulin (Skene, 1989), contains two arginine, four lysine, and no negatively charged residues. McLaughlin and his co-workers (Kim et al., 1994) have shown that the membrane affinity of this peptide increases as a function of the mole percent of acidic lipid. Upon serine phosphorylation by protein kinase C, the net charge of the peptide decreases from +6 to +2, and the measurements show that the membrane affinity of the phosphorylated peptide is significantly lower than that of the nonphosphorylated peptide.

Ben-Tal (1997, and unpublished data) used the continuum solvent model as described in the studies of short basic peptides (Section III.B.1) to calculate the binding of the neuromodulin peptide to lipid bilayers. A model of the peptide (in extended conformation) was constructed using InsightII (MSI, San Diego, CA), and its interaction free energy with atomic models of lipid bilayers containing different mole percent of acidic lipids at 100 mM salt concentration was calculated for different peptide–membrane configurations. Again, the most negative value of the interaction free energy was obtained when the van der Waals surfaces of the

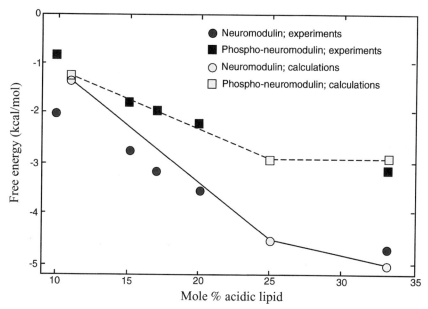

FIGURE 7 Neuromodulin and phospho-neuromodulin association with lipid bilayers: the experimentally determined standard Gibbs free energies of association of neuromodulin (39–55) (*circles*) and phospho-neuromodulin (39–55) (*squares*) to PC:PG unilamellar vesicles in 100 mM KCl (Kim *et al.,* 1994). The open circles and open squares represent the theoretical estimates for the association free energies of these peptides, respectively.

peptide and the bilayer were about 3 Å from each other. The association free energy ΔG_{tot} was approximated, and the value of the free energy at this minimum and the results are presented in Fig. 7 along with the measured values of McLaughlin and his co-workers (Kim *et al.,* 1994).

Overall, the calculated values of the association free energy were in good agreement with measured values. The calculations reproduced the dependence of the association free energy on the mole percent of acidic lipids in the membrane both for the phosphorylated and the nonphosphorylated peptides. There were deviations between the calculated and measured values at low concentration of acidic lipid (12 mol%); the calculated value was approximately 0.7 kcal/mol higher (i.e., less negative) than the measured value. This is presumably because of the oversimplified model used (discussed in Ben-Tal *et al.,* 1996b). The model does not take into account lipid motions and it may well be that at low acidic lipid concentrations, peptide adsorption onto the bilayer induces an accumulation of acidic lipids at the expense of zwitterionic lipids near the peptide (e.g., May *et al.,* 2000).

The model involves several arbitrary assumptions. First, the structure of the peptide is unknown. Extended conformation was used as an approximation of a random coil. Preliminary results showed that the membrane affinity of the peptide is significantly lower in an α-helical conformation. This is because the α-helix conformation prevents at least some of the basic residues from being in close proximity to the membrane plane. The main advantage of the extended form over other possible conformations is that it enables all the basic residues to be in the vicinity of the membrane plane, and thus it is likely to provide the maximal ΔG_{hg} contribution to the association free energy.

Second, the model is based on the assumption that the peptide does not penetrate the membrane. There is experimental evidence that this is the case for the tri-, penta-, and heptalysine peptides (Ben-Tal *et al.,* 1996b), and the very hydrophilic nature of the neuromodulin (39–55) peptide suggests that it behaves similarly. However, there are no experimental data in support of this assumption. In view of these shortcomings in the calculations, the agreement between the calculations and experiments with respect to the membrane affinity of the neuromodulin (39–55) peptide may have been fortuitous.

3. Conclusions

Several generalizations can be made about the association of basic peptides with bilayers containing acidic lipids. First, continuum solvent model calculations suggest that the long-range Coulomb attraction between the oppositely charged peptide and membrane is overly compensated for by the short-range image charge and Born repulsion. The balance between these two effects leads to a minimum in the electrostatic free energy when the van der Waals surfaces of the peptide and the membrane are about 3 Å (i.e., one water layer) from each other. Experimental data suggest that basic peptides are adsorbed onto membrane surfaces in very close proximity to the headgroups and yet they do not penetrate the bilayer (e.g., Ben-Tal *et al.,* 1996b, 1997a). However, it is unclear whether there is a layer of water between the peptide and the headgroup region.

Second, continuum solvent models based on atomic models of the peptide and the lipid bilayer are often successful in reproducing experimental partitioning coefficients of peptide–membrane association in cases where the peptides do not penetrate the membrane. A theoretical treatment of peptides that penetrate the membrane is likely to require replacing the static model of the lipid bilayer with one that incorporates lipid motions.

Third, the calculations and measurements suggest a rule of thumb to approximately account for the contributions of basic residues to the association of peptides and proteins to membranes. At physiological conditions, that is, about 25–33 mol% acidic lipid at a monovalent salt concentration of about 150 mM, each lysine or arginine residue in the vicinity of the acidic lipids contributes about 1 kcal/mol to the free energy of adsorption onto the cytoplasmic side of cellular membranes.

C. Category C: Peptides That Interact with Both the Hydrocarbon and Polar Headgroup Regions of the Bilayer

This category includes peptides and protein fragments that are in physical contact with the hydrocarbon and the polar headgroup regions simultaneously and whose interaction free energies with these two regions are of comparable magnitude. We will discuss two examples here: the N-terminal segment of the peripheral membrane protein Src, and the pore-forming domain of the insecticidal toxin δ-endotoxin produced by *Bacillus thuringiensis*.

1. The Src Peptide

The v-Src oncoprotein (pp60$^{\text{v-src}}$) and its normal cellular homologue, c-Src (pp60$^{\text{c-src}}$), are membrane-bound, nonreceptor tyrosine kinases (Resh, 1993). Their N-termini contain two motifs essential for membrane localization: a stretch of polar amino acids rich in positively charged residues and a myristate attached cotranslationally to the N-terminal glycine by *N*-myristoyltransferase (Murray *et al.*, 1997). Association of Src with the plasma membrane is inhibited by mutations that produce nontransforming phenotypes that either reduce the net positive charge in the N-terminal cluster of basic residues (Kaplan *et al.*, 1990; Sigal *et al.*, 1994) or prevent myristoylation (Buss *et al.*, 1986; Cross *et al.*, 1984; Kamps *et al.*, 1985).

Experimental data suggest that the cluster of basic residues at the N-terminus of Src interacts electrostatically with acidic lipids in the membrane, but does not penetrate the membrane (Murray *et al.*, 1997, and references therein), whereas the myristate partitions into the hydrocarbon region of the lipid bilayer (Peitzsch and McLaughlin, 1993). These key observations were later used by Murray *et al.* (1998) to experimentally and theoretically investigate the binding of myristoylated and nonmyristoylated peptides corresponding to residues 2–19 of avian c-Src, GSS**K**S**KPK**D**PSQRRRSL**E. They used the continuum solvent model presented in Section III.B and atomic models of the peptide and the bilayer to calculate the free energy of interaction of the peptides with lipid bilayers for different peptide–membrane configurations. They then used these free energy values to deduce the molar partition function of the peptides between the bilayer and the aqueous phase, carrying out statistical thermodynamic averaging to calculate the partition function of the free and membrane-adsorbed peptides.

Again, the results were in good agreement with the measured peptide partitioning onto phospholipid vesicles. Membrane association of the nonmyristoylated Src peptide increased as the mole percent of acidic lipid in the vesicles was increased, the ionic strength of the solution was decreased, or the net positive charge of the peptide was increased (due to phosphorylation). The theoretical model also correctly predicted the measured partitioning of the myristoylated peptide, myr-Src(2–19); for example, adding 33% acidic lipid to electrically neutral vesicles increased the partitioning of myr-Src(2–19) by a factor of 100. Phosphorylating either serine 12 or serine 17 decreased the partitioning of myr-Src(2–19) onto vesicles containing acidic lipid by a factor of 10.

The success of the theoretical model in this study was based on two facts. First, the coupling between the contributions of the myristate and the cluster of basic residues to membrane association is weak (Buser *et al.*, 1994; Murray *et al.*, 1997). Thus, the computational effort was dedicated to accurately calculating the electrostatic free energy of interaction of the cluster of positively charged residues with the negatively charged membrane. The contribution of the myristate was added at a later stage and the assumption was that it is, in essence, independent of changes in the membrane affinity of the cluster of positively charged residues, such as those due to peptide phosphorylation. Second, experimental data indicated that Src(2–19) is a random coil both in the aqueous phase and when membrane-associated, and that it does not penetrate the bilayer. Thus, approximate atomic models of the peptide (in extended form) and of a bilayer patch were built and the calculations involved a search in configurational space, considering the peptide and the bilayer as rigid bodies.

In summary, experimental data suggest that the theoretical treatment of the Src peptides requires explicitly taking into account only the peptide interactions with the polar headgroups. The experimentally derived contributions due to interactions of the myristate with the hydrocarbon region can be added separately.

2. The *Bacillus thuringiensis* δ-Endotoxin

δ-Endotoxins are highly potent insecticidal toxins produced by *Bacillus thuringiensis*. Research over the past three decades has attempted to elucidate the structure of the δ-endotoxins and its implications for their activity. The X-ray crystal structure of the *Bacillus thuringiensis* δ-endotoxin in its soluble form showed that the toxin comprises three domains (Gouaux, 1997; Li *et al.*, 1991). We will focus here on one of them: the pore-forming domain, which is an α-helix bundle consisting of seven helices.

The structure of the *Bacillus thuringiensis* δ-endotoxin in lipid bilayers was, until recently, unknown. In a combined experimental and theoretical effort, Gazit *et al.* (1998) elucidated the structure and organization of the pore-forming domain of δ-endotoxin in lipid bilayers. They used fluorescent and FTIR measurements to determine the affinity and orientation of seven synthetic peptides in small unilamellar vesicles, corresponding to each of the seven helices of the pore-forming domain. In addition, they used resonance energy transfer measurements of all possible combinatorial pairs of membrane-bound helices to map the network of interactions between helices in their membrane-associated state. The results are consistent with a situation in which helix 1 does not associate with the lipid bilayer, helices 4 and 5 insert into the membrane antiparallel to each other as a helical hairpin structure, and helices 2, 3, 6, and 7 lie on the membrane surface, radiating outward from their point of entry into the bilayer like the spokes of a wheel.

Gazit *et al.*'s study also reported a series of Monte Carlo simulations to probe low-energy configurations of peptides corresponding to helices 2–7 in the lipid bilayer. The energy function used in the simulations included the X-PLOR force field (Brunger, 1992) to account for ΔG_{con} contributions to ΔG_{tot}. The ΔG_{sol} and ΔG_{hg} contributions to ΔG_{tot} were taken into account using the "hydrophobic potential" approach (Biggin and Sansom, 1996; La Rocca and Sansom, 1998) described in Section III.A.5. The results of the simulations showed that each of the six peptides tends to adopt a helical conformation and to associate with the membrane. The calculated value of ΔG_{tot} was usually larger in magnitude than the measured value, in some cases by a factor of two, but the correlation between the calculated and measured values was good. However, the situation was not as good regarding the most likely orientations of the peptides in the bilayer. For example, the calculations suggested that all the helices, including helices 4 and 5, which have been experimentally shown to be transmembrane (Gazit *et al.*, 1998), are surface-bound in oblique orientations. The authors suggested that the reason for the discrepancy between the theoretical results and the measurements may be that membrane insertion of these two helices involves the formation of homodimers.

In summary, Gazit *et al.*'s study provides yet another example of the fact that current methodology usually cannot fully describe peptide–membrane systems, unless specific assumptions about the system at hand are included.

3. Conclusions

We have reviewed two different methods that have been used to elucidate thermodynamic aspects of peptide–membrane interactions for peptides of category C, that is, peptides that interact both with the hydrocarbon and the headgroup regions of the lipid bilayer. The first method is based on continuum solvent model; the peptide is represented in atomic detail and the membrane is represented as an all-atom model in one of them (Murray *et al.*, 1998). The other method (Gazit *et al.*, 1998) is based on MD simulations, in which the peptide is described in atomic detail and the bilayer is described as a structureless medium.

The two methods were successful in the reproduction of some, but not all, of the experimental data on the systems at hand. In cases where the structure of the peptide and its orientation in the bilayer are accurately predicted, the value of ΔG_{tot} is unreasonably high in magnitude, and in cases where the value of ΔG_{tot} is reasonable, the orientation of the peptide in the membrane is in conflict with experimental values. The most successful study of the above was on the interactions of the Src(2–19) peptide with lipid bilayers. However, the success in this case may be attributed to the fact that ample experimental data on the structure of the peptide and the nature of its interactions with the bilayer were implemented in the theoretical model. Overall, our theoretical understanding of the thermodynamics of association of peptides of category C with lipid bilayers is far from being satisfactory and requires that experimental data be incorporated into the theoretical model in one way or another.

IV. OPEN QUESTIONS

We have described a theoretical framework to clarify the energetics of peptide–membrane association (Fig. 2) and demonstrated its utility in mean-field-based computational studies of peptides that interact with the hydrocarbon region of the membrane, the polar headgroup region, or both. In many cases, the theoretical treatment is successful in reproducing experimental data (e.g., regarding the most likely conformation and orientation of peptides in the membrane), thus providing molecular interpretation of the experimental data. Nevertheless, the failure of the mean-field model to deal with certain peptide–membrane systems implies that our theoretical understanding of the thermodynamics of such systems is incomplete. Indeed, many peptide–membrane systems that have been studied experimentally are too complicated for a theoretical treatment, and for many cases in which theoretical models were applied, the experimental data could not be fully explained. For example, in most of the approaches described here it is assumed that the structure of the membrane does not change upon peptide association and in some of the approaches the three-dimensional structure of the peptide is taken as a given, based on prior experimental knowledge.

A successful theoretical treatment of the energetics of peptide–membrane systems requires that several fundamental questions about peptide–membrane interactions be answered. For example, the free energy of transfer of the peptide backbone hydrogen bond from the aqueous phase into the hydrocarbon region of the membrane is not yet known; estimates range from 0.6 (Roseman, 1988b) to 2.2 kcal/mol (Ben-Tal *et al.,* 1996a). As discussed by White and Wimley (1999), this value, which is of central importance to our understanding of peptide–membrane interactions, has major implications for folding and stability in membrane proteins. The free energy of transfer of the peptide backbone hydrogen bond from the aqueous phase into the polar headgroup region of the bilayer is also unknown (White and Wimley, 1999).

The contribution of the polar headgroup region to the energetics of peptide–membrane systems requires further clarification. For example, it is unclear why the aromatic residues Phe and Trp have high affinity to the water bilayer interface (e.g., Killian and von Heijne, 2000; Pilpel et al., 1999; White and Wimley, 1999; Yau et al., 1998). Most of the hydrophobicity scales indicate that these residues are highly hydrophobic, and yet they are much less commonly found at the hydrocarbon region of the bilayer than the aliphatic residues (Ala, Leu, Ile, and Val). The high affinity of the aromatic residues to the polar headgroup region has been variously attributed to amphipathic interactions of the imino group's hydrogen bonding, specific dipolar interactions, enhanced acyl-chain van der Waals interactions, and cation–π interactions (Yau et al., 1998). Another possible explanation is that the rigidity of the aromatic residues leads to yet another entropy penalty due to their confinement effect on the lipid chains. This qualitative explanation needs to be numerically tested. Overall, estimates of ΔG_{lip} and ΔG_{def} have been derived from models in which simplified geometry was used to represent the peptide (e.g., Fattal and Ben-Shaul, 1993). The assumption is that peptides can be approximated as cylinders and that their effective diameter is approximately constant. Continuum solvent model calculations have shown that the exact geometry of the peptide significantly affects the electrostatic and nonpolar contributions to the solvation free energy (Honig and Nicholls, 1995). It may well be that ΔG_{lip} and ΔG_{def} are similarly affected. This issue awaits further studies using detailed statistical thermodynamic models at the atomic level.

The effect of the polar headgroups on the pK_{a} of the titratable amino acids is well documented experimentally (e.g., Section II.A), but awaits a theoretical explanation. For the most part this effect can probably be explained based on desolvation, but the extent of desolvation and the contributions of the different chemical groups need to be elucidated. Specific interactions between polar residues and the polar headgroups need to be studied.

The value of the dielectric constant in the polar headgroup region is unknown, and, given the chemical complexity of this region, it is not even clear if assigning a single (average) value to this region is meaningful (Fig. 1; White and Wimley, 1999). Peptide and protein association with lipid bilayers may involve association with the phosphate groups of the lipid polar heads. This, in turn, may lead to full or at least partial desolvation of these groups. The energetic cost of this process is unknown (Ben-Tal et al., 1997a).

Answering these questions requires a combination of theoretical and experimental effort.

Acknowledgments

We thank Adam Ben-Shem, Stuart McLaughlin, Yechiel Shai, and Jefferey Skolnick for suggestions and comments on the manuscript. This work was supported by Israel Science Foundation grant 683/97 and fellowships from the Wolfson and Alon Foundations to N.B.-T.

References

Aderem, A. (1992). The MARCKS brothers: A family of protein kinase C substrates. *Cell* **71,** 713–716.

Arkin, I. T., Adams, P. D., Brunger, A. T., Smith, S. O., and Engelman, D. M. (1997). Structural perspectives of phospholamban, a helical transmembrane pentamer. *Annu. Rev. Biophys. Biomol. Struct.* **26,** 157–179.

Banerjee, U., and Chan, S. I. (1983). Structure of alamethicin in solution: Nuclear magnetic resonance relaxation studies. *Biochemistry* **22,** 3709–3713.

Barranger-Mathys, M., and Cafiso, D. S. (1996). Membrane structure of voltage-gated channel forming peptides by site-directed spin-labeling. *Biochemistry* **35,** 498–505.

Bechinger, B. (1997). Structure and functions of channel-forming peptides: Magainins, cecropins, melittin and alamethicin. *J. Membrane Biol.* **156,** 197–211.

Bechinger, B., Zasloff, M., and Opella, S. J. (1998). Structure and dynamics of the antibiotic peptide PGLa in membranes by solution and solid-state nuclear magnetic resonance spectroscopy. *Biophys. J.* **74,** 981–987.

Ben-Shaul, A., Ben-Tal, N., and Honig, B. (1996). Statistical thermodynamic analysis of peptide and protein insertion into lipid membranes. *Biophys. J.* **71,** 130–137.

Ben-Tal, N., Ben-Shaul, A., Nicholls, A., and Honig, B. (1996a). Free-energy determinants of alpha-helix insertion into lipid bilayers. *Biophys. J.* **70,** 1803–1812.

Ben-Tal, N., Honig, B., Peitzsch, R. M., Denisov, G., and McLaughlin, S. (1996b). Binding of small basic peptides to membranes containing acidic lipids: Theoretical models and experimental results. *Biophys. J.* **71,** 561–575.

Ben-Tal, N., Honig, B., Miller, C., and McLaughlin, S. (1997a). Electrostatic binding of proteins to membranes. Theoretical predictions and experimental results with charybdotoxin and phospholipid vesicles. *Biophys. J.* **73,** 1717–1727.

Ben-Tal, N., Sitkoff, D., Topol, I. A., Yang, A. S., Burt, S. K., and Honig, B. (1997b). Free energy of amide hydrogen bond formation in vacuum, in water, and in liquid alkane solution. *J. Phys. Chem. B* **101,** 450–457.

Ben-Tal, N., Sitkoff, D., Bransburg-Zabary, S., Nachliel, E., and Gutman, M. (2000a). Theoretical calculations of the permeability of monensin–cation complexes in model biomembranes. *Biochem. Biophys. Acta* **1466,** 221–233.

Ben-Tal, N., Honig, B., Bagdassarian, C. K., and Ben-Shaul, A. (2000b). Association entropy in adsorption processes. *Biophys. J.* **79,** 1180–1187.

Bentz, J. (1993). "Viral Fusion Mechanisms." CRC Press, Boca Raton, FL.

Berneche, S., Nina, M., and Roux, B. (1998). Molecular dynamics simulation of melittin in a dimyristoylphosphatidylcholine bilayer membrane. *Biophys. J.* **75,** 1603–1618.

Biggin, P. C., and Sansom, M. S. (1996). Simulation of voltage-dependent interactions of alpha-helical peptides with lipid bilayers. *Biophys. Chem.* **60,** 99–110.

Biggin, P. C., and Sansom, M. S. (1999). Interactions of alpha-helices with lipid bilayers: A review of simulation studies. *Biophys. Chem.* **76,** 161–183.

Biggin, P. C., Breed, J., Son, H. S., and Sansom, M. S. (1997). Simulation studies of alamethicin-bilayer interactions. *Biophys. J.* **72,** 627–636.

Blackshear, P. J. (1993). The MARCKS family of cellular protein kinase C substrates. *J. Biol. Chem.* **268,** 1501–1504.

Boguslavsky, V., Rebecchi, M., Morris, A. J., Jhon, D. Y., Rhee, S. G., and McLaughlin, S. (1994). Effect of monolayer surface pressure on the activities of phosphoinositide-specific phospholipase C-beta 1, -gamma 1, and -delta 1. *Biochemistry* **33**, 3032–3027.

Bowie, J. U. (1999). Helix-bundle membrane protein fold templates. *Protein Sci.* **8**, 2711–2719.

Brady, G. P., and Sharp, K. A. (1997). Entropy in protein folding and in protein–protein interactions. *Curr. Opin. Struct. Biol.* **7**, 215–221.

Brunger, A. T. (1992). "X-PLOR, a System for X-Ray Crystallography and NMR." Yale University Press, New Haven, CT.

Buser, C. A., Sigal, C. T., Resh, M. D., and McLaughlin, S. (1994). Membrane binding of myristylated peptides corresponding to the NH$_2$ terminus of Src. *Biochemistry* **33**, 13093–13101.

Buss, J. E., Kamps, M. P., Gould, K., and Sefton, B. M. (1986). The absence of myristic acid decreases membrane binding of p60src but does not affect tyrosine protein kinase activity. *J. Virol.* **58**, 468–474.

Cadwallader, K. A., Paterson, H., Macdonald, S. G., and Hancock, J. F. (1994). N-terminally myristoylated Ras proteins require palmitoylation or a polybasic domain for plasma membrane localization. *Mol. Cell. Biol.* **14**, 4722–4730.

Cafiso, D. (1994). Alamethicin: A peptide model for voltage gating and protein–membrane interactions. *Annu. Rev. Biophys. Biomol. Struct.* **23**, 141–165.

Cafiso, D. (1999). Interaction of natural and model peptides with membranes. *In* "Current Topics in Membranes." pp. 197–228. Academic Press, San Diego, CA.

Carr, C. M., and Kim, P. S. (1993). A spring-loaded mechanism for the conformational change of influenza hemagglutinin. *Cell* **73**, 823–832.

Carr, C. M., and Kim, P. S. (1994). Flu virus invasion: Halfway there. *Science* **266**, 234–236.

Clark, B. A., and Gray, D. M. (1989). A CD determination of the alpha-helix contents of the coat proteins of four filamentous bacteriophages: fd, IKe, Pf1, and Pf3. *Biopolymers* **28**, 1861–1873.

Cortese, J. D., McIntyre, J. O., Duncan, T. M., and Fleischer, S. (1989). Cooperativity in lipid activation of 3-hydroxybutyrate dehydrogenase: Role of lecithin as an essential allosteric activator. *Biochemistry* **28**, 3000–3008.

Cross, F. R., Garber, E. A., Pellman, D., and Hanafusa, H. (1984). A short sequence in the p60src N terminus is required for p60src myristylation and membrane association and for cell transformation. *Mol. Cell. Biol.* **4**, 1834–1842.

Deber, C. M., and Goto, N. K. (1996). Folding proteins into membranes. *Nat. Struct. Biol.* **3**, 815–818.

de Kroon, A. I. P. M., de Gier, J., and de Kruijff, B. (1993). Lipid–peptide interactions in model systems: Membrane insertion and translocation of peptides. *In* "Protein–Lipid Interactions" (A. Watts, ed.), pp. 107–126. Elsevier Science, Amsterdam.

Dell'Acqua, M. L., Faux, M. C., Thorburn, J., Thorburn, A., and Scott, J. D. (1998). Membrane-targeting sequences on AKAP79 bind phosphatidylinositol-4,5-bisphosphate. *EMBO J.* **17**, 2246–2260.

Denisov, G., Wanaski, S., Luan, P., Glaser, M., and McLaughlin, S. (1998). Binding of basic peptides to membranes produces lateral domains enriched in the acidic lipids phosphatidylserine and phosphatidylinositol 4,5-bisphosphate: An electrostatic model and experimental results. *Biophys. J.* **74**, 731–744.

Durell, S. R., Martin, I., Ruysschaert, J. M., Shai, Y., and Blumenthal, R. (1997). What studies of fusion peptides tell us about viral envelope glycoprotein-mediated membrane fusion (review). *Mol. Membrane Biol.* **14**, 97–112.

Edholm, O., and Jähnig, F. (1988). The structure of a membrane-spanning polypeptide studied by molecular dynamics. *Biophys. Chem.* **30**, 279–292.

Efremov, R. G., Nolde, D. E., Vergoten, G., and Arseniev, A. S. (1999a). A solvent model for simulations of peptides in bilayers. I. Membrane-promoting alpha-helix formation. *Biophys. J.* **76**, 2448–2459.

Efremov, R. G., Nolde, D. E., Vergoten, G., and Arseniev, A. S. (1999b). A solvent model for simulations of peptides in bilayers. II. Membrane-spanning alpha-helices. *Biophys. J.* **76,** 2460–2471.

Efremov, R. G., Nolde, D. E., Volynsky, P. E., Chernyavsky, A. A., Dubovskii, P. V., and Arseniev, A. S. (1999c). Factors important for fusogenic activity of peptides: Molecular modeling study of analogs of fusion peptide of influenza virus hemagglutinin. *FEBS Lett.* **462,** 205–210.

Eisenberg, D., and McLachlan, A. D. (1986). Solvation energy in protein folding and binding. *Nature* **319,** 199–203.

Engelman, D. M., and Steitz, T. A. (1981). The spontaneous insertion of proteins into and across membranes: The helical hairpin hypothesis. *Cell* **23,** 411–422.

Engelman, D. M., Steitz, T. A., and Goldman, A. (1986). Identifying nonpolar transbilayer helices in amino acid sequences of membrane proteins. *Annu. Rev. Biophys. Biophys. Chem.* **15,** 321–353.

Erickson, H. P. (1989). Co-operativity in protein–protein association: The structure and stability of the actin filament. *J. Mol. Biol.* **206,** 465–474.

Esmann, M., and Marsh, D. (1985). Spin-label studies on the origin of the specificity of lipid–protein interactions in Na^+,K^+-ATPase membranes from *Squalus acanthias*. *Biochemistry* **24,** 3572–3578.

Esposito, G., Carver, J. A., Boyd, J., and Campbell, I. D. (1987). High-resolution ^{1}H NMR study of the solution structure of alamethicin. *Biochemistry* **26,** 1043–1050.

Fasman, G. D., and Gilbert, W. A. (1990). The prediction of transmembrane protein sequences and their conformation: An evaluation. *Trends Biochem. Sci.* **15,** 89–92.

Fattal, D. R., and Ben-Shaul, A. (1993). A molecular model for lipid–protein interaction in membranes: The role of hydrophobic mismatch. *Biophys. J.* **65,** 1795–1809.

Finkelstein, A. V., and Janin, J. (1989). The price of lost freedom: Entropy of bimolecular complex formation. *Protein Eng.* **3,** 1–3.

Forrest, L. R., and Sansom, M. S. (2000). Membrane simulations: Bigger and better? *Curr. Opin. Struct. Biol.* **10,** 174–181.

Fox, R. O., Jr., and Richards, F. M. (1982). A voltage-gated ion channel model inferred from the crystal structure of alamethicin at 1.5-Å resolution. *Nature* **300,** 325–330.

Gazit, E., La Rocca, P., Sansom, M. S., and Shai, Y. (1998). The structure and organization within the membrane of the helices composing the pore-forming domain of *Bacillus thuringiensis* δ-endotoxin are consistent with an "umbrella-like" structure of the pore. *Proc. Natl. Acad. Sci. USA* **95,** 12289–12294.

Gersappe, D., Li, W., and Balazs, A. C. (1993). Computational studies of protein adsorption at bilayer interfaces. *J. Chem. Phys.* **99,** 7209–7213.

Gilson, M. K., Rashin, A., Fine, R., and Honig, B. (1985). On the calculation of electrostatic interactions in proteins. *J. Mol. Biol.* **184,** 503–516.

Gilson, M. K., Given, J. A., Bush, B. L., and McCammon, J. A. (1997). The statistical-thermodynamic basis for computation of binding affinities: A critical review. *Biophys. J.* **72,** 1047–1069.

Gouaux, E. (1997). Channel-forming toxins: Tales of transformation. *Curr. Opin. Struct. Biol.* **7,** 566–573.

Han, X., Bushwellar, J. H., Cafiso, D. S., and Tamm, L. K. (2001). Membrane structure and fusion-triggering conformational change of the fusion domain from influenza hemagglutinin. *Nature Struct. Biol.* **8,** 715–720.

Han, X., Steinhauer, D. A., Wharton, S. A., and Tamm, L. K. (1999). Interaction of mutant influenza virus hemagglutinin fusion peptides with lipid bilayers: Probing the role of hydrophobic residue size in the central region of the fusion peptide. *Biochemistry* **38,** 15052–15059.

Hancock, J. F., Paterson, H., and Marshall, C. J. (1990). A polybasic domain or palmitoylation is required in addition to the CAAX motif to localize p21ras to the plasma membrane. *Cell* **63**, 133–139.

He, K., Ludtke, S. J., Heller, W. T., and Huang, H. W. (1996). Mechanism of alamethicin insertion into lipid bilayers. *Biophys. J.* **71**, 2669–2679.

Heimburg, T., and Marsh, D. (1995). Protein surface-distribution and protein–protein interactions in the binding of peripheral proteins to charged lipid membranes. *Biophys. J.* **68**, 536–546.

Heimburg, T., and Marsh, D. (1996). Thermodynamics of the interaction of proteins with lipid membranes. *In* "Biological Membranes" (J. Merz and B. Roux, eds.), pp. 405–462. Birkhauser, Boston.

Heimburg, T., Angerstein, B., and Marsh, D. (1999). Binding of peripheral proteins to mixed lipid membranes: Effect of lipid demixing upon binding. *Biophys. J.* **76**, 2575–2586.

Herzyk, P., and Hubbard, R. E. (1998). Using experimental information to produce a model of the transmembrane domain of the ion channel phospholamban. *Biophys. J.* **74**, 1203–1214.

Hill, C. P., Worthylake, D., Bancroft, D. P., Christensen, A. M., and Sundquist, W. I. (1996). Crystal structures of the trimeric human immunodeficiency virus type 1 matrix protein: Implications for membrane association and assembly. *Proc. Natl. Acad. Sci. USA* **93**, 3099–3104.

Hill, T. L. (1985). Cooperativity Theory in Biochemistry: Steady-State and Equilibrium System (T. L. Hill, ed.), Springer-Verlag, New York.

Holtzer, A. (1994). Does Flory–Huggins theory help in interpreting solute partitioning experiments? *Biopolymers* **34**, 315–320.

Honig, B. H., and Hubbell, W. L. (1984). Stability of "salt bridges" in membrane proteins. *Proc. Natl. Acad. Sci. USA* **81**, 5412–5416.

Honig, B., and Nicholls, A. (1995). Classical electrostatics in biology and chemistry. *Science* **268**, 1144–1149.

Honig, B., Sharp, K., and Yang, A. S. (1993). Macroscopic models of aqueous solutions: Biological and chemical applications. *J. Phys. Chem.* **97**, 1101–1109.

Horton, N., and Lewis, M. (1992). Calculation of the free energy of association for protein complexes. *Protein Sci.* **1**, 169–181.

Horvath, L. I., Brophy, P. J., and Marsh, D. (1988). Influence of lipid headgroup on the specificity and exchange dynamics in lipid–protein interactions: A spin-label study of myelin proteolipid apoprotein–phospholipid complexes. *Biochemistry* **27**, 5296–5304.

Houbre, D., Duportail, G., Deloulme, J. C., and Baudier, J. (1991). The interactions of the brain-specific calmodulin-binding protein kinase C substrate, neuromodulin (GAP 43), with membrane phospholipids. *J. Biol. Chem.* **266**, 7121–7131.

Hristova, K., and White, S. H. (1998). Determination of the hydrocarbon core structure of fluid dioleoylphosphocholine (DOPC) bilayers by X-ray diffraction using specific bromination of the double-bonds: Effect of hydration. *Biophys. J.* **74**, 2419–2433.

Huang, H. W., and Wu, Y. (1991). Lipid–alamethicin interactions influence alamethicin orientation. *Biophys. J.* **60**, 1079–1087.

Hucho, F., Tsetlin, V. I., and Machold, J. (1996). The emerging three-dimensional structure of a receptor: The nicotinic acetylcholine receptor. *Eur. J. Biochem.* **239**, 539–557.

Jacobs, R. E., and White, S. H. (1989). The nature of the hydrophobic binding of small peptides at the bilayer interface: Implications for the insertion of transbilayer helices. *Biochemistry* **28**, 3421–3437.

Jähnig, F. (1983). Thermodynamics and kinetics of protein incorporation into membranes. *Proc. Natl. Acad. Sci. USA* **80**, 3691–3695.

Jähnig, F. (1990). Structure predictions of membrane proteins are not that bad. *Trends Biochem. Sci.* **15**, 93–95.

Janin, J. (1996). Quantifying biological specificity: The statistical mechanics of molecular recognition. *Proteins* **25**, 438–445.

Jayasinghe, S., Barranger-Mathys, M., Ellena, J. F., Franklin, C., and Cafiso, D. S. (1998). Structural features that modulate the transmembrane migration of a hydrophobic peptide in lipid vesicles. *Biophys. J.* **74**, 3023–3030.

Kamps, M. P., Buss, J. E., and Sefton, B. M. (1985). Mutation of NH_2-terminal glycine of p60src prevents both myristoylation and morphological transformation. *Proc. Natl. Acad. Sci. USA* **82**, 4625–4628.

Kaplan, J. M., Varmus, H. E., and Bishop, J. M. (1990). The src protein contains multiple domains for specific attachment to membranes. *Mol. Cell. Biol.* **10**, 1000–1009.

Keller, S. L., Bezrukov, S. M., Gruner, S. M., Tate, M. W., Vodyanoy, I., and Parsegian, V. A. (1993). Probability of alamethicin conductance states varies with nonlamellar tendency of bilayer phospholipids. *Biophys. J.* **65**, 23–27.

Kellis, J. T., Jr., Nyberg, K., Sali, D., and Fersht, A. R. (1988). Contribution of hydrophobic interactions to protein stability. *Nature* **333**, 784–786.

Kessel, A., Cafiso, D. S., and Ben-Tal, N. (2000a). Continuum solvent model calculations of alamethicin–membrane interactions: Thermodynamic aspects. *Biophys. J.* **78**, 571–583.

Kessel, A., Schulten, K., and Ben-Tal, N. (2000b). Calculations suggest a pathway for the transmembrane migration of a hydrophobic peptide. *Biophys. J.* **79**, 2322–2330.

Killian, J. A. (1998). Hydrophobic mismatch between proteins and lipids in membranes. *Biochim. Biophys. Acta* **1376**, 401–415.

Killian, J. A., and von Heijne, G. (2000). How proteins adapt to a membrane–water interface. *Trends Biochem. Sci.* **25**, 429–434.

Kim, J., Mosior, M., Chung, L. A., Wu, H., and McLaughlin, S. (1991). Binding of peptides with basic residues to membranes containing acidic phospholipids. *Biophys. J.* **60**, 135–148.

Kim, J., Blackshear, P. J., Johnson, J. D., and McLaughlin, S. (1994). Phosphorylation reverses the membrane association of peptides that correspond to the basic domains of MARCKS and neuromodulin. *Biophys. J.* **67**, 227–237.

Kurumbail, R. G., Stevens, A. M., Gierse, J. K., McDonald, J. J., Stegeman, R. A., Pak, J. Y., Gildehaus, D., Miyashiro, J. M., Penning, T. D., Seibert, K., Isakson, P. C., and Stallings, W. C. (1996). Structural basis for selective inhibition of cyclooxygenase-2 by anti-inflammatory agents. *Nature* **384**, 644–648 [Erratum, *Nature* (1997) **385**, 555].

Kyte, J., and Doolittle, R. F. (1982). A simple method for displaying the hydropathic character of a protein. *J. Mol. Biol.* **157**, 105–132.

La Rocca, P., and Sansom, M. S. (1998). Peptide–bilayer interactions: Simulation studies. *Biochem. Soc. Trans.* **26**, S302.

La Rocca, P., Shai, Y., and Sansom, M. S. (1999). Peptide–bilayer interactions: Simulations of dermaseptin B, an antimicrobial peptide. *Biophys. Chem.* **76**, 145–159.

Lemmon, M. A., Ferguson, K. M., O'Brien, R., Sigler, P. B., and Schlessinger, J. (1995). Specific and high-affinity binding of inositol phosphates to an isolated pleckstrin homology domain. *Proc. Natl. Acad. Sci. USA* **92**, 10472–10476.

Lewis, B. A., and Engelman, D. M. (1983a). Bacteriorhodopsin remains dispersed in fluid phospholipid bilayers over a wide range of bilayer thicknesses. *J. Mol. Biol.* **166**, 203–210.

Lewis, J. R., and Cafiso, D. S. (1999). Correlation between the free energy of a channel-forming voltage-gated peptide and the spontaneous curvature of bilayer lipids. *Biochemistry* **38**, 5932–5938.

Li, J. D., Carroll, J., and Ellar, D. J. (1991). Crystal structure of insecticidal delta-endotoxin from *Bacillus thuringiensis* at 2.5 Å resolution. *Nature* **353**, 815–821.

Liu, Y. C., and Storm, D. R. (1990). Regulation of free calmodulin levels by neuromodulin: Neuron growth and regeneration. *Trends Pharmacol. Sci.* **11**, 107–111.

MacKenzie, K. R., Prestegard, J. H., and Engelman, D. M. (1997). A transmembrane helix dimer: Structure and implications. *Science* **276**, 131–133.

MacNaughtan, W., Snook, K. A., Caspi, E., and Franks, N. P. (1985). An X-ray diffraction analysis of oriented lipid multilayers containing basic proteins. *Biochim. Biophys. Acta* **818**, 132–148.

Macosko, J. C., Kim, C. H., and Shin, Y. K. (1997). The membrane topology of the fusion peptide region of influenza hemagglutinin determined by spin-labeling EPR. *J. Mol. Biol.* **267**, 1139–1148.

Massiah, M. A., Starich, M. R., Paschall, C., Summers, M. F., Christensen, A. M., and Sundquist, W. I. (1994). Three-dimensional structure of the human immunodeficiency virus type 1 matrix protein. *J. Mol. Biol.* **244**, 198–223.

May, S., and Ben-Shaul, A. (1999). Molecular theory of lipid–protein interaction and the alpha-HII transition. *Biophys. J.* **76**, 751–767.

May, S., Harries, D., and Ben-Shaul, A. (2000). Lipid demixing and protein–protein interactions in the adsorption of charged proteins on mixed membranes. *Biophys. J.* **79**, 1747–1760.

McCammon, J. A. (1998). Theory of biomolecular recognition. *Curr. Opin. Struct. Biol.* **8**, 245–249.

McDonnell, P. A., Shon, K., Kim, Y., and Opella, S. J. (1993). fd coat protein structure in membrane environments. *J. Mol. Biol.* **233**, 447–463.

McLaughlin, S., and Aderem, A. (1995). The myristoyl–electrostatic switch: A modulator of reversible protein–membrane interactions. *Trends Biochem. Sci.* **20**, 272–276.

Milik, M., and Skolnick, J. (1993). Insertion of peptide chains into lipid membranes: An off-lattice Monte Carlo dynamics model. *Proteins* **15**, 10–25.

Milik, M., and Skolnick, J. (1995). A Monte Carlo model of fd and Pf1 coat proteins in lipid membranes. *Biophys. J.* **69**, 1382–1386.

Moll, T. S., and Thompson, T. E. (1994). Semisynthetic proteins: Model systems for the study of the insertion of hydrophobic peptides into preformed lipid bilayers. *Biochemistry* **33**, 15469–15482.

Mosior, M., and McLaughlin, S. (1992). Binding of basic peptides to acidic lipids in membranes: Effects of inserting alanine(s) between the basic residues. *Biochemistry* **31**, 1767–1773.

Mouritsen, O. G., and Bloom, M. (1984). Mattress model of lipid–protein interactions in membranes. *Biophys. J.* **46**, 141–153.

Murphy, K. P., Xie, D., Thompson, K. S., Amzel, L. M., and Freire, E. (1994). Entropy in biological binding processes: Estimation of translational entropy loss. *Proteins* **18**, 63–67.

Murray, D., Ben-Tal, N., Honig, B., and McLaughlin, S. (1997). Electrostatic interaction of myristoylated proteins with membranes: Simple physics, complicated biology. *Structure* **5**, 985–989.

Murray, D., Hermida-Matsumoto, L., Buser, C. A., Tsang, J., Sigal, C. T., Ben-Tal, N., Honig, B., Resh, M. D., and McLaughlin, S. (1998). Electrostatics and the membrane association of Src: Theory and experiment. *Biochemistry* **37**, 2145–2159.

Murray, D., Arbuzova, A., Hangyas-Mihalyne, G., Gambhir, A., Ben-Tal, N., Honig, B., and McLaughlin, S. (1999). Electrostatic properties of membranes containing acidic lipids and adsorbed basic peptides: Theory and experiment. *Biophys. J.* **77**, 3176–3188.

Murray, D., Ben-Tal, N., Petrey, D., Honig, B., and McLaughlin, S. (2001). Structure-based calculation of the electrostatic free energy required to concentrate acidic lipids into a localized region of membrane. *Biophys. J.* Submitted.

Nakashima, Y., Wiseman, R. L., Konigsberg, W., and Marvin, D. A. (1975). Primary structure and sidechain interactions of PFL filamentous bacterial virus coat protein. *Nature* **253**, 68–71.

Newton, A. C. (1993). Interaction of proteins with lipid headgroups: Lessons from protein kinase C. *Annu. Rev. Biophys. Biomol. Struct.* **22**, 1–25.

Nicholls, A., and Honig, B. (1991). A rapid finit difference algorithem utilizing successive over-relaxation to solve Poisson–Boltzmann equation. *J. Comp. Chem.* **12**, 435–445.

Nicholls, A., Sharp, K. A., and Honig, B. (1991). Protein folding and association: Insights from the interfacial and thermodynamic properties of hydrocarbons. *Proteins* **11**, 281–296.

North, C. L., Barranger-Mathys, M., and Cafiso, D. S. (1995). Membrane orientation of the N-terminal segment of alamethicin determined by solid-state ^{15}N NMR. *Biophys. J.* **69**, 2392–2397.

Novotny, J., Bruccoleri, R. E., and Saul, F. A. (1989). On the attribution of binding energy in antigen–antibody complexes McPC 603, D1.3, and HyHEL-5. *Biochemistry* **28**, 4735–4749.

Opella, S. J., Marassi, F. M., Gesell, J. J., Valente, A. P., Kim, Y., Oblatt-Montal, M., and Montal, M. (1999). Structures of the M2 channel-lining segments from nicotinic acetylcholine and NMDA receptors by NMR spectroscopy. *Nat. Struct. Biol.* **6**, 374–379.

Page, M. I., and Jencks, W. P. (1971). Entropic contributions to rate accelerations in enzymic and intramolecular reactions and the chelate effect. *Proc. Natl. Acad. Sci. USA* **68**, 1678–1683.

Palczewski, K., Kumasaka, T., Hori, T., Behnke, C. A., Motoshima, H., Fox, B. A., Trong, I. L., Teller, D. C., Okada, T., Stenkamp, R. E., Yamamoto, M., and Miyano, M. (2000). Crystal structure of rhodopsin: A G protein-coupled receptor. *Science* **289**, 739–745.

Pastor, R. W. (1994). Molecular dynamics and Monte Carlo simulations of lipid bilayers. *Curr. Opin. Struct. Biol.* **4**, 486–492.

Peitzsch, R. M., and McLaughlin, S. (1993). Binding of acylated peptides and fatty acids to phospholipid vesicles: Pertinence to myristoylated proteins. *Biochemistry* **32**, 10436–10443.

Picot, D., Loll, P. J., and Garavito, R. M. (1994). The X-ray crystal structure of the membrane protein prostaglandin H2 synthase-1. *Nature* **367**, 243–249.

Pilpel, Y., Ben-Tal, N., and Lancet, D. (1999). kPROT: A knowledge-based scale for the propensity of residue orientation in transmembrane segments. Application to membrane protein structure prediction. *J. Mol. Biol.* **294**, 921–935.

Pinheiro, T. J. (1994). The interaction of horse heart cytochrome c with phospholipid bilayers. Structural and dynamic effects. *Biochimie* **76**, 489–500.

Pinheiro, T. J., and Watts, A. (1994a). Lipid specificity in the interaction of cytochrome c with anionic phospholipid bilayers revealed by solid-state ^{31}P NMR. *Biochemistry* **33**, 2451–2458.

Pinheiro, T. J., and Watts, A. (1994b). Resolution of individual lipids in mixed phospholipid membranes and specific lipid–cytochrome c interactions by magic-angle spinning solid-state phosphorus-31 NMR. *Biochemistry* **33**, 2459–2467.

Pool, C. T., and Thompson, T. E. (1998). Chain length and temperature dependence of the reversible association of model acylated proteins with lipid bilayers. *Biochemistry* **37**, 10246–10255.

Popot, J. L. (1993). Integral membrane protein structure: Transmembrane alpha-helices as autonomous folding domains. *Curr. Opin. Struct. Biol.* **3**, 532–540.

Popot, J. L., and Engelman, D. M. (1990). Membrane protein folding and oligomerization: The two-stage model. *Biochemistry* **29**, 4031–4037.

Popot, J. L., Gerchman, S. E., and Engelman, D. M. (1987). Refolding of bacteriorhodopsin in lipid bilayers: A thermodynamically controlled two-stage process. *J. Mol. Biol.* **198**, 655–676.

Pouny, Y., Rapaport, D., Mor, A., Nicolas, P., and Shai, Y. (1992). Interaction of antimicrobial dermaseptin and its fluorescently labeled analogues with phospholipid membranes. *Biochemistry* **31**, 12416–12423.

Resh, M. D. (1993). Interaction of tyrosine kinase oncoproteins with cellular membranes. *Biochim. Biophys. Acta* **1155**, 307–322.

Resh, M. D. (1994). Myristylation and palmitylation of Src family members: The fats of the matter. *Cell* **76**, 411–413.

Roberts, M. F. (1996). Phospholipases: Structural and functional motifs for working at an interface. *FASEB J.* **10**, 1159–1172.

Rose, G. D. (1987). Protein hydrophobicity: Is it the sum of its parts? *Proteins* **2**, 79–80 .

Roseman, M. A. (1988a). Hydrophilicity of polar amino acid side-chains is markedly reduced by flanking peptide bonds. *J. Mol. Biol.* **200**, 513–522.

Roseman, M. A. (1988b). Hydrophobicity of the peptide C=O· · ·H—N hydrogen-bonded group. *J. Mol. Biol.* **201**, 621–623.

Roux, B., and Woolf, T. B. (1996). Molecular dynamics of Pf1 coat protein in a phospholipid bilayer. *In* "Biological Membranes: A Molecular Perspective from Computation and Experiment" (K. J. Merz and B. Roux, eds.), pp. 555–587. Birkhauser, Boston.

Saberwal, G., and Nagaraj, R. (1994). Cell-lytic and antibacterial peptides that act by perturbing the barrier function of membranes: Facets of their conformational features, structure-function correlations and membrane-perturbing abilities. *Biochim. Biophys. Acta* **1197**, 109–131 [Erratum, *Biochim. Biophys. Acta* (1995) **1235**, 159].

Sansom, M. S. (1998). Models and simulations of ion channels and related membrane proteins. *Curr. Opin. Struct. Biol.* **8**, 237–244.

Schlenkrich, M., Brickmann, J., MacKerell, A. D. J., and Karplus, M. (1996). An empirical potential energy function for phospholipids: Criteria for parameter optimization and applications. *In* "Biological Membranes: A Molecular Perspective from Computation and Experiment" (K. J. Merz and B. Roux, eds.), pp. 31–82. Birkhauser, Boston.

Seagraves, C., and Reinhardt, W. P. (1995). A two-lattice model of membrane proteins: Configuration as a function of sequence. *J. Chem. Phys.* **103**, 5091–5101.

Searle, M. S., and Williams, D. H. (1992). The cost of conformational order: Entropy changes in molecular associations. *J. Am. Chem. Soc.* **114**, 10690–10697.

Searle, M. S., Williams, D. H., and Gerhard, U. (1992). Partitioning of free energy contributions in the estimation of binding constants: Residual motions and consequences of amide–amide hydrogen bond strengths. *J. Am. Chem. Soc.* **114**, 10697–10704.

Shai, Y. (1995). Molecular recognition between membrane-spanning polypeptides. *Trends Biochem. Sci.* **20**, 460–464.

Shai, Y. (1999). Mechanism of the binding, insertion and destabilization of phospholipid bilayer membranes by alpha-helical antimicrobial and cell non-selective membrane-lytic peptides. *Biochim. Biophys. Acta* **1462**, 55–70.

Sheinerman, F. B., Norel, R., and Honig, B. (2000). Electrostatic aspects of protein–protein interactions. *Curr. Opin. Struct. Biol.* **10**, 153–159.

Shon, K. J., Kim, Y., Colnago, L. A., and Opella, S. J. (1991). NMR studies of the structure and dynamics of membrane-bound bacteriophage Pf1 coat protein. *Science* **252**, 1303–1305.

Sigal, C. T., Zhou, W., Buser, C. A., McLaughlin, S., and Resh, M. D. (1994). Amino-terminal basic residues of Src mediate membrane binding through electrostatic interaction with acidic phospholipids. *Proc. Natl. Acad. Sci. USA* **91**, 12253–12257.

Sitkoff, D., Sharp, K., and Honig, B. (1994). Accurate calculation of hydration free energies using macroscopic solvent models. *J. Phys. Chem.* **98**, 1978–1988.

Sitkoff, D., Ben-Tal, N., and Honig, B. (1996). Calculation of alkane to water solvation free energies using continuum solvent models. *J. Phys. Chem.* **100**, 2744–2752.

Skene, J. H. (1989). Axonal growth-associated proteins. *Annu. Rev. Neurosci.* **12**, 127–156.

Skene, J. H., and Virag, I. (1989). Posttranslational membrane attachment and dynamic fatty acylation of a neuronal growth cone protein, GAP-43. *J. Cell. Biol.* **108**, 613–624.

Skolnick, J., and Milik, M. (1996). Monte Carlo models of spontaneous insertion of peptides into lipid membranes. *In* "Biological Membranes: A Molecular Perspective from Computation and Experiment" (K. J. Merz and B. Roux, eds.), pp. 535–554. Birkhauser, Boston.

Stams, T., Nair, S. K., Okuyama, T., Waheed, A., Sly, W. S., and Christianson, D. W. (1996). Crystal structure of the secretory form of membrane-associated human carbonic anhydrase IV at 2.8-Å resolution. *Proc. Natl. Acad. Sci. USA* **93,** 13589–13594.

Tanford, C. (1991). "The Hydrophobic Effect: Formation of Micelles and Biological Membranes," 2nd ed. Krieger, Malabar, FL.

Thorgeirsson, T. E., Russell, C. J., King, D. S., and Shin, Y. K. (1996). Direct determination of the membrane affinities of individual amino acids. *Biochemistry* **35,** 1803–1809.

Tidor, B., and Karplus, M. (1994). The contribution of vibrational entropy to molecular association: The dimerization of insulin. *J. Mol. Biol.* **238,** 405–414.

Vajda, S., Weng, Z., Rosenfeld, R., and DeLisi, C. (1994). Effect of conformational flexibility and solvation on receptor–ligand binding free energies. *Biochemistry* **33,** 13977–13988.

von Heijne, G. (1981a). Membrane proteins: The amino acid composition of membrane-penetrating segments. *Eur. J. Biochem.* **120,** 275–278.

von Heijne, G. (1981b). On the hydrophobic nature of signal sequences. *Eur. J. Biochem.* **116,** 419–22.

von Heijne, G. (1998). Life and death of a signal peptide. *Nature* **396,** 111–113.

Wallace, B. A. (1990). Gramicidin channels and pores. *Annu. Rev. Biophys. Biophys. Chem.* **19,** 127–157.

Warshel, A. (1991). "Computer Modeling of Chemical Reactions in Enzymes and Solutions." Wiley, New York.

Weng, Z., Vajda, S., and Delisi, C. (1996). Prediction of protein complexes using empirical free energy functions. *Protein Sci.* **5,** 614–626.

White, S. H., and Wimley, W. C. (1999). Membrane protein folding and stability: Physical principles. *Annu. Rev. Biophys. Biomol. Struct.* **28,** 319–365.

White, S. H., Wimley, W. C., and Selsted, M. E. (1995). Structure, function, and membrane integration of defensins. *Curr. Opin. Struct. Biol.* **5,** 521–527.

Wiener, M. C., and White, S. H. (1992). Structure of a fluid dioleoylphosphatidylcholine bilayer determined by joint refinement of X-ray and neutron diffraction data. III. Complete structure. *Biophys. J.* **61,** 437–447.

Wimley, W. C., and White, S. H. (1996). Experimentally determined hydrophobicity scale for proteins at membrane interfaces. *Nat. Struct. Biol.* **3,** 842–848.

Wojcik, J., Altmann, K. H., and Scheraga, H. A. (1990). Helix–coil stability constants for the naturally occurring amino acids in water. XXIV. Half-cysteine parameters from random poly (hydroxybutylglutamine-co-S-methylthio-L-cysteine). *Biopolymers* **30,** 121–134.

Wu, Y., He, K., Ludtke, S. J., and Huang, H. W. (1995). X-ray diffraction study of lipid bilayer membranes interacting with amphiphilic helical peptides: Diphytanoyl phosphatidylcholine with alamethicin at low concentrations. *Biophys J.* **68,** 2361–2369.

Yang, A. S., and Honig, B. (1995). Free energy determinants of secondary structure formation. I. alpha-Helices. *J. Mol. Biol.* **252,** 351–365.

Yau, W. M., Wimley, W. C., Gawrisch, K., and White, S. H. (1998). The preference of tryptophan for membrane interfaces. *Biochemistry* **37,** 14713–14718.

Yee, A. A., and O'Neil, J. D. (1992). Uniform 15N labeling of a fungal peptide: The structure and dynamics of an alamethicin by 15N and 1H NMR spectroscopy. *Biochemistry* **31,** 3135–3143.

Yu, Y. G., Thorgeirsson, T. E., and Shin, Y. K. (1994). Topology of an amphiphilic mitochondrial signal sequence in the membrane-inserted state: A spin labeling study. *Biochemistry* **33,** 14221–14226.

Yue, K., and Dill, K. A. (1996). Folding proteins with a simple energy function and extensive conformational searching. *Protein Sci.* **5,** 254–261.

Zheng, N., and Gierasch, L. M. (1996). Signal sequences: The same yet different. *Cell* **86,** 849–852.

Zhou, W., Parent, L. J., Wills, J. W., and Resh, M. D. (1994). Identification of a membrane-binding domain within the amino-terminal region of human immunodeficiency virus type 1 Gag protein which interacts with acidic phospholipids. *J. Virol.* **68,** 2556–2569.

CHAPTER 9

Investigating Ion Channels Using Computational Methods

Eric Jakobsson,[*,†,‡,§] **R. Jay Mashl,**[*,†] **and Tsai-Tien Tseng**[‖]

*Beckman Institute for Advanced Science and Technology, †National Center for Supercomputing Applications, ‡Department of Molecular and Integrative Physiology, §Bioengineering Program, ‖Center for Biophysics and Computational Biology Molecular Biophysics Training Program, University of Illinois at Urbana-Champaign, Urbana, Illinois 61801

I. Introduction
II. Understanding Channel Function from Structure
    A. Water in Gramicidin Channels
    B. Potassium Ions in the KcsA Channel
    C. Beyond Permeation: Can Computation Help Us Understand Gating?
III. Understanding Channel Function from Structural Homology
    A. Bioinformatics Studies of Ion Channels
    B. The Transport Classification System
    C. Phylogenetics of the VIC Family
    D. Genome Analyses of the VIC Family
IV. Future Prospects
    References

Ion channels are of enormous biological importance and are also susceptible to extraordinarily precise measures of function and mechanism, based on to patch clamp technology. The richness of functional data makes ion channels excellent prototype proteins for the development of computational methods for understanding the relationships between macromolecular sequence, structure, and function. In this chapter we outline the overall strategy that is emerging for analyzing and simulating channel structure and function. The essence of the strategy is to describe the channel in a hierarchy of levels of coarse graining, with the appropriate simulation method at each level, and to connect the levels by statistical mechanical theory. We illustrate this strategy with two examples from our laboratory in which we are able to successfully predict the channel permeability directly from the structure: water permeation in gramicidin and potassium permeability in the KcsA ion channel. Remarkably, the detailed simulations of the permeant motions in the channel reveal knock-on, knock-off mechanisms for permeation that resemble simplified models used by channel biophysicists for years in the absence of structural data on the channels.

In addition to simulation and statistical mechanics, bioinformatics is also emerging as a useful computational approach for understanding ion channels. The first task in this effort is to develop a sound classification system for channels, including channel-specific rules for inferring similarities and differences in channel sequences. Once this is done, the classification scheme can be used as a context for inferring functional attributes and structural motifs from channel sequence data.

I. INTRODUCTION

From the time of Hodgkin and Huxley (1952), ion channels, mathematical models, and computational studies have been well suited to one another. The mathematical model built by Hodgkin and Huxley, at a time when there was not much access to computers and no clear idea of the molecular basis of bioelectricity, that is, ion channels, prefigured the rise of biomolecular computations with almost eerie prescience. Despite their phenomenological origin, the Hodgkin–Huxley concepts can now be understood through their molecular bases. We now know that the mathematical variables in the model, originally conceived as mathematical abstractions, correspond closely to particular molecular segments of the channel proteins. For example, the activation variables m and n correspond mainly to the S4 helices of sodium and potassium channels. The inactivation variable h corresponds to a portion of the channel protein, the N-terminus for potassium channels and an intracellular linker in sodium channels. Also, the reversal potential is determined by the selectivity filter in the narrowest part of the channel, a few residues in the S5–S6 linker. Furthermore, the maximum single-channel conductance is determined by the properties of the permeation pathway, consisting largely of the S5–S6 linker and the S6 helix. Although the first solutions of the Hodgkin–Huxley equations were achieved by heroic efforts with an electromechanical calculator, the form of the equations cried out for the aid of a programmed digital computer, work which was initiated by Cole *et al.* (1955). Even though today much more molecular knowledge of ion permeation is available, ion channels remain especially suitable biological entities for computational studies. The reason is the same as it was four decades ago, that the experimentally measurable signal from the functioning of ion channels is amplified by the capacitance of the membrane in which they reside, so that precise quantifiable functional information is the hallmark of experimental ion channel studies. Computational investigations facilitate the connection of this information to its molecular basis.

Today the functional information is augmented by high-resolution structural models of the pore-containing regions of several ion channels, including gramicidin (Ketchem *et al.,* 1993), several porins (Dutzler *et al.,* 1999; Fu *et al.,* 2000; Meyer *et al.,* 1997; Murata *et al.,* 2000; for a review see Jap and Walian, 1996), the large-conductance stretch-activated ion channel from the inner membrane of bacteria (Chang *et al.,* 1998), and the bacterial potassium channel KcsA (Doyle *et al.,* 1998; Perozo *et al.,* 1999). In combination with the precise functional data, these structures provide a scaffold for developing methodologies to relate protein structure to function.

In addition to a few high-resolution ion channel structures, at this writing approximately 1000 native sequences have been determined by homology to be likely members of the voltage-gated channel superfamily, consisting primarily of the potassium channels, the sodium channels, and the calcium channels. This number is growing continually. A rough measure may be readily made at any time by taking the amino acid sequence of any known member of this family and BLASTing it against the nonredundant protein sequence database of the San Diego Supercomputer Center (http://workbench.sdsc.edu). The massive amount of sequence data, combined with the extensive quantitative experimental data (including data on chimeras and site-directed mutants as well as native channels), provides a unique computational laboratory for the development of detailed function prediction from homology for newly sequenced channel genes.

The remainder of this chapter is divided into two parts. The first deals with relating channel structure to function, and the second deals with function prediction from structural homology.

II. UNDERSTANDING CHANNEL FUNCTION FROM STRUCTURE

The key to predicting fluxes in ion channels is to divide the problem into hierarchies. The results at each level serve to provide parameters for descriptions at coarser grained levels. In order from fine-grained to coarse-grained, the levels of analysis are as follows:

1. Quantum mechanical description. At this level one establishes the appropriate force fields that govern the interactions among the atoms of the system, including charge distributions, bond properties, and the nature of nonbonded interactions.

2. Atomically detailed pseudo-classical description. At this level one approximates atoms as charged, somewhat polar, somewhat elastic balls connected by springs (bonds). The parameters in these calculations are established by quantum mechanical considerations (level 1), in particular by a combination of interpretation of spectroscopic experiments on small molecules that are substituents of biomolecules and direct quantum mechanical calculation of charge distributions. The most important calculations at level 2 are molecular dynamics (to ascertain

Ion Channel Computational Hierarchy

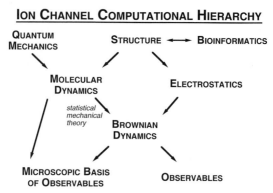

FIGURE 1   The ion channel computational hierarchy, showing the primary flow of information from the atomic level to experimentally observable quantities.

mobilities and effective potentials governing close interactions between components of the system), electrostatics (to map out electrical potentials governing ion–protein and ion–ion interactions), and Monte Carlo (to create equilibrium ensembles of system conformational states). The products of this description are mobilities and potentials governing the motions of ions through channels and (in principle, but not yet demonstrated) the gating motions of channels.

3. Diffusive or mesoscopic level of description. At this level, one utilizes the potentials and mobilities from the finer grained calculations as inputs to Brownian dynamics or continuum calculations of fluxes and (potentially, but not yet demonstrated) kinetic descriptions of channel gating. These descriptions correspond numerically to observations and experiments.

The entire ion channel computational hierarchy is shown in Fig. 1.

Our laboratory's most seminal contribution to the computational methodology of ion channels was to do the first Brownian dynamics simulations of flux through ion channels (Cooper *et al.*, 1985). This work built directly on the insight, first elucidated by Einstein in 1905, that Brownian motion is the particulate manifestation of the phenomenon of diffusion. Brownian dynamics, in this view, is another way of solving the Nernst–Planck equation. An advantage of Brownian dynamics is that it automatically accounts for complex patterns of ion–ion interactions simply by accumulating statistics on the ions' trajectories. It thus replaces very complicated mathematical techniques for accounting for many-body ion–ion interactions in a way that explicitly accounts for the particulate nature of charge carriers. Our early work was a demonstration in principle that, given the mobility and ion/channel/water interaction potentials governing motion in the channel, fluxes could be predicted. Remarkably, we found that much of the behavior of potassium channels could be replicated by postulating an ion/channel potential profile as a simple potential well (Bek and Jakobsson, 1994). Now, of course, potentials can be computed from the true high-resolution structure of potassium channels (Doyle *et al.*, 1998).

A particularly interesting theoretical aspect of this work lies in the connection between level 2 (molecular dynamics and electrostatics) and level 3 (Brownian dynamics). Molecular dynamics simulations are interesting and visually appealing, and can reveal microscopic details of mechanism, but are of such short duration (tens of nanoseconds for the most heroic) for proteins that it is impossible to compute ionic fluxes directly from them. Very large single-channel currents are about 10 pA, which is about one ion every 10 ns. So a very long calculation with a large applied electric field for a high conductance channel might get one ion across the channel. The accurate calculation of even a single point on a current–voltage curve, let alone a family of current–voltage curves for multiple concentrations, is thus beyond the scope of molecular dynamics simulations. In addition to the unavailability of sufficient computing power, there is an unsolved methodological problem. At this writing, nobody has used a boundary condition representing a bulk electrolyte concentration in a molecular dynamics calculation in a satisfactory way. Even if we had a solution to this boundary value problem, it would be more clever to use statistical mechanics than to use brute-force Newtonian simulations to calculate the properties of electrolytes and the kinetics of reactions involving electrolytes.

It is important to emphasize that the ions undergo motion in response to their environment according to the model of atomic interactions established at levels 1 and 2. The bridge from level 2 to level 3 is then established by inserting ion mobilities and free energy profiles for the effective ion/channel interaction into more coarse-grained theories, such as Brownian dynamics calculations. These mobilities and profiles from molecular dynamics simulations are calculated using fundamental statistical mechanical theory and sampling techniques, which we now describe.

In the hierarchy at level 2, ion mobility, or equivalently, the ion diffusion coefficient, in each region of the channel may be obtained by analyzing the ion position and velocity trajectories from molecular dynamics. The diffusion coefficient can then be considered from two points of view. On the one hand, the short-term fluctuations in ion trajectories give rise to the *fluid dynamic diffusion coefficient.* The fluid dynamic diffusion coefficient results from the systematic local friction between the ion and its environment and can be obtained by using the velocity autocorrelation function (VACF) from statistical mechanics. The VACF measures the average change in direction and speed of an ion between two points in time and is given by the general expression $\langle \mathbf{v}(t_1) \cdot \mathbf{v}(t_2) \rangle$, where $\mathbf{v}(t)$ is the velocity vector of the ion at the time t, and the angle brackets denote the average. For simple Brownian motion [as seen, for example, in simpler fluids such as liquid argon (Rahman, 1964)], the VACF is proportional to a single exponential function. In ion channels we focus on only the component of the velocity along the permeation pathway, aligned in the z direction. Under steady-state conditions, the VACF then depends only on the time interval Δt according to

$$\langle v_z(0)v_z(\Delta t)\rangle = A[v_z(0)]^2 \exp(-\Delta t/\tau), \tag{1}$$

where the time constant τ, the reciprocal of the local friction, is related to the fluid dynamic diffusion coefficient D_f by the expression

$$\tau = \frac{m\,D_f}{k_B T}, \tag{2}$$

where m is the mass of the ion, T is the temperature, and k_B is Boltzmann's constant.

Long time differences in ion trajectories, on the other hand, give rise to the *effective diffusion coefficient*. One computes from the fluctuating positions of the ion in the molecular dynamics (MD) calculations the mean-square displacement (MSD) of the ion along the permeation pathway in the time interval Δt, that is,

$$\mathrm{MSD}(\Delta t) = \langle [z(\Delta t) - z(0)]^2\rangle. \tag{3}$$

The diffusion coefficient D for the region is just half the slope of the MSD versus the time lag:

$$D = \frac{1}{2}\frac{d(\mathrm{MSD}(\Delta t))}{d(\Delta t)}. \tag{4}$$

It is generally expected that the fluid dynamic and effective diffusion coefficients vary in magnitude with the ion's position in the channel. In our laboratory we have computed both diffusion coefficients for ions in gramicidin and potassium channels. In these channels the local friction of ions moving within the channel was similar to that of ions in bulk water. Molecular dynamics simulations of the potassium channel revealed that potassium ions in the selectivity filter can be effectively immobilized by the negative structural charges of carbonyl oxygens lining the selectivity filter lumen. This partial confinement of ions in the selectivity filter lowers the *effective* mobility of ions in the selectivity filter (Allen *et al.*, 1999; Chiu *et al.*, 1993). However, the fluid dynamic diffusion coefficient for ions in the selectivity filter is very similar to those values obtained for ions elsewhere in the channel.

As the ion moves within the channel, we expect that an ion will see a series of free energy potential wells and barriers due to the protein along the permeation pathway. As mentioned, the negatively charged amino acid residues in the selectivity filter of potassium channels would provide potential wells for cations along the permeation pathway. In the gramicidin channel, carbonyl oxygens periodically disposed from the protein backbone provide the potential wells. The rate of permeation is defined by how frequently the ions cross the free energy barriers, but, as indicated by the "raw" molecular dynamics trajectories, the ions have a natural tendency to spend practically all their time in the potential wells. Thus, the important rate-limiting regions of the free energy profile are not well-sampled. The key to circumventing this central problem in molecular dynamics simulations is to incorporate the notion of *importance sampling*. In importance sampling, the ion is forced to spend roughly equal time near each position z_0 along the permeation pathway. This is accomplished by first applying to the ion an artificial potential $w(z, z_0)$, which restrains the ion to the position z_0, and then performing molecular dynamics to obtain new, biased trajectories. Then in each small region in which the ion is restrained, the shape of the free energy curve $\mathcal{G}(z, z_0)$ is derived by applying Boltzmann statistics to the biased probability distribution $p(z, z_0)$ of ion positions in that region, according to the equation

$$\mathcal{G}(z, z_0) = -RT \ln p(z, z_0) - w(z, z_0) + C(z_0), \qquad (5)$$

where R is the ideal gas constant, T is temperature, and $C(z_0)$ is a constant to be determined for a given small region. It is in this manner that the artificial restraint compensates for the biased probability distribution. [This procedure and analysis is described in detail in the context of ion permeation in Roux and Karplus (1991).] The entire free energy curve $\mathcal{G}(z)$ is then pieced together from the results from different regions. However, the sampling in a given small region often needs to be improved, for example, by varying the strength or the form of the restraining potential $w(z, z_0)$. The statistics accumulated from all simulations may then be combined using an approach developed by Kumar and co-workers called the weighted histogram analysis method (WHAM). This method has the added benefit of finding an optimal matching of the free energy curves $\mathcal{G}(z, z_0)$ from the individual regions. The result is a profile for the potential of mean force $\mathcal{G}(z)$ along the permeation pathway of the channel.

Although the WHAM approach is rigorous, in practice a simplified electrostatics approach may be used to efficiently approximate the free energy profile of an ion along the permeation pathway of a channel of fixed conformation. For these calculations a realistic dielectric environment is provided for the channel by placing around it a layer of neutral nonpolar spheres of dielectric constant 20, postulated to be situated between the aromatic residues at the lipid–water interfaces (Cowan *et al.*, 1992), to emulate the low-dielectric region of the membrane. The dielectric constant of the protein is set to 20, which appears to be the optimum for computing the ionization states of residues in many proteins (Antosiewicz *et al.*, 1994). Outside these regions the dielectric constant is set to 80, the dielectric constant of water. The program UHBD (Madura *et al.*, 1995), based on the widely used Poisson–Boltzmann continuum electrostatics model for charge interactions in proteins, may then be used to calculate the ionization states of all titratable residues. Most importantly, this program provides for a better approximation to the bulk electrolyte concentration boundary condition than can be achieved using pure molecular dynamics. The total electrostatic force acting on an ion as a function of position along the permeation pathway is calculated directly using UHBD, and the axial component of this force is integrated over the length of the channel to give the potential profile of the ion. In practice, an asymmetry of charge in the channel protein combined with the absence of mobile shielding charges in the electrolyte in this reduced system may be compensated for by adjusting the computed transmembrane potential to zero by the addition of a constant field. The potential-of-mean-force profile thus obtained from molecular dynamics or from electrostatics using a fixed structure can serve as input into the next level of the hierarchy of calculations.

At this time, our laboratory has achieved two successes in accurately predicting ion channel function from structure, with no arbitrary parameters. One is for water permeation through the gramicidin channel (Chiu *et al.,* 1999) and the other is for potassium ions through the KcsA potassium channel (Mashl *et al.* manuscript submitted). Both of these calculations are notable for the ability to stretch the domain of inference from molecular dynamics calculations to well beyond what actually happens in the calculations themselves.

A. Water in Gramicidin Channels

The computations in Chiu *et al.* (1999) are totally straightforward by today's standards. A gramicidin channel is embedded in a phospholipid bilayer and solvated and equilibrated, and the system is permitted to undergo thermal motions for several nanoseconds. The motion of the water in the channel, as revealed by the molecular dynamics simulations, is simple. The water resides in an obligatory single file in the long, narrow channel (see Fig. 2) and shakes. The water motions are much more restricted inside the channel than outside, by a combination of water–water and water–channel hydrogen bonds along the water chain. Every once in a while (in this case, a "while" is a couple of hundred picoseconds), a water molecule breaks free from one end of the chain of channel waters and drifts off into the surrounding electrolyte. Equally often, a water molecule happens to impinge on an end of the chain of waters and attaches itself to the end. In fact, not only do these two events happen with the same frequency, but in our observation (admittedly, of only a few events), these two events occur in mandatory alternation. In our experience, the channel always contains either eight or nine water molecules. The end of the "8-state" is always marked by a water molecule appending itself to one end of the channel to create a "9-state." Similarly, the end of the "9-state" is always marked by a water molecule leaving one end of the channel to create an "8-state." Interestingly, solving all the equations of motion for a system with thousands of degrees of freedom results in a system which can be represented by a trivially simple description. The water motion through the channel is effectively a one-dimensional random walk with the size of each step being one-half of the water–water spacing

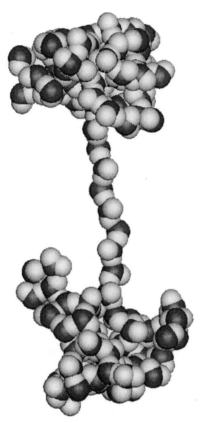

FIGURE 2 Snapshot of a chain of waters in the gramicidin channel during the course of a molecular dynamics simulation. The clusters of waters on the two ends of the channel are in the headgroup region of the surrounding phospholipid membrane (not shown). Net translocations are built up of single events in which a water molecule enters one end of the channel and another water molecule leaves the other end. Hydrogen atoms in the water molecules that are projecting from the side of the water chain are hydrogen-bonded to carbonyl oxygens in the gramicidin peptide (not shown).

in the channel, or about 1.3 Å. Einstein (1905, in Einstein, 1956) showed that this kind of microscopic random walk leads to a macroscopic diffusion, where the diffusion coefficient D is related to the frequency τ and distance λ of the random walk by the expression

$$D = \frac{\lambda^2}{2\tau}.$$

(6)

So, in terms of describing permeation, the entire molecular dynamics calculation gives us two numbers, the frequency and the magnitude of the effective random step of the chain of waters in the channel. From this, the diffusion coefficient is calculated from Eq. (6), and the permeability is derived from the relationship that the permeability is the diffusion coefficient divided by the length of the channel. Remarkably, this calculation replicates within the expected random error the experimentally determined water permeability of the gramicidin channel. Equally remarkably, whereas the experimental determination involves the motion of many thousands of molecules across the channel, in the molecular dynamics simulation that confirmed that number no single water molecule in the channel moved more than about 9 Å within the channel in the course of the simulation. Statistical mechanics analysis of molecular dynamics ion channel trajectories effectively multiplies the effective computer power manyfold.

In addition to being an ion channel and a water channel, the gramicidin channel is the best proton channel known (DeCoursey and Cherny, 2000). The visualization of Fig. 2 shows clearly why this should be so, as there is an excellent pathway for proton transfer from one end of the channel to another.

B. Potassium Ions in the KcsA Channel

The (closed) X-ray structure of the KcsA potassium channel was determined in 1998 (Doyle *et al.,* 1998) and the general form of the motions involved in the channel opening were determined by spin resonance studies (Perozo *et al.,* 1999). Figure 3 shows a very small sampling of ion motions in the channel as simulated by Brownian dynamics, revealing the mechanism of ion translocation across the channel. This figure shows just about 20 ns in the life of the channel. The horizontal axis is time, and the vertical axis is distance along the channel, with the extracellular end of the channel toward the top of the figure. The K channel structure is superimposed on the graph, with the carbonyl groups visualized as sticks pointing in toward the lumen of the narrow selectivity filter. A striking feature of Fig. 3 is that the motion is quite stereotyped. It is seen that most of the time the selectivity filter is occupied by two ions. Permeation through the channel is essentially by a three-ion "carom shot." Every once in a while (now the magnitude of a "while" depends on the bath concentration of permeant ions and the electric field across the membrane) an ion approaches the occupied selectivity filter (here, from the top of the figure) and succeeds in pushing ions out of the other end of the selectivity filter to the other end of the channel. In short, the full calculation produces a picture of stereotyped motions with preferred positions and movements between those positions based on knock-on and knock-off motions, much as has been postulated for years (Hille, 1992, Chapters 9 and 20). But now the rates of knock-on and knock-off are derived directly from detailed physics and the structure of the channel, as opposed to being postulated and built into the model.

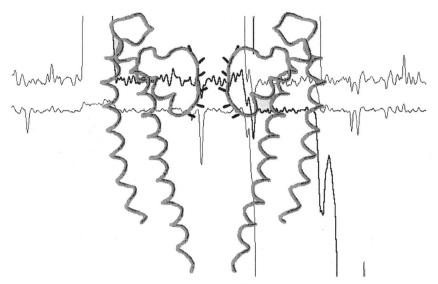

FIGURE 3 Potassium ion trajectories as computed by Brownian dynamics ion trajectories in a model of the open KcsA channel (duration ∼20 ns, under low pH conditions at 0 mV membrane potential in 250 mM K$^+$ solution). Vertical axis is position along the channel axis. Time increases from left to right along the horizontal axis. For ease of visualization the data were smoothed using a window of 120 ps, effectively filtering out oscillations greater than ∼10 GHz. An ion that has successfully traversed the channel is shown in bold relief. Ribbon representations of the backbones (*gray*) from two opposing subunits are superposed, showing the negatively charged carbonyl groups (*black*) projecting into the selectivity filter lumen. The unitary event of ion translocation is a "knock-off" mechanism, in which an ion approaching the doubly occupied selectivity filter from one end of the channel causes the ion in the closer binding site to move to the farther binding site and the ion in the farther binding site to leave the channel from the other end.

The full details of the calculation on which Fig. 3 is based are in a specialized paper (Mashl *et al.,* 2001). In outline, the channel was model-built based on published data from Doyle *et al.* (1998) and Perozo *et al.* (1999), the diffusion coefficient for ions in the channel was determined by molecular dynamics simulations, and the potential profile for ion–channel interactions was calculated by electrostatics. The potentials and the diffusion coefficients were put into a Brownian dynamics model for the channel (Bek and Jakobsson, 1994). Again, the total amount of molecular dynamics that underlies this calculation is very small (about 2 ns) compared to what would be required to directly compute the full electrophysiology of a channel from molecular dynamics (many *micro*seconds for the permeation, orders of magnitude more for the gating).

Although these are early days for working out the details for predicting the ion channel fluxes from structures, we have had enough proof-of-principle that we can be rather confident about how the computations will be structured and that the structure will be according to the chart in Fig. 1. We are now in the stage of refining the algorithms and theory and improving the software. For example, we are working on integrating the different levels of software with each other. It should become possible for the investigator to fill in a Web form with details of the experiments to be simulated for a given channel structure and have the underlying program automatically do both the level 2 (molecular dynamics and electrostatics) and level 3 (Brownian dynamics) calculations and subsequent analysis (e.g., current–voltage and conductance–concentration curves) automatically.

C. Beyond Permeation: Can Computation Help Us Understand Gating?

With regard to computational studies of gating, we are in a much more preliminary state than for permeation. Gating motions are in three dimensions, as opposed to the essentially one-dimensional motion of the permeating ion or water molecule. However, major experimental insights are being generated on gating by insights from such techniques as spin resonance (Barranger-Mathys and Cafiso, 1996; Gross *et al.*, 1999; Jiang *et al.*, 1997; Perozo *et al.*, 1999) and fluorescence resonance energy transfer (Glauner *et al.*, 1999; Selvin, 2000; for a recent review see Bezanilla, 2000). These studies enormously reduce the universe of possible conformations of closed and open channels and the possible motions that may lead from one to the other. As hard as the gating problem is, it should be easier than the overall protein folding problem. Since gating involves relatively subtle conformational changes, it may be somewhat akin to what is called the "endgame" in protein folding, that is, the last stage in packing the protein into its most favorable configuration (Levitt *et al.*, 1997). In gating, however, there are presumably several functionally identifiable, structurally similar favorable configurations, that is, the closed state and the various conducting substates of a channel.

III. UNDERSTANDING CHANNEL FUNCTION FROM STRUCTURAL HOMOLOGY

A. Bioinformatics Studies of Ion Channels

With an increasing amount of sequence data, we are entering a new and exciting era of basic research and applied biotechnology. Hundreds of completely sequenced genomes will be at our disposal in the foreseeable future. Bioinformatics is a multidisciplinary approach to analyzing genomic sequences in an efficient manner. With its roots in molecular evolution, bioinformatics successfully combines biology, mathematics, and computer science into an essentially new discipline. As an emerging method for studying ion channels, bioinformatics will allow us to elucidate the physiological importance of ion channels in many organisms, deduce function relationships, and understand the modular design of ion channels.

B. The Transport Classification System

Beyond the extraction of information from genomic data, bioinformatics also addresses the overall organization of knowledge. The universal classification system for living organisms introduced by Linnaeus allowed the rationalization of complex biological relationships into an evolutionary framework. The Enzyme Commission developed the EC system as a standardized classification system for enzymes in the context of functional relationships. However, the EC system contains no classification for enzymes catalyzing vectorial reactions across cell membranes. Furthermore, the EC system classifies enzymes solely on the basis of function. In order to add the benefits of phylogeny to the classification, the transport classification (TC) system was established (Saier, 2000b). A continuously updated description of the TC system is available on the World Wide Web (http://www-biology.ucsd.edu/~msaier/transport/). A short summary and a table of representative members as well as literature references are available for each permease family.

Unlike the EC system, the TC system classifies permeases on the basis of both function and phylogeny. According to early studies, transporter proteins can be grouped into families exclusively on the basis of degrees of similarity observed for their amino acid sequences. Previous studies in transmembrane transport suggested that molecular phylogeny could serve as a guide for substrate specificity, protein structure, transport mechanism, and mode of energy coupling (Saier, 2000b). Therefore, phylogeny will be a rational basis for the classification of ion channels as well as other transmembrane transporters.

In the TC system, families were established based on the exhibited sequence similarity of their members. In order for two proteins to belong to the same family, they must exhibit a region of 60 amino acids or more in comparable portions of the two proteins with a binary comparison score in excess of nine standard deviations. Therefore, the probability that the degree of sequence similarity between the two proteins occurred by chance is less than 10^{-19}. Binary comparison scores can be calculated using several programs, such as RDF2 (Pearson and Lipman, 1988) or GAP (Devereux *et al.,* 1984) in the GCG package, with proteins assigned to different families if they do not exhibit a sufficient degree of sequence similarity (Saier, 1994). Total similarity is augmented with motif recognition (e.g., the selectivity filter in voltage-gated channels) to resolve ambiguities.

There are five basic criteria governing the nomenclature of the TC systems rationalized by function and phylogeny: transporter class, subclass within the class, family, subfamily, and substrate specificity and polarity. For the KcsA potassium channel that we use for simulation the first three criteria are class 1 (channels), subclass A (α-helical transmembrane subunits), and family 1 (VIC or voltage-gated ion channels). The designation "voltage-gated channels" is historical, but in fact not all of the members of this family are voltage-gated. So the overriding classification criterion is phylogeny, not function. The members of this class are defined by sequence similarity rather than voltage-gating. Again historically, the origin of study of this class of transporters was in the context of neurophysiology, but we now know that this class is ubiquitous in all forms of life including many that do not have nervous systems—microbes, plants, fungi, nematodes, metazoa, etc.

C. Phylogenetics of the VIC Family

The VIC family is one of the 43 members of the α-helical channel subclass (see Table I). Within this subclass, the VIC family is unique in its extreme variety of topological types. A single polypeptide from the VIC family can have 2, 4, 6, 8, 12, or 24 transmembrane segments (TMSs). Members with two TMSs often contain the basic structure necessary to carry out the transport activity. The basic pore-forming monomeric unit consists of a single TMS with a second TMS providing structural support immediately behind the channel-lining helix. The structurally elucidated KcsA channel of *Streptomyces lividans* contains only the minimal channel-forming unit. In this type of two-TMS polypeptide channel of the VIC family, no other structural elements are required for transport activity, but four subunits are necessary to form the channel. The four-TMS polypeptides were the result of an intragenic tandem duplication of the element encoding the two-TMS subunit, and consequently, two, rather than four polypeptides make up the intact channel complex. Many potassium channels of the VIC family include a TMS (TMS 1) involved in the insertion of the protein into the membrane, TMSs 2–4 involved in regulation, and two final TMSs responsible for the formation of a channel that is homologous to the two-TMS channels. On the other hand, calcium channels were found with either 12 or 24 transmembrane segments. In general, the topology of the sodium channels is 24 TMSs. It is proposed that both calcium and sodium channels arose by tandem intragenic duplication (Hille, 1992a,b). As opposed to the homotetrameric structure of the potassium channels, calcium and sodium channels have an overall heterotetrameric structure.

Besides the unique multiple topology of the voltage-gated ion channels, preliminary sequence analyses and phylogenetic studies have suggested modular design of these proteins. Phylogenetic trees constructed by Nelson *et al.* (1999) indicated that a two-TMS pore module of potassium channels emerged early in evolution. This was indicated by the fact that bacterial KcsA protein, the inward-rectifying KGIRK from *Mus musculus,* and outward-rectifying pore modules stem from points near the center of the phylogenetic tree and on distinct branches. All of

TABLE I
Alpha-Type Channels from the TC System

1.A. Alpha-type channels

1.A.1 Voltage-gated ion channel (VIC) superfamily
1.A.2 Animal inward rectifier K^+ channel (IRK-C) family
1.A.3 Ryanodine-inositol 1,4,5-triphosphate receptor Ca^{2+} channel (RIR-CaC) family
1.A.4 Transient receptor potential Ca^{2+} channel (TRP-CC) family
1.A.5 Polycystin cation channel (PCC) family
1.A.6 Epithelial Na^+ channel (ENaC) family
1.A.7 ATP-gated cation channel (ACC) family
1.A.8 Major intrinsic protein (MIP) family
1.A.9 Ligand-gated ion channel (LIC) family of neurotransmitter receptors
1.A.10 Glutamate-gated ion channel (GIC) family of neurotransmitter receptors
1.A.11 Chloride channel (ClC) family
1.A.12 Organellar chloride channel (O-ClC) family
1.A.13 Epithelial chloride channel (E-ClC) family
1.A.14 Nonselective cation channel-1 (NSCC1) family
1.A.15 Nonselective cation channel-2 (NSCC2) family
1.A.16 Yeast stretch-activated, cation-selective, Ca^{2+} channel, Mid1 (Mid1) family
1.A.17 Chloroplast outer envelope solute channel (CSC) family
1.A.18 Chloroplast envelope anion channel-forming Tic110 (Tic110) family
1.A.19 Influenza virus matrix-2 channel (IVC) family
1.A.20 gp91phox phagocyte NADPH oxidase-associated cytochrome b558
    (CybB) H^+-channel family
1.A.21 Bcl-2 (Bcl-2) family
1.A.22 Large conductance mechanosensitive ion channel (MscL) family
1.A.23 Small conductance mechanosensitive ion channel (MscS) family
1.A.24 Gap junction-forming connexin (connexin) family
1.A.25 Gap Junction-forming innexin (innexin) family
1.A.26 Plant plasmodesmata (PPD) family
1.A.27 Phospholemman (PLM) family
1.A.28 P21 holin S (P21 holin) family
1.A.29 l holin S (l holin) family
1.A.30 P2 holin TM (P2 holin) family
1.A.31 LydA holin (LydA holin) family
1.A.32 PRD1 holin M (PRD1 holin) family
1.A.33 T7 holin (T7 holin) family
1.A.34 HP1 holin (HP1 holin) family
1.A.35 T4 holin (T4 holin) family
1.A.36 T4 immunity holin (T4 immunity holin) family
1.A.37 f29 holin (f29 holin) family
1.A.38 f11 holin (f11 holin) family
1.A.39 fAdh holin (fAdh holin) family
1.A.40 fU53 holin (fU53 holin) family
1.A.41 LrgA holin (LrgA holin) family
1.A.42 ArpQ holin (ArpQ holin) family
1.A.43 Cph1 holin (Cph1 holin) family
1.A.44 Urea transporter (UT) family
1.A.45 Urea/amide channel (UAC) family
1.A.46 H^+- or Na^+-translocating bacterial flagellar motor (Mot) family

the six-TMS voltage-gated channels, such as K1.6 from *Rattus norvegicus,* cluster together, as do the six-TMS cyclic-nucleotide-gated (CNG) channels. The configuration of CNG channels with other six-TMS channels also suggested that these two groups of proteins arose independently of each other.

As opposed to the homotetrameric topology of potassium channels, calcium and sodium channels have heterotetrameric structures. Repeats I–IV from representative members of this family were used to study their evolutionary relationships with one another. Several points are noteworthy, according to the analyses of Nelson *et al.* (1999). First, all homologous repeats I cluster together, as do repeats II–IV. This suggests that an intragenic quadruplication event occurred before the gene duplication event that resulted in the individual divergent proteins. Second, within each of the four clusters, sodium- and calcium-specific repeats cluster separately. This suggests that a single evolutionary event gave rise to two different substrate specificities. Third, repeats I and II cluster more closely with each other than they do than with either III or IV. Conversely, III and IV cluster more closely with each other than they do with either I or II. This suggests that quadruplication occurred in two distinct steps in the process of evolution.

The above results suggest in a preliminary way support for the concept of the highly modular design of the VIC family. The basic two-TMS subunit in the homotetrameric formation is sufficient to carry out necessary transport activity in many prokaryotic organisms. Further additions of other modules allow specific cellular targeting and regulatory mechanisms. Exploiting this concept of modularity may make it possible to do much more detailed sequence-based functional annotation and function prediction than has yet been done.

D. Genome Analyses of the VIC Family

Although the voltage-gated ion channel (VIC) family is present in all three kingdoms of life, a search of the complete microbial genomes revealed that just 10 of the 18 (at that time) contained VICs (Paulsen *et al.*, 2000). None of the 18 contained more than two copies of VIC genes. *Escherichia coli, Synechocystis, Aquifex aeolicus, Methanobacterium thermoautotrophicum,* and *Archaeoglobus fulgidus* each was found to contain two genes homologous to members of the VIC superfamily. The first high-resolution structure of this family, KcsA, was from a bacterium, *Synechocystis.* As discussed previously, the X-ray crystallography structure demonstrated that KcsA contains only the basic two-TMS topology (Doyle *et al.*, 1998). A recent analysis of the *Pseudomonas* genome revealed two VIC family channels with possible specificity for potassium (Stover *et al.*, 2000).

Similar whole-genome analyses were also conducted on two eukaryotic genomes. Two members of the VIC family were found in yeast, *Saccharomyces cerevisiae.* According to homology searches, one of these two members found in yeast transports potassium. Another member exhibits homology toward the α subunit of the calcium channels. More than 80 genes encoding VIC superfamily homologues were found in the genome of *Caenorhabditis elegans*. A recently completed genome of *Xylella fastidiosa* also contains a possible potassium channel with homology to the VIC superfamily.

Currently, there is only one reported potassium channel from virus. Described by Plugge *et al.* (2000), a *Chlorella* virus, PBCV-1, encodes a potassium channel. This potassium channel, Kcv1, has an open reading frame (ORF) of 94 codons. It was found to have the classic potassium channel motif, TXGXG. Furthermore, phylogenetic analyses have confirmed that this protein is distantly related to other known potassium channels.

IV. FUTURE PROSPECTS

With respect to the study of voltage-gated ion channels, there are several new and exciting challenges. The TC system needs to be integrated into various software programs and databases to improve understanding on the genome level. An accurate and comprehensive database based on the modules of this family needs to be established. This database should include the very precise functional information that patch clamp techniques have provided for VIC family members. Finally, refined motif definitions and position-specific scoring matrices specific to the VIC family are needed to facilitate detailed functional annotation and modular analysis of VIC sequences.

The state of various techniques in experimental and theoretical biology ensures that in the foreseeable future, we will acquire sequence data on VICs much more quickly than we will structural or functional data. By developing bioinformatics studies to go hand in hand with simulation and experiment, the range of applicability of the experimental and simulation studies will be extended.

Acknowledgments
T.T.T. acknowledges valuable discussions with Dr. Milton H. Saier, Jr., and Dr. Ian T. Paulsen and is also grateful to Mathew Cicero for his assistance in the preparation of this manuscript. This work was supported by grants to E.J. from the National Science Foundation, the National Institutes of Health, and a research fund established from royalties for the Biology Workbench. T.T.T. is a National Institutes of Health Molecular Biophysics trainee.

References
Allen, T. W., Kuyucak, S., and Chung, S.-H. (1999). Molecular dynamics study of the KcsA potassium channel. *Biophys. J.* **77,** 2502–2516.
Antosiewicz, J., McCammon, J. A., and Gilson, M. K. (1994). Prediction of pH-dependent properties of proteins. *J. Mol. Biol.* **238,** 415–436.

Barranger-Mathys, M., and Cafiso, D. S. (1996). Membrane structure of voltage-gated channel forming peptides by site-directed spin-labeling. *Biochemistry* **35**, 498–505.

Bek, S., and Jakobsson, E. (1994). Brownian dynamics study of a multiply-occupied cation channel: Application to understanding permeation in potassium channels. *Biophys. J.* **66**, 1028–1038.

Bezanilla, F. (2000). The voltage sensor in voltage-dependent ion channels. *Physiol. Rev.* **80**, 555–592.

Chang, G., Spencer, R. J., Lee, A. R., Barclay, M. T., and Rees, D. C. (1998). Structure of the MscL homolog from Mycobacterium tuberculosis: A gated mechanosensitive ion channel. *Science* **282**, 2220–2226.

Chiu, S.-W., Novotny, J. A., and Jakobsson, E. (1993). The nature of ion and water barrier crossings in a simulated ion channel. *Biophys. J.* **64**, 98–109.

Chiu, S.-W., Subramaniam, S., and Jakobsson, E. (1999). Simulation study of a gramicidin/lipid bilayer system in excess water and lipid. II. Rates and mechanisms of water transport. *Biophys. J.* **76**, 1939–1950.

Cole, K. S., Antosiewicz, H. A., and Rabinowitz, P. (1955). Automatic computation of nerve excitation. *J. Soc. Indust. Appl. Math.* **3**, 153–172.

Cooper, K., Jakobsson, E., and Wolynes, P. (1985). The theory of ion transport through membrane channels. *Prog. Biophys. Mol. Biol.* **46**, 51–96.

Cowan, S. W., Schirmer, T., Rummel, G., Steiert, M., Ghosh, R., Pauptit, R. A., Jansonius, J. N., and Rosenbusch, J. P. (1992). Crystal structures explain functional properties of two *E. coli* porins. *Nature* **358**, 727–733.

DeCoursey, T. E., and Cherny, V. V. (2000). Common themes and problems of bioenergetics and voltage-gated proton channels. *Biochim. Biophys. Acta* **1458**, 104–119.

Devereux, J., Haeberli, P., and Smithies, O. (1984). A comprehensive set of sequence analysis programs for the VAX. *Nucleic Acids Res.* **12**, 387–395.

Doyle, D. A., Cabral, J. M., Pfuetzner, R. A., Kuo, A. L., Gulbis, J. M., Cohen, S. L., Chait, B. T., and MacKinnon, R. (1998). The structure of the potassium channel: Molecular basis of K^+ conduction and selectivity. *Science* **280**, 69–77.

Dutzler, R., Rummel, G., Alberti, S., Hernandez-Alles, S., Phale, P., Rosenbusch, J., Denedi, V., and Schirmer, T. (1999). Crystal structure and functional characterization of OmpK36, the osmoporin of *Klebsiella pneumoniae*. *Structure* **7**, 425–434.

Einstein, A. (1956). *Investigations on the Theory of the Brownian Motion*. Dover, Mineola, NY.

Fu, D., Libson, A., Miercke, L. J., Weitzman, C., Nollert, P., Krucinski, J., and Stroud, R. M. (2000). Structure of a glycerol-conducting channel and the basis for its selectivity. *Science* **290**, 481–486.

Glauner, K. S., Mannuzzu, L. M., Gandhi, C. S., and Isacoff, E. Y. (1999). Spectroscopic mapping of voltage sensor movement in the Shaker potassium channel. *Nature* **402**, 813–817.

Gross, A., Columbus, L., Hideg, K., Altenbach, C., and Hubbell, W. L. (1999). Structure of the KcsA potassium channel from Streptomyces lividans: A site-directed spin labeling study of the second transmembrane segment. *Biochemistry* **38**, 10324–10335.

Hille, B. (1992). "Ionic Channels of Excitable Membranes," 2nd ed. Sinauer, Sunderland, MA.

Hodgkin, A. L., and Huxley, A. F. (1952). A quantitative description of membrane current and its application to conduction and excitation in nerve. *J. Physiol.* **117**, 500–544.

Jap, B. K., and Walian, P. J. (1996). Structure and functional mechanism of porins. *Physiol. Rev.* **76**, 1073–1088.

Jiang, X., Payne, M. A., Cao, Z., Foster, S. B., Feix, J. B., Newton, S. M., and Klebba, P. E. (1997). Ligand-specific opening of a gated-porin channel in the outer membrane of living bacteria. *Science* **276**, 1261–1264.

Ketchem, R. R., Hu, W., and Cross, T. A. (1993). High-resolution conformation of gramicidin A in a lipid bilayer by solid-state NMR. *Science* **261**, 1457–1460.

Kumar, S., Bouzida, D., Swendsen, R. H., Kollman, P. A., and Rosenberg, J. M. (1992). The weighted histogram analysis method for free-energy calculations on biomolecules. *J. Comp. Chem.* **13**, 1011–1021.

Levitt, M., Gerstein, M., Huang, E., Subbiah, S., and Tsai, J. (1997). Protein folding: The endgame. *Annu. Rev. Biochem.* **66,** 549–579.

Madura, J. D., Briggs, J. M., Wade, R. C., Davis, M. E., Luty, B. A., Ilin, A., Antosiewicz, J., Gilson, M. K., Bagheri, B., Scott, L. R., and McCammon, J. A. (1995). Electrostatics and diffusion of molecules in solution: Simulations with the University of Houston Brownian dynamics program. *Comput. Phys. Commun.* **91,** 57–95.

Mashl, R. J., Tang, Y., Schnitzer, J., and E. Jakobsson (2001). Heirarchical approach to predicting permeation in ion channels. *Biophys. J.* **81,** 3005–3015.

Meyer, J. E. W., Hofnung, M., and Schulz, G. E. (1997). Structure of maltoporin from *Salmonella typhimurium* ligated with a nitrophenyl-maltotrioside. *J. Mol. Biol.* **266,** 761–775.

Murata, K., Mitsuoka, K., Hirai, T., Walz, T., Agre, P., Heymann, J. B., Engel, A., and Fujiyoshi, Y. (2000). Structural determinants of water permeation through aquaporin-1. *Nature* **407,** 599–605.

Nelson, R. D., Kuan, G., Saier, M. H. Jr., and Montal, M. (1999). Modular assembly of voltage-gated channel proteins: A sequence analysis and phylogenetic study. *J. Mol. Microbiol. Biotechnol.* **1,** 281–287.

Paulsen, I. T., Nguyen, L., Sliwinski, M. K., Rabus, R., and Saier, M. H. Jr. (2000). Microbial genome analyses: Comparative transport capabilities in eighteen prokaryotes. *J. Mol. Biol.* **301,** 75–101.

Pearson, W. R., and Lipman, D. J. (1988). Improved tools for biological sequence comparison. *Proc. Natl. Acad. Sci. USA* **85,** 2444–2448.

Perozo, E., Cortes, D. M., and Cuello, L. G. (1999). Structural rearrangements underlying K^+ channel activation gating. *Science* **285,** 73–78.

Plugge, B., Gazzarrini, S., Nelson, M., Cerana, R., Van Etten, J. L., Derst, C., DiFrancesco, D., Moroni, A., and Thiel, G. (2000). A potassium channel protein encoded by Chlorella virus PBCV-1. *Science* **287,** 1641.

Rahman, A. (1964). Correlations in the motion of atoms in liquid argon. *Phys. Rev.* **136,** A405–A411.

Roux, B., and Karplus, M. (1991). Ion transport in a model gramicidin channel: Structure and thermodynamics. *Biophys. J.* **59,** 961–981.

Saier, M. H., Jr. (1994). Computer-aided analyses of transport protein sequences: Gleaning evidence concerning function, structure, biogenesis, and evolution. *Microbiol. Rev.* **58,** 71–93.

Saier, M. H., Jr. (2000a). Families of proteins forming transmembrane channels. *J. Membr. Biol.* **175,** 165–180.

Saier, M. H., Jr. (2000b). A functional–phylogenetic classification system for transmembrane solute transporters. *Microbiol. Mol. Biol. Rev.* **64,** 354–411.

Saier, M. H., Jr., and Tseng, T.-T. (1999). Evolutionary origins of transmembrane transport systems. *In* "Transport of Molecules Across Microbial Membranes". (J. K. Broome-Smith, S. Baumberg, C. J. Stirling, F. B. Ward, eds.), pp. 252–274. Cambridge University Press, Cambridge.

Selvin, P. R. (2000). The renaissance of fluorescence resonance energy transfer. *Nat. Struct. Biol.* **7,** 730–734.

Stover, C. K., Pham, X. Q., Erwin, A. L., Mizoguchi, S. D., Warrener, P., Hickey, M. J., Brinkman, F. S., Hufnagle, W. O., Kowalik, D. J., Lagrou, M., Garber, R. L., Goltry, L., Tolentino, E., Westbrock-Wadman, S., Yuan, Y., Brody, L. L., Coulter, S. N., Folger, K. R., Kas, A., Larbig, K., Lim, R., Smith, K., Spencer, D., Wong, G. K., Wu, Z., Paulsen, I. T., Reizer, J., Saier, M. H. Jr., Hancock, R. E. W., Lory, S., and Olson, M. V. (2000). Complete genome sequence of *Pseudomonas aeruginosa* PAO1, an opportunistic pathogen. *Nature* **406,** 959–964.

PART III

Experimental Investigations of
Peptide-Lipid Interactions

CHAPTER 10

The Role of Electrostatic and Nonpolar Interactions in the Association of Peripheral Proteins with Membranes

Diana Murray,[*,1,2] **Anna Arbuzova,**[†] **Barry Honig,**[*]
and Stuart McLaughlin[†,1]

*Department of Biochemistry and Molecular Biophysics, Columbia University, New York, New York 10032; †Department of Physiology and Biophysics, Health Sciences Center, SUNY, Stony Brook, New York 11794

I. Introduction
II. Electrostatic and Hydrophobic Interactions Mediate the Membrane Association of Important Biological Proteins
III. The Membrane Interaction of Simple Basic Peptides: Theory and Experiment
IV. The Combination of Experimental and Theoretical Approaches Provides a Detailed Comprehensive Picture of the Membrane Association of Myristoylated Proteins
    A. Src: Myristate plus Basic Cluster
    B. MARCKS
V. Future Directions
    A. Membrane Targeting
    B. Computational Studies
    References

[1]Address correspondence to Diana Murray (*Dim2007@med.cornell.edu*) or Stuart McLaughlin (*smcl@epo.som.sunysb.edu*).
[2]Current address: Diana Murray, Dept. of Microbiology and Immunology, Weill Medical College of Cornell University, 1300 York Avenue, Box 62, New York, NY 10021.

Current Topics in Membranes, Volume 52

The membrane association of acylated and prenylated peripheral proteins, such as Src, the myristoylated alanine-rich C kinase substrate (MARCKS), and K-ras4B, plays an important role in cellular signal transduction. This chapter reviews experimental and computational studies of the membrane partitioning of peptides corresponding to the membrane-interacting regions of these proteins. The computational model partitions the membrane interaction free energies into three components: electrostatic attraction between basic groups on the protein and acidic phospholipids in the membrane, desolvation of the protein and membrane as they associate, and nonpolar burial of aromatic side chains into the membrane interface. The electrostatic components of the binding free energy are calculated by solving the Poisson–Boltzmann equation for protein/membrane systems represented in atomic detail (finite-difference Poisson–Boltzmann [FDPB] method). The nonpolar component is calculated as the product of an interfacial hydrophobicity coefficient and the change in solvent-accessible surface area of aromatic side chains as they penetrate the interface. The model predicts how membrane association changes as a function of the electrostatic properties of the system and how different combinations of electrostatic and nonpolar forces dictate a wide range of membrane-binding properties. The success of the FDPB methodology in describing experimentally characterized biophysical systems establishes the applicability of classical electrostatics and the continuum approach to protein/membrane systems and justifies its extension to predicting the structural origins of the interfacial association of proteins of known structure. The biological implications of recent experimental measurements of the partitioning of MARCKS onto membranes containing phosphatidylinositol 4,5-bisphosphate are discussed.

I. INTRODUCTION

The binding of peripheral proteins to membranes is crucial to many biological processes, such as signal transduction, vesicle trafficking, and retroviral assembly, and is often accomplished through protein motifs that interact with membrane lipids. Many peripheral proteins use domains of well-defined structure to target specific lipids (Hurley and Misra, 2000). For example, the pleckstrin homology (PH) domain of phospholipase C-δ (PLC-δ) anchors the protein to phosphatidylinositol 4,5-bisphosphate (PIP$_2$) in the plasma membrane, the C1 domain of protein kinase C (PKC) binds to diacylglycerol in nuclear and plasma membranes, and the FYVE domain of the early endosome associated protein (EEA1) binds specifically to phosphatidylinositol 3-phosphate, a lipid often found localized to endosomal membranes. Other proteins use unstructured motifs that bind membranes through nonspecific electrostatic (Murray *et al.,* 1997) and nonpolar interactions (Bhatnagar and Gordon, 1997; Johnson and Cornell, 1999; Resh, 1999). For example, a substrate of PKC, MARCKS, binds to the plasma membrane through hydrophobic insertion of an N-terminal myristate into the membrane hydrocarbon and electrostatic interaction of a cluster of basic amino acids with acidic phospholipid headgroups; the nonreceptor tyrosine kinase Src binds in a similar manner (McLaughlin and Aderem, 1995). In this chapter, we focus on the nonspecific electrostatic and nonpolar interactions responsible for the membrane localization of proteins like MARCKS and Src. We stress the interplay of experiment and computation, which provides a powerful approach to describing the underlying molecular and physical basis of these interactions.

II. ELECTROSTATIC AND HYDROPHOBIC INTERACTIONS MEDIATE THE MEMBRANE ASSOCIATION OF IMPORTANT BIOLOGICAL PROTEINS

In vivo studies have shown that a number of proteins, for example, K-ras4B (Hancock *et al.,* 1990), Src (Sigal *et al.,* 1994), MARCKS (Seykora *et al.,* 1996; Swierczynski and Blackshear, 1996), and human immunodeficiency virus type 1 Gag polyprotein (HIV-1 Gag) (Rein *et al.,* 1986; Zhou and Resh, 1996), require both a lipophilic attachment (either an acyl chain or a prenyl group) and a cluster of basic residues to bind to the plasma membrane and that membrane association is crucial for their biological functions. For example, the small GTPase K-ras4B has a CAAX motif at its C-terminus, which is sequentially farnesylated, AAX-proteolyzed, and, finally, methylated (Magee and Marshall, 1999). The farnesyl chain (a 15-carbon isoprenoid) partitions hydrophobically into the membrane hydrocarbon, but does not provide sufficient energy to anchor the protein to membranes (Ghomashchi *et al.,* 1995; Silvius and l'Heureux, 1994). K-ras4B has a second membrane-binding motif, a C-terminal cluster of basic residues (see Table I). Hancock *et al.* (1990) showed that the plasma membrane targeting of K-ras4B requires both the CAAX motif and the adjacent basic cluster. Biophysical studies with peptides corresponding to the C-terminus of K-ras4B show that

TABLE I

Clusters of Basic Residues That Interact with Acidic Phospholipids[a]

Src	myristate-GSS**KSKPK***D*PSQ**RRR**
MARCKS	myristate-... **KKKKKR**FSF**KK**SF**K**LSGFSF**KKNKK**
HIV-1 Gag	myristate-GA**R**ASVLSGG*EL*D**R**W*E***K**I**RLR**PGG**KKK**Y**K**L
K-ras4B	G**KKKKKK**S**K**TSC-farnesyl
Rap1-A	PV*E***KKK**P**KKK**SC-geranylgeranyl
Rac-1	CPPPV**KKRKRK**C-geranylgeranyl
Rho-A	LQA**RR**G**KKK**SGC-geranylgeranyl
BASP-1	myristate-GG**K**LS**KKKK**GY

[a]Basic residues are boldface, acid residues are italic.

partitioning to membranes containing 20 mol% acidic phospholipid is 300-fold stronger than the partitioning to pure zwitterionic phospholipid membranes, suggesting the basic sequence may contribute favorable electrostatic interactions to the membrane association of the intact protein (Ghomashchi *et al.*, 1995; Leventis and Silvius, 1998). Other small GTPases, for example, Rap1-A, Rac-1, and Rho-A, also contain an isoprenyl group and an adjacent cluster of basic residues (see Table I).

The role of hydrophobic and electrostatic interactions in the membrane association of myristylated proteins (Src, MARCKS, HIV-1 Gag; see Table I) is also well established. Like the farnesyl group of K-ras4B, myristate (a 14-carbon fatty acid) partitions hydrophobically into the membrane hydrocarbon (Bhatnagar and Gordon, 1997; Ghomashchi *et al.*, 1995; Peitzsch and McLaughlin, 1993; Resh, 1999; Silvius and l'Heureux, 1994). Src and MARCKS are discussed in more detail in the sections below. HIV-1 Gag plays a central role in the assembly of new virions in HIV-infected cells (Garnier *et al.*, 1998; Wills and Craven, 1991). Studies from many labs have shown that plasma membrane association of Gag is crucial for virion formation and that myristoylation is required for this association (Bryant and Ratner, 1990; Rein *et al.*, 1986; Zhou *et al.*, 1994). As for farnesyl, myristate alone is not sufficient to anchor proteins at the plasma membrane, and like K-ras4B, HIV-1 Gag has a cluster of basic residues adjacent to the site of lipid modification (Table I) that contributes to membrane association through electrostatic interactions with acidic lipids (Zhou *et al.*, 1994). The structure of the N-terminal matrix domain of HIV-1 Gag revealed that these basic residues form a flat, positively charged basic surface adjacent to the site of myristoylation (Hill *et al.*, 1996).

What is the purpose of nonspecific association to membrane surfaces? The translocation of a protein from the cytoplasm to a membrane facilitates its interactions with other membrane-bound or embedded molecules by a "reduction-of-dimensionality" mechanism. Kinetic analyses suggest that the most important feature of this phenomenon is a simple enhancement of the equilibrium concentrations of the molecules at the membrane surface (Kholodenko *et al.*, 2000; McCloskey and Poo, 1986). Membrane association effectively increases the concentration of a peripheral protein in a thin ($d \approx 1$ nm) surface layer adjacent to the membrane. Assuming the cell is a sphere of radius R equal to a few micrometers, one finds that the volume of the surface phase ($V = 4\pi R^2 d$) is about 1/1000 the volume of the cell ($V = 4\pi R^3/3$). Thus, anchoring a protein to a cell's plasma membrane increases its concentration 1000-fold, greatly enhancing its ability (or the ability of proteins it may recruit to the plasma membrane) to interact with effectors and substrates. For example, Ras-GTP mediates the plasma membrane association of the serine/threonine kinase Raf1; this recruitment facilitates a number of post-translational modifications to Raf1 which activate the kinase (Morrison and Cutler, 1997). Similarly, plasma membrane association of HIV-1 Gag polyproteins facilitates the homo-oligomerization required for virion assembly and budding (Garnier *et al.*, 1998), and translocation of PKC from the cytoplasm to the membrane (Oancea and Meyer, 1998) facilitates its ability to phosphorylate its membrane-bound substrates, such as MARCKS.

A number of recent reviews describe the different "switch" mechanisms by which the membrane association of lipid-modified proteins may be regulated (Johnson and Cornell, 1999; McLaughlin and Aderem, 1995; Resh, 1999). The availability of myristate for membrane partitioning is regulated in a number of proteins. The myristate moieties of the inactive forms of both the small GTPase Arf and the visual transduction protein recoverin are sequestered in hydrophobic cavities in the proteins, keeping the proteins cytosolic. Activation of Arf and recoverin by GTP and calcium, respectively, causes protein conformational changes that result in the extrusion of myristate and concomitant membrane targeting. Recent work indicates that the membrane association of the HIV-1 matrix domain is regulated by proteolysis: In the context of a newly formed virion, the Gag polyprotein is cleaved by the viral protease, exposing an as-yet-unknown "signal" on the matrix domain that sequesters the myristate chain (Spearman *et al.,* 1997; Zhou and Resh, 1996). Since the hydrophobic energy due to myristate partitioning alone is not enough to keep a protein adsorbed to the plasma membrane (Peitzsch and McLaughlin, 1993), mechanisms that interfere with other membrane-binding motifs may result in significant translocation of the protein to the cytosol. As discussed in more detail below, the membrane association of MARCKS is regulated by a myristoyl electrostatic switch (McLaughlin and Aderem, 1995). Phosphorylation of serines within the basic cluster by PKC weakens the electrostatic attraction of the basic effector region to acidic lipids and results in the desorption of MARCKS from plasma membrane to cytosol in many cell types. The use of multiple membrane-interacting motifs by peripheral proteins allows for a wide variety of signal-specific membrane recruitment and translocation mechanisms.

Proteins in the same family may use different motifs to bind to membranes (Bhatnagar and Gordon, 1997; Resh, 1999). Src appears to be the only member of the Src family of nonreceptor tyrosine kinases that uses an acyl chain plus basic cluster. Other family members are also N-myristylated, but have an adjacent palmitate (16-carbon fatty acid) moiety instead of a basic cluster. The Ras isoforms H-Ras and N-Ras lack the C-terminal basic cluster of K-ras4B, but are palmitylated on C-terminal cysteines. The different types of lipid modifications seem to be correlated with localization. For example, the dually acylated Src family members are localized in detergent-resistant, cholesterol-enriched fractions of membranes known as rafts, as are other molecules with multiple saturated acyl chains, such as GAP43 (D. A. Brown and London, 2000; Simons and Ikonen, 1997).

Other peripheral proteins, such as the type II β-phosphatidylinositol 3-kinase, AKAP79, mylein basic protein, and a number of proteins containing C2 domains, use basic groups to bind to membrane surfaces, but have no lipophilic modifications. Although the focus here is on proteins that use both a basic cluster and a lipid group to bind to membranes, the principles that are described should be of general applicability to a wide range of peripheral proteins.

III. THE MEMBRANE INTERACTION OF SIMPLE BASIC PEPTIDES: THEORY AND EXPERIMENT

Traditionally, the electrostatic properties of membranes and peptide/membrane systems have been described by "smeared-charge" models based on Gouy–Chapman (GC) theory (McLaughlin, 1989). This theory assumes the charges due to the acidic lipids are smeared uniformly over a planar membrane surface and can be described by a uniform surface charge density. Studies from a number of labs have established that the electrostatic properties of phospholipid membranes can be described adequately by smeared-charge theory, at least for membranes containing physiological concentrations (>10%) of monovalent acidic lipids (Peitzsch *et al.*, 1995). As illustrated in Fig. 1 A, the electrostatic equipotential profiles above a bilayer containing 33 mol% acidic phospholipid in 100 mM KCl are essentially flat and uniform. When the membrane contains a low concentration of acidic lipid, it can be shown that the essentially hemispherical equipotential profiles around each lipid are given approximately by twice the Debye–Hückel expression; the factor of two arises because of the "image charge" effect (McLaughlin, 1989). Figure 1B illustrates the −25 mV electrostatic equipotential profile of a 33 mol% acidic lipid membrane that contains a single PIP$_2$ molecule; the multivalent acidic lipid significantly enhances the negative potential of the membrane in its vicinity.

GC theory has also been applied to describe the membrane binding of multivalent peptides (Ben-Tal *et al.*, 1996; Heimburg and Marsh, 1996; Mosior and McLaughlin, 1991, 1992). The charges on the bound peptide are treated as point charges that do not perturb the electrostatic potential of the membrane. Their positive charges, like the charges on acidic lipids, are smeared uniformly on the planar membrane surface and contribute to a net surface charge density. In order to reproduce quantitatively the experimentally determined membrane partitioning, GC theory typically has to invoke an effective charge for the peptide that is significantly less than its net charge (Heimburg and Marsh, 1996; Mosior and McLaughlin, 1991, 1992). In addition, smeared-charge models cannot account for complex patterns of electrostatic potential, which depend upon a protein's shape as well as the specific location of its charged and polar groups (Honig and Nicholls, 1995). Recent computational studies show that in order to describe the observed electrostatic properties of membranes with adsorbed basic peptides, it is crucial to account for the molecular detail of the peptides (Murray *et al.*, 1999). As shown in Fig. 1C, an adsorbed basic peptide perturbs the electrostatic potential in a highly localized manner and produces a strong positive potential in its vicinity.

Work from the Honig lab has applied the Poisson–Boltzmann (PB) equation to describe the electrostatic properties of proteins and nucleic acids (Honig and Nicholls, 1995). The PB equation is solved numerically in the finite-difference approximation (FDPB) for static, atomic-level models of the biological macromolecules. The surrounding aqueous phase is represented implicitly as a

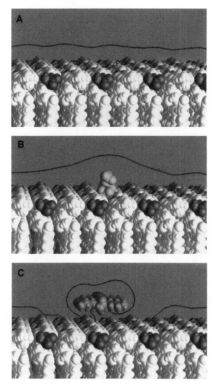

FIGURE 1   The negative electrostatic potential of a 2:1 PC/PS membrane is significantly perturbed by multivalent molecules. The red (blue) lines represent the two-dimensional -25 ($+25$) mV equipotential contours obtained from FDPB calculations. In each panel, the membrane composition is 2:1 PC/PS and the ionic strength, [KCl], is 100 mM. (A) The -25 mV electrostatic equipotential contour is essentially flat and located ~ 10 Å above the membrane surface. (B) A trivalent PIP_2 lipid (headgroup colored yellow) greatly increases the negative potential of the membrane in its vicinity. (C) A membrane-adsorbed basic peptide (heptalysine; blue) introduces a significant positive potential at the membrane surface. (See color plate.)

homogeneous medium of dielectric constant 80. Monovalent salt ions are treated in the mean-field approximation, so that ion–ion interactions and the desolvation of ions near molecular surfaces are ignored. In the FDPB implementation, an ion exclusion layer of 2 Å thickness is applied to the molecular surfaces of all molecules (see Fig. 3 of Ben Tal *et al.,* 1996). As recently reviewed (Vlachy, 1999), PB theory has been tested extensively against computer simulations that treat ions explicitly, and for many situations, its description is adequate. Indeed, it has been highly successful in predicting salt effects observed for macromolecular systems (Honig and Nicholls, 1995) (see below).

Recently the FDPB methodology has been successfully applied to describe the electrostatic properties of membranes and protein/membrane systems (Ben-Tal *et al.*, 1996; Murray *et al.*, 1997; Peitzsch *et al.*, 1995). The assumption that water molecules adjacent to a membrane surface can be treated theoretically as a dielectric continuum is supported by surface force, X-ray diffraction, and other experiments, as reviewed elsewhere (McLaughlin, 1989). The FDPB approach sacrifices accuracy in the representation of molecular details and motions, but retains the ability to accurately describe long-range electrostatic interactions, which are important for protein/membrane interactions. Molecular dynamics (MD) simulations, on the other hand, sacrifice accuracy in the physical model (e.g., through truncation or approximation of long-range electrostatic interactions) to achieve accuracy in molecular representation. Hence, MD simulations have been extremely valuable in depicting the short-time (less than microseconds) detailed motions of lipids in a membrane environment (Merz and Roux, 1996), but have been limited in their capacity to describe the electrostatic properties of membranes and protein/membrane systems (Jakobsson, 1997). As illustrated by the examples below, the FDPB approach provides a computational, structure-based framework for (1) testing detailed physical models, (2) predicting the electrostatic properties of protein/membrane systems, and (3) dissecting the various energetic components of protein/membrane interactions and assessing quantitatively their importance.

The FDPB methodology is described in detail elsewhere (Ben-Tal *et al.*, 1996; Gilson and Honig, 1988; Misra *et al.*, 1994). Here an overview is given that focuses on issues relevant to membrane and protein/membrane systems. The application of the methodology is illustrated by describing the interaction of a heptalysine peptide ($Z = +7$) with a membrane composed of zwitterionic (phosphatidylcholine, PC) and acidic (phosphatidylserine, PS) lipids (Fig. 1C). This type of system is a good model for how the N-terminal basic cluster on Src and the C-terminal basic cluster on K-ras4B interact with membranes (Murray *et al.*, 1997). The membrane models are built as described in Peitzsch *et al.* (1995): Atomic-level models of PC and PS lipids are arranged on a hexagonal lattice in which each lipid occupies approximately 68 $\mathring{A}^2$ in the plane of the membrane (McIntosh and Simon, 1986). The peptide in Fig. 1C is docked in the aqueous solution above the membrane in its minimum electrostatic free energy orientation. The electrostatic interaction free energy can be calculated as a function of the orientation of the peptide at the membrane surface, the mole percent acidic lipid in the membrane, and the ionic strength of the solution.

For the numerical solution of the PB equation, each atom of the peptide and membrane is assigned a radius and a partial charge, which is located at its nucleus. The peptide/membrane model is then mapped onto a three-dimensional lattice of points, each of which represents a small region of the peptide, membrane, or solvent. Smooth molecular surfaces for the peptide and membrane are generated by rolling a spherical probe with the radius of a water molecule (1.4 $\mathring{A}$) over the surfaces defined by the van der Waals radii of the constituent atoms. Lattice points that lie within the molecular surfaces of the peptides and bilayer are assigned a dielectric constant of 2, and lattice points outside the molecular surfaces, corresponding to the aqueous phase, are assigned a dielectric constant of 80. Salt ions are excluded from a region that extends 2 Å (the radius of a Na^+ ion) beyond the van der Waals surfaces of the peptide and membrane, as illustrated in Fig. 3 of Ben Tal *et al.* (1996). The electrostatic potential and the mean distribution of the monovalent salt ions at each lattice point are calculated by solving the nonlinear Poisson–Boltzmann equation,

$$\nabla[\epsilon(r)\nabla\phi(r)] - \epsilon_r\kappa(r)^2 \sinh[\phi(r)] + \frac{e^2}{\epsilon_0 kT}\rho^f(r) = 0,$$

where $\epsilon(r)$ is the dielectric constant, $\phi(r)$ is the electrostatic potential, and $\rho^f(r)$ is the charge density of the fixed charges. The PB equation is mapped onto the cubic lattice and solved for $\phi(r)$ using the finite-difference approximation and the quasi-Newton method (Holst, 1993) combined with three levels of multigriddings (Holst and Saied, 1993).

A sequence of focusing runs of increasing resolution is employed to accurately calculate the electrostatic potentials. In a typical initial calculation, the peptide/membrane model fills a small percentage of the lattice ($\sim 10\%$) and the potentials at the boundary points of the lattice are approximately zero. This procedure ensures that the system is electroneutral. Subsequent calculations employ successively finer resolutions and extract the potentials at boundary points from a preceding, coarser grain calculation. The solutions $\phi(r)$ to the Poisson–Boltzmann equation are used to calculate the electrostatic free energy of the peptide/membrane system $G_{el}(P \cdot M)$, using an equation given by Sharp and Honig (1990). The electrostatic free energy of interaction ΔG_{el} is calculated as the difference between the electrostatic free energy of the peptide and membrane when they are close together, $G_{el}(P \cdot M)$, and that when they are far apart, $G_{el}(P)$ and $G_{el}(M)$:

$$\Delta G_{el} = G_{el}(P \cdot M) - \{G_{el}(P) + G_{el}(M)\}.$$

The curve given by the triangles in Fig. 2 illustrates the calculated electrostatic free energy of interaction as a function of the distance R between the van der Waals surfaces of the peptide and the membrane for an orientation in which the peptide is parallel to the membrane surface as in Fig. 1C. When the peptide is far from the membrane surface, it experiences an electrostatic attraction that drives it toward the negatively charged membrane. This "Coulombic" attraction increases as the peptide approaches the membrane. Close to the membrane surface, charged and polar groups on both the peptide and membrane are desolvated, or stripped of water molecules, which is energetically unfavorable and results in a repulsion at small distances. The minimum electrostatic free energy of interaction for heptalysine is

Lys7 on 2:1 PC:PS, 100 mM KCl

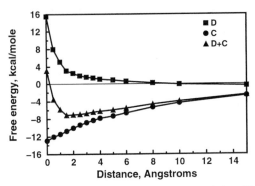

FIGURE 2 FDPB calculations of the peripheral membrane association of heptalysine. The electrostatic interaction free energies are plotted as a function of distance between the van der Waals surfaces of a heptalysine peptide and a 2:1 PC/PS bilayer in 100 m*M* KCl. The circles illustrate the long-range electrostatic attraction between basic groups on the peptide and the acidic phospholipid headgroups. The squares illustrate the unfavorable desolvation energy. The balance between the long-range attraction and short-range desolvation repulsion results in an electrostatic free energy minimum (triangles) when the peptide and membrane are separated by a layer of water. See Murray *et al.* (1999) for details of the calculations.

predicted to occur where the van der Waals surfaces of the peptide and membrane are separated by a distance of about the thickness of a layer of water. At this distance the Coulombic attractive force is balanced by the repulsive desolvation force.

The components of the electrostatic free energy of interaction, namely the Coulombic attraction and the desolvation repulsion, can be calculated explicitly. The desolvation penalty ΔG_{desolv} is determined at each R by calculating to what extent the presence of the peptide and the membrane shield the others from favorable interactions with the solvent (Misra and Honig, 1995; Misra *et al.*, 1994). This is done by discharging the peptide and membrane in turn and calculating the electrostatic free energy of the system:

$$\Delta G_{\text{desolv}} = \Delta G_{\text{el}}(P(Q = 0) \cdot M) + \Delta G_{\text{el}}(P \cdot M(Q = 0)).$$

The Coulombic component ΔG_{Coul} is simply the difference between the total electrostatic free energy of interaction and the desolvation penalty:

$$\Delta G_{\text{Coul}} = \Delta G_{\text{el}} - \Delta G_{\text{desolv}}.$$

Both the Coulombic and desolvation components are plotted in Fig. 2. The circles illustrate the long-range electrostatic attraction between basic groups on the peptide and the acidic lipid headgroups; this attraction increases as the peptide approaches the membrane surface. The squares illustrate the shorter range desolvation repulsion; as the peptide approaches the membrane surface, both molecules are stripped of water molecules, resulting in the transfer of charged and polar groups from a region of high dielectric constant in the aqueous phase to a region of low dielectric constant. As the peptide comes very close to the membrane, the repulsive desolvation force becomes greater than the attractive Coulombic force, resulting in an electrostatic minimum free energy orientation in which the peptide adsorbs outside the envelope of the polar headgroup region of the membrane. This prediction is consistent with nuclear magnetic resonance (NMR; Roux *et al.,* 1988), high-pressure fluorescence (Montich *et al.,* 1993), and monolayer surface pressure measurements (Ben-Tal *et al.,* 1996), which indicate that simple basic peptides do not penetrate the membrane interface. As shown below, favorable nonpolar interactions, mediated by hydrophobic groups, may contribute significantly to the membrane association and tip the force balance toward interfacial penetration.

In the FDPB calculations of the electrostatic component of peptide/membrane interactions to date, it has been assumed that the lipids are frozen, both vertically and laterally. Specifically, the calculations assume the lipids do not demix laterally at the membrane surface as the peptide adsorbs, that the orientation of the lipid headgroups is constant, and that there are no membrane undulations or lipid bobbing motions. Although these approximations are extreme (see, e.g., McIntosh and Simon, 1994), the FDPB calculations apparently capture the essence of the equilibrium binding of peripheral peptides. Experiments showing that pentalysine partitions similarly onto dimyristoylphosphatidylcholine/dimyristoylphosphatidylglycerol (DMPC/DMPG) membranes in the liquid crystalline and gel phases suggest that, at least for this type of peripheral membrane association, lipid motions and demixing can be ignored (Kim *et al.,* 1991). In addition, electron paramagnetic resonance (EPR) experiments showed that spin-labeled acidic lipids do not accumulate around pentalysine when it adsorbs to the membrane surface (Kleinschmidt and Marsh, 1997). A recent theoretical paper addresses the issue of lipid demixing produced by the adsorption of an idealized basic protein and predicts that under physiological conditions the demixing effect should not be large for monovalent acidic lipids (May *et al.,* 2000). Below, experimental results are discussed that suggest demixing effects can be very large for the multivalent acidic lipid PIP_2. Aside from the greater computational expense, there is no reason why lateral diffusion of lipids and lipid demixing cannot be incorporated into FDPB calculations.

The concentration of peptide at each distance from the membrane surface is proportional to the exponent of its interaction free energy. As suggested by the electrostatic free energy curve (Fig. 2, triangles), the bound peptides, which associate with the membrane through long-range electrostatic interactions, can be located at an appreciable distance from the membrane surface. For example, the peptide concentration at $R = 6$ Å is predicted to be only 10-fold lower than that at $R = 3$ Å and 5000-fold greater than the concentration in the bulk. This is the hallmark of a "nonspecific" electrostatic interaction and is more accurately described by a partition coefficient, rather than a binding constant, which assumes the formation of a 1 : 1 complex between a peptide and a lipid (White *et al.*, 1998). The "excess" peptide concentration at a distance R from the membrane is defined as the difference between the peptide concentration at R and the bulk peptide concentration. Integrating the excess peptide concentration over R gives the Gibbs surface excess (see Fig. 2 in Murray *et al.*, 1997), that is, moles of peptide bound per unit area of membrane surface. The Gibbs surface excess is simply related to the molar partition coefficient that is measured experimentally.

In practice, the Gibbs surface excess is calculated by considering many randomly chosen orientations of the peptide with respect to the membrane at each R (Ben-Tal *et al.*, 1996; Murray *et al.*, 1998). Although the peptide model is rigid throughout the calculations, the incorporation of many different orientations approximates a complete sampling over different peptide structures. For simple basic peptides like heptalysine, the molar partition coefficient calculated in this way consistently gives a binding energy that underestimates the experimentally determined value by 1.5 kcal/mol. This suggests that the model ignores and/or misrepresents some interactions. Nevertheless, an important result obtained from the comparison of computational predictions and experimental results for simple model systems is that the theoretical model correctly predicts how the membrane binding depends on factors that affect the electrostatic properties of the system. [Indeed, for predicting relative binding effects, it has been found that consideration of a single orientation of the peptide with respect to the membrane is sufficient (Ben-Tal *et al.*, 1996, 1997; Murray *et al.*, 1998)]. For example, the model correctly predicts that the binding of charybdotoxin ($Z = +4$) decreases by five orders of magnitude when the monovalent ion concentration in solution is increased from 10 to 150 mM (Ben-Tal *et al.*, 1997). The model also adequately describes the observation that the binding free energy of Lys$_3$, Lys$_5$, and Lys$_7$ onto 2 : 1 PC/PS in 100 mM KCl is -3, -5 and -7 kcal/mol, respectively, that is, the binding free energy under physiological conditions is about -1 kcal/mol per basic residue (Ben-Tal *et al.*, 1996). Computational studies on experimentally well characterized systems such as these show that the FDPB methodology can account for how the membrane binding increases as the ionic strength of the solution decreases, as the mole percent acidic lipid increases, and as the net charge of the peptide increases.

Experimental studies show that the binding of model peptides to membranes depends only weakly on the chemical nature of either the basic residues (Arg vs. Lys) or the monovalent acidic lipid (PS vs. PG, phosphatidylglycerol) (Kim *et al.*, 1991; Mosior and McLaughlin, 1992). Taken together, these observations support a model in which long-range, nonspecific electrostatic interactions drive the membrane association and that simple basic sequences of the form (Lys)$_N$ and (Arg)$_N$ ($N \geq 2$) can contribute significant membrane binding energy under physiological conditions (Ben-Tal *et al.*, 1996; Kim *et al.*, 1991; Mosior and McLaughlin, 1991, 1992). For example, the molar partition coefficient of pentalysine onto 2 : 1 PC/PS membranes in 100 mM KCl is 10^3 M^{-1}. Although not strong enough by itself to anchor a protein at the plasma membrane surface, a basic cluster may act in conjunction with other membrane-binding motifs to effect stable plasma membrane association, for example, a myristate (as for Src) or farnesyl group (as for K-ras4B).

The agreement of computational predictions based on solutions to the Poisson–Boltzmann equation with results from experimental studies on simple model systems supports the idea that nonspecific electrostatic interactions drive the association of basic clusters with membranes containing acidic lipids. The FDPB methodology provides a framework for dissecting protein/membrane interactions and for examining in detail the balance between the attractive and repulsive forces involved, and it suggests ways to examine the effect of additional forces, for example, the hydrophobic partitioning of lipophilic groups into the membrane interior or of aromatic groups into the polar headgroup region. The next section describes applications to more biological systems as well as the incorporation of nonpolar interactions.

IV. THE COMBINATION OF EXPERIMENTAL AND THEORETICAL APPROACHES PROVIDES A DETAILED COMPREHENSIVE PICTURE OF THE MEMBRANE ASSOCIATION OF MYRISTOYLATED PROTEINS

A. Src: Myristate plus Basic Cluster

Src belongs to a family of nonreceptor protein tyrosine kinases that are found associated principally with cellular membranes (M. T. Brown and Cooper, 1996; Cooper, 1990; Parsons and Parsons, 1997). Membrane binding is required for its function, and myristate plays a key role in its membrane localization (Resh, 1999). Nonmyristoylated v-Src mutants are found in the cytoplasm and do not transform cells, even though the kinase activity is the same as that of wild-type protein (Buss et al., 1986; Kamps et al., 1985). Although myristate is required for Src membrane binding, it alone is not sufficient (Buser et al., 1994; Sigal et al., 1994). Src contains a second membrane-binding motif, an N-terminal cluster of basic residues (Kaplan et al., 1990). Mutating away these basic residues also produces nontransforming phenotypes (Sigal et al., 1994). This suggests that Src uses a combination of motifs—a myristate plus a cluster of basic residues—to effect membrane localization.

Biophysical studies of the isolated protein show that Src binds to zwitterionic (PC) membranes with a molar partition coefficient of 10^3 M^{-1} (Sigal et al., 1994). Simple myristoylated peptides bind to PC membranes with a molar partition coefficient of 10^4 M^{-1} (Peitzsch and McLaughlin, 1993). Measurements of the membrane partitioning of acylated peptides show that each CH_2 group contributes about 0.8 kcal/mol to the membrane binding energy, in agreement with Tanford's observations on the hydrophobic partitioning of fatty acids into oil from water. Silvius (this volume, Chapter 13) considers the membrane partitioning of acylated and prenylated peptides in more detail. Myristoylated peptides corresponding to the N-terminus of Src bind to PC membranes with a partition coefficient similar to that of other myristoylated peptides ($K = 10^4 M^{-1}$). The partitioning of the intact Src protein onto PC membranes is about 10-fold lower for reasons that are not well understood. Incorporating acidic lipids into the membrane increases the binding of both the myristoylated peptide and the intact Src protein by exactly the same factor. Specifically, the partitioning of both the isolated Src protein and myristoylated peptides corresponding to its N-terminus is 1000-fold higher when the membrane contains 33 mol% acidic lipid (Buser *et al.*, 1994). There is good evidence that the N-terminal cluster of basic residues is responsible for the electrostatic interaction between the intact protein and membranes, because mutating away the basic residues decreases the partitioning of the protein onto PC/PS vesicles (Sigal *et al.*, 1994). These observations provide strong evidence that the N-terminal basic residues do indeed contribute significantly to the membrane association of Src by interacting electrostatically with acidic lipids. Computational predictions obtained by applying the FDPB methodology to models of the Src N-terminal peptide (Fig. 3) agree with these experimental observations and provide further support for the role of nonspecific electrostatic interactions in the membrane association of Src. Of course, these interactions do not rule out protein–protein interactions (e.g., utilizing the SH2 and SH3 domains of Src) that could help direct Src to different membranes or affect the lateral organization of Src within a given membrane.

For peptides corresponding to the N-terminus of Src, the individual hydrophobic and electrostatic contributions to the binding energy are additive (Buser *et al.*, 1994). If K_A is the "hydrophobic" molar partition coefficient for the myristolyated peptide onto PC vesicles and K_B is the "electrostatic" molar partition coefficient for the nonmyristoylated peptide into PC/PS vesicles, the overall binding of the myristoylated peptide to PC/PS vesicles is $K = \alpha K_A K_B$, where α is dependent on the "distance" between the two binding sites, as can be seen from models that consider the myristate and basic cluster as points connected by a flexible string of length L. Partitioning of the myristate into the membrane confines the basic cluster to a hemisphere of radius L above the membrane surface and facilitates its adsorption to the membrane; the shorter the string, the stronger the "synergism" between the two membrane-binding motifs. For the short "string" of Src, $\alpha \approx 1$. This "ball-and-string" model applies to the membrane partitioning of peptides corresponding to the C-terminus of K-ras4B as well (Table I) (Ghomashchi *et al.*, 1995). Although the ball-and-string model accounts for the additivity of electrostatic and hydrophobic interactions in the membrane partitioning of Src, it is descriptive rather than predictive. Figure 3 shows the predicted

Myr–Src(2–19)

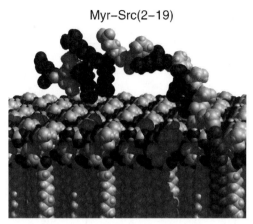

FIGURE 3 Molecular model of the N-terminal region of Src interacting with the upper leaflet of a 2 : 1 PC/PS bilayer. The front portion of the bilayer has been removed to expose the myristate (green). In agreement with ESR and CD measurements, the 18-residue peptide (myristate-GSSKSKPKDPSQRRRSLE-NH$_2$) is shown in an extended conformation. The six basic residues are blue and the two acidic residues are red. In the bilayer, PS is identified by its exposed nitrogen (blue); oxygen is red, phosphorus is yellow, and carbon and hydrogen are gray. Reprinted from Murray *et al.* (1997), copyright 1997, with permission from Elsevier Science. (See color plate.)

minimum free energy orientation of the myristoylated Src(2–19) peptide above a 2 : 1 PC/PS membrane in 100 mM KCl obtained using the FDPB methodology (Murray *et al.,* 1998). The conformation is consistent with monolayer, circular dichroism, and electron paramagnetic resonance (EPR) measurements (Buser *et al.,* 1994; Victor and Cafiso, 1998). The Gibbs surface excess and molar partition coefficient were determined by calculating and averaging over the electrostatic free energies of interaction for many different sampled orientations of the peptide anchored at its N-terminus to the membrane surface (Murray *et al.,* 1998). The prediction underestimates the absolute binding energy by 1 kcal/mol, but, as shown in Fig. 4, the relative binding, as functions of the mole percent acidic lipid in the membrane (Fig. 4, left) and the ionic strength of the solution (Fig. 4, right), is accurately predicted.

The three-pronged approach to the problem of Src membrane association— (1) *in vivo* transformation studies of wild-type and mutant v-Src, (2) *in vitro* biophysical studies of the intact proteins and peptide models of its membrane-interacting regions, and (3) computational studies with molecular models—is a powerful one and provides a detailed molecular and mechanistic picture of how Src binds to membranes. The approach also can be applied to obtain insight into the membrane-binding mechanisms of other peripheral proteins, for example, MARCKS, K-ras4B, and HIV-1 Gag.

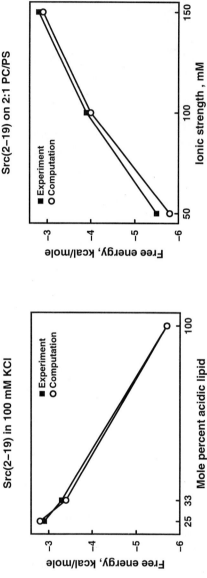

FIGURE 4 The membrane binding of nonmyristoylated peptides corresponding to residues 2–19 of Src is driven by electrostatic interactions. The molar partition coefficients determined experimentally and predicted by the FDPB methodology are plotted as a function of the mole percent of acidic lipid (*left*) and the ionic strength of the solution (*right*). The filled symbols are the experimentally determined values and the open symbols are approximate values for the partition coefficient calculated by considering only the orientation of maximum interaction. Data from Murray *et al.* (1998), which contains the experimental and computational details.

B. MARCKS

MARCKS is a widely distributed major PKC substrate implicated in phago-cytosis, secretion, and membrane recycling (Allen and Aderem, 1996). Although separated by 150 residues (Figure 5, left), both an N-terminal myristate and a cluster of basic residues [the effector region, MARCKS(151–175); Table I] are required for the plasma membrane association of MARCKS.

Unlike those of K-ras4B and Src, MARCKS's basic cluster contains aromatic residues that penetrate the membrane interface. EPR measurements of spin-labeled peptides corresponding to the effector region of MARCKS showed that the peptide lies at the membrane interface in a nonhelical conformation with its basic residues in the aqueous phase and its five Phe penetrating the polar headgroup region (Fig. 5, right) (Qin and Cafiso, 1996). This conformation agrees with independent circular dichroism (CD) and monolayer penetration experiments (Kim *et al.,* 1994a). As for Src, there is good evidence that the membrane-binding properties measured *in vitro* with intact MARCKS and peptides corresponding to its effector region reflect the membrane association of MARCKS in cells. Figure 5 (left) illustrates the mechanisms by which MARCKS binds to the plasma membrane: The N-terminal myristate partitions into the membrane hydrocarbon core, the basic residues in the effector region interact electrostatically with acidic lipids, the aromatic residues in the effector region partition into the membrane interface, and the rest of the protein, which has regions that are highly acidic, should be repelled from the plasma membrane and remain in the cytoplasm.

The MARCKS basic effector region contains 13 basic residues and 5 aromatic Phe residues (Table I). Experimental studies show that peptides corresponding to this region bind strongly to membranes containing acidic lipids (K $\sim 10^6 \, M^{-1}$ for 5:1 PC/PS in 100 mM KCl) (Arbuzova *et al.,* 2000; Kim *et al.,* 1994b). Unlike heptalysine and Src(2–19), the MARCKS peptide penetrates the membrane when it binds. Experimental studies show that the area of a monolayer kept at constant pressure increases upon addition of MARCKS peptide to the subphase. In addition, the MARCKS peptide binds weakly, but detectably, to neutral PC vesicles, indicating that forces other than electrostatic ones are involved in the membrane association (Arbuzova *et al.,* 2000).

Because of its high net charge ($Z = +13$), the MARCKS peptide experiences very strong electrostatic interactions with the membrane surface. FDPB calcu-lations with MARCKS(151–175) provide insight into the physical factors that drive membrane association (Arbuzova *et al.,* 2000). The circles and squares in Fig. 6 (left) show, respectively, the Coulombic attraction to and desolvation re-pulsion from a 5:1 PC/PS membrane in 100 mM KCl. Note that the magnitudes of these interactions are much larger than those for heptalysine (Fig. 2), even though the mole percent acidic lipid is actually lower in this case. Most pro-teins that use electrostatic interactions to bind peripherally to membranes also

MARCKS

MARCKS(151-175)

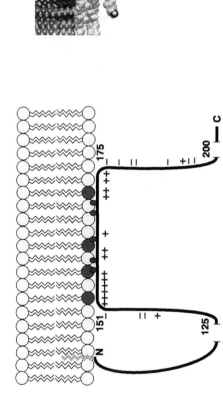

N 151 175 200 — C

125

FIGURE 5 Models for the membrane association of MARCKS. **Left:** A cartoon model of the interaction of the intact MARCKS protein with a PC (white headgroups)/PIP$_2$ (red headgroups) membrane. The myristate is colored yellow, the location of basic and acidic residues throughout region 125–200 are represented schematically by blue plus signs and red minus signs, respectively, and the aromatic residues in the basic effector domain are represented as green ovals penetrating the membrane interface. **Right:** An atomic level model for the interfacial association of MARCKS(151–175) based on the EPR measurements of spin-labeled peptides by Qin and Cafiso (1996). Reprinted from Arbuzova et al. (1998), copyright 1998, with permission from Elsevier Science. (See color plate.)

contain aromatic residues that partition into the membrane interface. Work from Steve White's lab has shown that aromatic residues (Trp, Tyr, and Phe) partition very favorably into the interface of palmitoyloleoylphosphatidylcholine (POPC) bilayers (White and Wimley, 1999; Wimley and White, 1996). Their experimentally derived interfacial hydrophobicity scale was used to incorporate the aromatic interactions into the FDPB calculations. The aromatic component ΔG_{arom} is given by

$$\Delta G_{arom} = \gamma_{if}\Delta A,$$

where γ_{if} is the interfacial aromatic surface tension coefficient ($\gamma_{if} = 0.13$ kcal/mol/Å^2) derived from the partitioning data of Wimley and White (1996) and ΔA is the change in solvent-accessible surface area of the aromatic side chains as the peptide associates with the membrane interface. In addition to the long-range electrostatic attraction and short-range desolvation penalty, there is a favorable short-range aromatic contribution (diamonds, Fig. 6, left) when the peptide is close enough to the membrane surface so that the Phe side chains can partition into the membrane interface. The favorable Coulombic and aromatic contributions combine to outcompete the desolvation penalty and allow the peptide to partially penetrate the membrane interface. Indeed, the calculations predict that a peptide corresponding to the MARCKS effector region has a shorter average distance ($<R> = 1.5$ Å, less than a layer of water) from the membrane surface than either heptalysine or a MARCKS peptide in which all five Phe are mutated to Ala ($<R> = 2.5$ Å). Figure 6 (right) illustrates the total "electrostatic plus aromatic" interaction free energy of the MARCKS peptide with PC (triangles) and 5:1 PC/PS (upside-down triangles) membranes. In agreement with experiment (Arbuzova *et al.*, 2000), the calculations predict that the peptide binds weakly to PC membranes and that the partitioning of the peptide increases 10^4-fold when 17 mol% acidic lipid is incorporated into the membrane. The computational methodology provides a physical model for how different magnitudes of these forces (Coulomb attraction, desolvation repulsion, aromatic attraction), which are reflected in the residue character and structure of proteins, combine to produce different membrane-binding behaviors.

The extrapolation from basic peptide to protein is not as straightforward for MARCKS as for Src. In the case of Src, the membrane binding energies of the myristate and basic cluster are additive for both the myristoylated peptide and the intact protein (Buser *et al.*, 1994; Resh, 1999; Sigal *et al.*, 1994). As discussed above, incorporating ~30% monovalent acidic lipid into a PC vesicle increases the binding of both the myristoylated Src peptide and the intact Src protein about 10^3-fold. In the case of MARCKS, incorporating ~20% monovalent acidic lipid into a PC vesicle increases the binding of the effector domain peptide 10^4-fold, as predicted theoretically (Arbuzova *et al.*, 2000), but increases the binding of the intact MARCKS protein only 10^2-fold (Kim *et al.*, 1994b). This can be rationalized by considering the physical character of the entire MARCKS protein,

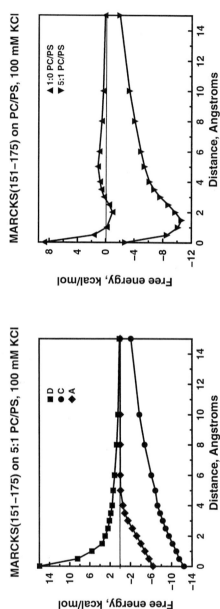

FIGURE 6 FDPB calculations of the interfacial association of MARCKS(151–175). **Left:** The electrostatic and aromatic interaction free energies are plotted as a function of distance between the van der Waals surfaces of the backbone of the MARCKS peptide and a 5 : 1 PC/PS bilayer in 100 m*M* KCl. The squares illustrate the unfavorable energy of desolvating the peptide and the membrane. The circles illustrate the long-range electrostatic attraction between basic groups on the peptide and the acidic phospholipid headgroups. The diamonds represent the short-range aromatic contribution due to insertion of the peptide's five Phe residues into the membrane interface. Adapted with permission from Arbuzova *et al.* (2000). Copyright 2000 American Chemical Society. **Right:** The total (electrostatic plus nonpolar) interaction free energy is plotted as a function of distance between the van der Waals surfaces of the backbone of the MARCKS peptide and 1 : 0 (triangles) and 5 : 1 (upside-down triangles) PC/PS bilayers in 100 m*M* KCl.

depicted in Fig. 5 (left). Several acidic residues (red minus signs) on either side of the basic effector domain are confined within a few Debye lengths ($1/\kappa = 1$ nm for 0.1 M salt) of the negatively charged membrane surface and thus are expected to contribute a significant electrostatic repulsion to the membrane interaction of the protein. In addition, the length of the "string" (150 amino acids) that connects the myristate and basic cluster should decrease the coupling between the two membrane-binding motifs (myristate and basic cluster), and hence the total binding energy should be significantly less that the sum of the individual binding energies.

In many other respects, though, the basic effector domain peptide, MARCKS (151–175), faithfully mimics how the intact protein interacts with membranes, as discussed in detail elsewhere (Arbuzova *et al.*, 1998). In brief, the peptide encapsulates three important aspects of the MARCKS/membrane interaction. First, binding of calcium–calmodulin (Ca–CaM) to the peptide and the protein occurs with the same high affinity (1–10 nM K_d). Increasing the level of Ca–CaM produces translocation of both the peptide and purified protein from phospholipid vesicles to the surrounding aqueous solution; in many cell types it also produces translocation of the intact protein from membrane to cytoplasm. The mechanism by which Ca–CaM rapidly dissociates the bound effector domain peptide from the membrane–solution interface is described in detail elsewhere (Arbuzova *et al.*, 1997). Second, phosphorylation of the effector region by PKC introduces three phosphates (six negative charges) into the cluster of basic residues. This weakens the binding of the effector domain peptide to negatively charged membranes and also produces translocation of the intact protein from phospholipid vesicles (Kim *et al.*, 1994a,b). The experimental evidence suggests that phosphorylation exerts its effects in both cases through a simple electrostatic mechanism; FDPB calculations support this model (Murray, unpublished calculation). In many cell types, PKC phosphorylation of MARCKS produces translocation from membrane to cytoplasm (Allen and Aderem, 1995; Guadagno *et al.*, 1992; Rosen *et al.*, 1990; Swierczynski and Blackshear, 1995). Third, recent results show that MARCKS (151–175) can bind PIP$_2$ with high affinity (Arbuzova *et al.*, 2000). For example, introducing 1% PIP$_2$ into a PC vesicle increases the binding of MARCKS(151–175) by a factor of 10^4. The binding has been measured with a centrifugation technique using a radioactive peptide and with a fluorescent technique using acrylodan-labeled peptide. Several lines of evidence suggest that one MARCKS(151–175) binds 3–4 PIP$_2$ to form an electroneutral complex (Wang *et al.*, 2001). For example, the fluorescence from PC/PIP$_2$ vesicles containing NBD-labeled PIP$_2$ is quenched by addition of MARCKS(151–175), which suggests the peptide induces a strong demixing of PIP$_2$ in the plane of the membrane (Wang *et al.*, 2001).

The interaction of PIP$_2$ with MARCKS(151–175) may be compared to its interaction with neomycin and the PH domain of PLC-δ_1, two other well-characterized molecules that bind PIP$_2$ with high affinity. Addition of $\sim 10^{-5}$, 10^{-6}, and 10^{-8} M PIP$_2$ in the form of PC/PIP$_2$ vesicles binds 50% of neomycin (Gabev *et al.*, 1989), the PH domain (Garcia *et al.*, 1995), and MARCKS effector domain (Arbuzova *et al.*, 2000), respectively, that is, MARCKS(151–175) partitions most strongly onto the PC/PIP$_2$ membranes. Both MARCKS(151–175) and neomycin interact as strongly with PI(3,4)P$_2$ as with PI(4,5)P$_2$, whereas the PH domain of PLC-δ is specific for PI(4,5)P$_2$. Neomycin and the PH domain form 1 : 1 complexes with PIP$_2$ (Ferguson *et al.*, 1996; Gabev *et al.*, 1989; Garcia *et al.*, 1995; Lemmon *et al.*, 1995), whereas MARCKS(151–175) interacts with several PIP$_2$ (Wang *et al.*, 2001). Electrostatic interactions are important, because increasing the salt concentration from 100 to 500 mM decreases the binding of MARCKS(151–175) to 99.9 : 0.1 PC/PIP$_2$ vesicles one hundred-fold. The role of other, more specific lipid–protein interactions remains to be determined.

Studies of the inhibition of PLC-mediated hydrolysis of PIP$_2$ suggest that the strong interaction between PIP$_2$ and MARCKS(151–175) occurs with the intact MARCKS protein as well. Both the protein and the peptide (see Fig. 7) decrease the PLC-induced hydrolysis of PIP$_2$ in phospholipid vesicles (Glaser *et al.*, 1996). The simplest interpretation of these results is that both the effector domain in the intact protein and the effector domain peptide bind PIP$_2$ with high affinity and compete successfully with the catalytic domain of PLC for this lipid. Experiments with monolayers (Wang *et al.*, 2001) show that these observations are not artifacts due to peptide-induced aggregation of vesicles, a problem which unfortunately complicates many experiments with basic peptides and lipid vesicles (Murray *et al.*, 1999).

Can the strong interaction between the effector domain of MARCKS and PIP$_2$ observed with *in vitro* model systems be extrapolated to living cells? This remains to be determined, but the hypothesis that MARCKS can bind a significant fraction of the PIP$_2$ in a typical cell suggests three testable corollaries. First, MARCKS and PIP$_2$ should have comparable cellular concentrations; they do in many cell types. In nerve cells, for example, both MARCKS and PIP$_2$ are present at about 10 μM (Albert *et al.*, 1987). Second, overexpression of MARCKS should induce enhanced production of PIP$_2$ if MARCKS buffers a significant fraction of the PIP$_2$. It does, at least in one cell type (Laux *et al.*, 2000). Third, MARCKS and PIP$_2$ should be colocalized in cells. MARCKS is not uniformly distributed in the plasma membrane of some cell types. In fibroblasts, for example, it is concentrated in ruffles (Myat *et al.*, 1997). If it binds a significant fraction of PIP$_2$, then PIP$_2$ should be colocalized with MARCKS in ruffles, and it is (Honda *et al.*, 1999; Tall *et al.*, 2000). Of course, there are many other, unrelated mechanisms by which PIP$_2$ and PIP$_3$ (Czech, 2000; Martin, 1998) could be sequestered in ruffles, for example, through localized synthesis.

Although the functional role of the sequestration of PIP$_2$ by MARCKS can only be determined in cells or organisms, experiments on model systems and theoretical calculations can help guide cell biologists in designing and interpreting *in vivo* studies. Phosphoinositides are important for the production of second

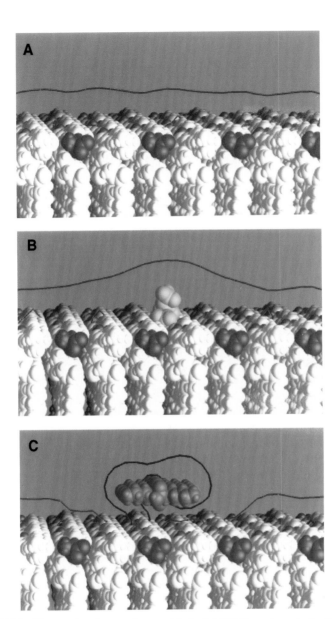

FIGURE 1   The negative electrostatic potential of a 2:1 PC/PS membrane is significantly perturbed by multivalent molecules. The red (blue) lines represent the two-dimensional –25 (+25) mV equipotential contours obtained from FDPB calculations. In each panel, the membrane composition is 2:1 PC/PS and the ionic strength [KCl], is 100 mM. (A) The –25 mV electrostatic equipotential contour is essentially flat and located ~10 Å above the membrane surface. (B) A trivalent PIP_2 lipid (denoted by its yellow headgroup) greatly increases the negative potential of the membrane in its vicinity. (C) A membrane-adsorbed basic peptide (heptalysine; blue) introduces a significant positive potential at the membrane surface.

Myr–Src(2–19)

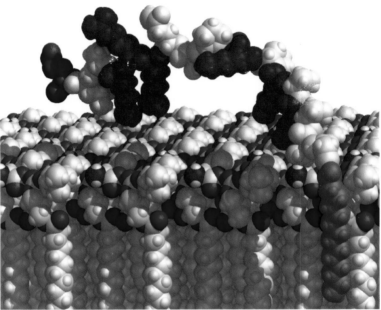

FIGURE 3 Molecular model of the N-terminal region of Src interacting with the upper leaflet of a 2:1 PC/PS bilayer. The front portion of the bilayer has been removed to expose the myristate (green).   In agreement with ESR and CD measurements, the 18-residue peptide (myristate-GSSKSKPKDPSQRRRSLE-NH$_2$) is shown in an extended conformation. The six basic residues are blue and the two acidic residues are red. In the bilayer, PS is identified by its exposed nitrogen (blue); oxygen is red, phosphorus is yellow, and carbon and hydrogen are gray. Reprinted from Murray *et al.,* (1997), copyright 1997, with permission from Elsevier Science.

MARCKS

MARCKS(151-175)

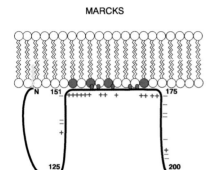

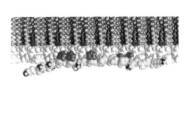

FIGURE 5   Models for the membrane association of MARCKS. **Left:** A cartoon model of the interaction of the intact MARCKS protein with a PC (white headgroups)/PIP$_2$ (red headgroups) membrane. The myristate is colored yellow, the location of basic and acidic residues throughout region 125–200 are represented schematically by blue plus signs and red minus signs, respectively, and the aromatic residues in the basic effector domain are represented as green ovals penetrating the membrane interface. **Right:** An atomic level model for the interfacial association of MARCKS(151–175) based on the EPR measurements of spin-labeled peptides by Qin and Cafiso (1996). Reprinted from Arbuzova *et al.,* (1998), copyright 1998, with permission from Elsevier Science.

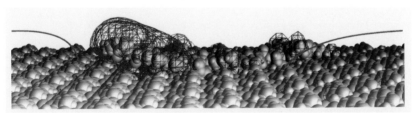

FIGURE 8   FDPB calculations of the electrostatic properties of the effector region of MARCKS, MARCKS(151–175), adsorbed to the surface of a 2:1 PC/PS membrane in 100 m*M* KCl. The blue mesh and red line represent, respectively, +25-mV and –25-mV equipotential contours. Reprinted from Arbuzova *et al.,* (1998), copyright 1998, with permission from Elsevier Science.

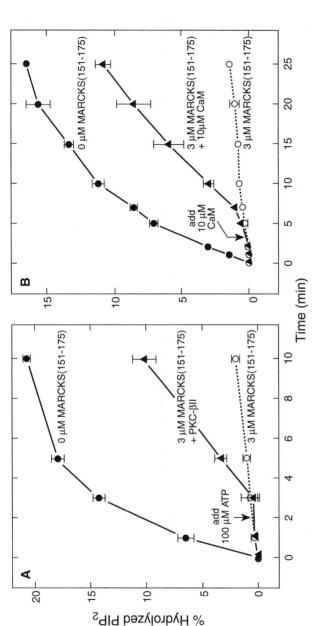

FIGURE 7 (A) PKC phosphorylation of MARCKS(151–175) reverses the MARCKS(151–175)-induced inhibition of PLC-catalyzed hydrolysis of PIP_2. The filled circles illustrate that PLC, added at time 0 min, rapidly hydrolyzes PIP_2 in large unilamellar vesicles. The open circles show that the addition of MARCKS(151–175) prior to PLC addition inhibits the activity of PLC >10-fold (slope of line decreases). The filled triangles illustrate that when PKC is present, the addition of ATP at time 2 min (arrowhead) increases the activity of PLC significantly. The vesicles are made up of 1% PIP_2 and a mixture of zwitterionic and monovalent acidic lipids that mimics the composition of the cytoplasmic surface of the plasma membrane. (B) Ca–calmodulin reverses the inhibition of PLC by MARCKS(151–175). The open circles show that adding MARCKS(151–175) inhibits PLC activity >10-fold. The filled triangles show that adding calmodulin (CaM) at time 3 min rapidly reverses the inhibition. Glaser *et al.* (1996) contains the experimental details. Reprinted with permission from Glaser *et al.* (1996). Copyright 1996 The American Society for Biochemistry and Molecular Biology.

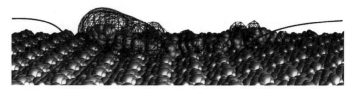

FIGURE 8 FDPB calculations of the electrostatic properties of the effector region of MARCKS, MARCKS(151–175), adsorbed to the surface of a 2 : 1 PC/PS membrane in 100 mM KCl. The blue mesh and red line represent, respectively, +25 mV and −25 mV equipotential contours. Reprinted from Arbuzova *et al.* (1998), copyright 1998, with permission from Elsevier Science. (See color plate.)

messengers, exocytosis, and endocytosis, the regulation of ion channels, and the recruitment and anchoring of specific proteins to membrane surfaces (Cockcroft, 2000; Czech, 2000; Martin, 1998). Because they play a role in so many cellular functions, it would be surprising if their distribution and accessibility were not carefully regulated by the cell. Experiments on model systems suggest mechanisms by which this could occur. For example, as described above and depicted in Fig. 7, MARCKS and MARCKS(151–175) inhibit the PLC-induced hydrolysis of PIP$_2$ in phospholipid vesicles. Figure 7 also illustrates mechanisms by which the MARCKS-induced inhibition of PLC hydrolysis is relieved. Both PKC phosphorylation of MARCKS(151–175) and the binding of Ca–CaM to MARCKS-(151–175) release the bound PIP$_2$ by causing desorption of the effector domain from the membrane. It remains to be determined if these phenomena also occur to a significant degree in the living cell and if the cell can thus produce localized bursts of PIP$_2$ in response to these two signals. FDPB calculations of the electrostatic properties of MARCKS(151–175)/membrane systems help visualize the mechanisms of PIP$_2$ sequestration and release. Figure 8 depicts the +25 mV (blue mesh) and −25 mV (red line) equipotential contours around a membrane-adsorbed MARCKS peptide. The peptide (net charge +13) dramatically perturbs the negative potential of the membrane and provides a strong positive potential at the membrane surface that could serve as a highly localized sink for the multivalent acidic PIP$_2$.

V. FUTURE DIRECTIONS

A. Membrane Targeting

In summary, the function of MARCKS is not well understood and could involve both its ability to bind actin (Aderem, 1992; Hartwig *et al.,* 1992; Nairn and Aderem, 1992) and its ability to bind PIP$_2$ as discussed here. In contrast, the biological functions of Src, K-ras4B, and HIV-1 Gag have been well established. However, the mechanisms by which any of these proteins are targeted to the plasma membrane are not known. Recent work from the Hancock and Philips labs has shown that H- and N-ras are targeted to the plasma membrane by a mechanism that involves their palmitoylation on internal membranes and subsequent transfer to the plasma membrane by exocytosis (Apolloni *et al.*, 2000; Choy *et al.*, 1999). Silvius and co-workers (Leventis and Silvius, 1998; Roy *et al.*, 2000) have proposed a simple biophysical mechanism by which K-ras4B may be targeted to the plasma membrane: Stronger electrostatic attraction drives the preferential binding to plasma membrane. This proposal relies on the premise that the electrostatic potential of the cytoplasmic surface of the plasma membrane is more negative than the cytoplasmic surfaces of other intracellular membranes. Because the surface area of internal membranes is about 10-fold greater than the surface area of the plasma membrane in a typical cell, Silvius and co-workers estimate that the surface potential of the plasma membrane would have to be about 20 mV more negative than the surface potential of internal membranes in order for an electrostatic mechanism to drive plasma membrane targeting. This could occur if, for example, the "flippase" for PS is more active in the plasma membrane (Daleke and Lyles, 2000), resulting in a higher negative surface charge density. Although the magnitude of the surface potentials of the plasma and internal membranes is not well established, studies of the cellular distribution of K-ras4B/GFP constructs suggest a nonspecific electrostatic mechanism is plausible. The fact that the fluorescent constructs are targeted to the plasma membrane even when the C-terminal sequence of basic residues is scrambled argues, at least, against a targeting mechanism that involves specific protein–protein recognition. If the nonspecific electrostatic targeting mechanism is operative, it could also act to target Src, MARCKS, HIV-1 Gag, and other proteins with basic residues to the plasma membrane.

It is important to note that even if electrostatics produces a bias for the plasma membrane, other interactions can override this mechanism and direct a protein to a different subcellular localization. For example, v- and c-Src have the same N-terminal sequences (myristate plus basic cluster, see Table I), but are targeted to different membranes. v-Src and activated c-Src localize to the plasma membrane, whereas the inactive form of c-Src is localized primarily to endosomal membranes. These observations suggest that other factors, for example, protein–protein interactions, contribute to their subcellular localization. In its inactive state, Src forms a compact folded structure through a series of intramolecular interactions mediated by its SH2 and SH3 domains. Activation of the enzyme leads to the disruption of these interactions and concomitant availability of the SH2 and SH3 domains to interact with other ligands. Resh and co-workers found that the SH2 domain mediates cytoskeletal association of v-Src, demonstrating that domains other than the membrane-interacting motifs contribute to the subcellular localization of Src. Interestingly, a "Src-like" mutant construct of MARCKS, one in which the long stretch of residues between the myristate and basic effector domain is removed, is targeted to the nucleus rather than the plasma membrane as one might have expected (Seykora *et al.*, 1996). The targeting of MARCKS is complicated by at least two additional factors. First, it is possible that MARCKS is targeted to the plasma membrane through its interaction with PIP_2. The plasma membrane has a high relative concentration of PIP_2 and this directs the plasma membrane targeting of PLC-δ: Its PH domain anchors the protein to the plasma membrane through its high-affinity interaction with PIP_2 (Lemmon and Ferguson, 2000). The high affinity of the effector domain of MARCKS for PIP2 could act in a similar manner. Second, MARCKS can bind actin, and protein–protein interactions could both target it to different membranes and modify the lateral distribution of MARCKS in the plasma membrane. The role of phosphoinositides in regulating membrane–cytoskeletal interactions has long been recognized and was elegantly demonstrated by recent experiments using both GFP constructs and laser tweezers (Raucher *et al.*, 2000). In conclusion, the cellular mechanisms for plasma membrane targeting of K-ras4B, Src, MARCKS, and HIV-1 Gag remain to be elucidated, as do the mechanisms by which the lateral distribution is modified; this area of investigation will surely continue to benefit from the new cell biology approaches being developed.

B. Computational Studies

The FDPB approach provides a framework for (1) describing in quantitative terms the physical forces that drive nonspecific protein–lipid interactions, (2) developing structure-based models for peptides and proteins at the membrane interface, and (3) identifying protein sequence and structure patterns that are predictive of membrane binding potential. The methodology described here for calculating the membrane interaction of peptides can be applied to proteins of known structure as well. The next step toward gaining both a deeper and broader understanding of the mechanisms proteins use to bind to membranes involves the integration of detailed computer calculations of the underlying physical interactions, as described here, with tools for exploiting and analyzing the vast amounts of genomic data becoming available. These tools include facilities for sequence analysis, structure comparison and prediction, and surface property analysis. The synthesis of these computational approaches can be used to describe similarities and differences within and across whole protein families involved in membrane association (e.g., for the C2 domains; Murray and Honig, 2001). Representative structures for peripheral proteins have recently become available, which allows for this type of computational analysis. Calculation of the interaction of these proteins with realistic models of membranes will provide information unobtainable through structural analysis of the proteins alone. Building on the examples described in this chapter, computational results can be used in new ways to continue to aid in the design and interpretation of experiments, leading to new insights into the molecular basis of protein–membrane association.

Acknowledgments

This work was supported by National Institutes of Health Grant GM24971 and National Science Foundation Grant MCB9729538 to S.M., National Science Foundation Grant MCB9808902 to B.H. and Helen Hay Whitney and Sloan/U.S. Department of Energy Computational Biology postdoctoral fellowships to D.M.

References

Aderem, A. (1992). Signal transduction and the actin cytoskeleton: The roles of MARCKS and profilin. *Trends Biochem. Sci.* **17,** 438–443.

Albert, K. A., Nairn, A. C., and Greengard, P. (1987). The 87-kDa protein, a major specific substrate for protein kinase C: Purification from bovine brain and characterization. *Proc. Natl. Acad. Sci. USA* **84,** 7046–7050.

Allen, L. A., and Aderem, A. (1995). Protein kinase C regulates MARCKS cycling between the plasma membrane and lysosomes in fibroblasts. *EMBO J.* **14,** 1109–1120.

Allen, L. A., and Aderem, A. (1996). Mechanisms of phagocytosis. *Curr. Opin. Immunol.* **8,** 36–40.

Apolloni, A., Prior, I. A., Lindsay, M., Parton, R. G., and Hancock, J. F. (2000). H-ras but not K-ras traffics to the plasma membrane through the exocytic pathway. *Mol. Cell. Biol.* **20,** 2475–2487.

Arbuzova, A., Wang, J., Murray, D., Jacob, J., Cafiso, D. S., and McLaughlin, S. (1997). MARCKS, membranes, and calmodulin: Kinetics of interaction. *J. Biol. Chem.* **272,** 27167–27177.

Arbuzova, A., Murray, D., and McLaughlin, S. (1998). MARCKS, membranes and calmodulin: Kinetics of interaction. *Biochim. Biophys. Acta* **1376,** 369–379.

Arbuzova, A., Wang, L., Wang, J., Hangyas-Mihalyne, G., Murray, D., Honig, B., and McLaughlin, S. (2000). Membrane binding of peptides containing both basic and aromatic residues. Experimental studies with peptides corresponding to the scaffolding region of caveolin and the effector region of MARCKS. *Biochemistry* **39,** 10330–10339.

Ben-Tal, N., Honig, B., Peitzsch, R. M., Denisov, G., and McLaughlin, S. (1996). Binding of small basic peptides to membranes containing acidic lipids: Theoretical models and experimental results. *Biophys. J.* **71,** 561–575.

Ben-Tal, N., Honig, B., Miller, C., and McLaughlin, S. (1997). Electrostatic binding of proteins to membranes: Theoretical prediction and experimental results with charybdotoxin and phospholipid vesicles. *Biophys. J.* **73,** 1717–1727.

Bhatnagar, R. S., and Gordon, J. I. (1997). Understanding covalent modifications of proteins by lipids: Where cell biology and biophysics mingle. *Trends Cell Biol.* **7,** 14–21.

Brown, D. A., and London, E. (2000). Structure and function of sphingolipid- and cholesterol-rich membrane rafts. *J. Biol. Chem.* **275,** 17221–17224.

Brown, M. T., and Cooper, J. A. (1996). Regulation, substrates and functions of Src. *Biochim. Biophys. Acta* **1287,** 121–149.

Bryant, M., and Ratner, L. (1990). Myristoylation-dependent replication and assembly of human immunodeficiency virus 1. *Proc. Natl. Acad. Sci. USA* **87,** 523–527.

Buser, C. A., Sigal, C. T., Resh, M. D., and McLaughlin, S. (1994). Membrane binding of myristylated peptides corresponding to the NH_2-terminus of Src. *Biochemistry* **33,** 13093–13101.

Buss, J. E., Kamps, M. P., Gould, K., and Sefton, B. M. (1986). The absence of myristic acid decreases membrane binding of p60src but does not affect tyrosine protein kinase activity. *J. Virol.* **58,** 468–474.

Choy, E., Chiu, V. K., Silletti, J., Feoktistov, M., Morimoto, T., Michaelson, D., Ivanov, I. E., and Philips, M. R. (1999). Endomembrane trafficking of ras: The CAAX motif targets proteins to the ER and Golgi. *Cell* **98,** 69–80.

Cockcroft, S. (2000). "Biology of Phosphoinositides." Oxford University Press, Oxford.

Cooper, J. A. (1990). The Src family of protein-tyrosine kinases. *In* "Peptides and Protein Phosphorylation" (B. E. Kemp, ed.), pp. 85–113. CRC Press, Boca Raton, FL.

Czech, M. P. (2000). PIP$_2$ and PIP$_3$: Complex roles at the cell surface. *Cell* **100**, 603–606.

Daleke, D. L., and Lyles, J. V. (2000). Identification and purification of aminophospholipid flippases. *Biochim. Biophys. Acta* **1486**, 108–127.

Ferguson, K. M., Lemmon, M. A., Schlessinger, J., and Sigler, P. B. (1996). Structure of the high affinity complex of inositol trisphosphate with a phospholipase C pleckstrin homology domain. *Cell* **83**, 1037–1046.

Gabev, E., Kasianowicz, J., Abbott, T., and McLaughlin, S. (1989). Binding of neomycin to phosphatidylinositol 4,5-bisphosphate (PIP$_2$). *Biochim. Biophys. Acta* **979**, 105–112.

Garcia, P., Gupta, R., Shah, S., Morris, A. J., Rudge, S. A., Scarlata, S., Petrova, V., McLaughlin, S., and Rebecchi, M. J. (1995). The pleckstrin homology domain of phospholipase C-δ_1 binds with high affinity to phosphatidylinositol 4,5-bisphosphate in bilayer membranes. *Biochemistry* **34**, 16228–16234.

Garnier, L., Bowzard, B., and Wills, J. W. (1998). Recent advances and remaining problems in HIV assembly. *AIDS* **12**, S5–S16.

Ghomashchi, F., Zhang, X., Liu, L., and Gelb, M. H. (1995). Binding of prenylated and polybasic peptides to membranes: Affinities and intervesicle exchange. *Biochemistry* **34**, 11910–11918.

Gilson, M., and Honig, B. (1988). Calculation of the total electrostatic energy of a macromolecular system: Solvation energies, binding energies, and conformational analysis. *Proteins* **4**, 7–18.

Glaser, M., Wanaski, S., Buser, C. A., Boguslavsky, V., Rashidzada, W., Morris, A., Rebecchi, M., Scarlata, S. F., Runnels, L. W., Prestwich, G. D., Chen, J., Aderem, A., Ahn, J., and McLaughlin, S. (1996). MARCKS produces reversible inhibition of phospholipase C by sequestering phosphatidylinositol 4,5-bisphosphate in lateral domains. *J. Biol. Chem.* **271**, 26187–26193.

Guadagno, S. N., Borner, C., and Weinstein, I. B. (1992). Altered regulation of a major substrate of protein kinase C in rat 6 fibroblasts overproducing PKC beta I. *J. Biol. Chem.* **267**, 2697–2707.

Hancock, J. F., Paterson, H., and Marshall, C. J. (1990). A polybasic domain or palmitoylation is required in addition to the CAAX motif to localize p21ras to the plasma membrane. *Cell* **63**, 133–139.

Hartwig, J. H., Thelen, M., Rosen, A., Janmey, P. A., Nairn, A. C., and Aderem, A. (1992). MARCKS is an actin filament crosslinking protein regulated by protein kinase C and calcium-calmodulin. *Nature* **356**, 618–622.

Heimburg, T., and Marsh, D. (1996). Thermodynamics of the interaction of proteins with lipid membranes. *In* "Biological Membranes" (K. Merz, Jr., and B. Roux, eds.), pp. 405–464. Birkhauser, Boston.

Hill, C. P., Bancroft, D. P., Christensen, A. M., and Sundquist, W. I. (1996). Crystal structures of the trimeric human immunodeficiency virus type I matrix protein: Implications for membrane binding and assembly. *Proc. Natl. Acad. Sci. USA* **93**, 3099.

Holst, M. (1993). Numerical solutions to the finite-difference Poisson–Boltzmann equation. Thesis/Dissertation, University of Illinois, Champaign-Urbana, IL.

Holst, M., and Saied, F. (1993). Multigrid solution of the Poisson–Boltzmann equation. *J. Comp. Chem.* **14**, 105–113.

Honda, A., Nogami, M., Yokozeki, T., Yamazaki, M., Nakamura, H., Watanabe, H., Kawamoto, K., Nakayama, K., Morris, A. J., Frohman, M. A., and Kanaho, Y. (1999). Phosphatidylinositol 4-phosphate 5-kinase alpha is a downstream effector of the small G protein ARF6 in membrane ruffle formation. *Cell* **99**, 521–532.

Honig, B. H., and Nicholls, A. (1995). Classical electrostatics in biology and chemistry. *Science* **268**, 1144–1149.

Hurley, J. H., and Misra, S. (2000). Signaling and subcellular targeting by membrane-binding domains. *Annu. Rev. Biophys. Biophys. Chem.* **29**, 49–79.

Jakobsson, E. (1997). Computer simulation studies of biological membranes: Progress, promise and pitfalls. *Trends Biochem. Sci.* **22**, 339–344.

Johnson, J. E., and Cornell, R. B. (1999). Amphitropic proteins: Regulation by reversible membrane interactions. *Mol. Membrane Biol.* **16**, 217–235.

Kamps, M. P., Buss, J. E., and Sefton, B. M. (1985). Mutation of NH2-terminal glycine of p60src prevents both myristoylation and morphological transformation. *Proc. Natl. Acad. Sci. USA* **82**, 4625–4628.

Kaplan, J. M., Varmus, H. E., and Bishop, J. M. (1990). The Src protein contains multiple domains for specific attachment to membranes. *Mol. Cell. Biol.* **10**, 1000–1009.

Kholodenko, B. N., Hoek, J. B., and Westerhoff, H. V. (2000). Why cytoplasmic signaling proteins should be recruited to cell membranes. *Trends Cell Biol.* **10**, 173–178.

Kim, J., Mosior, M., Chung, L. A., Wu, H., and McLaughlin, S. (1991). Binding of peptides with basic residues to membranes containing acidic phospholipids. *Biophys. J.* **60**, 135–148.

Kim, J., Blackshear, P. J., Johnson, J. D., and McLaughlin, S. (1994a). Phosphorylation reverses the membrane association of peptides that correspond to the basic domains of MARCKS and neuromodulin. *Biophys. J.* **67**, 227–237.

Kim, J., Shishido, T., Jiang, X., Aderem, A., and McLaughlin, S. (1994b). Phosphorylation, high ionic strength, and calmodulin reverse the binding of MARCKS to phospholipid vesicles. *J. Biol. Chem.* **269**, 28214–28219.

Kleinschmidt, J. H., and Marsh, D. (1997). Spin-label electron spin resonance studies on the interactions of lysine peptides with phospholipid membranes. *Biophys. J.* **73**, 2546–2555.

Laux, T., Fukami, K., Thelen, M., Golub, T., Frey, D., and Caroni, P. (2000). GAP43, MARCKS, CAP23 modulate PI(4,5)P2 at plasmalemmal rafts, and regulate cell cortex actin dynamics through a common mechanism. *J. Cell Biol.* **149**, 1455–1472.

Lemmon, M. A., and Ferguson, K. M. (2000). Pleckstrin homology domains: Phosphoinositide-regulated membrane tether. *In* "Biology of Phosphoinositides." (S. Cockcroft, ed.), pp. 131–165. Oxford University Press, Oxford.

Lemmon, M. A., Ferguson, K. M., O'Brien, R., Sigler, P. B., and Schlessinger, J. (1995). Specific and high-affinity binding of inositol phosphates to an isolated pleckstrin homology domain. *Proc. Natl. Acad. Sci. USA* **92**, 10472–10476.

Leventis, R., and Silvius, J. R. (1998). Lipid-binding characteristics of the polybasic carboxy-terminal sequence of K-ras4B. *Biochemistry* **37**, 7640–7648.

Magee, T., and Marshall, C. J. (1999). New insights into the interaction of ras with the plasma membrane. *Cell* **98**, 9–12.

Martin, T. F. J. (1998). Phosphoinositide lipids as signaling molecules: Common themes for signal transduction, cytoskeletal regulation, and membrane trafficking. *Annu. Rev. Cell Dev. Biol.* **14**, 231–264.

May, S., Harries, D., and Ben-Shaul, A. (2000). Lipid demixing and protein–protein interactions in the adsorption of charged proteins on mixed membranes. *Biophys. J.* **79**, 1747–1760.

McCloskey, M. A., and Poo, M. M. (1986). Rates of membrane-associated reactions: Reduction of dimensionality revisited. *J. Cell Biol.* **102**, 88–96.

McIntosh, T. J., and Simon, S. A. (1986). Area per molecule and distribution of water in fully hydrated dilauroylphosphatidylethanolamine bilayers. *Biochemistry* **25**, 4948–4952.

McIntosh, T. J., and Simon, S. A. (1994). Hydration and steric pressures between phospholipid bilayers. *Annu. Rev. Biophys. Biomol. Struct.* **23**, 27–51.

McLaughlin, S. (1989). The electrostatic properties of membranes. *Annu. Rev. Biophys. Biophys. Chem.* **18**, 113–136.

McLaughlin, S., and Aderem, A. (1995). The myristoyl–electrostatic switch: A modulator of reversible protein–membrane interactions. *Trends Biochem. Sci.* **20**, 272–276.

Merz, K., and Roux, B. (1996). "Biological Membranes: A Molecular Perspective from Computation and Experiment." Birkhauser, Boston.

Misra, V., and Honig, B. (1995). On the magnitude of the electrostatic contribution to ligand–DNA interactions. *Proc. Natl. Acad. Sci. USA* **92**, 4691–4695.

Misra, V., Sharp, K. A., Friedman, R. A., and Honig, B. (1994). Salt effects on ligand–DNA binding: Minor groove binding antibiotics. *J. Mol. Biol.* **238**, 245–263.

Montich, G., Scarlata, S., McLaughlin, S., Lehrmann, R., and Seelig, J. (1993). Thermodynamic characterization of the association of small basic peptides with membranes containing acidic lipids. *Biochim. Biophys. Acta* **1146**, 17–24.

Morrison, D. K., and Cutler, R. E., Jr. (1997). The complexity of Raf-1 regulation. *Curr. Opin. Cell Biol.* **9**, 174–179.

Mosior, M., and McLaughlin, S. (1991). Peptides that mimic the pseudosubstrate region of protein kinase C bind to acidic lipids in membranes. *Biophys. J.* **60**, 149–159.

Mosior, M., and McLaughlin, S. (1992). Binding of basic peptides to acidic lipids in membranes: Effects of inserting alanine(s) between the basic residues. *Biochemistry* **31**, 1767–1773.

Murray, D., Ben-Tal, N., Honig, B., and McLaughlin, S. (1997). Electrostatic interaction of myristoylated proteins with membranes: Simple physics, complicated biology. *Structure* **5**, 985–989.

Murray, D., Hermida-Matsumoto, L., Buser, C. A., Tsang, J., Sigal, C., Ben-Tal, N., Honig, B., Resh, M. D., and McLaughlin, S. (1998). Electrostatics and the membrane association of Src: Theory and experiment. *Biochemistry* **37**, 2145–2159.

Murray, D., Arbuzova, A., Mihaly, G., Ghambir, A., Ben-Tal, N., Honig, B., and McLaughlin, S. (1999). Electrostatic properties of membranes containing acidic lipids and adsorbed basic peptides: Theory and experiment. *Biophys. J.* **77**, 3176–3188.

Murray, D., and Honig, B. (2001). Electrostatic control of the membrane targeting of C2 domains. *Molecular Cell.* In press.

Myat, M. M., Anderson, S., Allen, L. H., and Aderem, A. (1997). Identification of the basolateral targeting determinant of a peripheral membrane protein, MacMARCKS, in polarized cells. *Curr. Biol.* **7**, 611–614.

Nairn, A. C., and Aderem, A. (1992). Calmodulin and protein kinase C cross-talk: The MARCKS protein is an actin filament and plasma membrane cross-linking protein regulated by protein kinase C phosphorylation and by calmodulin. *Ciba Found. Symp.* **164**, 145–154.

Oancea, E., and Meyer, T. (1998). Protein kinase C as a molecular machine for decoding calcium and diacylglycerol signals. *Cell* **95**, 307–318.

Parsons, J. T., and Parsons, S. J. (1997). Src family protein tyrosine kinases: Cooperating with growth factor and adhesion signaling pathways. *Curr. Opin. Cell Biol.* **9**, 187–192.

Peitzsch, R. M., and McLaughlin, S. (1993). Binding of acylated peptides and fatty acids to phospholipid vesicles: Pertinence to myristoylated proteins. *Biochemistry* **32**, 10436–10443.

Peitzsch, R. M., Eisenberg, M., Sharp, K. A., and McLaughlin, S. (1995). Calculations of the electrostatic potential adjacent to model phospholipid bilayers. *Biophys. J.* **68**, 729–738.

Qin, Z., and Cafiso, D. S. (1996). Membrane structure of the protein kinase C and calmodulin binding domain of myristoylated alanine rich C kinase substrate determined by site-directed spin labeling. *Biochemistry* **35**, 2917–2925.

Raucher, D., Stauffer, T., Chen, W., Shen, K., Guo, S., York, J. D., Sheetz, M. P., and Meyer, T. (2000). Phosphatidylinositol 4,5-bisphosphate functions as a second messenger that regulates cytoskeleton–plasma membrane adhesion. *Cell* **100**, 221–228.

Rein, A., McClure, M. R., Rice, N. R., Luftig, R. B., and Schultz, A. M. (1986). Myristylation site in Pr65gag is essential for virus particle formation by Moloney murine leukemia virus. *Proc. Natl. Acad. Sci. USA* **83**, 7246–7250.

Resh, M. D. (1999). Fatty acylation of proteins: New insights into membrane targeting of myristoylated and palmitoylated proteins. *Biochim. Biophys. Acta* **1451**, 1–16.

Rosen, A., Keenan, K. F., Thelen, M., Nairn, A. C., and Aderem, A. (1990). Activation of protein kinase C results in the displacement of its myristoylated, alanine-rich substrate from punctate structures in macrophage filopodia. *J. Exp. Med.* **172**, 1211–1215.

Roux, M., Neumann, J. M., Bloom, M., and Devaux, P. F. (1988). ^{2}H and ^{31}P NMR study of pentalysine interaction with headgroup deuterated phosphatidylcholine and phosphatidylserine. *Eur. Biophys. J.* **16**, 267–273.

Roy, M., Leventis, R., and Silvius, J. R. (2000). Mutational and biochemical analysis of plasma membrane targeting mediated by the farnesylated, polybasic carboxy terminus of K-ras4B. *Biochemistry* **39**, 8298–8307.

Seykora, J. T., Myat, M. M., Allen, L. A., Ravetch, J. V., and Aderem, A. (1996). Molecular determinants of the myristoyl–electrostatic switch of MARCKS. *J. Biol. Chem.* **271**, 18797–18802.

Sharp, K. A., and Honig, B. H. (1990). Calculating total electrostatic energies with the nonlinear Poisson–Boltzmann equation. *J. Phys. Chem.* **94**, 7684–7692.

Sigal, C. T., Zhou, W., Buser, C. A., McLaughlin, S., and Resh, M. D. (1994). The amino terminal basic residues of Src mediate membrane binding through electrostatic interaction with acidic phospholipids. *Proc. Natl. Acad. Sci. USA* **91**, 12253–12257.

Silvius, J. R., and l'Heureux, F. (1994). Fluorimetric evaluation of the affinities of isoprenylated peptides for lipid bilayers. *Biochemistry* **33**, 3014–3022.

Simons, K., and Ikonen, E. (1997). Functional rafts in cell membranes. *Nature* **387**, 569–572.

Spearman, P., Horton, R., Ratner, L., and Kuli-zade, I. (1997). Membrane binding of human immunodeficiency virus type 1 matrix protein *in vivo* supports a conformational myristyl switch mechanism. *J. Virol.* **71**, 6582–6592.

Swierczynski, S. L., and Blackshear, P. J. (1995). Membrane association of the myristoylated alanine-rich C kinase substrate (MARCKS) protein. Mutational analysis provides evidence for complex interactions. *J. Biol. Chem.* **270**, 13436–13445.

Swierczynski, S. L., and Blackshear, P. J. (1996). Myristoylation-dependent and electrostatic interactions exert independent effects on the membrane association of the myristoylated alanine-rich C-kinase substrate. *J. Biol. Chem.* **271**, 23424–23430.

Tall, E. G., Spector, I., Pentyala, S. N., Bitter, I., and Rebecchi, M. J. (2000). Dynamics of phosphatidylinositol 4,5-bisphosphate in actin-supported structures. *Curr. Biol.* **10**, 743–746.

Victor, K., and Cafiso, D. (1998). Structure and position of the N-terminal membrane-binding domain of pp60src at the membrane interface. *Biochemistry* **37**, 3402–3410.

Vlachy, V. (1999). Ionic effects beyond Poisson–Boltzmann theory. *Annu. Rev. Phys. Chem.* **50**, 145–165.

Wang, J., Arbuzova, A., Hangyas-Mihalyne, G., and McLaughlin, S. (2001). The effector domain of myristoylated alanine-rich C kinase substrate binds strongly to phosphatidylinositol 4,5-bisphosphate. *J. Biol. Chem.* **276**, 5012–5019.

White, S. H., and Wimley, W. C. (1999). Membrane protein folding and stability: Physical principles. *Annu. Rev. Biophys. Biomol. Struct.* **28**, 319–365.

White, S. H., Wimley, W. C., Ladokhin., A. S., and Hristova, K. (1998). Protein folding in membranes: Determining the energetics of peptide–bilayer interactions. *Meth. Enzymol.* **295**, 62–87.

Wills, J. W., and Craven, R. C. (1991). Form, function, and use of retroviral Gag proteins. *AIDS* **5**, 639–654.

Wimley, W. C., and White, S. H. (1996). Experimentally determined hydrophobicity scale for proteins at membrane interfaces. *Nat. Struct. Biol.* **3**, 842–848.

Zhou, W., and Resh, M. D. (1996). Differential membrane binding of the human immunodeficiency virus type 1 matrix protein. *J. Virol.* **70**, 8540–8548.

Zhou, W., Parent, L. J., Wills, J. W., and Resh, M. D. (1994). Identification of a membrane-binding domain within the amino-terminal region of human immunodeficiency virus type 1 Gag protein which interacts with acidic phospholipids. *J. Virol.* **68**, 2556–2569.

CHAPTER 11

The Energetics of Peptide-Lipid Interactions: Modulation by Interfacial Dipoles and Cholesterol

Thomas J. McIntosh, * **Adriana Vidal,** * **and Sidney A. Simon**†

*Department of Cell Biology and †Department of Neurobiology, Duke University Medical Center Durham, North Carolina 27710

I. Introduction
 A. Energetics of Partitioning
 B. Contributions to the Partitioning Process
II. Results
 A. Measurements of Free Energy of Partitioning
 B. Location of Signal Peptide in the Bilayer
III. Discussion
 A. Roles of Electrostatics, Peptide Conformational Changes, and "Classical" Hydrophobic Effect
 B. Role of Dipoles in the Headgroup and Hydrocarbon Regions of the Bilayer
 C. Roles of Area per Molecule and Isothermal Compressibility Modulus
IV. Conclusions
 References

The goal of this work is to determine how specific compositional, structural, and mechanical properties of the lipid bilayer modulate the membrane binding and conformation of a biologically significant peptide, the signal sequence of the bacterial protein LamB. The combination of circular dichroism, microelectrophoresis, and fluorescence quenching data showed that this peptide converted from a random coil in the aqueous phase to a "hammock" configuration in bilayers, with both termini exposed to the aqueous phase on the outside of the lipid vesicle and a central hydrophobic α-helical segment located near the hydrocarbon–water interface. Although the free energy of binding ΔG depended on the charge of the lipid,

Current Topics in Membranes, Volume 52

indicating the importance of electrostatics, the peptide bound to zwitterionic phosphatidylcholine (PC) bilayers with $\Delta G = -6.3$ kcal/mol. Analysis of isothermal titration calorimetry (ITC) data indicated that both conformational changes in the peptide and the hydrophobic effect contributed to this binding to PC bilayers. Peptide binding and α-helix formation were essentially eliminated by adding lipids containing large, oriented dipoles located a few angstroms into the bilayer hydrocarbon region, pointing to an important role for interfacial dipoles. That is, for PC bilayers containing either 6-ketocholestanol or PCs with nitroxide moieties at the 7-position in one of their acyl chains, ΔG was reduced to -3 kcal/mol. Binding experiments with polyunsaturated PCs showed that ΔG was not strongly dependent on molecular area. However, incorporation of cholesterol into the PC bilayer, which increases the area expansion or isothermal compressibility modulus, modified the energetics of binding. Increasing concentrations of cholesterol reduced ΔG to about -5 kcal/mol at equimolar cholesterol, but even more markedly reduced the enthalpy ΔH of binding from -8 kcal/mol in the absence of cholesterol to near 0 kcal/mol with equimolar cholesterol. Thus, peptide binding was enthalpically driven for PC bilayers, but entropically driven for equimolar PC : cholesterol bilayers. We conclude that variations in lipid composition that alter bilayer properties, including fixed charges, interfacial dipoles, and isothermal compressibility modulus, can strongly influence peptide binding.

I. INTRODUCTION

A. Energetics of Partitioning

The partitioning of peptides from an aqueous phase to a lipid bilayer involves several energetic contributions to the changes in the free energy of transfer ΔG. These contributions involve electrostatic, ion–dipole, dipole–dipole, and hydrogen-bond interactions of the peptide with the lipid headgroup, image charge and Born energy repulsion arising from the insertion of charges or dipoles into low-dielectric-constant regions, hydrophobic interactions, van der Waals attraction, and conformational changes in both the peptide and the bilayer (Ben-Shaul *et al.,* 1996; Ben-Tal *et al.,* 1996; Hunt *et al.,* 1997a; Jones and Gierasch, 1994b; Wimley and White, 1993). As described in this chapter, as well as in other chapters in this book, fundamental thermodynamic information about the different components of the transfer process can be obtained by investigating these interactions in well-characterized systems. In peptide–lipid interaction studies in which ΔG has been measured, many researchers have selectively modified the peptide and have chosen a particular type of lipid bilayer (or systematically changed the bilayer's charge density). This is a reasonable approach, because electrostatic interactions have proven to be among the strongest interactions in determining the partitioning of solutes into bilayers (see Chapters 8 and 10). However, many of the other contributions listed above can significantly modulate the partitioning of peptides into bilayers, as described in several chapters in this book. In this chapter we focus on the changes in ΔG that occur for a particular peptide (the signal sequence of the bacterial protein LamB) when selected chemicophysical properties of the lipid bilayer are modified. These properties include the presence of charges and dipoles in the lipid headgroup or acyl chain regions, the dipole potential, the bilayer area compressibility modulus, and the number of double bonds in the acyl chains. Of course, as will be detailed below, these various properties are to some extent coupled, so that changing one parameter may alter others.

The change in free energy of transfer ΔG of a peptide from water to a membrane is made up of several terms (Ben-Shaul *et al.*, 1996; Hunt *et al.*, 1997a; Russell *et al.*, 1999; White and Wimley, 1999) and can be expressed as

$$\Delta G = -RT \ln K_p = \Delta G_{qE} + \Delta G_{solv} + \Delta G_{bilayer}, \tag{1}$$

where RT is the thermal energy, K_p is the mole fraction bilayer–water partition coefficient, ΔG_{qE} is the electrostatic contribution of binding due to fixed charges, ΔG_{solv} is the contribution due to the changes in solvation of both the peptide (Born energies) and the bilayer upon partitioning and is made up of the classical hydrophobic free energy and the energy difference arising from placing charges (or partial charges) in lower dielectric constant regions, and $\Delta G_{bilayer}$ represents the effects of the bilayer on the partitioning process.

B. Contributions to the Partitioning Process

1. Electrostatic and Hydrophobic Interactions

In most investigations of solute binding to bilayers, two attractive ($\Delta G < 0$) interactions have been considered: electrostatic and hydrophobic. An attractive electrostatic component of binding can arise when the fixed charges on the peptide and bilayer are of opposite signs. Experimentally this interaction has been determined and quantified by changing the charge density of the lipids and/or by increasing the ionic strength and using the nonlinear Poisson–Boltzmann equation to calculate the interactions (see Chapters 2, 8, and 10). Attractive electrostatic interactions usually dominate the partitioning when small groups of highly positively charged amino acids in peptides (or proteins) bind to bilayers containing significant amounts of negatively charged lipids, such as phosphatidylserine (PS) or phosphatidylglycerol (PG). Increasing the charge on either the peptide or the bilayer increases the binding in a quantitatively predictable manner. [Once the peptide is bound, the proximity of a positively charged group on the peptide to a negatively charged group in the bilayer depends on the topology of the charge distribution, the net charges on the peptide and bilayer, and the dehydration energy of these groups (Honig and Nichols, 1995).] However, ΔG_{qE} is small for peptides that interact with electrically neutral or zwitterionic bilayers, such as those composed solely of the membrane lipid phosphatidylcholine (PC).

Classical hydrophobic interactions have also been explored. These hydrophobic interactions are entropically driven, so that $\Delta G \sim T \Delta S$, where T is the temperature and ΔS is the change in entropy, and arise predominantly from the entropic changes associated with the insertion of nonpolar groups in water. For amino acid side chains, the hydrophobic interactions are proportional to the exposed hydrophobic solute area. This value is usually taken to be -20 to -25 cal/mol/Å^2, but depends on the position where the molecule partitions into the bilayer. Therefore it may be different depending on whether a hydrophobic residue partitions into the interfacial or the acyl chain region (Thorgeirsson *et al.*, 1996).

2. Bilayer Effects

In considering the partitioning of peptides into lipid bilayers one must appreciate the presence of three separate environments, each with distinct solvent properties: the aqueous phase, the bilayer hydrocarbon region, and the water–bilayer interfacial region (the location of the lipid polar headgroups). The aqueous phase is a hydrogen-bonding solvent characterized by a high dielectric constant, the hydrocarbon chain region is apolar, having a low dielectric constant, whereas the interfacial region can have a wide variability in solvent properties due to its charges, dipoles, hydrophobic groups (such as choline), H-bond acceptors, and H-bond donors and the fact that the volume fraction of interfacial water decreases toward the hydrocarbon region.

$\Delta G_{bilayer}$ has several contributing factors, including conformational changes of the peptide caused by the bilayer, perturbation of the lipid structure, and a variety of interactions due to specific properties of both the interfacial and the hydrocarbon regions of the bilayer (Ben-Shaul *et al.*, 1996; Hunt *et al.*, 1997a; White and Wimley, 1999). Specific properties of the lipid bilayer's headgroup and hydrocarbon region can influence $\Delta G_{bilayer}$. Modifications in the bilayer interfacial region are of particular importance because (1) the interfacial region is where water-soluble peptides first come into contact with the bilayer, (2) the partitioning of peptides into the interface is important in a number of biological processes, and (3) the chemical composition of the interfacial region differs among biological membranes as well as between the extracellular and cytoplasmic sides of plasma membranes.

a. Conformational Changes in the Peptide Caused by the Bilayer. The monomeric conformation of peptide in water is often quite different than when it is associated with the bilayer headgroup and/or the hydrocarbon region. Many peptides, such as amphipathic peptides and, most importantly to this chapter, signal peptides, form random coils in water that convert to α helices in hydrophobic environments such as sodium dodecylsulfate (SDS) micelles or lipid bilayers (Hammen *et al.*, 1996; McKnight *et al.*, 1989; Wieprecht *et al.*, 1997, 1999a). This change from a random coil to an α helix produces a large favorable free energy change, which promotes partitioning into the bilayer. In a clever series of experiments Ladokhin and White (1999) and Wieprecht *et al.* (1999a, 2000) used several double-D isomers of amphipathic peptides to determine the free energy gained per amino acid residue in forming an α helix upon binding. By combining circular dichroism (CD) with partitioning measurements, Ladokhin and White (1999) found for the amphipathic peptide melittin that ΔG per residue is about -0.4 kcal/mol. By combining CD and isothermal titration calorimetry (ITC) measurements, Wieprecht *et al.* (1999a) found a value of $\Delta G = -0.14$ kcal/mol per residue for the amphipathic peptide magainin and Wieprecht *et al.* (2000) measured a value of $\Delta G = -0.2$ kcal/mol/residue for the presequence of the mitochondrial protein rhodanese. These ITC studies (Wieprecht *et al.*, 1999a, 2000) also yielded the changes in enthalpy ΔH and entropy ΔS per residue of the interaction and showed that, at least for these two small (about 25 amino acids) amphipathic peptides, (1) α-helix formation accounted for about one-half of the total free energy of binding and (2) the peptide binding to phospholipid bilayers was driven by enthalpy and opposed by the changes in entropy.

b. Hydrocarbon Thickness of Bilayer: Hydrophobic Mismatch. As described in detail in Chapters 8 and 12, the hydrophobic thickness of the bilayer can influence the interactions between bilayers and transmembrane peptides or proteins. When the hydrophobic length of the peptide does not match the hydrophobic thickness of the lipid, there is a large cost of free energy, which can be reduced either by the bilayer locally deforming around the peptide or by the peptide tilting (when the peptide is longer than the bilayer hydrophobic thickness). Nielsen *et al.* (1998) recently described the energetic costs of peptide-induced bilayer deformation due to hydrophobic mismatch (originally called the "mattress model"; Mouritsen and Bloom, 1984).

To critically evaluate this energetic term for membranes, it is important to know the hydrophobic thickness of the lipid bilayer and how it is modified by the number of methylene groups and double bonds in the phospholipid hydrocarbon chains as well as by the incorporation of cholesterol. Several methods have been used to determine the hydrophobic thickness of the bilayer, including gravimetric analysis of X-ray diffraction data (Lecuyer and Dervichian, 1969; Luzzati, 1968; Rand *et al.*, 1988), Fourier analysis of lamellar diffraction (McIntosh and Simon, 1986b; Petrache *et al.*, 1998), X-ray solution scattering (B. A. Lewis and Engelman, 1983), neutron diffraction (Buldt *et al.*, 1979; Zaccai *et al.*, 1975), and a combination of capacitance and X-ray diffraction data (Simon *et al.*, 1982). These studies found that (1) in fluid bilayers the thickness increases linearly with the length of the acyl chain with an increment of about 0.8 Å/CH$_2$ group (B. A. Lewis and Engelman, 1983), (2) increasing concentrations of cholesterol in PC bilayers monotonically increases the bilayer thickness, by up to approximately 5 Å with equimolar cholesterol (Lecuyer and Dervichian, 1969), and (3) the presence of cholesterol decreases the depth of water penetration into the hydrocarbon region, thereby increasing the width of the low-dielectric-constant region of the bilayer by about 5 Å (Simon *et al.,* 1982).

c. Roles of Area per Lipid Molecule and Isothermal Compressibility Modulus K_T.   For small molecules that completely partition into the nonpolar region (e.g., hexane), the properties of the hydrocarbon region can influence the partition. For example, adding cholesterol usually decreases the partitioning of small hydrophobic molecules into the acyl chain region (De Young and Dill, 1988). Presumably, similar effects will also be true for peptides. Two possible reasons for cholesterol's effects on partitioning are that the incorporation of cholesterol decreases the area per phospholipid molecule (Lecuyer and Dervichian, 1969) and increases the bilayer's cohesive interactions (the work to separate or condense the acyl chains of the bilayer). The increases in cohesive interactions are demonstrated by measured increases in the area expansion of the isothermal compressibility modulus, K_T [dyn/cm] $= (\delta G/\delta A)_T$, which is the slope of the linear region of the relation between the applied tension T (in dyn/cm) and the change in area α. The modulus K_T has been measured for many bilayers (Kwok and Evans, 1981; Needham and Nunn, 1990; Rawicz *et al.,* 2000), and increases significantly with increasing cholesterol concentration (Needham and Nunn, 1990). In addition, the compressibility modulus is a critical parameter in determining the energy of the hydrophobic mismatch between peptide and bilayer (Section I.B.2b), because the energy to deform a bilayer is proportional to K_T (Nielsen *et al.,* 1998).

d. Dipole Potential.   For bilayers, the potential arising from the sum of the perpendicular components of the dipoles making up the headgroup region and the associated polarized water molecules is called the dipole potential, ΔVd. For phospholipids or sphingolipids, ΔVd is positive toward the bilayer interior by several hundred millivolts (Brockman, 1994; Smaby and Brockman, 1990). This short-range potential has been shown to influence the partitioning of charged hydrophobic ions (Franklin and Cafiso, 1993). Since amphipathic helices that partition into the interface have a large dipole moment, their location and orientation might also be influenced by this large interfacial potential (Cafiso, 1995). To date, few experiments have been designed to critically test this effect. As detailed below, we (Voglino *et al.,* 1998) have investigated the role of dipole potential on peptide binding by determining how the LamB peptide binding depends on ΔVd. This was achieved by comparing the binding of peptides to bilayer lipids containing cholesterol and its analogue 6-ketocholestanol (which is similar to cholesterol, but has a ketone in the 6-position; Fig. 1). Both cholesterol and 6-ketocholestanol increase the dipole potential, but the increase is much larger for 6-ketocholestanol, because of the position and orientation of the dipolar ketone group (Simon *et al.*, 1992; Voglino *et al.*, 1998).

e. Lateral Pressure Gradients. Unlike an isotropic solvent, a lipid bilayer has a unique property, namely that the stresses throughout the acyl chain region are anisotropic and vary with the depth within the bilayer. Although the total lateral pressure in the bilayer is approximately 0 dyn/cm, the stresses are distributed throughout the bilayer as a consequence of the negative excess free energy arising from the surface tension at the hydrocarbon–water interfaces, which must be balanced by positive pressures distributed throughout the acyl chain region. At the two hydrocarbon–water interfaces the compressive lateral pressure is concentrated over a thin region and is on the order of 10^8–10^9 dyn/cm (100–1000 atm). In this region this very large pressure can be altered by changes in lipid composition. Cantor's calculations (Cantor, 1997, 1999a,b) indicate that the incorporation of cholesterol, the distribution of double bonds, and the presence of small amphipathic molecules cause large redistributions in lateral pressure, which could modulate the conformation of membrane-associated proteins. Because conformational energy is involved in the free energy of transfer (see Section I.B.2a), this could, in principle, alter the membrane–water partition coefficient.

f. Spontaneous Curvature or Curvature–Elastic Effects. Various polar lipids can form lamellar (bilayer) as well as a variety of nonlamellar phases (Luzzati *et al.*, 1968). Lipids, such as diacyl chain phosphatidylcholine (PC), that have approximately the same projected headgroup and acyl chain areas, form lamellar phases. Lyso PCs, whose headgroup area is much larger than the acyl chain area, tend to form micelles, and diacyl chain phosphatidylethanolamine (PE), whose headgroup area is smaller than the area of its acyl chains, forms inverted hexagonal phases (Israelachvili *et al.*, 1980; Kirk *et al.*, 1984). The molecules in micellar and hexagonal phases exhibit spontaneous curvature. Spontaneous curvature represents the curvature that one monolayer of a bilayer would have if it were unconstrained by being in a bilayer (Gruner, 1985). The work or free energy associated with this tendency to bend is $W_B = B/2(1/R - 1/R_0)^2$, where B is the bilayer bending modulus, R is the radius of curvature of the lipid phase, and R_0 is the spontaneous radius of curvature (Gruner *et al.*, 1986). Changing the curvature–elastic energy can alter K_p, because the free energy of transfer $\Delta G = -RT \ln(K_p)$ represents the difference in chemical potential of the peptide in water and the peptide associated with the bilayer. It follows that changing the chemical potential of the bilayer by changing the curvature–elastic energy will alter ΔG. Thus, from the above equation, it is predicted that the changes in free energy of binding ΔW_B should be directly proportional to the bending modulus and inversely proportional to the curvature $C = 1/R_0$. Although there have been studies showing that changes in curvature alter conductances of ion channels formed from alamethicin (Keller *et al.*, 1993) or gramicidin A (Lundbaek *et al.*, 1997), there have been very few tests to determine how changing the spontaneous curvature effects peptide binding. The most thorough and convincing study of the effect of the curvature–elastic energy on peptide binding was done by J. R. Lewis and Cafiso (1999). They measured the binding of alamethicin to dioleoylphosphatidylcholine (DOPC) bilayers that contained different mole fractions of dioleoylphosphatidylethanolamine (DOPE) or mono and dimethyl DOPEs which had different values of R_0. Plots of the change in free energy versus $C(= 1/R_0)$ gave a straight line, indicating that curvature of elastic energy is a critical factor in modulating the changes in the partition coefficient of transmembrane peptides.

g. Solvent Properties of the Interfacial Region and Steric Barriers. The interfacial region exhibits a wide variability in solvent properties for different classes of lipid. In part this is because the dielectric constant ε varies from that of bulk water ($\varepsilon = 78$) to that of hydrocarbon ($\varepsilon = 2$) over a relatively short distance that depends on the lipid composition. For example, the headgroup thickness can vary from about 8 Å for PE bilayers (McIntosh and Simon, 1986a), to 10 Å for PC bilayers (McIntosh and Simon, 1986a; Nagle *et al.*, 1996; Petrache *et al.*, 1998, Small, 1967) to about 25 Å for the membrane glycolipid ganglioside GM1 (McIntosh and Simon, 1994), to over 60 Å for specific lipopolysaccharides (LPSs), the major lipids on the surface of Gram-negative bacteria (Snyder *et al.*, 1999). To put these dimensions in perspective for lipid–peptide interactions, the diameter of the core of an α helix is approximately 10 Å. Therefore, an α-helical peptide oriented parallel to the bilayer surface could be almost completely embedded in the interfacial region of a PC bilayer (White and Wimley, 1994, 1999; White *et al.*, 1998) or a GM1 bilayer, but not a PE bilayer. The rapid change in dielectric constant across the interface reflects, in essence, the change in volume fraction of interfacial water. Since the activity of interfacial water is dependent on the chemical composition of the interface, it follows that the work to remove water upon the binding of peptides may differ among these different lipid interfaces and that K_p may depend on the interfacial environment.

The wide interfacial regions of particular lipids, such as the bacterial LPS and the ganglioside GM1, produce a steric barrier to partitioning of hydrophobic peptides into the hydrophobic interior of membranes. The range and magnitude of this barrier have been measured for both LPS (Snyder *et al.*, 1999) and GM1 (McIntosh and Simon, 1994). In the case of LPS, this headgroup permeability barrier is thought to be critical for the resistance of Gram-negative bacteria to various antibiotics, such as the amphipathic peptides magainin and melittin (Banemann *et al.*, 1998; Macias *et al.*, 1990; Rana and Blazyk, 1991); mutant bacteria containing LPSs with smaller numbers of saccharides have greater susceptibilities to these peptides. In terms of GM1, it has been shown that the introduction into liposomes of either GM1 or synthetic lipids with covalently attached polyethyleneglycol headgroups (PEG–lipids) changes the bilayer partitioning of large proteins into the membrane interface (Chonn and Cullis, 1992; Du and Chandaroy, 1997; Senior *et al.,* 1991). This barrier property has an important practical application for *in vivo* drug delivery, because the introduction of GM1 or PEG-lipids into liposomes is thought to prevent binding of serum proteins (opsonization) when the liposomes are injected into the blood stream. This effect decreases phagocytosis of liposomes by macrophages in the liver and spleen, thereby increasing the blood circulation time of the liposomes and allowing targeting to solid tumors. The steric barrier produced by PEG–lipids was measured as a function of both PEG size and concentration of PEG–lipid in the bilayer and shown to correlate closely with *in vivo* blood circulation times (Kenworthy *et al.,* 1995a,b). Although these wide interfacial regions can prevent protein binding, smaller molecules, such as lysolipids, can diffuse through the PEG polymer network and reach the hydrocarbon–water interface (Needham *et al.,* 1997) (see Chapter 15).

Other differences in interfacial properties between the two most common phospholipids in biological membranes, PC and PE, are in their H-bonding capabilities (Boggs, 1987; Nagle, 1980) and strength of the interactions with π bonds on aromatic molecules. In this regard, we found that the binding of polyphenols to the monolayers was greater for PC than PE interfaces, mostly as a consequence of the strong π-choline bonds in PC bilayers (Huh *et al.,* 1996; McIntosh *et al.,* 1999; Simon *et al.,* 1994). Consistent with this interpretation, Huh *et al.* (1996) found that decreasing the surface density of PC headgroups by the addition of cholesterol (which acts as a spacer molecule for these purposes) also decreased the binding of polyphenols. Although it has not been demonstrated experimentally, it seems likely these same interactions occur with peptide aromatic groups, such as tryptophan (White and Wimley, 1999; Yau *et al.,* 1998).

h. Presence of Dipoles in the Interfacial or Hydrocarbon Regions of the Bilayer. Although the roles of fixed charges in modulating peptide binding have been analyzed both experimentally and theoretically (see Chapter 10), the effects of fixed dipoles in modulating peptide binding have not been throughly explored. This is a potentially important consideration, because the interfacial region of lipid bilayer contains dipoles from the lipid headgroups and polarized water distributed throughout the interface. Moreover, molecules adsorbed to the interface or incorporated into the bilayer, such as anesthetics or proteins, also can contribute oriented dipoles while replacing interfacial water molecules (Cafiso, 1995; Qin *et al.,* 1995).

The hydrocarbon region generally contains acyl chains with various numbers of double bonds, which give this region a slightly more polar character than that of an *n*-alkane. The insertion of dipoles in the hydrocarbon region is energetically unfavorable, being proportional to the square of the permanent dipole moment (Israelachvili, 1985). Nevertheless, in most biological membranes, dipoles are in the hydrocarbon region due to the presence of the amino acid residues in transmembrane peptides. If proteins are present in sufficiently high concentrations, as in the case of the inner mitochondrial membrane, they can increase the dielectric constant and hence increase the partition coefficient of charged or dipolar molecules into this region (Dilger *et al.,* 1979). Dipoles can also be inserted at any depth into the hydrocarbon region of bilayers by covalently linking them to the lipid chains. Common examples are nitroxide-labeled phospholipids, which contain covalently linked doxyl groups (see Fig. 1), and are used in spin-label experiments and fluorescence quenching experiments designed to determine the depth of penetration of peptides in the hydrocarbon region (Abrams and London, 1992; Chattopadhyay and London, 1987). These dipole layers can interact with specific groups on membrane peptides through dipole–dipole interactions and/or can change the dielectric profile or energy barrier in the bilayer (Flewelling and Hubbell, 1986).

II. RESULTS

To elucidate the role of bilayer properties on peptide binding, we chose to study the signal sequence of the bacterial protein LamB, because its interactions with negatively charged bilayers have been well characterized (Hoyt *et al.,* 1991; Jones and Gierasch, 1994a,b; McKnight *et al.,* 1991a,b; Wang *et al.,* 1993) and because of its biological significance. In both bacteria and eukaryotes, proteins that are to be transported out of the cytoplasm contain N-terminal extension sequences, called signal or leader sequences, which increase the efficiency of protein transport across membranes. A typical signal sequence (von Heijne, 1995; Wickner and Leonard, 1996), such as the signal sequence of LamB (Fig. 1), has a net positive charge on its N-terminus, a hydrophobic portion that adopts an α-helical conformation in hydrophobic environments, and a more polar cleavage region. Small changes in leader sequences, especially in the hydrophobic regions, can markedly affect export (Jones and Gierasch, 1994b). A central issue in protein trafficking is to understand how signal sequences interact with biological membranes to facilitate protein transport across them. In most cases there is a requirement of specific proteinaceous components for proper export (Joly and Wickner, 1993; Schatz and Beckwith, 1990; Wickner and Leonard, 1996), although there is evidence (Martoglio *et al.,* 1995) indicating that during membrane translocation the signal sequence is in direct contact with lipids. These data, together with the fact that the conformation of signal peptides is sensitive to the lipid environment (Bruch *et al.,* 1989; Chaloin *et al.,* 1997; McKnight *et al.,* 1991a), suggest a possible role of membrane lipids in the translocation process.

A. Measurements of Free Energy of Partitioning

By the combined use of an ultrafiltration binding assay (Sophianopoulos *et al.,* 1978), fluorescence spectroscopy, CD, and microelectrophoresis, we (Voglino

LIPIDS:

6–Ketocholestanol   7-Doxyl PC

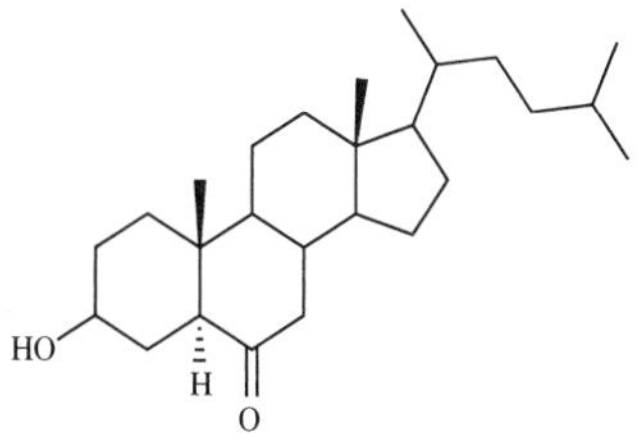

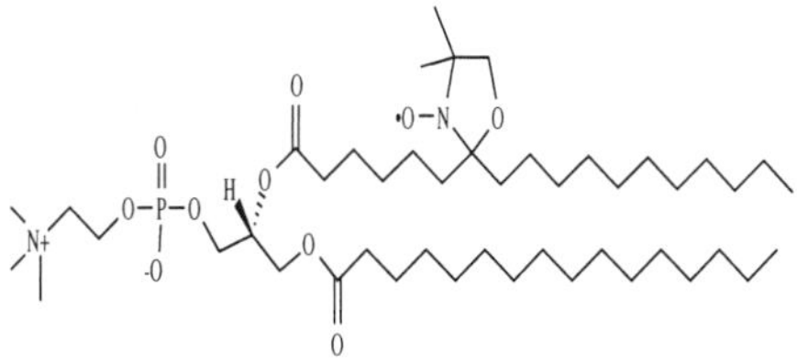

PEPTIDES:

LamB   Net Charge

```
+            + +                              –
NH3·MMITLRKLPLAVAVAAGVMSAQAMA·C00          (+2)
```

LamB-W

```
+            + +                              –
NH3·MMITLRKLPLAVAVAAGWMSAQAMA·C00          (+2)
```

FIGURE 1 Structures of 6-ketocholestanol and 1-palmitoyl-2-stearoyl (7-doxyl)-*sn*-glycerol-3-phosphocholine, and amino acid sequences of the leader sequence of the protein LamB and its analogue LamB-W. The net charges for these peptides are given under the assumption that at pH 7 both the C- and N-termini are fully charged. Taken with permission from Voglino *et al.* (1998).

et al., 1998, 1999) studied the partitioning into bilayers of LamB-W (Fig. 1). The signal sequence of LamB was modified for fluorescence spectroscopy by the substitution of a tryptophan (W) residue at position 18 near the center of the hydrophobic region of the peptide (Jones and Gierasch, 1994a). The partition coefficient of LamB-W was measured with lipid bilayers that were modified by changing (1) the surface charge density (by varying the concentration of negatively charged lipids), (2) the distribution of exogenously added dipoles in and near the interfacial region (by using bilayers containing 6-ketocholestanol and phospholipids with nitroxide groups in their acyl chains), and (3) the area per lipid molecule and area compressibility modulus (by using PCs with multiple double bonds and by adding cholesterol).

1. Role of Electrostatics and Classical Hydrophobic Effect

In agreement with previous results (Jones and Gierasch, 1994a,b), we (Voglino *et al.,* 1998) found that the positively charged LamB-W bound avidly to negatively

TABLE I

Free Energy of Binding for LamB-W to Different
Bilayers[a]

Lipid	ΔG (kcal/mol)
8 : 2 EPC : PS	−8.0
EPC	−6.3
DAPC	−6.2
1 : 1 EPC : 6-ketocholestanol	−3.9
7-Doxyl PC	−3.0
9 : 1 EPC : 7-doxyl PC	−5.6
9 : 1 EPC : 12-doxyl PC	−6.1
8 : 2 EPC : cholesterol	−6.0
6 : 4 EPC : cholesterol	−4.8
1 : 1 EPC : cholesterol	−4.7
1 : 1 SM : cholesterol	−4.6
6,7-Br PC	−6.1
9,10-Br PC	−6.2
11,12-Br PC	−6.1

[a]Experimental results, except those for EPC : cholesterol and
SM : cholesterol, are taken from Voglino *et al.* (1998, 1999).

charged bilayers containing a mixture of the zwitterionic phospholipid egg
phosphatidylcholine (EPC) and the negatively charged phospholipid phosphatidylserine (PS). For 4 : 1 EPC : PS our direct (ultrafiltration) binding assay showed
that at a phospholipid concentration of 1.5 mM and a LamB-W concentration of
5 μM, greater than 90% of the peptide bound to the lipid, so that the free energy
of transfer was $\Delta G = -8.0$ kcal/mol (Table I).

We next investigated, using small unilamellar vesicles (SUVs), how selected
properties of the lipid bilayer modified LamB-W partitioning and conformation.
First, we studied the role of surface charge by comparing the LamB-W binding
to 4 : 1 EPC : PS with binding data obtained with zwitterionic EPC bilayers. As
expected, the positively charged LamB-W peptide bound less to EPC bilayers
($\Delta G = -6.3$ kcal/mol) than to the EPC : PS bilayers (Table I). Measurements
of the heat of binding obtained using ITC gave for EPC an enthalpy of binding
$\Delta H = -7.8$ kcal/mol (Fig. 2). Since $\Delta G = \Delta H - T\Delta S$, where ΔH and $T\Delta S$
are the enthalpic and entropic contributions, respectively, it follows that $-T\Delta S =$
1.5 kcal/mol (at 298°K), and therefore the binding of LamB-W to PC vesicles is
enthalpically driven.

The penetration depth and conformation of LamB-W in EPC bilayers were
analyzed by fluorescence spectroscopy and CD. Measurements of the changes in
tryptophan fluorescence showed the presence of a 10-nm blue shift in the maximum wavelength (Fig. 3). This relatively small blue shift is consistent with the

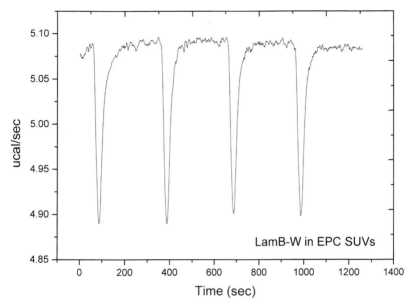

FIGURE 2 Titration calorimetric traces of the binding of LamB-W with SUVs of EPC. The scan shows the results from four 10-μl injections of 120 μM LamB-W solution into a reaction cell containing 20 mM EPC. The buffer was 20 mM KCl, 5 mM Hepes adjusted to pH 7.4.

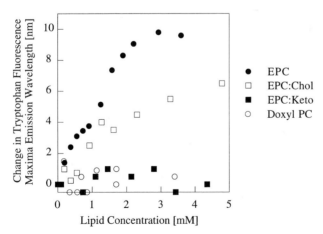

FIGURE 3 Change in LamB-W tryptophan fluorescence maxima emission wavelength (blue shift) plotted as a function of lipid concentration for EPC, 1:1 EPC:cholesterol, 1:1 EPC:6-ketocholestanol, and 7-doxyl PC single-walled vesicles. Taken with permission from Voglino *et al.* (1998).

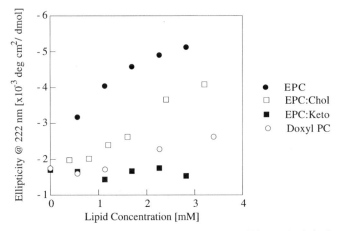

FIGURE 4 Circular dichroism measurements of ellipticity at 222 nm (ϑ_{222}) for LamB-W as a function of lipid concentration for EPC, 1 : 1 EPC : cholesterol, 1 : 1 EPC : 6-ketocholestanol, and 7-doxyl PC single-walled vesicles. The magnitude of the negative peak in ellipticity at 222 nm is a measure of the percentage of α-helix formation. Taken with permission from Voglino *et al.* (1998).

tryptophan-18 being located in the interfacial region rather than deep in the acyl chain region (Jones and Gierasch, 1994a). This is because larger (20–30 nm) blue shifts are usually observed when tryptophans are buried deeper in the hydrocarbon core of the bilayer (Jones and Gierasch, 1994a). Circular dichroism experiments (Fig. 4) showed that the conformation of LamB-W changed from a random coil in the aqueous phase to an α helix in the presence of EPC (Voglino *et al.*, 1998), similar to the conformation change found for LamB-W in the presence of negatively charged bilayers (Jones and Gierasch, 1994a).

2. Role of Dipoles in the Headgroup and Hydrocarbon Regions of the Bilayer

We also tested the effects of interfacial dipoles on LamB-W partitioning by measuring peptide binding to bilayers composed of equimolar EPC and 6-ketocholestanol (an analogue of cholesterol with a ketone group at the 6-position) and the spin probe 1-palmitoyl-2-stearoyl(7-doxyl)-*sn*-glycero-3-phosphocholine (7-doxyl PC), which has a nitroxide group at the 7-position on one acyl chain (Fig. 1). Both 6-ketocholestanol and 7-doxyl PC have large dipoles (the ketone and nitroxide groups, respectively) located at approximately the same depth in the bilayer (4–5 Å from the hydrocarbon–water interface) (Voglino *et al.*, 1998). Despite these similarities, the dipole potentials of monolayers of 7-doxyl PC and EPC : 6-ketocholestanol (1 : 1) were quite different, 368 and 703 mV, respectively (Voglino *et al.*, 1998), compared to the dipole potential of 415 mV of EPC (Simon and McIntosh, 1989). That is, 7-doxyl PC had a smaller dipole potential than

EPC, whereas 1 : 1 EPC : 6-ketocholestanol had a much larger dipole potential than EPC. The free energy of binding of LamB-W was only $\Delta G = -3.9$ kcal/mol to 1 : 1 EPC : 6-ketocholestanol and $\Delta G = -3.0$ kcal/mol to bilayers of 7-doxyl PC (Table I). Thus, compared to EPC, LamB-W binding was considerably reduced for both these bilayers with dipoles near the hydrocarbon–water interface in a manner not related to differences in dipole potential. CD experiments (Fig. 4) showed that, compared to EPC, the peptide random-coil to α-helix conformational change was practically eliminated for both 1 : 1 EPC : 6-ketocholestanol and 7-doxyl PC. Moreover, there was no appreciable blue shift in tryptophan fluorescence for both 1 : 1 EPC : 6-ketocholestanol and 7-doxyl PC bilayers (Fig. 3). These results show, for the first time, that the binding and conformation of interfacial peptides is quite sensitive to exogenously added dipoles located near the interfacial region (Voglino *et al.*, 1998).

Because 7-doxyl PC is often used as a spin probe or as a fluorescent quencher, we analyzed the effects on peptide binding of the incorporation of this and other spin probes and quenchers into bilayers at the concentrations typically used in structural studies. We (Voglino *et al.*, 1999) found that both the concentration of spin probe and the location of the nitroxide group on the lipid chain modified LamB-W binding. The binding (Table I), the blue shift (Fig. 5), and the ellipticity (Fig. 6) were reduced by the presence of 7-doxyl PC in a concentration-dependent manner (Voglino *et al.*, 1999). Interestingly, the binding was decreased at 7-doxyl PC concentrations (10 mol%) typically used in quenching experiments. However,

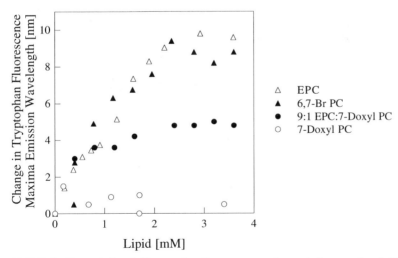

FIGURE 5 Change in LamB-W tryptophan fluorescence maxima emission wave-length (blue shift) plotted as a function of lipid concentration for EPC, 6,7-Br PC, 7-doxyl PC, and 9 : 1 EPC : 7-doxyl PC. Taken with permission from Voglino *et al.* (1999).

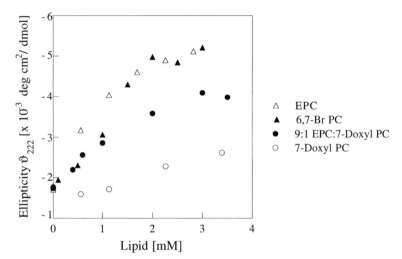

FIGURE 6   Circular dichroism measurements of ellipticity at 222 nm (ϑ_{222}) for LamB-W as a function of lipid concentration for EPC, 6,7-Br PC, 7-doxyl PC, and 9 : 1 EPC : 7-doxyl PC. Taken with permission from Voglino *et al.* (1999).

binding was not appreciably affected by the presence of 10 mol% 12-doxyl PC (Table I), a lipid probe where the nitroxide group is located at the 12-position on the acyl chain. These results indicate that the position of the nitroxide probe affected the amount of peptide bound; the probe had a much larger effect when it was located closer to the interface. This means that the location of these dipoles in the bilayer is critical to LamB-W partitioning, with dipoles closer to the interfacial region decreasing binding to a larger extent than dipoles located deeper within the hydrocarbon core of the bilayer. Moreover, because fluorescence quenching depends both on the proximity of the probe to the tryptophan residue and the amount of peptide in the bilayer, this binding difference makes the use of these nitroxide probes problematic for parallax analysis of peptide tryptophan depth (Abrams and London, 1992; Chattopadhyay and London, 1987) for the LamB-W peptide (Voglino *et al.*, 1999).

3. Roles of Area per Molecule and Isothermal Compressibility Modulus

Two related factors that could potentially modulate peptide–lipid interactions are the area per lipid molecule and the work necessary to separate acyl chains when a peptide partitions into the lipid bilayer. As described in Section I.B.2c, a parameter that characterizes the latter energy term is the isothermal compressibility modulus K_T. To test the roles of area per molecule and K_T, we measured binding with lipids with known areas and compressibility moduli. For example, bilayers containing lipids with unsaturated hydrocarbon chains have larger areas per molecule than EPC, but similar compressibility moduli (Rawicz *et al.,* 2000), whereas the addition of cholesterol lowers the area per phospholipid molecule (Lecuyer and Dervichian, 1969) and also increases K_T, particularly for lipids with saturated chains, such as sphingomyelin (SM) (Needham and Nunn, 1990). For example, $K_T = 192$ dyn/cm for EPC, $K_T = 660$ dyn/cm for equimolar EPC:cholesterol, and $K_T = 1800$ dyn/cm for equimolar SM:cholesterol bilayers (Needham and Nunn, 1990).

As shown in Table I, the free energies of binding were quite similar for SUVs composed of EPC (area per lipid molecule of 64 Å^2) (McIntosh *et al.,* 1995) and diarchidonyl phosphatidylcholine (DAPC) (a PC with four double bonds per acyl chain and an area per molecule of 73 Å^2) (McIntosh *et al.,* 1995). This indicates that, at least over this area range, the area per molecule is not a significant factor in determining binding energy. However, the addition of cholesterol to EPC decreased ΔG in a concentration-dependent manner (Table I) and ΔG was quite low for 1 : 1 SM:cholesterol bilayers. These results indicate that K_T plays an important role in LamB-W binding to bilayers.

To obtain a better understanding of the thermodynamic changes associated with increasing the cholesterol concentration in the bilayer, we measured the heat of binding of LamB-W in EPC at 20, 40, and 50 mol% cholesterol (see Fig. 7). The

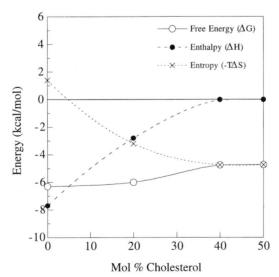

FIGURE 7 Thermodynamic parameters of binding of LamB-W to EPC bilayers obtained as a function of cholesterol content of the bilayer.

ITC measurements revealed that the enthalpy of binding ΔH decreased monotonically with mole percent cholesterol, essentially disappearing at 40 and 50 mol%.
By subtracting ΔH from ΔG we obtained the entropic contribution $(-T\Delta S)$
to the free energy of binding. As shown in Fig. 7, in the absence of cholesterol
the free energy of binding of LamB-W to EPC bilayers was favored by enthalpy,
but opposed by changes in entropy. However, as the concentration of cholesterol
increased, the binding was dominated by entropic changes. A similar phenomenon
has been reported by Wieprecht *et al.,* (1999b) for the antibacterial peptide
magainin.

B. Location of Signal Peptide in the Bilayer

Since nitroxide probes may affect peptide binding and hence give misleading
information as to the depth of penetration of the probe, we sought to find probes
that did not alter peptide binding. In this regard we found that brominated phosphatidylcholines (BrPCs) with bromines located along the acyl chains had no
appreciable effect on peptide–lipid interactions, irrespective of the position of the
bromines (Voglino *et al.,* 1999). That is, the binding of LamB-W (Table I) was
quite similar for EPC and BrPC bilayers with the bromines located at carbons
6 and 7, carbons 9 and 10, and carbons 11 and 12 along the acyl chains. In addition,
the fluorescence blue shift (Fig. 6) and ellipticity (Fig. 7) were similar for EPC and
6,7-BRPC. Therefore, since the partitioning was the same for LamB-W to each of
the BrPCs, these brominated probes (unlike the nitroxide probes) could be used for
the fluorescence quenching analysis of the bilayer depth of the tryptophan residue
in LamB-W.

Parallax analysis (Chattopadhyay and London, 1987) of fluorescence quenching
with the brominated probes indicated that the tryptophan residue near the middle
of the hydrophobic portion of the peptide (Fig. 1) was located in the hydrocarbon
region, but only 4 Å from the hydrocarbon–water interface (Voglino *et al.,* 1999).
This result was consistent with the relatively small (10 nm) blue shift in tryptophan
fluorescence noted in Section II.A.1. To obtain additional data on the orientation
of LamB-W in the bilayer, we changed the charges near the N- and C-termini and
found that LamB-W peptides with either positive or negative charges at the N- or
C-termini bound to a similar extent to PC bilayers. Microelectrophoresis experiments showed that both the N- and C-termini were on the outside of the bilayer
(Voglino *et al.,* 1998). As noted in Section II.A.1, in agreement with previous
studies (Jones and Gierasch, 1994a; McKnight *et al.,* 1991a,b), our CD measurements (Voglino *et al.,* 1998) showed that the peptide conformation changed from
random coil to α helix when it bound to the bilayer. Taken together, these fluorescence quenching, fluorescence blue shift, CD, and microelectrophoresis data
indicated that LamB-W adopted a "hammock" configuration in the bilayer, with

both termini exposed to the aqueous phase on the same side of the bilayer and the central hydrophobic region of the peptide forming an α helix located near the hydrocarbon–water interface (Voglino *et al.*, 1998). This result was different than the model of Jones and Gierasch (1994a), who, based in part on fluorescence quenching results with nitroxide probes, depicted the C-terminus partially inserting into the bilayer hydrocarbon region.

III. DISCUSSION

The partitioning of water-soluble peptides or proteins into lipid bilayers is of direct biological significance, because several proteins are inserted into membranes post-translationally (Basaran *et al.*, 1999), some cytosolic enzymes are activated upon binding to the plasma membrane (Hannun and Bell, 1987; Majerus *et al.*, 1986), mitochondrial signal sequences may bind to the lipid bilayer before interacting with the proteinaceous import machinery in the mitochondrion (Wang and Wiener, 1994), and there are several types of water-soluble, amphipathic peptides that partition into the bacterial membrane and act as antibiotics (Hristova *et al.*, 1997; Jo *et al.*, 1998; Lohner and Prenner, 1999; Lohner *et al.*, 1997; Ludtke *et al.*, 1996; Vaara, 1992; Wieprecht *et al.*, 1999b). Moreover, knowledge of the energetics of peptide partitioning should provide detailed information concerning the lipid-mediated interactions that determine peptide structure, stability, and function in the membrane (Mobashery *et al.*, 1997; Mouritsen, 1998; White and Wimley, 1998, 1999). In this chapter we have studied the leader sequence of the LamB protein, which is a random coil in buffer and forms an α helix in the presence of lipid bilayers (Jones and Gierasch, 1994a,b; McKnight *et al.*, 1991a) and partitions into the bilayer interfacial region (Voglino *et al.*, 1998, 1999). We have explored how the partitioning of this interfacial peptide is modulated by several of the specific properties of the lipid bilayer discussed in Section I.B.2: (1) surface charge, (2) presence of exogenously added dipoles in the interfacial region, (3) dipole potential, (4) area per lipid molecule, and (5) isothermal compressibility modulus.

A. Roles of Electrostatics, Peptide Conformational Changes, and "Classical" Hydrophobic Effect

Our binding data (Table I) show that, like many other small peptides with a net positive charge (Dathe *et al.*, 1996; Hammen *et al.*, 1996; Kim *et al.*, 1994; Mosior and McLaughlin, 1991; Wang and Wiener, 1994), LamB-W avidly partitioned into zwitterionic lipid vesicles. However, we also found that LamB-W partitioned into electrically neutral bilayers, such as EPC, although with a lower value of ΔG (Table I, Figs. 2–4). These binding, fluorescence, and CD data indicate that, although electrostatics plays an important role in the binding of LamB-W to negatively charged bilayers (Jones and Gierasch, 1994a,b), other interactions, such as peptide conformational changes and the hydrophobic effect, must also play a significant role in LamB-W–bilayer interactions, especially to zwitterionic lipids (Voglino *et al.*, 1998).

To gain insights into the relative magnitudes of both peptide conformational changes and the hydrophobic effect in LamB-W binding, we used ITC results from LamB-W (Fig. 7) and two other peptides that form interfacial helices upon binding to bilayers, magainin (Wieprecht *et al.*, 1999b) and the presequence of the protein rhodanese (Wieprecht *et al.*, 2000) (see Section I.B.2a). The thermodynamic parameters of LamB-W binding to SUVs of EPC were $\Delta G = -6.3$ kcal/mol, $\Delta H = -7.8$ kcal/mol, and $-T \Delta S = 1.5$ kcal/mol, indicating that the partitioning was enthalpically driven. Free energies per amino acid residue for the random coil-to-α helix transition (hel) at 298 K were as follows: for magainin, $\Delta G_{hel} = -0.14$ kcal/mol/residue, $\Delta H_{hel} = -0.70$ kcal/mol/residue, and $-T \Delta S_{hel} = 0.56$ kcal/mol/residue (Wieprecht *et al.*, 1999a), and for the presequence of rhodanese, $\Delta G_{hel} = -0.20$ kcal/mol/residue, $\Delta H_{hel} = -0.53$ kcal/mol/residue, and $-T \Delta S_{hel} = 0.33$ kcal/mol/residue (Wieprecht *et al.*, 2000). Here we assume that these values are also appropriate for the random coil–helix change in LamB-W. Therefore, using the average value for these two sets of experiments and estimates from CD data (Voglino *et al.*, 1998) that, on average, about 10 amino acid residues of LamB-W become α-helical on binding, we calculate that for LamB-W the energies due to peptide conformation changes upon binding are approximately $\Delta G_{hel} = -1.7$ kcal/mol, $\Delta H_{hel} = -6.2$ kcal/mol, and $-T \Delta S_{hel} = 4.5$ kcal/mol. Thus, for LamB-W binding to EPC the fraction of the free energy change accompanying the random coil-helix transition is about 27% of the total binding energy $(-1.7$ kcal/mol)/$(-6.3$ kcal/mol), a lower value than obtained by Wieprecht *et al.* (1999b) for magainin (50–60%). Such a difference is reasonable, because the helical content is lower in lipid-bound magainin (Wieprecht *et al.*, 1999b) than in EPC-bound LamB-W (Voglino *et al.*, 1998). About 79% of the enthalpy of binding $[(-6.2$ kcal/mol)/$(-7.8$ kcal/mol)] can be accounted for by this random coil-α helix transition, which clearly dominates the total free energy of binding. As would be expected, the conversion from a random coil to the more regular α helix is entropically unfavorable. The nonhelical (nh) contributions of LamB-W binding to EPC can be estimated by subtracting the total energies from these estimated contributions of the coil–helix transition. This calculation gives $\Delta G_{nh} = -4.6$ kcal/mol, $\Delta H_{nh} = -1.6$ kcal/mol, and $-T \Delta S_{nh} = -3.0$ kcal/mol. These values, summarized in Table II, imply that the nonhelical interactions associated with LamB-W binding have a large entropic component, which likely arise, at least in part, from hydrophobic interactions (Jones and Gierasch, 1994a,b). In summary, the major driving interaction of LamB-W with EPC SUVs is the enthalpy gained

TABLE II

Energetics of Binding of LamB-W to EPC Bilayers[a]

	ΔG (kcal/mol)	ΔH (kcal/mol)	$-T\Delta S$ (kcal/mol)
Total	−6.3	−7.8	1.5
Coil–helix conformation change	−1.7	−6.2	4.5
Nonhelical contribution	−4.6	−1.6	−3.0

[a]Calculated from energies per residue for the random coil–helix transition for two amphipathic peptides (Wieprecht *et al.,* 1999a, 2000) and assuming 10 residues are in the α helix in LamB-W.

by the coil–helix transition. The loss of entropy associated with this conformational transition is to a large extent compensated by the classical hydrophobic entropy gained by removing nonpolar amino acids from the aqueous phase (Jones and Gierasch, 1994a,b; Tanford, 1980).

B. Role of Dipoles in the Headgroup and Hydrocarbon Regions of the Bilayer

We also explored the effects on binding of dipoles inserted near the interfacial region of the bilayer. The addition of equimolar 6-ketocholestanol to EPC had a dramatic effect on LamB-W binding, reducing the free energy of binding from −6.3 to −3.9 kcal/mol (Table I). These binding data are completely consistent with fluorescence measurements, which show that, in the presence of EPC:6-ketocholestanol bilayers, the tryptophan at position 18 is not in a very low dielectric constant environment (Fig. 3), and CD spectra, which indicate that there is little α-helix formation (Fig. 4). The generality of these observations to other types of peptides is not known at present. However, in one recent study Cladera and O'Shea (1998) used fluorescent probe measurements to determine the interaction of the mitochondrial amphipathic signal sequence p25 with EPC in the presence and the absence of 6-ketocholestanol. They found that 15 mol% 6-ketocholestanol had little effect on the dissociation constant, but increased the p25 binding "capacity" and modified the conformation of the peptide. Thus, even small concentrations of 6-ketocholestanol can effect the interaction of peptides with bilayers.

A central issue in understanding the role of interfacial dipoles in peptide–bilayer interactions is to determine the mechanisms by which 6-ketocholestanol destabilizes the interaction between LamB-W and the bilayer. Several possible factors can be eliminated. First, because EPC:6-ketocholestanol bilayers had the same electrophoretic mobility and hence surface charge density as EPC bilayers (Voglino *et al.,* 1998), the difference in binding properties of these bilayers cannot be due to electrostatic interactions. Second, effects of changes in bilayer structure (e.g., area per molecule) can be eliminated, because the incorporation of 6-ketocholestanol into EPC bilayers had little effect on bilayer organization or the bilayer compressibility modulus (Simon *et al.,* 1992; Voglino *et al.,* 1998). In this regard, the addition of 6-ketocholestanol differed from that of cholesterol (although both decreased the binding), because cholesterol increased bilayer thickness and markedly increased the compressibility modulus. Third, changes in dipole potential could not explain the small binding to bilayers of both EPC : 6-ketocholestanol and 7-doxyl PC, because 1 : 1 EPC : ketocholesterol had a higher dipole potential (703 mV) than EPC (415 mV), whereas 7-doxyl PC had a lower dipole potential (369 mV).

Another factor to consider in LamB-W partitioning into the bilayer is the location of the dipolar ketone group in EPC : 6-ketocholestanol bilayers. As noted in Section II, the ketone group is located about 4 Å into the bilayer's hydrocarbon region (see Fig. 1) and energetic calculations suggest that the ketone group is hydrated (Cladera and O'Shea, 1998; Simon *et al.,* 1992). Therefore one possible effect of 6-ketocholestanol is to increase the hydrophilic width (polarity) of the interfacial region by about 4 Å in the direction of the acyl chains. As noted in Section I.B.2g, the hydration properties and width of the interfacial region are critical to the interaction of water-soluble peptides with bilayers. With the "hammock" model described above, much of the helical region of the LamB-W peptide must be located near the interfacial region of the bilayer. As noted by White and Wimley (1994), the width of the polar headgroup region of PC bilayers is approximately the same as the diameter of the core of an α helix. Thus, a 4-Å increase in the width of the polar interfacial region would have the effect of increasing the width of the potential energy barrier to the insertion of hydrophobic peptides into the bilayer. Given the configurational constraints for LamB-W outlined in Section II.B, such an increase in interfacial width could mean that fewer of the hydrophobic amino acids in the α helix would be exposed to the low-dielectric acyl chain region of the bilayer. This would decrease the contribution of the hydrophobic energy to the free energy of binding and tilt the energy balance in such a manner as to markedly reduce peptide binding to the bilayer. It follows that increasing the polarity of the acyl chain region near the hydrocarbon–water interface by the addition of dipoles, such as the ketone group in 6-ketocholestanol, could cause a reduction in the hydrophobic free energy, which, as shown in Section III.A.1 and by Jones and Gierasch (1994b), is a major factor driving LamB-W binding to neutral bilayers. The addition of equimolar 6-ketocholestanol reduced the free energy of transfer of LamB-W to EPC bilayers by about 2 kcal/mol (Table I), a value that is comparable to the free energy of transfer (-1.9 kcal/mol) of a tryptophan residue from water to a PC bilayer (Wimley *et al.,* 1996). Therefore, just a few nonpolar amino acid residues failing to become dehydrated as a result of an increase in the interfacial width could energetically account for the inhibitory effect of 6-ketocholestanol.

The above arguments indicate that a key factor involved in 6-ketocholestanol's effect on partitioning is not only the presence of its ketone dipole, but the dipole's location in the hydrocarbon region of the bilayer near the interface. The model described is consistent with the observation that bilayers formed from 7-doxyl PC (Fig. 1) also did not appreciably bind LamB-W (Table I, Fig. 4). In this lipid, the dipolar nitroxide group would be expected to be located in the hydrocarbon region of the bilayer at about the same depth in the hydrocarbon region as the ketone in 6-ketocholestanol (Voglino *et al.*, 1998). Thus, bilayers formed from pure 7-doxyl PC, like bilayers of EPC:6-ketocholestanol, should have a relatively wide interfacial region and thus reduce the free energy of transfer of the hydrophobic helix. Moreover, 12-doxyl PC, which has a nitroxide moiety located deeper in the bilayer, had little effect on LamB-W binding (Table I). Therefore we argue that a key factor in binding of LamB to bilayers is the location of interfacial dipoles.

The properties and width of the interfacial region of bilayers should also be important in biological membranes, where several types of molecules with polar moieties could modify the effective width of the interfacial region. For example, transmembrane proteins or drugs that partition into the interfacial region could also introduce polar groups (dipoles) into the bilayer near the water–hydrocarbon interface.

C. Roles of Area per Molecule and Isothermal Compressibility Modulus

Experiments with bilayers composed of the polyunsaturated lipid DAPC and with EPC : cholesterol mixtures (Table I) allowed us to examine the relative effects of both area per lipid molecule and K_T on binding. The area per molecule is significantly larger for DAPC (76 Å^2) than for EPC (63 Å^2) (McIntosh *et al.*, 1995), and the area per EPC molecule decreases with increasing cholesterol concentration (Lecuyer and Dervichian, 1969). However, K_T is similar for DAPC and EPC (Rawicz *et al.*, 2000), whereas K_T increases with increasing cholesterol concentration (Needham and Nunn, 1990). Therefore, our binding data (Table I), which show similar values of ΔG for DAPC and EPC, but a decrease in ΔG with increasing cholesterol concentration, appear to indicate that the compressibility modulus is a more important factor than area per molecule in modulating partitioning. That is, these data point to the important effect that cholesterol has on increasing the work necessary to separate the acyl chains in the bilayer to allow peptide binding. In other words, bilayers with equimolar cholesterol tend to "squeeze out" the peptide from the more hydrophobic regions of the bilayer. One less interesting possibility that could rationalize the effect of cholesterol on decreasing peptide binding is that the binding depends on the size of the vesicles (Wieprecht *et al.*, 1999b) and that the SUV vesicles formed in the presence of cholesterol are larger than EPC SUVs. We note that our previous experiments (Voglino *et al.*, 1998), done at one concentration of cholesterol, found a smaller effect of cholesterol on binding. However, these more extensive binding (Table I) and ITC experiments (Fig. 7) clearly indicate an important influence of cholesterol.

The thermodynamic parameters for the partitioning of LamB-W into 1 : 1 EPC : cholesterol SUVs indicate that this interaction is entropy-driven in the sense that $\Delta G = -4.7$ kcal/mol, $\Delta H \approx 0$ kcal/mol, and $-T\Delta S = -4.7$ kcal/mol (Fig. 7). Comparisons of these values with those obtained from the EPC measurements (Tables I and II) show that in the presence of cholesterol there is a much smaller enthalpic contribution to the binding and a much larger entropic contribution. Part of the reduced enthalpy (and free energy) of binding in the presence of cholesterol likely arises from the smaller conversion of random coil to α helix of the peptide, as indicated by the CD measurements (see Fig. 4). As noted in Section III.A.1, the random coil-to-α helix conformation change provides a major fraction of the enthalpic change in binding, but is entropically unfavorable. For 1 : 1 EPC : cholesterol the value of $-T\Delta S = -4.7$ kcal/mol is somewhat larger than the entropy calculated for the nonhelical contribution of binding to EPC ($T\Delta S_{nh} = -3.0$ kcal/mol, see Table II), indicating that at least some of this energy arises from the hydrophobic interactions. That is, although the binding of LamB-W to EPC is driven primarily by the enthalpic gain in the conversion of the random coil to the α helix upon lipid binding, for 1 : 1 EPC : cholesterol the LamB-W binding is to a large extent entropically driven by the classical hydrophobic effect. The entropic energy may also be larger for bilayers containing cholesterol, because cholesterol orders the hydrocarbon chains in the bilayer (Lecuyer and Dervichian, 1969), so that peptide partitioning might have a larger bilayer disordering effect for cholesterol-containing bilayers.

IV. CONCLUSIONS

The composition, structure, and mechanical properties of the lipid bilayer can all potentially modify the partitioning of a water-soluble peptide into a membrane. The experiments described in this chapter emphasize the significant role that the bilayer plays in the partitioning of LamB-W, a surface-active peptide. Increasing the fixed negative charge on the bilayer increases the binding of the positively charged peptide. Moreover, a relatively small change in the composition of the interfacial region caused by the presence of oriented dipoles on the lipid molecules disrupts the binding of a signal peptide by changing the hydration properties of the interface. Thus, the presence of both fixed charges (which control electrostatic interactions) and fixed dipoles (which change the width and hydration properties of the interfacial region) can markedly effect the energetics of peptide binding. The addition of cholesterol, by virtue of its ability to increase the work to separate the acyl chains, was found to decrease the binding by decreasing the enthalpic contributions.

Acknowledgments

This work was supported by NIH grant GM27278. We thank Laura Voglino for all of her experimental work on this project.

References

Abrams, F. S., and London, E. (1992). Calibration of the parallax fluorescence quenching method of determining membrane penetration depth: Refinement and comparison of quenching by spin-labeled and brominated lipids. *Biochemistry* **31**, 5312–5322.

Banemann, A., Deppisch, H., and Gross, R. (1998). The lipopolysaccharide of *Bordetella bronchisepticsa* acts as a protective shield against antimicrobial peptides. *Infection Immunity* **66**, 5607–5612.

Basaran, N., Doebler, R. W., Goldston, H., and Holloway, P. W. (1999). Effect of lipid unsaturation on the binding of native and a mutant form of cytochrome b5 to membranes. *Biochemistry* **38**, 15245–15252.

Ben-Shaul, A., Ben-Tal, N., and Honig, B. (1996). Statistical thermodynamic analysis of peptide and protein insertion into lipid membranes. *Biophys. J.* **71**, 130–137.

Ben-Tal, N., Honig, B., Peitzsch, R. M., Denisov, G., and McLaughlin, S. (1996). Binding of small basic peptides to membranes containing acidic lipids: Theoretical models and experimental results. *Biophys. J.* **71**, 561–575.

Boggs, J. M. (1987). Lipid intermolecular hydrogen bonding: Influence on structural organization and membrane function. *Biochim. Biophys. Acta* **906**, 353–404.

Brockman, H. L. (1994). Dipole potentials of lipid membranes. *Chem. Phys. Lipids* **73**, 57–79.

Bruch, M. D., Mcknight, C. J., and Gierasch, L. M. (1989). Helix formation and stability in a signal sequence. *Biochemistry* **28**, 8554–8561.

Buldt, G., Gally, H. U., Seelig, J., and Zaccai, G. (1979). Neutron diffraction studies on phophatidyl-choline model membranes I. Head group conformation. *J. Mol. Biol.* **134**, 673–691.

Cafiso, D. S. (1995). Influence of charges and dipoles on macromolecular adsorption and permeability. *In* "Permeability and Stability of Lipid Bilayers." (E. A. Dosalvo, and S. A. Simon, eds.), pp. 179–195. CRC Press, Boca Raton, FL.

Cantor, R. S. (1997). The lateral pressure profile in membranes: A physical mechanism of general anesthesia. *Biochemistry* **36**, 2340–2343.

Cantor, R. S. (1999a). Lipid composition and the lateral pressure profile in bilayers. *Biophys. J.* **76**, 2625–2639.

Cantor, R. S. (1999b). Solute modulation of conformational equilibria in intrinsic membrane proteins: Apparent "cooperativity" without binding. *Biophys. J.* **77**, 2643–2647.

Chaloin, L., Vidal, P., Heitz, A., Van Mau, N., Mery, J., Divita, G., and Heitz, F. (1997). Conformations of primary amphipathic carrier peptides in membrane mimicking environments. *Biochemistry* **36**, 11179–11187.

Chattopadhyay, A., and London, E. (1987). Parallax method for direct measurement of membrane penetration depth using fluorescence quenching by spin-labeled phospholipids. *Biochemistry* **26**, 39–45.

Chonn, A., and Cullis, P. (1992). Ganglioside GM1 and hydrophilic polymers increase liposome circulation times by inhibiting the association of blood proteins. *J. Liposome Res.* **2**, 397–410.

Cladera, J., and O'Shea, P. (1998). Intramembrane molecular dipoles affect the membrane insertion and folding of a model amphiphilic peptide. *Biophys. J.* **74**, 2434–2442.

Dathe, M., Schumann, M., Wieprecht, T., Winkler, A., Beyermann, M., Krause, E., Matsuzaki, K., Murase, O., and Bienert, M. (1996). Peptide helicity and membrane surface charge modulate the balance of electrostatic and hydrophobic interactions with lipid bilayers and biological membranes. *Biochemistry* **35**, 12612–12622.

DeYoung, L. R., and Dill, K. A. (1988). Solute partitioning into lipid bilayer membranes. *Biochemistry* **27**, 5281–5289.

Dilger, J. P., McLaughlin, S. G. A., McIntosh, T. J., and Simon, S. A. (1979). The dielectric constant of phospholipid bilayer and the permeability of membranes to ions. *Science* **206**, 1196–1198.

Du, H., and Chandaroy, H. S. W. (1997). Grafted poly-(ethylene glycol) on lipid surfaces inhibits protein adsorption and cell adhesion. *Biochim. Biophys. Acta* **1326**, 236–248.

Flewelling, R. F., and Hubbell, W. L. (1986). The membrane dipole potential in a total membrane potential model. Applications to hydrophobic ion interactions with membranes. *Biophys. J.* **49,** 541–552.

Franklin, J. C., and Cafiso, D. S. (1993). Internal electrostatic potentials in bilayers: Measuring and controlling dipole potentials in lipid vesicles. *Biophys. J.* **65,** 289–299.

Gruner, S. M. (1985). Intrinsic curvature hypothesis for biomembrane lipid composition: A role for non-bilayer lipids. *Proc. Natl. Acad. Sci. USA* **82,** 3665–3669.

Gruner, S. M., Parsegian, V. A., and Rand, R. P. (1986). Directly measured deformation energy of phospholipid HII hexagonal phases. *Faraday Disc. Chem. Soc.* **81,** 29–37.

Hammen, P. K., Gorenstein, D. G., and Weiner, H. (1996). Amphilicity determines binding properties of three mitochondrial presequences to lipid surfaces. *Biochemistry* **35,** 3772–3781.

Hannun, Y. A., and Bell, R. M. (1987). Lysophingolipids inhibit protein kinase C: Implications for the sphingolipidoses. *Science* **235,** 670–674.

Honig, B., and Nichols, A. (1995). Classical electrostatics in biology and chemistry. *Science* **268,** 1144–1148.

Hoyt, D. W., Cyr, D. M., Gierasch, L. M., and Douglas, M. G. (1991). Interaction of peptides corresponding to mitochondrial presequences with membranes. *J. Biol. Chem.* **266,** 21693–21699.

Hristova, K., Selsted, M. E., and White, S. H. (1997). Critical role of lipid compostion in membrane permeabilization by rabbit neutrophil defensins. *J. Biol. Chem.* **272,** 24224–24233.

Huh, N.-W., Porter, N. A., McIntosh, T. J., and Simon, S. A. (1996). The interaction of polyphenols with bilayers: Conditions for increasing bilayer adhesion. *Biophys. J.* **71,** 3261–3277.

Hunt, J., Earnest, T. N., Bousche, O., Kalgatgi, K., Reilly, K., Rothschild, K. J., and Engelman, D. M. (1997a). A biophysical study of integral membrane protein folding. *Biochemistry* **36,** 15156–15176.

Hunt, J. F., Rath, P., Rothschild, K. J., and Engleman, D. M. (1997b). Spontaneous, pH-dependent membrane insertion of a transbilayer α-helix. *Bichemistry* **36,** 15177–15192.

Israelachvili, J. N. (1985). "Intermolecular and Surface Forces," Academic Press, London.

Israelachvili, J. N., Marcelja, S., and Horn, R. G. (1980). Physical principles of membrane organization. *Q. Rev. Biophys.* **13,** 121–200.

Jo, E., Blazyk, J., and Boggs, J. M. (1998). Insertion of magainin into the lipid bilayer detected using lipid photolabels. *Biochemistry* **37,** 13791–13799.

Joly, J. C., and Wickner, W. (1993). The SecA and SecY subunits of translocase are the nearest neighbors of a transocating protein, shielding it from phospholipids. *EMBO J.* **12,** 255–263.

Jones, J. D., and Gierasch, L. M. (1994a). Effect of charged residue substitutions on the membrane-interactive properties of signal sequences of the *Escherichia coli* LamB protein. *Biophys. J.* **67,** 1534–1545.

Jones, J. D., and Gierasch, L. M. (1994b). Effect of charged residue substitutions on the thermodynamics of signal peptide–lipid interactions for *Escherichia coli* LamB signal sequence. *Biophys. J.* **67,** 1546–1561.

Keller, S. L., Bezrukov, S. M., Gruner, S. M., Tate, M. W., Vodyanoy, I., and Parsegian, V. A. (1993). Probability of alamethicin conductance states varies with nonlamellar tendency of bilayer phospholipids. *Biophys. J.* **65,** 23–27.

Kenworthy, A. K., Hristova, K., Needham, D., and McIntosh, T. J. (1995a). Range and magnitude of the steric pressure between bilayers containing phospholipids with covalently attached poly(ethyleneglycol). *Biophys. J.* **68,** 1921–1936.

Kenworthy, A. K., Simon, S. A., and McIntosh, T. J. (1995b). Structure and phase behavior of lipid suspensions containing phospholipids with covalently attached poly(ethylene glycol). *Biophys. J.* **68,** 1903–1920.

Kim, J., Blackshear, P. J., Johnson, J. D., and McLaughlin, S. (1994). Phosphorylation reverses the membrane association of peptides that correspond to the basic domains of MARCKS and neuromodulin. *Biophys. J.* **67,** 227–237.

Kirk, G. L., Gruner, S. M., and Stein, D. L. (1984). A thermodynamic model of the lamellar to inverse hexagonal phase transition of lipid membrane–water systems. *Biochemistry* **23**, 1093–1102.

Kwok, R., and Evans, E. (1981). Thermoelasticity of large lecithin vesicles. *Biophys. J.* **35**, 637–652.

Ladokhin, A. S., and White, S. H. (1999). Folding of amphipathic alpha-helices on membranes: Energetics of helix formation by melittin. *J. Mol. Biol.* **285**, 1363–1369.

Lecuyer, H., and Dervichian, D. G. (1969). Structure of aqueous mixtures of lecithin and cholesterol. *J. Mol. Biol.* **45**, 39–57.

Lewis, B. A., and Engelman, D. M. (1983). Lipid bilayer thickness varies linearly with acyl chain length in fluid phosphatidylcholine vesicles. *J. Mol. Biol.* **166**, 211–217.

Lewis, J. R., and Cafiso, D. S. (1999). Correlation of the free energy of a channel-forming voltage-gated peptide and the spontaneous curvature of bilayer lipids. *Biochemistry* **38**, 5932–5938.

Lohner, K., and Prenner, E. J. (1999). Differential scanning calorimetry and X-ray diffraction studies of the specificity of the interaction of antimicrobial peptides with membrane-mimetic systems. *Biochim. Biophys. Acta* **1462**, 141–156.

Lohner, K., Latal, A., Leher, R. I., and Ganz, T. (1997). Differential scanning microcalorimetry indicates that human defensin HNP-2 interacts specifically with biomembrane mimetic systems. *Biochemistry* **36**, 1525–1531.

Ludtke, S., He, K., and Huang, H. (1996). Membrane thinning caused by magainin 2. *Biochemistry* **34**, 16764–16769.

Lundbaek, J. A., Maer, A. M., and Andersen, O. S. (1997). Lipid bilayer electrostatic energy, curvature stress, and assembly of gramicidin channels. *Biochemistry* **36**, 5695–5701.

Luzzati, V. (1968). X-ray diffraction studies of lipid–water systems. *In* "Biological Membranes" (D. Chapman, ed.), pp. 71–123. Academic Press, New York.

Luzzati, V., Gulik-Krzywicki, T., and Tardieu, A. (1968). Polymorphism of Lecithins. *Nature* **218**, 1031–1034.

Macias, E. A., Rana, F., Blazyk, J., and Modrzakowski, M. C. (1990). Bactericidal activity of magainin 2: Use of lipopolysaccharide mutants. *Can. J. Microbiol.* **36**, 582–584.

Majerus, P. W., Connolly, T. M., Deckmyn, H., Ross, T. S., Bross, T. E., Ishii, H., Bansal, V. S., and Wilson, D. B. (1986). The metabolism of phosphoinositide-derived messenger molecules. *Science* **234**, 1519–1526.

Martoglio, B., Hofmann, M. W., Brunner, J., and Dobberstein, B. (1995). The protein–conducting channel in the membrane of the endoplasmic reticulum is open laterally toward the lipid bilayer. *Cell* **81**, 207–214.

McIntosh, T. J., and Simon, S. A. (1986a). Area per molecule and distribution of water in fully hydrated dilauroylohosphatidylethanolamine bilayers. *Biochemistry* **25**, 4948–4952.

McIntosh, T. J., and Simon, S. A. (1986b). The hydration force and bilayer deformation: A reevaluation. *Biochemistry* **25**, 4058–4066.

McIntosh, T. J., and Simon, S. A. (1994). Long- and short-range interactions between phospholipid/ganglioside GM1 bilayers. *Biochemistry* **33**, 10477–10486.

McIntosh, T. J., Advani, S., Burton, R. E., Zhelev, D. V., Needham, D., and Simon, S. A. (1995). Experimental tests for protrusion and undulation pressures in phospholipid bilayers. *Biochemistry* **34**, 8520–8532.

McIntosh, T. J., Pollastri, M. P., Porter, N. A., and Simon, S. A. (1999). Tannic acid and polyphenolic analogs increase adhesion between lipid bilayers by forming interbilayer bridges. *In* "Plant Polyphenols. 2. Chemistry, Biology, Pharmacology, Ecology" (R. W. Hemmingway, ed.), pp. 451–470. Plenum Press, New York.

McKnight, C. J., Briggs, M. S., and Gierasch, L. M. (1989). Functional and nonfunctional LamB signal sequences can be distinguished by their biophysical properties. *J. Biol. Chem.* **264**, 17293–17297.

McKnight, C. J., Rafalski, M., and Gierasch, L. M. (1991a). Fluorescence analysis of tryptophan-containing variants of the LamB signal sequence upon insertion into a lipid bilayer. *Biochemistry* **30**, 6241–6246.

McKnight, C. J., Stradley, S. J., Jones, J. F., and Gierasch, L. M. (1991b). Conformational and membrane-binding properties of a signal sequence are largely unaltered by its adjacent mature region. *Proc. Natl. Acad. Sci. USA* **88,** 5799–5803.

Mobashery, N., Nielsen, C., and Andersen, O. S. (1997). The conformational preference of gramicidin channels is a function of lipid bilayer thickness. *FEBS Lett.* **412,** 15–20.

Mosior, M., and McLaughlin, S. (1991). Peptides that mimic the pseudosubstrate region of protein kinase C bind to acidic lipids in membranes. *Biophys. J.* **60,** 149–160.

Mouritsen, O. G. (1998). Self-assembly and organization of lipid–protein membranes. *Curr. Opin. Colloid Interface Sci.* **3,** 78–87.

Mouritsen, O. G., and Bloom, M. (1984). Mattress model of lipid-protein interactions in membranes. *Biophys. J.* **46,** 141–153.

Nagle, J. F. (1980). Theory of the main lipid bilayer phase transition. *Annu. Rev. Phys. Chem.* **31,** 157–195.

Nagle, J. F., Zhang, R., Tristram-Nagle, S., Sun, W., Petrache, H. I., and Suter, R. M. (1996). X-ray structure determination of fully hydrated Lα phase dipalmitoylphosphatidylcholine bilayers. *Biophys. J.* **70,** 1419–1431.

Needham, D., and Nunn, R. S. (1990). Elastic deformation and failure of lipid bilayer membranes containing cholesterol. *Biophys. J.* **58,** 997–1009.

Needham, D., Stoicheva, N., and Zhelev, D. V. (1997). Exchange of monooleoylphosphatidylcholine as monomer and micelle with membranes containing poly(ethylene glycol)–lipid. *Biophys. J.* **73,** 2615–2629.

Nielsen, C., Goulian, M., and Andersen, O. S. (1998). Energetics of inclusion-induced bilayer deformations. *Biophys. J.* **74,** 1966–1983.

Petrache, H. I., Tristram-Nagle, S., and Nagle, J. F. (1998). Fluid phase structure of EPC and DMPC bilayers. *Chem. Phys. Lipids* **95,** 83–94.

Qin, Z., Szabo, G., and Cafiso, D. (1995). Anesthetics reduce the magnitude of the membrane dipole potential. Measurements in lipid vesicles using voltage-sensitve spin probes. *Biochemistry* **34,** 5536–5543.

Rana, F. R., and Blazyk, J. (1991). Interactions between the antimicrobial peptide, magainin 2, and *Salmonella typhimurium* lipopolysaccharides. *FEBS Lett.* **293,** 11–15.

Rand, R. P., Fuller, N., Parsegian, V. A., and Rau, D. C. (1988). Variation in hydration forces between neutral phospholipid bilayers: Evidence for hydration attraction. *Biochemistry* **27,** 7711–7722.

Rawicz, W., Olbrich, K. C., McIntosh, T. J., Needham, D., and Evans, E. (2000). Effect of chain length and unsaturation on elasticity of lipid bilayers. *Biophys. J.* **79,** 328–339.

Russell, C. J., Thorgeirsson, T. E., and Shin, Y.-K. (1999). The membrane affinities of the aliphatic amino acid side chains in an alpha-helical context are independent of membrane immersion depth. *Biochemistry* **38,** 337–346.

Schatz, P., and Beckwith, J. (1990). Genetic analysis of protein export in *Escherichia coli. Annu. Rev. Genet.* **24,** 215–248.

Senior, J., Delgado, C., Fisher, D., Tilcock, C., and G., G. (1991). Influence of surface hydrophilicity of liposomes on their interaction with plasma protein and clearance from the circulation: Studies with poly(ethylene glycol)-coated vesicles. *Biochim. Biophys. Acta* **1062,** 77–82.

Simon, S. A., and McIntosh, T. J. (1989). Magnitude of the solvation pressure depends on dipole potential. *Proc. Natl. Acad. Sci. USA* **86,** 9263–9267.

Simon, S. A., McIntosh, T. J., and Latorre, R. (1982). Influence of cholesterol on water penetration into bilayers. *Science* **216,** 65–67.

Simon, S. A., McIntosh, T. J., Magid, A. D., and Needham, D. (1992). Modulation of the interbilayer hydration pressure by the addition of dipoles at the hydrocarbon/water interface. *Biophys. J.* **61,** 786–799.

Simon, S. A., Disalvo, E. A., Gawrisch, K., Borovyagin, V., Toone, E., Schiffman, S. S., Needham, D., and McIntosh, T. J. (1994). Increased adhesion between neutral lipid bilayers: Interbilayer bridges fromed by tannic acid. *Biophys. J.* **66**, 1943–1958.

Smaby, J. M., and Brockman, H. L. (1990). Surface dipole moments of lipids at the argon–water interface: Similarities among glycerol-ester-based lipids. *Biophys. J.* **58**, 195–204.

Small, D. M. (1967). Phase equilibria and structure of dry and hydrated egg lecithin. *J. Lipid Res.* **8**, 551–557.

Snyder, S., Kim, D., and McIntosh, T. J. (1999). Lipopolysacccharide bilayer structure: Effect of chemotype, core mutations, divalent cations, and temperature. *Biochemistry* **38**, 10758–10767.

Sophianopoulos, J. A., Durham, S. J., Sophianopoulos, A. J., Ragsdale, H. L., and Cropper, W. P. (1978). Ultrafiltration is theoretically equivalent to equilibrium dialysis but much simpler to carry out. *Arch. Biochem. Biophys.* **187**, 132–137.

Tanford, C. (1980). "The Hydrophobic Effect: Formation of Micelles and Biological Membranes," 2nd ed. Wiley, New York.

Thorgeirsson, T. E., Russell, C. J., King, D. S., and Shin, Y.-K. (1996). Direct determination of the membrane affinities of individual amino acids. *Biochemistry* **35**, 1803–1809.

Vaara, M. (1992). Agents that increase the permeability of the outer membrane. *Microbiol. Rev.* **56**, 395–411.

Voglino, L., McIntosh, T. J., and Simon, S. A. (1998). Modulation of the binding of signal peptides to lipid bilayers by dipoles near the hydrocarbon–water interface. *Biochemistry* **37**, 12241–12252.

Voglino, L., Simon, S. A., and McIntosh, T. J. (1999). Orientation of LamB signal peptides in bilayers: Influence of lipid probes on peptide binding and interpretation of fluorescence quenching data. *Biochemistry* **38**, 7509–7516.

von Heijne, G. (1995). Membrane protein assembly: Rules of the game. *BioEssays* **17**, 25–30.

Wang, Y., and Wiener, H. (1994). Evaluation of electrostatic and hydrophobic effects on the interaction of mitochondrial signal sequences with phospholipid bilayers. *Biochemistry* **33**, 12860–12867.

Wang, Z., Jones, J. D., Rizo, J., and Gierasch, L. M. (1993). Membrane-bound conformation of a signal peptide: A nuclear Overhauser effect analysis. *Biochemistry* **32**, 13991–13999.

White, S. H., and Wimley, W. C. (1994). Peptides in lipid bilayers: Structural and thermodynamic basis for partioning and folding. *Curr. Opin. Struct. Biol.* **4**, 79–86.

White, S. H., and Wimley, W. C. (1998). Hydrophobic interactions of peptides with membrane interfaces. *Biochim. Biophys. Acta* **1376**, 339–352.

White, S. H., and Wimley, W. C. (1999). Membrane protein folding and stability: Physical principles. *Annu. Rev. Biophys. Struct.* **28**, 319–365.

White, S. H., Wimley, W. C., Lasokhin, A. S., and Histrova, K. (1998). Protein folding in membranes: Determining the energetics of peptide–bilayer interactions. *Meth. Enzymol.* **295**, 62–88.

Wickner, W., and Leonard, M. R. (1996). *Escherichia coli* preprotein translocase. *J. Biol. Chem.* **271**, 29514–29516.

Wieprecht, T., Dathe, M., Krause, E., Beyermann, M., Maloy, W. L., MacDonald, D. L., and, Bienert, M. (1997). Modulation of membrane activity of amphipathic, antibacterial peptides by slight modification of the hydrophobic moment. *FEBS Lett.* **417**, 135–140.

Wieprecht, T., Apostolov, O., Beyermann, M., and Seelig, J. (1999a). Thermodynamics of the alpha-helix–coil transition of amphipathic peptides in a membrane environment: Implications for the peptide–membrane binding equilibrium. *J. Mol. Biol.* **294**, 785–794.

Wieprecht, T., Beyermann, M., and Seelig, J. (1999b). Binding of antibacterial magainin peptides to electrically neutral membrane: Thermodynamics and structure. *Biochemistry* **38**, 10377–10387.

Wieprecht, T., Apostolov, O., Beyermann, M., and Seelig, J. (2000). Interaction of a mitochondrial presequence with lipid membranes: Role of helix formation for membrane binding and perturbation. *Biochemistry* **39**, 15297–15303.

Wimley, W. C., and White, S. H. (1993). Membrane partitioning: Distinguishing bilayer effects from the hydrophobic effect. *Biochemistry* **32,** 6307–6312.

Wimley, W. C., Creamer, T. P., and White, S. H. (1996). Solvation energies of amino acid side chains and backbone in a family of host–guest pentapeptides. *Biochemistry* **35,** 5109–5124.

Yau, W.-M., Wimley, W. C., Gawrisch, K., and White, S. H. (1998). The preference of tryptophan for membrane interfaces. *Biochemistry* **37,** 14713–14718.

Zaccai, G., Blasie, J. K., and Schoenborn, B. P. (1975). Neutron diffraction studies on the location of water in lecithin bilayer model membranes. *Proc. Natl. Acad. Sci. USA* **72,** 376–380.

CHAPTER 12

Transmembrane α Helices

Sanjay Mall, J. Malcolm East, and Anthony G. Lee

Division of Biochemistry and Molecular Biology, School of Biological Sciences, University of Southampton, Southampton, SO16 7PX, United Kingdom

  I. Intrinsic Membrane Proteins
 II. Effects of Intrinsic Membrane Proteins on Lipid Bilayers
III. Model Transmembrane α Helices
    A. Lipid–Peptide Interactions
    B. Effects of Hydrophobic Mismatch
    C. Interactions with Anionic Phospholipids
    D. Effects of Phospholipid Phase
    E. Effects of Cholesterol
IV. Biological Consequences of Hydrophobic Mismatch
    References

I. INTRINSIC MEMBRANE PROTEINS

Intrinsic membrane proteins span their membranes as α helices or as β sheets, driven by the requirement to maximize the number of hydrogen bonds. The cost of transferring a peptide bond from water into a nonpolar environment is about 25 kJ mol^{-1} when not hydrogen-bonded, but only about 2.4 kJ mol^{-1} when hydrogen-bonded (Roseman, 1988). Breaking the hydrogen bonds in a 20-residue α helix in a nonpolar environment would cost about 400 kJ mol^{-1}. This high cost can be avoided if the membrane-spanning region of a membrane protein is organized as an α helix or as a β sheet, because in these structures the peptide backbone can form hydrogen bonds internally.

Formation of β sheets is seen in bacterial outer membrane proteins, which span the membrane as β barrels. Proteins in other membranes span the lipid bilayer as right-handed α helices. An ideal α helix has 3.6 residues per turn and a translation per residue of 1.50 Å. With a thickness for the hydrocarbon core of a lipid

Current Topics in Membranes, Volume 52

bilayer of 30 Å, about 20 residues will be required to span the core of the bilayer. The residues in the transmembrane region will be predominantly hydrophobic. However, for membrane proteins such as transporters and ion channels the transmembrane α helices will also have to contain polar groups; the transmembrane α helices will then be amphipathic rather than just hydrophobic. An extreme case could be a transmembrane α helix totally surrounded by other α helices in the center of a helical bundle: Such a helix would not need to be hydrophobic at all, because it is not in direct contact with the lipid bilayer. However, although small in number, the available high-resolution structures for membrane proteins show no evidence for the presence of such purely hydrophilic transmembrane α helices. It may be that the process of insertion of membrane proteins into the membrane during biogenesis requires all the transmembrane α helices to be relatively hydrophobic.

Analysis of the compositions of a large number of membrane proteins predicted to contain single transmembrane α helices has shown that the amino acid composition of a transmembrane α helix is distinctly different from that of hydrophobic α helices in water-soluble proteins. As expected, hydrophobic residues make up the bulk of the residues, the most common being Leu (Landolt-Marticorena *et al.*, 1993; Wallin *et al.*, 1997). Amino acids essentially excluded are the basic (Arg and Lys) and acidic (Asp and Glu) amino acids and their amide counterparts (Asn and Gln). Transmembrane α helices are, however, relatively rich in bulky residues, such as Ile, Val, and Thr, which, in water-soluble proteins, are classed as membrane destabilizers (their bulky side chains interfere sterically with the carbonyl oxygen in the preceding turn of the α helix and thus destabilize the helical conformation). Thus, factors such as residue volume and packing, which are important in determining helix stability in water-soluble proteins, are not so important for transmembrane α helices, at least for membrane proteins containing single transmembrane α helices; effects of large residue volume will, in the membrane, be balanced by the favorable hydrophobic interactions of a large side chain with the fatty acyl chains. In water-soluble proteins, the conformationally flexible Gly residue is also classed as a helix breaker, because it is an intrinsically flexible residue with the potential to adopt most of the dihedral angles available in a Ramachandran plot. The observation that Gly is quite common in transmembrane α helices suggests that its potential flexibility is constrained in the bilayer environment. Because Gly possesses the smallest of all the side chains, it may play a role in mediating helix–helix interactions and packing in the membrane. The polar amino acids most commonly found within transmembrane α helices are Cys, Thr, and Ser. These residues can be stabilized within a hydrophobic environment by hydrogen bonding between their polar side chains and the peptide backbone at positions $i - 3$ and $i - 4$ (Eilers *et al.*, 2000).

Figure 1 shows the positional preferences of the residues in the transmembrane α helices of type I membrane proteins with a single transmembrane α helix oriented

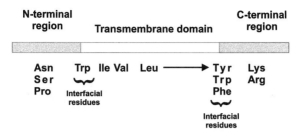

FIGURE 1 Positional preferences for amino acids in the transmembrane domains of human type I membrane proteins with single transmembrane α helices. Modified from Landolt-Marticorena *et al.* (1993).

with its C-terminus on the cytoplasmic side of the membrane (Landolt-Marticoreno *et al.*, 1993). The amino-terminal end of the transmembrane domain contains an Ile-rich region followed by a Val-enriched region. The carboxyl-terminal half of the transmembrane α helix is Leu-rich. Ala is found randomly distributed throughout the transmembrane domain. Aromatic residues are found located preferentially in the boundary regions, with Trp at either end of the transmembrane domain, but with Tyr and Phe only at the carboxyl-terminal boundary. Unlike the other aromatic amino acids, Phe, is also found in the hydrophobic segment as well as in the boundary region.

The polar regions flanking the transmembrane domain are enriched in Arg and Lys on the C-terminal side; Asn, Ser, and Pro are enriched in the N-terminal flanking region. The presence of a positively charged C-terminus (cytoplasmic) could play a role in the process of insertion into the membrane, according to the inside positive rule of von Heijne (1996). The presence of particular residues at the N- and C-terminal ends of the helices could also be important in meeting the requirement to "cap" the ends of the α helices; the initial four −NH and final four −C=O groups of an α helix have no hydrogen-bonding partners provided by the peptide backbone of the α helix itself, and so suitable hydrogen-bonding partners have to be provided in some other way. One way is to extend the helix by three or four residues at each end with polar residues containing suitable hydrogen-bonding partners such as Pro and Asn. Alternatively, if the hydrophobic, nonpolar residues in the transmembrane α helix extend into the headgroup region, hydrogen bonds could form with suitable groups in the glycerol backbone and headgroup regions of the lipid bilayer. Either way, if about 20 residues are required to span the hydrophobic core of the bilayer, the total helix length could be up to 28 residues. The result is that there is a degree of indeterminacy in where the ends of transmembrane α helices should be drawn; the precise ends of transmembrane α helices are often not known.

The observed preference for Trp and Tyr residues for the ends of transmembrane α helices agrees with measurements of the binding of small peptides at the lipid–water interface, which show that aromatic residues have a preference for the interface (Wimley and White, 1996). Further, a number of small tryptophan analogues have been shown to bind in the glycerol backbone and lipid headgroup region of the bilayer, stabilized partly by location of the aromatic ring in the electrostatically complex environment provided by this region of the bilayer, and partly by exclusion of the flat, rigid ring system from the hydrocarbon core of the bilayer for entropic reasons (Yau *et al.*, 1998). Thus, although it is agreed that aromatic residues at the ends of transmembrane α helices probably act as "floats" at the interface serving to fix the helix within the lipid bilayer, it is unclear whether the aromatic rings are located in the hydrocarbon or the headgroup region of the bilayer. This uncertainty is also apparent in the crystal structures of a number of membrane proteins. For example, the Trp residues in the bacterial potassium channel KcsA (Doyle *et al.*, 1998) are found clustered at the ends of the transmembrane α helices, forming clear bands on the two sides of the membrane, as shown in Fig. 2. However, the Tyr residues clearly form a band on the periplasmic side of the membrane above the band formed by the Trp residues. Similarly, in the bacterial photosynthetic reaction center (Rees *et al.*, 1994) the majority of the Trp residues are found near the periplasmic side of the protein near the ends of the transmembrane α helices, as shown in Fig. 3. However, the band of Trp residues is more diffuse than in KcsA, and some Trp residues are likely to be located in the hydrocarbon core and some in the headgroup region. The average number of

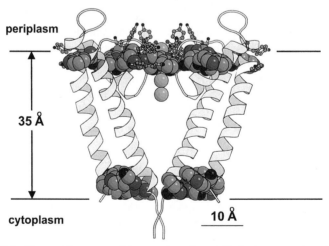

FIGURE 2   The crystal structure of the potassium channel KcsA. A cross section with just two of the four identical subunits is shown. Trp residues are shown in space-fill representation and Tyr residues are shown in ball-and-stick representation. Two potassium ions in space-fill representation are shown moving through the channel. The separation between the two planes representing the outer edges of the Trp residues is 35 Å. (Protein Data Bank [PDB] file 1bl8.)

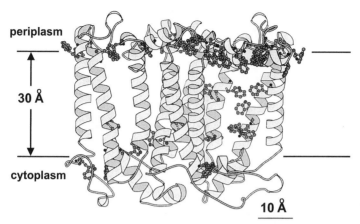

periplasm

30 Å

cytoplasm

10 Å

FIGURE 3 The structure of the L and M subunits of the photosynthetic reaction center of *Rhodobacter sphaeroides*. Trp residues are shown in ball-and-stick representation. An approximate location for the hydrophobic core of the bilayer of thickness 30 Å is shown, as defined by the surface covered by detergent. (PDB file 1aij.)

residues in the transmembrane α helices of the bacterial photoreaction center is 26, corresponding to a length of about 39 Å. The stretch of hydrophobic residues in these helices is, however, only about 19 amino acids or about 28.5 Å long (Ermler *et al.*, 1994; Michel and Deisenhofer, 1990). This matches the thickness of the nonpolar region of the complex (about 30 Å) as defined experimentally as the part covered by detergent in the crystal (Roth *et al.*, 1989, 1991). Detergent is seen to cover some of the Trp residues on the periplasmic side of the membrane, but not others (Roth *et al.*, 1991). The distribution of Trp residues on the cytoplasmic side of the complex is much less distinct than on the periplasmic side (Fig. 3). If the hydrocarbon core of the bilayer around the complex does have a thickness of 30 Å, then again the Trp residues on the cytoplasmic side of the membrane will be located in both the hydrocarbon core and the headgroup regions of the bilayer (Fig. 3).

In the Ca^{2+}-ATPase of skeletal muscle sarcoplasmic reticulum the situation is more complex, as shown in Fig. 4 (Toyoshima *et al.*, 2000). Many of the transmembrane α helices extend above the membrane surface to form a central stalk linking the transmembrane region to the cytoplasmic head of the protein. As a consequence some of the helices are very long; helix M5, for example, contains 41 residues. A ring of Trp residues can be seen on the cytoplasmic side of the membrane helping to define the location of the membrane surface (Fig. 4). A Lys residue (Lys-262) in transmembrane α helix M3 can be seen pointing up from the hydrophobic core of the bilayer like a snorkel. Because the cost of burying a charged residue in the hydrophobic core of a bilayer is very high (about

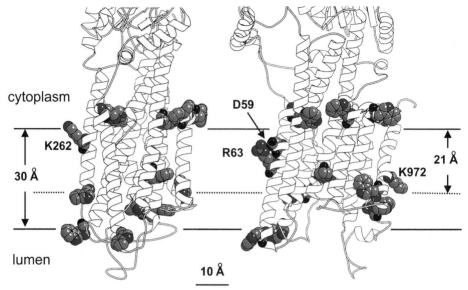

FIGURE 4   The transmembrane region of the Ca^{2+}-ATPase of skeletal muscle sarcoplasmic reticulum. Two views are shown. The view on the left shows the location of Lys-262 in transmembrane α helix M3 pointing up toward the cytoplasmic membrane surface. The view on the right shows Lys-972 in M10 pointing down toward the luminal membrane surface and Asp-59 stacked against Arg-63 to form an ion pair. A possible location for the cytoplasmic surface and two possible locations for the luminal surface are shown. (PDB file 1eul.)

37 kJ mol^{-1} for a Lys residue; Engelman *et al.,* 1986), it is likely that the amino group on Lys-262 will be located at the interface; the Trp residues in the Ca^{2+}-ATPase will then be located in the headgroup region of the bilayer. The structure of the Ca^{2+}-ATPase is also unusual in that the first transmembrane α helix contains two polar residues, Asp-59 and Arg-63, pointing out into the hydrocarbon core; presumably, stacking of Asp-59 against Arg-63 allows formation of an ion pair.

The distribution of Trp residues on the lumenal face of the Ca^{2+}-ATPase is much more diffuse than on the cytoplasmic side. The hydrophobic thickness of the Ca^{2+}-ATPase could be expected to be about 30 Å, because that is the hydrophobic thickness of a bilayer of di(C18:1)PC,[1] the phospholipid that supports highest activity for the ATPase (East and Lee, 1982). However, as shown in Fig. 4, this definition locates the lumenal loops between transmembrane α helices M5 and M6

---

[1]Phospholipid designations are PC, PS, and PA for phosphatidylcholine, phosphatidylserine, and phosphatidic acid, respectively. Fatty acyl chains are given in the format *m:n*, where *m* is the number of carbon atoms and *n* is the number of double bonds. Thus, for example, dioleoylphosphatidylcholine is di(C18:1)PC.

and between M9 and M10 within the hydrocarbon core. Further, it locates a Lys residue (Lys-972) totally within the hydrocarbon core, which seems unlikely. The hydrophobic thickness of the bilayer would have to be about 21 Å to locate the two interhelical loops and Lys-972 at the lumenal surface (Fig. 4); this is close to the thickness of a bilayer of di(C14:1)PC (22 Å). The crystal structure shown in Fig. 4 corresponds to the Ca^{2+}-bound, E1 conformation of the ATPase. It has been shown that the E1 conformation of the ATPase is favored by di(C14:1)PC, whereas di(C18:1)PC favors the other major conformation of the ATPase, E2 (Starling *et al.,* 1994). Thus, it is possible that conformational changes within the transmembrane region of the Ca^{2+}-ATPase lead to changes in the interhelical loops and thus to changes in the effective hydrophobic thickness of the ATPase (Lee and East, 2001).

II. EFFECTS OF INTRINSIC MEMBRANE PROTEINS ON LIPID BILAYERS

Incorporation of a protein into a lipid bilayer can be expected to have significant effects on the properties of the bilayer. The rough surface presented by a protein to the surrounding lipid bilayer will tend to produce poor packing unless the lipid fatty acyl chains distort to match the surface of the protein. In a liquid crystalline bilayer the lipid fatty acyl chains are disordered, because the chains undergo extensive wobbling fluctuations. The presence of a rigid protein surface would be expected to reduce the extent of these motional fluctuations. However, the chains will have to tilt and become conformationally disordered to maximize contact with the rough surface of the protein. The net result is that the presence of a protein will lead to decreased order for the chains, with a wide range of chain orientations relative to the bilayer normal, but with reduced extent and rate of motion. Because of the reduced motion, lipids adjacent to membrane proteins are often referred to as being motionally restricted.

It is clear, therefore, that the reasons for the disorder of the bulk lipids and the disorder of the lipids adjacent to the protein (the boundary or annular lipids) are different; for the bulk phospholipids the disorder is dynamic, whereas for the boundary lipids the disorder is static. An example is provided by the bacterio-rhodopsin trimer, whose crystal structure is unusual in showing a few well-defined lipid molecules (Belrhali *et al.,* 1999; Luecke *et al.,* 1999). Figure 5 shows some of the lipids located at the surface of the trimer. The electron densities for the chains are well defined, but the headgroups are disordered, so that the headgroups could not be identified; the lipids were therefore modeled simply as 2,3-di-*O*-phytanyl-*sn*-propane (Belrhali *et al.,* 1999). The considerable static disorder of the chains is clear in Fig. 5, the rotational disorder of the chains being necessary to obtain good van der Waals contacts with the molecularly rough surface of the bacteriorhodopsin trimer. Lipids on the extracellular side of the membrane are better resolved than

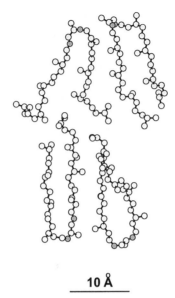

10 Å

FIGURE 5 Structures of four phospholipid molecules identified in X-ray crystallographic studies of bacteriorhodopsin (Belrhali *et al.*, 1999). The lipids have been modeled as 2,3-di-*O*-phytanyl-*sn*-propane. (PDB file 1qhj.)

those on the cytoplasmic side; the degree of order of the lipids parallels that of the protein, which is also greater on the extracellular side (Grigorieff *et al.*, 1996). The average distance between the glycerol backbone oxygens for phospholipids on the two sides of the membrane was 31.6 Å (Mitsuoka *et al.*, 1999). As expected, this closely matches the hydrophobic length of the transmembrane helices of bacteriorhodopsin; the mean helix length is 23 residues, corresponding to a length of about 35 Å.

The crystal structure also makes clear the very different conformations adopted by the various lipid molecules located on the surface of the trimer. For example, one lipid molecule forms a hydrogen bond from its ether oxygens to a tyrosine —OH group at the end of a transmembrane α helix (Fig. 6; Belrhali *et al.*, 1999; Essen *et al.*, 1998). The result is that the strength of the interactions of individual boundary lipid molecules with the protein will be different.

The disorder of the chains seen in Fig. 6 is consistent with the results of molecular dynamics simulations of the bacteriorhodopsin trimer in a bilayer of diphytanyl phosphatidylglycerophosphate (Edholm *et al.*, 1995). The molecular dynamics simulation agrees with experiment in predicting higher order for both the lipids and the protein on the extracellular side of the membrane; fluctuations in the loops and the ends of helices on the cytoplasmic side of the membrane are greater than on the extracellular side. This is also seen in fluctuations of the lipids, with lipids on

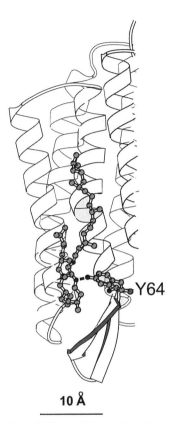

FIGURE 6   Hydrogen bonding in lipid–protein interactions. Tyr-64 at the end of a transmembrane α helix of bacteriorhodopsin is hydrogen-bonded to an ether oxygen of 2,3-di-O-phytanyl-sn-propane. The hydrogen bond is shown by the dash. (PDB file 1qhj.)

the cytoplasmic side of the membrane fluctuating more strongly than those on the extracellular side (Edholm *et al.,* 1995). The calculated order parameters for the chains are low, mainly due to a static tilt of the chains necessary to allow them to nestle against the rough surface of the protein. The chains in the purple membrane behave more like parts of the protein than parts of a fluid lipid phase, consistent with the idea of boundary lipids (Edholm *et al.,* 1995).

The boundary lipids and the bulk lipids in a membrane can be distinguished experimentally in many systems, because of the static disorder of the boundary lipids and the dynamic disorder of the bulk lipids. Static and dynamic disorder give rise to very different electron spin resonance (ESR) spectra for spin-labeled lipids, and ESR spectra for membrane protein systems usually show two components, one "immobilized," corresponding to boundary lipid, and one relatively mobile, corresponding to bulk lipid (Devaux and Seigneuret, 1985; Marsh, 1995). Studies with oriented samples have confirmed a wide range of orientational distributions for the boundary lipid, in contrast to the bulk lipid phase, where motion of the lipid long axis is about the bilayer normal (Jost *et al.,* 1973; Pates and Marsh, 1987).

A particularly important feature of a membrane protein as far as the lipid bilayer is concerned is the thickness of the transmembrane region of the protein. The cost of exposing hydrophobic fatty acyl chains or protein residues to water is such that the hydrophobic thickness of the protein should match that of the bilayer. The question then is how the system responds when these do not match. Most models of hydrophobic mismatch assume that the lipid chains in the vicinity of the protein adjust their length to that of the protein, with the protein acting as a rigid body. When the thickness of the bilayer is less than the hydrophobic length of the peptide the lipid chains must be stretched. Conversely, when the thickness of the bilayer is greater than the hydrophobic length of the peptide the lipid chains must be compressed (Fig. 7). Stretching the fatty acyl chains will effectively decrease the surface area they occupy in the membrane surface, and, conversely, compressing the chains will increase the effective area occupied in the surface (Fig. 7). Thus,

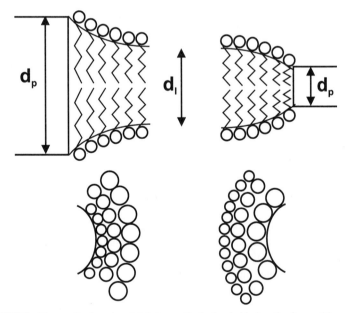

FIGURE 7 The result of a mismatch between the hydrophobic length of a peptide and the hydrophobic thickness of a lipid bilayer. Left: Positive hydrophobic mismatch ($d_P > d_L$). Right: Negative mismatch ($d_P < d_L$). The top shows a "side" view of the chain packing around the peptide and the bottom shows a top view, illustrating the variation in chain cross-sectional area with distance from the peptide. Figure based on Fattal and Ben-Shaul (1993).

changes in fatty acyl chain order are linked to changes in average interfacial areas per lipid molecule.

A number of terms have been suggested to contribute to the total free energy cost of deforming a lipid bilayer around a protein molecule (Fattal and Ben-Shaul, 1993; Nielsen *et al.*, 1998):

1. Loss of conformational entropy of the chains imposed by the presence of the rigid protein wall
2. Bilayer compression/expansion energy due to changes in the membrane thickness
3. Surface energy changes due to changes in the area of the bilayer–water interface
4. Splay energy due to changes in the cross-sectional energy available to the chains along their length, resulting from curvature of the monolayer surface near the protein

A number of models have been proposed to estimate these terms. Fattal and Ben-Shaul (1993) calculated the total lipid–protein interaction free energy as the sum of chain and headgroup terms. For the chains the loss of conformational entropy imposed by the rigid protein wall was positive (unfavorable) even for perfect hydrophobic matching. The other contribution to the chain term arose from the requirement for hydrophobic matching and the consequent stretching or compression of the chain. The term due to the headgroup region was treated as an interfacial free energy, including an attractive term associated with exposure of the hydrocarbon core to the aqueous medium and a repulsive term due to electrostatic and excluded-volume interactions between the headgroups (excluded-volume interactions signify that no two atoms can occupy the same position in space). The resulting profile of energy of interaction as a function of hydrophobic mismatch was fairly symmetrical about the point of zero mismatch. The lipid perturbation energy F (in units of kT per angstrom of protein circumference) calculated by Ben-Shaul (1995) fits to the equation

$$F = 0.37 + 0.005(d_P - d_L)^2, \tag{1}$$

where d_P and d_L are the hydrophobic thicknesses of the protein and lipid bilayer, respectively, and the unperturbed bilayer thickness is 24.5 Å. The hydrophobic thickness of a bilayer of phosphatidylcholine in the liquid crystalline phase is given by

$$d_L = 1.75(n - 1), \tag{2}$$

where n is the number of carbon atoms in the fatty acyl chain (Lewis and Engelman, 1983; Sperotto and Mouritsen, 1993). A simple, but crude calculation gives an idea of the size of the effect that can be expected from hydrophobic mismatch. It is assumed that the protein is very large and so appears flat to a lipid molecule. It is also assumed that all the lipid perturbation energy is concentrated in the first shell of lipids around the protein. Equation (1) then shows that if a lipid occupies 6 Å of the protein circumference, the lipid–protein interaction energy would change by 3.6 kJ mol^{-1} for a hydrophobic mismatch of 7 Å, corresponding to an increase in fatty acyl chain length of 4 carbons, and by 22.8 kJ mol^{-1} for a hydrophobic mismatch of 17.5 Å, corresponding to an increase in acyl chain length of 10 carbons. Changes in interaction energies of 3.6 and 22.8 kJ mol^{-1} correspond to decreases in the lipid–protein binding constant by factors of 4.3 and 10^4, respectively. If the change in lipid–protein interaction energy were to propagate out from the protein surface to affect more than the first shell of lipids, effects of hydrophobic mismatch would be reduced. For example, if effects were averaged over three shells of lipids, changes in fatty acyl chain lengths by 4 and 10 carbons from that giving optimal interaction would decrease lipid–protein binding constants by factors of 1.6 and 21, respectively.

Energies of the magnitude calculated by Ben-Shaul (1995) are easily sufficient to result in conformational changes on a protein. If a protein conformational change results in a change in the hydrophobic thickness of the protein, the change will result in a deformation of the adjacent bilayer. Because the equilibrium constant describing the equilibrium between two conformational states of a protein is determined by the total free energy difference between the two states, the energetic cost of the membrane deformation will contribute toward the equilibrium constant.

The approach adopted by Nielsen *et al.* (1998) came to rather similar conclusions. The most important energy terms were found to be the splay energy and the compression–expansion term, the splay energy term being most important close to the protein, with the compression–expansion term being more important further from the protein. Even though the bilayer deformation was calculated to extend some 30 Å from the protein, most of the deformation was found concentrated in the component immediately adjacent to the protein.

An alternative model for mismatch is the mattress model of Mouritsen and Bloom (1984, 1993). This again expresses mismatch as two terms. The first is an excess hydrophobic free energy associated with exposing either lipid chains or the protein surface to the aqueous medium. The second is proportional to the contact area between the lipid chains and the hydrophobic surface of the protein. The calculations showed that, for a protein of hydrophobic thickness 20 Å, which matches a bilayer of di(C14:0)PC in the liquid crystalline state, the binding constant for di(C14:0)PC is about a factor of 2.5 greater than for a bilayer of di(C18:0)PC, which will give a bilayer too thick by 7 Å (Sperotto and Mouritsen, 1993). Thus, effects of mismatch calculated in this way are similar to those calculated using the approach of Fattal and Ben-Shaul (1993).

The importance of hydrophobic matching has been confirmed in a number of experimental studies (Dumas *et al.*, 1999; Killian, 1998). For example, although bacteriorhodopsin has relatively little effect on the phase transition temperatures of di(C14:0)PC or di(C16:0)PC (Alonso *et al.,* 1982), it increases the transition temperature of di(C12:0)PC and decreases that of di(C18:0)PC (Piknova *et al.,* 1993). This is consistent with hydrophobic matching models; because di(C12:0)PC gives a too thin bilayer in the liquid crystalline phase, bacteriorhodopsin favors the gel phase, whereas di(C18:0)PC gives a too thick bilayer in the gel phase, so that bacteriorhodopsin favors the liquid crystalline phase. Hydrophobic matching could also be the explanation for the unexpected observation that, in mixtures of di(C14:0)PC and di(C18:0)PC at temperatures where the mixture contains both gel and liquid crystalline phases, bacteriorhodopsin partitions equally between the two phases (Piknova *et al.,* 1997; Schram and Thompson, 1997). This contrasts with the observed exclusion of bacteriorhodopsin from gel-phase lipid in mixtures containing single species of phospholipid (Alonso *et al.,* 1982; Cherry *et al.,* 1978). It has been suggested that this shows that the requirements of hydrophobic matching are of prime importance; the hydrophobic thickness of bacteriorhodopsin is intermediate between the hydrophobic thickness of di(C14:0)PC in the gel phase and di(C18:0)PC in the liquid crystalline phase, so that bacteriorhodopsin shows little preference between the two. In mixtures of di(C12:0)PC and di(C18:0)PC, where, at low temperatures, two separate gel phases are formed, one enriched in di(C12:0)PC and one enriched in di(C18:0)PC, bacteriorhodopsin partitions very strongly into the di(C12:0)PC-enriched domains; this could be because the hydrophobic thickness of bacteriorhodopsin is better matched to gel-state di(C12:0)PC than to gel-state di(C18:0)PC (Dumas *et al.,* 1997). However, in studies of the Ca^{2+}-ATPase of sarcoplasmic reticulum using either spin-labeled (London and Feigenson, 1981) or brominated phospholipids (East and Lee, 1982), strengths of binding of liquid crystalline-phase phospholipids to the ATPase were found to be independent of fatty acyl chain length. These results are not consistent with the expectations of hydrophobic matching theory and suggest that, in liquid crystalline bilayers, α-helical membrane proteins are not rigid, but, in fact, can distort to match the thickness of the bilayer. Such a distortion could explain why bilayer thickness affects the activity of membrane proteins such as Ca^{2+}-ATPase (Lee, 1998). The structural distortion could take the form of a change in the tilt of the transmembrane α helices with respect to the bilayer normal or could be a change in the packing of the transmembrane α helices.

III. MODEL TRANSMEMBRANE α HELICES

A. Lipid-Peptide Interactions

An alternative approach to these questions, avoiding the complexity of real membrane proteins, is to use simple model transmembrane α helices, which can be synthesized chemically in large quantity. A number of studies have used peptides of the type Ac-K_2-G-L_n-K_2-A-amide (P_n) consisting of a long sequence of hydrophobic Leu residues capped at both the N- and C-terminal ends with a pair of charged lysine residues. The poly(Leu) region forms a maximally stable α helix, particularly in the hydrophobic environment of the lipid bilayer. The charged Lys caps were chosen both to anchor the ends of the peptides in the lipid headgroup region and to inhibit the aggregation of the peptides in the membrane. The peptide has been shown to adopt the expected α-helical structure in both liquid crystalline- and gel-phase bilayers (Davis *et al.*, 1983; Huschilt *et al.*, 1989; Zhang *et al.*, 1992a). Rates of hydrogen/deuterium exchange for the peptide P_{16} in lipid bilayers suggest that at least 80% of the peptide is in an α-helical conformation in the bilayer, meaning that the whole of the poly(Leu) core must be α-helical (Zhang *et al.*, 1992b). Rates of hydrogen/deuterium exchange were greater at the N- and C-termini of the peptides than in the middle, suggesting some unraveling of the peptide at its ends (Zhang *et al.*, 1992b).

Experiments with the peptides of this type suggest that about 15 lipid molecules are required for complete incorporation of the peptide into a bilayer of the appropriate thickness (Webb *et al.*, 1998). This agrees with estimates from molecular modeling that about 16–18 lipid molecules will be required to form a complete bilayer shell around the peptide. At molar ratios of lipid less than this, non-bilayer phases can be induced, particularly when the hydrophobic length of the peptide is less than the hydrophobic thickness of the bilayer and when the peptide contains interfacial aromatic groups (de Planque *et al.*, 1999; Killian *et al.*, 1996; Morein *et al.*, 1997).

Effects of single transmembrane α helices on lipid bilayers are likely to be less than those of a protein containing a bundle of transmembrane α helices. The cross-sectional area of a single transmembrane α helix is not much greater than that of a phospholipid molecule in the liquid crystalline phase, so that the hydrophobic surface presented to the lipid molecules is rather small. The structure of the lipid bound to bacteriorhodopsin shown in Fig. 6 shows the two chains interacting predominantly with two different transmembrane α helices. This kind of interaction will obviously not be possible with a single transmembrane α helix. Less extensive interactions between lipids and single transmembrane α helices than between lipids and membrane proteins is suggested by ESR experiments. Whereas ESR spectra of spin-labeled lipids in the presence of membrane proteins typically show two-component spectra, as described above, ESR spectra for lipid bilayers containing the peptide L_{24} and for a tryptophan-containing peptide of the type $AW_2(LA)_n W_2A$ are single-component (de Planque *et al.*, 1998; Subczynski *et al.*, 1998). This means either that the lipid fatty acyl chains are not "immobilized" on the peptide surface or that the rate of exchange between bulk and boundary lipid is fast on the ESR time scale (i.e., exchange is faster than 10^7 s^{-1}).

Effects of peptides on chain order in di(C14:0)PC or di(C16:0)PC measured using deuterium nuclear magnetic resonance (NMR) methods are small in both the liquid crystalline and the gel phases (Davis *et al.*, 1983; de Planque *et al.*, 1998, 1999; Roux *et al.*, 1989). Thus, addition of a peptide $AW_2(LA)_7W_2A$ to bilayers of di(C12:0)PC in the liquid crystalline phase resulted in only a 1.4-Å increase in thickness (de Planque *et al.*, 1998), whereas about an 11-Å increase would be necessary for the bilayer thickness to match the hydrophobic length of the peptide. Similarly, Nezil and Bloom (1992) estimated that the peptide P_{24} increased the thickness of a bilayer of (C16:0,C18:1)PC by just 0.6 Å, despite the hydrophobic mismatch between the peptide and the lipid bilayer being ca. 10 Å. Increases in chain order caused by P_{24} in (C16:0,C18:1)PC in the liquid crystalline phase were detected using ESR, but again the effects were small (Subczynski *et al.*, 1998).

Effects of peptides on chain order will depend on the relative hydrophobic length of the peptide compared to the hydrophobic thickness of the bilayer, with long peptides decreasing order and short peptides increasing order, and such effects have been detected using infrared (IR) spectroscopy, but again effects were small (Zhang *et al.*, 1992a). Thus it appears that lipids will distort slightly to improve the match between the hydrophobic length of the peptide and the hydrophobic thickness of the bilayer, but the extent of these modifications is very limited and much less than required to produce full matching.

A number of studies have been published on the effects of these peptides on the phase transition properties of lipid bilayers. Addition of the peptide P_{16} to bilayers of di(C16:0)PC both broadens the main gel-to-liquid crystalline phase transition and decreases the enthalpy of the transition (Morrow *et al.*, 1985). Similar effects have been seen on incorporation of membrane proteins such as bacteriorhodopsin and Ca^{2+}-ATPase (Alonso *et al.*, 1982; Gomez-Fernandez *et al.*, 1980). The decrease in enthalpy of the transition has often been taken to mean that the lipids adjacent to the protein (the boundary lipids) are very strongly perturbed by the peptide and so are unable to take part in the normal phase transition: They are effectively withdrawn from the transition. However, deuterium NMR spectra of mixtures of P_{24} and di(C16:0)PC above and below the phase transition are typical of liquid crystalline- and gel-phase lipids, respectively, with no evidence for any "special" lipid unable to take part in the phase transition (Huschilt *et al.*, 1985). Similarly, as already described, ESR spectra of spin-labeled lipids show the presence of a single type of lipid in the system, not separate bulk and boundary lipids in slow exchange (Subczynski *et al.*, 1998). Thus the peptides (or proteins) do not remove lipid from the main transition, but, rather, perturb the whole lipid bilayer. The peptide decreases the enthalpy difference between the liquid crystalline and gel phases, whereas the lipids in the bilayer remain recognizably liquid crystalline or gel (Morrow *et al.*, 1985).

Morrow *et al.* (1985) showed that mixtures of lipids and peptides can be modeled in terms of regular solution theory (Lee, 1978). Unfortunately, the number of free parameters in fitting to regular solution theory is high, so that little useful information is obtained from the analysis, apart from showing that the data are consistent with regular solution theory. Effects of peptides or proteins on phase transition properties have therefore been interpreted qualitatively in terms of a two-component model in which one component is more or less unperturbed bulk lipid and the other is highly perturbed boundary lipid, which undergoes a broad phase transition of low enthalpy; this approach works for peptides of the poly(Leu) type, but, for some reason, does not work with peptides of the $K_2(LA)_n K_2$ type (Zhang *et al.*, 1992a, 1995). Differential scanning calorimetry (DSC) thermograms for mixtures with the poly(Leu) peptides have been fitted to two components, attributed to the phase transitions of peptide-free and boundary lipid, respectively. The phase transition temperature for the peptide-free component is slightly less than that for pure lipid; this is probably due to the normal colligative effects that will follow from mixing the "pure" lipid phase with the boundary lipid, the latter acting as an "impurity." The phase transition temperature for the boundary lipid is higher than the bulk transition temperature for short-chain lipids, but is lower for long-chain lipids (Zhang *et al.*, 1992a). The same observation has been made for membrane proteins (Dumas *et al.*, 1999; Piknova *et al.*, 1993). This is consistent with the idea that short fatty acyl chains have to stretch to match the hydrophobic thickness of the membrane protein, whereas long fatty acyl chains have to compress. However, the experimental changes are much smaller than expected from models of hydrophobic matching.

Further insights into how transmembrane α helices might interact with lipid molecules in a bilayer have come from molecular dynamics simulations. One study was of a transmembrane α helix of 32 alanine residues in a bilayer of di(C14:0)PC in the liquid crystalline state (Shen *et al.*, 1997). The peptide is not an ideal model for a transmembrane α helix, because it lacks charged groups at each end to interact with the polar headgroups of the phospholipids. Nevertheless many features of the simulation are informative. The simulation was started with the peptide as a pure α helix. The central 15 residues (Ala-12–26), which interacted just with the lipid fatty acyl chains, remained as a stable α helix. The N- and C-terminal regions of the α helix, located in the lipid headgroup region, were less stable and fluctuated more, because of transient hydrogen bonding between the peptide bond amide hydrogen and the phosphate or fatty acyl ester oxygen atoms and the water; as a result, the ends of the helices become frayed. The length of the central helical region oscillated slightly about a 22-Å average expected for an α helix, varying between 20 and 23 Å.

The helix was tilted up to $30°$ with respect to the bilayer normal. Because the helix contains no resides that would make strong contacts in the headgroup region or with water, there is no reason for it not to tilt (Shen *et al.*, 1997). Tilting in fact allows more hydrophobic contact by allowing more of the Ala residues to be located in the core of the bilayer. The presence of the peptide had little effect on the calculated properties of the bilayer. The average bilayer thickness was not significantly changed, although the average order parameter for the CH_2 groups in the chains decreased in the presence of the peptide.

Many different lipid molecules contributed to the immediate surroundings of the peptide (Shen *et al.*, 1997). Even if the fatty acyl chain of a particular lipid was immediately adjacent to the peptide, the headgroup of the lipid could be a substantial distance away. Given the diameter of the helix and the size of the phosphate group, a phosphate immediately adjacent to the helix would be between 8 and 10 Å away from the center of the helix (Shen *et al.*, 1997). On average, five lipid molecules having their chains adjacent to the helix also had their headgroups adjacent, using this definition. However, the headgroup, as given by the position of the phosphorus atom, could be up to 16–18 Å away. Since the average distance between lipid headgroups is 8 Å, this puts these lipids in the "second" shell around the peptide. Other lipids existed between these extremes, suggesting a very diffuse environment around the protein rather than a discrete set of well-ordered shells of lipid. Only rarely did an entire lipid molecule pack tightly around the helix. The shell around the helix contains contributions from a large number of lipid molecules, each contributing a small number of atoms (Shen *et al.*, 1997). Thus there is no evidence for a distinct shell of lipids around the peptide and any perturbation of the lipids extends out just a few Ångstroms.

The lack of a clear shell of lipids around the poly(Ala) peptide contrasts with the boundary lipids observed for membrane proteins and illustrated for bacteriorhodopsin in Figs. 5 and 6. In part this could be an intrinsic feature of single transmembrane α helices, which will not be able to present a large surface area on which fatty acyl chains could be immobilized. The regular structure of a poly(Ala) peptide compared to the rough surface of a typical α-helical peptide might also contribute to the lack of an immobilization shell of lipids. However, a significant factor is also likely to be the lack of polar groups at the ends of the peptide able to interact with the phospholipid headgroups. The importance of polar groups at the ends of the peptides has been shown in a simulation of isolated helices from bacteriorhodopsin (Woolf, 1998). The simulations show that a small number of the lipids surrounding a helix interact with it much more strongly than other lipids, due to a combination of van der Waals (mediated by chains) and charge interactions (mediated by headgroups) (Woolf, 1998).

A molecular dynamics simulation of the peptide L_{16} in a bilayer of di(C14:0)PC in the liquid crystalline phase showed that the peptide tilted with an average angle of 15.3° with respect to the bilayer normal, even though the thickness of the hydrophobic region of the bilayer (22 Å) was a good match to the length of the α helix, 24 Å (Belohorcova *et al.*, 1997). A molecular dynamics simulation has been reported for PF1 coat protein in di(C16:0)PC (Roux and Woolf, 1996). The coat protein contains an amphipathic α helix on the bilayer surface and a hydrophobic transmembrane α helix. Fatty acyl chains next to the transmembrane α helix were slightly more ordered than bulk lipids (Roux and Woolf, 1996). Similarly, in a simulation of the seventh transmembrane α helix of the serotonin 5HT receptor in di(C14:0)PC, lipids in contact with the peptide had slightly higher order parameters than bulk lipids (Duong *et al.*, 1999).

B. Effects of Hydrophobic Mismatch

The theories described above show that there will be an energetic cost associated with any change in the thickness of a bilayer. This would be reflected in values for the equilibrium constant describing the binding of lipids to the protein. A lipid that can bind to a protein without a change in bilayer thickness would bind more strongly to the protein than one for which binding required a change in bilayer thickness. The strength of interaction between a peptide and a phospholipid in a bilayer can be measured using a fluorescence quenching method (Webb *et al.*, 1998). Peptides used are of the type $KKGL_7WL_9KKA$ (L_{16}) and $KKGL_{10}WL_{12}KKA$ (L_{22}) containing a central Trp residue as a fluorescence reporter group. The peptide is incorporated into bilayers containing the brominated phospholipid dibromostearoylphosphatidylcholine (di($Br_2C18:0$)PC); di($Br_2C18:0$)PC behaves much like a conventional phospholipid with unsaturated fatty acyl chains, because the bulky bromine atoms have effects on lipid packing similar to those of a *cis* double bond (East and Lee, 1982). Contact between the bromine atoms in the lipid and the Trp residue in the peptide leads to fluorescence quenching. In mixtures of brominated and nonbrominated phospholipids, the degree of quenching of the fluorescence of the tryptophan residue is related to the fraction of the surrounding (boundary) phospholipid molecules that are brominated, and thus to the strength of binding of the nonbrominated lipid to the peptide. An example of the method is shown in Fig. 8. The fluorescence intensity for the peptide L_{16} incorporated into bilayers of di($Br_2C18:0$)PC at a molar ratio of lipid to peptide of $100:1$ is 5% of that in di(C18:1)PC, demonstrating highly efficient quenching of the tryptophan by the bromine-containing fatty acyl chains (Fig. 8). The fluorescence intensity in mixtures of di($Br_2C18:0$)PC and di(C18:1)PC decreases with increasing content of di($Br_2C18:0$)PC, reflecting the increasing number of boundary lipids that will be di($Br_2C18:0$)PC. As shown in Fig. 8, fluorescence quenching curves for L_{16} in mixtures of di($Br_2C18:0$)PC and (C14:1)PC show more fluorescence quenching at intermediate mole fractions of di($Br_2C18:0$)PC than in mixtures with di(C18:1)PC. This shows that di(C18:1)PC binds more strongly to the peptide than does di(C14:1)PC. The results can be analyzed to give relative lipid-binding constants, as described in Webb *et al.* (1998). These lipid-binding constants for L_{16} and L_{22} are given in Table I.

For L_{22} strongest binding is seen with di(C22:1)PC, for which the relative binding constant is about double that for di(C18:1)PC (Table I). The hydrophobic length of the peptide L_{22} is about 36 Å, calculated for a stretch of 24 hydrophobic residues in total, with a helix translation of 1.5 Å per residue. Thus, strongest binding is seen when the hydrophobic length of the peptide matches the hydrophobic thickness of the bilayer, as expected from theories of hydrophobic mismatch. However, relative binding constants do not continue to decrease with decreasing chain length from di(C18:1)PC to di(C14:1)PC as would have been predicted (Table I). An even

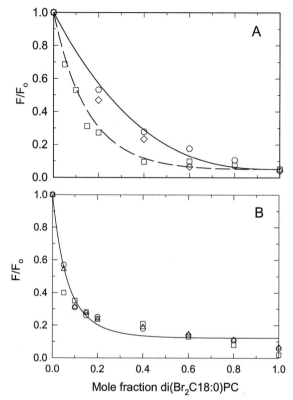

FIGURE 8 Fluorescence quenching method for determining lipid binding constants. (A) Fluorescence intensities for L_{16} in mixtures containing di(Br_2C18:0)PC. The L_{16} was incorporated into a mixture of di(Br_2 C18:0)PC and either di(C18:1)PC (○) or di(C14:1)PC (□), or a mixture of di(C14:1)PC and cholesterol at a 1:1 molar ratio (◇), at a molar ratio of peptide:phospholipid of 1:100. Fluorescence intensities are expressed as a fraction of that recorded for L_{16} in di(C18:1)PC. (B) Fluorescence intensities in mixtures of di(Br_2C18:0)PC and gel-phase lipid. Fluorescence intensities are shown for L_{16}(○) and Y_2L_{14} (△) in mixtures of di(Br_2C18:0)PC and di(C16:0)PC and for L_{16} in mixtures of di(Br_2C18:0)PC and sphingomyelin (□), at 25°C.

larger deviation from theoretical predictions is observed with the short peptide L_{16} (Table I). In this case strongest binding is observed with di(C18:1)PC, with binding decreasing with decreasing chain length to di(C14:1)PC, as expected. However, the peptide was found not to incorporate at all into bilayers of di(C24:1)PC, instead forming aggregates of peptide separate from the bilayer. Ren *et al.* (1999) obtained very similar results, except that under their conditions, unincorporated peptide bound to the surface of the lipid bilayer, with the long axis of the peptide parallel to the surface. Similarly, L_{16} was found to be only partly incorporated into

TABLE I

Relative Phospholipid-Binding Constants for Peptides L_{16} and L_{22}

Lipid	Bilayer thickness[a] (Å)	Relative binding constant[b]	
		L_{16}	L_{22}
di(C14:1)PC	22.8	0.4	0.9
di(C16:1)PC	26.3	0.8	0.7
di(C18:1)PC	29.8	1.0	1.0
di(C20:1)PC	33.3	0.7	1.8
di(C22:1)PC	36.8	—	2.0
di(C24:1)PC	40.3	—	1.5

[a] Bilayer hydrophobic thickness d calculated from $d = 1.75(n - 1)$, where n is the number of carbon atoms in the fatty acyl chain (Sperotto and Mouritsen, 1988).
[b] Estimated hydrophobic length is 27 Å for L_{16} and 36 Å for L_{22}.

bilayers of di(C22:1)PC (Webb *et al.*, 1998). Thus, a short peptide cannot incorporate into a too-thick bilayer. It is suggested that a too-thin bilayer can match to a too-long peptide both by a slight stretching of the lipid and by tilting of the long axis of the helix with respect to the bilayer normal so that its effective length across the bilayer is reduced. However, a too-thick bilayer can only match a too-thin peptide by compression of the lipid, which becomes energetically unfavorable when the difference between the bilayer thickness and the peptide length exceeds about 6 Å (Webb *et al.*, 1998).

Possible effects of aromatic residues at the ends of transmembrane α helices have been studied using peptides $K_2GFL_6WL_8FK_2A$ (F_2L_{14}) and $K_2GYL_6WL_8YK_2A$ (Y_2L_{14}), in which one Leu residue at each end of the poly(Leu) stretch is replaced by either a Phe or a Tyr (Mall *et al.*, 2000). In contrast to the results with L_{16}, peptide F_2L_{14} incorporated fully into bilayers of di(C24:1)PC, and Y_2L_{14} partitioned partially into di(C24:1)PC. The fluorescence quenching method was again used to obtain binding constants for phosphatidylcholines to the peptides, measured relative to the binding constant for di(C18:1)PC (Table II). The effective hydrophobic length of the peptide Y_2L_{14} might be expected to be somewhat greater than that of L_{16}; if the peptide is modeled as an α helix with the two Tyr residues oriented to be roughly parallel to the long axis of the α helix, the distance between the two Tyr—OH groups is ca. 33 Å, about 6 Å greater than the hydrophobic length of L_{16}. The hydrophobic length of Y_2L_{14} calculated in this way matches the hydrophobic thickness of a bilayer of di(C20:1)PC, whereas the relative lipid-binding constants increase from di(C14:1)PC to di(C22:1)PC (Table II). Similarly, relative binding constants for F_2L_{14} increase with increasing chain length from di(C14:1)PC to di(C24:1)PC. The results with Y_2L_{14} and F_2L_{14} show that introduction of aromatic

TABLE II

Relative Lipid-Binding Constants for Peptides Containing Aromatic Groups

| Lipid | d^a (Å) | Relative binding constant | | |
		Y_2L_{14}	F_2L_{14}	Y_2L_{20}
di(C14:1)PC	22.8	0.46	0.50	0.47
di(C16:1)PC	26.1	0.74	0.83	0.45
di(C18:1)PC	29.8	1	1	1
di(C20:1)PC	33.3	1.02	1.16	1.22
di(C22:1)PC	36.8	1.37	1.62	0.96
di(C24:1)PC	40.3	—	2.33	1.19

^{a}Hydrophobic thickness of the bilayer calculated from $d = 1.75(n - 1)$, where n is the number of carbon atoms in the fatty acyl chain (Sperotto and Mouritsen, 1988).

residues at the two ends of the hydrophobic sequence increases the ability of the short peptides to partition into thick lipid bilayers. The observation that the highest relative binding constant is obtained with bilayers considerably thicker than the calculated hydrophobic length of the peptides suggests that the presence of aromatic residues at the ends of the helices could lead to marked thinning of the bilayer around the peptides (Mall *et al.*, 2000).

The chain length dependence of lipid binding to Y_2L_{20} is much less marked than for the shorter peptides; the relative binding constant increases from di(C14:1)PC to di(C18:1)PC, but then hardly changes with increasing chain length between di(C18:1)PC and di(C24:)PC (Table II). This contrasts with L_{22}, which shows markedly stronger interaction with di(C22:1)PC than with phospholipids with shorter or longer chains. This again suggests that the introduction of the two Tyr residues leads to an increase in the thickness of the bilayer with which optimal interaction of the peptide is observed.

C. Interactions with Anionic Phospholipids

Interactions between transmembrane α helices and the phospholipid headgroups also have to be considered. Using the fluorescence quenching method, it was shown that a small number of anionic phospholipid molecules (possibly just one) bound strongly to the peptide L_{16}, the remaining molecules binding with an affinity equal to that of phosphatidylcholine. The binding constant for the strongly bound phosphatidic acid molecule relative to phosphatidylcholine in a medium of low ionic strength was 8.6 (in mole fraction units), corresponding to a difference in unitary binding energies of -5.3 kJ mol^{-1} (Mall *et al.*, 1998). At pH 7.2, phosphatidic acid bears a single negative charge (Cevc, 1990). The binding constant for phosphatidic acid changed little with ionic strength, suggesting that the interaction with the positively charged peptide did not follow simply from a high positive potential in the vicinity of the positively charged Lys residues on the peptide, increasing the local concentration of anionic phospholipid. The energy of interaction between two ions U is given by

$$U = z_1 z_2 e^2 / 4\pi \varepsilon_0 \varepsilon_r r, \tag{3}$$

where z_1 and z_2 are the charges on the two ions ε_0 is the permittivity of a vacuum, ε_r is the relative permittivity (dielectric constant) of the medium, and r is the distance between the two ions. Assuming a dielectric constant of 78.5 (water), we find that an energy of interaction of 5.3 kJ mol^{-1} corresponds to a distance of separation between two monovalent ions of 3.3 Å. This therefore suggests that strong interaction requires the anionic headgroup of phosphatidic acid to be in close contact with one of the Lys residues on the peptide. Once this strong interaction with a single phosphatidic acid molecule has been made, other phosphatidic acid molecules will then interact with L_{16} relatively nonspecifically, with a binding constant relative to phosphatidylcholine close to 1.

The relative binding constants for phosphatidylserine were less than for phosphatidic acid and are more sensitive to ionic strength (Mall *et al.*, 1998). For phosphatidylserine, the presence of the positively charged ammonium group as well as the negatively charged carboxyl group in the headgroup region may reduce interaction with the positively charged peptide. In contrast to L_{16}, the binding constants for anionic phospholipids to L_{22} are very similar to those for zwitterionic phospholipids, with a relative binding constant close to 1. It could be that tilting of L_{22} in the bilayer, necessary to match the hydrophobic length of L_{22} to the hydrophobic thickness of a bilayer of di(C18:1)PC, locates the Lys residues on the peptide too far from the lipid headgroup region to allow a strong interaction between the anionic phospholipid and the peptide.

Both phosphatidylserine and phosphatidic acid bind more strongly to the peptides $Y_2 L_{14}$ and $Y_2 L_{20}$ than does phosphatidylcholine, the effect of anionic phospholipid decreasing slightly with increasing ionic strength. However, in this case the experiments are consistent with a model in which the binding constants for all the anionic phospholipid molecules binding to the peptide are increased slightly (Mall *et al.*, 2000). This suggests that the presence of the Tyr residues prevents close association of the anionic phospholipid group with the cationic Lys residues.

These results suggest that the effects of charge on the interactions between phospholipids and transmembrane α helices will often be rather small and will be strongly dependent on the detailed structure of the peptide and its orientation in the membrane. This picture is consistent with the results of the molecular dynamics simulations of individual α helices of bacteriorhodopsin in bilayers of di(C14:0)PC, which showed that a small proportion of the lipid molecules interacted with the α helices much more strongly than the others, and that these strong interactions were dominated by electrostatic terms rather than van der Waals terms (Woolf, 1998).

In general, binding constants for phospholipids to membrane proteins also show relatively little selectivity for anionic phospholipids. For example, binding constants for phosphatidic acid and phosphatidylserine relative to phosphatidylcholine are close to 1 for the Ca^{2+}-ATPase (Dalton et al., 1998), and binding constants for phosphatidic acid and phosphatidylserine for the $(Na^{+}-K^{+})$-ATPase are about twice those for phosphatidylcholine (Esmann and Marsh, 1985). However, there is evidence for the presence of a small number of "special" anionic phospholipids binding to some membrane proteins, acting as "cofactors." An example is provided by cytochrome c oxidase, whose crystal structure shows the presence of a lipid molecule bound between the transmembrane α helices (Iwata et al., 1995). Interaction between an anionic phospholipid and a binding site on a membrane protein would be specific if strong binding requires close interaction between the anionic headgroup and a positively charged residue on the protein, as suggested by the results presented here.

D. Effects of Phospholipid Phase

The phase of the phospholipid is important in determining interactions with transmembrane α helices. As shown in Fig. 8, fluorescence quenching is much more marked in mixtures of di(Br_2C18:0)PC and di(C16:0)PC at temperatures where both liquid crystalline and gel phases are present than in mixtures of di(Br_2C18:0)PC and di(C18:1)PC (Mall et al., 2000). Thus, L_{16} is excluded from regions of lipid in the gel phase and accumulates in regions in the liquid crystalline phase. The binding constants of L_{16} and L_{22} for di(C16:0)PC in the gel phase relative to di(C18:1)PC in the liquid crystalline phase are ca. 0.15 (Mall et al., 2000). This is consistent with the expectation that van der Waals contacts between an all-trans fatty acyl chain and the molecularly rough surface of a peptide will be poor; one way that this poor packing can be overcome is by exclusion of the peptide from the gel phase. Quenching plots for Y_2L_{14} are very similar to those of L_{16} (Fig. 8), showing that the presence of bulky aromatic residues does not have any significant effect on the selectivity for liquid crystalline- over gel-phase lipid (Mall et al., 2000). Further, since Y_2L_{14} shows a preference for longer chain phospholipids than L_{16}, Y_2L_{14} might have been expected to show a greater preference for gel-phase lipid than L_{16}, because phospholipid in the gel phase gives a thicker bilayer than the corresponding lipid in the liquid crystalline phase. Because Y_2L_{14} and L_{16} show equal preferences for liquid crystalline- over gel-phase lipid, any effects of hydrophobic matching between the peptide and the bilayer must be small compared to effects of lipid phase on the interaction energies between the peptides and the lipid (Mall *et al.*, 2000). Preferential partitioning of proteins from domains in the gel phase into domains of liquid crystalline lipid has been demonstrated for a variety of membrane proteins, including bacteriorhodopsin (Cherry *et al.*, 1978) and Ca^{2+}-ATPase (East and Lee, 1982; Kleeman and McConnell, 1976).

Effects of sphingomyelin at 25°C are very similar to the effects of gel-phase di(C16:0)PC (Fig. 8; Mall *et al.*, 2000). Mixtures of bovine brain sphingomyelin and di(C18:1)PC are in a two-phase region at 25°C, with gel-phase domains enriched in sphingomyelin (Untracht and Shipley, 1977). Thus, partitioning of the peptides between gel- and liquid crystalline-phase lipid shows little dependence on the structure of the phospholipid. It has been suggested that plasma membranes of mammalian cells contain domains or "rafts" enriched in sphingomyelin and that particular enzymes, particularly those associated with cell signaling, are concentrated within the rafts (Simons and Ikonen, 1997). The results presented here suggest that membrane proteins containing transmembrane α helices will tend to be excluded from these rafts, and it may therefore be significant that many of the signaling proteins suggested to be contained within the rafts are anchored to the membrane by glycosylphosphatidylinositol anchors (Harder and Simons, 1997).

E. Effects of Cholesterol

The presence of cholesterol has a marked effect on incorporation of the peptides into phospholipid bilayers (Webb *et al.*, 1998). Incorporation of cholesterol at a 1:1 molar ratio to phospholipid leads to a general reduction in incorporation of the peptides L_{16} and L_{22}, but superimposed on this effect is a chain length effect. In the presence of cholesterol, the binding constant of P_{16} for di(C14:1)PC relative to di(C18:1)PC increased from 0.4 to about 1, as expected if the presence of cholesterol increases the effective chain length of the C14 chain so that it more nearly matches the hydrophobic length of the peptide (Fig. 8). Consistent with this interpretation, the presence of high molar ratios of cholesterol prevented the incorporation of P_{16} into bilayers of di(C18:1)PC. Nezil and Bloom (1992) showed that incorporation of cholesterol at 33 mol% increases bilayer thickness by about 4 Å.

Studies with brominated analogues of cholesterol showed that cholesterol binds to the peptides with a binding constant only a factor of about two less strongly than di(C18:1)PC (Mall *et al.*, 1998). This is rather surprising, given the relatively rigid structure of the steroid ring of cholesterol and the molecularly rough surface of the peptide. In other studies, it has been shown that cholesterol binds relatively weakly at the lipid–protein interface of the ATPase (Simmonds *et al.,* 1982, 1984); comparison with the peptide studies reported here suggests that weak binding of cholesterol to the ATPase involves interactions in the lipid headgroup region rather than interactions between the sterol ring and the hydrophobic transmembrane α helices.

IV. BIOLOGICAL CONSEQUENCES OF HYDROPHOBIC MISMATCH

The requirement to match the hydrophobic thickness of a membrane protein to that of the surrounding lipid bilayer could be important in a number of ways. Targeting of proteins to their correct final destinations in a cell is essential in maintaining cell integrity. In the bulk flow model, the vast majority of proteins synthesized in the endoplasmic reticulum (ER) are believed to leave the ER by default and flow along the exocytic pathway until they reach the plasma membrane (Nilsson and Warren, 1994). Some proteins, however, have to be retained at particular points along the exocytic pathway. Compartmental localization could be achieved in one of two ways. The first involves a retention signal in the protein, which, at the appropriate point in the exocytic pathway, prevents forward movement of the protein by denying it access to budding transport vesicles of the onward pathway. The second involves a retrieval signal, leading to recapture of the protein after it has left the compartment in which it resides. The classical retrieval signal is the KDEL sequence found in many ER-resident proteins; the situation appears to be different for Golgi-resident proteins, where membrane-spanning domains act as retention signals (Nilsson and Warren, 1994).

Despite the extensive flux of proteins through the Golgi, the Golgi maintains its own distinctive population of resident proteins. Furthermore, the distribution of enzymes within the Golgi is organized according to function, so that, for example, the distributions of glycosyltransferases and glycosidases, although overlapping, are distinct (Colley, 1997; Roth, 1987). Many of the proteins in the Golgi membrane are predicted to contain a single transmembrane α helix, oriented with the N- and C-termini on the inner and outer faces of the membrane, respectively. The Golgi retention signal in such proteins has been shown to involve the membrane spanning domain (Munro, 1991, 1995; Nilsson *et al.,* 1991; Swift and Machamer, 1991). However, the membrane-spanning domains show no sequence homology, and it has not been possible to identify any particular motif leading to retention (Bretscher and Munro, 1993; Colley *et al.,* 1992; Munro, 1991). Thus, sialyltransferase remains localized in the Golgi even when its 17-amino-acid transmembrane domain is replaced by 17 Leu residues (Munro, 1991). However, a longer stretch of 23 Leu residues did not provide an efficient retention signal (Munro, 1991). Similarly, a 4-residue insertion into the transmembrane domain of galactosyltransferase reduced its retention in the Golgi (Masibay *et al.,* 1993). The reverse effect has been shown with the influenza virus neuraminidase, which shifted from the plasma membrane to the Golgi and ER when the number of residues in the transmembrane domain was reduced (Sivasubramanian and Nayak, 1987).

The lack of a clear retention motif, together with the inability to saturate the mechanism for Golgi retention by overexpression, suggests that retention is not a receptor-mediated event (Nilsson and Warren, 1994). One possible model is then retention by preferential interaction with membranes of optimal thickness (Nilsson and Warren, 1994). Both Bretscher and Munro (1993) and Masibay *et al.* (1993) showed that transmembrane domains of Golgi proteins are shorter (average 15 residues) than transmembrane domains of plasma membrane proteins (average 20 residues). It has therefore been suggested that if the Golgi membrane is thinner than the plasma membrane, membrane proteins with short transmembrane domains will interact "more strongly" with the lipid bilayer of the Golgi than with that of the plasma membrane, leading to retention in the Golgi (Bretscher and Munro, 1993; Masibay *et al.*, 1993). The studies with model peptides described above show that a protein containing a transmembrane α helix with a hydrophobic length greater than the hydrophobic thickness of the Golgi membrane will be able to move out of the Golgi into the plasma membrane. However, a protein whose transmembrane α helix has a hydrophobic length less than the hydrophobic thickness of a particular membrane will not be able to enter that membrane, and such a protein would then be retained in the Golgi (Webb *et al.*, 1998).

Studies of targeting of proteins in yeast are also consistent with a lipid-based model (Rayner and Pelham, 1997). The length of the transmembrane domain is important in targeting with long helices (24 residues), ensuring transport to the plasma membrane. However, for proteins with shorter transmembrane domains, the relative hydrophobicity of the transmembrane domain has been suggested to be important as well as its length, this determining targeting to the Golgi and the vacuole (Rayner and Pelham, 1997).

Retention of some membrane proteins in the ER could also depend on the length of the transmembrane domain of the protein. An important class of ER membrane proteins are those with an N-terminal catalytic domain exposed to the cytoplasm and a C-terminal membrane anchor. Such proteins are inserted into the ER membrane post-translationally by a signal-recognition-particle-independent pathway. No ER retrieval signals have been identified in these proteins. Instead, it has been observed that the hydrophobic domain is rather short. For example, cytochrome b_5, a protein of this type, has a transmembrane domain containing just 17 hydrophobic amino acid residues (Pedrazzini *et al.*, 1996). If the length of the hydrophobic stretch is increased to 22 residues, the protein is transported out of the ER along the secretory pathway (Pedrazzini *et al.*, 1996). It could therefore be that matching of the thickness of the lipid bilayer and the transmembrane length of the protein is important in retention in ER, as was suggested for the Golgi complex. Although the length of the transmembrane domain appears to be the most important factor, the structure of the C-terminal, lumenal, region has also been shown to contribute to retention (Honsho *et al.*, 1998). Experiments with another C-terminal-anchored protein, the ubiquitin-conjugating enzyme UBC6 from yeast, suggest that the thickness requirements of the ER and Golgi membranes may be different, explaining targeting between these two organelles (Yang *et al.*, 1997). Whereas UBC6 containing the wild-type 17-residue transmembrane domain targets to the ER, increasing the length of the transmembrane domain to 21 residues results in movement to the Golgi, and increasing the length further to 26 residues allows movement to the plasma membrane.

These experiments show that the length of the transmembrane α helix is often an important factor in targeting, although it is likely to be only one of a number of important factors. The lengths of the transmembrane α helices are also likely to be important in the proper function of membrane proteins containing multiple trans-membrane α helices. An example already described is that of the Ca^{2+}-ATPase, which shows highest ATPase activity in di(C18:1)PC and lower activities in bilay-ers of phospholipids with longer or shorter fatty acyl chains (Lee, 1998). Changes in the ATPase underlying these changes in ATPase activity are complex (Lee, 1998), but all must be mediated by the transmembrane α helices, because these are the parts of the ATPase that can "sense" the change in bilayer thickness. In the case of the Ca^{2+}-ATPase it seems that, as described above, the two major confor-mational states of the ATPase (E1 and E2) have different preferences for bilayer thickness, the E1 conformation favoring thin bilayers and the E2 conformation favoring thick bilayers (Lee, 1998). Changing the bilayer thickness could change the tilt of the transmembrane α helices in the Ca^{2+}-ATPase, it could change the packing of the helices, and, possibly, it could lead to changes in the structures of the loops connecting the helices, changing the effective lengths of the helices. All these changes could be linked to changes in the phosphorylation domain of the Ca^{2+}-ATPase, located well above the surface of the membrane.

If, as seems likely, the various membranes in a cell have different thicknesses because of their different lipid compositions, the structure of each membrane protein will have evolved to match the thickness of the membrane in which it resides.

Acknowledgment

We thank the Biotechnology and Biological Sciences Research Council (BBSRC) for financial support of the original studies reported here.

References

Alonso, A., Restall, C. J., Turner, M., Gomez-Fernandez, J. C., Goni, F. M., and Chapman, D. (1982). Protein–lipid interactions and differential scanning calorimetric studies of bacteriorhodopsin re-constituted lipid–water systems. *Biochim. Biophys. Acta* **689**, 283–289.

Belohorcova, K., Davis, J. H., Woolf, T. B., and Roux, B. (1997). Structure and dynamics of an amphiphilic peptide in a lipid bilayer: A molecular dynamics study. *Biophys. J.* **73**, 3039–3055.

Belrhali, H., Nollert, P., Royant, A., Menzel, C., Rosenbusch, J. P., Landau, E. M., and Pebay-Peyroula, E. (1999). Protein, lipid and water organization in bacteriorhodopsin crystals: A molecular view of the purple membrane at 1.9 Å resolution. *Structure* **7,** 909–917.

Ben-Shaul, A. (1995). Molecular theory of chain packing, elasticity and lipid–protein interaction in lipid bilayers. *In* "Handbook of Biological Physics. Volume 1A. Structure and Dynamics of Membranes" (R. Lipowsky and E. Sackmann, eds.), pp. 359–401. Elsevier, Amsterdam.

Bretscher, M. S., and Munro, S. (1993). Cholesterol and the Golgi apparatus. *Science* **261,** 1280–1281.

Cevc, G. (1990). Membrane electrostatics. *Biochim. Biophys. Acta* **1031,** 311–382.

Cherry, R. J., Muller, U., Henderson, R., and Heyn, M. P. (1978). Temperature-dependent aggregation of bacteriorhodopsin in dipalmitoyl- and dimyristoyl-phosphatidylcholine vesicles. *J. Mol. Biol.* **121,** 283–298.

Colley, K. J. (1997). Golgi localization of glycosyltransferases: More questions than answers. *Glycobiology* **7,** 1–13.

Colley, K. J., Lee, E. U., and Paulson, J. P. (1992). The signal anchor and stem regions of the β-galactoside α2,6-sialyltransferase may each act to localize the enzyme to the Golgi apparatus. *J. Biol. Chem.* **267,** 7784–7793.

Dalton, K. A., East, J. M., Mall, S., Oliver, S., Starling, A. P., and Lee, A. G. (1998). Interaction of phosphatidic acid and phosphatidylserine with the Ca^{2+}-ATPase of sarcoplasmic reticulum and the mechanism of inhibition. *Biochem. J.* **329,** 637–646.

Davis, J. H., Clare, D. M., Hodges, R. S., and Bloom, M. (1983). Interaction of a synthetic amphiphilic polypeptide and lipids in a bilayer structure. *Biochemistry* **22,** 5298–5305.

de Planque, M. R. R., Greathouse, D. V., Koeppe, R. E., Schafer, H., Marsh, D., and Killian, J. A. (1998). Influence of lipid/peptide hydrophobic mismatch on the thickness of diacylphosphatidylcholine bilayers: A ^{2}H NMR and ESR study using designed transmembrane alpha-helical peptides and gramicidin A. *Biochemistry* **37,** 9333–9345.

de Planque, M. R. R., Kruijtzer, J. A. W., Liskamp, R. M. J., Marsh, D., Greathouse, D. V., Koeppe, R. E., de Kruijff, B., and Killian, J. A. (1999). Different membrane anchoring positions of tryptophan and lysine in synthetic transmembrane alpha-helical peptides. *J. Biol. Chem.* **274,** 20839–20846.

Devaux, P. F., and Seigneuret, M. (1985). Specificity of lipid-protein interactions as determined by spectroscopic techniques. *Biochim. Biophys. Acta* **822,** 63–125.

Doyle, D. A., Cabral, J. M., Pfuetzner, R. A., Kuo, A., Gulbis, J. M., Cohen, S. L., Chait, B. T., and Mackinnon, R. (1998). The structure of the potassium channel: Molecular basis of K^+ conduction and selectivity. *Science* **280,** 69–77.

Dumas, F., Sperotto, M. M., Lebrun, M. C., Tocanne, J. F., and Mouritsen, O. G. (1997). Molecular sorting of lipids by bacteriorhodopsin in dilauroylphophatidylcholine/distearoylphosphatidylcholine lipid bilayers. *Biophys. J.* **73,** 1940–1953.

Dumas, F., Lebrun, M. C., and Tocanne, J. F. (1999). Is the protein/lipid hydrophobic matching principle relevant to membrane organization and functions? *FEBS Lett.* **458,** 271–277.

Duong, T. H., Mehler, E. L., and Weinstein, H. (1999). Molecular dynamics simulation of membranes and a transmembrane helix. *J. Comput. Phys.* **151,** 358–387.

East, J. M., and Lee, A. G. (1982). Lipid selectivity of the calcium and magnesium ion dependent adenosinetriphosphatase, studied with fluorescence quenching by a brominated phospholipid. *Biochemistry* **21,** 4144–4151.

Edholm, O., Berger, O., and Jahnig, F. (1995). Structure and fluctuations of bacteriorhodopsin in the purple membrane: A molecular dynamics study. *J. Mol. Biol.* **250,** 94–111.

Eilers, M., Shekar, S. C., Shieh, T., Smith, S. O., and Fleming, P. J. (2000). Internal packing of helical membrane proteins. *Proc. Natl. Acad. Sci. USA* **97,** 5796–5801.

Engelman, D. M., Steitz, T. A., and Goldman, A. (1986). Identifying nonpolar transbilayer helices in amino acid sequences of membrane proteins. *Annu. Rev. Biophys. Biophys. Chem.* **15,** 321–353.

Ermler, U., Michel, H., and Schiffer, M. (1994). Structure and function of the photosynthetic reaction center from *Rhodobacter sphaeroides*. *J. Bioenerget. Biomembr.* **26,** 5–15.

Esmann, M., and Marsh, D. (1985). Spin label studies on the origin of the specificity of lipid–protein interactions in Na^+,K^+-ATPase membranes from *Squalus acanthias*. *Biochemistry* **24**, 3572–3578.

Essen, L. O., Siegert, R., Lehmannn, W. D., and Oesterhelti, D. (1998). Lipid patches in membrane protein oligomers: Crystal structure of the bacteriorhodopsin–lipid complex. *Proc. Natl. Acad. Sci. USA* **95**, 11673–11678.

Fattal, D. R., and Ben-Shaul, A. (1993). A molecular model for lipid–protein interaction in membranes: The role of hydrophobic mismatch. *Biophys. J.* **65**, 1795–1809.

Gomez-Fernandez, J. C., Goni, F. M., Bach, D., Restall, C. J., and Chapman, D. (1980). Protein–lipid interaction. Biophysical studies of $(Ca^{2+} + Mg^{2+})$-ATPase reconstituted systems. *Biochim. Biophys. Acta* **598**, 502–516.

Grigorieff, N., Cesta, T. A., Downing, K. H., Baldwin, J. M., and Henderson, R. (1996). Electron-crystallographic refinement of the structure of bacteriorhodopsin. *J. Mol. Biol.* **259**, 393–421.

Harder, T., and Simons, K. (1997). Caveolae, DIGs, and the dynamics of sphingolipid–cholesterol microdomains. *Curr. Opin. Cell Biol.* **9**, 534–542.

Honsho, M., Mitoma, J. Y., and Ito, A. (1998). Retention of cytochrome b_5 in the endoplasmic reticulum is transmembrane and luminal domain-dependent. *J. Biol. Chem.* **273**, 20860–20866.

Huschilt, J. C., Hodges, R. S., and Davis, J. H. (1985). Phase equilibria in an amphiphilic peptide–phospholipid model membrane by deuterium nuclear magnetic resonance difference spectroscopy. *Biochemistry* **24**, 1377–1386.

Huschilt, J. C., Millman, B. M., and Davis, J. H. (1989). Orientation of α-helical peptides in a lipid bilayer. *Biochim. Biophys. Acta* **979**, 139–141.

Iwata, S., Ostermeier, C., Ludwig, B., and Michel, H. (1995). Structure at 2.8 Å resolution of cytochrome c oxidase from *Paracoccus denitrificans*. *Nature* **376**, 660–669.

Jost, P. C., Griffith, O. H., Capaldi, R. A., and Vanderkooi, G. A. (1973). Identification and extent of fluid bilayer regions in membranous cytochrome oxidase. *Biochim. Biophys. Acta* **311**, 141–152.

Killian, J. A. (1998). Hydrophobic mismatch between proteins and lipids in membranes. *Biochim. Biophys. Acta* **1376**, 401–416.

Killian, J. A., Salemink, I., dePlanque, M. R. R., Lindblom, G., Koeppe, R. E., and Greathouse, D. V. (1996). Induction of nonbilayer structures in diacylphosphatidylcholine model membranes by transmembrane alpha-helical peptides: Importance of hydrophobic mismatch and proposed role of tryptophans. *Biochemistry* **35**, 1037–1045.

Kleeman, W., and McConnell, H. M. (1976). Interactions of proteins and cholesterol with lipids in bilayer membranes. *Biochim. Biophys. Acta* **419**, 206–222.

Landolt-Marticorena, C., Williams, K. A., Deber, C. M., and Reithmeier, R. A. F. (1993). Non-random distribution of amino acids in the transmembrane segments of type 1 single span membrane proteins. *J. Mol. Biol.* **229**, 602–608.

Lee, A. G. (1978). Calculation of phase diagrams for non-ideal mixtures of lipids, and a possible non-random distribution of lipids in lipid mixtures in the liquid crystalline phase. *Biochim. Biophys. Acta* **507**, 433–444.

Lee, A. G. (1998). How lipids interact with an intrinsic membrane protein: The case of the calcium pump. *Biochim. Biophys. Acta* **1376**, 381–390.

Lee, A. G., and East, J. M. (2001). What the structure of a calcium pump tells us about its mechanism. *Biochem. J.* **356**, 665–683.

Lewis, B. A., and Engelman, D. M. (1983). Lipid bilayer thickness varies linearly with acyl chain length in fluid phosphatidylcholine vesicles. *J. Mol. Biol.* **166**, 211–217.

London, E., and Feigenson, G. W. (1981). Fluorescence quenching in model membranes. 2. Determination of local lipid environment of the calcium adenosinetriphosphatase from sarcoplasmic reticulum. *Biochemistry* **20**, 1939–1948.

Luecke, H., Schobert, B., Richter, H. T., Cartailler, J. P., and Lanyi, J. K. (1999). Structure of bacteriorhodopsin at 1.55 Å resolution. *J. Mol. Biol.* **291**, 899–911.

Mall, S., Sharma, R. P., East, J. M., and Lee, A. G. (1998). Lipid–protein interactions in the membrane: Studies with model peptides. *Faraday Disc.* **111,** 127–136.

Mall, S., Broadbridge, R., Sharma, R. P., Lee, A. G., and East, J. M. (2000). Effects of aromatic residues at the ends of transmembrane alpha-helices on helix interactions with lipid bilayers. *Biochemistry* **39,** 2071–2078.

Marsh, D. (1995). Specificity of lipid–protein interactions. *In* "Biomembranes. Volume 1. General Principles" (A. G. Lee, ed.), pp. 137–186. JAI Press, Greenwich, CT.

Masibay, A. S., Balaji, P. V., Boeggeman, E. F., and Qasba, P. K. (1993). Mutational analysis of the Golgi retention signal of bovine β-1,4-galactosyl transferase. *J. Biol. Chem.* **268,** 9908–9916.

Michel, H., and Deisenhofer, J. (1990). The photosynthetic reaction center from the purple bacterium *Rhodopseudomonas viridis:* Aspects of membrane protein structure. *Curr. Top. Memb. Transp.* **36,** 53–69.

Mitsuoka, K., Hiral, T., Murata, K., Miyazawa, A., Kidera, A., Kimura, Y., and Fujiyoshi, Y. (1999). The structure of bacteriorhodopsin at 3.0 Å resolution based on electron crystallography: Implication of the charge distribution. *J. Mol. Biol.* **286,** 861–882.

Morein, S., Strandberg, E., Killian, J. A., Persson, S., Arvidson, G., Koeppe, R. E., and Lindblom, G. (1997). Influence of membrane-spanning alpha-helical peptides on the phase behavior of the dioleoylphosphatidylcholine/water system. *Biophys. J.* **73,** 3078–3088.

Morrow, M. R., Huschilt, J. C., and Davis, J. H. (1985). Simultaneous modeling of phase and calorimetric behavior in an amphiphilic peptide/phospholipid model membrane. *Biochemistry* **24,** 5396–5406.

Mouritsen, O. G., and Bloom, M. (1984). Mattress model of lipid–protein interactions in membranes. *Biophys. J.* **46,** 141–153.

Mouritsen, O. G., and Bloom, M. (1993). Models of lipid–protein interactions in membranes. *Annu. Rev. Biophys. Bioeng.* **22,** 145–171.

Munro, S. (1991). Sequences within and adjacent to the transmembrane segment of alpha-2,6-sialyltransferase specify Golgi retention. *EMBO J.* **10,** 3577–3588.

Munro, S. (1995). An investigation of the role of transmembrane domains in Golgi protein retention. *EMBO J.* **14,** 4695–4704.

Nezil, F. A., and Bloom, M. (1992). Combined influence of cholesterol and synthetic amphiphilic peptides upon bilayer thickness in model membranes. *Biophys. J.* **61,** 1176–1183.

Nielsen, C., Goulian, M., and Andersen, O. S. (1998). Energetics of inclusion-induced bilayer deformations. *Biophys. J.* **74,** 1966–1983.

Nilsson, T., and Warren, G. (1994). Retention and retrieval in the endoplasmic reticulum and the Golgi apparatus. *Curr. Opin. Cell Biol.* **6,** 517–521.

Nilsson, T., Lucocq, J. M., Mackay, D., and Warren, G. (1991). The membrane spanning domain of β-1,4-galactosyl transferase specifies *trans* Golgi localization. *EMBO J.* **10,** 3567–3575.

Pates, R. D., and Marsh, D. (1987). Lipid mobility and order in bovine rod outer segment disk membranes:. A spin-label study of lipid–protein interactions. *Biochemistry* **26,** 29–39.

Pedrazzini, E., Villa, A., and Borgese, N. (1996). A mutant cytochrome b$_5$ with a lengthened membrane anchor escapes from the endoplasmic reticulum and reaches the plasma membrane. *Proc. Natl. Acad. Sci. USA* **93,** 4207–4212.

Piknova, B., Perochon, E., and Tocanne, J. F. (1993). Hydrophobic mismatch and long-range protein/lipid interactions in bacteriorhodopsin/phosphatidylcholine vesicles. *Eur. J. Biochem.* **218,** 385–396.

Piknova, B., Marsh, D., and Thompson, T. E. (1997). Fluorescence quenching and electron spin resonance study of percolation in a two-phase lipid bilayer containing bacteriorhodopsin. *Biophys. J.* **72,** 2660–2668.

Rayner, J. C., and Pelham, H. R. B. (1997). Transmembrane domain-dependent sorting of proteins to the ER and plasma membrane in yeast. *EMBO J.* **16,** 1832–1841.

Rees, D. C., Chirino, A. J., Kim, K. H., and Komiya, H. (1994). Membrane protein structure and stability: Implications of the first crystallographic analyses. *In* "Membrane Protein Structure" (S. H. White, ed.), pp. 3–26. Oxford University Press, Oxford.

Ren, J. H., Lew, S., Wang, J. Y., and London, E. (1999). Control of the transmembrane orientation and interhelical interactions within membranes by hydrophobic helix length. *Biochemistry* **38**, 5905–5912.

Roseman, M. A. (1988). Hydrophobicity of the peptide C=O··H—N hydrogen bonded group. *J. Mol. Biol.* **201**, 621–623.

Roth, J. (1987). Subcellular organization of glycosylation in mammalian cells. *Biochim. Biophys. Acta* **906**, 405.

Roth, M., Lewit-Bentley, A., Michel, H., Deisenhofer, J., Huber, R., and Oesterhelt, D. (1989). Detergent structure in crystals of a bacterial photosynthetic reaction centre. *Nature* **340**, 659–662.

Roth, M., Arnoux, B., and Reiss-Husson, F. (1991). Structure of the detergent phase and protein–detergent interactions in crystals of the wild-type (Strain Y) *Rhodobacter sphaeroides* photochemical reaction center. *Biochemistry* **30**, 9403–9413.

Roux, B., and Woolf, T. B. (1996). Molecular dynamics of Pf1 coat protein in a phospholipid bilayer. *In* "Biological Membranes" (K. M. Merz and B. Roux, eds.), pp. 555–587. Birkhauser, Boston.

Roux, M., Neumann, J. M., Hodges, R. S., Devaux, P. F., and Bloom, M. (1989). Conformational changes of phospholipid headgroups induced by a cationic integral membrane peptide as seen by deuterium magnetic resonance. *Biochemistry* **28**, 2313–2321.

Schram, V., and Thompson, T. E. (1997). Influence of the intrinsic membrane protein bacteriorhodopsin on gel-phase domain topology in two-component phase-separated bilayers. *Biophys. J.* **72**, 2217–2225.

Shen, L. Y., Bassolino, D., and Stouch, T. (1997). Transmembrane helix structure, dynamics, and interactions: Multi-nanosecond molecular dynamics simulations. *Biophys. J.* **73**, 3–20.

Simmonds, A. C., East, J. M., Jones, O. T., Rooney, E. K., McWhirter, J., and Lee, A. G. (1982). Annular and non-annular binding sites on the (Ca^{2+} + Mg^{2+})-ATPase. *Biochim. Biophys. Acta* **693**, 398–406.

Simmonds, A. C., Rooney, E. K., and Lee, A. G. (1984). Interactions of cholesterol hemisuccinate with phospholipids and (Ca^{2+}–Mg^{2+})-ATPase. *Biochemistry* **23**, 1432–1441.

Simons, K., and Ikonen, E. (1997). Functional rafts in cell membranes. *Nature* **387**, 569–572.

Sivasubramanian, N., and Nayak, D. P. (1987). Mutational analysis of the signal-anchor domain of influenza virus neuraminidase. *Proc. Natl. Acad. Sci. USA* **84**, 1–5.

Sperotto, M. M., and Mouritsen, O. G. (1988). Dependence of lipid membrane phase transition temperature on the mismatch of protein and lipid hydrophobic thickness. *Eur. Biophys. J.* **16**, 1–10.

Sperotto, M. M., and Mouritsen, O. G. (1993). Lipid enrichment and selectivity of integral membrane proteins in two-component lipid bilayers. *Eur. Biophys. J.* **22**, 323–328.

Starling, A. P., Khan, Y. M., East, J. M., and Lee, A. G. (1994). Characterization of the single Ca^{2+} binding site on the Ca^{2+}-ATPase reconstituted with short and long chain phosphatidylcholines. *Biochem. J.* **304**, 569–575.

Subczynski, W. K., Lewis, R. N. A. H., McElhaney, R. N., Hodges, R. S., Hyde, J. S., and Kusumi, A. (1998). Molecular organization and dynamics of 1-palmitoyl-2-oleoylphosphatidyl-choline bilayers containing a transmembrane alpha-helical peptide. *Biochemistry* **37**, 3156–3164.

Swift, A. M., and Machamer, C. E. (1991). A Golgi retention signal in a membrane-spanning domain of coronavirus E1 protein. *J. Cell Biol.* **115**, 19–30.

Toyoshima, C., Nakasako, M., Nomura, H., and Ogawa, H. (2000). Crystal structure of the calcium pump of sarcoplasmic reticulum at 2.6 Å resolution. *Nature* **405**, 647–655.

Untracht, S. M., and Shipley, G. G. (1977). Molecular interactions between lecithin and sphingomyelin. *J. Biol. Chem.* **252**, 4449–4457.

von Heijne, G. (1996). Principles of membrane protein assembly and structure. *Prog. Biophys. Mol. Biol.* **66,** 113–139.

Wallin, E., Tsukihara, T., Yoshikawa, S., vonHeijne, G., and Elofsson, A. (1997). Architecture of helix bundle membrane proteins: An analysis of cytochrome *c* oxidase from bovine mitochondria. *Protein. Sci.* **6,** 808–815.

Webb, R. J., East, J. M., Sharma, R. P., and Lee, A. G. (1998). Hydrophobic mismatch and the incorporation of peptides into lipid bilayers: A possible mechanism for retention in the Golgi. *Biochemistry* **37,** 673–679.

Wimley, W. C., and White, S. H. (1996). Experimentally determined hydrophobicity scale for proteins at membrane interfaces. *Nat. Struct. Biol.* **3,** 842–848.

Woolf, T. B. (1998). Molecular dynamics simulations of individual alpha-helices of bacteriorhodopsin in dimyristoylphosphatidylcholine. II. Interaction energy analysis. *Biophys. J.* **74,** 115–131.

Yang, M., Ellenberg, J., Bonifacino, J. S., and Weissman, A. M. (1997). The transmembrane domain of a carboxyl-terminal anchored protein determines localization to the endoplasmic reticulum. *J. Biol. Chem.* **272,** 1970–1975.

Yau, W. M., Wimley, W. C., Gawrisch, K., and White, S. H. (1998). The preference of tryptophan for membrane interfaces. *Biochemistry* **37,** 14713–14718.

Zhang, Y. P., Lewis, R. N. A. H., Hodges, R. S., and McElhaney, R. N. (1992a). Interaction of a peptide model of a hydrophobic transmembrane α-helical segment of a membrane protein with phosphatidylcholine bilayers: Differential scanning calorimetric and FTIR spectroscopic studies. *Biochemistry* **31,** 11579–11588.

Zhang, Y. P., Lewis, R. N. A. H., Hodges, R. S., and McElhaney, R. N. (1992b). FTIR spectroscopic studies of the conformation and amide hydrogen exchange of a peptide model of the hydrophobic transmembrane α-helices of membrane proteins. *Biochemistry* **31,** 11572–11578.

Zhang, Y. P., Lewis, R. N. A. H., Hodges, R. S., and McElhaney, R. N. (1995). Peptide models of helical hydrophobic transmembrane segments of membrane proteins. 2. Differential scanning calorimetric and FTIR spectrooscopic studies of the interaction of Ac-K_2-(LA)$_{12}$-K_2-amide with phosphatidylcholine bilayers. *Biochemistry* **34,** 2362–2371.

CHAPTER 13

Lipidated Peptides as Tools for Understanding the Membrane Interactions of Lipid-Modified Proteins

John R. Silvius

Department of Biochemistry, McGill University, Montreal, Quebec, Canada H3G 1Y6

I. Introduction
II. Structures of Lipid Modifications of Proteins
III. Chemistry of Peptide Lipidation
IV. Thermodynamics of Association of Lipidated Peptides with Lipid Bilayers
V. Kinetics of Dissociation of Lipidated Sequences from Bilayers: Protein Targeting by Kinetic Trapping
VI. Partitioning of Lipidated Peptides into Liquid-Ordered Lipid Domains
VII. Conclusion
References

A number of intracellular and cell-surface proteins are modified with one or more lipidic residues, which in many cases play very important roles in the biological function of the protein. In many cases lipidation appears to be important for localization of a protein to the correct membrane or submembrane compartment, which in turn is essential to the proper biological functioning of the protein. Studies of synthetic lipidated peptides can provide useful information for understanding the interactions of lipidated proteins with membranes, the distribution of lipidated proteins between different membrane domains, and the effects of lipidation on protein conformation. This chapter summarizes insights that such studies have provided into biological questions such as the mechanisms of subcellular targeting of lipid-modified proteins and the physical/structural bases of their biological function.

I. INTRODUCTION

A number of intracellular and cell-surface proteins are now known to be co-valently linked to one or more lipidic groups. For many such proteins these lipid modifications have been shown to be of clear, and sometimes of critical, importance to the protein's cellular functions.

One of the more common biological roles of protein lipidation is to influence the subcellular distributions of proteins, for example, by promoting membrane association of a protein that otherwise would behave as a soluble species or by targeting a protein to a particular membrane compartment or submembrane domain. This chapter reviews the contributions that studies of lipid-modified peptides have made toward elucidating the physical and structural factors governing the membrane interactions, and through such interactions, the subcellular targeting, of lipidated proteins. As discussed previously (Epand, 1997), understanding these factors can in turn provide important insights into the functioning of many lipidated proteins, for which correct targeting is often critical for normal and/or pathological functioning.

II. STRUCTURES OF LIPID MODIFICATIONS OF PROTEINS

A variety of types of lipid modifications of proteins have been identified in both bacterial and eukaryotic systems (Fig. 1). Bacterial toxins from species such as *Bordetella pertussis* and pathogenic *Escherichia coli* have been shown to be N-acylated on lysine residues (Stanley *et al.,* 1994), and a number of proteins exposed on the surfaces of Gram-negative and Gram-positive bacteria are anchored via their amino-termini to *N*-palmitoyl-*S*-(diacylglyceryl)-cysteinyl residues (Mizushima, 1984; Nielsen and Lampen, 1982).

Five major classes of lipid modifications have been observed for eukaryotic cell proteins, and new types of modifications continue to be discovered (see, e.g., Kojima *et al.,* 1999). Diverse proteins of the cell surface are anchored via their carboxy-termini to glycosylphosphatidylinositol (GPI) residues or similar glycosylated inositol-lipid anchors (McConville and Ferguson, 1993). Much progress has been made in understanding both the synthesis and the subcellular routing of GPI-anchored proteins, although a variety of mechanistic issues remain to be addressed in both areas. A more recently discovered lipid modification of cell-surface proteins is the anchoring to cholesterol of several members of the Hedgehog family of morphogenetic factors (Porter *et al.,* 1996). Through an autocatalyzed cleavage reaction of the precursor form of these proteins, the amino-terminal domain becomes ester-linked to cholesterol via the newly generated carboxy-terminus. Interestingly, at least one member of the Hedgehog family is additionally modified with a palmitoyl group which is amide-linked to the amino-terminal cysteine residue (Pepinsky *et al.,* 1998).

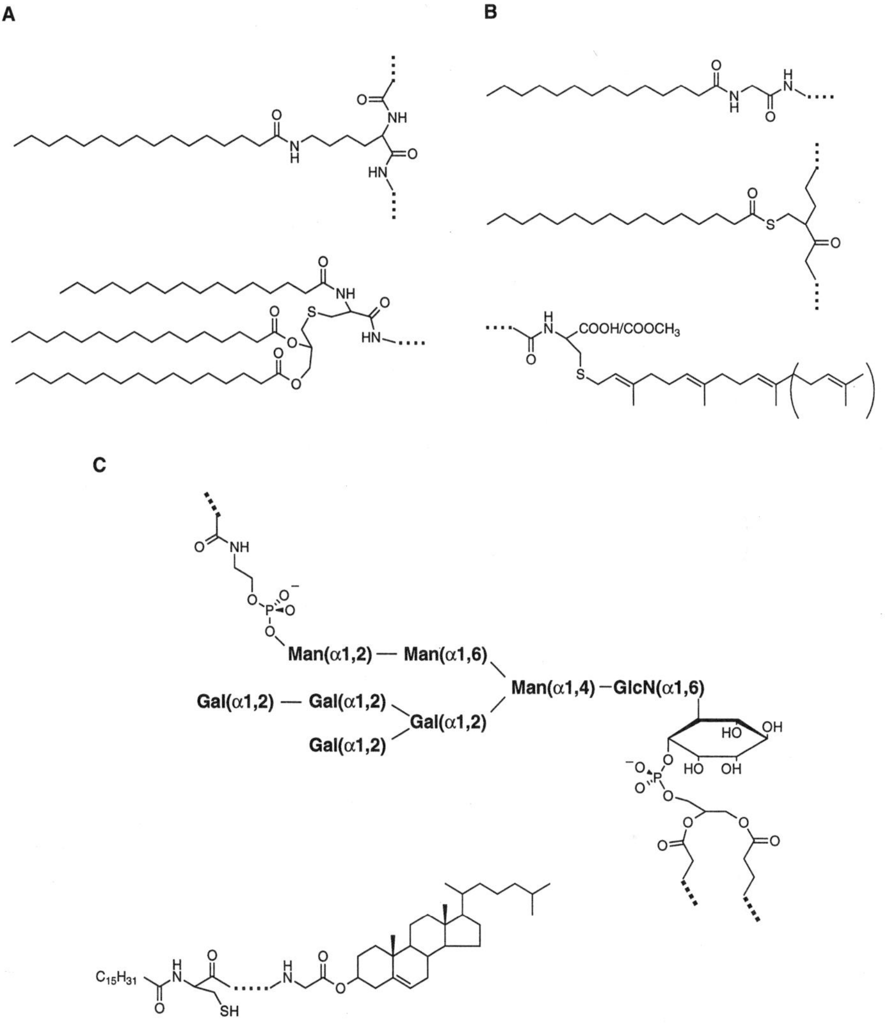

FIGURE 1 Structures of some common lipid modifications of proteins. (A) Two lipid modifications found in bacterial proteins: N-ε-acylation of a lysine residue (upper structure) and N-palmitoylation accompanied by (diacylglyceryl)-thioetherification of an amino-terminal cysteine residue (lower structure). (B) Lipid modifications of intracellular proteins of eukaryotic cells. Top to bottom: N-myristoylation of an amino-terminal glycine residue, S-acylation of an internal cysteine residue, and S-prenylation (with/without accompanying O-methylation) of a carboxy-terminal cysteine residue. Proteins of the Rab/Ypt family are often modified with a second S-prenyl group coupled to a cysteine residue at the (ω-1) or (ω-2) position. (C) Lipid modifications of cell-surface-associated proteins of eukaryotic cells: the glycosylphosphatidylinositol (GPI) "anchor" attached to the carboxy-terminus of the trypanosomal variant surface glycoprotein (*upper structure;* acyl chains are shown truncated) and carboxy-terminal cholesterol-esterification, accompanied by amino-terminal N-palmitoylation, of the morphogenetic factor Sonic hedgehog (*lower structure*). GPI-linked proteins exhibit considerable diversity in both the glycosyl residues and the acyl/alkyl chains of their "anchor" structures. In yeast, phosphoceramides replace the phosphatidylinositol moieties found in mammalian cells.

Intracellular proteins can be modified by amino-terminal myristoylation, carboxy-terminal prenylation, or S-acylation of cysteine residues; this last modification may occur in combination with myristoylation or prenylation. Cotranslational N-myristoylation occurs only on amino-terminal glycine residues; both the relevant protein N-myristoyltransferases and a well-defined consensus sequence for this modification have been identified (Johnson *et al.*, 1994a; Taniguchi, 1999). In most tissues N-myristoyl groups are incorporated with high specificity, but in retinal cells unsaturated 14-carbon chains as well as saturated lauroyl groups may also be incorporated (Johnson *et al.*, 1994b). A distinct N-terminal modification, an N-palmitoyl residue, has been found for the α subunit of the heterotrimeric G protein subunit α_s (Christiane Kleuss, personal communication).

Prenylation of eukaryotic proteins occurs on carboxy-terminal cysteine residues, whose carboxyl groups are often O-methylated as well. A variety of intracellular proteins, including many monomeric G proteins and heterotrimeric G protein γ subunits, are modified with farnesyl (C_{15}) or geranylgeranyl (C_{20}) groups. The precursor forms of these proteins terminate in $-CAAX$ sequences, which are sequentially processed by prenylation (by cytoplasmic farnesyltransferase or geranylgeranyltransferase I), proteolytic removal of the three C-terminal amino acid residues, and, in many cases, final O-methylation of the cysteine carboxyl group (Fu and Casey, 1999; Zhang and Casey, 1996). Monomeric G proteins of the Rab/Ypt family commonly terminate in $-CC$ or $-CXC$ sequences, both of whose cysteine residues are prenylated by geranylgeranyltransferase II acting in concert with a "Rab escort protein" (REP). Whereas almost all known prenylated proteins are synthesized as soluble precursors, prenylation of the prostacyclin receptor, an integral membrane protein, has also recently been reported (Hayes *et al.*, 1999).

S-Acylation of intracellular and integral membrane proteins is potentially reversible, a fact that may be important in the function of proteins such as the β_2-adrenergic and other G protein-coupled receptors and the heterotrimeric G proteins, such as G_s (Morello and Bouvier, 1996; Mumby *et al.*, 1994). The biochemistry of protein S-acylation is much less well elucidated than is that of the other lipid modifications discussed above. Evidence has been reported for enzymic S-acylating activities in eukaryotic cells and isolated cellular membranes (Berthiaume and Resh, 1995; Dunphy and Linder, 1998; Dunphy *et al.*, 1996; Veit *et al.*, 1999). Nonetheless, the protein S-acyltransferase responsible for modification of proteins like those noted above remains to be definitively identified, although both lysosomal and cytoplasmic acylprotein thioesterases have been cloned (Camp *et al.*, 1994; Duncan and Gilman, 1998; Soyombo and Hofmann, 1997). A further complicating factor is the finding that various proteins and even simple peptides can undergo nonenzymic S-acylation *in vitro* in the presence of long-chain acyl-CoAs, raising the possibility that physiological S-acylation of at least some cellular proteins could be nonenzymic (reviewed in Bano *et al.*, 1998;

Bizzozero, 1997; Dunphy and Linder, 1998). The physiological importance of nonenzymic S-acylation, however, remains unclear, in part because within eukaryotic cells the concentrations of long-chain acyl-CoAs are maintained at very low levels by the ubiquitous cytoplasmic acyl-CoA-binding protein (ACBP). Accordingly, in *in vitro* assays ACBP strongly inhibits the nonenzymic S-acylation of various peptides and proteins, but exerts much more modest effects on enzymically mediated S-acylation (Dunphy *et al.*, 2000; Leventis *et al.*, 1997).

III. CHEMISTRY OF PEPTIDE LIPIDATION

As discussed in this and other chapters in this volume, lipidated peptides have been employed to elucidate diverse aspects of the physical chemistry, biochemistry, and biology of lipidated proteins. Below I present a brief summary of the methods most commonly used to create peptides modified with lipidic residues of the types found in intracellular proteins (N- and S-acyl groups and isoprenyl residues). Straightforward modifications of existing methods also allow preparation of peptides ester-linked to cholesterol at their carboxy-termini (Wang *et al.*, 2001). Chemical approaches have not yet been described for synthesis of the complex glycolipid anchors of GPI-linked proteins. Diverse methods have been described to link peptides (and proteins) to simpler phospholipid derivatives, for example, for liposomal incorporation, but will not be discussed here. Methods have also been described to couple peptides via their amino termini to *N*-palmitoyl-*S*-(diacylglyceryl)-cysteinyl residues like those which anchor diverse hydrophilic proteins to bacterial membranes (Metzger *et al.*, 1991).

Of the physiological lipid modifications found in proteins, N-acyl groups are usually the simplest to incorporate into peptides. Acyl groups can be coupled to peptide amino-termini as a final step in solid-phase synthesis, using the same basic chemistries as are used to add protected amino acyl residues to the growing peptide chain. In principle, peptides bearing N-ε-acylated lysine residues could also be prepared by solid-phase methods, incorporating at the appropriate step(s) in the synthesis either an N-ε-acyllysine derivative or a lysine derivative whose ε-amino group could subsequently be selectively deprotected and N-ε-acylated. Acyl groups can be coupled to free amino residues of deprotected peptides, using, for example, succinimidyl esters, although selective modification of particular amino groups can be difficult to achieve.

The choice of procedures for incorporation of isoprenyl residues into peptides is constrained by the relative lability of these groups to both strongly reducing and strongly acidic conditions. Consequently, prenyl groups are normally incorporated into protected or deprotected peptides after cleavage of the latter from the support resin (Liu *et al.*, 1995; Naider and Becker, 1997). Simple cysteine derivatives can be readily prenylated using the appropriate prenyl bromide or chloride under mildly basic conditions. This method has also been used to acylate cysteinyl-peptides in which other potentially nucleophilic groups (amino or carboxyl) are blocked or absent (Liu *et al.,* 1995). However, these conditions can be of limited usefulness for prenylating many peptides, due to problems of solubility and/or regioselectivity. Such problems can often be overcome by using the prenyl halide in combination with the cysteinyl-peptide, employing as catalyst either KF · 2H$_2$O (Shahinian and Silvius, 1995; Silvius and l'Heureux, 1994; Xue *et al.,* 1990, 1991) or zinc(II) acetate under acidic conditions (Dawe *et al.,* 1997; Ghomashchi *et al.,* 1995; Xue *et al.,* 1992). The latter reaction conditions allow cysteine residues to be prenylated with high selectivity in the presence of unprotected amino or car-boxyl groups. Peptides (or proteins) terminating in suitable amino acid sequences −CX$_1$X$_2$X$_3$ can be enzymatically prenylated on the indicated cysteine residue using purified protein farnesyltransferase or geranylgeranyltransferase I (Thissen and Casey, 1993).

Strategies for peptide S-acylation must be chosen in the light of the relative lability of the thioester linkage and the relatively low regioselectivity of most acti-vated derivatives of carboxylic acids. Allyl-ester protection of carboxyl groups, and protection of amino groups with either allyoxycarbonyl or enzymically cleavable residues, allows preparation of protected S-acyl peptides, which can subsequently be deprotected without cleaving the thioester linkage (Cotte *et al.,* 1999; Nägele *et al.,* 1998; Schmittberger and Waldman, 1999). Selective thioacylation of pep-tides bearing unprotected amino as well as sulfhydryl groups has been reported using acyl chlorides in trifluoroacetic acid, although under these conditions hy-droxyl as well as thiol residues become acylated (Yousefi-Salakdeh *et al.,* 1999).

One of the most selective methods reported for chemical S-acylation of certain peptides (and proteins) involves the incubation of the peptide at near-neutral pH in a micellar codispersion with an acyl thioester. The success of this reaction requires that the peptide bind to suitably constructed micelles via hydrophobic and/or electrostatic interactions, such that the thiol residue(s) to be modified are exposed at the micelle surface. The usual S-acyl donor in such reactions has been the appropriate acyl-CoA (Duncan and Gilman, 1996; Quesnel and Silvius, 1994), although other, simpler thioesters may support much faster rates of peptide S-acylation in some bilayer or micellar systems (Dunphy *et al.,* 2000; J. Silvius and R. Leventis, unpublished data). At neutral pH, peptide hydroxyl and amino residues do not become detectably acylated using these acyl donors.

A particularly interesting recent development is the coupling of lipidated pep-tides to soluble proteins to obtain "neolipidated" protein species (Bader *et al.,* 2000). Although the reported semisynthesis of lipidated H-Ras by this method utilized a nonpeptide linker to join lipidated C-teminal peptides to the remainder of the Ras protein, in principle, intein-based chemistry could be used in an analo-gous manner to achieve a completely "native" protein structure (Iakovenko *et al.,* 2000).

IV. THERMODYNAMICS OF ASSOCIATION OF LIPIDATED PEPTIDES WITH LIPID BILAYERS

The subcellular targeting of a given lipidated (or other) protein has multiple facets, including its overall distribution between the aqueous phase and membranes (or other insoluble cellular structures), its association with particular membranes or other structures, and even potentially its distribution between different domains within a specific membrane. There is evidence that protein-coupled lipid groups can play a role in determining protein targeting at all these levels.

This and the following sections will discuss current knowledge of the ways in which lipidic groups of proteins contribute to their subcellular targeting, with particular emphasis on the contributions that studies of lipidated peptides have made to understanding the physical and structural bases of these phenomena. Such studies can be particularly informative for proteins whose targeting is primarily dictated by a small region of sequence including the lipidated amino acid residue(s), as is the case for species including the monomeric G proteins H-, N-, and K-Ras and the Src-homologous nonreceptor tyrosine kinases, such as p56[lck] and p59[fyn] (Choy *et al.*, 1999; McCabe and Berthiaume, 1999; Wolven *et al.*, 1997; Zlatkine *et al.*, 1997). Lipidated peptides have been most extensively studied with regard to their interactions with artificial (i.e., lipid) membranes. It must also be noted, however, that lipidic groups can participate in protein–protein as well as protein–lipid interactions. This fact has been most clearly established for interactions in the soluble phase (Gosser *et al.*, 1997; Hoffman *et al.*, 2000; Hogle *et al.*, 1985; Loew *et al.*, 1998; Mondal *et al.*, 2000), but may also be important in binding of certain lipidated proteins to membrane proteins (Sinensky, 2000). In the latter regard, knowledge of the nature of interactions between protein-bound (or peptide-bound) lipid groups and the membrane lipid bilayer can still provide important information on the "background level" of protein–membrane association against which other, specific interactions must be detected.

N-Acylated peptides, the first lipidated peptides whose interactions with lipids were characterized in detail, have provided very useful information for understanding the nature and the affinity of the interaction between myristoylated proteins and the membrane lipid bilayer. Nuclear magnetic resonance (NMR) studies of the myristoylated amino-terminal peptide of the protein kinase A catalytic subunit bound to lipid bicelles (Struppe *et al.*, 1998) have provided direct confirmation that the N-myristoyl group intercalates into the bilayer with its long axis parallel to the lipid acyl chains (see also Vergères *et al.*, 1995). Peitzsch and McLaughlin (1993) measured the partition coefficients for association of a series of N-terminally acylated glycines and short peptides with lipid vesicles, and concluded that a glycine-coupled N-myristoyl group associates with the lipid bilayer with a free energy of roughly -8 kcal mol^{-1}. This value agrees well with those obtained subsequently from measurements of the lipid-binding affinity of the myristoylated proteins transducin and hisactophilins I and II (Hanakam *et al.,* 1996; Seitz *et al.,* 1999).

The thermodynamic data just discussed lead to three important conclusions about the potential role of myristoylation in protein targeting to membranes. First, for a mammalian cell, where the concentration of membrane bilayer lipids exposed to the cytoplasm lies in the low-millimolar range, a protein will need to bind to the membrane bilayer with a free energy of partitioning of ca. -10 kcal mol^{-1} or more to ensure $>99\%$ binding to cellular membranes (Fig. 2). It is thus evident that a myristoylated protein will bind efficiently to cellular membranes only when the free energy of intercalation of the myristoyl group into the bilayer is augmented by that of other protein–membrane interactions. Second, however, it is clear that the free energy of these latter interactions may be rather modest, and that the membrane/soluble distribution of a myristoylated protein may be sensitively adjusted by modulating these additional interactions of the protein with the membrane. Finally, the free energy of intercalation of an N-myristoyl group into

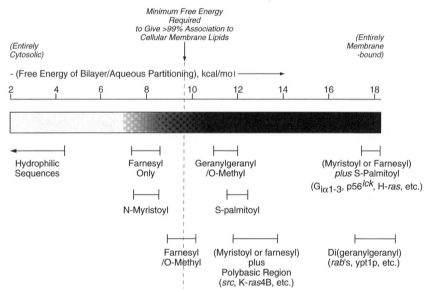

FIGURE 2   Free energies of association of different lipid-modified protein groups with lipid bilayers, as estimated from measurements of the affinities of association of lipid-modified peptides with lipid vesicles. The vertical dashed line represents the calculated minimum free energy of membrane binding required to give $>99\%$ membrane association of an intracellular protein if the concentration of membrane lipids exposed to the cytoplasm lies in the low millimolar range, as has been estimated for a typical mammalian cell. Shading schematically illustrates the predicted distribution of a protein between the aqueous and membrane phases under intracellular conditions (*white,* entirely aqueous; *black,* entirely membrane-bound) if the protein binds to membranes with the indicated free energy.

the membrane lipid bilayer is sufficiently large that a protein bearing a fully exposed N-myristoyl residue would be expected to be partially (though by no means completely) membrane-associated even if the protein showed no other favorable interaction with membrane components. In order to ensure that under a given condition a myristoylated protein is completely localized to the cytoplasm, it may thus be necessary to shield the myristoyl group at least partially from the aqueous phase through intra- or intermolecular interactions.

The conclusions just noted have been borne out experimentally in studies of various myristoylated proteins. As illustrated in Fig. 3, studies of diverse myristoylated proteins have revealed several types of interactions that can act synergistically with membrane intercalation of the myristoyl residue to promote efficient membrane binding. Moreover, these interactions often appear to be modulated physiologically

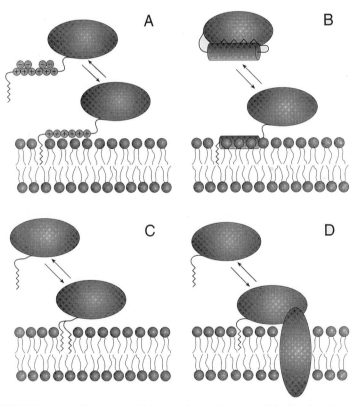

FIGURE 3 Four possible ways in which intercalation of a protein-linked myristoyl group into a membrane lipid bilayer may act synergistically with other protein–membrane interactions to promote efficient membrane binding. The models shown are discussed further in the text.

in order to regulate the intracellular distribution of the lipidated protein. Studies of lipidated peptides have contributed significantly toward elucidating the physical bases of these synergistic interactions and their regulation.

As illustrated in Fig. 3A (and as discussed in more detail by Murray *et al.* in Chapter 10), electrostatic interactions between clusters of basic amino acid residues and the negatively charged membrane interface serve to enhance the membrane-binding affinity of certain myristoylated proteins, such as myristoylated alanine-rich C kinase substrate (MARCKS) and $pp60^{src}$. Magnetic resonance methods have provided a fairly complete picture of the disposition at the lipid bilayer interface of peptides corresponding to the polybasic regions of these proteins (Victor and Cafiso, 1998; Victor *et al.*, 1999). The electrostatic interactions between the protein and anionic membrane lipids may be attenuated by phosphorylation within the polybasic region or by sequestration of this region by binding to another protein, as studies with model peptides have demonstrated (Arbuzova *et al.*, 1998; Kim *et al.*, 1994).

Interactions of hydrophobic amino acid residues with the membrane interface can provide a second means of enhancing the membrane-binding affinity of myristoylated proteins (Fig. 3B). A particularly instructive example of this phenomenon is seen with the monomeric GTP-binding protein Arf1. This protein possesses an amphiphilic α helix, adjacent to the myristoylated glycine residue, which interacts with the bilayer interface when the myristoyl group intercalates into the bilayer, augmenting the latter interaction to enhance protein–membrane binding (Antonny *et al.*, 1997; Losonczi *et al.*, 2000). Arf1 interconverts between a membrane-bound and a cytoplasmic form by a switching mechanism that is controlled by the binding and hydrolysis of GTP, both of which are also highly regulated processes (Chavrier and Goud, 1999). Interestingly, in the cytoplasmic form of the protein, the amino-terminal amphipathic helix appears to form part of a hydrophobic pocket within which the myristoyl group becomes intramolecularly sequestered (Goldberg, 1998). As noted earlier, such sequestration may be necessary to shift the membrane/cytoplasmic distribution of the protein entirely to the latter state. Intramolecular sequestration of the myristoyl chain has been observed in the soluble forms of certain other myristoylated proteins as well, such as retinal recoverin (Hughes *et al.*, 1995; Tanaka *et al.*, 1998).

A third potential mechanism for enhancing the membrane-binding affinity of a myristoylated protein is to couple additional lipidic groups to it, as illustrated in Fig. 3C (Dunphy and Linder, 1998; Resh, 1999). A variety of myristoylated proteins, including Src-homologous nonreceptor tyrosine kinases such as $p56^{lck}$ and $p59^{fyn}$ and heterotrimeric G protein α subunits of the $G_{i/o}$ family, undergo post-translational S-acylation, which, as expected, substantially increases the affinity of membrane association. However, S-acylation of such proteins can have additional physical consequences, which may play important further roles in protein function. First, as discussed below, conversion of a singly to a multiply lipidated protein (e.g., S-acylation of an already N-myristoylated protein) essentially abolishes the protein's ability to diffuse spontaneously between different membranes, such that certain myristoylated proteins may become "kinetically trapped" by S-acylation in particular cellular compartments. Second, modification of proteins with multiple saturated acyl chains appears to enhance their tendency to partition into "lipid raft" domains within membranes, a property that, as discussed below, can be very important to the proper functioning of these proteins in a cellular context (Brown and London, 1998; Simons and Ikonen, 1997).

A final, and obvious, manner in which the affinity of a myristoylated protein for membranes can be enhanced is by the direct interaction of the myristoylated species with an already membrane-associated protein (Fig. 3D). Specific protein–protein interactions can play crucially important roles in the membrane recruitment of various myristoylated proteins. No clear examples have been reported in which modulation of such protein–protein interactions directly modulates the membrane/cytoplasmic distribution of a myristoylated protein. However, it has been shown that the interaction of the farnesylated rhodopsin kinase (GRK-1) with rhodopsin-containing rod outer segment membranes is strongly enhanced by rhodopsin activation (Inglese *et al.,* 1992). Targeting of nascent myristoylated p56[lck] and of $G_{\alpha z}$ to specific cellular membranes has been shown to be facilitated by coexpression of CD4 and of $G_{\beta\gamma}$, respectively, although proper membrane targeting of p56[lck] can be achieved even in the absence of CD4 (Bijlmakers and Marsh, 1999; Bijlmakers *et al.,* 1997; Fishburn *et al.,* 1999).

The affinities of protein-coupled S-isoprenyl groups for partitioning into lipid bilayers have also been evaluated, using lipidated peptides based on the carboxy-terminal structures of various prenylated proteins (Ghomaschi *et al.,* 1995; Leventis and Silvius, 1998; Shahinian and Silvius, 1995; Silvius and l'Heureux, 1994). As indicated in Fig. 2, peptides bearing a farnesylated, but unmethylated carboxy-terminal cysteine residue associate with lipid bilayers with affinities which are similar to those observed for myristoylated peptides and which therefore are (marginally) too weak to ensure efficient membrane binding under intracellular conditions (Silvius and l'Heureux, 1994). However, methylation of the farnesylated cysteine enhances the affinity of prenylated peptides for neutral bilayers by up to 25-fold and for negatively charged bilayers by over 50-fold, such that under intracellular conditions a protein terminating in a farnesylated/O-methylated cysteine residue could be efficiently (though reversibly) anchored to cellular membranes purely through the intercalation of the prenyl group into the membrane bilayer. This prediction agrees with the observation that normally soluble proteins, such as the green fluorescent protein, when linked to a simple ($-$CAAX) farnesylation motif become predominantly (albeit reversibly) membrane-associated in intact mammalian cells (Choy *et al.,* 1999). The prediction from peptide-binding studies that O-methylation strongly enhances the membrane-binding affinity of a farnesylated cysteine residue agrees well with the observation that K-Ras4B binds much more weakly to membranes in its unmethylated [−C(farnesyl)−OH] form than in its mature, O-methylated form (Hancock *et al.,* 1991).

Lipid bilayer-partitioning measurements have revealed that geranylgeranylated peptides typically exhibit a 30- to 60-fold greater affinity for lipid bilayers than do analogous farnesylated peptides under the same conditions, and that methylation of the prenylated cysteine residue enhances bilayer partitioning by factors similar to those noted above for farnesylated peptides (Silvius and l'Heureux, 1994). As a result, it is predicted that a geranylgeranylated protein will bind efficiently to cellular membranes via hydrophobic interactions, independent of its methylation status, unless the geranylgeranyl group is shielded from the aqueous phase by intermolecular (or, potentially, intramolecular) interactions. This prediction agrees well with the observation that geranylgeranylated monomeric G proteins occur in the soluble fraction of mammalian cells only in the form of protein complexes, within which the prenyl groups are sequestered from the aqueous phase (Regazzi *et al.,* 1992). Sequestration of peptide- and protein-conjugated prenyl groups within hydrophobic pockets in proteins has been demonstrated by structural studies of Rho guanine nucleotide dissociation inhibitor (RhoGDI; Gosser *et al.,* 1997; Hoffman *et al.,* 2000) and is likely to occur also for complexes of prenylated proteins with RabGDI (Schalk *et al.,* 1996) and small G protein GDP dissociation stimulator (smgGDS; Kawamura *et al.,* 1991).

As discussed above for myristoylated proteins, prenylated proteins may bind to membranes through simultaneous interactions of both the prenyl group and other regions of the protein sequence with the target membrane. In K-Ras4B and several other monomeric G proteins, in proximity to the prenylated carboxy-terminus is found a region rich in basic amino acid residues, which markedly enhances the affinity of the protein carboxy-terminus for negatively charged lipid bilayers (Ghomaschi *et al.,* 1995; Hancock *et al.,* 1991; Leventis and Silvius, 1998). For yeast **a**-factor, a farnesylated dodecapeptide [YIIKGVFWDPAC(farnesyl)−OMe], hydrophobic amino acid side chains and the carboxy-terminal farnesyl group combine to confer a high affinity for lipid bilayers (Epand *et al.,* 1993). While **a**-factor is known to act by binding to a proteinaceous receptor on its target cell (Ste3p; Davis and Davey, 1997), the relatively strong binding of the lipidated peptide to lipid bilayers may enhance the initial encounter of secreted **a**-factor molecules with the surface of a target cell, following which peptide molecules may diffuse laterally to bind to their receptors.

Lipidated peptides have been used to study not only peptide/lipid bilayer interactions, but also the nature of the binding sites for lipidated protein sequences in biological membranes. Thissen and Casey (1993) showed that peptides incorporating specifically the −C(prenyl)AAX motif [but not a fully processed −C(prenyl)−OMe structure] bound specifically to a receptor on microsomal membranes; intriguingly, further studies showed that for partially processed K-Ras4B the responsible binding protein was tubulin (Thissen *et al.,* 1997). Siddiqui *et al.*

(1998) identified a putative receptor that binds specifically diverse prenylated proteins and peptides in plasma membranes of mammalian fibroblasts. A question that frequently arises concerning the interactions of lipidated proteins with membrane-bound effector, modulatory, and other proteins is whether during such interactions the lipidic group participates directly in the protein–protein interaction or instead simply aids to anchor the lipidated protein to the membrane via intercalation in the lipid bilayer. Studies of different functions of specific prenylated proteins, such as H- and K-Ras and the prenylated γ subunits of heterotrimeric G proteins, have provided evidence both for and against the possibility of specific interactions between protein-bound lipid groups and membrane-bound effector or other proteins (reviewed in Parish and Rando, 1996; Sinensky, 2000; see also Dudler and Gelb, 1997; Kisselev *et al.*, 1995; McGeady *et al.*, 1995). Lipidated peptides may prove particularly useful for addressing such questions, because of the potential to vary the structure of the lipidic group and thereby to determine this aspect of the structural specificity of the interaction (Kilic *et al.*, 1997; Kisselev *et al.*, 1995; Sinensky, 2000).

The interactions of a limited number of S-acylated peptides with lipid bilayers have also been characterized. S-Acylated hydrophilic peptides partition from the aqueous phase into lipid bilayers with significantly greater affinity than do terminally N-acylated peptides with similar sequences (Shahinian and Silvius, 1995), suggesting that a thioester linkage may permit deeper intercalation of the acyl chain into the lipid bilayer than does an amide linkage to glycine. Palmitoylation of a peptide corresponding to the transmembrane sequence of the vesicular stomatitis virus G protein was shown to produce a modest change in conformation (slightly increased helicity) (Joseph and Nagaraj, 1995). An NMR study of O-palmitoylated gramicidin A in detergent micelles similarly found that acylation caused modest, localized perturbations in side-chain conformation and interactions (Koeppe *et al.*, 1995).

V. KINETICS OF DISSOCIATION OF LIPIDATED SEQUENCES FROM BILAYERS: PROTEIN TARGETING BY KINETIC TRAPPING

Lipidated peptides have been used to evaluate the kinetics as well as the thermodynamics of interaction of lipidated protein sequences with membrane lipids. The results of a series of such studies are summarized in Fig. 4. Lipidated peptides have proven particularly useful for these studies, because they can be easily and specifically fluorescence-labeled, permitting their diffusion between different lipid vesicles to be conveniently monitored in real time using resonance energy transfer-based approaches (Ghomashchi *et al.*, 1995; Shahinian and Silvius, 1995). Peptides modified with a single prenyl, N-acyl, or S-acyl group typically associate with and dissociate from membrane lipid bilayers with rapid half-times (seconds

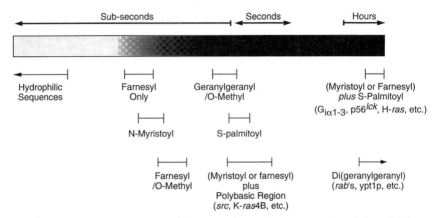

FIGURE 4 Schematic summary of the expected rates of spontaneous dissociation of different lipid-modified protein sequences from lipid bilayers, estimated from measurements of the kinetics of dissociation of lipid-modified peptides from lipid vesicles. As in Fig. 2, shading schematically illustrates the predicted aqueous/membrane distribution of a protein bearing the indicated lipidic groups (*white,* entirely aqueous; *black,* entirely membrane-bound) under intracellular conditions if association of the lipidic group with the lipid bilayer constitutes the predominant mode of protein/membrane interaction.

or less). This is generally true even for peptides bearing hydrophobic or polybasic amino acid sequences adjacent to the lipidated amino acid residue (Leventis and Silvius, 1998; Schroeder *et al.,* 1997), although the geranylated/polybasic carboxy-terminal peptide of the G25K small G protein has been reported to dissociate very slowly from negatively charged bilayers (Ghomashchi *et al.,* 1995).

In contrast to the behavior of singly lipidated peptides, peptides bearing dual lipid modifications of the types found in intracellular proteins (N/S-diacylation, digeranylgeranylation, S-acylation combined with prenylation, and di-S-acylation) show very slow dissociation from lipid bilayers, with half-times typically of the order of tens of hours (Shahinian and Silvius, 1995). Lipid-anchored macromolecules may dissociate from bilayers at moderately (severalfold) faster rates than would the isolated "anchor" structure alone (Shahinian and Silvius, 1995; Silvius and Zuckermann, 1993). One can nonetheless estimate that a multiply lipidated protein will dissociate from cellular membranes with extremely slow kinetics unless the process is accelerated by specific cellular factors. In principle, this physical behavior would permit multiply lipidated proteins to be targeted to particular membranes by processes based on "kinetic targeting." In such a process a protein, once incorporated into a particular membrane in a multiply lipidated form through the action of a specific cellular factor(s), will remain localized to that membrane (and/or to "downstream" membrane compartments, if transferred there via membrane traffic) until the same or other cellular factors subsequently promote the protein's release from the membrane.

The physical principles just described suggested that certain myristoylated or monoprenylated proteins that were also S-acylation substrates could be "kinetically targeted" to a particular cellular membrane by undergoing S-acylation specifically at that membrane (or, in principle, in a membrane compartment lying "upstream" along a vesicular membrane-trafficking pathway). Direct evidence to support this possibility was obtained through experiments in which cultured cells were incubated with fluorescent lipidated peptides that undergo intracellular S-acylation (Schroeder *et al.*, 1996, 1997; Waldmann *et al.*, 1997). These peptides were based on the lipidated termini of proteins such as H- and N-Ras, Src-family nonreceptor tyrosine kinases such as p56[lck] and p59[fyn], and the α subunits of heterotrimeric G proteins of the $G_{i/o}$ family. The strategy employed in these studies is illustrated in Fig. 5, along with structures of some representative fluorescent peptides employed.

Because uncharged monolipidated (myristoylated or prenylated) peptides can rapidly diffuse to, between, and across cellular membranes, the peptides can be readily incorporated into cultured cells from the external medium. By virtue of these same properties, the peptides can subsequently be removed efficiently from mammalian cells (by washing with serum albumin) so long as the peptides remain in their original monolipidated (i.e., non-S-acylated) forms. By contrast, after intracellular S-acylation the peptides become in essence irreversibly anchored to the membranes, where they become S-acylated (and/or in membranes to which they may be subsequently transferred by membrane traffic). As a result, cells that are incubated first with a fluorescent monolipidated peptide that is an S-acylation substrate, then with serum albumin, will show fluorescence labeling specifically in those compartments in which the S-acylated form of the peptide accumulates. Using this strategy and, in some cases, treatments to suppress intracellular membrane trafficking pathways, it was possible to demonstrate an S-acylating activity with broad specificity toward both myristoylated and prenylated peptides in the plasma membrane of intact mammalian cells and a second S-acylating activity with a much different (and possibly more restricted) specificity in the Golgi (Schroeder *et al.*, 1996).

Subsequent studies of the intracellular trajectory followed by proteins such as p59[fyn] (van't Hof and Resh, 1997) and the α subunit of G_z (Fishburn *et al.*, 1999) have suggested that, consistent with the results discussed above, after synthesis and contranslational myristoylation the nascent proteins diffuse to the plasma membrane, where they become S-acylated and thereby "trapped" at this locus. Other proteins, including p56[lck] (in CD4-expressing cells) and H- and N-Ras, after synthesis appear to undergo initial S-acylation in the Golgi (or endoplasmic reticulum) followed by vesicular trafficking to the plasma membrane (Bijlmakers

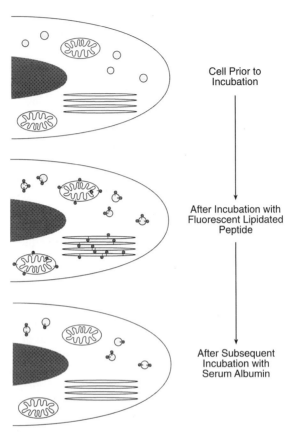

FIGURE 5   Schematic illustration of the general approach used to identify subcellular loci of S-acylation of fluorescent singly lipidated peptides. In the monoacylated or monoprenylated form administered to the cells (shown as shaded circles with a single "tail"), the fluorescent lipopeptides readily transfer across and between membranes, thereby gaining access to the various membrane compartments in the cell. In one or more of these compartments the lipopeptide substrate may acquire a second lipid chain through S-acylation. In this case, after subsequent treatment with serum albumin to remove the singly lipidated form of the fluorescent peptide, membranes in which peptide S-acylation occurs (or potentially, membranes to which the S-acylated peptide is transferred by membrane traffic) will incorporate the extraction-resistant S-acylated peptide. These membranes can then be selectively visualized by fluorescence microscopy.

and Marsh, 1999; Choy *et al.,* 1999). In all of these processes "kinetic trapping" of the protein by S-acylation appears to be an important element in the mechanism of protein targeting, although the detailed nature of the targeting process can vary depending on the participation of other factors, such as specific receptors for the nascent protein (Bijlmakers and Marsh, 1999; Fishburn *et al.,* 2000) or vesicular transport processes, as just noted.

VI. PARTITIONING OF LIPIDATED PEPTIDES INTO LIQUID-ORDERED LIPID DOMAINS

A novel recent application of lipidated peptides is to monitor the distribution of these species between liquid-ordered and liquid-disordered regions in mixed-lipid membranes containing cholesterol (Wang *et al.,* 2000, 2001). The plasma and certain other membranes of eukaryotic cells appear to incorporate special sphingolipid- and cholesterol-enriched domains known as "lipid rafts" (Simons and Ikonen, 1997). Considerable evidence suggests that the lipid component of these domains exists in a liquid-ordered state, which is distinct from the liquid-disordered state formed by typical unsaturated membrane phospholipids (Brown and London, 1998). Certain types of lipid modifications of proteins, notably acylation with multiple saturated acyl chains, appear to promote protein localization to rafts even when such motifs are transplanted to heterologous proteins (Brown and London, 1998; Resh, 1999; Zlatkine *et al.,* 1997). By contrast, other modifications (such as prenylation or S-acylation with polyunsaturated fatty acyl chains) have been reported to antagonize raft association (Brown and London, 1998; Melkonian *et al.,* 1999; Moffett *et al.,* 2000; Resh, 1999; Webb *et al.,* 2000).

To date, the conclusions just noted have been derived largely from measurements of the association of lipidated proteins with the low-density insoluble fraction obtained upon treating membranes with certain nonionic detergents at low temperatures (Brown and London, 1998). This procedure entails the risk that the distribution of proteins into rafts could be perturbed by either the presence of detergent or the unphysiologically low temperatures employed. Therefore, the availability of complementary assays of raft association based on other criteria would be desirable. One such approach, which can directly monitor the distribution of fluorescent-labeled lipidated peptides between liquid-ordered (raftlike) and liquid-disordered domains in lipid bilayers (vesicles), is illustrated schematically in Fig. 6. In this approach lipid vesicles are prepared by combining multiple

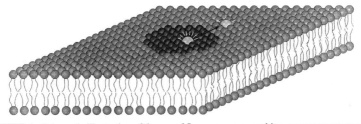

FIGURE 6 Schematic illustration of the use of fluorescence-quenching measurements to monitor the distribution of a fluorescent probe between coexisting liquid-ordered domains (*central darker region*) and liquid-disordered regions (*surrounding lighter area*) in lipid bilayers. The bilayers contain a fluorescence quencher that is enriched in the liquid-disordered domains. As a result, the probe emits stronger fluorescence when associated with liquid-ordered than with liquid-disordered domains.

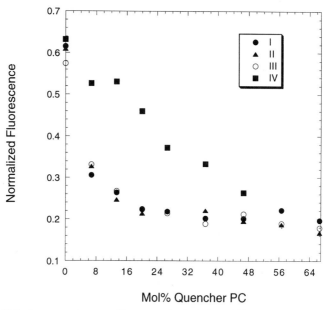

I R_1 = farnesyl, R_2 = palmitoyl
II R_1 = geranylgeranyl, R_2 = palmitoyl
III R_1 = farnesyl, R_2 = myristoleoyl
IV R_1 = hexadecyl, R_2 = stearoyl

FIGURE 7 Representative quenching curves for the indicated bimane-labeled lipidated peptides bound to bilayers (lipid vesicles) composed of mixtures of dipalmitoylphosphatidylcholine (DPPC), the quencher species 1-palmitoyl-2-(12-doxylstearoyl) phosphatidylcholine (12-doxyl PC), and cholesterol. To generate the data shown, the normalized fluorescence of the indicated peptides was measured as a function of the relative proportions of the two phospholipid components while keeping the proportion of cholesterol constant at 33 mol%. In this system the DPPC and 12-doxyl PC components become preferentially enriched in liquid-ordered and liquid-disordered domains, respectively. The comparatively efficient quenching of the fluorescence of species I–III thus indicates the comparatively high affinity of these species for the (12-doxyl PC-enriched) liquid-disordered phase. By contrast, the less efficient quenching of the fluorescence of species IV indicates that it shows a relatively greater affinity for the (DPPC-enriched) liquid-ordered domains.

phospho- or sphingolipid components, one of which is a quencher of peptide fluorescence, with physiological proportions of cholesterol. The lipids are chosen such that the vesicle bilayers are made up of a mixture of liquid-ordered (l_o) and liquid-disordered (l_d) domains, with the spin-labeled species accumulating preferentially in the latter. In such a system the distribution of a bilayer-intercalated peptide between the l_o and l_d domains will be directly reflected in the efficiency with which the peptide's fluorescence is quenched by the spin-labeled species.

Representative results from quenching experiments like those just outlined are illustrated in Fig. 7 for a series of doubly lipidated peptides based on a sequence resembling that of the S-acylated/farnesylated carboxy-terminus of H-ras. The data are presented in the form of a quenching curve, where the normalized fluorescence intensity measured for each peptide is plotted as a function of the content of the spin-labeled quencher lipid 1-palmitoyl-2-(12-doxylstearoyl) phosphatidylcholine (12-doxyl PC) in bilayers prepared from dipalmitoylphosphatidylcholine, 12-doxyl PC, and 33 mol% cholesterol. In this system the spin-labeled PC is enriched in the l_d domains, and therefore the fluorescence of a given lipidated peptide will be more strongly quenched, the greater is the extent to which it partitions into the l_d phase. It can be seen from Fig. 7 that the peptides doubly modified with either a farnesyl or a geranylgeranyl group plus a palmitoyl residue show strong fluorescence quenching, which is not further enhanced if the saturated palmitoyl chain is replaced by an unsaturated myristoleoyl (14 : 1c) chain. Since a myristoleoyl chain itself confers substantially higher relative affinity for the l_d phase than does a saturated palmitoyl chain (Wang *et al.,* 2000), we can conclude that in this system the prenylated peptides partition essentially completely into the l_d phase regardless of whether they bear a saturated or an unsaturated S-acyl chain. By contrast, when the same peptide is doubly modified with an S-stearoyl chain and with a saturated hexadecyl group on the terminal cysteine residue, the resulting species shows significant partitioning into the l_o phase, as evidenced by the significantly reduced efficiency of fluorescence quenching in the range of bilayer compositions where the quencher-depleted l_o and the quencher-enriched l_d phases coexist (Fig. 7). The conclusion from these results that farnesyl and geranylgeranyl groups show a strong preference for l_d-phase over l_o-phase domains in membranes agrees with evidence derived from detergent-solubilization assays that prenylated proteins are depleted from membrane lipid rafts (Melkonian *et al.,* 1998; Moffet *et al.,* 2000).

VII. CONCLUSION

Lipidated proteins have provided a very useful tool for understanding the physical characteristics of the interactions between lipidated proteins and the membrane lipid bilayer. Understanding these properties in turn has led to useful insights into the biology of lipidated proteins, particularly the basis of their subcellular and, in some cases, submembrane localization. Lipidated peptides have also provided useful insights into the ways in which lipid chains can participate directly in protein–protein interactions in the soluble phase. For many lipidated proteins that interact with membrane-bound effectors, regulatory proteins, etc., it remains to be established whether the lipidic group participates directly in such protein–protein interactions or simply intercalates into the lipid bilayer while purely peptidic residues from the two proteins interact directly. Such questions are often of more than purely fundamental interest, for example, in efforts to design drugs that can disrupt interactions between a given lipidated signaling protein and a membrane-bound effector. Important questions also remain to be addressed concerning the effects of lipid modifications on protein conformation and on the lateral distributions of proteins into different membrane domains, such as "lipid rafts." Synthetic lipidated peptides can contribute importantly in the investigation of such questions by virtue of the ease with which they can be prepared and manipulated, the ability to vary freely the structure of both the lipidic and the peptidic moieties, and the potential to couple diverse probe residues to the peptide in a straightforward manner. For these reasons lipidated peptides species should continue to constitute useful tools for elucidating the structure and physical bases of the biological functions of lipid-modified proteins.

Acknowledgments

I thank Dr. Maurine Linder (Washington University, St. Louis) for useful discussions. My laboratory is supported by operating grants from the Canadian Institutes of Health Research and les Fonds FCAR du Québec.

References

Antonny, B., Beraud-Dufour, S., Chardin, P., and Chabre, M. (1997). N-terminal hydrophobic residues of the G-protein ADP-ribosylation factor-1 insert into membrane phospholipids upon GDP to GTP exchange. *Biochemistry* **36,** 4675–4684.

Arbuzova, A., Murray, D., and McLaughlin, S. (1998). MARCKS, membranes, and calmodulin: Kinetics of their interaction. *Biochim. Biophys. Acta* **1376,** 369–379.

Bader, B., Kuhn, K., Owen, D. J., Waldmann, H., Wittinghofer, A., and Kuhlmann, J. (2000). Bioorganic synthesis of lipid-modified proteins for the study of signal transduction. *Nature* **403,** 223–226.

Bano, M. C., Jackson, C. S., and Magee, A. I. (1998). Pseudo-enzymatic S-acylation of a myristoylated *yes* protein tyrosine kinase peptide *in vitro* may reflect non-enzymatic S-acylation *in vivo*. *Biochem. J.* **330,** 723–731.

Berthiaume, L., and Resh, M. D. (1995). Biochemical characterization of a palmitoyl acyltransferase activity that palmitoylates myristoylated proteins. *J. Biol. Chem.* **270,** 22399–22405.

Bijlmakers, M. J., and Marsh, M. (1999). Trafficking of an acylated cytosolic protein: Newly synthesized p56[lck] travels to the plasma membrane via the exocytic pathway. *J. Cell Biol.* **145,** 457–468.

Bijlmakers, M. J., Isobe-Nakamura, M., Ruddock, L. J., and Marsh, M. (1997). Intrinsic signals in the unique domain target p56[lck] to the plasma membrane independently of CD4. *J. Cell Biol.* **137,** 1029–1040.

Bizzozero, O. A. (1997). The mechanism and functional roles of protein palmitoylation in the nervous system. *Neuropediatrics* **28,** 23–26.

Brown, D. A., and London, E. (1998). Functions of lipid rafts in biological membranes. *Annu. Rev. Cell Dev. Biol.* **14,** 111–136.

Camp, L. A., Verkruyse, L. A., Afendis, S. J., Slaughter, C. A., and Hofmann, S. L. (1994). Molecular cloning and expression of palmitoyl-protein thioesterase. *J. Biol. Chem.* **269,** 23212–23219.

Chavrier, P., and Goud, B. (1999). The role of ARF and Rab GTPases in membrane transport. *Curr. Opin. Cell Biol.* **11,** 466–475.

Choy, E., Chiu, V. K., Silletti, J., Feoktistov, M., Morimoto, T., Michaelson, D., Ivanov, I. E., and Philips, M. R. (1999). Endomembrane trafficking of ras: The CAAX motif targets proteins to the ER and Golgi. *Cell* **98,** 69–80.

Cotte, A., Bader, B., Kuhlmann, J., and Waldmann, H. (1999). Synthesis of the N-terminal lipohexapeptide of human $G_{\alpha o}$-protein and fluorescent-labeled analogues for biological studies. *Chem. Eur. J.* **5,** 922–936.

Davis, K., and Davey, J. (1997). G-protein-coupled receptors for peptide hormones in yeast. *Biochem. Soc. Trans.* **25,** 1015–1021.

Dawe, A. L., Becker, J. M., Jiang, Y., Naider, F., Eummer, J. T., Mu, Y. Q., and Gibbs, R. A. (1997). Novel modifications to the farnesyl moiety of the **a**-factor lipopeptide pheromone from *Saccharomyces cerevisiae:* A role for isoprene modifications in ligand presentation. *Biochemistry* **36,** 12036–12044.

Dudler, T., and Gelb, M. H. (1997). Replacement of the H-Ras farnesyl group by lipid analogues: Implications for downstream processing and effector activation in *Xenopus* oocytes. *Biochemistry* **36,** 12434–12441.

Duncan, J. A., and Gilman, A. G. (1996). Autoacylation of G protein alpha subunits. *J. Biol. Chem.* **271,** 23594–23600.

Duncan, J. A., and Gilman, A. G. (1998). A cytoplasmic acyl-protein thioesterase that removes palmitate from G protein alpha subunits and p21[ras]. *J. Biol. Chem.* **273,** 15830–15837.

Dunphy, J. T., and Linder, M. E. (1998). Signalling functions of protein palmitoylation. *Biochim. Biophys. Acta* **1436,** 245–261.

Dunphy, J. T., Greentree, W. K., Manahan, C. L., and Linder, M. E. (1996). G-protein palmitoyltransferase activity is enriched in plasma membranes. *J. Biol. Chem.* **271,** 7154–7159.

Dunphy, J. T., Schroeder, H., Leventis, R., Greentree, W. K., Knudsen, J. K., Silvius, J. R., and Linder, M. (2000). Differential effects of acyl-CoA binding protein on enzymatic and nonenzymatic thioacylation of protein and peptide substrates. *Biochim. Biophys. Acta* **1485,** 185–198.

Epand, R. M. (1997). Biophysical studies of lipopeptide–membrane interactions. *Biopolymers* **43,** 15–24.

Epand, R. F., Xue, C. B., Wang, S. H., Naider, F., Becker, J. M., and Epand, R. M. (1993). Role of prenylation in the interaction of the **a**-factor mating pheromone with phospholipid bilayers. *Biochemistry* **32,** 8368–8373.

Fishburn, C. S., Herzmark, P., Morales, J., and Bourne, H. R. (1999). $G_{\beta\gamma}$ and palmitate target newly synthesized $G_{\alpha z}$ to the plasma membrane. *J. Biol. Chem.* **274,** 18793–18800.

Fishburn, C. S., Pollitt, S. K., and Bourne, H. R. (2000). Localization of a peripheral membrane protein: $G_{\beta\gamma}$ targets $G_{\alpha z}$. *Proc. Natl. Acad. Sci. USA* **97,** 1085–1090.

Fu, H. W., and Casey, P. J. (1999). Enzymology and biology of CaaX protein prenylation. *Rec. Prog. Hormone Res.* **54,** 315–342.

Ghomashchi, F., Zhang, X., Liu, L., and Gelb, M. H. (1995). Binding of prenylated and polybasic peptides to membranes: Affinities and intervesicle exchange. *Biochemistry* **34,** 11910–11918.

Goldberg, J. (1998). Structural basis for activation of ARF GTPase: Mechanisms of guanine nucleotide exchange and GTP-myristoyl switching. *Cell* **95,** 237–248.

Gosser, Y. Q., Nomanbhoy, T. K., Aghazadeh, B., Manor, D., Combs, C., Cerione, R. A., and Rosen, M. K. (1997). C-terminal binding domain of Rho GDP-dissociation inhibitor directs N-terminal inhibitory peptide to GTPases. *Nature* **387**, 814–819.

Hanakam, F., Gerisch, G., Lotz, S., Alt, T., and Seelig, A. (1996). Binding of hisactophilin I and II to lipid membranes is controlled by a pH-dependent myristoyl-histidine switch. *Biochemistry* **35**, 11036–11044.

Hancock, J. F., Cadwallader, K., and Marshall, C. J. (1991). Methylation and proteolysis are essential for efficient membrane binding of prenylated p21$^{K\text{-ras(B)}}$. *EMBO J.* **10**, 641–646.

Hayes, J. S., Lawler, O. A., Walsh, M. T., and Kinsella, B. T. (1999). The prostacyclin receptor is isoprenylated. Isoprenylation is required for efficient receptor-effector coupling. *J. Biol. Chem.* **274**, 23707–23718.

Hoffman, G. R., Nassar, N., and Cerione, R. A. (2000). Structure of the Rho family GTP-binding protein Cdc42 in complex with the multifunctional regulator RhoGDI. *Cell* **100**, 345–356.

Hogle, J. M., Chow, M., and Filman, D. J. (1985). Three-dimensional structure of poliovirus at 2.9 Å resolution. *Science* **229**, 1358–1365.

Hughes, R. E., Brzovic, P. S., Klevit, R. E., and Hurley, J. B. (1995). Calcium-dependent solvation of the myristoyl group of recoverin. *Biochemistry* **34**, 11410–11416.

Iakovenko, A., Rostkova, E., Merzlyak, E., Hillebrand, A. M., Thoma, N. H., Goody, R. S., and Alexandrov, K. (2000). Semi-synthetic Rab proteins as tools for studying intermolecular interactions. *FEBS Lett.* **468**, 155–158.

Inglese, J., Koch, W. J., Caron, M. G., and Lefkowitz, R. J. (1992). Isoprenylation in regulation of signal transduction by G-protein-coupled receptor kinases. *Nature* **359**, 147–150.

Johnson, D. R., Bhatnagar, R. S., Knoll, L. J., and Gordon, J. I. (1994a). Genetic and biochemical studies of protein N-myristoylation. *Annu. Rev. Biochem.* **63**, 869–914.

Johnson, R. S., Ohguro, H., Palczewski, K., Hurley, J. B., Walsh, K. A., and Neubert, T. A. (1994b). Heterogeneous N-acylation is a tissue- and species-specific posttranslational modification. *J. Biol. Chem.* **269**, 21067–21071.

Joseph, M., and Nagaraj, R. (1995). Conformations of peptides corresponding to fatty acylation sites in proteins. A circular dichroism study. *J. Biol. Chem.* **270**, 19439–19445.

Kawamura, S., Kaibuchi, K., Hiroyoshi, M., Hata, Y., and Takai, Y. (1991). Stoichiometric interaction of smg p21 with its GDP/GTP exchange protein and its novel action to regulate the translocation of smg p21 between membrane and cytoplasm. *Biochem. Biophys. Res. Commun.* **174**, 1095–1102.

Kilic, F., Dalton, M. B., Burrell, S. K., Mayer, J. P., Patterson, S. D., and Sinensky, M. (1997). *In vitro* assay and characterization of the farnesylation-dependent prelamin A endoprotease. *J. Biol. Chem.* **272**, 5298–5304.

Kim, J., Blackshear, P. J., Johnson, J. D., and McLaughlin, S. (1994). Phosphorylation reverses the membrane association of peptides that correspond to the basic domains of MARCKS and neuromodulin. *Biophys. J.* **67**, 227–237.

Kisselev, O., Ermolaeva, M., and Gautam, N. (1995). Efficient interaction with a receptor requires a specific type of prenyl group on the G protein gamma subunit. *J. Biol. Chem.* **270**, 25356–25358.

Kleuss, C., and Gilman, A. G. (1997). $G_{\alpha s}$ contains an unidentified covalent modification that increases its affinity for adenylyl cyclase. *Proc. Natl. Acad. Sci. USA* **94**, 6116–6120.

Koeppe, R. E., Killian, J. A., Vogt, T. C., de Kruijff, B., Taylor, M. J., Mattice, G. L., and Greathouse, D. V. (1995). Palmitoylation-induced conformational changes of specific side chains in the gramicidin transmembrane channel. *Biochemistry* **34**, 9299–9306.

Kojima, M., Hosoda, H., Date, Y., Nakazato, M., Matsuo, H., and Kangawa, K. (1999). Ghrelin is a growth-hormone-releasing acylated peptide from stomach. *Nature* **402**, 656–659.

Leventis, R., and Silvius, J. R. (1998). Lipid-binding characteristics of the polybasic carboxy-terminal sequence of K-Ras4B. *Biochemistry* **37**, 7640–7648.

Leventis, R., Juel, G., Knudsen, J. K., and Silvius, J. R. (1997). Acyl-CoA binding proteins inhibit the nonenzymic S-acylation of cysteinyl-containing peptide sequences by long-chain acyl-CoAs. *Biochemistry* **36,** 5546–5553.

Liu, L., Jang, G.-F., Farnsworth, C., Yokoyama, K., Glomset, J. A., and Gelb, M. H. (1995). Synthetic prenylated peptides: Studying prenyl protein-specific endoprotease and other aspects of protein prenylation. *Meth. Enzymol.* **250,** 189–206.

Loew, A., Ho, Y.-K., Blundell, T., and Bax, B. (1998). Phosducin induces a structural change in transducin $\beta\gamma$. *Structure* **6,** 1007–1019.

Losonczi, J. A., Tian, F., and Prestegard, J. H. (2000). Nuclear magnetic resonance studies of the N-terminal fragment of adenosine diphosphate ribosylation factor 1 in micelles and bicelles: Influence of N-myristoylation. *Biochemistry* **39,** 3804–3816.

McCabe, J. B., and Berthiaume, L. G. (1999). Functional roles for fatty acylated amino-terminal domains in subcellular localization. *Mol. Biol. Cell* **10,** 3771–3786.

McConville, M. J., and Ferguson, M. A. (1993). The structure, biosynthesis and function of glycosylated phosphatidylinositols in the parasitic protozoa and higher eukaryotes. *Biochem. J.* **294,** 305–324.

McGeady, P., Kuroda, S., Shimizu, K., Takai, Y., and Gelb, M. H. (1995). The farnesyl group of H-Ras facilitates the activation of a soluble upstream activator of mitogen-activated protein kinase. *J. Biol. Chem.* **270,** 26347–26351.

Melkonian, K. A., Ostermeyer, A. G., Chen, J. Z., Roth, M. G., and Brown, D. A. (1999). Role of lipid modifications in targeting proteins to detergent-resistant membrane rafts. Many raft proteins are acylated, while few are prenylated. *J. Biol. Chem.* **274,** 3910–3917.

Metzger, J., Wiesmuller, K. H., Schaude, R., Bessler, W. G., and Jung, G. (1991). Synthesis of novel immunologically active tripalmitoyl-S-glycerylcysteinyl lipopeptides as useful intermediates for immunogen preparations. *Int. J. Pept. Prot. Res.* **37,** 46–57.

Mizushima, S. (1984). Post-translational modification and processing of outer membrane prolipoproteins in *Escherichia coli. Mol. Cell. Biochem.* **60,** 5–15.

Moffett, S., Brown, D. A., and Linder, M. E. (2000). Lipid-dependent targeting of G proteins into rafts. *J. Biol. Chem.* **275,** 2191–2198.

Mondal, M. S., Wang, Z., Seeds, A. M., and Rando, R. R. (2000). The specific binding of small molecule isoprenoids to RhoGDP dissociation inhibitor (RhoGDI). *Biochemistry* **39,** 406–412.

Morello, J. P., and Bouvier, M. (1996). Palmitoylation: A post-translational modification that regulates signalling from G-protein coupled receptors. *Biochem. Cell Biol.* **74,** 449–457.

Mumby, S. M., Kleuss, C., and Gilman, A. G. (1994). Receptor regulation of G-protein palmitoylation. *Proc. Natl. Acad. Sci. USA* **91,** 2800–2804.

Nägele, E., Schelhaas, M., Kuder, N., and Waldmann, H. (1998). Chemoenzymatic synthesis of N-Ras lipopeptides. *J. Am. Chem. Soc.* **120,** 6889–6902.

Naider, F., and Becker, J. M. (1997). Synthesis of prenylated peptides and peptide esters. *Biopolymers* **43,** 3–14.

Nielsen, J. B., and Lampen, J. O. (1982). Glyceride-cysteine lipoproteins and secretion by Gram-positive bacteria. *J. Bacteriol.* **152,** 315–322.

Parish, C. A., and Rando, R. R. (1996). Isoprenylation/methylation of proteins enhances membrane association by a hydrophobic mechanism. *Biochemistry* **35,** 8473–8477.

Peitzsch, R. M., and McLaughlin, S. (1993). Binding of acylated peptides and fatty acids to phospholipid vesicles: Pertinence to myristoylated proteins. *Biochemistry* **32,** 10436–10443.

Pepinsky, R. B., Zeng, C., Wen, D., Rayhorn, P., Baker, D. P., Williams, K. P., Bixler, S. A., Ambrose, C. M., Garber, E. A., Miatkowski, K., Taylor, F. R., Wang, E. A., and Galdes, A. (1998). Identification of a palmitic acid-modified form of human Sonic hedgehog. *J. Biol. Chem.* **273,** 14037–14045.

Porter, J. A., Young, K. E., and Beachy, P. A. (1996). Cholesterol modification of Hedgehog signaling proteins in animal development. *Science* **274,** 255–259.

Quesnel, S., and Silvius, J. R. (1994). Cysteine-containing peptide sequences exhibit facile uncatalyzed transacylation and acyl-CoA-dependent acylation at the lipid bilayer interface. *Biochemistry* **33,** 13340–13348.

Regazzi, R., Kikuchi, A., Takai, Y., and Wollheim, C. B. (1992). The small GTP-binding proteins in the cytosol of insulin-secreting cells are complexed to GDP dissociation inhibitor proteins. *J. Biol. Chem.* **267,** 17512–17519.

Resh, M. D. (1999). Fatty acylation of proteins: New insights into membrane targeting of myristoylated and palmitoylated proteins. *Biochim. Biophys. Acta* **1451,** 1–16.

Schalk, I., Zeng, K., Wu, S. K., Stura, E. A., Matteson, J., Huang, M., Tandon, A., Wilson, I. A., and Balch, W. E. (1996). Structure and mutational analysis of Rab GDP-dissociation inhibitor. *Nature* **381,** 42–48.

Schmittberger, T., and Waldmann, H. (1999). Synthesis of the palmitoylated and prenylated C-terminal lipopeptides of the human R- and N-Ras proteins. *Bioorg. Med. Chem.* **7,** 749–762.

Schroeder, H., Leventis, R., Shahinian, S., Walton, P. A., and Silvius, J. R. (1996). Lipid-modified, cysteinyl-containing peptides of diverse structures are efficiently S-acylated at the plasma membrane of mammalian cells. *J. Cell Biol.* **134,** 647–660.

Schroeder, H., Leventis, R., Rex, S., Schelhaas, M., Nägele, E., Waldmann, H., and Silvius, J. R. (1997). S-Acylation and plasma membrane targeting of the farnesylated carboxyl-terminal peptide of N-Ras in mammalian fibroblasts. *Biochemistry* **36,** 13102–13109.

Seitz, H. R., Heck, M., Hofmann, K. P., Alt, T., Pellaud, J., and Seelig, A. (1999). Molecular determinants of the reversible membrane anchorage of the G-protein transducin. *Biochemistry* **38,** 7950–7960.

Shahinian, S., and Silvius, J. R. (1995). Doubly-lipid-modified protein sequence motifs exhibit long-lived anchorage to lipid bilayer membranes. *Biochemistry* **34,** 3813–3822.

Siddiqui, A. A., Garland, J. R., Dalton, M. B., and Sinensky, M. (1998). Evidence for a high affinity, saturable, prenylation-dependent p21^{Ha-ras} binding site in plasma membranes. *J. Biol. Chem.* **273,** 3712–3717.

Silvius, J. R., and l'Heureux, F. (1994). Fluorimetric evaluation of the affinities of isoprenylated peptides for lipid bilayers. *Biochemistry* **33,** 3014–3022.

Silvius, J. R., and Zuckermann, M. J. (1993). Interbilayer transfer of phospholipid-anchored macromolecules via monomer diffusion. *Biochemistry* **32,** 3153–3161.

Simons, K., and Ikonen, E. (1997). Functional rafts in cell membranes. *Nature* **387,** 569–572.

Sinensky, M. (2000). Recent advances in the study of prenylated proteins. *Biochim. Biophys. Acta* **1484,** 93–106.

Soyombo, A. A., and Hofmann, S. L. (1997). Molecular cloning and expression of palmitoyl-protein thioesterase 2 (PPT2), a homolog of lysosomal palmitoyl-protein thioesterase with a distinct substrate specificity. *J. Biol. Chem.* **272,** 27456–27463.

Stanley, P., Packman, L. C., Koronakis, V., and Hughes, C. (1994). Fatty acylation of two internal lysine residues required for the toxic activity of *Escherichia coli* hemolysin. *Science* **266,** 1992–1996.

Struppe, J., Komives, E. A., Taylor, S. S., and Vold, R. R. (1998). ^{2}H NMR studies of a myristoylated peptide in neutral and acidic phospholipid bicelles. *Biochemistry* **37,** 15523–15527.

Tanaka, T., Ames, J. B., Kainosho, M., Stryer, L., and Ikura, M. (1998). Differential isotope labeling strategy for determining the structure of myristoylated recoverin by NMR spectroscopy. *J. Biomol. NMR* **11,** 135–152.

Taniguchi, H. (1999). Protein myristoylation in protein–lipid and protein–protein interactions. *Biophys. Chem.* **82,** 129–137.

Thissen, J. A., and Casey, P. J. (1993). Microsomal membranes contain a high affinity binding site for prenylated peptides. *J. Biol. Chem.* **268,** 13780–13783.

Thissen, J. A., Gross, J. M., Subramanian, K., Meyer, T., and Casey, P. J. (1997). Prenylation-dependent association of Ki-Ras with microtubules. Evidence for a role in subcellular trafficking. *J. Biol. Chem.* **272**, 30362–30370.

van't Hof, W., and Resh, M. D. (1997). Rapid plasma membrane anchoring of newly synthesized p59^{fyn}: Selective requirement for NH$_2$-terminal myristoylation and palmitoylation at cysteine-3. *J. Cell Biol.* **136**, 1023–1035.

Veit, M., Sachs, K., Heckelmann, M., Maretzki, D., Hofmann, K. P., and Schmidt, M. F. (1999). Palmitoylation of rhodopsin with S-protein acyltransferase: Enzyme catalyzed reaction versus autocatalytic acylation. *Biochim. Biophys. Acta* **1394**, 90–98.

Vergères, G., Manenti, S., Weber, T., and Sturzinger, C. (1995). The myristoyl moiety of myristoylated alanine-rich C kinase substrate (MARCKS) and MARCKS-related protein is embedded in the membrane. *J. Biol. Chem.* **270**, 19879–19887.

Victor, K., and Cafiso, D. S. (1998). Structure and position of the N-terminal membrane-binding domain of pp60^{src} at the membrane interface. *Biochemistry* **37**, 3402–3410.

Victor, K., Jacob, J., and Cafiso, D. S. (1999). Interactions controlling the membrane binding of basic protein domains: Phenylalanine and the attachment of the myristoylated alanine-rich C-kinase substrate protein to interfaces. *Biochemistry* **38**, 12527–12536.

Waldmann, H., Schelhaas, M., Nägele, E., Kuhlmann, J., Wittinghofer, A., Schroeder, H., and Silvius, J. R. (1997). Chemoenzymatic synthesis of fluorescent N-ras lipopeptides and their use in membrane localization studies *in vivo. Ang. Chem. Int. Ed. Engl.* **36**, 2238–2241.

Wang, T., Leventis, R., and Silvius, J. R. (2000). Fluorescence-based evaluation of the partitioning of lipids and lipidated peptides into liquid-ordered lipid microdomains: A model for molecular partitioning into "lipid rafts." *Biophys. J.* **79**, 919–933.

Wang, T.-Y., Leventis, R., and Silvius, J. R. (2001). Partitioning of lipidated peptide sequences into liquid-ordered lipid domains in model and biological membranes. *Biochemistry* **40**, 13031–13040.

Webb, Y., Hermida-Matsumoto, L., and Resh, M. D. (2000). Inhibition of protein palmitoylation, raft localization, and T cell signaling by 2-bromopalmitate and polyunsaturated fatty acids. *J. Biol. Chem.* **275**, 261–270.

Wolven, A., Okamura, H., Rosenblatt, Y., and Resh, M. D. (1997). Palmitoylation of p59^{fyn} is reversible and sufficient for plasma membrane association. *Mol. Biol. Cell* **8**, 1159–1173.

Xue, C. B., Ewenson, A., Becker, J. M., and Naider, F. (1990). Solution phase synthesis of *Saccharomyces cerevisiae* **a**-mating factor and its analogs. *Int. J. Pept. Prot. Res.* **36**, 362–373.

Xue, C. B., Becker, J. M., and Naider, F. (1991). Synthesis of S-alkyl and C-terminal analogs of the *Saccharomyces cerevisiae* **a**-factor. Influence of temperature on the stability of Fmoc and OFm groups toward HF. *Int. J. Pept. Prot. Res.* **37**, 476–486.

Xue, C.B., Becker, J., and Naider, F. (1992). Efficient regioselective isoprenylation of peptides in acidic aqueous solution using zinc acetate as catalyst. *Tetrahedron Lett.* **33**, 1435–1438.

Yousefi-Salakdeh, E., Johansson, J., and Stromberg, R. (1999). A method for S- and O-palmitoylation of peptides: Synthesis of pulmonary surfactant protein-C models. *Biochem. J.* **343**, 3557–3562.

Zhang, F. L., and Casey, P. J. (1996). Protein prenylation: Molecular mechanisms and functional consequences. *Annu. Rev. Biochem.* **65**, 241–269.

Zlatkine, P., Mehul, B., and Magee, A. I. (1997). Retargeting of cytosolic proteins to the plasma membrane by the Lck protein tyrosine kinase dual acylation motif. *J. Cell Sci.* **110**, 673–679.

CHAPTER 14

Experimental and Computational Studies of the Interactions of Amphipathic Peptides with Lipid Surfaces

Jere P. Segrest,*'† Martin K. Jones,* Vinod K. Mishra,*
and G. M. Anantharamaiah*'†

*Department of Medicine and the Atherosclerosis Research Unit, and †Department of Biochemistry
and Molecular Genetics, UAB Medical Center, Birmingham, Alabama 35294

I. Introduction
II. Amphipathic α Helixes
   A. Background
   B. Programs for Computational Analysis of Amphipathic α Helixes
   C. Experimental and Computational Studies of Amphipathic α Helixes in
      Apolipoprotein A-I
   D. Experimental and Computational Studies of Model Amphipathic α-Helical Peptides
III. Amphipathic β Strands/Sheets
   A. Background
   B. Computational Analysis of Amphipathic β Strands
   C. Experimental Studies of Peptide Analogues of Amphipathic β Strands from
      Apolipoprotein B
   References

Amphipathic α helixes and amphipathic β sheets are important surface lipid-associating motifs in plasma lipoproteins and membrane proteins. Amphipathic α helixes provide readily reversible lipid association, act as peptide detergents, and have optimal interactions with phospholipid bilayers. Amphipathic β sheets, on the other hand, interact with lipids in an essentially irreversible manner, completely lack detergent properties, and, in plasma lipoproteins, are likely to interact more readily with phospholipid monolayer:neutral core lipid emulsion particles than with phospholipid bilayers. Peptide mimics of these two surface-associating motifs have proven useful in establishing both the physical chemical and the biological properties of these motifs as analogues for the intact surface-associating proteins of which they are an integral part. To further aid in our understanding of these motifs, computational methods for simulations of interactions of these motifs with lipid surfaces have been developed and are undergoing refinement.

I. INTRODUCTION

Proteins that insert into lipid surfaces are fundamental to biology and medicine. Lipid–water interfaces leave a signature in the amino acid sequence of those proteins evolutionarily adapted to interact with lipid. This laboratory been particularly interested in the properties of two of these "lipid signatures," the amphipathic α helix (Segrest *et al.*, 1974, 1994a) and the amphipathic β strand (Segrest *et al.*, 1994b; 1998). One approach we and others have used is synthesis and analysis of peptide analogues of these lipid-associating motifs. A complementary approach has been development and use of windows-based computer programs for simulations of interactions of these motifs with lipid surfaces.

In this chapter we survey the current state of knowledge of the amphipathic α helix and the amphipathic β strand. A major focus will be a discussion of the state of development of computer algorithms for study of these lipid-associating motifs. The amphipathic α helix and the amphipathic β strand can be membrane-spanning and/or surface-associating. Although our topic in this chapter is the surface-associating forms of these two motifs, as a point of reference, the membrane-spanning protein porin will also be discussed.

II. AMPHIPATHIC α HELIXES

A. Background

The amphipathic α helix, defined as an α helix with opposing polar and nonpolar faces oriented along its long axis (Segrest *et al.*, 1974, 1990), is a common secondary structural motif in biologically active peptides and proteins. In a review article from this laboratory (Segrest *et al.*, 1990), naturally occurring amphipathic α helixes were grouped into seven distinct classes: A, apolipoproteins; H, polypeptide hormones: L, lytic polypeptides; G, globular proteins; K, calmodulin-regulated protein kinases; C, coiled-coil proteins; and M, transmembrane proteins. These groupings were based upon a detailed analysis of physical chemical and structural properties using helical wheel projections. The primary determinants of class were found to be characteristics of the polar face: charge, charge density, charge distribution, and angle subtended.

Class A amphipathic α helixes represent the lipid-associating amphipathic α-helical domains of the exchangeable apolipoproteins (Segrest *et al.*, 1990). The exchangeable apolipoproteins are defined as those apolipoproteins soluble in aqueous solutions, and the structural motif, the amphipathic α helix, responsible for their lipid association has been extensively studied (Segrest *et al.*, 1974, 1992, 1994a). The most distinctive feature of the class A amphipathic α helix is a unique clustering of positively charged residues at the polar–nonpolar interface and negatively charged amino acid residues at the center of the polar face. We have suggested that the amphipathic basic residues, when associated with phospholipid, extend ("snorkel") toward the polar face of the helix to insert their charged moieties into the aqueous milieu (Fig. 1). The snorkel hypothesis is supported by the results of a number of experimental studies from our laboratory (Anantharamaiah *et al.*, 1985; Kanellis *et al.*, 1980; Mishra *et al.*, 1994; Palgunachari *et al.*, 1996; Segrest *et al.*, 1983, 1990, 1992, 1994a), as well as from others (Rozek *et al.*, 1995).

An additional class of amphipathic α helix, termed G*, was also found in the exchangeable apolipoproteins (Jones *et al.*, 1992; Segrest *et al.*, 1992). This class is distinguished by a random radial arrangement of positively and negatively charged residues. Class G* amphipathic α helixes differ from those of class G in having both a greater hydrophobic moment and a greater nonpolar-face hydrophobicity. The class G* amphipathic α helixes are postulated to prefer protein–protein interactions over protein–lipid interactions (Segrest *et al.*, 1992, 1994a). The best-characterized example of this class is the N-terminal four-helix-bundle globular domain of apolipoprotein E (Wilson *et al.*, 1991), which consists of four class G* amphipathic α helixes and, depending upon local conditions, folds as a four-helix-bundle protein or associates with lipid (Wilson *et al.*, 1991).

Class L amphipathic α helixes include venoms such as bombolitins and mastoparan from hymenoptera, which are hemolytic (Argiolas and Pisano, 1983); antibiotics, such as the magainins isolated from *Xenopus laevis* skin (Zasloff *et al.*, 1988); and seminalplasmin from semen (Sitaram and Nagaraj, 1990). As the name implies, these peptides disrupt artificial phospholipid bilayers, although magainin and seminalplasmin are not hemolytic. Unlike the apolipoproteins and peptide hormones (class H), each peptide of this class consists entirely of an amphipathic α helix.

Class L and class H amphipathic helixes have several similar properties. Both types have large mean hydrophobic moments and are highly positively charged. In addition, both have intermediate charge densities and have polar faces that subtend an average angle of $100°$ or less perpendicular to the long axis of the helix. There are three significant differences between the two classes: (i) Most strikingly, the peptide hormones have a mean Lys/Arg ratio of 0.7, whereas class L peptides have a mean Lys/Arg ratio of 46. (ii) The class H peptides have a higher hydrophobic moment, but a lower nonpolar-face hydrophobicity than the class L helixes. (iii) Class L peptides have a bimodal cluster of positively charged amino acid residues, whereas class H peptides have only a single cluster.

B. Programs for Computational Analysis of Amphipathic α Helixes

1. Algorithm for Calculating Lipid Affinity

The rationale for derivation of the algorithm for calculating the lipid affinity of amphipathic α helixes, Λα, is as follows: The snorkeling of basic residues allows for greater penetration of class A amphipathic helixes into the hydrophobic interior of phospholipid monolayers than would otherwise be possible; the greater is the angle of the snorkel wedge (Fig. 1), the greater is the lipid penetration (Palgunachari *et al.*, 1996). Using neutron diffraction, Jacobs and White (1989) measured the gradient that H_2O forms from the outside to the inside of a phospholipid monolayer. Because the free energy of the hydrophobic effect decreases with a decrease in concentration of H_2O, the deeper the penetration of an amphipathic helix into the interior of a phospholipid monolayer, the more effective is the hydrophobicity (i.e., the lower is the free energy) of its nonpolar face. Therefore, the overall lipid affinity of an amphipathic helix will partially depend upon its depth of lipid penetration (Palgunachari *et al.*, 1996).

Combining the water gradient determined by Jacobs and White with the free energy of transfer of an amino acid residue from H_2O to varying H_2O:organic solvent mixtures (see Palgunachari *et al.*, 1996), we derived a free energy gradient δ_t (Å), which is a function of depth of penetration in angstroms. The δ_t for each amino acid residue in a given amphipathic α helix is multiplied by the free energy ΔG of transfer from hydrocarbon to water in kcal/mol for that amino acid residue. The ΔG of transfer represents a modification of the Goldman–Engelman–Steitz (GES) scale of hydrophobicity (Engelman *et al.*, 1986) as described elsewhere (Mishra *et al.*, 1998; Palgunachari *et al.*, 1996).

Calculation of $\Lambda\alpha$ is based upon a three-dimensional helical cylinder analysis. The program considers each amphipathic helix to be a cylinder 14 Å in diameter with a pitch of *n* residues per turn, where *n* is equal to 3.6 for an ideal helix, but may be any number. The $\Lambda\alpha$ algorithm then determines the orientation of the amphipathic helix relative to the plane of the hydrated phospholipids that produces the maximum lipid affinity. Rules for determination of orientation are based upon three charged residues, with positively charged residues allowed to snorkel, determining the plane. Because the maximum value of $\Lambda\alpha$ does not always occur for a plane that passes through three charged residues, the plane is allowed to pivot through two charged residues, and even one, so long as the pivoting charged groups remain in the water; the plane orientation is adjusted at 1° increments to give multiple rotations and tilts.

The present $\Lambda\alpha$ algorithm provides the following output: (i) a calculated $\Lambda\alpha$, (ii) a calculated angle of tilt, (iii) a diagrammatic representation of the plane of

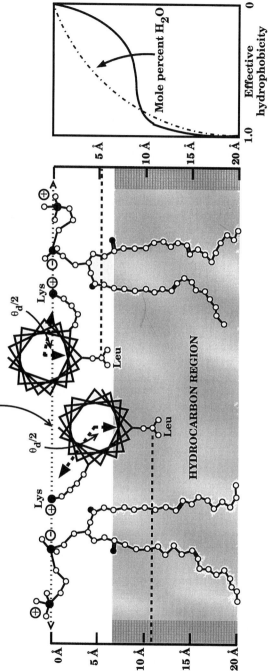

FIGURE 1 Schematic diagram of the theory behind the derivation of the $\Lambda\alpha$ algorithm. Left: Ball-and-stick molecular model (to scale) illustrating the snorkel principle. Dimensions are based on neutron scattering studies of Jacobs and White (1989). The two phospholipid molecules represent dioleoylphosphatidylcholine. Amphipathic helixes are represented in a helical wheel projection, showing a single Lys and Leu in each. Leu is at the center of the nonpolar face and Lys is shown in snorkel orientation for two different snorkel wedge angles (see Mishra $et\ al.$, 1994, for definition). The hydrocarbon region is assumed to start $\sim$7 Å from the plane of the lipid–water interface (Mishra $et\ al.$, 1994; Palgunachari $et\ al.$, 1996). Right: Plot of water gradient (Jacobs and White, 1989) and derived free energy gradient (Mishra $et\ al.$, 1994) δ_t (Å), which is a function of depth of penetration in angstroms. (See text for additional information).

the hydrated phospholipid relative to the N- and C-terminal ends of the amphipathic helix (displayed on the helical wheel output), (iv) a depth of penetration (in angstroms) for each residue, and (v) a depth-of-penetration contour map plotted on the helical net output.

2. Identification and Computational Analysis within a Given Amino Acid Sequence

The program αLOCATE identifies potential amphipathic α helixes within an amino acid sequence using termination rules described elsewhere (Segrest *et al.*, 1994a). To be able to identify the presence of a class A amphipathic helix in a given amino acid sequence, a mathematically defined motif for the particular charge distribution associated with various classes of amphipathic helixes was derived (Hazelrig *et al.*, 1993) and implemented in αLOCATE.

The program creates either a linear or a two-dimensional plot. The two-dimensional plot displays the location of each sequence with the desired motif on the y axis and a measure of $\Lambda\alpha$ on the x axis (see Fig. 3). Available options include minimum allowable cutoff values for amphipathic motif parameters, such as sequence length and $\Lambda\alpha$.

3. Identification and Computational Analysis within Databases

αLOCATEdb is programmed to search the Swiss Protein database, using a range of selection parameters. Additionally, once proteins possessing the amphipathic α-helical motif with the properties selected are identified, they can be sorted by number or density of amphipathic α helixes per protein (e.g., number per 100 residues) and/or species.

C. Experimental and Computational Studies of Amphipathic α Helixes in Apolipoprotein A-I

1. Background

Apolipoprotein (apo) A-I is a 243-residue protein containing a globular (possibly lipid-binding) N-terminal domain (residues 1–43) and a lipid-binding C-terminal domain (residues 44–243) (Segrest *et al.*, 1994a). On the basis of analyses of protein and gene sequences, it is known that there are eight 22-mer and two 11-mer tandem amino acid sequence repeats (called helixes 1–10), often punctuated by prolines, encoded in exon 4 of the apo A-I gene (residues 44–241); each helix has the periodicity of an amphipathic α helix (Segrest *et al.*, 1974, 1994a).

Apo A-I is an integral component of both spheroidal circulating high-density lipoprotein (HDL) particles and the geometrically simpler discoidal nascent HDL particles. HDL discs are small unilamellar bilayers surrounded by apo A-I monomers (Atkinson *et al.*, 1980; Wlodawer *et al.*, 1979). Two general discoidal models have been proposed for apo A-I on the disc rim: (i) Molecules of apo A-I form essentially continuous amphipathic α helixes parallel to the plane of the disc (the "belt" model) (Segrest, 1977); (ii) the 22-mer amphipathic α helical repeats of apo A-I form tandem antiparallel helixes perpendicular to the plane of the disc (the "picket-fence" model) (Tall *et al.*, 1977).

In an exciting development, a 4-Å-resolution, solution-phase X-ray structure for residues 44–243 of apo A-I was determined (Borhani *et al.*, 1997). Previous studies had shown that this deletion mutant retains the lipid-bound conformation of intact apo A-I in both the presence and the absence of lipid (Rogers *et al.*, 1997). The structure determined for this fragment of apo A-I was that of an almost continuous amphipathic α-helical nonplanar horseshoe-shaped object, a structure suggested by the authors to support the double-belt model for discoidal HDL (Borhani *et al.*, 1997). However, because this structure for apo A-I was determined from aqueous-phase crystals, its relevance to lipid-associated apo A-I was uncertain. Since a lipid environment has an overriding effect on apolipoprotein structure (Segrest *et al.*, 1992), a detailed understanding of the lipid-associated structure of apo A-I is crucial.

2. Computer Model

We recently proposed an atomic-resolution molecular model for the lipid-associated C-terminal domain of apo A-I (Segrest *et al.*, 1999b). The atomic details of this model follow *a priori* from the starting assumption that lipid imposes profound constraints on the conformation and orientation of lipid-associated proteins (Segrest *et al.*, 1992). In the resulting model, two apo A-I molecules are wrapped beltwise around a small discoidal patch of bilayer containing 160 lipid molecules (Fig. 2A). The C-terminal domain of each apo A-I monomer (residues 44–243) forms a curved, planar, amphipathic α-helical ring with 11/3 (approximately 3.67) residues per turn (termed an α11/3 helix) in which the hydrophobic surface faces inward toward the lipid disc (Figs. 2B and 2C).

The first step in the derivation of the double-belt model was the observation that tandem helixes 1–10 of apo A-I are evolutionarily adapted to form a continuous amphipathic α helix, suggesting a belt model. Figure 3A (center) is a continuous α-helical net display of apo A-I plotted, using the $\Lambda\alpha$ function for calculating lipid affinity, with the pitch of an idealized α helix, 3.6 (18/5) residues per turn. In this display, hydrophobic residues are represented by closed black circles and prolines by larger closed gray circles; it can be seen that the hydrophobic face of helixes 1–10 forms one complete turn of a continuous right-handed spiral (pitch = 22/3.6 = approximately 6.11 residues per 22-mer repeat). If closed into a circle, this hydrophobic face would twist around the resulting torus, rather than lying on the inside as required for the belt model.

Examination of alternative possibilities demonstrated that helixes 1–10 plotted with a pitch of 3 turns per 11 residues suggested by the 22-mer/11-mer tandem periodicity creates a 198-residue α helix, termed an α11/3 helix, with a straight (planar) hydrophobic face (Fig. 3B, center); this helix pitch is essentially indistinguishable

A

B

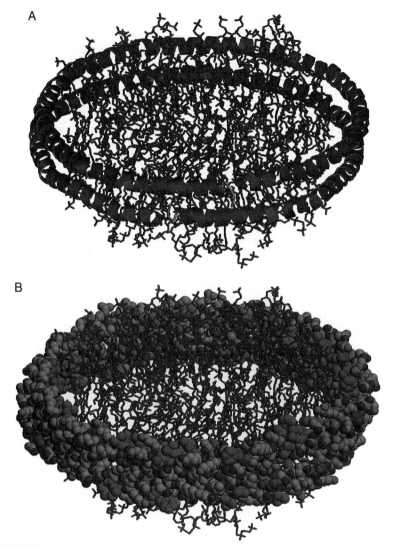

FIGURE 2 Three views of the detailed molecular model proposed for discoidal human apo A-I : phospholipid complexes. The molecular models (Segrest *et al.,* 1999b) were generated with Rasmol (Sayle and Milner-White, 1995). (A) Phospholipid bilayer disc displayed in molecular stick mode and protein displayed in cartoon mode. Color code: phospholipid, gold; helixes 1–10 (residues 44–243) of apo A-I, red; residues 33–43 (one of three G* amphipathic α-helical repeats at the N-terminal end), blue; prolines, green. (B) Lipid displayed in molecular stick mode and protein displayed in space-filling mode. Color code: phospholipid, gold; hydrophobic residues (L, F, M, V, W, Y, I), green; prolines, magenta; other residues, CPK. (C) Lipid and protein displayed in space-filling modes. Color code: phospholipid, gold; hydrophobic residues (L, F, M, V, W, Y, I), green; prolines, magenta; other residues, CPK. (See color plate.)

C

FIGURE 2   (*Continued*)

from an idealized α helix (Segrest *et al.*, 1999b). Helixes have been shown to curve away from environments with higher, and toward those with lower, dielectric constants. Thus, if associated with a lipid bilayer disc, a continuous amphipathic α helix like apo A-I would spontaneously wrap around the disc edge, hydrophobic face inward (Fig. 2B).

The general concept of a double-belt model for discoidal HDL has been confirmed independently by several laboratories (Borhani *et al.*, 1997; Klon *et al.*, 2000; Koppaka *et al.*, 1999; Li *et al.*, 2000; Maiorano and Davidson, 2000; Segrest *et al.*, 1999b, 2000). At issue now are molecular details, in particular whether the detailed model we developed (Segrest *et al.*, 1999b, 2000) is correct. One feature of our model is the presence of a number of perfectly aligned interhelical salt bridges between the two apo A-I monomers in the belt (Segrest *et al.*, 1999b). We are using site-directed mutagenesis to test this prediction.

3. Computational Analysis

A feature of our model that is particularly unusual is the proposed $\alpha 11/3$ helix conformation. An alternate model might invoke the existence of localized kinks, perhaps in the region of the prolines, in an otherwise continuous amphipathic α helix, kinks that might keep the hydrophobic face of the belt in the plane of the disc. In an effort to distinguish between these possibilities, we hypothesized that each 11-mer/22-mer repeat with an 11/3 α-helical pitch will have a more favorable $\Lambda\alpha$ when analyzed as an $\alpha 11/3$ helix than when analyzed as an idealized α helix.

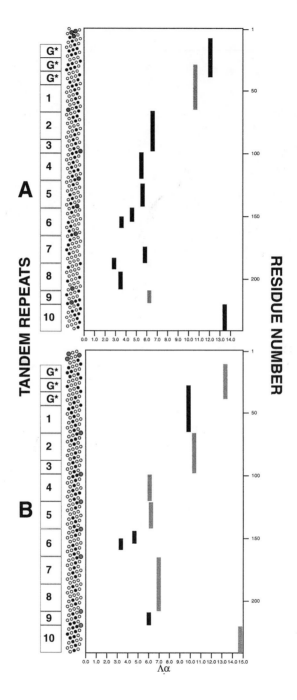

As a test, apo A-I was analyzed by the program αLOCATE using two different choices of helical pitch, 3.6 and 11/3 residues per turn.

Compare the αLOCATE analyses of apo A-I as a continuous idealized α helix (Fig. 3A, right) and as a continuous α11/3 helix (Fig. 3B, right). Amphipathic α-helical regions predicted to have the more favorable $\Lambda\alpha$ for a given conformation are highlighted in gray. Note that helixes 2–8 and 10 all have a more favorable $\Lambda\alpha$ when analyzed as α11/3 helixes, whereas helix 1 (joined with the adjacent 11-mer G* helix in exon 3) and helix 9 have the more favorable $\Lambda\alpha$ when analyzed as idealized α helixes. The current αLOCATE algorithm fails to identify helix 6 as a continuous helix, but, when analyzed separately, this helix also has the more favorable $\Lambda\alpha$ when analyzed as an α11/3 helix.

As we noted in our proposal of an atomic-resolution molecular model of discoidal HDL (Segrest *et al.*, 1999b), striking 11-mer/22-mer periodicity in perfectly aligned interhelical salt bridges was found encoded into the 11-mer tandem sequence repeats of apo A-I. We suggested that this periodicity reflected the geometric relationship of an α11/3 helix to the conformation of apo A-I on the edge of discoidal HDL. In our model of the 10 tandem helical repeats, only helixes 1 and 9 do not form interhelical salt bridges (Segrest *et al.*, 1999b). Consideration of our model in light of the results of Fig. 3 suggests that helixes 1 and 9 are the only helical repeats of the 10 encoded by exon 4 whose function does not require the α11/3 helical conformation. All in all, the results of Fig. 3 are suggestive of the presence of the α11/3 helical conformation in most portions of apo A-I monomers associated with phospholipid discs.

4. Peptide Analogues

Another approach to delineating the overall lipid-associating properties of apo A-I is to study the lipid-associating properties of individual and overlapping tandem amphipathic helical domains of this apolipoprotein. To achieve this goal, we studied an extensive set of synthetic peptide analogues of apo A-I (Mishra *et al.*, 1998; Palgunachari *et al.*, 1996). Figure 4 is a diagrammatic summary of these studies. The lipid affinity (ΔG) of each peptide studied is represented by the apex of a triangle, the base of which represents the location of that peptide in the sequence of apo A-I. The length of individual peptide analogues is coded by degree of shading: the shorter the peptide, the darker the shading.

Figure 4 illustrates clearly that lipid affinity is greatest at the N- and C-terminal ends of the sequence encoded by exon 4, residues 44–241, results supported by

---

FIGURE 3 Helical net and αLOCATE analyses of human apo A-I. (A) Analysis as an idealized α helix. (B) Analysis as an α11/3 helix. The y axes represent residue number. Left: Location of tandem helical repeats. Center: Helical net diagram with hydrophobic residues in black and prolines in gray. Right: Two-dimensional αLOCATE analysis with $\Lambda\alpha$ plotted on the x axis. The sequence position and $\Lambda\alpha$ of each selected amphipathic α helix is indicated by a black bar. Gray bars indicate the helix with the higher $\Lambda\alpha$ compared to its counterpart in the other frame.

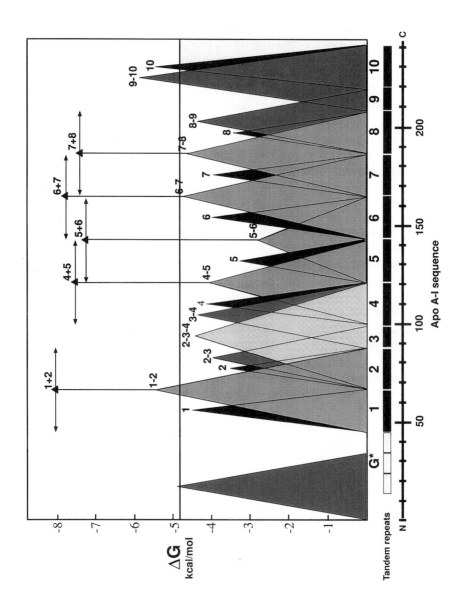

site-directed mutagenesis studies from other laboratories (Burgess *et al.,* 1999; Laccotripe *et al.,* 1997; Sorci-Thomas *et al.,* 1993). In particular, Fig. 4 shows a U-shaped affinity curve similar to the lipid affinities calculated in Fig. 3. Further, if individual peptide analogues form idealized α helixes when associated with the surface of phospholipid vesicles, as opposed to a phospholipid disc edge, then Fig. 3A should more accurately predict lipid affinities than Fig. 3B. Consistent with this assumption, from Fig. 4, the 33-mer representing helixes 2 and 3 has a lower measured lipid affinity than helix 1, a result predicted by Fig. 3A, but not Fig. 3B.

Figure 5 shows $\Lambda\alpha$ analyses of helixes 5 and 10 displayed as helical wheel and helical net diagrams. Helixes 5 and 10 are shown as idealized and 11/3 α-helical nets in Figs. 5A and 5B, respectively. The calculated center of the hydrophobic face for each helix is centered in the net. The contour lines on the nets indicate the depth of penetration for each residue of the helix. The thin contour line is the calculated membrane surface and the thicker line is the 5-Å contour line; not shown in either of the helical net representations is a third, thicker, 10-Å contour line. The distinguishing difference between these two $\Lambda\alpha$ computations is that helix 10 is inserted into the membrane parallel to its surface, without the tilt associated with helix 5. Helix 5 is shown as an idealized and an 11/3 α helical wheel in Figs. 5C and 5D, respectively. The hydrophobic face for each wheel is up. The lines in front and back of the wheels mark the plane of the membrane relative to the helix; the idealized α helix is tilted more than the α11/3 helix and for both, the N-terminal end is inserted less deeply than the C-terminal end.

D. Experimental and Computational Studies of Model Amphipathic α-Helical Peptides

1. Class A Amphipathic α-Helical Peptides

a. Physical Chemical Studies. We have designed and synthesized a series of model class A amphipathic α-helical peptides, whose lipid-associating properties have been extensively studied. The peptide 18A, DWLKAFYDKVAEKLKEAF, its N- and C-terminally blocked version, Ac-18A-NH$_2$, and their analogues have been shown to act as peptide detergents, interacting with phospholipid to convert it

---

FIGURE 4 Plot of the experimentally derived free energy of lipid association for synthetic peptide homologues of single and multiple tandem repeating units of human apo A-I. The *x* axis, residue position and position of tandem helical repeats; the *y* axis, experimentally measured free energy of lipid association for each peptide (Mishra and Palgunachari, 1996; Palgunachari *et al.,* 1996). Each peptide is represented by a triangle; the apex plots the measured free energy of lipid association and the base indicates the peptide position in the apo A-I amino acid sequence. The darker is the shading of the triangle, the smaller is the peptide. Peptide lengths are 22, 33, 44, and 55 residues. The small, closed triangles bisected by horizontal double-headed arrows indicate the expected free energy of lipid association for 44-mers if the free energies of individual 22-mers sum in a linear fashion. The horizontal line at approximately 5 kcal/mol is the experimentally determined free energy of lipid association of native apo A-I.

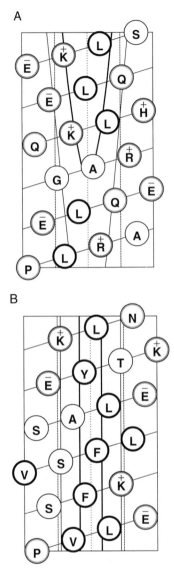

FIGURE 5 Helical wheel and helical net $\Lambda\alpha$ diagrams of helixes 5 and 10 from human apo A-I displayed as idealized α and α 11/3 helixes. (A, B) $\Lambda\alpha$ analyses of helixes 5 and 10 displayed as idealized and 11/3 α helical nets, respectively. The calculated center of the hydrophobic face for each helix is positioned in the center of each net. The contour lines on the nets indicate the depth of membrane penetration for each residue of the helix. The thin contour line is the calculated membrane surface and the thicker line is the 5-Å contour line. (C, D) $\Lambda\alpha$ analyses of helix 5 displayed as idealized and 11/3 α helical wheels, respectively. The hydrophobic face for each wheel is up. The lines in front (thicker gray) and back (thinner black) of the wheels mark the tilt of the plane of the membrane relative to the helix.

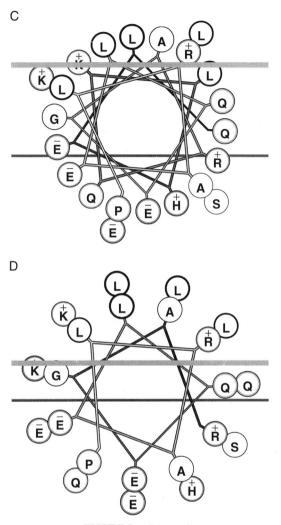

FIGURE 5 (*Continued*)

to discoidal (mixed micelle) bilayer complexes (Anantharamaiah *et al.,* 1985; Brouillette *et al.,*1984; Epand *et al.,*1987; Mishra *et al.,*1998; Srinivas *et al.,*1992; Wlodawer *et al.,*1979). Details of the lipid interactions of these peptides have been studied by a number of investigators using a variety of physical chemical techniques, including nuclear magnetic resonance (NMR) (Anantharamaiah *et al.,* 1985; Brouillette *et al.,* 1984; Lund-Katz *et al.,* 1990, 1995; Spuhler *et al.,* 1994), calorimetry (Brouillette *et al.,* 1984; Epand *et al.,* 1990; Gazzara *et al.,* 1997a,b; Mishra *et al.,*1994, 1995, 1998; Palgunachari *et al.,* 1996; Polozov *et al.*, 1994, 1997, 1998; Spuhler *et al.*, 1994; Venkatachalapathi *et al.*, 1993), circular dichroic spectroscopy (Anantharamaiah *et al.*, 1985; Epand *et al.*, 1987, 1989; Mishra *et al.*, 1994, 1995, 1998; Palgunachari *et al.*, 1996; Venkatachalapathi *et al.*, 1993), fluorescence spectroscopy (Epand *et al.*, 1987, 1989; Gazzara *et al.*, 1997a,b; Mishra *et al.*, 1994, 1995; Polozov *et al.*, 1998; Venkatachalapathi *et al.*, 1993), surface pressure measurements (Epand *et al.*, 1995; Gillotte *et al.*, 1999; Mishra *et al.*, 1995, 1998; Palgunachari *et al.*, 1996), negative stain electron microscopy (Anantharamaiah *et al.*, 1985; Brouillette *et al.*, 1984; Epand *et al.*, 1987; Mishra *et al.*, 1998; Srinivas *et al.*, 1992; Wlodawer *et al.*, 1979), vesicle leakage (Kanellis *et al.*, 1980; Polozov *et al.*, 1994, 1997; Segrest *et al.*, 1983; Srinivas *et al.*, 1992), and nondenaturing gradient gel electrophoresis (Anantharamaiah *et al.*, 1985; Brouillette *et al.*, 1984; Epand *et al.*, 1987, 1989).

As one detailed example, the transbilayer position and orientation of the peptide Ac-18A-NH_2 was determined by a novel X-ray diffraction method in oriented fluid-state dioleoylphosphatidylcholine (DOPC) bilayers (Hristova *et al.*, 1999). The diffraction results (Fig. 6) showed that Ac-18A-NH_2 was located in the bilayer interface and that its transbilayer distribution could be described by a Gaussian function with a $1/e$ half-width of 4.5 (± 0.3) Å located 17.1 (± 0.3) Å from the bilayer center, close to the glycerol moiety. Molecular modeling suggested that Ac-18A-NH_2 is helical and oriented with an angle of 0–6° to the bilayer plane. The width of the Gaussian distribution indicated that the Ac-18A-NH_2 helix penetrated the hydrocarbon core to the level of the DOPC double bonds. Bilayer perturbations caused by Ac-18A-NH_2 were surprisingly modest, consisting of a 2.6-Å decrease in bilayer thickness (Hristova *et al.*, 1999).

The amphipathic peptide Ac-18A-NH_2 is displayed above the X-ray scattering density plot of Fig. 6 in the form of a $\Lambda\alpha$ helical wheel diagram. The $\Lambda\alpha$ algorithm calculates that this amphipathic α helix would penetrate to within approximately 10–11 Å of the bilayer center and have a tilt to the bilayer plane of 4.6°, results compatible with the experimental findings.

b. Biological Studies. The peptide 18A and its analogues possess a number of biological properties that mimic apo A-I, including (i) activation of the plasma enzyme lecithin:cholesteryl acyl transferase (Anantharamaiah *et al.*, 1990), (ii) inhibition of human immunodeficiency virus type 1 (HIV-1) : GP1-41-mediated HIV fusion (Owens *et al.*, 1990), (iii) inhibition of neutrophil activation (Blackburn *et al.*, 1991), (iv) stimulation of human placental lactogen synthesis (Jorgensen *et al.*, 1989), and (v) stimulation of cholesterol efflux from cholesterol-loaded cells (Yancey *et al.*, 1995).

In recent studies from our laboratory, the chronic injection of two different synthetic amphipathic α-helical peptides into animal models susceptible to development of atherosclerosis resulted in inhibition of atherosclerosis. Inhibition by the first peptide was lipoprotein-independent; inhibition by the second peptide was lipoprotein-dependent.

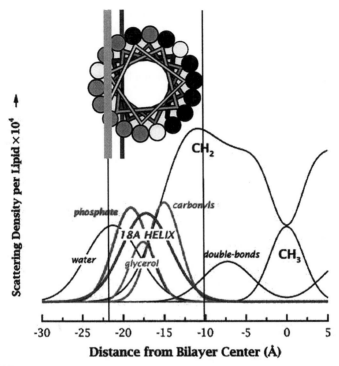

FIGURE 6 Transbilayer position and orientation of the peptide Ac-18A-NH$_2$ as determined by a novel X-ray diffraction method in oriented fluid-state dioleoylphosphatidylcholine bilayers. Summary of the results of the structural refinement of Ac-18A-NH$_2$ in DOPC bilayers showing the transbilayer distribution of the amphipathic α helix in the context of the structure of the fluid DOPC bilayer. The inset shows a helical wheel plot of Ac-18A-NH$_2$ produced by the $\Lambda\alpha$ algorithm, which is aligned parallel to the bilayer plane with the nonpolar surface facing the hydrocarbon core of the bilayer (to scale). Color code for the amino acid residues: hydrophobic (L, F, W, Y), black; basic, blue; acidic, red; other, white. The Gaussian distribution of the helix indicates that the thermally disordered surface of the helix penetrates the hydrocarbon core to the level of the double bonds. The helix axis, located 17.1 Å from the bilayer center, coincides closely with the mean position of the DOPC glycerol groups, located about 17.6 Å from the center. The thin vertical lines indicate the calculated position of the water/lipid interface and the most deeply buried residue. Figure modified from Hristova *et al.* (1999) with permission. (See color plate.)

In the first study, an analogue of 18A possessing three additional phenylalanine residues, thus an increased lipid affinity, inhibited atherosclerosis in a lipoprotein-independent manner. Chronic administration of this peptide inhibited atherosclerosis lesion formation in C57BL6 mice administered a high-fat diet, but had no effect on plasma cholesterol levels (Garber *et al.*, 2001).

In a second study, another peptide, Ac-hE-18A-NH$_2$, in which hE is an 11-residue, highly basic analogue of the LDL-receptor-binding domain of apo E, associated with the surface of human LDL and very low density lipoprotein

(VLDL), and enhanced the uptake and degradation of these lipoproteins in HepG2 cells (Datta *et al.*, 2000). Acute injection of this peptide into apo E(−/−) mice rapidly resulted in the clearance of atherogenic lipoproteins; chronic administration resulted in reversal of atherosclerosis (Datta *et al.*, 2001). Thus this amphipathic α-helical peptide inhibited atherosclerosis in a lipoprotein-dependent manner.

2. Class L Amphipathic α-Helical Peptides

Magainins and mastoparans are examples of peptide antibiotics and peptide venoms, respectively. They have been grouped together as class L amphipathic α helices (Segrest *et al.*, 1990) because of similarities in the distribution of Lys residues along the polar face of the helix. Class A amphipathic α helices are postulated to act as detergents by virtue of their wedge-shaped cross section (Fig. 7.) Using computer analyses of naturally occurring class A and class L amphipathic helixes, we designed two archetypical model peptides. Analogues of these two peptides, incorporating substitutions or modifications of interfacial or basic residues, had the following effects (Tytler *et al.*, 1993): Class A peptides stabilized bilayer structure, reduced leakage from large unilamellar vesicles and erythrocytes, and inhibited lysis induced by class L peptides. Class L peptides destabilized bilayer structure in model membranes and increased binding of class A peptides to erythrocytes. The ability of class L analogues to lyse membranes and induce inverted lipid phases was reduced by either decreasing the bulk of an interfacial residue, increasing the angle subtended by the polar face, or increasing the bulk of the basic residues. The ability of the class A analogue to stabilize bilayer structure and inhibit erythrocyte lysis by class L peptides was enhanced by methylating the Lys residues. These results can be explained by a model that we termed the "reciprocal-wedge hypothesis" (Fig. 7). By analogy to the reciprocal effects of phospholipid shapes on membrane structure, we proposed that the wedge shape of class A helixes stabilizes membrane bilayers, whereas the inverted-wedge shape

Class A   **Class L**

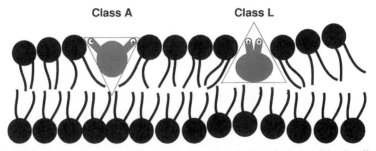

FIGURE 7   Schematic representation of the reciprocal-wedge hypothesis describing the effects of class A and class L amphipathic α helixes on phospholipid bilayers. The helixes are shown schematically in cross section. The approximate radial positions of the basic residues in each helix are indicated by two arms per helix. The triangle surrounding each helix indicates its average cross-sectional shape.

of class L helixes destabilizes membrane bilayers, and thus one class will neutralize the effect of the other class on membranes (Tytler *et al.,* 1993).

Class L venoms, such as mastoparans, lyse both eukaryotic and prokaryotic cells, whereas class L antibiotics, such as magainins, specifically lyse bacteria. Sequence analysis showed that class L antibiotics have a Glu residue on the nonpolar face of the amphipathic helix; this is absent from class L venoms. We synthesized three model class L peptides with or without Glu on the nonpolar face. Hemolysis, bacteriolysis, and bacteriostasis studies using these peptides showed that the specificity of lysis is due to both the presence of a Glu residue on the nonpolar face of the helix and the bulk of the nonpolar face (Tytler *et al.,* 1995). Studies using large unilamellar phospholipid vesicles showed that the inclusion of cholesterol greatly inhibited leakage by the two Glu-containing peptides. These results cannot be attributed to changes in the phase behavior of the lipids caused by the inclusion of cholesterol or to differences in the secondary structure of the peptides. These results suggested that eukaryotic cells are resistant to lysis by magainins due to peptide–cholesterol interactions in their membranes, which inhibit the formation of peptide structures capable of lysis, perhaps by hydrogen bonding between Glu and cholesterol (Tytler *et al.,* 1995). Bacterial membranes, lacking cholesterol, are susceptible to lysis by magainins.

III. AMPHIPATHIC β STRANDS/SHEETS

A. Background

The B apolipoproteins, which are present in chylomicrons, very low density lipoproteins (VLDL), intermediate-density lipoproteins (IDL), and low-density lipoproteins (LDL), are highly insoluble in aqueous solution, and thus remain with the lipoprotein particle throughout its metabolism. Because of the size and insoluble nature of apo B, it has been difficult to confirm the structural motifs responsible for the lipid-associating properties of this nonexchangeable apolipoprotein.

Several different research groups have postulated that amphipathic β strands contribute to the high affinity of apo B-100 for the lipid surface of VLDL and LDL (Cladaras *et al.,* 1986; Goormaghtigh *et al.,* 1989, 1993; Lins *et al.,* 1994; Nolte, 1994; Olofsson *et al.,* 1987; Segrest *et al.,* 1994b; Yang *et al.,* 1986). We initially developed the LOCATE program to examine the lipid-associating domains of apo B-100 in more detail. Analysis with this program confirmed the presence of two regions of amphipathic β strands alternating with two regions of amphipathic α helixes (De Loof *et al.,* 1987; Nolte, 1994) and suggested the presence of a third N-terminal, mixed amphipathic α-helical and β-strand domain, giving apo B-100 a pentapartite structure: NH_2-$\beta\alpha_1$-β_1-α_2-β_2-α_3-COOH (Segrest *et al.,* 1994b). In this report, we suggested that the two amphipathic β-strand domains in apo B-100

(β_1 and β_2) represented the irreversibly lipid-associated regions of this irreversibly associated apolipoprotein.

In a second publication, we compared the complete sequence of human apolipoprotein B-100 with partial sequences from eight additional species of vertebrates (Segrest *et al.*, 1998). We showed that amphipathic β strands cluster in two domains in all species for which these regions have been sequenced, with apparent conservation of several individual amphipathic β strands.

Although the amphipathic β strand is generally accepted as a major lipid-associating motif in apo B, little direct evidence for the motif exists and little is known about its mode of association with phospholipid monolayers and its biological functions. Small and Atkinson (1997) analyzed the amphipathic β strands in the N-terminal half of apo B. Their analysis located 57 amphipathic β strands of 11 residues or more in this region of apo B; 41 were between residues 968 and 1882. These investigators proposed that this region of apo B forms a nearly continuous amphipathic β sheet. When they modeled the 41 amphipathic β strands in apo B as a continuous sheet, they calculated the total ΔG of lipid association for the sheet to be on the order of -500 kcal/mol, an extremely high lipid affinity (Small and Atkinson, 1997). This proposal is similar to the one made by this laboratory, that the two amphipathic β-strand domains (β_1 and β_2), with, for all practical purposes, infinite lipid affinity (Segrest *et al.*, 1994b, 1998), represent the irreversibly lipid-associated regions of apo B-100.

B. Computational Analysis of Amphipathic β Strands

1. Algorithms for Identification and Characterization of Amphipathic β Strands

The rationale for derivation of the $\Lambda\beta$ algorithm for calculating the lipid affinity of amphipathic β strands and the termination rules for the program βLOCATE, which identifies potential amphipathic β strands within a given amino acid sequence or an amino acid database, have been described elsewhere (Segrest *et al.*, 1994b, 1998, 1999a).

2. LOCATE Analyses of Amphipathic Motifs in Apolipoprotein B

βLOCATE and αLOCATE plots of human apo B are shown in Fig. 8. The proposed pentapartite organization of apo B-100 is indicated by large boxes

---

FIGURE 8 βLOCATE and αLOCATE analyses of human apo B. The *y* axis represents the amino acid sequence position for each identified amphipathic motif and the *x* axis shows the minimal lipid affinity used to select by each row of amphipathic motifs. The gray rectangles in each row indicate the sequence length and position of each amphipathic motif selected. The large boxes denote the three α domains and the two β domains. (A) βLOCATE analysis with minimal sequence length of 6 residues and minimal hydrophobic moment of 1.0. (B) αLOCATE analysis with minimal sequence length of 10 residues.

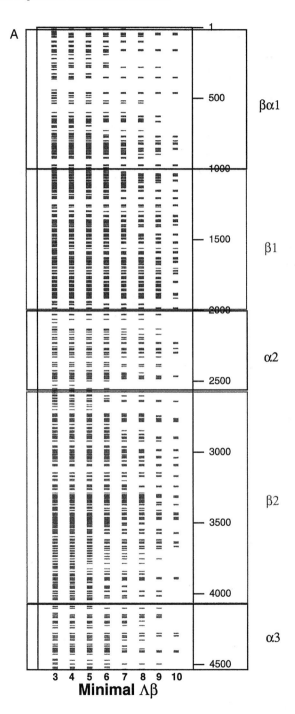

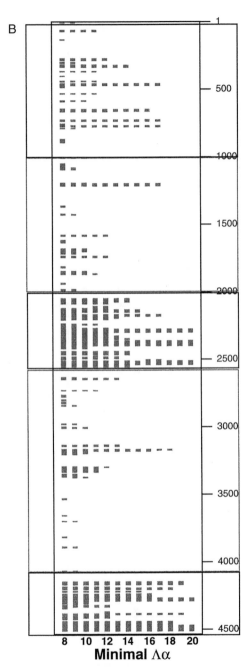

FIGURE 8 (*Continued*)

for the α and β domains. Figure 8A is a plot of amphipathic β strands in which the cutoff for $\Lambda\beta$ increases on the x axis from 5 to 10 kcal/mol. Figure 8B is a plot of amphipathic α helixes in which the cutoff for $\Lambda\alpha$ increases on the x axis from 8 to 20 kcal/mol. This type of lipid affinity gradient plot is a useful schematic for illustrating the wide variation in lipid affinity of amphipathic β strands and α helixes within the apo B sequence.

The five domains of apo B are not entirely homogeneous; there appears to be some admixture of amphipathic α helixes within the β domains and vice versa (Segrest *et al.*, 1998). This is especially true of the globular $\beta\alpha_1$ domain, residues 1–1000, whose heterogeneity in secondary structural elements relates to its structural homology to lamprey lipovitellin (Bradbury *et al.*, 1999; Mann *et al.*, 1999; Perez *et al.*, 1991; Segrest *et al.*, 1999a; Steyrer *et al.*, 1990).

3. Comparison of Amphipathic β Strands/Sheets from apo B with Those from Other Proteins

Amphipathic β strands are found in at least two other two classes of proteins in addition to apo B. Examples of amphipathic β strands are the hollow β-barrel porins found among the outer membrane proteins from Gram-negative bacteria. These transmembrane proteins contain mixed polar and nonpolar sequences arranged in amphipathic antiparallel β barrels, which are inclined to the plane of the membrane (Buchanan, 1999; Koebnik *et al.*, 2000). The polar residues of the amphipathic β strands that form the porins face the channel interior on the concave side of the β sheets, whereas the majority of the hydrophobic residues face the surrounding membrane lipid on the convex side of the β sheets. A seven-stranded amphipathic β sheet from the X-ray structure of porin from *Rhodopseudomonas blastica* (Kreusch *et al.*, 1994) is shown in Figs. 9A–9C . In Figs. 9A and 9B, the hydrophobic (convex) face of the seven-stranded sheet is shown in cartoon and space-filling forms, respectively, while Fig. 9C shows the hydrophilic (concave) face in space-filling form.

A second example of amphipathic β strands is found in a nonmammalian form of lipoprotein. During a βLOCATEdb search for protein sequences that contained similar amphipathic β strands to apo B, four vitellogenins, the precursor form of lipovitellin, from chicken, frog, lamprey, and *Caenorhabditis elegans* appeared on the list of candidate proteins (Segrest *et al.*, 1999a). Lipovitellin is a lipoprotein found in the egg yolk of oocytes of many egg-laying animals. Lipovitellins are 15% lipid by weight and thus are proteolipids rather than classical lipoproteins, which are 50% or greater lipid by weight. The X-ray crystal structure for lamprey lipovitellin has been determined at 2.8-Å resolution (Anderson *et al.*, 1998; Banaszak *et al.*, 1991; Raag *et al.*, 1988; Sharrock *et al.*, 1992). The X-ray crystal structure shows that lamprey lipovitellin contains a "lipid pocket" lined by three antiparallel amphipathic β sheets, designated βB, βA, and βD. A seven-stranded amphipathic β sheet from the antiparallel amphipathic β sheet, βD, is

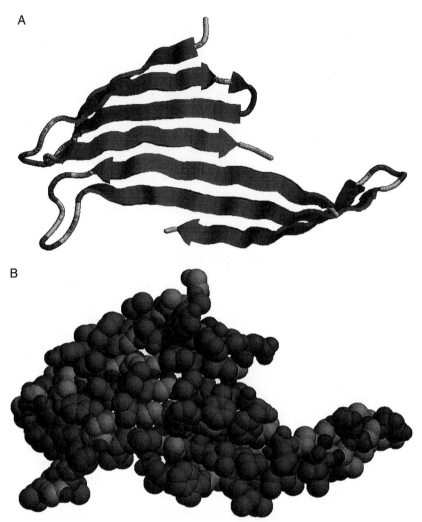

FIGURE 9 X-ray structures of a selected seven-stranded amphipathic β sheet from porin (*Rhodopseudomonas blastica*) and from lipovitellin (lamprey). The molecular models were generated with Rasmol (Sayle and Milner-White, 1995). Color code for the space-filling models: hydrophobic residues (L, F, M, V, W, Y, A, I), gold; basic residues (L, R, H), blue; acidic residues (D, E), red; other, CPK. The X-ray structure for porin is from Kreusch *et al.* (1994) and that for lipovitellin is from Anderson *et al.* (1998). (A) A 113-residue-long, seven-stranded amphipathic β sheet, residues 2–57 and 233–289, from porin displayed in cartoon mode and viewed from the hydrophobic (convex) face. (B) A seven-stranded amphipathic β sheet from porin displayed in space-filling mode and viewed from the hydrophobic (convex) face. (C) A seven-stranded amphipathic β sheet from porin displayed in space-filling mode and viewed from the hyrdrophilic (concave) face. (D) A 96-residue-long, seven-stranded amphipathic β sheet, residues 1374–1458 and 1493–1503, from lipovitellin displayed in cartoon mode and viewed from the hydrophobic (concave) face. (E) A seven-stranded amphipathic β sheet from lipovitellin displayed in space-filling mode and viewed from the hydrophobic (concave) face. (F) A seven-stranded amphipathic β sheet from lipovitellin displayed in space-filling mode and viewed from the hydrophilic (convex) face. (See color plate.)

C

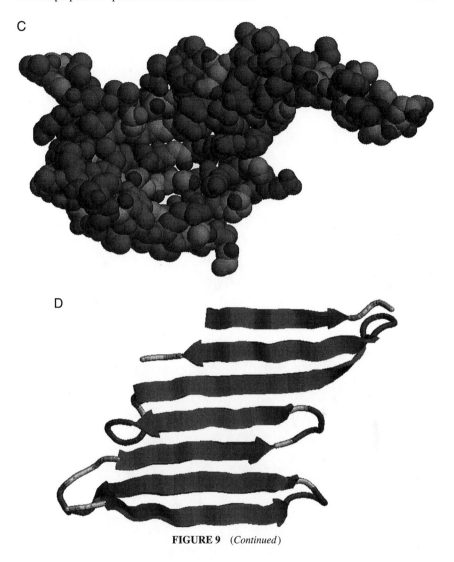

D

FIGURE 9 *(Continued)*

shown in Figs. 9D–9F. In Figs. 9D and 9E, the hydrophobic (concave) face of the seven-stranded sheet is shown in cartoon and space-filling forms, respectively; in Fig. 9F, the hydrophilic (convex) face is displayed in space-filling form.

The two seven-stranded amphipathic β sheets shown in Fig. 9 provided an opportunity to test the robustness of the βLOCATE algorithm. In Fig. 10 the locations of amphipathic β strands predicted by the current version of βLOCATE are shown on stereo models of the two seven-stranded amphipathic β sheets. This algorithm, using a cutoff of $\Lambda\beta \geq 5.0$, predicts amphipathic β strands well for the

E

F

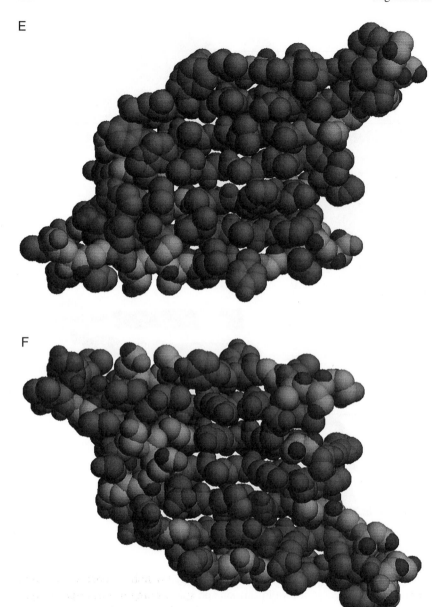

FIGURE 9   (*Continued*)

A

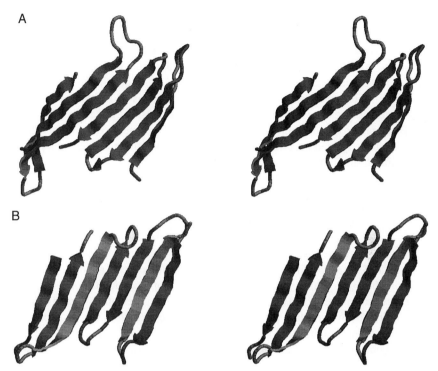

B

FIGURE 10   Stereo models of the seven-stranded amphipathic β sheets from porin and lipovitellin showing the location of amphipathic β strands predicted by the current version of βLOCATE. The two cross-eyed stereo models were created by Rasmol (Sayle and Milner-White, 1995) and are displayed in cartoon mode. Amphipathic β strands selected by the current version of βLOCATE for a $\Lambda\beta \geq 5$ are highlighted in gold. The X-ray structure for porin is from Kreusch *et al.* (1994) and that for lipovitellin is from Anderson *et al.* (1998). (A) A 113-residue-long, seven-stranded amphipathic β sheet, residues 2–57 and 233–289, from porin (*Rhodopseudomonas blastica*) displayed in cartoon mode and viewed from the hydrophobic (convex) face. (B) A 96-residue-long, seven-stranded amphipathic β sheet, residues 1374–1458 and 1493–1503, from lipovitellin (lamprey) displayed in cartoon mode and viewed from the hydrophilic (convex) face. Two amphipathic β strands selected by the current version of βLOCATE for $\Lambda\beta$ <5.0 but $\geq$2.5 are highlighted in yellow. (See color plate.)

porin amphipathic β sheet (highlighted in orange), missing only one short stretch of the penultimate strand on the left edge of the sheet. However, the prediction of βLOCATE for the lipovitellin amphipathic β sheet is less robust; two of the seven strands are predicted only if a lower $\Lambda\beta$ cutoff is used (highlighted in yellow) and only a portion of either is completely predicted.

The seven-stranded amphipathic β sheets shown for porin and lipovitellin in Fig. 9 support the plausibility of similar structures in apo B. βLOCATE analyses of the seven-stranded amphipathic β sheets from porin (113 residues) and lipovitellin (96 residues) were compared with βLOCATE analysis of the first

100 residues of the β_1 domain of apo B (residues 1001–1100). The results are shown in Fig. 11. In all three examples, βLOCATE identifies seven potential amphipathic β strands. The mean $\Lambda\beta$ is 11.0, 9.9, and 7.4 and the mean residue length per strand is 12.6, 10.9, and 9.6 for porin, apo B, and lipovitellin, respectively. In these three sequences, the highest $\Lambda\beta$ identified is 14.4 (porin) and the lowest is 2.5 (lipovitellin), whereas the longest strand identified is 17 residues (porin) and the shortest is 7 residues (lipovitellin). Thus, on the basis of this limited analysis, apo B appears to contain amphipathic β strands that are intermediate in robustness between those formed by a transmembrane protein and those formed by a protein containing a small lipid pocket.

C. Experimental Studies of Peptide Analogues of Amphipathic β Strands from Apolipoprotein B

1. Peptides Synthesized

βLOCATE predicts three amphipathic β strands between residues 1456 and 1496 in the β_1 domain of apo B. These three amphipathic β strands and the intervening turns are shown in Fig. 12A in the form of a schematic β-strand diagram with a pitch of two residues per turn. Three peptide analogues that progressively overlap the three amphipathic β strands (see Fig. 11A) were synthesized and their physical chemical properties, including lipid interactions, studied.

2. Studies of Secondary Structure

Circular dichroism (CD) spectra of B:1479–499, B:1466–1499, and B:1456–1499 were examined in buffer and in sodium dodecylsulfate. In buffer, all three peptides were random coil. In sodium dodecylsulfate, peptides B:1479–1499 and B:1466–1499 were random coil, but B:1456–1499, with a single minimum at 218 nm, adopted a β-sheet conformation (Fig. 12B).

The secondary structure of the longest of the peptides, B:1456–1499, was studied in the presence of POPC multilamellar vesicles (MLV) using Fourier transform infrared (FTIR) spectroscopy. The amide I' region of the FTIR spectrum was deconvoluted and curve-fitted (Fig. 12C). The results show that B:1456–1499 adopted a conformation that was 0.51 mole fraction β sheet and 0.24 mole fraction

---

FIGURE 11 Two-dimensional βLOCATE analyses of the seven-stranded amphipathic β sheets from porin and lipovitellin and the first 100 residues of the β_1 domain of human apolipoprotein. The *y* axis represents residue number and the *x* axis represents $\Lambda\beta$. The sequence position and $\Lambda\alpha\beta$ of each selected amphipathic β strand is indicated by a gray bar. (A) Analysis of the 113-residue-long, seven-stranded amphipathic β sheet, residues 2–57 and 233–289, from porin (*Rhodopseudomonas blastica*). (B) Analysis of the 96-residue-long, seven-stranded amphipathic β sheet, residues 1374–1458 and 1493–1503, from lipovitellin (lamprey). (C) Analysis of residues 1001–1100 of apo B.

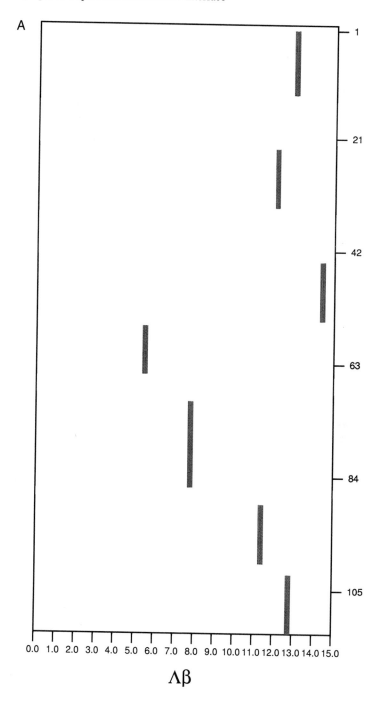

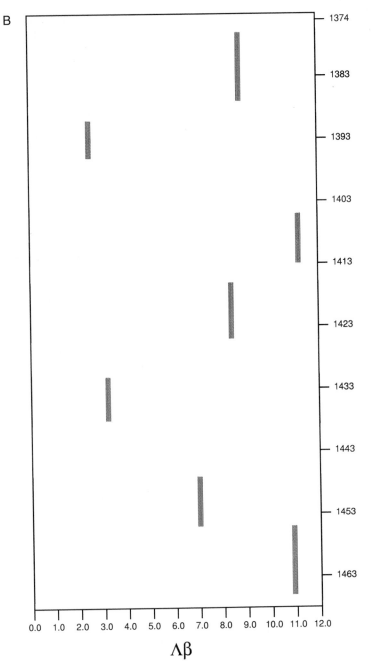

FIGURE 11 (*Continued*)

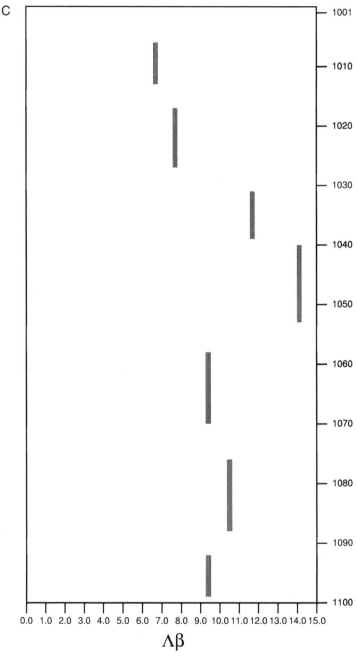

FIGURE 11 (*Continued*)

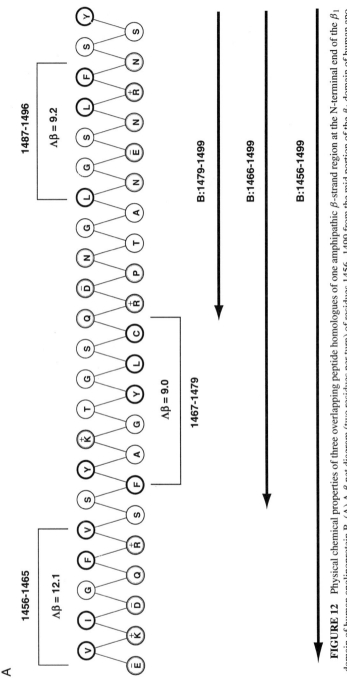

FIGURE 12 Physical chemical properties of three overlapping peptide homologues of one amphipathic β-strand region at the N-terminal end of the β₁ domain of human apolipoprotein B. (A) A β net diagram (two residues per turn) of residues 1456–1499 from the mid portion of the β₁ domain of human apo B. Hydrophobic residues, black; prolines, basic residues, acidic residues, and polar residues shaded. The three amphipathic β strands predicted by βLOCATE are indicated by brackets. The sequences of the three overlapping peptide homologues studied are indicated by arrows. (B) CD spectroscopy of the peptides B:1479–1499, B:1466–1499, and B:1456–1499 examined in sodium dodecylsulfate (V. K. Mishra *et al.*, unpublished data). (C) FTIR spectroscopy of the longest of the peptides, B:1456–1499, in the presence of POPC multilamellar vesicles (V. K. Mishra *et al.*, unpublished data). Peptide in D₂O was mixed with POPC MLV in D₂O at a lipid : peptide ratio of 2 : 1, w : w. The mixture was incubated overnight at room temperature in the dark under helium. The amide I′ region of the FTIR spectrum was deconvoluted and curve-fitted.

B

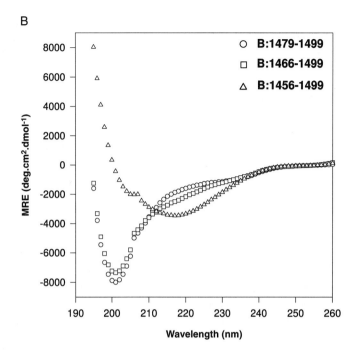

C

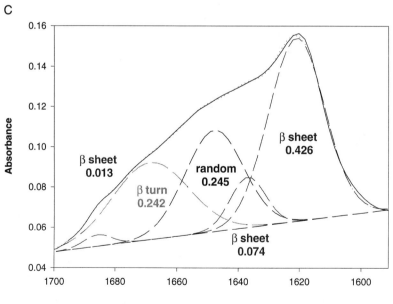

FIGURE 12 *(Continued)*

β turn when associated with the surface of POPC MLV. If this peptide forms a three-stranded amphipathic β sheet with the dimensions predicted by Fig. 12A on the surface of the POPC MLV, one can calculate that 12 residues (25%) of B:1456–1499 should be involved in turns.

3. Studies of Lipid Affinity

The lipid affinity of peptides B:1479–1499 and B:1466–1499 were determined by measuring their partition coefficients into POPC MLV as described elsewhere (Palgunachari *et al.*, 1996). Based upon the partition coefficients determined, the free energy of binding, ΔG, to POPC MLV was −4.6 and −5.3 kcal/mol for B:1479–1499 and B:1466–1499, respectively.

To summarize, we have shown that (i) an apo B domain containing a minimum of three amphipathic β strands adopts a β-sheet conformation on the surface of phospholipid, (ii) a single amphipathic β strand of apo B has measurable lipid affinity, and (iii) the lipid affinity of an apo B peptide analogue containing two amphipathic β strands is higher than that of an apo B peptide analogue containing only one of the two amphipathic β strands.

Based upon studies of the interactions of small peptides with water : phospholipid interfaces by Jacobs and White (1989), it seems likely that unfavorable energy costs associated with the lack of hydrogen bonds between amphipathic β strands will drive individual amphipathic β strands toward β sheets on lipid surfaces whenever possible. We proposed elsewhere (Segrest *et al.*, 2001) a direct contact of the hydrophobic face of amphipathic β sheets with the neutral lipid core of apo B-containing lipoproteins and suggested that this contact will play a role in core lipid organization and metabolism. Further, the creation of neutral lipid core ridges beneath amphipathic β sheets would contribute additional favorable free energy not available to individual strands. Thus, phospholipid vesicles would seem not to be the ideal lipid surfaces for study of lipid interactions of amphipathic β sheets. Such studies might better be done in the presence of lipid emulsions in which a neutral lipid core underlies the phospholipid surface monolayer, thus allowing direct contact of amphipathic β sheets with core lipid.

References

Anantharamaiah, G. M., Jones, J. L., Brouillette, C. G., Schmidt, C. F., Chung, B. H., Hughes, T. A., Bhown, A. S., and Segrest, J. P. (1985). Studies of synthetic peptide analogs of the amphipathic helix: Structure of complexes with dimyristoyl phosphatidylcholine. *J. Biol. Chem.* **260**, 10248–10255.

Anantharamaiah, G. M., Venkatachalapathi, Y. V., Brouillette, C. G., and Segrest, J. P. (1990). Use of synthetic peptide analogues to localize lecithin : cholesterol acyltransferase activating domain in apolipoprotein A-I. *Arteriosclerosis* **10**, 95–105.

Anderson, T. A., Levitt, D. G., and Banaszak, L. J. (1998). The structural basis of lipid interactions in lipovitellin, a soluble lipoprotein. *Structure* **6**, 895–909.

Argiolas, A., and Pisano, J. J. (1983). Facilitation of phospholipase A2 activity by mastoparans, a new class of mast cell degranulating peptides from wasp venom. *J. Biol. Chem.* **258**, 13697–13702.

Atkinson, D., Small, D. M., and Shipley, G. G. (1980). X-ray and neutron scattering studies of plasma lipoproteins. *Ann. N.Y. Acad. Sci.* **348,** 284–298.

Banaszak, L., Sharrock, W., and Timmins, P. (1991). Structure and function of a lipoprotein: Lipovitellin. *Annu. Rev. Biophys. Biophys. Chem.* **20,** 221–246.

Blackburn, W. D., Jr., Dohlman, J. G., Venkatachalapathi, Y. V., Pillion, D. J., Koopman, W. J., Segrest, J. P., and Anantharamaiah, G. M. (1991). Apolipoprotein A-I decreases neutrophil degranulation and superoxide production. *J. Lipid. Res.* **32,** 1911–1918.

Borhani, D. W., Rogers, D. P., Engler, J. A., and Brouillette, C. G. (1997). Crystal structure of truncated human apolipoprotein A-I suggests a lipid-bound conformation. *Proc. Natl. Acad. Sci. USA* **94,** 12291–12296.

Bradbury, P., Mann, C. J., Kochl, S., Anderson, T. A., Chester, S. A., Hancock, J. M., Ritchie, P. J., Amey, J., Harrison, G. B., Levitt, D. G., Banaszak, L. J., Scott, J., and Shoulders, C. C. (1999). A common binding site on the microsomal triglyceride transfer protein for apolipoprotein B and protein disulfide isomerase. *J. Biol. Chem.* **274,** 3159–3164.

Brouillette, C. G., Jones, J. L., Ng, T. C., Kercret, H., Chung, B. H., and Segrest, J. P. (1984). Structural studies of apolipoprotein A-I/phosphatidylcholine recombinants by high-field proton NMR, nondenaturing gradient gel electrophoresis, and electron microscopy. *Biochemistry* **23,** 359–367.

Buchanan, S. K. (1999). Beta-barrel proteins from bacterial outer membranes: Structure, function and refolding. *Curr. Opin. Struct. Biol.* **9,** 455–461.

Burgess, J. W., Frank, P. G., Franklin, V., Liang, P., McManus, D. C., Desforges, M., Rassart, E., and Marcel, Y. L. (1999). Deletion of the C-terminal domain of apolipoprotein A-I impairs cell surface binding and lipid efflux in macrophage. *Biochemistry* **38,** 14524–14533.

Cladaras, C., Hadzopoulou-Cladaras, M., Nolte, R. T., Atkinson, D., and Zannis, V. I. (1986). The complete sequence and structural analysis of human apolipoprotein B-100: Relationship between apoB-100 and apoB-48 forms. *EMBO J.* **5,** 3495–3507.

Datta, G., Chaddha, M., Garber, D. W., Chung, B. H., Tytler, E. M., Dashti, N., Bradley, W. A., Gianturco, S. H., and Anantharamaiah, G. M. (2000). The receptor binding domain of apolipoprotein E, linked to a model class A amphipathic helix, enhances internalization and degradation of LDL by fibroblasts. *Biochemistry* **39,** 213–220.

Datta, G., Garber, D. W., Chung, B. H., Chaddha, M., Dashti, N., Bradley, W. A., Gianturco, S. H., and Anantharamaiah, G. M. (2001). Cationic domain 141-150 of apoE covalently linked to a class A amphipathic helix enhances atherogenic lipoprotein metabolism *in vitro* and *in vivo J. Lipid Res.* **42,** 959–966.

De Loof, H., Rosseneu, M., Yang, C. Y., Li, W. H., Gotto, A. M., Jr., and Chan, L. (1987). Human apolipoprotein B: Analysis of internal repeats and homology with other apolipoproteins. *J. Lipid Res.* **28,** 1455–1465.

Engelman, D. M., Steitz, T. A., and Goldman, A. (1986). Identifying nonpolar transbilayer helices in amino acid sequences of membrane proteins. *Annu. Rev. Biophys. Biophys. Chem.* **15,** 321–353.

Epand, R. M., Gawish, A., Iqbal, M., Gupta, K. B., Chen, C. H., Segrest, J. P., and Anantharamaiah, G. M. (1987). Studies of synthetic peptide analogs of the amphipathic helix: Effect of charge distribution, hydrophobicity, and secondary structure on lipid association and lecithin:cholesterol acyltransferase activation. *J. Biol. Chem.* **262,** 9389–9396.

Epand, R. M., Surewicz, W. K., Hughes, D. W., Mantsch, H., Segrest, J. P., Allen, T. M., and Anantharamaiah, G. M. (1989). Properties of lipid complexes with amphipathic helix-forming peptides: Role of distribution of peptide charges. *J. Biol. Chem.* **264,** 4628–4635.

Epand, R. M., Segrest, J. P., and Anantharamaiah, G. M. (1990). Thermodynamics of the binding of human apolipoprotein A-I to dimyristoylphosphatidylglycerol. *J. Biol. Chem.* **265,** 20829–20832.

Epand, R. M., Shai, Y., Segrest, J. P., and Anantharamaiah, G. M. (1995). Mechanisms for the modulation of membrane bilayer properties by amphipathic helical peptides. *Biopolymers* **37,** 319–338.

Garber, D. W., Datta. G., Chaddha, M., Palgunachari, M. N., Hama, S. Y., Navab, M., Fogelman, A. M., Segrest, J. P., and Anantharamaiah, G. M. (2001). A new synthetic class A amphipathic peptide analogue protects mice from diet-induced atherosclerosis. *J. Lipid Res.* **42,** 545–552.

Gazzara, J. A., Phillips, M. C., Lund-Katz, S., Palgunachari, M. N., Segrest, J. P., Anantharamaiah, G. M., Rodrigueza, W. V., and Snow, J. W. (1997a). Effect of vesicle size on their interaction with class A amphipathic helical peptides. *J. Lipid Res.* **38,** 2147–2154.

Gazzara, J. A., Phillips, M. C., Lund-Katz, S., Palgunachari, M. N., Segrest, J. P., Anantharamaiah, G. M., and Snow, J. W. (1997b). Interaction of class A amphipathic helical peptides with phospholipid unilamellar vesicles. *J. Lipid Res.* **38,** 2134–2146.

Gillotte, K. L., Zaiou, M., Lund-Katz, S., Anantharamaiah, G. M., Holvoet, P., Dhoest, A., Palgunachari, M. N., Segrest, J. P., Weisgraber, K. H., Rothblat, G. H., and Phillips, M. C. (1999). Apolipoprotein-mediated plasma membrane microsolubilization: Role of lipid affinity and membrane penetration in the efflux of cellular cholesterol and phospholipid. *J. Biol. Chem.* **274,** 2021–2028.

Goormaghtigh, E., De Meutter, J., Vanloo, B., Brasseur, R., Rosseneu, M., and Ruysschaert, J. M. (1989). Evaluation of the secondary structure of apo B-100 in low-density lipoprotein (LDL) by infrared spectroscopy. *Biochim. Biophys. Acta* **1006,** 147–150.

Goormaghtigh, E., Cabiaux, V., De Meutter, J., Rosseneu, M., and Ruysschaert, J. M. (1993). Secondary structure of the particle associating domain of apolipoprotein B-100 in low-density lipoprotein by attenuated total reflection infrared spectroscopy. *Biochemistry* **32,** 6104–6110.

Hazelrig, J. B., Jones, M. K., and Segrest, J. P. (1993). A mathematically defined motif for the radial distribution of charged residues on apolipoprotein amphipathic alpha helixes. *Biophys. J.* **64,** 1827–1832.

Hristova, K., Wimley, W. C., Mishra, V. K., Anantharamiah, G. M., Segrest, J. P., and White, S. H. (1999). An amphipathic alpha-helix at a membrane interface: A structural study using a novel X-ray diffraction method. *J. Mol. Biol.* **290,** 99–117.

Jacobs, R. E., and White, S. H. (1989). The nature of the hydrophobic binding of small peptides at the bilayer interface: Implications for the insertion of transbilayer helices. *Biochemistry* **28,** 3421–3437.

Jones, M. K., Anantharamaiah, G. M., and Segrest, J. P. (1992). Computer programs to identify and classify amphipathic alpha helical domains. *J. Lipid Res.* **33,** 287–296.

Jorgensen, E. V., Anantharamaiah, G. M., Segrest, J. P., Gwynne, J. T., and Handwerger, S. (1989). Synthetic amphipathic peptides resembling apolipoproteins stimulate the release of human placental lactogen. *J. Biol. Chem.* **264,** 9215–9219.

Kanellis, P., Romans, A. Y., Johnson, B. J., Kercret, H., Chiovetti, R., Jr., Allen, T. M., and Segrest, J. P. (1980). Studies of synthetic peptide analogs of the amphipathic helix: Effect of charged amino acid residue topography on lipid affinity. *J. Biol. Chem.* **255,** 11464–11472.

Klon, A. E., Jones, M. K., Segrest, J. P., and Harvey, S. C. (2000). Molecular belt models for the apolipoprotein A-I paris and milano mutations. *Biophys. J.* **79,** 1679–1685.

Koebnik, R., Locher, K. P., and Van Gelder, P. (2000). Structure and function of bacterial outer membrane proteins: Barrels in a nutshell. *Mol. Microbiol.* **37,** 239–253.

Koppaka, V., Silvestro, L., Engler, J. A., Brouillette, C. G., and Axelsen, P. H. (1999). The structure of human lipoprotein A-I: Evidence for the "belt" model. *J. Biol. Chem.* **274,** 14541–14544.

Kreusch, A., Neubuser, A., Schiltz, E., Weckesser, J., and Schulz, G. E. (1994). Structure of the membrane channel porin from *Rhodopseudomonas blastica* at 2.0 Å resolution. *Protein Sci.* **3,** 58–63.

Laccotripe, M., Makrides, S. C., Jonas, A., and Zannis, V. I. (1997). The carboxyl-terminal hydrophobic residues of apolipoprotein A-I affect its rate of phospholipid binding and its association with high density lipoprotein. *J. Biol. Chem.* **272,** 17511–17522. [Erratum, *J. Biol. Chem.* (1998) **273,** 15878].

Li, H., Lyles, D. S., Thomas, M. J., Pan, W., and Sorci-Thomas, M. G. (2000). Structural determination of lipid-bound apo A-I using fluorescence resonance energy transfer. *J. Biol. Chem.* **275,** 37048–37054.

Lins, L., Brasseur, R., Rosseneu, M., Yang, C. Y., Sparrow, D. A., Sparrow, J. T., Gotto, A. M., Jr., and Ruysschaert, J. M. (1994). Structure and orientation of apo B-100 peptides into a lipid bilayer. *J. Protein Chem.* **13,** 77–88.

Lund-Katz, S., Anantharamaiah, G. M., Venkatachalapathi, Y. V., Segrest, J. P., and Phillips, M. C. (1990). Nuclear magnetic resonance investigation of the interactions with phospholipid of an amphipathic alpha-helix-forming peptide of the apolipoprotein class. *J. Biol. Chem.* **265,** 12217–12223.

Lund-Katz, S., Phillips, M. C., Mishra, V. K., Segrest, J. P., and Anantharamaiah, G. M. (1995). Microenvironments of basic amino acids in amphipathic alpha-helices bound to phospholipid: ^{13}C NMR studies using selectively labeled peptides. *Biochemistry* **34,** 9219–9226.

Maiorano, J. N., and Davidson, W. S. (2000). The orientation of helix 4 in apolipoprotein A-I-containing reconstituted high density lipoproteins. *J. Biol. Chem.* **275,** 17374–17380.

Mann, C. J., Anderson, T. A., Read, J., Chester, S. A., Harrison, G. B., Kochl, S., Ritchie, P. J., Bradbury, P., Hussain, F. S., Amey, J., Vanloo, B., Rosseneu, M., Infante, R., Hancock, J. M., Levitt, D. G., Banaszak, L. J., Scott, J., and Shoulders, C. C. (1999). The structure of vitellogenin provides a molecular model for the assembly and secretion of atherogenic lipoproteins. *J. Mol. Biol.* **285,** 391–408.

Mishra, V. K., and Palgunachari, M. N. (1996). Interaction of model class A1, class A2, and class Y amphipathic helical peptides with membranes. *Biochemistry* **35,** 11210–11220.

Mishra, V. K., Palgunachari, M. N., Segrest, J. P., and Anantharamaiah, G. M. (1994). Interactions of synthetic peptide analogs of the class A amphipathic helix with lipids: Evidence for the snorkel hypothesis. *J. Biol. Chem.* **269,** 7185–7191.

Mishra, V. K., Palgunachari, M. N., Lund-Katz, S., Phillips, M. C., Segrest, J. P., and Anantharamaiah, G. M. (1995). Effect of the arrangement of tandem repeating units of class A amphipathic alpha-helixes on lipid interaction. *J. Biol. Chem.* **270,** 1602–1611.

Mishra, V. K., Palgunachari, M. N., Datta, G., Phillips, M. C., Lund-Katz, S., Adeyeye, S. O., Segrest, J. P., and Anantharamaiah, G. M. (1998). Studies of synthetic peptides of human apolipoprotein A-I containing tandem amphipathic alpha-helixes. *Biochemistry* **37,** 10313–10324.

Nolte, R. T. (1994). Structural analysis of the human apolipoproteins: An integrated approach utilizing physical and computational methods. *Thesis, Department of Biophysics.* Boston University, Boston, MA.

Olofsson, S. O., Bjursell, G., Bostrom, K., Carlsson, P., Elovson, J., Protter, A. A., Reuben, M. A., and Bondjers, G. (1987). Apolipoprotein B: Structure, biosynthesis and role in the lipoprotein assembly process. *Atherosclerosis* **68,** 1–17.

Owens, B. J., Anantharamaiah, G. M., Kahlon, J. B., Srinivas, R. V., Compans, R. W., and Segrest, J. P. (1990). Apolipoprotein A-I and its amphipathic helix peptide analogues inhibit human immunodeficiency virus-induced syncytium formation. *J. Clin. Invest.* **86,** 1142–1150.

Palgunachari, M. N., Mishra, V. K., Lund-Katz, S., Phillips, M. C., Adeyeye, S. O., Alluri, S., Anantharamaiah, G. M., and Segrest, J. P. (1996). Only the two end helixes of eight tandem amphipathic helical domains of human apo A-I have significant lipid affinity: Implications for HDL assembly. *Arterioscler. Thromb. Vasc. Biol.* **16,** 328–338.

Perez, L. E., Fenton, M. J., and Callard, I. P. (1991). Vitellogenin: Homologs of mammalian apolipoproteins? *Comp. Biochem. Physiol. B* **100,** 821–826.

Polozov, I. V., Polozova, A. I., Anantharamaiah, G. M., Segrest, J. P., and Epand, R. M. (1994). Mixing rates can markedly affect the kinetics of peptide-induced leakage from liposomes. *Biochem. Mol. Biol. Int.* **33,** 1073–1079.

Polozov, I. V., Polozova, A. I., Tytler, E. M., Anantharamaiah, G. M., Segrest, J. P., Woolley, G. A., and Epand, R. M. (1997). Role of lipids in the permeabilization of membranes by class L amphipathic helical peptides. *Biochemistry* **36**, 9237–9245.

Polozov, I. V., Polozova, A. I., Mishra, V. K., Anantharamaiah, G. M., Segrest, J. P., and Epand, R. M. (1998). Studies of kinetics and equilibrium membrane binding of class A and class L model amphipathic peptides. *Biochim. Biophys. Acta* **1368**, 343–354.

Raag, R., Appelt, K., Xuong, N. H., and Banaszak, L. (1988). Structure of the lamprey yolk lipid–protein complex lipovitellin–phosvitin at 2.8 Å resolution. *J. Mol. Biol.* **200**, 553–569.

Rogers, D. P., Brouillette, C. G., Engler, J. A., Tendian, S. W., Roberts, L., Mishra, V. K., Anantharamaiah, G. M., Lund-Katz, S., Phillips, M. C., and Ray, M. J. (1997). Truncation of the amino terminus of human apolipoprotein A-I substantially alters only the lipid-free conformation. *Biochemistry* **36**, 288–300.

Rozek, A., Buchko, G. W., and Cushley, R. J. (1995). Conformation of two peptides corresponding to human apolipoprotein C-I residues 7–24 and 35–53 in the presence of sodium dodecyl sulfate by CD and NMR spectroscopy. *Biochemistry* **34**, 7401–7408.

Sayle, R. A., and Milner-White, E. J. (1995). RASMOL: Biomolecular graphics for all. *Trends Biochem. Sci.* **20**, 374.

Segrest, J. P. (1977). Amphipathic helixes and plasma lipoproteins: Thermodynamic and geometric considerations. *Chem. Phys. Lipids* **18**, 7–22.

Segrest, J. P., Jackson, R. L., Morrisett, J. D., and Gotto, A. M., Jr. (1974). A molecular theory of lipid–protein interactions in the plasma lipoproteins. *FEBS Lett.* **38**, 247–258.

Segrest, J. P., Chung, B. H., Brouillette, C. G., Kanellis, P., and McGahan, R. (1983). Studies of synthetic peptide analogs of the amphipathic helix: Competitive displacement of exchangeable apolipoproteins from native lipoproteins. *J. Biol. Chem.* **258**, 2290–2295.

Segrest, J. P., De Loof, H., Dohlman, J. G., Brouillette, C. G., and Anantharamaiah, G. M. (1990). Amphipathic helix motif: Classes and properties. *Proteins* **8**, 103–117 [Erratum, *Proteins* (1991) **9**, 79].

Segrest, J. P., Jones, M. K., De Loof, H., Brouillette, C. G., Venkatachalapathi, Y. V., and Anantharamaiah, G. M. (1992). The amphipathic helix in the exchangeable apolipoproteins: A review of secondary structure and function. *J. Lipid Res.* **33**, 141–166.

Segrest, J. P., Garber, D. W., Brouillette, C. G., Harvey, S. C., and Anantharamaiah, G. M. (1994a). The amphipathic alpha helix: A multifunctional structural motif in plasma apolipoproteins. *Adv. Protein Chem.* **45**, 303–369.

Segrest, J. P., Jones, M. K., Mishra, V. K., Anantharamaiah, G. M., and Garber, D. W. (1994b). apoB-100 has a pentapartite structure composed of three amphipathic alpha-helical domains alternating with two amphipathic beta-strand domains: Detection by the computer program LOCATE. *Arterioscler. Thromb.* **14**, 1674–1685.

Segrest, J. P., Jones, M. K., Mishra, V. K., Pierotti, V., Young, S. H., Boren, J., Innerarity, T. L., and Dashti, N. (1998). Apolipoprotein B-100: Conservation of lipid-associating amphipathic secondary structural motifs in nine species of vertebrates. *J. Lipid Res.* **39**, 85–102.

Segrest, J. P., Jones, M. K., and Dashti, N. (1999a). N-terminal domain of apolipoprotein B has structural homology to lipovitellin and microsomal triglyceride transfer protein: A "lipid pocket" model for self-assembly of apo B-containing lipoprotein particles. *J. Lipid Res.* **40**, 1401–1416.

Segrest, J. P., Jones, M. K., Klon, A. E., Sheldahl, C. J., Hellinger, M., De Loof, H., and Harvey, S. C. (1999b). A detailed molecular belt model for apolipoprotein A-I in discoidal high density lipoprotein. *J. Biol. Chem.* **274**, 31755–31758.

Segrest, J. P., Li, L., Anantharamaiah, G. M., Harvey, S. C., Liadaki, K. N., and Zannis, V. (2000). Structure and function of apolipoprotein A-I and high-density lipoprotein. *Curr. Opin. Lipidol.* **11**, 105–115.

Segrest, J. P., Jones, M. K., De Loof, H., and Dashti, N. (2001). Structure of apolipoprotein B-100 in low density lipoproteins. *J. Lipid Res.* **42**, 1346–1367.

Sharrock, W. J., Rosenwasser, T. A., Gould, J., Knott, J., Hussey, D., Gordon, J. I., and, Banaszak, L. (1992). Sequence of lamprey vitellogenin. *Implications for the lipovitellin crystal structure. J. Mol. Biol.* **226**, 903–907.

Sitaram, N., and Nagaraj, R. (1990). A synthetic 13-residue peptide corresponding to the hydrophobic region of bovine seminalplasmin has antibacterial activity and also causes lysis of red blood cells. *J. Biol. Chem.* **265**, 10438–10442.

Small, D. M., and Atkinson, D. (1997). The first beta sheet region of apoB (apoB21-41) is a amphipathic ribbon 50–60 Å wide and 200 Å long which initiates triglyceride binding and assembly of nascent lipoproteins. *Circulation* **96**, 1.

Sorci-Thomas, M., Kearns, M. W., and Lee, J. P. (1993). Apolipoprotein A-I domains involved in lecithin–cholesterol acyltransferase activation: Structure : function relationships. *J. Biol. Chem.* **268**, 21403–21409.

Spuhler, P., Anantharamaiah, G. M., Segrest, J. P., and Seelig, J. (1994). Binding of apolipoprotein A-I model peptides to lipid bilayers: Measurement of binding isotherms and peptide–lipid headgroup interactions. *J. Biol. Chem.* **269**, 23904–23910.

Srinivas, S. K., Srinivas, R. V., Anantharamaiah, G. M., Segrest, J. P., and Compans, R. W. (1992). Membrane interactions of synthetic peptides corresponding to amphipathic helical segments of the human immunodeficiency virus type-1 envelope glycoprotein. *J. Biol. Chem.* **267**, 7121–7127.

Steyrer, E., Barber, D. L., and Schneider, W. J. (1990). Evolution of lipoprotein receptors: The chicken oocyte receptor for very low density lipoprotein and vitellogenin binds the mammalian ligand apolipoprotein E. *J. Biol. Chem.* **265**, 19575–19581.

Tall, A. R., Small, D. M., Deckelbaum, R. J., and Shipley, G. G. (1977). Structure and thermodynamic properties of high density lipoprotein recombinants. *J. Biol. Chem.* **252**, 4701–4711.

Tytler, E. M., Segrest, J. P., Epand, R. M., Nie, S. Q., Epand, R. F., Mishra, V. K., Venkatachalapathi, Y. V., and Anantharamaiah, G. M. (1993). Reciprocal effects of apolipoprotein and lytic peptide analogs on membranes. Cross-sectional molecular shapes of amphipathic alpha helixes control membrane stability. *J. Biol. Chem.* **268**, 22112–22118.

Tytler, E. M., Anantharamaiah, G. M., Walker, D. E., Mishra, V. K., Palgunachari, M. N., and Segrest, J. P. (1995). Molecular basis for prokaryotic specificity of magainin-induced lysis. *Biochemistry* **34**, 4393–4401.

Venkatachalapathi, Y. V., Phillips, M. C., Epand, R. M., Epand, R. F., Tytler, E. M., Segrest, J. P., and Anantharamaiah, G. M. (1993). Effect of end group blockage on the properties of a class A amphipathic helical peptide. *Proteins* **15**, 349–359.

Wilson, C., Wardell, M. R., Weisgraber, K. H., Mahley, R. W., and Agard, D. A. (1991). Three-dimensional structure of the LDL receptor-binding domain of human apolipoprotein E. *Science* **252**, 1817–1822.

Wlodawer, A., Segrest, J. P., Chung, B. H., Chiovetti, R., Jr., and Weinstein, J. N. (1979). High-density lipoprotein recombinants: Evidence for a bicycle tire micelle structure obtained by neutron scattering and electron microscopy. *FEBS Lett.* **104**, 231–235.

Yancey, P. G., Bielicki, J. K., Johnson, W. J., Lund-Katz, S., Palgunachari, M. N., Anantharamaiah, G. M., Segrest, J. P., Phillips, M. C., and Rothblat, G. H. (1995). Efflux of cellular cholesterol and phospholipid to lipid-free apolipoproteins and class A amphipathic peptides. *Biochemistry* **34**, 7955–7965.

Yang, C. Y., Chen, S. H., Gianturco, S. H., Bradley, W. A., Sparrow, J. T., Tanimura, M., Li, W. H., Sparrow, D. A., DeLoof, H., and Rosseneu, M., et al. (1986). Sequence, structure, receptor-binding domains and internal repeats of human apolipoprotein B-100. *Nature* **323**, 738–742.

Zasloff, M., Martin, B., and Chen, H. C. (1988). Antimicrobial activity of synthetic magainin peptides and several analogues. *Proc. Natl. Acad. Sci. USA* **85**, 910–913.

CHAPTER 15

Interactions of pH-Sensitive Peptides and Polymers with Lipid Bilayers: Binding and Membrane Stability

Doncho V. Zhelev and David Needham

Department of Mechanical Engineering and Materials Science, Duke University, Durham, North Carolina 27708

I. Introduction
II. Interaction of Viral Fusion Peptides with the Membrane
III. Interaction of AcE4K with Lipid Bilayer Membranes (SOPC)
IV. Effect of Peptide Binding on Membrane Stability
V. Membrane Instability Induced by the pH-Sensitive Polymer Poly(2-Ethylacrylic Acid) (PEAA)
VI. Conclusion
References

The controlled release of content from liposomes is critical for their successful use as drug delivery systems. The most commonly used triggering mechanisms for content release are changes in the environment, such as a change in temperature or pH. In this case, the structure of the liposome membrane changes either by itself or by the presence of environmentally sensitive molecules, which leads to a change of membrane permeability. In this chapter we discuss the interaction of pH-sensitive molecules, such as the influenza fusion peptide mutant AcE4K and the polymer poly(2-ethylacrylic acid), with 1-stearoyl-2-oleoyl-*sn*-glycero-3-phosphocholine membranes. We emphasize the ability of these molecules to rearrange the structure of the membrane. This ability is derived from the strength of molecular binding to the membrane interface as well as from the cohesiveness of the membrane itself. The binding of molecules to the membrane has been studied extensively; however, the coupling between binding and membrane cohesiveness is a relatively uncharted area. The binding of pH-sensitive molecules to the membrane is coupled with their protonation, whereas membrane cohesiveness depends on membrane composition. A simple model is introduced where the binding to the membrane and the work for displacing the membrane lipids and creating a vacancy for the binding molecules determine the outcome of the event. The work for creating a vacancy is dependent on membrane cohesiveness, and can be increased by adding cholesterol. The model predicts that the addition of cholesterol will increase the threshold pH for content release. The experimental data fit very well to the predictions of the model. This effect provides an opportunity for designing systems to release drugs in environments of unique pH, such as tumor tissues and endosome interiors.

I. INTRODUCTION

Despite recent successes in using liposomes to retain agents, evade the body's defenses, and target, both passively and specifically, certain diseased tissues (Gabizon and Martin, 1997; Goren et al., 1996; Lee and Low, 1995; Suzuki et al., 1995), one of the major challenges to the use of liposomes as effective drug delivery systems is the release of encapsulated agents. This feature requires the design of appropriate triggering mechanisms to cause the release of encapsulated drugs in response to environmental conditions. pH is a promising trigger for release from liposomes, especially for acidic environments, such as tumor interstitial spaces and endosomes (Conner et al., 1984; Seki and Tirrell, 1984; Wang and Huang, 1987).

One strategy that has been investigated for pH-sensitive liposomes is the use of phase-changing lipids that undergo a liquid crystalline to hexagonal phase transition at an acidic pH. This transition allows the liposomes to fuse with biological membranes and release entrapped contents (Conner et al., 1984; Straubinger et al., 1985). Other strategies involve pH-sensitive macromolecules such as polymers and peptides (Bailey et al., 1997; Pecheur et al., 1997; Plank et al., 1994; Thomas and Tirrell, 1992). Interactions between these molecules and liposomes can also lead to membrane fusion. In addition, these types of interactions can cause an increase in permeability of synthetic membranes or even disruption and solubilization of liposomes (Thomas and Tirrell, 1992). The specific effects that such triggering, pH-sensitive molecules have on the bilayer membrane depend on their concentration relative to the lipid concentration. Studies of lipid structures for different ratios of the pH-sensitive polymer poly(2-ethylacrylic acid) (PEAA) and the lipid dipalmitoylphosphatidylcholine (DPPC) showed that at pH 8.5, liposomes were still present for a lipid to polymer ratio of 1 : 1. With a decrease in pH, the diameter of the liposomes decreased from approximately 90 nm in basic solutions to 16 nm due to a transition occurring near pH 6.5 (Borden et al., 1988). These results indicate that the polymer induces a vesicle-to-micelle transition that corresponds to the hydrophobic transition of the polymer (Borden *et al.*, 1988). At lower polymer : lipid ratios (near 1 : 100) PEAA acts on the bilayer to induce the formation of pores in diphytanoyl phosphatidylcholine bilayers rather than the complete destruction and micelle formation seen at higher polymer concentrations. These pores were found to be cation-selective and often stable for many seconds (Chung *et al.*, 1996). In these transitions, the pH-sensitive polymer and the lipid are mixed in solution. In nature, the pH-sensitive molecules that trigger the release of contents of cells or invade cells are attached to membrane structures. Examples are the fusion peptides of pH-sensitive viruses (White *et al.*, 1983). This feature of the viral fusion peptides is mimicked by attaching pH-sensitive molecules to membrane lipids of liposomes. Maeda *et al.* (1988) attached the polymer PEAA to the surface of phosphatidylcholine liposomes. With this design, the group found that only slight variations in pH, within the physiological range, caused leakage of an encapsulated fluorescent dye. A variation from pH 7 to 6.5 showed a complete release of vesicle contents within minutes (Maeda *et al.*, 1988). A similar construct was made by attaching a mutant of the influenza virus fusion peptide to distearoylglycerol (Bailey *et al.*, 1997). The resulting lipopeptide was capable of inducing pH-dependent leakage and (monolayer) fusion of vesicle bilayer membranes. The leakage of content induced by pH-sensitive molecules depends on the interaction of these molecules with the membrane. In this regard different membranes may interact differently with the trigger molecules.

The pH-dependent triggering of contents release by pH-sensitive polymers and peptides has been fairly well characterized in the literature for the usual pure lipid compositions, such as egg phosphatidylcholine (PC) and DPPC. It has been established that the destabilization of the lipid membrane by these molecules is driven by their hydrophobic nature, which depends on the degree of protonation as a result of the surrounding pH. However, there are no studies showing the role of the membrane mechanical properties in this process. The focus of our present work is on detemining the mechanism for membrane destabilization by pH-sensitive peptides and polymers. We emphasize the role of the membrane elasticity and the work needed to expand membranes in this process. Our rationale is that the rearrangement of the membrane lipids during membrane destabilization requires work, which is obtained from the decrease in free energy after protonation and binding to the membrane. The work for rearranging the membrane components is dependent on the cohesiveness of the membrane, which is characterized by the membrane area expansion modulus K and degree of area strain α. This rearrangement leads to membrane instability, usually measured by the amount of a marker dye released from the vesicle sample, which occurs at a fairly sharp and well-characterized pH for pure lipid membranes. An increase in the area expansion modulus produced, for example, by including cholesterol in the membrane increases membrane cohesiveness, and a higher degree of protonation of the trigger molecule is required in order to achieve the same degree of membrane instability (contents release). In this chapter, we discuss the mechanisms involved in these instability phenomena by considering two molecules, the peptide AcE4K and the pH-sensitive polymer poly(2-ethylacrylic acid) (PEAA). The peptide represents a group of fusion peptides related to the influenza fusion peptide, and the polymer represents a synthetic alternative to biological molecules.

II. INTERACTION OF VIRAL FUSION PEPTIDES WITH THE MEMBRANE

Fusion peptides are essential parts of viral fusion proteins. These peptides provide the attachments of the virus to the target membrane and are involved in the process of pore formation in the target membrane. The importance of fusion peptides for viral fusion is better understood by considering the overall process of fusion and the role of the different regions of fusion proteins in this process. Chapter 18 in this volume provides a comprehensive discussion of protein-mediated membrane fusion and more specifically the fusion mediated by the influenza hemagglutinin (HA). In our discussion we will mention only the major stages of the fusion process facilitated by HA and the role of the fusion peptide and will focus our attention on the changes that the fusion peptides undergo in this process.

The influenza fusion protein has two disulfide-linked subunits, HA1 and HA2 (Bullough *et al.,* 1994), one of which (HA1) contains the binding site for sialic acid (Sauter *et al.,* 1989), whereas the other (HA2) contains a highly conserved apolar sequence of about 22 amino acids at its N-terminal, called the fusion peptide (Durrer *et al.,* 1996; Gething *et al.,* 1986; Harter *et al.,* 1989). The fusion facilitated by HA2 does not require receptors on the host membrane, and thus it appears to involve only the lipid component of the membrane. The fusion process has several stages. In the first stage, the exposure of the virus to mildly acidic conditions (~pH 5) (Gaudin *et al.,* 1995; White and Wilson, 1987) results in a conformational change of the HA ectodomain (Brunner *et al.,* 1991; Shanguan *et al.,* 1998) and extension of the fusion peptide 100 Å away from its original position (Bullough *et al.,* 1994; Carr and Kim, 1993; Carr *et al.,* 1997). In the second stage of fusion, the exposed fusion peptide inserts into the host membrane (Ruigrok *et al.,* 1988) and creates a physical link between the viral envelope and the host membrane. The exposed fusion peptide can also insert in the viral envelope (Weber *et al.,* 1994), suggesting that the HA ectodomain is flexible. The third stage of viral fusion is the formation of the fusion pore. This stage is critical for the fusion event and is considered its climax. The formation of the fusion pore has two intermediates. In the first intermediate, the viral envelope and the host membrane are linked and the formation of "flickering" or reversible pores is observed (Chernomordik *et al.,* 1995). The linkage between the two membranes is confirmed by the exchange of lipids between their outside monolayers (Chernomordik *et al.,* 1998; Tse *et al.,* 1993). In the second intermediate the fusion pore increases in size, which is observed from an increase of the overall capacitance in patch clamp measurements (Chernomordik *et al.*, 1998). This intermediate is irreversible and results in complete fusion of the viral envelope with the host membrane.

The means of attachment of the fusion protein to the viral envelope plays an important role in the fusion process. Point mutations in the transmembrane region of HA2 or its complete substitution with glycosylphosphatidylinositol (GPI-HA) arrests fusion, even though it does not affect the hemifusion between the viral envelope and the host membrane and lipid transfer between them (Hussler *et al.*, 1997; Kemble *et al.*, 1994; Melikyan *et al.*, 1995). The labeling of the inner and outer membrane monolayers with different fluorescent markers shows that GPI-HA induces hemifusion, in which both the inside and the outside membrane monolayers exchange lipid, but there is no contents mixing (Hussler *et al.*, 1997). Other important conditions for pore formation are the presence of at least three HA2 trimers in the pore region (Bentz, 2000; Bentz *et al.*, 1990; Danieli *et al.*, 1996) and the tightness of the trimer aggregates (Bentz, 2000; Chernomordik *et al.*, 1998). Finally, the substitution of the transmembrane domain of HA with the transmembrane domain of other viral fusion proteins recovers the fusion capability of the protein (Schroth-Diez *et al.*, 1998). This final finding suggests that the role of the transmembrane domain is physical, and other means of attachment of the fusion peptide to the membrane can be explored in order to mimic the viral fusion event (Bailey *et al.*, 1997).

From the above discussion it is clear that the fusion peptide is essential for fusion. Thus, the study of its interaction with the membrane and the mimicking of this interaction by model systems are critical in order to more fully understand its role in the fusion event. The importance of the fusion peptide for fusion is demonstrated by the fact that its removal from the otherwise intact HA protein results in complete abolition of fusion (Ruigrok *et al.*, 1988). Furthermore, mutations of HA2 in the fusion peptide region affect significantly the ability of HA2 to fuse membranes. In this regard, the glycine at the amino-terminal is critical because its deletion abolishes fusion completely (Gething *et al.*, 1986). Several studies have focused on the ability of HA2 fusion peptide mutants to insert into membranes (Gray *et al.*, 1996; Lear and DeGrado, 1987; Rafalski *et al.*, 1991) or to initiate content leakage (Alford *et al.*, 1994; Stegmann *et al.*, 1985). These studies indicate that the ability of HA2 mutants to insert into membranes depends on their forming α-helical structures (Gray *et al.*, 1996). However, the ability to form α-helical structures is not enough to promote fusion (Burger *et al.*, 1991). The ability of the fusion peptides to insert into membranes has been demonstrated directly by their insertion into monolayers (Burger *et al.*, 1991; Rafalski *et al.*, 1991) as well as in vesicle membranes (Longo *et al.*, 1997). Recent Fourier-transform infrared (FTIR) measurements and electron paramagnetic resonance (EPR) measurements have shown that the position of the soluble wild-type fusion peptide in the membrane is at the membrane interface (Macoscko *et al.*, 1997; Zhou *et al.*, 2000) and is oriented parallel to the interface (Ishiguro *et al.*, 1996; Zhou *et al.*, 2000).

In a continuing effort to reveal the role of fusion peptides in membrane fusion, Bailey *et al.* (1997) synthesized the lipopeptide lipo-AcE4K and studied its ability to promote membrane hemifusion and fusion. The peptide AcE4K (Ac-GLFEAI AGFIENGWEGMIDGK) from the lipo-AcE4K is similar to the HA2 fusion peptide mutant E4 introduced by Rafalski *et al.* (1991), which, in turn, is similar to the wild-type fusion peptide w-20 from the X31 strain of the influenza virus. E4 is preferred over w-20, because it has better solubility in water and its binding to the membrane is similar to the binding of w-20 (Bailey *et al.*, 1997). The measurement of the secondary structure of the lipo-AcE4K in buffer showed that the peptide is in an α-helical conformation independent of pH (Bailey *et al.*, 1997), whereas the fluorescence from the tryptophan present in the peptide showed a blue shift with a decrease of pH from 7 to 5. This behavior is different from that of the wild-type HA2 peptide and its mutants, for which both the tryptophan fluorescence blue shift and the formation of α-helical structures are dependent on pH. To resolve this discrepancy, we have characterized the interaction of AcE4K with phophatidylcholine membranes (Zhelev *et al.*, 2001).

III. INTERACTION OF AcE4K WITH LIPID BILAYER MEMBRANES (SOPC)

Although AcE4K is similar to the E4 peptide studied by Rafalski *et al.* (1991), there are two important differences between the two peptides: the N-terminal of AcE4K is acylated, whereas that of E4 is not, and AcE4K has a lysine at its C-terminal. At pH 5, the insertion of E4 in the membrane is coupled to a blue shift of the fluorescence of its tryptophan and with the formation of α-helical secondary structure (Rafalski *et al.*, 1991). Similarly, AcE4K showed a pH-dependent fluorescence of its tryptophan maximum fluorescence and a pH-dependent formation of α-helical secondary structures (Fig. 1). The data in Fig. 1 show that AcE4K has two distinct secondary structures: random coil and α-helical. The peptide is primarily in a random coil conformation at high pH and forms an α-helical structure at low pH.

The measurement of both tryptophan fluorescence and circular dicroism spectra in the absence of lipid showed that there was no blue shift of the tryptophan fluorescence or the formation of α-helical structures with change in the pH, showing that both the measured blue shift and the formation of the α-helical structure are a direct result of peptide binding to the membrane.

The inspection of the helical wheel of AcE4K (Fig. 2) suggests that the folded (α-helical) peptide is amphiphilic and most likely inserts at the membrane interface. This hypothesis was tested by measuring the depth of insertion of AcE4K

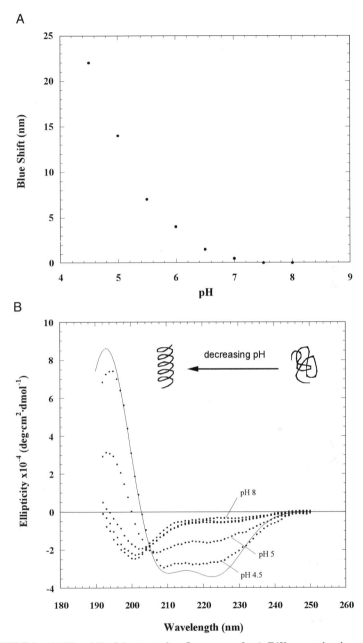

FIGURE 1 (A) Blue shift of the tryptophan fluorescence for AcE4K versus the change of the buffer pH. The peptide concentration is 10 μM and the SOPC concentration is 0.38 mM. (B) Circular dichroism spectra for AcE4K in solutions with different pH. The peptide concentration is 10 μM and the lipid concentration is 0.36 mM. The solid curve represents the calculated spectra for 100% ellipticity using the model of Chang *et al.* (1978).

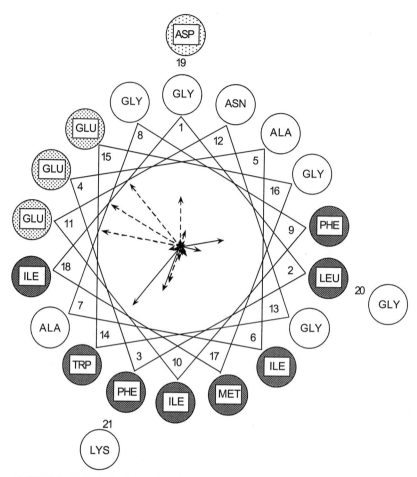

FIGURE 2   Helical wheel depiction of AcE4K. The black amino acids are hydrophobic and the white amino acids are hydrophilic. The gray amino acids have pK 4.7, and are expected to determine the protonation of AcE4K for pH close to 5. The arrows represent the Gibbs energy of amino acid transfer from solution to the PC membrane, measured by Wimley and White (1996). Solid arrows represent the negative Gibbs energies and the dashed arrows represent the positive energies at pH 8.

in the membrane. AcE4K has one tryptophan, the fluorescence of which is quenched by bromine. We studied the quenching of the tryptophan fluorescence by the dibromo lipids 1-palmitoyl-2-stearoyl(6–7)dibromo-*sn*-glycero-3-phosphocholine [PSPC(6–7)Br] and 1-palmitoyl-2-stearoyl(9–10)dibromo-*sn*-glycero-3-phosphocholine [PSPC(9–10)Br] as well as the fluorescence of tryptophan for AcE4K bound to 1-stearoyl-2-oleoyl-*sn*-glycero-3-phosphocholine (SOPC) membranes, to measure the depth of peptide insertion (Fig. 3). The tryptophan is close to the

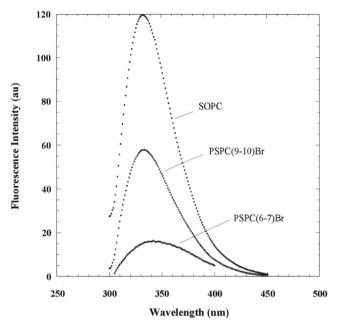

FIGURE 3 Fluorescence intensity of tryptophan from AcE4K inserted in SOPC membranes and membranes made of di-bromo lipids.

middle of the α helix, and therefore the depth of its insertion is a good estimate for the depth of insertion of AcE4K in the membrane. The depth of insertion of the tryptophan side chain was calculated using the parallax method (Chattopadhyay and London, 1987),

$$z_{F1} = \pm \frac{L_{12}^2 + (1/\pi C)\,\ln(F_1/F_2)}{2L_{12}}, \tag{1}$$

where z_{F1} is the distance of the tryptophan side chain from the shallower quencher PSPC(6–7)Br, L_{12} is the distance between the two quenchers, C is the quencher concentration per unit membrane area (in this case, the area occupied by the two quenchers was equal to the area of one lipid molecule), and the ratio F_1/F_2 is the ratio of the intensity of the tryptophan fluorescence from the vesicles made of PSPC(6–7)Br to the tryptophan fluorescence from the vesicles made of PSPC (9–10)Br. The measured depth of insertion of the tryptophan was 8 Å from the lipid's carbonyl group. (The position of the lipid carbonyl groups determines the dividing surface between the hydrophilic and the hydrophobic region of the membrane, which represents the membrane interface.) This distance is very close to the distance (5 Å) for the same side chain of the wild-type HA2 fusion peptide w-2 from the lipid's carbonyl group measured by EPR (Zhou *et al.*, 2000). The measurement of tryptophan quenching was performed at pH 5 and 6. In both cases the depth of insertion of the tryptophan side chain in the membrane was the same. This result shows that the depth of insertion of AcE4K in the membrane is independent of pH. This is similar to the depth of insertion of the wild-type HA2 fusion peptide (Lunenberg *et al.*, 1995; Zhou *et al.*, 2000), which also is independent of pH.

The apparent discrepancy between the dependence of the tryptophan blue shift and the formation of α-helical structure on pH and the independence of the depth of peptide insertion of pH is resolved by measuring the pH dependence of AcE4K partitioning in the membrane (Fig. 4). The partitioning of AcE4K in the SOPC membrane depends strongly on pH. Therefore, the measured pH dependences of the tryptophan blue shift and the formation of α-helical structures in Fig. 1 are a direct result of the increased amount of bound peptide as the pH decreases from 7.5 to 4.5. The partition coefficient allows us to determine the Gibbs energy of binding ΔG^0, which is calculated from

$$\Delta G^0 = -RT \ln K_{\mathrm{p}}, \qquad (2)$$

where R is the gas constant, T is the temperature, and K_{p} is the partition coefficient.

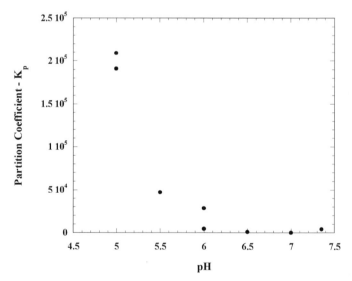

FIGURE 4   Partition coefficient K_{p} for AcE4K in SOPC membranes, determined with the filtration method. The vesicles and the peptide are added to a centrifuge tube with a filter that allows the solution with the peptide to pass, but retains the lipid. The partition coefficient is calculated by measuring the peptide concentration when half of the solution is filtered and comparing this concentration to the peptide concentration in a similar sample, but without the lipid. The partition coefficient for pH 4.5, which is not shown in the graph, is 1×10^6.

For pH 7, when almost all the peptide is in solution and the binding to the membrane is at a minimum, the Gibbs energy of binding is on the order of 3.4 kcal/mol. For pH 4.5, when almost all peptide is bound to the membrane, the Gibbs energy is on the order of 8.9 kcal/mol. The transition pH for the change of the Gibbs energy from minimum to maximum is close to pH 4.7, which is the pK for glutamatic acid and aspartic asid. AcE4K has three glutamatic acids and one aspartic acid, which suggests that the dependence of AcE4K binding on pH is coupled with amino acid protonation. The Gibbs energies of binding to the membrane of amino acids incorporated in small peptide chains have been measured by Wimley and White (1996). The Gibbs energy of binding for the glutamatic acid changes from −2.02 kcal/mol at pH 8 to 0.01 kcal/mol at pH 2, whereas the same Gibbs energy for the aspartic acid changes from −1.23 kcal/mol at pH 8 to 0.07 kcal/mol at pH 2. Thus, the total change of the Gibbs energy of binding for the three glutamatic acids and the aspartic acid of AcE4K is −7.4 kcal/mol when the pH changes from 8 to 2. This is close to the change, on the order of 5.5 kcal/mol, of the Gibbs energy of binding of AcE4K when the pH changes from 7.5 to 4.5. This similarity of the change of the Gibbs energy of binding of the four amino acids and of the whole fusion peptide supports the hypothesis that the dependence of AcE4K partitioning on pH is due primary to amino acid protonation and that the four amino acids are critical to this event.

If the protonation of AcE4K is the major factor governing the dependence of the Gibbs energy of binding on pH, the binding of AcE4K to the membrane will be driven by hydrophobic interactions. In this case the hydrophobic energy of the peptide will be reduced when it is inserted at the membrane interface, because the side chains of the hydrophobic amino acids will be in contact with the membrane's hydrocarbon region. At the same time, the insertion of the peptide in the membrane requires the creation of a vacancy among the membrane lipids, where the peptide is inserted. The creation of such a vacancy will require work of membrane expansion, which will decrease the gain of free energy due to the reduced exposed hydrophobic surface after peptide insertion. It is important to know how the work for creating a vacancy in the membrane interface compares to the gain of free energy during peptide binding.

We developed an assay for measuring the work for creating a vacancy for peptide insertion based on our assay for characterizing molecular exchange (Needham and Zhelev, 1995). In this assay (Fig. 5A), a single vesicle is held by a pipet and two other pipets deliver either peptide solution or peptide-free solution, respectively. Because peptide insertion is expected to increase the total area of the lipid vesicle membrane, the kinetics of both peptide adsorption and desorption can be determined by measuring the vesicle projection length in the holding pipet, which is proportional to the total vesicle area. The measured increase of the vesicle projection length in the presence of peptide and the decrease of the projection length

A

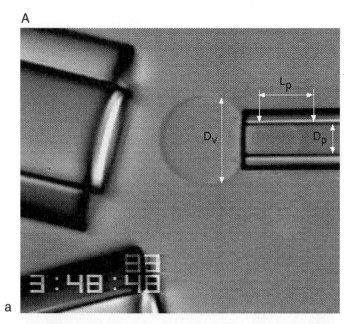

a

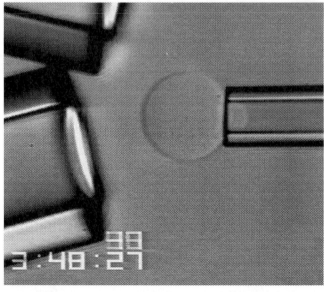

b

FIGURE 5 (A) Videomicrograph showing the arrangement of the holding pipet and the two flow pipets that deliver different solutions at a controlled flow rate to the test vesicle. (a) Initial vesicle projection length L_p in flow conditions is established when the lower pipet is used to flow peptide-free buffer over the vesicle. (b) The lower pipet is replaced by the upper pipet, which delivers buffer containing peptide. The projection length inside the holding pipet increases as the peptide intercalates into the vesicle membrane.

B

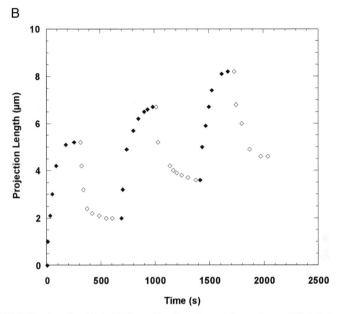

FIGURE 5 (*Continued*)   (B) AcE4K-peptide adsorption and desorption at pH 5.4. A single vesicle is exposed continuously to peptide buffer for 7 min (*closed diamonds*) and then the peptide is "washed" (*open diamonds*) with peptide-free buffer at the same pH. The adsorption and the desorption of the peptide is repeated three times. The projection length change after every "washing" of the adsorbed peptide is similar, and represents the area occupied by the adsorbed peptide before the desorption started. The projection length change during peptide desorption corresponds to a 6% change of the vesicle area.

when the vesicle is "washed" with peptide-free solution (Fig. 5B) correspond to peptide adsorption and desorption, respectively. The initial increase in vesicle projection length likely includes both increase in membrane area and loss of vesicle volume due to compromised membrane permeability that occurs when peptide is in the membrane. Thus, the decrease of the vesicle projection length during peptide desorption is used to calculate the area occupied by a single peptide molecule in the membrane (Zhelev *et al.*, 2001). The measured area of AcE4K in the SOPC membrane is 226 Å^2. The work for creating such a vacancy for peptide insertion W_v^0 is dependent on the membrane elasticity and the area of the lipids associated with the inserted peptide (Zhelev, 1998),

$$W_v^0 = K \frac{A_p^2}{A_{lp}} N_a, \tag{3}$$

where K is the area expansion modulus, which characterizes membrane elasticity, A_p is the area occupied by a single peptide molecule in the membrane, A_{lp} is the initial area occupied by the lipids "associated" with the inserted peptide, and N_a is Avogadro's number.

The area expansion modulus for the SOPC membranes is equal to 190 mN/m (Needham and Nunn, 1990) and the initial area occupied by the lipids "associated" with the inserted peptide is equal to 3770 $Å^2$ (Zhelev et al., 2001). Then, the work for insertion of 1 mol of AcE4K in the SOPC membrane is -3.7 kcal/mol. This value is almost half of the value of the Gibbs energy for peptide binding at pH 5.5, which is -6.4 kcal/mol. Thus, the maximum Gibbs energy for peptide binding is equal to the sum $\Delta G^0 + W_v^0$, and for pH 5.5 is on the order of -10.1 kcal/mol. The ratio of the maximum Gibbs energy to the area occupied by 1 mol of peptide is equal to the peptide apparent interfacial tension. Interestingly, for pH 5.5 this apparent interfacial tension is on the order of 31 mN/m, and for pH 4.5 it is on the order of 38.6 mN/m. These values are very close to the value of the apparent interfacial tension for the membrane hydrocarbon region, on the order of 33 mN/m (Zhelev, 1998). The calculation of the latter apparent interfacial tension is based on the measured activation energy for transfer of a single hydrocarbon to water, which is on the order of 0.68 kcal/mol (Israelashvili, 1991). Another interfacial tension of interest is the alkane–water interfacial tension, which is on the order of 50 mN/m (Small, 1986). The alkane–water interfacial tension represents the maximum possible hydrocarbon–water interfacial tension. The comparison of the above interfacial tensions suggests that the major driving force for AcE4K binding is the gain in free energy due to removing the amino acid side chains from their water environment and bringing them into a hydrophobic environment.

The comparison of the work for creating a vacancy for AcE4K with the gain in free energy after the binding of the peptide shows that the work for creating a vacancy represents a major portion of the total energy gain. This work is strongly dependent on the tension of the inside and the outside membrane monolayers, which is accounted for by the dependence between the work to create a vacancy and the area expansion modulus [Eq. (3)]. We continue this discussion below, when we consider the leakage across the membrane induced by the insertion of the pH-sensitive PEAA polymer.

IV. EFFECT OF PEPTIDE BINDING ON MEMBRANE STABILITY

In addition to creating a link between the viral envelope and the host membrane, the bound peptide rearranges the structure of the host membrane, which ultimately results in the formation of a membrane pore. As discussed above, the fusion pore involves pore formation in both the host membrane and the viral envelope. The transmembrane portion of HA2 plays an important role in the overall fusion dynamics, especially for the formation of the pore in the viral envelope.

However, the pore region in the host membrane is equally important, and its characterization is necessary in order to understand the overall fusion event. We will start with a discussion of the size of the pore region and the apparent peptide-to-lipid ratio in it.

The size of the pore region has been measured from electron micrographs of influenza virions fusing with glycophorin-containing liposomes (Kanaseki et al., 1997). The radius of the pore region is on the order of 10 nm, which corresponds to 7850 $Å^2$. The area of one lipid molecule is 65 $Å^2$ (McIntosh and Simon, 1986); therefore, there are 120 lipids in one of the membrane monolayers in the pore region. The analysis of the kinetics of influenza virus fusion with cell membranes shows that the fusion site contains at least three HA trimers (Bentz, 2000; Bentz et al., 1990; Danieli et al., 1996). On the other hand, the fusion peptide can partition into both the viral envelope and the host membrane (Weber et al., 1994). Then, the lipid-to-peptide ratio in the pore region ranges from 30, when half of the fusion peptides partition into the host membrane, to 13, when all fusion peptides are inserted into the host membrane. These values are higher than the same lipid-to-peptide ratio for AcE4K determined from titration calorimetry measurements (Zhelev et al., 2001), which is 58. Taking into account the fact that the area of the fusion peptide in the membrane is almost 3.5 times larger than the area of a lipid molecule, we find that the presence of the fusion peptide in the pore region represents a significant change of the membrane composition in this region. This change of composition can affect both the strength of the membrane and its overall elasticity.

The critical membrane tension, which is a measure of the membrane strength, for membranes containing peptide and for peptide-free membranes at pH 5.5 is shown in Fig. 6A for three bathing solutions, glucose, ionic buffer, and ionic buffer plus peptide. It is seen that the presence of inserted peptide decreases the membrane strength; however, this decrease is not dramatic. The measurement of the area expansion modulus requires the determination of the area change as function of applied membrane tension, and the slope of the change of the applied membrane tension versus the relative area change is equal to the area expansion modulus (Fig. 6B). It is seen from the data in Fig. 6B that the area expansion modulus can be measured before peptide adsorption and after peptide desorption. However, the area expansion modulus cannot be measured in the presence of adsorbed peptide, because, under these conditions, the membrane is leaky and the measured vesicle projection length in the holding pipet is dependent on both membrane area change and change of vesicle volume. This behavior is consistent with the initial stage of viral fusion, when the conductivity of the membrane in the pore region changes spontaneously in either direction ("flickering pores") and lipid mixing between the host membrane and the viral envelope is still not observed (Chernomordik et al., 1998; Tse et al., 1993). Comparison of the data from Figs. 6A and 6B shows that whereas the membrane in the presence of adsorbed peptide is leaky, its

A

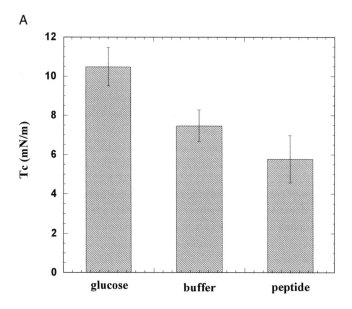

B

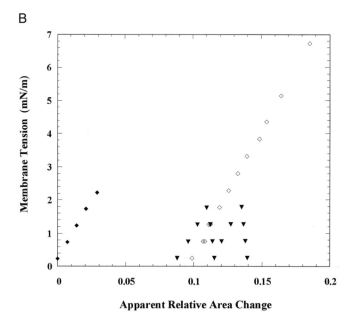

overall global stability is not significantly compromised. Indeed, our micropipet measurements show that under these conditions the membrane is stable for more that $\frac{1}{2}$ h. A decrease of the pH below pH 5 results in a pronounced decrease of membrane stability, and for pH below 4.8 the membrane breaks under low holding tension within seconds after the initial exposure to peptide solution. The threshold pH for membrane breakdown is close to the pH 5 where the influenza viral fusion takes place. Therefore it is important to characterize the peptide–membrane interactions for this pH.

In characterizing membrane instability induced by the AcE4K peptide, we tried to answer two questions: (1) Is the membrane instability a result of the binding of single peptide molecules or of peptide aggregates? (2) Is the binding of the fusion peptide in the pore region reversible? The first question was prompted by the closeness of the threshold pH for membrane instability to the pK for the protonated glutamatic acid and aspartic acid of AcE4K, whereas the second question was aimed at allowing us to come to a conclusion about the peptide–peptide interactions in the pore region. To answer the first question we compared the lifetime of vesicle membranes directly exposed to peptide solution at pH 4.7 to that of membranes that had been "preloaded" with peptide at pH 5.5 and then exposed to peptide solution at pH 4.7 (Fig. 7). The lifetime of the membranes already containing saturating amount of peptide is almost four times larger than the lifetime of the membranes directly exposed to peptide solution (Zhelev $et\ al.$ 2001). Since the position and the depth of insertion of the peptide are independent of pH, the lowering of the pH is not expected to affect the peptides already bound to the membrane. (The exposure of vesicle membranes "preloaded" with peptide at pH 5.5 to peptide-free

---

FIGURE 6 (A) Critical membrane tension for SOPC membranes. The critical membrane tension is measured using single vesicles made in sucrose. The solution in the vesicle exterior is glucose, peptide-free buffer, or buffer with 10 μM AcE4K peptide. The critical membrane tension in the buffer solutions is measured 5 min after exposing the vesicle to the buffer. The critical membrane tension in the peptide solution is measured 10 min after exposure to peptide solution. The temperature is 25°C and the pH is 5.5. (B) Measurement of the area expansion modulus from the dependence of the applied membrane tension on the relative change of the vesicle area (all data are for a single vesicle). The solid diamonds represent the data before the exposure of the vesicle to peptide solution. The solid triangles represent measurements in the presence of peptide and the open diamonds represent the data after peptide "washing". For each set of data the membrane tension is increased and then decreased. For the data in the absence of peptide the area changes corresponding to the same membrane tensions are the same and the measured data during the decrease of the membrane tension retrace the same data when the membrane tension is increased. The data for the decrease of the membrane tension in the presence of peptide do not retrace the same data when the membrane tension is increased. This is a result of the presence of adsorbed peptide, which induces membrane leakage and vesicle volume loss. The volume loss affects the measured vesicle projection length in the holding pipet, and is calculated as an apparent increase of the membrane area. The area expansion modulus can be measured only when the measurement of the change of the vesicle projection length during the increase of the membrane tension is retraced by the change of the projection length during the decrease of this tension.

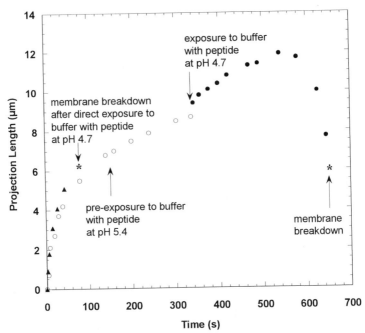

FIGURE 7 Change of the vesicle projection length after direct exposure to AcE4K peptide solution at pH 4.7 (*triangles*) and initial exposure to peptide solution at pH 5.4 (*open circles*) and then exposure to peptide solution at pH 4.7 (*closed circles*). The direct exposure of the vesicle to peptide solution at pH 4.7 results in vesicle breakdown during adsorption. The exposure of the vesicle to peptide solution at pH 5.4 leads to increase of the projection length. The subsequent exposure to peptide solution at pH 4.7 results in additional increase of the projection length and then fast decrease of the projection length. The increase of the projection length could be a result of additional area change and/or volume loss. The fast decrease of the projection length is a result of pore formation and vesicle swelling. The peptide concentration is 10 μM.

solution at pH 4.7 resulted in peptide desorption similar to the desorption at pH 5.5.) Therefore, the only difference between the two experiments is the presence of the peptide already inserted in the membrane. The inserted peptide apparently decreases the available area for peptide binding; therefore, the overall rate of peptide transfer to the membrane will be decreased. Thus, the increased membrane lifetime in the presence of preadsorbed peptide means that the species destabilizing the membrane was likely transferred to the membrane from the bulk solution. This observation does not prove directly that peptide aggregates are involved in destabilizing the membrane; however, because only two secondary structures were measured for AcE4K (Fig. 1B), random coil in the bulk and α-helical in the membrane, it is likely that the instability of the membrane is induced by peptide aggregates formed in the solution.

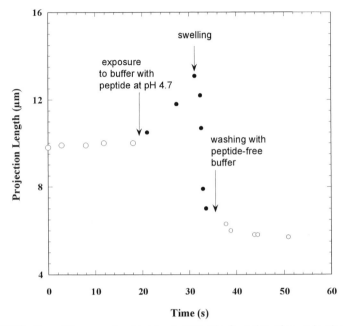

FIGURE 8 Reversible pores stimulated by AcE4K. The formation of pores is stimulated by exposing a single vesicle to peptide solution with reduced concentration at pH 4.7. After pore formation, the vesicle starts to swell and its projection length in the holding pipet starts to decrease. In the process of swelling, the vesicle is exposed to peptide-free buffer at pH 4.7. The exposure to peptide-free buffer stops the swelling almost immediately, and the projection length in the holding pipet approaches a steady state.

The reversibility of peptide binding is tested by "washing" the vesicles with peptide-free solution after the formation of a large pores. The data in Fig. 7 show that the vesicle with preadsorbed peptide did not break instantaneously, but first swelled; then the membrane failed because of the increase of the interior osmotic pressure. The conditions for swelling in these experiments were provided by using larger molecules (sucrose) in the vesicle interior compared to the molecules in the bathing medium surrounding the vesicle (glucose, sodium chloride, etc.) (Zhelev *et al.*, 2001).

The data in Fig. 8 represent an experiment in which the vesicle was exposed to peptide that induced vesicle swelling and then, during the swelling, was washed with peptide-free solution. Immediately after exposing the vesicle to the peptide-free solution, the swelling stopped and the peptide that was adsorbed to the vesicle membrane desorbed. This immediate cessation of vesicle swelling after exposure to peptide-free solution shows that the peptide molecules in the pore region do not interact strongly with each other and do not form oligomers. Therefore, the number of peptides in the pore region is not fixed and the size of the pore will depend on the number of involved peptide molecules. This lack of peptide oligomerization at the pore region explains the existence of the "flickering" pores, where the pores are formed and disappear depending on the apparent association and dissociation of the fusion peptides from the host membrane.

Finally, it is instructive to consider the formation of a pore in the host membrane as an energy-driven process and to compare the involved energies and work terms in order to determine the role of peptide binding in viral fusion. The measured Gibbs energy of binding at pH 5 is 30 kJ/mol, which corresponds to the Gibbs energy of binding for one molecule, 5×10^{-20} J, or about $12kT$. There are at least 4–5 bound peptides in the pore region, which have a total Gibbs energy from $48kT$ to $60kT$. The pore region in the membrane can be viewed as a dimple with a diameter on the order of 10 nm, whereas the size of the actual pore is on the order of 3–4 nm (Kanaseki *et al.*, 1997). Thus, the pore region can be modeled as a dome with an opening at its top (Kozlov and Chernomordik, 1997). This structure has a curvature, and therefore a bending energy associated with the deformation of the membrane. The bending energy of a pore region with diameter 10 nm and an opening at the top of 3.5 nm is $60kT$ (Kozlov and Chernomordik, 1997). This is similar to the energy of binding of five peptides. On the other hand, the work for breakdown (Zhelev, 1998) of the pore region is used as an estimate for the work to rearrange the membrane structure and form an opening in the membrane. For a pore region of 10 nm and for a membrane with a tensile strength of 10 mN/m and an area expansion modulus of 190 mN/m, this work is $2.6kT$. The value of the work for breakdown in the pore region is much smaller than the energy of binding. Therefore, the major role of the fusion peptides in viral fusion is to provide the anchoring between the viral envelope and the host membrane by forming the membrane dimple in the pore region.

V. MEMBRANE INSTABILITY INDUCED BY THE pH-SENSITIVE POLYMER POLY(2-ETHYLACRYLIC ACID) (PEAA)

In the above discussion of the action of the influenza virus fusion peptide in viral fusion we concluded that the protonation of the glutamatic acids and the aspartic acid of AcE4K is the major event initiating peptide binding and eventually leading to membrane instability and pore formation. Also, we concluded that the work for creating a vacancy in the membrane is significant compared to the gain of the free energy after peptide binding. In support of these arguments we have explored the role of membrane elasticity in the ability to create membrane defects by characterizing the binding of the pH-sensitive polymer poly(2-ethylacrylic acid) (PEAA) to SOPC/cholesterol membranes. PEAA is similar to AcE4K, in that its binding to membranes is pH-dependent. In aqueous solution, at pH above 6.2, PEAA is an extended random coil, whereas below this threshold pH it forms a globular hydrophobic coil (Borden *et al.*, 1987). A similar transition is observed

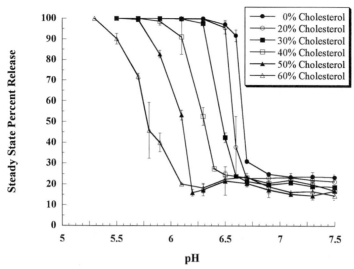

FIGURE 9 Plot of steady-state release of 6-CF (60 min after PEAA polymer addition) versus pH for 100-nm liposomes composed of SOPC and containing 0, 20, 30, 40, 50, and 60 mol% cholesterol.

for concentrated solutions of AcE4K, which below pH 4.5 form peptide aggregates and above pH 4.5 exist as a soluble random coil. The pH of transition of PEAA from an extended coil to a globular form increases with decreasing ionic strength and decreases with decreasing polymer molecular weight (Thomas and Tirrell, 1992). Interestingly, in the presence of lipid membranes the midpoint of the transition is shifted from pH 6.2 to 6.5 (Borden *et al.,* 1987), which is consistent with the increase of pH observed for AcE4K binding to the membrane relative to the pH of peptide self-aggregation. The relative amount of entrapped 6-carboxyfluorescein (6-CF) released from the vesicle after 1 h was used as a measure of the ability of PEAA to destabilize the membrane (Mills *et al.,* 1999). Figure 9 shows the dependence of the amount of released 6-CF on pH. The experimental data show that the pH for 6-CF release decreases with increasing cholesterol concentration in the membrane. The addition of cholesterol to the SOPC membrane increases the area expansion modulus from 190 to 1250 mN/m (Needham and Nunn, 1990). This is a roughly 6.5 times increase in the area expansion modulus, which according to Eq. (2), corresponds to the same increase for the work to create a vacancy. The work for creating a vacancy is compared to the Gibbs energy of polymer binding. The Gibbs energy of PEAA binding is dependent on the polymer partition coefficient in a similar way as the Gibbs energy of AcE4K binding is dependent on the peptide partition coefficient. The partition coefficient of PEAA binding is directly related to the equilibrium constant of polymer binding, which in turn is dependent on the

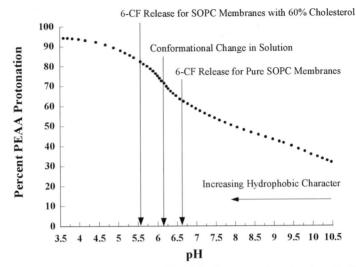

FIGURE 10   Plot showing percentage of PEAA that is protonated as a function of buffer pH.

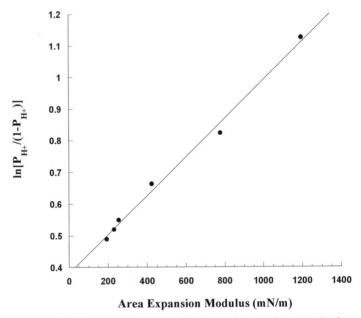

FIGURE 11   Plot of $\ln[P_{H+}/(1 - P_{H+})]$, where P_{H+} is the fraction of protonated polymer, versus the area expansion modulus for membranes containing different mole percent cholesterol. Numbers in parentheses represent mole percent cholesterol in the bilayer. The dependence of $\ln[P_{H+}/(1 - P_{H+})]$ on the area expansion modulus is almost linear, which is in agreement with Eq. (4).

fraction of the protonated polymer (Fig. 10) according to

$$\Delta G^0 = \text{const} - RT \ln\left(\frac{P_{H^+}}{1 - P_{H^+}}\right), \tag{4}$$

where P_{H^+} is the fraction of the protonated polymer.

The Gibbs energy of PEAA binding depends also on the area expansion modulus, and this dependence is given by Eq. (3). Therefore if the same percentage of released 6-CF for membranes with different mole percent of cholesterol corresponds to the same Gibbs energy of polymer binding, the area expansion modulus and the fraction of the protonated polymer will be related by

$$\ln\left(\frac{P_{H^+}}{1 - P_{H^+}}\right) = \text{const} + \frac{A_p^2 N_a}{RTA_{lp}} K, \tag{5}$$

where A_p is the area of the polymer in the membrane and A_{lp} is the initial area of the lipids associated with one polymer molecule. The area of the polymer in the membrane and the number of lipids per bound polymer are unknown; however, Eq. (5) predicts that there will be a linear relation between $\ln[P_{H^+}/(1 - P_{H^+})]$ and the area expansion modulus K. Figure 11 shows that this dependence indeed is linear.

The dependence of the work for vacancy formation on the area expansion modulus provides a useful tool for the controlled release of drugs from liposomes. Figure 12 shows the pH at onset of release for each cholesterol composition, along

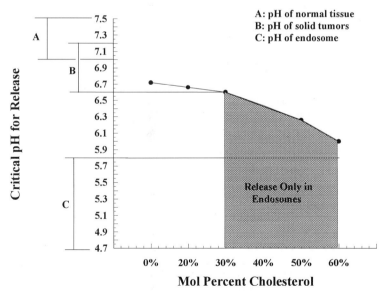

FIGURE 12 Plot of the critical pH at which onset of release of contents occurs for each of the cholesterol–lipid systems, along with the pH ranges for normal and tumor tissues (Lee *et al.*, 1996) and endosomal compartments (Severin *et al.*, 1994).

with the pH ranges of normal and tumor tissues and of endosomal compartments (Lee *et al.,* 1996; Severin *et al.,* 1994). It is seen from this figure that, with a pH slightly less than normal for tissues and blood, vesicles with less that 30% cholesterol will release their content into the surrounding tumor tissues, but vesicles containing more than 30% cholesterol will release their content only into the endosome, with no release in normal tissue. This type of delivery system is useful for genetic drugs and cytotoxic agents that require direct access to the cytosol or the nucleus. Although liposomal delivery agents are a long way from reaching the specificity and control of biological delivery systems, the combination of a pH-sensitive polymer and engineered liposomes could be an initial step in this direction.

VI. CONCLUSION

We have demonstrated in this chapter the importance of the membrane mechanical properties for molecular insertion. The insertion of molecules in the membrane interior requires both change of the Gibbs energy of the bound molecule relative to the Gibbs energy of the free molecule and work for creating a vacancy in the membrane for the inserted molecule. The Gibbs energy of binding represents the overall change of the free energy of the system. The comparison of this energy to the work for creating a vacancy calculated from the measured area per molecule and intermembrane tension shows that the value of the work for creating a vacancy is significant compared to the value of the Gibbs energy. This shows that the Gibbs energy of binding is substantially reduced compared to the maximum Gibbs energy of binding, where the same molecule binds at zero work for insertion. This simple model for molecular binding describes very well the pH-dependent content leakage from SOPC/cholesterol vesicles induced by PEAA protonation. Understanding the relationship between molecular binding and membrane cohesiveness, characterized by the area expansion modulus, provides a powerful tool for modulating the threshold for membrane instability, which is a necessary feature for the design of environmentally sensitive drug delivery systems.

Acknowledgments
D. Z. acknowledges support from NIH grant HL57629 and D. N. acknowledges support from GM40162. We also thank Dr. Natalia Stoicheva, Jeff Mills, and Dr. Gary Eichenbaum for contributions to this work.

References
Alford, D., Ellens, H., and Bentz, J. (1994). Fusion of influenza virus with sialic acid-bearing target membranes. *Biochemistry* **33,** 1977–1987.
Bailey, A. L., Monk, M. A., and Cullis, P. R. (1997). pH-Induced destabilization of lipid bilayers by a lipopetide derived from influenza hemagglutinin. *Biochim. Biophys. Acta* **1324,** 232–244.

Bentz, J. (2000). Membrane fusion mediated by coiled coils: A hypothesis. *Biophys J.* **78,** 886–900.

Bentz, J., Ellens, H., and Alford, D. (1990). An architecture for the fusion site of influenza hemagglutinin. *FEBS Lett.* **276,** 1–5.

Borden, K. A., Eum, K. M., Langley, K. H., and Tirrell, D. A. (1987). On the mechanism of polyelectrolyte-induced structural reorganization in thin molecular films. *Macromolecules* **20,** 454–456.

Borden, K. A., Eum, K. M., Langley, K. H., Tan, J. S., Tirrell, D. A., and Voycheck, C. L. (1988). pH-dependent vesicle-to-micelle transition in an aqueous mixture of dipalmitoylphosphatidylcholine and a hydrophobic polyelectrolyte. *Macromolecules* **21,** 2649–2651.

Brunner, J., Zugliani, C., and Mischler, R. (1991). Fusion activity of influenza virus PR8/34 correlates with a temperature-induced conformational change within the hemagglutinin ectodomain detected by photochemical labeling. *Biochemistry* **30,** 2432–2438.

Bullough, P. A., Hughson, F. M., Skehel, J. J., and Wiley, D. C. (1994). Structure of influenza haemagglutinin at the pH of membrane fusion. *Nature* **371,** 37–43.

Burger, K. N. J., Wharton, S. A., Demel, R. A., and Verkleij, A. J. (1991). The interaction of synthetic analogs of the N-terminal fusion sequence of influenza virus with a lipid monolayer. Comparison of fusion-active and fusion defective analogs. *Biochim. Biophys. Acta* **1065,** 121–129.

Carr, C. M., and Kim, P. S. (1993). A spring-loaded mechanism for the conformational change of influenza hemagglutinin. *Cell* **73,** 823–832.

Carr, C. M., Chaundry, C., and Kim, P. S. (1997). Influenza hemagglutinin is spring-loaded by a metastable native conformation. *Proc. Natl. Acad. Sci. USA* **94,** 14306–14313.

Chang, C. T., Wu, C.-S. C., and Yang, J. T. (1978). Circular dichroic analysis of protein conformation: Inclusion of the β-turns. *Analyt. Biochem.* **91,** 13–31.

Chattopadhyay, A., and London, E. (1987). Parallax method for direct measurement of membrane penetration depth utilizing fluorescence quenching by spin-labeled phospholipids. *Biochemistry* **26,** 39–45.

Chernomordik, L., Chanturya, A., Green, J., and Zimmerberg, J. (1995). The hemifusion intermediate and its conversion to complete fusion: Regulation by membrane composition. *Biophys. J.* **69,** 922–929.

Chernomordik, L., Frolov, V. A., Leikina, E., Bronk, P., and Zimmerberg, J. (1998). The pathway of membrane fusion catalyzed by influenza hemagglutinin: Restriction of lipids, hemifusion, and lipidic fusion pore formation. *J. Cell Biol.* **140,** 1369–1382.

Chung, J. C., Gross, D. J., Thomas, J. L., Tirrell, D. A., and Opsahl-Ong, L. R. (1996). pH-sensitive, cation-selective channels formed by a simple synthetic polyelectrolyte in artificial bilayer membranes. *Macromolecules* **29,** 4636–4641.

Conner, J., Yatvin, M. B., and Huang, L. (1984). pH-sensitive liposomes: Acid-induced liposome fusion. *Proc. Natl. Acad. Sci. USA* **81,** 1751–1781.

Danieli, T., Pelletier, S. L., Henis, Y. I., and White, J. M. (1996). Membrane fusion mediated by the influenza virus hemagglutinin requires the concerted action of at least three hemagglutinin trimers. *J. Cell Biol.* **133,** 559–569.

Durrer, P., Galli, C., Hoenke, S., Corti, C., Gluck, R., Vorherr, T., and Brunner, J. (1996). H^+-induced membrane insertion of influenza virus hemagglutinin involves the HA2 amino-terminal fusion peptide but not the coiled coil region. *J. Biol. Chem.* **271,** 13417–13421.

Gabizon, A., and Martin, F. (1997). Polyethylene glycol-coated (pegylated) liposomal doxorubicin: Rationale for use in solid tumours. *Drugs* **54**(Suppl. 4), 15–21.

Gaudin, Y., Ruigrok, R., and Brunner, J. (1995). Low-pH induced conformational changes in viral fusion proteins: Implications for the fusion mechanism. *J. Gen. Virol.* **76,** 1541–1556.

Gething, M.-J., Doms, R. W., York, D., and White, J. M. (1986). Studies on the mechanism of membrane fusion: Site-specific mutagenesis of the hemagglutinin of influenza virus. *J. Cell Biol.* **107,** 11–23.

Goren, D., Horowitz, A. T., Zalipsky, S., Woodle, M. C., Yarden, Y., and Gabizon, A. (1996). Targeting of stealth liposomes to erbB-2 (Her/2) Receptor: *In vitro* and *in vivo* studies. *Br. J. Cancer* **74,** 1749–1765.

Gray, C., Tatulian, S. A., Wharton, S. A., and Tamm, L. K. (1996). Effect of N-terminal glycine on the secondary structure, orientation, and interaction of the influenza hemagglutinin fusion peptide with lipid bilayers. *Biophys. J.* **70,** 2275–2286.

Harter, C., James, P., Bachi, T., Semenza, G., and Brunner, J. (1989). Hydrophobic binding of the ectodomain of influenza hemagglutinin to membranes occurs through the "fusion peptide." *J. Biol. Chem.* **264,** 6459–6464.

Hussler, F., Claque, M. J., and Herrmann, A. (1997). Meta-stability of the hemifusion intermediate induced by glycosylphosphatidylinositol-anchored influenza hemagglutinin. *Biophys. J.* **73,** 2280–2291.

Ishiguro, R. N., Matsumoto, M., and Takahashi, S. (1996). Interaction of fusogenic synthetic peptide with phspholipid bilayers: Orientation of the peptide α-helix and binding isotherm. *Biochemistry* **35,** 4976–4983.

Israelashvili, J. N. (1991). "Intermolecular and Surface Forces." Academic Press, San Diego, CA.

Kanaseki, T., Kawasaki, K., Murata, M., Ikeuchi, Y., and Ohnishi, S. (1997). Structural features of membrane fusion between influenza virus and liposome as revealed by quick-freezing electron microscopy. *J. Cell Biol.* **137,** 1041–1056.

Kemble, G. W., Danieli, T., and White, J. M. (1994). Lipid-anchored influenza hemagglutinin promotes hemifusion, not complete fusion. *Cell* **76,** 383–391.

Kozlov, M. M., and Chernomordik, L. V. (1997). A mechanizm of protein-mediated fusion: Coupling between refolding of the influenza hemagglutinin and lipid rearrangements. *Biophys. J.* **75,** 1384–1396.

Lear, J. D., and DeGrado, W. F. (1987). Membrane binding and conformational properties of peptides representing the N-terminus of influenza HA2. *J. Biol. Chem.* **262,** 6500–6504.

Lee, R. J., and Low, P. S. (1995). Folate-mediated tumor cell targeting of liposome-entrapped doxorubicin *in vitro. Biochim. Biophys. Acta* **1233,** 134–144.

Lee, R. J., Wang, S., and Low, P. S. (1996). Measurement of endosome pH following folate receptor-mediated endocytosis. *Biochim. Biophys. Acta* **1312,** 237–242.

Longo, M. L., Waring, A. J., and Hammer, D. A. (1997). Interaction of the hemagglutinin fusion peptide with lipid bilayers: Area expansion and permeation. *Biophys. J.* **73,** 1430–1439.

Lunenberg, J., Martin, I., Nussler, F., Ruysschaert, J.-M., and Herrmann, A. (1995). Structure and topology of the influenza virus fusion peptide in lipid bilayers. *J. Biol. Chem.* **270,** 27606–27614.

Macoscko, J. C., Kim, C.-H., and Shin, Y.-S. (1997). The membrane topology of the fusion peptide region of influenza hemagglutinin determined by spin-labeling EPR. *J. Mol. Biol.* **267,** 1139–1148.

Maeda, M., Kumano, A., and Tirrell, D. A. (1988). H^+-induced release of contents of phosphatidylcholine vesicles bearing surface-bound polyelectrolyte chains. *J. Am. Chem. Soc.* **110,** 7455–7459.

McIntosh, T. J., and Simon, S. A. (1986). Hydration force and bilayer deformation: A reevaluation. *Biochemistry* **25,** 4058–4066.

Melikyan, G. B., White, J. M., and Cohen, F. S. (1995). GPI-anchored influenza hemagglutinin induces hemifusion to both red blood cell and planar bilayer membranes. *J. Cell Biol.* **131,** 679–691.

Mills, J., Eichenbaum, G., and Needham, D. (1999). Effect of bilayer cholesterol and surface-grafted poly(ethyleneglycol) pH-induced release of contents from liposomes by polyethylacrylic acid. *J. Liposome Res.* **9,** 275–290.

Needham, D., and Nunn, R. S. (1990). Elastic deformation and failure of lipid membranes containing cholesterol. *Biophys. J.* **58,** 997–1009.

Needham, D., and Zhelev, D. V. (1995). Lysolipid exchange with lipid vesicle membranes. *Ann. Biomed. Eng.* **23,** 287–298.

Pecheur, E.-I., Hoekstra, D., Sainte-Marie, J., Maurin, L., Bienvenue, A., and Philippot, J. R. (1997). Membrane anchorage brings about fusogenic properties in a short synthetic peptide. *Biochemistry* **36,** 3773–3781.

Plank, C., Oberhauser, B., Mechtler, K., Koch, C., and Wagner, E. (1994). The influence of endosome-disruptive peptides on gene transfer using synthetic virus-like gene transfer system. *J. Biol. Chem.* **269,** 12918–12924.

Rafalski, M., Ortiz, A., Rockwell, A., van Ginkel, L. C., Lear, J. D., DeGrado, W. F., and Wilshut, J. (1991). Membrane fusion activity of the influenza virus hemagglutinin: Interaction of HA2 N-terminal peptides with phospholipid vesicles. *Biochemistry* **30,** 10211–10220.

Ruigrok, R. W. H., Aitken, A., Calder, L. J., Martin, S. R., Skehel, J. J., Wharton, S. A., Weis, W., and Wiley, D. C. (1988). Studies on the structure of the influenza virus hemagglutinin at the pH of membrane fusion. *J. Gen. Virol.* **69,** 2785–2795.

Sauter, N. K., Bernarski, M. D., Wutzburg, B. A., Hanson, J. E., Whitesides, G. M., Skehel, J. J., and Wiley, D. C. (1989). Hemagglutinins from two influenza virus variants bind to sialic acid derivatives with millimolar dissociation constants: A 500-MHz proton nuclear magnetic resonance study. *Biochemistry* **28,** 8388–8396.

Schroth-Diez, B., Ponimaskin, E., Reverey, H., Schmidt, M. F., and Herrmann, A. (1998). Fusion activity of transmembrane and cytoplasmic domain chimeras of the influenza virus glycoprotein hemagglutinin. *J. Virol.* **72,** 133–141.

Seki, K., and Tirrell, D. A. (1984). pH-depepndent complexation of poly(acrylic acid) derivatives with phospholipid vesicle membranes. *Macromolecules* **17,** 1692–1698.

Severin, T., Müller, B., Giese, G., Uhl, B., Wolf, B., Hauschildt, S., and Kreutz, W. (1994). pH-dependent LAK cell cytotoxicity. *Tumor Biol.* **15,** 304–310.

Shanguan, T., Siegel, D. P., Lear, J. D., Axelsen, P. H., Alford, D., and Bentz, J. (1998). Morphological changes and fusogenic activity of influenza virus hemagglutinin. *Biophys. J.* **74,** 54–62.

Small, D. M. (1986). "The Physical Chemistry of Lipids: From Alkanes to Phospholipids. Handbook of Lipid Research," Vol. 4. Plenum Press, New York.

Stegmann, T., Hoekstra, D., Scherphof, G., and Wilshut, J. (1985). Kinetics of pH dependent fusion between influenza virus and liposomes. *Biochemistry* **24,** 3107–3113.

Straubinger, R. M., Duzgunes, N., and Papahadjopoulos, D. (1985). pH-sensitive liposomes mediate cytoplasmic delivery of encapsulated macromolecules. *FEBS Lett.* **179,** 148–154.

Suzuki, S., Wantanabe, S., Masuko, T., and Hashimoto, Y. (1995). Preparation of long-circulating immunoliposomes containing adriamycin by a novel method to coat immunoliposomes with poly(ethylene glycol). *Biochim. Biophys. Acta* **1245,** 9–16.

Thomas, J. L., and Tirrell, D. A. (1992). Polyelectrolyte-sensitized phospholipid vesicles. *Acc. Chem. Res.* **25,** 336–342.

Tse, F. W., Iwata, A., and Almers, W. (1993). Membrane flux through the pore formed by a fusogenic viral envelope protein during cell fusion. *J. Cell Biol.* **121,** 543–552.

Wang, C.-Y., and Huang, L. (1987). pH-sensitive immunoliposomes mediate target-cell-specific delivery and contolled expression of a foreign gene in mouse. *Proc. Natl. Acad. Sci. USA* **84,** 7581–7855.

Weber, T., Paesold, G., Galli, C., Mischler, R., Semenza, G., and Brunner, J. (1994). Evidence for H^+-induced insertion of influenza hemagglutinin HA2 N-terminal segment into viral membrane. *J. Biol. Chem.* **269,** 18353–18358.

White, J. M., and Wilson, I. A. (1987). Antipeptide antibodies detect steps in a protein conformational change. Low pH activation of the influenza virus hemagglutinin. *J. Cell Biol.* **105,** 2887–2896.

White, J. M., Kielian, M., and Helenius, A. (1983). Membrane fusion proteins of enveloped animal viruses. *Q. Rev. Biophys.* **16,** 151–195.

Wimley, W. C., and White, S. H. (1996). Experimentally determined hydrophobicity scale for proteins at membrane interfaces. *Nat. Struct. Biol.* **3,** 842–848.

Zhelev, D. V. (1998). Material property characteristics for lipid bilayers containing lysolipid. *Biophys. J.* **75,** 321–330.

Zhelev, D. V., Stoicheva, N., Scherrer, P., and Needham, D. (2001). Interaction of synthetic HA2 influenza fusion peptide analog with model mebranes. *Biophys. J.* **81,** 285–304.

Zhou, Zh., Macosko, J. C., Hughes, D. W., Sayer, B. G., Hawes, J., and Epand, R. M. (2000). ^{15}N NMR study of the ionization properties of the influenza virus fusion peptide in zwitterionic phospholipid dispersions. *Biophys. J.* **78,** 2418–2425.

CHAPTER 16

The Hydrophobicity Threshold for Peptide Insertion into Membranes

Charles M. Deber,*† **Li-Ping Liu,***† **Chen Wang,***† **Natalie K. Goto,***†
and Reinhart A. F. Reithmeier†‡

*Structural Biology and Biochemistry, Research Institute, Hospital for Sick Children,
Toronto M5G 1X8, Ontario, Canada; Departments of †Biochemistry and ‡Medicine,
University of Toronto, Toronto M5S 1A8, Ontario, Canada

I. Introduction
II. Results
    A. Peptides Designed to Mimic Transmembrane Segments
    B. Quantifying "Threshold Hydrophobicity"
    C. Statistical Analysis of Protein Transmembrane Segments
    D. Construction of TM Helix and Non-TM Helix Databases
    E. Biological Correlation of "Threshold Hydrophobicity"
III. Discussion
    A. HPLC-Determined Hydrophobicity Scale
    B. Comparison with Other Hydropathy Scales
    C. Protein TM Segments Are Intrinsically Competent for Membrane Insertion
    D. Relationship of Primary Sequence to Partitioning of TM Segments
IV. Conclusions
    References

Peptides designed as model transmembrane segments are shown to insert spontaneously from water into micellar membranes only when their mean residue hydrophobicity is equivalent to or greater than a polyalanine segment. By using circular dichroism-derived peptide helicity as a probe of membrane insertion and correlating high-performance liquid chromatography-derived peptide hydrophobicity with helicity, we were able to determine quantitatively the hydrophobicity threshold. This analysis allows (1) assignment of hydropathy indices to the 20 commonly occurring amino acids and (2) assessment of the membrane insertion

Current Topics in Membranes, Volume 52

potential of a given transmembrane segment as "all-or-nothing," depending upon whether its segmental hydrophobicity, calculated from the indices of its component amino acids, surpasses or fails to meet the minimum hydrophobicity requirement for integration into the membrane. The minimum hydrophobicity threshold was found to be satisfied by >96% of over 5000 transmembrane segments derived from a database of single- and multispanning intrinsic membrane proteins. When applied *in vivo,* the notion of "threshold hydrophobicity" would allow the selective incorporation of transmembrane segments into the lipid bilayer during the biosynthetic translocation process without the requirement of any additional expenditure of energy.

I. INTRODUCTION

Minimal hydrophobicity of transmembrane (TM) segments is required during membrane protein biosynthesis for stop-transfer activity and for their effective release from the translocon into the endoplasmic reticulum (ER) membrane (Andrews and Johnson, 1996; Davis and Model, 1985; Johnson and van Waes, 1999; Spiess *et al.,* 1989). The notion of such a "threshold hydrophobicity" has also been considered important in translocon-independent spontaneous insertion of membrane proteins (Lee *et al.,* 1997; Shore *et al.,* 1995; Whitley *et al.,* 1996). By fine tuning hydrophobicity around a threshold, a membrane-interactive protein/peptide can be converted into a soluble form or vice versa (Chen and Kendall, 1995; Chung and Thompson, 1996). Accordingly, the localization of polypeptide segments in biological systems can be characterized as "all-or-nothing," where the hydrophobicity of a protein segment either surpasses or fails to meet the physical requirements for integration into the membrane. Yet, a unifying model for quantitation of "threshold hydrophobicity," especially as it applies to the step of membrane partitioning (Deber and Goto, 1996; Li and Deber, 1994; Wimley and White, 1996), remains to be presented.

By using a *de novo*-designed peptide library of alanine-based model TM peptides, we were able to detect the existence of a "threshold hydrophobicity" requirement (Liu *et al.,* 1996). This initial study was expanded to include all 20 commonly occurring amino acids, to allow a quantitative measure of the hydrophobicity threshold, which governs the fundamental interaction of polypeptide segments with membranes (Liu and Deber, 1998a). When considered in the context of a survey on a transmembrane segment database derived from single-spanning and multispanning membrane proteins, we find that over 96% of all TM helices satisfy the "threshold hydrophobicity" requirement. Here we provide an overview of these findings, which lead to the proposal that native transmembrane segments are intrinsically competent for spontaneous insertion into a lipid bilayer during the biosynthetic translocation process.

II. RESULTS

A. Peptides Designed to Mimic Transmembrane Segments

Hydrophobic residues such as Ala, Val, Ile, and Phe typically dominate the natural TM segments of proteins (Table I). To mimic their properties in a simplifying manner, we prepared a set of *de novo*-designed Ala-based peptides, KKAAA XAAAAAXAAWAAXAAAKKKK–amide, where X = each of the 20 commonly occurring amino acids. Peptides of this type, where X was substituted by each of the 20 commonly occurring amino acids, were synthesized by the continuous-flow 9-fluorenylmethoxycarbonyl (Fmoc) solid-phase method (Liu and Deber, 1997). When folded into an α-helical conformation, the 19-residue hydrophobic core of the peptide is of sufficient length to span a phospholipid bilayer (Reithmeier, 1995). Distributions of the three X residues have been designed to preserve both angular and longitudinal symmetry around the helix, thereby minimizing bias from the amphipathic character of the peptide that may arise when X is substituted by polar residues. This feature assures that all peptides can approach lipid environments in a uniform manner.

Peptides remained soluble and monomeric in both aqueous and membrane-mimetic media over a wide concentration range (5–250 μM), based on circular dichroism (CD) measurements and size exclusion high-performance liquid chromatography (HPLC). These advantages were realized through the inclusion of multiple lysine residues at both termini. In aqueous buffer, peptides formed a range of helical conformations, in accordance with the intrinsic α-helical propensities of the X residues (shown for selected peptides in Fig. 1) (Blaber *et al.,* 1993; Chou and

TABLE I

Selected Transmembrane Segments of Single-Spanning and Multispanning Membrane Proteins[a]

	TM Sequence	References
Single-spanning		
Glycophorin A (92–114)	ITLIIFGVMAGVIGTILLISYGI	MacKenzie *et al.,* 1997
M13 coat protein (21–39)	YIGYAWAMVVVIVGATIGI	Papavoine *et al.,* 1997
Insulin receptor (957–979)	IIIGPLIFVFLFSVVIGSIYLFL	Ullrich *et al.,* 1985
Multispanning		
Bacteriorhodopsin (F167–188)	VASTFKLRNVTVVLWSAYPVV	Pebay-Peyroula *et al.,* 1997
Photosynthetic reaction center (L33–53)	FFGVSAIFFIFLGVSLIGYAA	Deisenhofer *et al.,* 1985
Light-harvesting complex (C123–143)	SILAIWATQVILMGAVEGYRI	Savage *et al.,* 1996
Cystic fibrosis transmembrane conductance regulator (human) (195–215)	LALAHFVWIAPLQVALLMGLI	Riordan *et al.,* 1989

[a]Numbers in parentheses indicate the protein residue positions of the segments given. β-Branched residues (Val, Ile, Thr) are underlined; Gly residues are in boldface (see text).

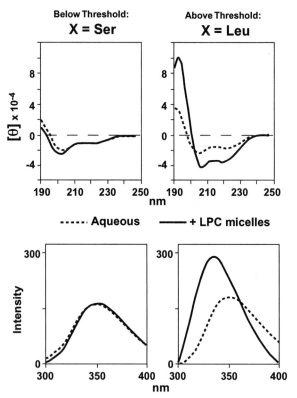

FIGURE 1 Spectra of model peptides KKAAAXAAAAAXAAWAAXAAAKKKK–amide, where X = Ser (*left*) and X = Leu (*right*). Top: Circular dichroism spectra. Bottom: Fluorescence spectra. Spectra are compared in aqueous buffer (*dashed lines*) and in lysophosphatidylcholine (LPC) micelles (*solid lines*). The peptides shown represent opposing examples of X-residue hydrophobicity and side-chain chemistry. CD spectra: Peptide concentrations were typically 30 μM in aqueous buffer and in various lipid micellar solutions. The aqueous buffer was prepared from 10 mM Tris–HCl, 10 mM NaCl, pH 7.0. Lipid micelle concentration: 10 mM lysophospholipid in 10 mM Tris–HCl, 10 mM NaCl, pH 7.0. Curves reported are based on triplicate measurements; standard deviation is ±1%. Fluorescence spectra: Peptide concentration was 4 μM in aqueous buffer (10 mM Tris–HCl, 10 mM NaCl, pH 7.0), in 4 mM lipid micelles. The spectra were recorded at 37°C with subtraction of the background of buffer or lipids in the absence of peptide. In aqueous buffer, the emission maximum for all peptides is centered around 350 nm and blue-shifted ca. 10–15 nm upon interaction with LPC micelles for peptides that assume full α-helical conformations.

Fasman, 1974; King and Sternberg, 1996). The peptides were then exposed to lipid micelles, because these are systems widely adopted as mimics of biological membranes (Kaiser and Kezdy, 1983; MacKenzie *et al.*, 1997; Papavoine *et al.*, 1997) and used to reconstitute membrane proteins *in vitro* to facilitate their subsequent insertion into lipid bilayers (Casey and Reithmeier, 1993; Tanford,

1980). In zwitterionic lysophosphatidylcholine micelles (LPC; net charge $= 0$ at pH 7.0) (Fig. 1, top), peptides either adopted fully helical conformations (X = Leu) or retained the conformation observed in aqueous buffer (X = Ser). The fully helical peptides inserted into the micelles while the other peptides remained free in solution. Peptide insertion into membranes by this subgroup was confirmed by observation of lipid-induced fully helical conformations and peptide Trp fluorescence emission spectra (Fig. 1, bottom) (10- to 15-nm blue shifts with concomitant enhancement in intensity).

B. Quantifying "Threshold Hydrophobicity"

From the combined HPLC (Liu and Deber, 1998a), CD, and fluorescence experiments described above, the requirement for a hydrophobicity threshold became apparent from the dramatic segregation of the 20 peptides into two distinct groups: fully membrane-inserted "hydrophobes" (where X = Ala, Cys, Ile, Leu, Met, Phe, Val, Trp, and Tyr peptides) and noninserted "hydrophiles" (X = Asn, Arg, Asp, Gln, Glu, Gly, His, Lys, Pro, Ser, and Thr peptides) (Fig. 2). The absence of partially inserted peptides suggests that a defined threshold must exist for successful insertion of the peptides. This group of hydrophobic amino acids (except Gly) corresponds to the amino acids found to be enriched in native TM segments (Landolt-Marticorena *et al.*, 1993). To quantitate this minimum hydrophobicity requirement, procedures similar to those of Sereda *et al.* 1994 were used to construct a hydrophobicity

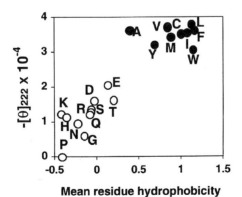

Mean residue hydrophobicity

FIGURE 2 Experimentally determined peptide helicity ($\theta_{222\,nm}$) in LPC micelles versus mean residue hydropathy of peptide hydrophobic segments AAAXAAAAAXAAWAAXAAA. Mean residue hydropathy is derived from the data in the "hydrophobicity" column of Table II by summing the values for the 19 residues in a given segment and then dividing by 19. Single letters refer to the guest X residues within the corresponding peptides. Peptides fully inserted into LPC micelles are shown as filled circles; noninserted peptides are shown as open circles. [Reprinted with permission from Liu and Deber (1998a). Copyright © 1998 John Wiley & Sons, Inc.]

TABLE II

Hydrophobicity^a and Helicity^b Scales Determined Experimentally from the Properties of
KKAAAXAAAAAXAAWAAXAAAKKKK–NH$_2$ Peptides^a

Amino acid	Hydrophobicity	Helicity	KD^c	GES^d	Eisenberg^e
F	5.00 (1)	1.26 (4)	2.8 (4)	3.7 (1)	1.19 (2)
W	4.88 (2)	1.07 (10)	−0.9 (11)	1.9 (7)	0.81 (5)
L	4.76 (3)	1.28 (2)	3.8 (3)	2.8 (4)	1.06 (4)
I	4.41 (4)	1.29 (1)	4.5 (1)	3.1 (3)	1.38 (1)
M	3.23 (5)	1.22 (6)	1.9 (6)	3.4 (2)	0.64 (6)
V	3.02 (6)	1.27 (3)	4.2 (2)	2.6 (5)	1.08 (3)
C	2.50 (7)	0.79 (19)	2.5 (5)	2 (6)	0.29 (9)
Y	2.00 (8)	1.11 (8)	−1.3 (12)	−0.7 (13)	0.26 (10)
A	−0.16 (9)	1.24 (5)	1.8 (7)	1.6 (8)	0.62 (7)
T	−1.08 (10)	1.09 (9)	−0.7 (9)	1.2 (9)	−0.05 (12)
E	−1.50 (11)	0.85 (18)	−3.5 (18)	−8.2 (17)	−0.74 (15)
D	−2.49 (12)	0.89 (16)	−3.5 (17)	−9.2 (19)	−0.9 (18)
Q	−2.76 (13)	0.96 (13)	−3.5 (15)	−4.1 (15)	−0.85 (17)
R	−2.77 (14)	0.95 (14)	−4.5 (20)	−12.3 (20)	−2.53 (20)
S	−2.85 (15)	1.00 (11)	−0.8 (10)	0.6 (11)	−0.18 (13)
G	−3.31 (16)	1.15 (7)	−0.4 (8)	1 (10)	0.48 (8)
N	−3.79 (17)	0.94 (15)	−3.5 (16)	−4.8 (16)	−0.78 (16)
H	−4.63 (18)	0.97 (12)	−3.2 (14)	−3 (14)	−0.4 (14)
P	−4.92 (19)	0.57 (20)	−1.6 (13)	−0.2 (12)	0.12 (11)
K	−5.00 (20)	0.88 (17)	−3.9 (19)	−8.8 (18)	−1.5 (19)

Adapted from Deber *et al.*, (2001).

^{a}Hydrophobicity of each guest **X** residue, scaled from HPLC retention times (Liu and Deber, 1998a).
The corresponding values of three known hydrophobicity scales are provided for comparison. The rank
of each amino acid with respect to each scale is indicated in parentheses.

^{b}Helical propensity of each guest **X** residue, scaled from circular dichroism measurements of peptides
in *n*-butanol (Liu and Deber, 1998b; Wang *et al.*, 1999).

^{c}Kyte and Doolittle (1982).

^{d}Engelman *et al.* (1986).

^{e}Eisenberg *et al.* (1984).

scale from relative retention times obtained from reversed-phase HPLC experiments (Table II). Based upon this scale, we defined a threshold hydrophobicity (at ca. 0.4, calculated as the mean residue hydrophobicity of the X = Ala peptide, which contains 18 Ala and one Trp in its hydrophobic core), above which peptide segments spontaneously integrate into LPC micelles.

C. Statistical Analysis of Protein Transmembrane Segments

Because the experimentally determined threshold hydrophobicity is derived from model peptides, we felt it of interest to use statistical analysis to demonstrate the existence of threshold hydrophobicity as a minimal requirement in natural

TM segments. However, due to the scarcity of solved membrane protein structures, it would be difficult to ascertain the statistical significance of an analysis on this handful of experimentally confirmed Protein Data Bank structures (Preusch, 1998). Therefore, we chose the TMbase as our target for analysis. Although TMbase includes bone fide TM segments from solved membrane protein structures, one must acknowledge that it contains mostly "putative" TM segments. Despite the impreciseness of this process, all the proteins included in TMbase are known to be membrane proteins, the segments chosen are demonstrably hydrophobic, and the segments are of membrane-spanning lengths. These three features insure that the overwhelming majority of the TMbase entries are indeed TM segments; in practice, it is the precise entry/exit residues that cannot be specified. Since the property we are assessing is the experimentally derived threshold hydrophobicity, there is no circularity involved in this approach.

D. Construction of TM Helix and Non-TM Helix Databases

The TM helix database was downloaded and developed from http://ulrec3. unil.ch/pub/tmbase or http://ncbi.nlm.nih.gov/repository/TMbase, which contains >8000 entries of TM segments (Hofmann and Stoffel, 1993). The database for non-TM helical segments (nontransmembrane α helices from globular and membrane proteins) was constructed by searching the SwissProt release 34 (April 10, 1997) under the keyword *helix*. The initial non-TM data-base contained 4649 entries. We further modified the databases as follows: (i) Duplication of TM segments from closely related proteins was avoided and (ii) segments of length shorter than 19 residues (length requirement for spanning a lipid bilayer) were also deleted. After these operations, 5444 entries for TM helices and 339 entries for non-TM helices remained. The TM helix database was then divided into single-spanners (876 helices) and multispanners (4568 helices), based on whether the segments were derived from single-spanning or multiple-spanning membrane proteins, respectively. The non-TM helix database was ultimately depleted after the second operation from 4649 entries to 339 entries, because most of the helical segments in the non-TM helix database were observed to be comparatively shorter (average length ca. 10 residues).

Values derived from the hydrophobicity scale developed in the present work identified ca. 96% of the segments in the TMbase to be at or above the "threshold." It is emphasized that this is not a result of the way the threshold was set: The demarcation point of threshold hydrophobicity set itself from the experimental observation that a stretch of Ala residues represents the approximate average minimum hydrophobicity requirement for spontaneous membrane insertion. In addition, the choice of numerical range is arbitrary, and the results are the same even if our scale (which uses limits of $+5$ and -5) is set up to be entirely positive or negative.

E. Biological Correlation of "Threshold Hydrophobicity"

The biological relevance to proteins of this peptide-based, experimentally determined threshold hydrophobicity was evaluated by examining mean residue hydrophobicity (calculated by the hydropathy scale in the present work; see Table II) for several thousand helices using databases derived from TM segments of single- and multispanning membrane proteins (Hofmann and Stoffel, 1993) and from non-TM helices (nontransmembrane α helices from globular and membrane proteins) (Fig. 3a). From the resulting distributions of mean residue hydrophobicities, we

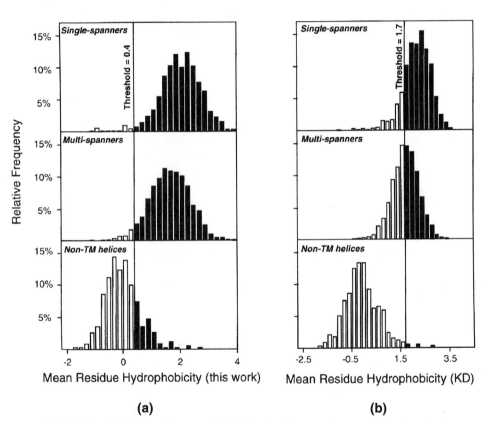

(a)   **(b)**

FIGURE 3 Distributions of mean residue hydropathy of transmembrane α helices from single-spanning integral membrane proteins ($n = 876$), TM α helices from multiple-spanning membrane proteins ($n = 4568$), and nontransmembrane α helices of length ≥ 19 residues ($n = 339$). (a) Hydropathy values obtained using the scale given in Table II, with threshold hydropathy indicated as a solid vertical line at the value of 0.4 as calculated for the X = Ala peptide. (b) Hydropathy values obtained using the Kyte–Doolittle scale (Kyte and Doolittle, 1982) with threshold hydropathy indicated as the value of 1.7 as calculated from the KD scale for the X = Ala peptide. [Reprinted with permission from Liu and Deber (1998a). Copyright © 1998 John Wiley & Sons, Inc.]

found that >97% of single-spanning TM segments and >96% of TM segments in multispanning (polytopic) membrane proteins are above the calculated threshold. Thus, the parameters for membrane insertion established for model peptides can be applied to natural TM segments contained in both single- and multispanning membrane proteins. When a comparative analysis was performed on 339 non-TM helices of length ≥ 19 residues, 22% were determined as above the threshold; thus, less than 2% of the total collected non-TM helices fulfill the requirements of both length ≥ 19 residues and hydrophobicity above the threshold. Hence, the length of a helix also appears to play a major role in determining the ability of the segment to partition into the membrane environment, a circumstance which likely results from the requirement for an appropriate match with the bilayer thickness (Kuroiwa *et al.*, 1991).

Parallel analysis was carried out using the Kyte–Doolittle (KD) hydropathy scale (Kyte and Doolittle, 1982), which is derived largely from considerations of buried versus exposed residues in a sample of crystalline globular proteins, using threshold hydropathy ($= 1.7$) as calculated by the KD scale for the X = Ala peptide. Results indicated that 84% of single-spanning TM segments and 52% of multispanning TM segments were calculated as above the threshold (Fig. 3b). In contrast, the KD scale correctly predicted greater than 98% of non-TM helices as below the threshold hydrophobicity required for a TM segment.

III. DISCUSSION

A. HPLC-Determined Hydrophobicity Scale

The manner in which the peptides interact with the hydrophobic surface of the column simulates the situation whereby hydrophobic segments in proteins approach a membrane surface. In effect, the HPLC-determined hydrophobicity scale with synthetic peptides should thus derive from essentially the same primary factors that determine the behavior of each side chain upon interaction with nonpolar environments. However, the measurement of peptide retention times by HPLC may in some instances reflect a combination of hydrophobicity and conformation (i.e., aqueous-based helical content prior to membrane insertion); the present work applies the method specifically to nonamphiphilic hydrophobic peptides. Thus, there arises the possibility that the measured HPLC retention times are determined simply by peptide helical content on the nonpolar surface of the HPLC column. This, however, is not borne out by the experimental results. For example, several peptides (X = F, W, L, I, M, V, C, Y, and A) possess virtually the same helical content in the nonpolar interior of micelles, yet their retention times range from 22.58 to 18.88 min (Liu and Deber, 1998a). If the HPLC measurements had indeed created a secondary structure propensity scale rather than a true hydrophobicity scale, one should observe a linear correlation between retention time and helicity; instead, we observe a *segregation* between two distinct clusters (Fig. 2), which is a direct manifestation of the threshold hydrophobicity requirement.

In light of these considerations, we proceeded to examine the "nonpolar phase" helical propensities of the 20 amino acids, by constructing a scale from CD measurements of the relative helical content of the 20 host-guest peptides in *n*-butanol (Liu and Deber, 1998b). Results are summarized in Table II. While the key result is that the most hydrophobic residues are essentially the most helix-supporting residues in nonpolar media, which explains fundamentally why β-formers in water, such as Val and Ile (Chou and Fasman, 1974), so readily support helical structure in membranes, not all residues were parallel in their hydrophobicity versus helicity. Most notably, Trp and Gly exchanged positions in the scales: Trp was ranked number 1 in hydrophobicity, but number 11 in helicity, whereas Gly was ranked number 16 in hydrophobicity, but was number 7 in its ability to support helices in nonpolar phases. These findings not only emphasize the importance of the dual phenomena of hydrophobicity and helicity, they also provide, in part, an explanation as to why Gly, the classic helix-breaker in aqueous phases, can be so prevalent in membranes.

B. Comparison with Other Hydropathy Scales

Studies on TM mimic peptides described herein led us to examine our findings in the context of other popular hydropathy scales. The observation that the KD scale, for example, underestimates the hydrophobicity threshold in a fraction of cases arises simply because several of its residues (notably Trp and Tyr, along with Glu, Asp, and Arg) have been assigned the much more hydrophilic positions versus the present HPLC-derived scale (Table II). Conceptually similar results are obtained from the Goldman–Engelman–Steitz (GES) scale (Engelman *et al.*, 1986) and the Eisenberg scale (Eisenberg *et al.*, 1984) (Table II). This interesting effect—that the "hydropathy" of a given residue is not a universal property, but must be considered a function of molecular environment—is most apparent in TM segments of multispanning proteins, which have a considerable total proportion of these five residues as required for functions such as channel formation. A further manifestation of the environmental dependence of hydropathy is that KD is essentially 100% successful in identifying non-TM helical segments (Fig. 3b), which demonstrates the great power of KD to address the folding of globular proteins, whereas the present HPLC scale overestimates about 25% of them. Thus, by using the present HPLC-derived hydrophobicity scale (which identifies TM segments) in conjunction with the KD scale (which identifies non-TM helices), one is able to identify a TM segment readily and accurately based simply upon its experimentally determined threshold hydrophobicity.

C. Protein TM Segments Are Intrinsically Competent for Membrane Insertion

Current hypotheses regarding structural/mechanistic details of TM segment insertion include opening of the translocon and lateral movement of the TM segment into the lipid bilayer (Do *et al.,* 1996; Martoglio *et al.,* 1995). The present observations of spontaneous peptide partitioning suggest that once the release of the nascent TM segment from the translocon has been triggered, there exists a universal threshold hydrophobicity requirement for movement of the segment into the lipid bilayer. Further, according to our delineation of threshold hydrophobicity, an overwhelming majority of all naturally occurring TM segments (ca. 96%) (Fig. 3a) are intrinsically competent for spontaneous insertion into the hydrophobic phase of the bilayer once the energy of passing any polar globular domain through the membrane has been circumvented. Since the translocon threads polar segments destined for the lumen through the membrane in a thermodynamically viable manner, the problem of TM segment insertion is reduced to an interaction between the TM segment and the lipids. This situation is simulated by a micellar suspension, because the hydrophobic core of a micelle can be penetrated isotropically; thus, outlying polar regions of the peptide do not need to pass through the thermodynamic barrier of the core. Thus, movement of TM segments into the lipid bilayer can be driven by the hydrophobic effect without the requirement for an additional expenditure of energy, a process regulated by the translocon through control of the timing of this release process (Borel and Simon, 1996; Thrift *et al.,* 1991).

D. Relationship of Primary Sequence to Partitioning of TM Segments

The observation that partitioning into lipid systems is essentially indifferent to primary sequences of model peptides as long as hydrophobicity is maintained is supported by the fact that lipid-interactive residues are not well-conserved across species (Taylor *et al.,* 1994) [e.g., 16% identity for the photosynthetic reaction center protein family (Rees *et al.,* 1989)]. The threshold value also correlates well with a minimum biosynthesis requirement that 19 Ala residues are sufficient for stop-transfer function (Kuroiwa *et al.,* 1991). In this context, it may be noted that although Ala-based peptides were studied in the present work, any segment of TM length with average hydropathy above the threshold, even a segment devoid of Ala, would behave similarly. This latter phenomenon is observed, for example, in several of the seven TM segments of the multispanning transport protein bacteriorhodopsin, where highly nonpolar residues compensate for charged residues (Henderson *et al.,* 1990). In a preliminary experimental test of this notion, we designed and synthesized the peptide KKATALVGAASLTAWVGLASAKKKK–amide, for which the sequence was randomized except that the Ser and Thr residues were suitably placed on the helical wheel to minimize amphipathic bias; this peptide, with average hydropathy ($\sim$0.6) just above the threshold hydrophobicity level, inserted spontaneously into LPC micelles.

However, hydrophobic effects alone do not provide a complete description of polypeptide–membrane interactions, that is, interhelical van der Waals packing interactions between TM helices will drive assembly of dimeric and multispanning membrane domains (Senes *et al.,* 2000; Wang and Deber, 2000). Glycine residues in TM segments may play a particular role in allowing intimate contacts between TM segments. Aromatic residues localized to the interfacial region of TM segments may assist in aligning them properly with respect to the lipid bilayer. In addition, TM segments with cytosolic Lys or Arg residues can be assisted into the membrane via electrostatic interactions with the headgroups of anionic lipids present in biological membranes (Kuroiwa *et al.,* 1991; Liu and Deber 1997).

IV. CONCLUSIONS

A quantitative threshold hydrophobicity for the spontaneous insertion of polypeptide segments into membranes was obtained through use of a stringently controlled TM-mimetic peptide system. Peptides inserted spontaneously from water into micellar membranes only when their hydrophobicity is equivalent to or greater than the threshold hydrophobicity, which can be approximated by a polyalanine segment of suitable length. The fact that the minimum hydrophobicity requirement is satisfied by >96% of over 5000 transmembrane segments derived from a database of single- and multispanning intrinsic membrane proteins provides direct evidence that threshold hydrophobicity is the rule followed by most TM helices. These findings, in turn, relate the concept of threshold hydrophobicity to the selective incorporation of transmembrane segments into the lipid bilayer during the biosynthetic translocation process without the requirement of any additional expenditure of metabolic energy. In conjunction with genome-wide analysis of the properties of integral membrane proteins from a variety of organisms (Wallin and von Heijne 1998), the present quantitation of threshold hydrophobicity can provide guidelines for the design of membrane-interactive peptides and proteins.

Acknowledgments

This work was supported, in part, by grants to C.M.D. and R.A.F.R. from the Canadian Institutes of Health Research, and by a grant to C.M.D. from the Natural Sciences and Engineering Research Council of Canada. L.-P.L. and C.W. acknowledge support from the Research Training Committee of the Hospital for Sick Children. N.K.G. held an NSERC postgraduate studentship.

References

Andrews, D. W., and Johnson, A. E. (1996). The translocon: More than a hole in the ER membrane. *Trends Biochem. Sci.* **21,** 365–369.

Blaber, M., Zhang, X.-J., and Matthews, B. W. (1993). Structural basis of amino acid alpha helix propensity. *Science* **260,** 1637–1640.

Borel, A. C., and Simon, S. M. (1996). Biogenesis of polytopic membrane proteins: Membrane segments assemble within translocation channels prior to membrane integration. *Cell* **85,** 379–389.

Casey, J. R., and Reithmeier, R. A. (1993). Detergent interaction with band 3, A model polytopic membrane protein. *Biochemistry* **32,** 1172–1179.

Chen, H., and Kendall, D. A. (1995). Artificial transmembrane segments. *J. Biol. Chem.* **270,** 14115–14122.

Chou, P. Y., and Fasman, G. D. (1974). Conformational parameters for amino acids in helical, β-sheet and random coil regions calculated from proteins. *Biochemistry* **13,** 211–222.

Chung, L. A., and Thompson, T. E. (1996). Design of membrane-inserting peptides: Spectroscopic characterization with and without lipid bilayers. *Biochemistry* **35,** 11343–11354.

Davis, N. G., and Model, P. (1985). An artificial anchor domain: Hydrophobicity suffices to stop transfer. *Cell* **41,** 607–614.

Deber, C. M., and Goto, N. K. (1996). Folding proteins into membranes. *Nat. Struct. Biol.* **3,** 815–818.

Deber, C. M., Wang, C., Liu, L.-P., Prior, A. S., Agrawal, S., Muskat, B. L., and Cuticchia, A. J. (2001). TM Finder: A prediction program for transmembrane protein segments using a combination of hydrophobicity and nonpolar phase helicity scales. *Protein Sci.* **10,** 212–219.

Deisenhofer, J., Epp, O., Miki, K., Huber, R., and Michel, H. (1985). Structure of the protein subunits in the photoreaction center of *Rhodopseudomonas viridis* and its implications for function. *Nature* **318,** 618–624.

Do, H., Falcone, D., Lin, J., Andrews, D. W., and Johnson, A. E. (1996). The cotranslational integration of membrane proteins into the phospholipid bilayer is a multistep process. *Cell* **85,** 369–378.

Eisenberg, D., Schwarz, E., Komaromy, M., and Wall, R. (1984). Analysis of membrane and surface protein sequences with the hydrophobic moment plot. *J. Mol. Biol.* **179,** 125–142.

Engelman, D. M., Steitz, T. A., and Goldman, A. (1986). Identifying nonpolar transbilayer helices in amino acid sequences of membrane proteins. *Annu. Rev. Biophys. Biophys. Chem.* **15,** 321–353.

Henderson, R., Baldwin, J. M., Ceska, T. A., Zemlin, F., Beckman, E., and Downing, K. H. (1990). Model for the structure of bacteriorhodopsin based on high-resolution electron cryo-microscopy. *J. Mol. Biol.* **213,** 899–929.

Hofmann, K., and Stoffel, W. (1993). A database of membrane-spanning protein segments. *Biol. Chem. Hoppe-Seyler* **374,** 166.

Johnson, A. E., and van Waes, M. A. (1999). The translocon: A dynamic gateway at the ER membrane. *Annu. Rev. Cell. Dev. Biol.* **15,** 799–842.

Kaiser, E. T., and Kezdy, F. J. (1983). Secondary structures of proteins and peptides in amphiphilic environments: A review. *Proc. Natl. Acad. Sci. USA* **80,** 1137–1143.

King, R. D., and Sternberg, M. J. (1996). Identification and application of the concepts important for accurate and reliable protein secondary structure prediction. *Protein Sci.* **5,** 2298–2310.

Kuroiwa, T., Sakaguchi, M., Mihara, K., and Omura, T. (1991). Systematic analysis of stop-transfer sequence for microsomal membrane. *J. Biol. Chem.* **266,** 9251–9255.

Kyte, J., and Doolittle, R. F. (1982). A simple method for displaying the hydropathic character of a protein. *J. Mol. Biol.* **157,** 105–132.

Landolt-Marticorena, C., Williams, K. A., Deber, C. M., and Reithmeier, R. A. F. (1993). Non-random distribution of amino acids in the transmembrane segments of human type I single span membrane proteins. *J. Mol. Biol.* **229,** 602–608.

Lee, S., Kiyota, T., Kunitake, T., Matsumoto, E., Yamashita, S., Anzai, K., and Sugihara, G. (1997). *De novo* design, synthesis and characterization of a pore-forming small globular protein and its insertion into lipid bilayers. *Biochemistry* **36**, 3782–3791.

Li, S.-C., and Deber, C. M. (1994). A measure of helical propensity for amino acids in membrane environments. *Nat. Struct. Biol.* **1**, 368–373.

Liu, L.-P., and Deber, C. M. (1997). Anionic phospholipids modulate peptide insertion into membranes. *Biochemistry* **36**, 5476–5482.

Liu, L. P., and Deber, C. M. (1998a). Guidelines for membrane protein engineering derived from *de novo* designed model peptides. *Biopolymers* **47**, 41–62.

Liu, L. P., and Deber, C. M. (1998b). Uncoupling protein hydrophobicity and helicity in nonpolar environments. *J. Biol. Chem.* **273**, 23645–23648.

Liu, L.-P., Li, S.-C., Goto, N. K., and Deber, C. M. (1996). Threshold hydrophobicity dictates helical conformations of peptides in membrane environments. *Biopolymers* **39**, 465–470.

MacKenzie, K. R., Prestegard, J. H., and Engelman, D. M. (1997). A transmembrane helix dimer: Structure and implications. *Science* **276**, 131–133.

Martoglio, B., Hofmann, M. W., Brunner, J., and Dobberstein, B. (1995). The protein-conducting channel in the membrane of the endoplasmic reticulum is open laterally toward the lipid bilayer. *Cell* **81**, 207–214.

Papavoine, C. H., Remerowski, M. L., Horstink, L. M., Konings, R. N., Hilbers, C. W., and van de Ven, F. J. (1997). Backbone dynamics of the major coat protein of bacteriophage M13 in detergent micelles by ^{15}N nuclear magnetic resonance relaxation measurements using the model-free approach and reduced spectral density mapping. *Biochemistry* **36**, 4015–4026.

Pebay-Peyroula, E., Rummel, G., Rosenbusch, J. P., and Landau, E. M. (1997). X-ray structure of bacteriorhodopsin at 2.5 angstroms from microcrystals grown in lipidic cubic phases. *Science* **277**, 1676–1681.

Preusch, P. C., Norvell, J. C., Cassatt, J. C., and Cassman, M. (1998). Progress away from 'no crystals, no grant.' *Nat. Struct. Biol.* **5**, 12–13.

Rees, D. C., Komiya, H., Yeates, T. O., Allen, J. P., and Feher, G. (1989). The bacterial photosynthetic reaction center as a model for membrane-proteins. *Annu. Rev. Biochem.* **58**, 607–633.

Reithmeier, R. A. F. (1995). Characterization and modeling of membrane proteins using sequence analysis. *Curr. Opin. Struct. Biol.* **5**, 491–500.

Riordan, J. R., Rommens, J. M., and Tsui, L.-C., *et al.* (1989). Identification of the cystic fibrosis gene: Cloning and characterization of complementary DNA. *Science* **245**, 1066–1073.

Savage, H., Cyrklaff, M., Montoya, G., Kuhlbrandt, W., and Sinning, I. (1996). Two-dimensional structure of light harvesting complex II (LHII) from the purple bacterium *R. sulfidophilum* and comparison with LHII from *R. acidophila*. *Structure* **4**, 243–252.

Senes, A., Gerstein, M., and Engelman, D. M. (2000). Statistical analysis of amino acid patterns in transmembrane helices: The GxxxG motif occurs frequently and in association with β-branched residues at neighboring positions. *J. Mol. Biol.* **296**, 921–936.

Sereda, T. J., Mant, C. T., Sonnichsen, F. D., and Hodges, R. S. (1994). Reversed-phase chromatography of synthetic amphiphatic alpha-helical peptides as a model for ligand/receptor Interactions: effect of changing hydrophobic environment of the relative hydrophilicity/hydrophobicity of amino-acid side-chains. *J. Chromatogr. A* **676**, 139–153.

Shore, G. C., McBride, H. M., Millar, D. G., Steenaart, N. A., and Nguyen, M. (1995). Import and insertion of proteins into the mitochondrial membrane. *Eur. J. Biochem.* **227**, 9–18.

Spiess, M., Handschin, C., and Baker, K. P. (1989). Stop-transfer activity of hydrophobic sequences depends on the translation system. *J. Biol. Chem.* **264**, 19117–19124.

Tanford, C. (1980). *"The Hydrophobic Effect: Formation of Micelles and Biological Membranes."* Wiley, New York.

Taylor, W. R., Jones, D. T., and Green, N. M. (1994). A method for α-helical integral membrane protein fold prediction. *Proteins* **18,** 281–294.

Thrift, R. N., Andrews, D. W., Walter, P., and Johnson, A. E. (1991). A nascent membrane protein is located adjacent to ER membrane proteins throughout its integration and translation. *J. Cell Biol.* **112,** 809–821.

Ullrich, A., Bell, J. R., Chen, E. Y., Herrera, R., Petruzzelli, L. M., Dull, T. J., Gray, A., Coussens, L., Liao, Y. C., and Tsubokawa, M., *et al.* (1985). Human insulin receptor and its relationship to the tyrosine kinase family of oncogenes. *Nature* **313,** 756–761.

Wallin, E., and von Heijne, G. (1998). Genome-wide analysis of integral membrane proteins from eubacterial, archaean, and eukaryotic organisms. *Protein Sci.* **7,** 1029–1038.

Wang, C., and Deber, C. M. (2000). Peptide mimics of transmembrane helices: Retention of an M13 coat protein dimerization motif. *J. Biol. Chem.* **275,** 16155–16159.

Wang, C., Liu, L.-P., and Deber, C. M. (1999). Helicity of hydrophobic peptides in polar vs. non-polar environments. *Phys. Chem. Chem. Phys.* **1,** 1539–1542.

Whitley, P., Grahn, E., Kutay, U., Rapoport, T. A., and von Heijne, G. (1996). A 12-residue-long polyleucine tail is sufficient to anchor synaptobrevin to the endoplasmic reticulum membrane. *J. Biol. Chem.* **271,** 7583–536.

Wimley, W. C., and White, S. H. (1996). Experimentally determined hydrophobicity scale for proteins at membrane interfaces. *Nat. Struct. Biol.* **3,** 842–848.

PART IV

Specialized Topics of Biological
Relevance

CHAPTER 17

Signal Sequence Function in the Mammalian Endoplasmic Reticulum: A Biological Perspective

Christopher V. Nicchitta

Department of Cell Biology, Duke University Medical Center, Durham, North Carolina 27710

I. Signal Sequences: Passage and Coronation
II. Signal Peptides: A Profile for Function
III. Signal Peptide Recognition: Where Biophysics Meets Biology
    A. The Membrane Bilayer as a Signal Sequence Receptor
    B. Protein-Based Recognition of Signal Peptides
IV. Yes, But What Happens at the Membrane?
V. Where to from Here?
VI. Conclusion
    References

The discovery of the signal peptide, an amino-terminal protein sequence that specifies targeting of newly synthesized polypeptides to the endoplasmic reticulum (ER), stands as one of the most significant in cell biology. The signal peptide performs a targeting function in the cell and serves as a paradigm for the processes by which proteins are targeted to other organelles of the eukaryotic cell, such as the nucleus, the mitochondria, and the peroxisome. Central to signal sequence function is its composite secondary structure, a conserved tripartite motif consisting of a positively charged amino terminus, a central hydrophobic core, and a carboxy-terminal polar domain. Of the three domains, it is the central hydrophobic core, a continuous stretch of 7–15 hydrophobic amino acids, that is functionally dominant. By virtue of its mean hydrophobicity, the central hydrophobic domain disposes the signal sequence to direct interactions with the lipid bilayer. Nonetheless, the predominant view in the cell biology community is that signal sequences function through direct protein–protein interactions to specify both targeting to the ER and regulation of the protein translocation machinery. This review focuses on signal sequence function, with critical emphasis on discerning lipid-dependent versus protein-dependent interactions with and within the ER membrane.

I. SIGNAL SEQUENCES: PASSAGE AND CORONATION

In 1999, the Nobel Assembly at the Karolinksa Institute awarded the Nobel Prize in Physiology or Medicine to Dr. Günter Blobel for the "signal hypothesis." In the signal hypothesis, Blobel proposed that proteins destined for secretion from the cell contain a 10- to 40-amino-acid N-terminal sequence that directs the protein, in the context of the ribosome, to the ER membrane. Once at the ER membrane, the signal sequence then serves an essential function(s) in initiating the vectorial transport of the nascent chain across the ER membrane (Blobel and Dobberstein, 1975). This proposal was later expanded to include, as paradigm, models for the role of such topogenic signals in the localization of proteins to any of the membrane compartments of the eukaryotic cell as well as the precise topological orientation of integral membrane proteins within biological membranes (Blobel, 1980). Decades of experimental study have established the validity of the signal hypothesis and thereby given topogenic signals a place among the scientific foundations of cell biology. It was appropriate that the fundamental significance of this hypothesis was ultimately recognized in science's highest award.

II. SIGNAL PEPTIDES: A PROFILE FOR FUNCTION

Signal peptides, also referred to as signal sequences, function in eukaryotes and prokaryotes to regulate access to the secretory and protein trafficking pathways of the cell (for review see Gierasch, 1989; also see Jones *et al.,* 1990; von Heinje, 1985; Walter and Johnson, 1994). Signal sequences are transient elements of a protein. They function early in synthesis to direct the nascent chain to the proper compartment in the cell, the ER, and to initiate the translocation event. Early in the translocation process, signal sequences are proteolytically removed by a heterooligomeric resident integral membrane protein complex, termed signal peptidase (Evans *et al.,* 1986).

Notably, signal sequences display little or no sequence conservation. Instead, signal sequences display a conserved physicochemical structure and display an overall length of 15–30 amino acids, a positively charged amino terminus (N domain), a central hydrophobic core (H domain), and a polar carboxy-terminal

Net Positive Charge   Hydrophobic Amino Acids   Polar Amino Acids

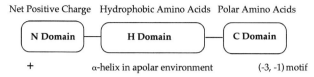

+   α-helix in apolar environment   (-3, -1) motif

FIGURE 1 Canonical signal peptide structure. Typically, signal peptides are between 15 and 30 amino acids long and have three regions that can be readily distinguished by their physicochemical properties. The N region, located at the amino terminus, is highly variable (1–20 amino acids) and is distinguished by the presence of a net positive charge. The H region is the functionally dominant region of the signal peptide and varies from 7 to 15 amino acids in length. It displays a high propensity for α-helix formation in nonpolar environments and contains highly hydrophobic residues, without charge interruption. The C region is polar and contains a specific motif, the (−3, −1 motif), which serves as the recognition site for signal peptidase, which cleaves the signal peptide from the nascent chain during translocation.

region (C domain) (Fig. 1). One can provide a structural summary as follows: The N domain is quite variable in length and always bears a positive charge; the H domain is enriched in hydrophobic residues (leucine, isoleucine, phenylalanine, methionine, and valine) and varies in length from 7 to 15 residues; and the C domain contains a motif, referred to as the (−3, −1) motif, which serves as the site of recognition and cleavage by the signal peptidase (von Heinje, 1984). The (−3, −1) motif reflects a convention in numbering of precursor proteins, such that the first residue in the mature protein is +1 and the adjacent residue of the signal sequence is −1. In the (−3, −1) motif, both residues are uncharged and relatively small (i.e., alanine, serine, glycine, cysteine).

The observation that the overall physicochemical structure of the signal sequence is conserved, in combination with the emergence of complete organism genomes, has fostered the development of predictive algorithms for the identification of signal peptides as well as signal peptide cleavage sites. In their most recent evolution, such algorithms have taken the form of artificial neural networks, which reliably perform both functions (Nielsen *et al.,* 1997). Discussions of the design of these artificial neural networks and the rationale used in algorithm optimization, as well as an interactive prediction display, are currently available at http://www.cbs.dtu.dk/services/SignalP/.

In reviewing the ever-expanding database of signal peptide sequences, one feature is overwhelmingly apparent. In analyzing the sequence content of the hydrophobic core, the signal peptides of eukaryotes can be readily distinguished from the signal peptides of Gram-negative and Gram-positive bacteria by the relative abundance of leucine residues (Fig. 2). The precise significance of this difference is not yet known; possible interpretations will be addressed in a later section.

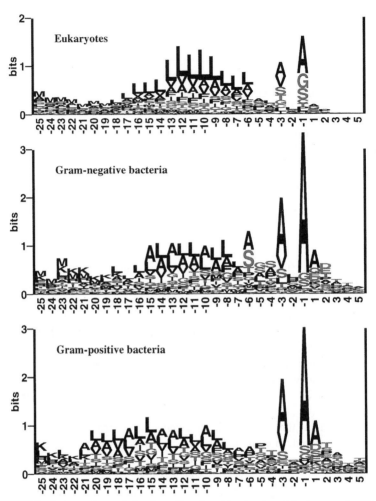

FIGURE 2   Sequence logos of signal peptides aligned by signal peptide cleavage site. The total height of the letter stack at each position is representative of the information content, whereas the relative amino acid abundance at that residue is displayed by the letter height. The information is defined as the difference between the maximal and the actual entropy. Charged residues are shown in blue and red, uncharged in green, and hydrophobic in black. Reprinted with permission from Nielsen *et al.* (1997). Copyright © 1997 Oxford University Press. (See color plate.)

III. SIGNAL PEPTIDE RECOGNITION: WHERE BIOPHYSICS MEETS BIOLOGY

A. The Membrane Bilayer as a Signal Sequence Receptor

The question of how a protein is translocated across a lipid bilayer has attracted the attention of scientific researchers whose interests span the subdisciplines of cell biology, biochemistry, and biophysics. That an amino-terminal signal sequence can satisfy the criteria of "necessary and sufficient" for protein translocation provides a superb opportunity for defining, from a reductionist point of view, the mechanism of signal peptide function. A fundamental and longstanding question regarding signal sequence function is the identification of the (a) signal sequence receptor. From one point of view, generally favored by cell biologists and biochemists, the signal sequence is predicted to participate in a series of protein-dependent membrane binding and translocation events that together define protein translocation as a protein-based process. To many in the biophysical community, the characteristic structure of the signal sequence, in particular the H domain, suggests that signal sequence function is dependent upon a capacity to spontaneously insert into the lipid bilayer. These differing views have spawned a decades-long debate tinged with acrimony and sparing of insight. In all fairness, though, these are very difficult questions. Is it solely the physicochemical properties of the signal peptide that are relevant to function? Or are the conserved physicochemical properties reflective of a degenerate protein–protein recognition process? It may be that the answer to both questions is yes.

We address each proposal in turn. Is it solely the physicochemical properties of the signal peptide that are relevant to function? The data in support of this view are derived from studies of environmental influences on signal peptide conformation as well as studies on signal peptide interactions with phospholipid monolayers and phospholipid bilayers. Early studies were focused on signal peptide structure and demonstrated that isolated signal peptides are capable of a high level of variability in secondary structure, as influenced by solution environment (reviewed in Gierasch, 1989). To identify a function correlated to signal peptide conformation, Gierasch and colleagues initiated a series of studies on the LamB signal peptide. Using a series of LamB signal peptide mutants and pseudorevertants, all of which had been previously analyzed for their capacity to support LamB export *in vivo* (Emr and Silhavy, 1982, 1983), Briggs and Gierasch (1984) demonstrated a clear correlation between the ability of the signal peptide to adopt a helical conformation in a lipid bilayer and its ability to support protein secretion, in agreement with the proposal of Emr and Silhavy (1983).

In subsequent studies, the conclusion became clear that although the capacity to assume a helical conformation in a hydrophobic environment was predictive of signal sequence function, the hydrophobicity of the central H domain was paramount. Because hydrophobicity has been determined to be an essential functional element of signal peptides, and given the finding that the overall hydrophobicity of the H domain bestows upon the signal sequence the capacity for spontaneous insertion into membrane bilayers, further studies were devoted to examining the interaction of signal peptides with model membrane systems. Two experimental approaches have proven especially valuable. In one approach, surface tensiometry was utilized to assess both signal peptide binding and signal peptide insertion into phospholipid monolayers. The primary and most substantial conclusion from these studies was that the capacity of signal peptides to bind to lipid membranes did not correlate with their ability to support secretion; instead, activity in secretion reflected the relative capacity to insert into the hydrocarbon region of the lipid monolayer (McKnight *et al.,* 1989). These studies were further extended in investigations into the topology of export-active and export-defective LamB signal peptides, in association with the membrane bilayer. As determined by fluorescence spectroscopy of a series of tryptophan variants, McKnight *et al.* (1991) conclusively demonstrated that the ability of a given LamB signal peptide to access the hydrocarbon domain of a lipid bilayer correlated strongly with protein export activity. It is clear, then, that signal peptides have an innate capacity to bind and insert into model membranes and the capacity to do so correlates strongly with *in vivo* function. Of high biological and biophysical interest, recent studies with the LamB peptide have demonstrated that physical properties of the bilayer, in particular membrane dipoles, can modulate rather dramatically the interaction of signal peptides with model membranes (Voglino *et al.,* 1998, 1999).

The results of biophysical analyses of signal peptide conformation and signal peptide interactions with lipid membranes provide strong support for the argument that signal peptides are likely to function through direct physical interaction with the lipid bilayer. Nonetheless, direct *in vivo* experimental support for this proposal is lacking. Recent studies have indicated, however, that signal peptides are in contact with the lipid bilayer under conditions where the nascent polypeptide chain resides in association with the ER membrane as a translocation intermediate (Martoglio *et al.,* 1995; Mothes *et al.,* 1998). These results derive from an experimental system that has proven to be accessible to biophysical measurements (Crowley *et al.,* 1993) and thus will likely prove suitable for future investigations into the biophysical mechanism of signal peptide function. In this system, a translocation-competent ER membrane fraction, obtained from canine pancreas, is used as a biologically active membrane fraction (Walter and Blobel, 1983). This membrane fraction can be used in a cotranslational translocation assay system consisting of the membrane fraction, a translationally competent cell lysate, usually obtained from rabbit reticulocytes or wheat germ, an energy-regenerating system, and mRNA. This system faithfully recapitulates the targeting, insertion, and translocation events of secretory protein translocation as well as membrane protein integration. In order to trap translocation intermediates in this system, artificially truncated mRNA molecules are used to program the translation reaction (Gilmore *et al.*, 1991; Perara *et al.*, 1986). To obtain truncated mRNA molecules, cDNA molecules can be cut at suitable sites within the coding sequence with restriction enzymes or, alternatively, truncations can be made at any point by use of the polymerase chain reaction. Subsequent transcription of the artificially truncated cDNA yields mRNA molecules that lack termination codons. During protein synthesis, then, translation proceeds until the ribosome reaches the 3′ end of the mRNA, at which point translation stalls to yield a stable ribosome/mRNA/nascent chain complex (RNC complex). By this approach the translation system can be programmed to yield a homogeneous population of nascent chains of defined length. This approach has proven invaluable in analyzing the environment of the nascent chain during the early stages of protein translocation (Connolly and Gilmore, 1986; Crowley *et al.*, 1994; Mothes *et al.*, 1994; Nicchitta *et al.*, 1995).

Additional experimental opportunities are afforded by taking advantage of the capacity of the ribosome to utilize chemically modified aminoacyl tRNA molecules during polypeptide synthesis. For example, the ε-amino group of lysyl-tRNA can be chemically modified to include photosensitive crosslinker or fluorescent dye moieties and subsequently utilized in an *in vitro* translation reaction (Crowley *et al.*, 1993, 1994; Krieg *et al.*, 1986; Martoglio *et al.*, 1995; Wiedmann *et al.*, 1986). By this technique, a given reporter moiety can be introduced into the nascent chain either at native codons or, as experimentally demonstrated by Mothes *et al.* (1994), at defined points in the nascent chain. In utilizing a suppressor tRNA-based variation of this approach, Martoglio *et al.* (1995) demonstrated by chemical crosslinking that the hydrophobic domain of the signal peptide resided in close physical proximity to membrane phospholipids during the early stages of the translocation reaction. This result was later confirmed in a detailed study demonstrating that the signal peptide progresses through a multistage insertion process wherein it initially resides in proximity to protein components of the translocation apparatus and subsequently, it has been proposed, binds to a site consisting of an interface between membrane protein and phospholipid components of the ER membrane (Mothes *et al.*, 1998).

The results of the studies by Martoglio *et al.* (1995) and Mothes *et al.* (1998) are intriguing, in that they clearly indicate that in the biological context of translocation across the ER membrane, the signal peptide can reside in immediate physical proximity to membrane phospholipids. However, though significant crosslinking of nascent translocation intermediates to membrane phospholipids was reported (Mothes *et al.*, 1998), these authors conclude that "all steps leading to nascent chain insertion, including signal sequence recognition in the membrane are likely based primarily, if not exclusively, on protein–protein interactions." The authors summarize their findings by concluding that "signal sequences are ultimately recognized by a proteinaceous binding site and [thus] exclude partitioning in the lipid bilayer as an exclusive mechanism of their recognition." As will be discussed below, the conclusion that signal sequence recognition is mediated by protein–protein inter-actions has strong precedent, at least when the phenomenon is examined in the aqueous cytoplasm.

B. Protein-Based Recognition of Signal Peptides

The results of the biophysical analyses of signal peptide/membrane interactions are compelling in their identification of membrane-driven structural changes in the signal peptide that correlate precisely with *in vivo* mutation and pseudorevertant studies of signal peptide function. Such studies certainly raise the specter of a lipid-dependent recognition event, but certainly do not exclude a protein-based recognition process. Furthermore, when the topic of signal peptide recognition is considered from a biological perspective, it can be concluded beyond reasonable doubt that the cell utilizes protein–protein interactions to identify signal peptides. At first glance, a protein-based recognition of a degenerate signal would seem implausible, but to the contrary, it not only occurs, it occurs in a manner that is conserved from prokaryotes to mammalian cells. The overall pathway for this process is schematically illustrated in Fig. 3 and begins in the aqueous cytosol.

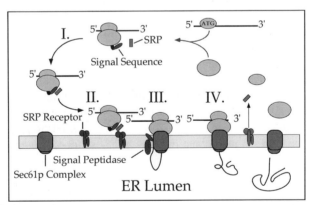

FIGURE 3   Schematic illustration of the protein translocation cycle in the mammalian endoplas-mic reticulum. In current views, the process of protein translocation begins in the cytosol. After the initiation of protein synthesis, the signal sequence is recognized in the context of the ribosome by the signal recognition particle (SRP), to yield the formation of a ribosome/nascent chain/SRP complex (stage I). This complex is targeted to the ER membrane through interaction with the SRP receptor (stage II). Subsequently, the ribosome/nascent chain complex binds to the translocon, the protein com-plex responsible for protein translocation, and translocation across the ER membrane ensues (stage III). The termination of protein synthesis is thought to result in the release of the ribosomal subunits back to the cytosol and the inactivation of the translocon (stage IV). (See color plate.)

Early in the synthesis of the nascent chain on a cytosolic ribosome, the signal sequence is recognized by the signal recognition particle (SRP) (stage I). This interaction leads to a pause in protein translation, thereby increasing the time window for the association of the ribosome/nascent chain/SRP complex with the ER membrane (Walter and Johnson, 1994). The process of ribosome targeting to the ER membrane is also ultimately dependent upon SRP, because there exists in the ER membrane a cognate receptor for SRP, the SRP receptor, also referred to as docking protein (Gilmore *et al.*, 1982; Meyer *et al.*, 1982). SRP thus serves three roles: (1) It recognizes the signal peptide, (2) it elicits, as a consequence of this recognition event, a kinetic slowing of protein translation, and (3) it directs, through interactions with its cognate receptor in the ER membrane, the ribosome/nascent chain complex to its appropriate location in the cell, the endoplasmic reticulum membrane (stage II).

The mammalian SRP is an 11S ribonucleoprotein complex of a 7S RNA and six polypeptides (Walter and Blobel, 1980, 1982). All components of SRP have been the subject of intensive study (for reviews see Lütcke, 1995; Walter and Johnson, 1994). In the context of this discussion, that is the recognition of signal peptides, the most critical component of SRP is the 54-kDa subunit, termed SRP54. SRP54 is a multidomain protein containing a GTP-binding domain (G domain) and a methionine-rich C-terminal domain, the M domain. That SPR54 is solely responsible for the recognition of the signal peptide was determined in coupled crosslinking proteolysis assays (Zopf *et al.*, 1990) and subdomain/subparticle reconstitution studies, which demonstrated that the M domain of SRP54 solely makes up the signal-peptide-binding site (Lütcke *et al.*, 1992).

The structural basis for signal sequence recognition by SRP54 was first inferred from secondary structure modeling studies and represents an intriguing motif (Bernstein *et al.*, 1989). In this analysis, modeling predictions of the M-domain secondary structure identified a series of methionine-rich amphipathic helices, with the methionine residues lining the hydrophobic face of the amphipathic helix and charged residues lining the opposite face. The proposed model for signal peptide recognition highlights the steric flexibility and overall hydrophobicity of the methionine side chains in the accommodation and recognition of the H domain of the signal peptide (Bernstein *et al.*, 1989). Direct structural validation of this model was obtained in crystal structure studies of the *Thermus aquaticus* SRP54 homologue, Ffh (Keenan *et al.*, 1998). From this structure it is apparent that four helices are arranged to form a hydrophobic groove (approximately 12 Å deep, 15 Å wide, and 25 Å long) with a hydrophobic surface area of approximately 1500 Å^2 (Keenan *et al.*, 1998). This groove can accommodate 15 amino acids in a helical conformation, which is approximately the size limit of natural variation in the H domain of the signal peptide (Keenan *et al.*, 1998). Interestingly, 11 of the methionine side chains line the hydrophobic face of the groove, as predicted by the modeling studies. Furthermore, the conservation of a methionine-rich signal-peptide-binding pocket across phylogeny suggests proof of principle for the protein-based recognition of a degenerate amino acid motif, as displayed by signal peptides.

IV. YES, BUT WHAT HAPPENS AT THE MEMBRANE?

The high-resolution crystal structure of Ffh, a *T. aquaticus* SRP54 homologue, combined with earlier secondary structure predictions provides a molecular model for protein-based recognition of signal peptides. These data, combined with the identification of SRP and SRP-receptor homologues in prokaryotes and eukaryotes, both simple and higher, indicate that in the aqueous cytoplasm signal sequence recognition occurs via protein–protein interactions. With the mammalian system as paradigm, recognition of the signal peptide by SRP in the cytosol yields the formation of a ribosome/nascent chain/SRP complex, which is then targeted to the ER membrane (Fig. 3, stages I–II). This targeting cycle occurs through the interaction of SRP with the SRP receptor. The subsequent formation of a translocation-competent ribosome–membrane junction (Fig. 3, stage III) requires the presence of GTP (Lütcke, 1995; Rapiejko and Gilmore, 1997; Stroud and Walter, 1999). SRP (SRP54) and both subunits of the SRP receptor are GTP-binding proteins, yet the signal sequence recognition reaction occurs with SRP54 and the SRP receptor in the nucleotide-free state (Rapiejko and Gilmore, 1997). GTP binding and hydrolysis are then coupled to the process of SRP release to the cytoplasm (Fig. 3, stage IV) (Rapiejko and Gilmore, 1997; Stroud and Walter, 1999). Operationally speaking, the process of targeting and GTP binding/exchange by the components of the SRP targeting pathway frees the signal peptide for interaction with components of the ER membrane and yields the signal-peptide-dependent formation of a translocation-competent junction between the ribosome and the translocation machinery (Fig. 3, stage III). The formation of such a junction yields a salt- and ethylene diaminetetraacetate-resistant association of the nascent chain with components of the ER membrane (Connolly and Gilmore, 1986; Jungnickel and Rapoport, 1995; Murphy *et al.*, 1997; Nicchitta and Blobel, 1989). Does this process formally require a signal sequence recognition event in the ER membrane, and if so, what ER membrane components (protein and/or lipid) serve as the signal sequence receptor? The answer to this question would contain the needed insights into the mechanistic elements of signal sequence function; it has not, however, been adequately resolved.

The existence of a proteinaceous signal sequence receptor has long been postulated and in early chemical crosslinking studies, an integral membrane glycoprotein was identified as residing in close proximity to the nascent chain (Wiedmann *et al.*, 1987). This protein, termed SSR (signal sequence receptor), was later determined to be dispensable for translocation (Migliaccio *et al.*, 1992). Following the report that the minimal translocation apparatus consisted of the SRP receptor and the heterooligomeric integral membrane protein complex Sec61, it became evident that should protein-dependent molecular recognition of the signal peptide occur in the ER membrane, the signal sequence receptor must then be a component of the Sec61 complex (Görlich and Rapoport, 1993; Mothes *et al.*, 1998). At present, however, no signal-peptide-binding site on any of the components of the Sec61 complex has been identified. Interestingly, and as noted previously, a highly conserved signal peptide recognition motif, the "methionine bristle," functions in the molecular recognition of signal peptides in the cytoplasm. No such motif exists, however, in any of the components of the Sec61 complex. By necessity, then, the postulated Sec61p-dependent molecular recognition of signal peptides in the ER membrane must occur through a mechanism distinct from that occurring in the cytoplasm.

The hypothesis that the molecular basis for signal peptide recognition in the ER membrane differs from that occurring in the cytoplasm is supported by a series of observations indicating that the structural requirements for signal sequence function in targeting differ from those necessary for translocation. In studies performed with canonical signal peptides, Belin *et al.* (1996) observed that the capacity of SRP to elicit elongation arrest did not correlate strongly with the activities of the signal peptide in translocation. Because elongation arrest requires the interaction of SRP with the signal peptide, these data suggest that the signal peptide recognition in the cytoplasm differs significantly from the signal peptide recognition in the ER membrane. To gain additional insights into the molecular basis for signal peptide recognition in the ER membrane, Zheng and Nicchitta (1999) prepared a chimera secretory protein consisting of the mature domain of prolactin, a well-established model secretory protein, and the signal peptide of LamB, for which a plethora of mutants and structural data were available. In their studies, it was reported that the wild-type (WT) LamB signal peptide was able to support SRP-dependent targeting to the ER membrane, yet did not support translocation (Zheng and Nicchitta, 1999). Thus, the studies of Zheng and Nicchitta (1999) and Belin *et al.* (1996), in which the activity of native signal peptides were evaluated in assays for SRP-dependent targeting and translocation, indicate that the structural requirements for signal peptide recognition in the cytosol differ from those necessary for signal peptide function in the translocation process per se.

In evaluating hypotheses for this apparent dichotomy in signal peptide function, the structural properties of prokaryotic and eukaryotic signal peptides were analyzed, whereupon it was apparent (Fig. 2) that eukaryotic signal peptides have a strong predominance of leucine residues in the H domain (Nielsen *et al.*, 1997). Prokaryotic signal peptides, in contrast, display a statistical prevalence of alanine and valine residues (Nielsen *et al.*, 1997). By available indices of hydrophobicity, valine, and to a lesser extent alanine, are of similar hydrophobicity to leucine (Engelman *et al.*, 1986). Nonetheless, natural selection has determined that an H domain containing equal proportions of leucine and alanine is ideal for translocation in prokaryotes, whereas leucine overwhelmingly dominates the H domain of eukaryotic signal peptides (Fig. 2) (Nielsen *et al.,* 1997). To test the hypothesis that a leucine-rich signal peptide is necessary for translocation in the eukaryotic system, point mutations were made in the LamB signal peptide and analyzed for gain-of-function activity in translocation assays (Zheng and Nicchitta, 1999). In these studies, a gain-of-function mutant was identified following conversion of two valine residues and one alanine residue of the LamB H domain to leucine (Zheng and Nicchitta, 1999). This mutant LamB signal peptide displayed efficient targeting and translocation activity in eukaryotic membranes (Zheng and Nicchitta, 1999). Further analyses of the mechanism underlying the gain-of-function mutations in the LamB signal peptide suggested that the gain of function yielded an enhanced interaction with SRP, thereby yielding more efficient formation of a translocation-competent ribosome-membrane junction (Zheng and Nicchitta, 1999). From these data, the authors proposed that the interaction of the signal peptide with SRP has significant structural consequences for the signal peptide. For example, under conditions in which the interaction between SRP and the signal peptide is weak, the signal peptide may assume a conformation refractory to subsequent translocation. This hypothesis emphasizes the role of signal peptide structure in the protein translocation and proposes that differences in signal peptide function in targeting and translocation are determined by the structure of the signal peptide. To test this prediction, short-chain forms of the wild-type and gain-of-function LamB–prolactin chimera were prepared and subjected to chemical denaturation prior to their exposure to ER membranes (Zheng and Nicchitta, 1999). Under these conditions, the signal peptide would be in an unstructured state, and thus, assumedly, able to replicate the behavior of isolated signal peptides. When presented to ER membranes in the denatured state, the wild-type and gain-of-function mutants bound to ER membranes with similar efficiencies and characteristics (Zheng and Nicchitta, 1999). To extrapolate from these data, we postulate that the interaction of the signal peptide with SRP maintains the signal peptide in an extended unstructured state suitable for interaction(s) with components of the ER membrane.

V. WHERE TO FROM HERE?

At present, the question of whether signal peptides function at the ER membrane through direct interaction with a proteinaceous receptor or by spontaneous insertion into the membrane bilayer remains unanswered. In many ways, the question remains unanswered because it is inappropriately general. By this it is meant that, if the mechanistic basis for signal peptide function is unknown, how can the question of whether function is displayed after spontaneous insertion into a membrane bilayer or binding to an integral membrane protein receptor be objectively answered? For example, if one believes that signal sequences act as ligands to gate an aqueous channel, and there is experimental evidence to support this (Simon and Blobel, 1992), then necessarily, any near-neighbor interactions between the signal peptide and integral membrane proteins would serve to support a protein–protein recognition model. Alternatively, if one believes that it is the act of insertion into a membrane, and perhaps the structural and topological consequences that such an insertion event may exert on the nascent chain, then indeed there is a preponderance of data indicating that signal peptides function by spontaneous, direct insertion into the membrane bilayer. In addition, though it is a notion not commonly expressed, it requires no great leap of faith to imagine that as the biological mechanism of action, signal peptides function by spontaneous insertion into the membrane and subsequent diffusion within the plane of the bilayer to interact with and regulate the activity of the protein translocation apparatus.

The difficulties confronting the direct experimental analysis of this question are manifold, but not uncommon to the study of membrane transport processes. Of particular relevance to the signal peptide, further studies of the topology of the membrane inserted form would be helpful, particularly if the topology could be determined in the context of a ribosome-associated nascent chain. Such an analysis would be most helpful if it could provide insight into the steady-state topological disposition of the N, H, and C domains of the signal peptide. Such knowledge would bring clarity to the apparently conflicting observations that in an assembled translocation intermediate, the signal peptide can be crosslinked to membrane phospholipids (Martoglio *et al.,* 1995; Mothes *et al.,* 1998), yet appears to reside in an aqueous environment (Crowley *et al.,* 1993). Rather than a leap, it would be but a short jump to consider that the N, H, and C domains of the signal peptide occupy different domains of the membrane during translocation. In this context it is important to note that the signal peptide displays different orientations at different stages of the translocation process, as proposed previously (Nicchitta and Zheng, 1997).

Another point to consider is the rather obvious conundrum presented by the observation that isolated signal peptides spontaneously insert into membranes and, upon their so doing, the H domain accesses the interior of the membrane. The conundrum is this: If it is the natural penchant for signal peptides to insert into membranes, and if the entirety of the signal peptide recognition process is mediated by protein–protein interactions (Mothes *et al.,* 1998), it would follow that upon the release of SRP, the signal peptide would have to be physically shielded from any interactions with the lipid component of the membrane. Is this a likely scenario? A model that invoked an initial, spontaneous insertion of the signal peptide into the membrane bilayer, with subsequent interactions with components of the translocation apparatus, would satisfyingly explain both datasets and alleviate this conundrum. Again, if it is considered that the signal can assume different topologies during the translocation reaction (Nicchitta and Zheng, 1997), a model in which very short nascent chains are primarily in the vicinity of membrane proteins (perhaps a consequence of a parallel disposition on the membrane), whereas in longer nascent chains the signal sequence would be fully inserted into the membrane in an antiparallel orientation, topologically competent for crosslinking to lipids, would provide a unified view of signal sequence function.

Once the precise details of signal peptide–membrane topology are determined, preferably in the context of a nascent chain rather than an isolated signal peptide, it could be determined whether the topology assumed by the signal peptide during translocation in native membranes is identical to that seen following spontaneous insertion into a liposome. If the two topologies are identical, it is reasonable to conclude that the initial interaction of the signal peptide with the membrane bilayer is indeed spontaneous and thus the known interactions of the nascent chain with components of the translocation apparatus are displayed at a later stage of the translocation event. Such events could include the gating of the lumenal side of the translocation pore (Hamman *et al.,* 1998) or alterations in the higher order structure of the translocation channel.

VI. CONCLUSION

In conclusion, the identification of the signal sequence and the elucidation of its primary role in protein secretion represents a discovery of enormous magnitude and biological significance. This discovery has yielded invaluable insights into the mechanism of membrane protein assembly as well as the basic blueprint for how proteins are trafficked in cells. Yet it remains to be determined how signal sequences function once presented to the ER membrane. It is somewhat ironic that the conceptual breakthrough afforded by viewing signal peptides and their variants as topogenic determinants has proven to be of great predictive power, yet so little is known concerning the topological and structural dynamics of the signal sequence itself. One can be comfortable in predicting that insights into these questions will provide the clues necessary for defining the molecular basis of signal sequence function.

References

Belin, D., Bost, S., Vassalli, J.-D., and Strub, K. (1996). A two-step recognition of signal sequences determines the translocation efficiency of proteins. *EMBO J.* **15,** 468–478.

Bernstein, H. D., Poritz, M. A., Strub, K., Hoben, P. J., Brenner, S., and Walter, P. (1989). Model for signal sequence recognition from amino acid sequence of 54K subunit of signal recognition particle. *Nature* **340,** 482–486.

Blobel, G. (1980). Intracellular protein topogenesis. *Proc. Natl. Acad. Sci. USA* **77,** 1496–1500.

Blobel, G., and Dobberstein, B. (1975). Transfer of proteins across membranes. I. Presence of proteolytically processed and unprocessed nascent immunoglobulin light chains on membrane-bound ribosomes of murine myeloma. *J. Cell Biol.* **67,** 835–851.

Briggs, M. S., and Gierasch, L. M. (1984). Exploring the conformational roles of signal sequences: Synthesis and conformational analysis of lambda receptor protein wild-type and mutant signal peptides. *Biochemistry* **23**, 3111–3114.

Connolly, T., and Gilmore, R. (1986). Formation of a functional ribosome–membrane junction during translocation requires the participation of a GTP-binding protein. *J. Cell Biol.* **103**, 2253–2261.

Crowley, K. S., Reinhart, G. D., and Johnson, A. E. (1993). The signal sequence moves through a ribosomal tunnel into a non-cytoplasmic aqueous environment at the ER membrane early in translocation. *Cell* **73**, 1101–1115.

Crowley, K. S., Liao, S., Worrell, V. E., Reinhart, G. D., and Johnson, A. E. (1994). Secretory proteins move through the endoplasmic reticulum membrane via an aqueous, gated pore. *Cell* **78**, 461–471.

Emr, S. D., and Silhavy, T. J. (1982). Molecular components of the signal sequence that function in the initiation of protein export. *J. Cell Biol.* **95**, 689–696.

Emr, S. D., and Silhavy, T. J. (1983). Importance of secondary structure in the signal sequence for protein secretion. *Proc. Natl. Acad. Sci. USA* **80**, 4599–4603.

Engelman, D. M., Steitz, T. A., and Goldman, A. (1986). Identifying nonpolar transbilayer helices in amino acid sequences of membrane proteins. *Annu. Rev. Biophys. Biophys. Chem.* **15**, 321–353.

Evans, E. A., Gilmore, R., and Blobel, G. (1986). Purification of microsomal signal peptidase as a complex. *Proc. Natl. Acad. Sci. USA* **83**, 581–585.

Gierasch, L. (1989). Signal sequences. *Biochemistry* **28**, 923–930.

Gilmore, R., Blobel, G., and Walter, P. (1982). Protein translocation across the endoplasmic reticulum. I. Detection in the microsomal membrane of a receptor for the signal recognition particle. *J. Cell Biol.* **95**, 461–469.

Gilmore, R., Collins, P., Johnson, J., Kellaris, K., and Rapiejko, P. (1991). Transcription of full-length and truncated mRNA transcripts to study protein translocation across the endoplasmic reticulum. *Meth. Cell Biol.* **34**, 223–239.

Görlich, D., and Rapoport, T. A. (1993). Protein translocation into proteoliposomes reconstituted from purified components of the endoplasmic reticulum membrane. *Cell* **75**, 615–630.

Hamman, B. D., Hendershot, L. M., and Johnson, A. E. (1998). BiP maintains the permeability barrier of the ER membrane by sealing the lumenal end of the translocon pore before and early in translocation. *Cell* **92**, 747–758.

Jones, J. D., McKnight, C. J., and Gierasch, L. M. (1990). Biophysical studies of signal peptides: Implications for signal sequence functions and the involvement of lipid in protein export. *J. Bioenerg. Biomembranes* **22**, 213–232.

Jungnickel, B., and Rapoport, T. A. (1995). A posttargeting signal sequence recognition event in the endoplasmic reticulum membrane. *Cell* **82**, 261–270.

Keenan, R. J., Freymann, D. M., Walter, P., and Stroud, R. M. (1998). Crystal structure of the signal sequence binding subunit of the signal recognition particle. *Cell* **94**, 181–191.

Krieg, U. C., Walter, P., and Johnson, A. E. (1986). Photocrosslinking of the signal sequence of nascent preprolactin to the 54-kilodalton polypeptide of the signal recognition particle. *Proc. Natl. Acad. Sci. USA* **83**, 8604–8608.

Lütcke, H. (1995). Signal recognition particle (SRP), a ubiquitous initiator of protein translocation. *Eur. J. Biochem.* **228**, 531–550.

Lütcke, H., High, S., Römisch, K., Ashford, A. J., and Dobberstein, B. (1992). The methionine-rich domain of the 54 kDa subunit of signal recognition particle is sufficient for the interaction with signal sequences. *EMBO J.* **11**, 1543–1551.

Martoglio, B., Hofmann, M. W., Brunner, J., and Dobberstein, B. (1995). The protein-conducting channel in the membrane of the endoplasmic reticulum is open laterally toward the lipid bilayer. *Cell* **81**, 201–214.

McKnight, C. J., Briggs, M. S., and Gierasch, L. M. (1989). Functional and nonfunctional LamB signal sequences can be distinguished by their biophysical properties. *J. Biol. Chem.* **264**, 17293–17307.

McKnight, C. J., Rafalski, M., and Gierasch, L. (1991). Fluorescence analysis of tryptophan-containing variants of the LamB signal sequence upon insertion into a lipid bilayer. *Biochemistry* **30,** 6241–6246.

Meyer, D. I., Krause, E., and Dobberstein, B. (1982). Secretory protein translocation across membranes: The role of the "docking protein." *Nature* **297,** 647–650.

Migliaccio, G. N., Nicchitta, C. V., and Blobel, G. (1992). The signal sequence receptor, unlike the signal recognition particle receptor, is not essential for protein translocation. *J. Cell Biol.* **117,** 15–25.

Mothes, W., Prehn, S., and Rapoport, T. A. (1994). Systematic probing of the environment of a translocating secretory protein during translocation through the ER membrane. *EMBO J.* **13,** 3973–3982.

Mothes, W., Jungnickel, B., Brunner, J., and Rapoport, T. A. (1998). Signal sequence recognition in cotranslational translocation by protein components of the endoplasmic reticulum membrane. *J. Cell Biol.* **142,** 355–364.

Murphy, E. C. I., Zheng, T., and Nicchitta, C. V. (1997). Identification of a novel stage of ribosome/nascent chain association with the endoplasmic reticulum membrane. *J. Cell Biol.* **136,** 1213–1226.

Nicchitta, C. V., and Blobel, G. (1989). Nascent chain binding and translocation are distinct processes: Differentiation by chemical alkylation. *J. Cell Biol.* **108,** 789–795.

Nicchitta, C. V., and Zheng, T. (1997). Regulation of the ribosome–membrane junction at early stages of presecretory protein translocation in the mammalian endoplasmic reticulum. *J. Cell Biol.* **139,** 1697–1708.

Nicchitta, C. V., Murphy, E. C., Haynes, R., and Shelness, G. S. (1995). Stage and ribosome-specific alterations in nascent chain–Sec61p interactions accompany translocation across the ER membrane. *J. Cell Biol.* **129,** 957–970.

Nielsen, H., Engelbrecht, J., Brunak, S., and von Heinje, G. (1997). Identification of prokaryotic and eukaryotic signal peptides and prediction of their cleavage sites. *Protein Eng.* **10,** 1–6.

Perara, E., Rothman, R., and Lingappa, V. (1986). Uncoupling translocation from translation: Implications for transport of proteins across membranes. *Science* **232,** 348–352.

Rapiejko, P. J., and Gilmore, R. (1997). Empty site forms of the SRP54 and SR alpha GTPases mediate targeting of ribosome–nascent chain complexes to the endoplasmic reticulum. *Cell* **89,** 703–713.

Simon, S. M., and Blobel, G. (1992). Signal peptides open protein-conducting channels in *E. coli*. *Cell* **69,** 677–684.

Stroud, R. M., and Walter, P. (1999). Signal sequence recognition and protein targeting. *Curr. Opin. Struct. Biol.* **9,** 754–759.

Voglino, L., McIntosh, T. J., and Simon, S. A. (1998). Modulation of the binding of signal peptides to lipid bilayers by dipoles near the hydrocarbon–water interface. *Biochemistry* **37,** 12241–12252.

Voglino, L., Simon, S. A., and McIntosh, T. J. (1999). Orientation of LamB signal peptides in bilayers: Influence of lipid probes on peptide binding and interpretation of fluorescence quenching data. *Biochemistry* **38,** 7509–7516.

von Heinje, G. (1984). How signal sequences maintain cleavage specificity. *J. Mol. Biol.* **173,** 243–251.

von Heinje, G. (1985). Signal sequences: The limits of variation. *J. Mol. Biol.* **184,** 99–105.

Walter, P., and Blobel, G. (1980). Purification of a membrane-associated protein complex required for protein translocation across the endoplasmic reticulum. *Proc. Natl. Acad. Sci. USA* **77,** 7112–7116.

Walter, P., and Blobel, G. (1982). Signal recognition particle contains a 7S RNA essential for protein translocation across the endoplasmic reticulum. *Nature* **299,** 691–698.

Walter, P., and Blobel, G. (1983). Preparation of microsomal membranes for cotranslational protein translocation. *Meth. Enzymol.* **96,** 84–93.

Walter, P., and Johnson, A. E. (1994). Signal sequence recognition and protein targeting to the endoplasmic reticulum membrane. *Annu. Rev. Cell Biol.* **10,** 87–119.

Wiedmann, M., Huth, A., and Rapoport, T. A. (1986). A signal sequence is required for the functions of the signal recognition particle. *Biochem. Biophys. Res. Commun.* **134,** 790–796.

Wiedmann, M., Kurzchalia, T. V., Hartmann, E., and Rapoport, T. A. (1987). A signal sequence receptor in the endoplasmic reticulum membrane. *Nature* **328,** 830–833.

Zheng, T., and Nicchitta, C. V. (1999). Structural determinants for signal sequence function in the mammalian endoplasmic reticulum. *J. Biol. Chem.* **274,** 36623–36630.

Zopf, D., Bernstein, H. D., Johnson, A. E., and Walter, P. (1990). The methionine-rich domain of the 54kd protein subunit of the signal recognition particle contains an RNA binding site and can be crosslinked to a signal sequence. *EMBO J.* **9,** 4511–4517.

CHAPTER 18

The Process of Membrane Fusion: Nipples, Hemifusion, Pores, and Pore Growth

Fredric S. Cohen, Ruben M. Markosyan, and Grigory B. Melikyan

Department of Molecular Biophysics and Physiology, Rush Medical College, Chicago, Illinois 60612

I. Introduction
II. Structural Features and Conformational Changes of Fusion Proteins
III. Hemifusion and Its Possible Role in the Fusion Process
IV. The Observed Temporal Sequence of Lipid Dye Spread and Fusion Pore Formation Depends upon Target Membrane and the Chosen Lipid Dye
V. The Dependence of Fusion on Spontaneous Curvature of Lipid Monolayers
VI. An Ectodomain of HA Anchored to a Membrane Can Yield Fusion or End-State Hemifusion
VII. Fusion Does Not Require a Precise Amino Acid Sequence of the TM Domain
VIII. Capturing Candidates of Transitional Hemifusion
IX. The Role of Fusion Proteins in Hemifusion and Pore Formation
X. The Effects of Lipid Composition on Pore Flickering and Growth
XI. Energy and Shape Considerations in Pore Growth
XII. Outlook
   References

In this chapter we discuss the current understanding of the process of membrane fusion and the contributions of proteins and lipids to this process. Many viral fusion proteins and proteins probably responsible for intracellular fusion fold into a bundle of coiled-coils. Formation of these coiled-coils is thought to be responsible for bringing membranes close together and may also cause the fusion event. Based on studies using the hemagglutinin (HA) of influenza virus, a prototypic fusion protein, the most likely sequence for fusion is as follows: Membranes make local contact by bending toward each other; the contacting lipid monolayers merge at these sites, yielding a state of hemifusion; with the subsequent merger

Current Topics in Membranes, Volume 52

of distal leaflets, fusion is completed. Many properties of lipids contribute to the fusion process. Important among these is spontaneous monolayer curvature. The contribution of spontaneous curvature to fusion is the best-understood aspect of the process in quantitative biophysical terms. Positive spontaneous curvature of contacting leaflets inhibits hemifusion, negative spontaneous curvature promotes it. Intermediate states that exhibit properties of hemifusion and can proceed on to full fusion have been captured and characterized. The transmembrane domain of HA greatly augments the transition from hemifusion to full fusion, but is not essential. The amino acid sequence of this domain is not critical, indicating that the conversion of hemifused to fused membranes occurs by a physical process. The growth of fusion pores is regulated by fusion proteins and affected by the lipid composition of the fused membranes.

I. INTRODUCTION

Fusion between membranes is a key event in diverse cellular processes. The proteins responsible for exocytosis and intracellular trafficking have, for the most part, been identified (Hanson *et al.*, 1997; Südhof, 1995). Because of the complexity of the systems, however, it has been difficult to determine which of these proteins specifically control membrane recognition, docking, and actual fusion. Viruses that infect cells by fusion, in contrast, are simpler systems and have envelopes that contain only a small number of proteins. This simplicity has allowed unambiguous identification of the proteins involved in viral binding and fusion (Hernandez *et al.*, 1996).

A viral envelope consists of a phospholipid bilayer incorporating multiple copies of a few integral membrane proteins. This bilayer is derived from the host cell: The virus buds after viral proteins and the viral nucleocapsid assemble; cell proteins are largely excluded. The virus initiates infection in another cell by depositing its genome into that cell's cytosol after membrane fusion. Depending on the virus, fusion occurs either directly with the cell plasma membrane or between envelope and endosomal membrane after the cell has internalized the virus by endocytosis. For internalized viruses, the low pH of endosomes triggers fusion by causing conformational changes in viral fusion proteins (White, 1994). For viruses that fuse directly to the plasma membrane, binding to the receptors of the target cell causes the conformational changes of the fusion protein.

Viral fusion is conveniently studied by expressing viral fusion proteins on cell surfaces and binding them to receptor-containing target cells. Such cell–cell fusion systems properly mimic the environment of natural viral fusion, because both occur in extracellular spaces. (The interior of an endosome is topologically equivalent to the extracellular space.) Intracellular fusion, on the other hand, would be difficult to study by cell–cell fusion even if the specific proteins were identified, because they would have to be engineered to expose and properly fold their intracellular portions in the extracellular environment. Fusion has been studied biophysically more extensively for the hemagglutinin (HA) of influenza virus than for any other virus. HA is responsible for both binding and fusion. Red blood cells (RBCs) are standardly used as the target cells. (Hemagglutinin even derives its name from the ability of influenza virus to agglutinate RBCs.) The hypotheses and experimental strategies for determining mechanisms of fusion in other systems, both viral (Hernandez *et al.,* 1996) and exocytotic (McNew *et al.,* 2000), are therefore often modeled on what is known of HA-mediated fusion. We therefore focus this review on HA-mediated fusion, with an emphasis on the work of our own laboratory. Some aspects of HA-induced fusion have been shown to generalize to other viral fusion proteins and the features discussed are, we believe, sufficiently general that they will apply to many of them.

II. STRUCTURAL FEATURES AND CONFORMATIONAL CHANGES OF FUSION PROTEINS

The HA glycoprotein is assembled from three identical monomers, each consisting of about 550 amino acid residues. The large, external portion of HA, about 510 residues for each monomer, is known as the ectodomain (Fig. 1). Each monomer is anchored to the membrane by a single transmembrane (TM) domain of about 25 residues; the cytoplasmic tail (CT) is short, about 10 amino acids. A monomer is synthesized as a single polypeptide chain, HA0, and is post-translationally cleaved into two subunits, HA1 and HA2. HA1 is responsible for binding to the target membranes; HA2 is the subunit that induces fusion. The N-terminus of HA2 is an apolar stretch of some 20 amino acids, known as the fusion peptide. The insertion of the fusion peptides into target membranes is critical for fusion.

The neutral-pH structure of the homotrimeric ectodomain of HA as well as a major portion of the ectodomain after it has reconfigured at low pH is known crystallographically (Bullough *et al.,* 1994; Chen *et al.,* 1999; Wilson *et al.,* 1981). The fusion peptide of each monomer is tucked away from the aqueous phase at neutral pH. This basic scheme holds for many viral fusion proteins (Hernandez *et al.,* 1996). For example, Env, the fusion protein of human immunodeficiency virus (HIV), is also homotrimeric, with each monomer synthesized as a single polypeptide chain, gp160, which is cleaved into two subunits, gp120 and gp41. Binding is through gp120, fusion through gp41. The N-terminus of gp41 constitutes the fusion peptide. For some viruses, fusion peptides are internal, rather than N-terminal.

The monomers of many viral fusion proteins besides HA and HIV Env are synthesized as single polypeptide chains which are post-translationally proteolytically cleaved. Postponing the generation of the active form safeguards against the protein

Influenza HA

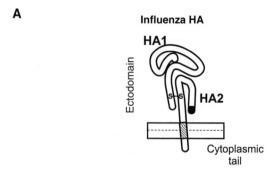

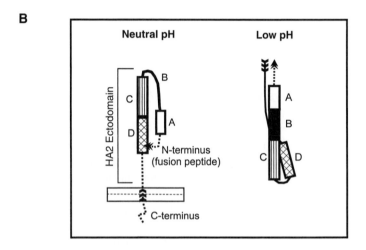

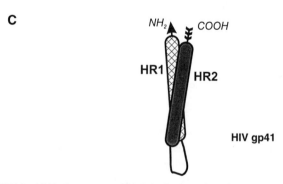

FIGURE 1 (A) Each monomer of the HA trimer consists of an HA1 and HA2 subunit linked by a single disulfide bond. Within the trimer, each fusion peptide (the filled section at the N-terminus of HA2) is sequestered from solution. The TM domain is shown striped. (B) The HA2 subunit of an individual monomer is shown. The orientation of HA2 relative to the viral membrane is indicated at neutral pH, but is unknown at low pH. When pH is lowered, the B loop becomes α-helical and A–C consequently

undergoing further conformational changes during intracellular transport. After cleaving, the protein gains additional configurational freedom. This results in some conformational changes before fusion is triggered. But when fusion is triggered the protein undergoes massive structural changes, changes impossible for a single chain. We thus envision that the gain of configurational freedom would be highly advantageous energetically for driving fusion: The energy difference between the initial state after cleavage and the final state of the protein induced by fusion conditions is very likely to be increased significantly. More of the energy stored in the protein could be available to be used for otherwise energetically unfavored steps of membrane rearrangements.

Each HA trimer is about 13.5 nm in length, consisting of a long stem and a globular head, and is oriented more or less perpendicular to the viral membrane (Wilson *et al.,* 1981). HA1 contributes to the stem and constitutes the entire head. Most of the stem is composed of HA2 (Fig. 1A). HA1 effectively clamps HA2 in its conformation. At neutral pH, segments C and D of each monomer are part of a trimeric central α-helical coiled-coil (Fig. 1B). Helix A is packed against the central portion of the continuous α helix of C and D. Helices A and C are connected by a loop (B). About 18 amino acids intervene between helix A and the fusion peptide. At low pH, HA1 dissociates from HA2 (but remains bound through a disulfide bond). The top of HA1 does not change conformation (Bizebard *et al.,* 1995), but the dissociation removes the clamp and HA2 undergoes a massive conformational change. Loop B converts into an α helix and becomes part of the central coiled-coil. This conversion of the B segment into an α helix projects the fusion peptide toward the target membrane, and is known as a "spring-coil" (Carr and Kim, 1993). Regions A–C thereby have become a continuous extended α helix. In addition, seven residues that had been part of the original continuous α helix of C and D refold into a loop, with the result that the D helices separate from each other, reverse their orientations, and pack against the triple-stranded coiled-coil of C helices. This antiparallel packing of D against C yields a structure known as a six-helix bundle. The three C helices form the central coiled-coil core and the three D helices are

---

form an extended α helix. The C helices of each monomer form a triple-stranded coiled-coil. Residues at the interface of helices C and D become a loop and each D helix inverts and packs into one of the three grooves created by the central triple-stranded coiled-coil. This creates a six-helix bundle with the D helix oriented antiparallel to the C helix. Bonds between residues adjacent to the fusion peptide (*arrowheads*) and the TM domain (*arrow tails*) cause the fusion peptides and TM domains to come into close contact. (C) Each monomer of HIV gp41 contains a heptad repeat (HR1) adjacent to the fusion peptide (*arrowhead*). Each HR1 of a trimer assembles into a central triple-stranded-coiled coil. A second heptad repeat (HR2) is separated from the TM domain (*arrow tails*) by a relatively short stretch of amino acids. HR1 and HR2 are connected by a long stretch of amino acids. HR2 packs into the grooves of the central coiled-coil. The formation of this six-helix bundle directly brings the TM domains toward the fusion peptides.

packed into the three grooves created by the coiled-coil. Although viral fusion proteins from some families of virus do not form coiled-coils (Rey *et al.*, 1995), many of them do. Six-helix bundles have been crystallographically observed in the final configuration of many viral fusion proteins (Skehel and Wiley, 1998).

HIV gp41, for example, exhibits a six-helix bundle in its final configuration (Fig. 1C). Although the TM domains and fusion peptides are not present in the crystallographically identified portion of the ectodomain of gp41, these two domains are adjacent to the C- and N-terminal helices of the bundle. Consequently, they should come into proximity with each other when the C-terminal helices pack antiparallel to the N-terminal α helices that make up the triple-stranded core, as the six-helix bundle is forming. In the case of HA, there is a long stretch of amino acids between the C-terminal α helices (i.e., the D helix) and the TM domains. Therefore, the formation of the bundle by itself would not force the TM domains and fusion peptides to approach each other. The crystallographic structure shows, however, that residues that end the α helices of the central core at the N-terminus (known as an N cap) bond with residues adjacent to the TM domain (Fig. 1B). Because the N cap is proximal to the fusion peptides, here, too, the fusion peptides should be close to the TM domains in the final structure. It may be generally true that in the final structure the TM domains and fusion peptides have come close to each other, and perhaps have even come into contact. Soluble *N*-ethylmaleimide-sensitive factor attachment protein receptors (SNAREs) are currently the best candidate proteins for mediating intracellular fusion; they form tetrameric coiled-coils (Poirier *et al.*, 1998; Sutton *et al.*, 1998). The finding that fusion proteins from different sources often fold as coiled-coils supports the expectation that the fusion mechanism will be similar in many different systems and indicates that the role of the coiled-coil is fundamental to the process of fusion.

III. HEMIFUSION AND ITS POSSIBLE ROLE IN THE FUSION PROCESS

Membrane fusion is a complex process in which a fusion pore—the structure that joins two membranes and allows aqueous continuity—forms and enlarges. It is clear that if one knew exactly how the pore formed, including the role of the fusion proteins and lipid rearrangements prior to pore formation, the problem of fusion would be largely solved. So little is definitively known about the events that result in fusion that even the specific composition of the early fusion pore is unresolved. There are only three possible compositional combinations for the initial pore, and all have been suggested. Either it is totally proteinaceous (Almers and Tse, 1990; Lindau and Almers, 1995), entirely lipidic (Siegel, 1993a), or a made up of a lipid–protein complex (Zimmerberg *et al.*, 1991). No matter what the role of proteins, it is clear that the lipids must temporarily leave their bilayer arrangement for a nonbilayer configuration as they rearrange from two separate contacting bilayers into one. It is the lipids that confer the membrane fluidity needed for fusion. The question as to the nature of the fusion pore revolves around an understanding of whether the lipids rearrange before or after the pore has formed.

When an aqueous continuity (two formerly separate compartments have become one) and a lipid continuity (two membranes have become one) have formed, fusion is complete. The routine method for assessing lipid continuity is to place fluorescent lipids into the outer lipid monolayer (outer leaflet) of one cell membrane and monitor their spread to the membrane of a bound cell. (In cell–cell fusion, the contacting leaflets are designated as the outer leaflets and the distal leaflets are the inner ones.) Aqueous continuity is monitored either by detecting transfer of dyes placed within the cytosol of one cell or, more sensitively, by measuring electrical capacitance to detect pore formation (Cohen and Melikyan, 1998).

Hemifusion is the merger of outer leaflets of membranes while the inner leaflets remain distinct and unaltered (Fig. 2). We describe fusion between cells and routinely use terminology appropriate to this configuration. If the merged outer leaflets clear from the region of contact, the inner leaflets can come into direct apposition and form a new single bilayer membrane, known as a hemifusion diaphragm, which continues to separate aqueous compartments. Hemifusion has been shown to occur in both model lipid bilayer (Chernomordik *et al.,* 1995a; Lee and Lentz, 1997) and plasma membrane systems that express HA (Chernomordik *et al.,* 1998; Kemble

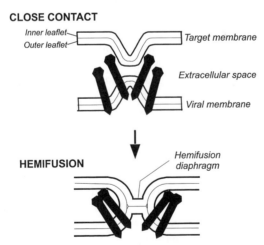

FIGURE 2 The bending of membranes into structures referred to as "nipples" allows membranes to locally come into close contact. It is not known whether a nipple in one membrane approaches a flat membrane or whether a nipple in each membrane contacts the other, as is illustrated. In the hemifusion process, each inner leaflet remains intact. Integral membrane proteins are excluded from the hemifusion diaphragm.

et al., 1994; Melikyan *et al.,* 1995c; Qiao *et al.,* 1999). On topological grounds, the hemifusion diaphragm should be a pure lipid bilayer membrane without integral membrane proteins. Portions of membrane proteins that are extracellular must remain extracellular (Fig. 2). If an integral membrane protein were to reside within the hemifusion diaphragm, the extracellular portion would become cytosolic, and this is not possible. An alternative way to understand why the hemifusion diaphragm is a pure lipid bilayer is to realize that the TM domain of a protein cannot span the inner leaflet of the other membrane in the hemifused state.

Hemifusion has been hypothesized as an intermediate state of full fusion for at least a quarter of a century (Palade, 1975). In biological fusion, the aqueous contents are transferred from one compartment to another in a conserved manner, without losses. If hemifusion is an intermediate state of full fusion, the inner leaflets are disrupted after the outer leaflets have merged: One set of leaflets always ensures that aqueous contents do not leak into extracellular spaces. Thus hemifusion would fulfill the biological expectation of conservation. The hemifusion model also has an elegance, in that it would naturally apply to any fusion system, independent of the structure of the fusion proteins, and it would maintain the sidedness of membranes. Only recently, however, has strong evidence emerged experimentally that hemifusion is in fact a precursor to full fusion.

IV. THE OBSERVED TEMPORAL SEQUENCE OF LIPID DYE SPREAD AND FUSION PORE FORMATION DEPENDS UPON TARGET MEMBRANE AND THE CHOSEN LIPID DYE

The principal alternative hypothesis to hemifusion as a precursor to fusion is the "gap-junction" model, in which the initial fusion pore is created solely by protein, in a manner analogous to gap junctions (Almers and Tse, 1990; Lindau and Almers, 1995). Multiple copies of a fusion protein would connect the aqueous compartments of two cells, and aqueous continuity would be created prior to lipid continuity, because the protein would form the walls of the initial pore, allowing aqueous contents to mix, without lipid merger. In contrast, if hemifusion occurs, the order is reversed: Lipid continuity of the outer leaflets would occur prior to aqueous continuity.

Because the temporal relation between lipid dye spread and pore formation is reversed in the two models, one would think that a straightforward way to resolve the question of which model is correct would be to simultaneously monitor both lipid dye spread and pore formation and see which occurs first. In practice, however, movement of lipid has proved to be more complicated and problematic than one would have anticipated. For RBCs as target, pore formation using HA-expressing cells is observed prior to spread of lipid dye and the lipid dye does not move well through pores until they enlarge (Qiao *et al.,* 1999; Tse *et al.,* 1993; Zimmerberg *et al.,* 1994). In fact, it has been claimed that lipid dye does not move at all through small pores (Tse *et al.,* 1993; Zimmerberg *et al.,* 1994). If lipid dye does not move through the wall of the initial pore, then hemifusion, it has been argued, could not have occurred as a prior step (Lindau and Almers, 1995). Lack of lipid movement has been the main evidence supporting the gap-junction hypothesis of fusion. More recent, extensive measurements, however, show that lipid dye might in fact move through the walls of small fusion pores, albeit at a slow rate (Markosyan *et al.,* 2000).

It has been shown that even for fused cells with enlarged pores (which must contain lipid), movement of lipid dye can be impeded to a surprisingly large degree (Melikyan *et al.,* 2000a). In the case of fusion of intact influenza virus to RBCs, the diffusion constant of lipid movement through enlarged pores is two orders of magnitude smaller than for diffusion in biological membranes (Georgiou *et al.,* 1989; Lowy *et al.,* 1990, 1995), illustrating the idiosyncrasies of lipid dye movement. In other cases, such as fusion of influenza to planar phospholipid bilayer membranes, movement of lipid dye is unimpeded, with a diffusion constant characteristic of that of bilayer membranes (Niles and Cohen, 1991). This disparity in lipid movement between virus and RBCs and virus and phospholipid membrane suggests that protein of the RBC, unrelated to fusion, may hinder lipid movement. The temporal sequence between lipid movement and pore formation between HA-expressing cells and planar phospholipid bilayer membranes has also been determined. The formation of pores was measured electrically.

Even though detecting pores by electrical means is intrinsically more sensitive than an observation of dye spread, lipid mixing can occur before pore formation with bilayer membranes as target (Razinkov *et al.,* 1999). Whether lipid dye spread is observed prior to or after, and, if the latter, how much after pore formation depends not only on the target membrane, but also on the lipid dye chosen as probe. When the fluorescent dye DiI was included in the planar membrane and its spread to cells as a result of hemifusion or fusion was monitored by video fluorescence microscopy, the lipid dye was always observed to spread before any pores were electrically detected. These cells clearly hemifused to bilayer membranes and then supported pore formation. When rhodamine (Rho-PE) was the fluorescent lipid probe, pore formation always occurred before dye was observed to spread. This indicates that the movement of Rho-PE is considerably more restricted than that of DiI. The ability of octadecylrhodamine B (R18), a popular dye used to follow lipid dye transfer, to spread was intermediate to that of DiI and Rho-PE. Thus it has not been possible to use movement of lipid dye as a definitive test to distinguish between the hemifusion and gap-junction hypotheses.

The hemifusion hypothesis is much more highly favored for a number of reasons. First, the very fact that fusion proteins can efficiently induce hemifusion strongly suggests that hemifusion is part of the natural process (Chernomordik *et al.,* 1998; Cleverley and Lenard, 1998; Kemble *et al.,* 1994; Melikyan *et al.,*

1995c, 1997a; Nüssler *et al.*, 1997; Qiao *et al.*, 1999). Fusion does not generally occur once lipid dye spread is observed, but the ease with which fusion proteins induce hemifusion does show that they readily induce merger of contacting leaflets. Second, the ability of membranes to fuse depends on their lipid composition (Chernomordik *et al.*, 1995b) and the initial pore conductances exhibit broad distributions (Lanzrein *et al.*, 1993; Plonsky and Zimmerberg, 1996; Spruce *et al.*, 1991). Neither of these findings is true for gap junctions (Wang and Veenstra, 1997). In addition, fusion does not depend on the amino acid sequence of the TM domain of fusion proteins (Melikyan *et al.*, 1999; Odell *et al.*, 1997; Wilk *et al.*, 1996); it does not require a full-length TM domain (Armstrong *et al.*, 2000) or even a TM domain at all, so long as the ectodomain is linked to the membrane (Frolov *et al.*, 2000; Markosyan *et al.*, 2000). Both of these findings are expected of a hemifusion mechanism, but are not easily accounted for by a gap-junction model. Furthermore, states intermediate to binding and lipid dye spread have been identified that can proceed to fusion and these states exhibit characteristics expected of hemifusion (Chernomordik *et al.*, 1998; Melikyan *et al.*, 2000b; Markosyan *et al.*, 2001).

V. THE DEPENDENCE OF FUSION ON SPONTANEOUS CURVATURE OF LIPID MONOLAYERS

The precise molecular rearrangements that must occur for two membranes to hemifuse are not known, but one could intuitively expect that the fewest possible number of lipid molecules would be disturbed to create the initial local connection. The structure that fulfills this criterion has been aptly termed a "stalk" (Gingell and Ginsberg, 1978; Markin *et al.*, 1984). For a stalk, or any other connecting structure, to form, the outer leaflets must bend toward each other. Calculations show that the minimum bending energy is indeed achieved by the stalk structure (Markin *et al.*, 1984; Siegel, 1993b).

When a thin structure such as a monolayer (or membrane) is bent, one side is stretched and the other is compressed. The neutral surface is located in the internal region; it is the mathematical surface that is not stressed when the structure bends (Safran, 1994). The energy required to bend monolayers or membranes is quantitatively accounted for by an equivalent of Hooke's law, $E_B = (B/2)(\kappa_1 + \kappa_2 - 2\kappa_0)^2$, where B is the bending modulus, in the range of 10–20 times $k_B T$, κ_1 and κ_2 are the two principal geometric curvatures of the neutral surface, and κ_0 is the spontaneous curvature of the monolayers or membranes (Helfrich, 1973). Spontaneous curvature can be understood on the basis of the effective shape of lipid molecules (Israelachvili *et al.*, 1976). For example, lysophosphatidylcholine (LPC) has a large polar portion, in terms of cross-sectional area, relative to its hydrophobic single acyl chain. The incorporation of LPC into a flat monolayer will thus displace the headgroup region of that monolayer more than the acyl chain region, and will cause the monolayer to curve toward the hydrocarbon core, defined as a positive curvature. The incorporation into a monolayer of a lipid, such as oleic acid or arachidonic acid, that has a small hydroxyl as headgroup will cause the monolayer to bend in the opposite direction, curving toward the aqueous phase, conferring a negative spontaneous curvature.

Both morphological (Frolov *et al.*, 2000) and functional evidence (Markosyan *et al.*, 1999) indicate that membranes approach each other by projecting themselves locally (Fig. 3). In cell–cell fusion, protuberances project outward from the cell as small "nipples." [In exocytotic fusion, these projections are called "dimples," because the plasma membrane bends inward toward the secretory granules (Chandler and Heuser, 1980; Ornberg and Reese, 1981).] An activation barrier probably exists between nipples and stalks. The process should be able to proceed not only from nipples to stalks, but in the opposite direction as well, so that the hemifused

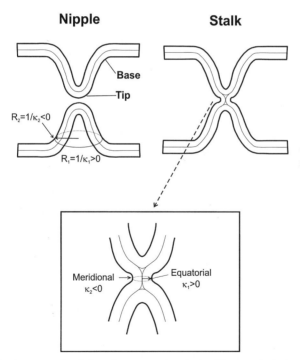

FIGURE 3   A cross-sectional view of two contacting nipples. The two principal curvatures of the tip of a nipple are positive. At the base, one curvature is positive (κ_1), the other is negative (κ_2). A stalk forms between the tips of the nipples, with the equatorial curvature positive and the meridional curvature negative. The meridional curvature of the unaltered portion of the nipple remains positive.

membranes of a stalk can separate back into two membranes. Because monolayers or membranes must continuously bend from the creation of nipples to the completion of pore enlargement, the energy required to bend them must be considered throughout the process. It has been established that for a wide variety of fusion systems, including those of HA, the addition of molecules with positive spontaneous curvature to outer leaflets inhibits fusion, whereas those with negative curvature tend to promote fusion (Chernomordik *et al.*, 1995b, 1997). The observed effects upon fusion of altering spontaneous curvature can be readily appreciated by noting the deviation of the geometric curvatures from spontaneous curvature at each step of the fusion process.

Nipples have curvatures that deviate from the spontaneous membrane curvature, and therefore their formation requires that energy be supplied by the fusion proteins. To simplify the discussion of membrane bending in fusion, we assume that the spontaneous curvatures of the monolayers are zero (i.e., $\kappa_0 = 0$). At the base of a nipple, the two curvatures have opposite signs and should have roughly equal magnitudes. Thus $\kappa_1 + \kappa_2 \approx 0$. Energy is not required to create this saddle structure (Kozlov and Chernomordik, 1998). Both curvatures at the tip of the nipple are positive, and thus energy must be supplied by fusion proteins to create the tip (Fig. 3). Less energy is required to create the tip of the nipple when positive-curvature lipids such as LPC are incorporated into outer leaflets; more energy is required when negative-curvature agents are incorporated into outer leaflets. The lowering of the energy of the nipple by incorporation of LPC would result in a larger barrier against stalk formation and thereby hinder the transition from nipples to stalks (Fig. 4).

Experimental evidence indicates that the reverse process, stalks converting back to nipples, does in fact also occur (Leikina and Chernomordik, 2000). For a stalk in an hourglass shape, a more positive spontaneous curvature of the outer leaflet would decrease the energy required to create the positive equatorial curvature and increase the energy required for the negative meridional curvature (Fig. 3). Calculations show that the negative meridional curvature is energetically much more consequential than the positive equatorial curvature (Kozlov *et al.*, 1989). The addition of LPC to outer leaflets should thus disfavor the dominant curvature and raise the energy of a toroidal stalk. In agreement with this, experimentally, the addition of LPC appears to augment the reversion (Leikina and Chernomordik, 2000; Melikyan *et al.*, 2000b). Changes in spontaneous curvature through altering lipid composition should also alter the absolute value of the activation barrier that separates the hemifused and nippled states. This alteration should therefore affect the transition between the two states in both the forward and reverse directions. Whether the incorporation of LPC into outer leaflets increases or decreases the absolute value of the activation barrier is not yet known. In any case, for a reversible transition between nipples and stalks, the net effect of increasing the curvature of outer leaflets to be more positive is to disfavor the transition from nipples to stalks and favor the transition in the opposite direction.

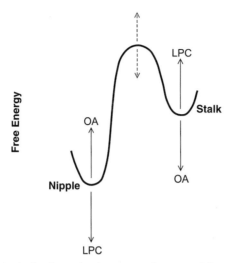

FIGURE 4 An activation barrier needs to be overcome for a state of close contact, created through nipples, to transit to a stalk. The addition of LPC, with its positive spontaneous curvature, to outer leaflets decreases the energy of a nipple and increases the energy of a stalk. The opposite pertains to the addition of agents with negative spontaneous curvature, such as oleic acid (OA), to outer leaflets. The changes in the absolute energy of the activation barrier upon altering spontaneous curvature are not yet known (*dashed arrows*).

The spontaneous curvature of inner leaflets also affects the fusion process, but in a different manner: it regulates the transition from hemifusion to pore formation. Positive spontaneous curvature of inner leaflets promotes pore formation, negative spontaneous curvature inhibits it (Chernomordik *et al.,* 1995b; Melikyan *et al.,* 1997a). The ability of spontaneous curvature to influence pore formation within phospholipid bilayer membranes is well known: LPC promotes the formation of lipidic pores (hence the name lyso) and agents such as oleic and arachidonic acid inhibit them. As would be expected, increasing the spontaneous curvature of inner leaflets to be more positive promotes the transition from hemifusion to fusion (Chernomordik *et al.,* 1998; Melikyan *et al.,* 1997a, 2000b). The transition from hemifusion to pore formation, however, is less well understood in many ways than is the transition from nipples to hemifusion, and the mechanism of pore formation requires much more experimental and theoretical investigation.

The addition of lipids and lipophilic agents affects a multitude of membrane properties; it is therefore not surprising that lipids affect fusion through other properties in addition to spontaneous curvature (Chanturiya *et al.,* 1999; Markosyan *et al.,* 2000). The varied mechanisms by which lipids affect fusion are complex and not understood in a systematic way, but the fact that lipids have marked effects on fusion is consistent with a mechanism in which lipids play a central role. This points much more toward hemifusion than toward an initial proteinaceous pore.

VI. AN ECTODOMAIN OF HA ANCHORED TO A MEMBRANE CAN YIELD FUSION OR END-STATE HEMIFUSION

The trimeric ectodomain of HA with each monomer linked to the lipid glyco-sylphosphatidylinositol (GPI) has been expressed on cell surfaces (Kemble *et al.*, 1993). That is, the TM domains and CTs of wild-type HA have been replaced by a GPI lipid. GPI–HA readily induces hemifusion, as evidenced by lipid dye spread without pore formation (Kemble *et al.*, 1994; Melikyan *et al.*, 1995c). This experimentally identified state of hemifusion does not proceed on to full fusion and therefore corresponds to a dead, end state. We refer to this state as end-state hemifusion. End-state hemifusion is experimentally much easier to achieve than is full fusion: Synthetic fusion peptides and large fragments of fusion proteins can induce lipid dye spread between membranes, but they do not induce fusion pores (Durell *et al.*, 1997; Epand *et al.*, 1999). It has not yet been shown that the isolated portions of the ectodomain induce hemifusion in the same manner as the intact ectodomain.

GPI–HA can also induce fusion pores prior to the observation of lipid dye spread (Frolov *et al.*, 2000; Markosyan *et al.*, 2000). The conditions required are much more limited than for HA and the pores do not enlarge: Many conditions that induce pores for HA induce only hemifusion for GPI-HA, but conditions that induce pores for GPI-HA also do so for HA. Lipid is undoubtedly a constituent of the GPI-HA fusion pore, because (1) TM domains are absent from GPI–HA and thus could not be lining the pore, (2) lipid readily passes through the initial GPI–HA pore, showing that a continuous lipid pathway is present in the pore, and (3) the ability of the pore to form is much more sensitive to changes in the lipid composition of the fusing cells than is the case for HA, indicative of the direct participation of lipid in the creation of the pore (Markosyan *et al.*, 2000).

The very fact that GPI–HA induces pores presents another strong argument that fusion does transit through hemifusion. In the gap-junction model, the TM domain would line the lumen of the pore within the HA-expressing membrane. But the TM domain is not present in GPI–HA and yet it induces pores. Therefore GPI–HA does not induce pores in a gap-junction manner. In light of the observation that an HA ectodomain linked to a membrane is capable of inducing a fusion pore at all, one has to assume that its manner of pore induction should bear similarities to pore induction via intact HA rather than by some totally different mechanism. Because the TM domain—the portion that must line the pore in a gap-junction model—is absent, lipid must be lining the GPI–HA pore. Lipidic pore formation follows naturally from merger of inner leaflets after hemifusion and accounts for the facile movement of lipid through the wall of the initial GPI–HA pore.

The finding that end-state hemifusion is more likely and pore formation less likely for GPI–HA than for HA shows that the ectodomain can cause fusion, but the TM domain greatly facilitates it. In addition, the TM domain is required for pore growth. Because GPI–HA always induces either pore formation or end-state hemifusion, the ectodomain alone should be responsible for hemifusion.

VII. FUSION DOES NOT REQUIRE A PRECISE AMINO ACID SEQUENCE OF THE TM DOMAIN

The existence of a TM domain clearly helps promote fusion, but the CT is not required. Fusion occurs as readily for HA with the CT deleted as it does for HA with an intact CT, in both virus infection (Simpson and Lamb, 1992) and cell–cell fusion systems (Melikyan *et al.*, 1997b). Chimeras in which the TM domain and/or CT of a protein unrelated to fusion were substituted for those of HA, but with the HA ectodomain always retained, yielded fusion that proceeded as well and under conditions as widespread as for HA (Melikyan *et al.*, 1999). In addition, these chimera pores enlarged as readily as those of HA. Thus, pore formation does not require a strict and limited amino acid sequence of the TM domain. This generalizes to other fusion proteins: HIV Env (Wilk *et al.*, 1996) and vesicular stomatitus virus G (Odell *et al.*, 1997) fusion proteins support fusion with TM domains derived from foreign proteins. In addition, TM domains derived from other fusion proteins effectively substitute for this domain in HA (Kozerski *et al.*, 2000; Schroth-Diez *et al.*, 1998). Furthermore, the TM domain of HA can be considerably shortened, from either the N- or the C-terminus, without impairing fusion (Armstrong *et al.*, 2000). The ability of a wide range of amino acid sequences for the TM domain to effectively promote pore formation runs contrary to the gap-junction theory, in which polar residues must line the water-filled lumen of the pore (Fig. 5). Such a pattern of polar residues would require amino acid sequences within the TM domain that are highly restricted.

VIII. CAPTURING CANDIDATES OF TRANSITIONAL HEMIFUSION

One can create conditions, usually through pH and temperature, that would induce fusion, except that the pH is not quite acidic enough and/or the temperature not quite high enough. Such conditions are said to be suboptimal. By manipulating these conditions in various ways, the process of fusion can be arrested in pre-fusion "intermediate" states. In order to be designated a true intermediate, the state must be capable of proceeding on to full fusion when the suboptimal condition is improved. For HA-mediated fusion, it is most convenient to lower pH to an optimal value for a given time, but at a suboptimal temperature, and then reneutralize and maintain cells at the suboptimal temperature. If fusion occurs when temperature is raised to the optimal 37°C at neutral pH, the captured state was an intermediate. Having reached the intermediate state, the pH-dependent steps of HA have occurred, but

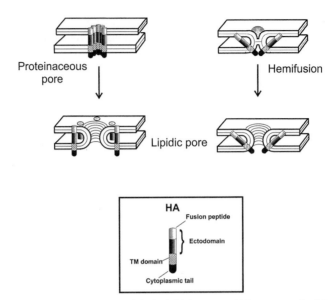

FIGURE 5 In the gap-junction model, the initial pore would be composed solely of protein. Because the TM domain would line the lumen of the pore within the viral membrane, pore formation should not occur for arbitrary amino acid sequences of this domain. In this model, lipids insinuate themselves between the individual fusion proteins after pore formation as part of pore growth. In a hemifusion model, lipids would line the initial pore lumen, whereas the TM domain would not.

temperature-sensitive steps remain to complete fusion. If full fusion does proceed through a state of hemifusion, one should be able to arrest fusion at the transitional hemifusion state.

How can one determine if the captured intermediate is a state of transitional hemifusion? Because fusion is not observed if lipid dye has spread (with RBCs as target), lipid dye is not expected to spread at the point of transitional hemifusion. Therefore dye spread cannot be employed as the primary indicator for merger of contacting leaflets. Indirect means must be used: aqueous dye spread as induced by both osmotic shock and addition of chlorpromazine (CPZ).

These assays have been shown to report the rupture of the hemifusion diaphragm of end-state hemifusion, readily measured as spread of aqueous dye. Osmotic swelling of cells that have hemifused leads to aqueous continuity (Melikyan *et al.*, 1995c), as does the addition of weak bases that are membrane-permeable and possess positive spontaneous curvature (Melikyan *et al.*, 1997a). When osmotic swelling ruptures the hemifusion diaphragm, aqueous continuity between cells obviously follows, whereas when rupture occurs at the nonhemifused portions of membranes, leakage (lysis) into the extracellular space is observed. Weak bases can gain access to inner leaflets, because the neutral form permeates the plasma membrane; the protonated, charged form accumulates in inner leaflets compared to outer leaflets (Deutike, 1968; Sheetz and Singer, 1974), because inner leaflets are negatively charged. Hence, the concentration of the weak base is higher within the hemifusion diaphragm, composed solely of inner leaflets, than in other parts of the membrane. Choosing a weak base with positive spontaneous curvature promotes the formation of lipidic pores within the hemifusion diaphragm. Several weak bases, including trifluoperazine, dibucaine, and CPZ, fulfill these criteria and form pores within the hemifusion diaphragm of end-state hemifusion (Melikyan *et al.,* 1997a). CPZ has become the standard weak base employed. Absence of lipid dye spread, advancement to fusion by raising temperature at neutral pH, and aqueous dye spread induced by osmotic swelling and addition of CPZ are the requirements for a candidate for a transitional hemifusion intermediate. By varying the value of the lowered pH and the temperature and time at which pH is lowered, several such intermediates have been identified (Chernomordik *et al.,* 1998; Leikina and Chernomordik, 2000; Markosyan *et al.,* 2001; Melikyan *et al.,* 2000b). A method has not been developed, however, that can definitively determine whether outer leaflets of the candidates have merged. We refer to the observed candidate states as "CPZ-sensitive," so as to not conflate them with the conjectured state of transitional hemifusion.

All CPZ-sensitive intermediates do not have the same properties. Some are resistant to further manipulations that would prevent fusion, whereas others are susceptible to such manipulations. States in the first category are termed "secure" and those in the latter category are termed "vulnerable" (Melikyan *et al.,* 2000b). In addition, some intermediates can remain kinetically stable over time (i.e., the kinetics of fusion upon raising temperature does not depend on how long the intermediate was maintained), whereas others do not. It has been discovered that secure intermediates are also stable intermediates and vulnerable intermediates are the ones that are not stable over time (Markosyan *et al.,* 2001). Based on energetic considerations, states that are secure should be more advanced in the fusion process and thus are the more promising candidates for transitional hemifusion than the vulnerable ones: Advanced states will have already passed through points in the process that are susceptible to manipulation and will have surmounted barriers that prevent them from falling back into states that are kinetically further from fusion.

IX. THE ROLE OF FUSION PROTEINS IN HEMIFUSION AND PORE FORMATION

The importance of the fusion peptide is beyond debate. Its role in hemifusion and pore formation, however, is far from understood on either the molecular or the physical level. Photolabeling experiments directly show that the fusion peptide inserts into target membranes, and this is the only portion of HA2 that does so (Harter *et al.,* 1989; Stegmann *et al.,* 1991). It is possible, particularly in view of the threefold symmetry of many viral fusion proteins, that one or two fusion peptides within a trimer insert into the target membrane and the remaining fusion peptides insert into the viral membrane (Wharton *et al.*, 1995). Insertion of fusion peptides into the viral membrane prior to the spring coil would be facilitated if HA bent at its base, adjacent to the viral membrane, soon after lowering of pH (Tatulian *et al.*, 1995). As a possible physical function, this tilting and insertion could generate torques that induce nipples (Kozlov and Chernomordik, 1998). Point mutations within the fusion peptide can result in end-state hemifusion or eliminate all fusion activity (Gething *et al.*, 1986; Qiao *et al.*, 1999; Schoch and Blumenthal, 1993; Steinhauer *et al.*, 1995). Mutations have been made in loop B that should hinder formation of its α helix and thereby affect the spring-coil transition; these mutation reduce the efficiency of fusion (Qiao *et al.*, 1998).

Multiple lines of evidence indicate that several fusion proteins act together to induce fusion (Bentz, 2000; Blumenthal *et al.*, 1996; Danieli *et al.*, 1996; Ellens *et al.*, 1990; Melikyan *et al.*, 1995a). Common sense dictates, and it is routinely imagined, that the fusion proteins form a ring around the initial nipple and the initial stalk (Chernomordik *et al.*, 1998; Hernandez *et al.*, 1996; Kozlov and Chernomordik, 1998; Melikyan *et al.*, 1999). At the point of hemifusion, the fusion peptides have inserted into the target membrane and the TM domains are located within the viral membrane. We envision that further conformational changes of the fusion protein must force the TM domains and fusion peptides closer together. To come together, they must pass through the local connection, whether stalk or hemifusion diaphragm, and in the process of doing so they rupture it, completing pore formation (Fig. 6). In the case of HIV Env, evidence indicates that the six-helix bundles and pores form simultaneously. That is, when the TM domain and fusion peptides are brought together by bundle formation, this directly induces the pores. Stated in energetic rather than mechanical terms, the free energy released by the formation of the bundle is directly utilized for pore formation (Melikyan *et al.*, 2000c). In the case of HA, the formation of bonds between the C-terminus and the N-terminus at the N cap which occurs along with bundle formation (Fig. 1) must force the TM domain and fusion peptides into close proximity (Chen *et al.*, 1999), inducing pores in the same manner. To summarize, in our view, HA of influenza virus and Env of HIV cause fusion by a variation on a common theme: The TM domains and fusion peptides, inserted in different membranes, are forced toward each other and, as a consequence, cause rupture.

X. THE EFFECTS OF LIPID COMPOSITION ON PORE FLICKERING AND GROWTH

It has been experimentally established that once a pore forms, its growth is dependent upon both fusion proteins and lipids. GPI–HA induces pores, but they do not enlarge (Frolov *et al.*, 2000; Markosyan *et al.*, 2000). Thus the TM domain

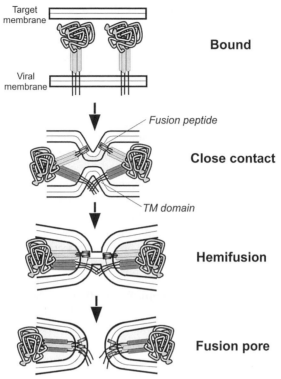

Target
membrane

Bound

Viral
membrane

Fusion peptide

Close contact

TM domain

Hemifusion

Fusion pore

FIGURE 6   The TM domains and fusion peptides have come into close contact in the final fusion protein structure (as illustrated in Fig. 1). If the final structure has not occurred at the point of hemifusion, the transition of fusion proteins into their final structure could drive the TM domain into the hemifusion diaphragm (or initial stalk) and disturb it, thereby creating pores.

must play a vital role in pore growth. Within the TM domain of HA itself, a point mutation has been found that leads to nonenlarging pores (Melikyan *et al.,* 2000b). Therefore, the TM domain is somehow intimately involved in pore growth, although the mechanism has yet to be elaborated. Point mutations within the fusion peptide can also impair pore growth (Gething *et al.,* 1986; Schoch and Blumenthal, 1993).

The effect of lipid composition on pore growth has been investigated with RBCs as target through the incorporation of fluorescent lipid probes into the RBCs. These probes are, of course, themselves lipids and so alter the composition when they are placed in the RBC membranes in the concentration range of a few mole percent. The probes influence the growth of HA-mediated pores, though their effects are not observed until somewhat high concentrations (about 5 mol%, depending on the probe). Fusion pores induced by GPI–HA do not enlarge when RBCs are the target. However, incorporating low concentrations of some probes (e.g., R18, DiI, or Rho-PE), but not others (e.g., PKH26), leads to pore growth. (The probes also increase the probability that GPI–HA induces pore formation rather than end-state hemifusion.) The altered lipid composition and augmentation of pore formation and growth does not correlate with any obvious physical parameter of the probe—spontaneous curvature, electrical charge, or flip-flop rate (Markosyan *et al.*, 2000). The very fact, however, that the formation and enlargement of GPI–HA pores depends strongly on lipid composition—significantly more so than for wild-type HA—suggests that the lipid is a more abundant component of a GPI–HA pore than an HA pore. This would be expected, because the TM domain of HA, thought to be part of the initial pore, is absent in GPI–HA.

Fusing HA-expressing to planar membranes has been experimentally useful in investigating the role of lipids in pore expansion—much more useful than RBCs as target—because lipid composition of the bilayer membrane can be conveniently controlled. The presence of particular lipids in bilayer membranes at concentrations that occur in biological systems can have profound effects on pore growth. It was found by voltage-clamping the planar membrane that fusion pores tend to open and close (a phenomenon known as "flicker") before one remains open and fully enlarges (Melikyan *et al.*, 1993, 1995b). When 30 mol% cholesterol was placed in a phospholipid target membrane, the first pore usually fully enlarged (Razinkov and Cohen, 2000). In contrast, including a few mole percent sphingomyelin in a target membrane without cholesterol led to a much longer period of flickering than in the case without sphingomyelin and the likelihood of full pore enlargement was greatly reduced. Cholesterol and sphingomyelin canceled out each other's effect when both were included in the target membrane: Flickering and full pore enlargement proceeded to the same extent as if neither was present. Cholesterol and sphingomyelin tend to associate with each other, perhaps to the point of forming small rafts (Simons and Ikonen, 1997). It is thus possible that this association is responsible for the cancellation of effects. This is substantiated by the findings that lactosylcerebroside (another sphingolipid), thought to associate with cholesterol, obviated the effect of cholesterol, whereas galactosylcerebroside, thought not to associate with cholesterol (Slotte *et al.*, 1993), did not cancel out cholesterol's effect on flickering (Razinkov and Cohen, 2000). Thus, properties of individual lipids and their ability to interact among themselves can strongly affect the evolution of fusion pores.

The ability to manipulate lipids allows the structure of the flickering pore to be inferred. There are two possible structures for the flickering pore. It could be composed of only inner leaflets, as an opening in a large "two-dimensional" hemifusion diaphragm. Alternatively, the pore could form in either a stalk or a nonexpanded hemifusion diaphragm, in which case the walls would consist of both inner and outer lipid leaflets. Such a pore would be "three-dimensional" and roughly hourglass-shaped (Fig. 7). A flickering pore composed of only inner

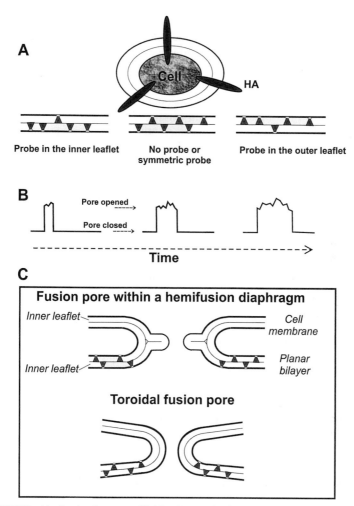

FIGURE 7   Altering the placement of lipid probes (A) that flip-flop in response to voltage indicates that flickering pores are not simply holes in hemifusion diaphragms (C). Pore flickers (B) have short open times when the probes are preferentially in the *trans* (corresponding to an inner) leaflet of a planar lipid bilayer. Open times are long when probe is flipped to the outer leaflet. Open times are intermediate for probe symmetrically placed. The absence of probe (in which case the lipid composition is identical in the two monolayers) gives the same open times as probes placed symmetrically. For a pore formed within a hemifusion diaphragm, the effect on open times should be greater when there is more probe within the inner leaflet. Experimentally, however, open times depend on the asymmetry of lipid composition of the two monolayers of the planar membrane. The walls of flickering pores should therefore be composed of both inner and outer leaflets, that is, the pore is roughly toroidal in shape.

leaflets might be considered particularly likely, because, as discussed, when planar membranes are the target, lipid dye spread can precede pore formation (Razinkov *et al.,* 1999) and thus an extended hemifusion diaphragm may have formed. The evidence, however, shows that the flickering pore wall is composed of fully assembled membranes (Razinkov *et al.,* 1998).

It was reasoned that if the pore is an opening within the hemifusion diaphragm, only lipids within inner leaflets should contribute to pore growth, but if the pore were hourglass-shaped, lipids of both leaflets should contribute. The cationic fluorescent lipid dyes R18 and DiI flip-flop across membranes in a voltage-dependent manner (Melikyan *et al.,* 1996). When placed in the planar membrane, their concentrations were made asymmetric to the same extent, but in opposite directions within each monolayer, by switching the polarity of the clamped membrane potential. If a pore were a hole within a hemifusion diaphragm, the open times should depend on the amount of probe within the inner leaflet of the planar membrane. Instead, the open times were the same with probe symmetrically placed as in the absence of probe. That is, open times depended on the direction of the asymmetry, and not on the amount of probe (Fig. 7). This is consistent with, and expected of, an hourglass-shaped (i.e., three-dimensional) pore. This finding leads to the conclusion that the stalk (or hemifusion diaphragm) in which the pore forms is a small, localized structure as opposed to an extended diaphragm. These experiments also demonstrated that with negative-curvature probes (i.e., R18 or DiI) in the outer leaflet, pore growth was inhibited; with probe in the inner leaflet, pore enlargement was facilitated. Thus, even small asymmetries in lipid composition in two monolayers of a membrane can greatly influence pore flickering and expansion. The negative meridional curvature of the toroidal pore is energetically more important than the positive equatorial one in controlling pore evolution when a planar bilayer membrane is the target.

XI. ENERGY AND SHAPE CONSIDERATIONS IN PORE GROWTH

Although the molecular mechanisms by which proteins and lipids control pore enlargement are not known, the elastic properties of membranes must be involved. For a pore to enlarge, the area of the pore wall must increase and the additional membrane must bend. The energy of bending has the form $(B/2)(\kappa_1 + \kappa_2 - 2\kappa_0)^2$ integrated over the area of the pore. The smaller the pore, the more the fusion proteins should constrain pore geometry. As the pore enlarges, the geometry should become freer.

We consider what would happen if the proteins constrain the geometry of the pore wall, by assuming that it connects parallel flat membranes and that the hourglass-shaped pore must remain toroidal (i.e., a hemisphere of revolution) with the length of the pore fixed (Chizmadzhev *et al.,* 1995, 2000). For $\kappa_0 > 0$, the pore resides

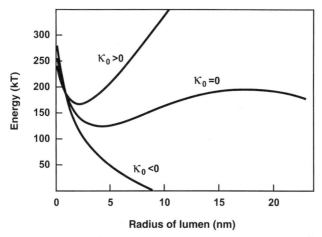

FIGURE 8 Energy of a toroidal pore connecting two parallel flat membranes. For $\kappa_0 > 0$, pore energy exhibits a minimum at a finite radius with energy increasing with increasing radius: The pore would never enlarge. For $\kappa_0 = 0$, the energy minimum is flanked by a finite barrier: The pore would enlarge if it could overcome the barrier. For $\kappa_0 < 0$, the pore energy decreases with increasing radius: The pore spontaneously enlarges. Pore formation at the tips of positive-curvature nipples is equivalent to making $\kappa_0 < 0$. If pore enlargement were nonexistent or gradual, it would indicate that the fusion proteins were constraining the pore.

in an energy minimum, and the energy continues to increase with greater pore radius (Fig. 8). The pore would remain at its minimum energy, never enlarging. For $\kappa_0 = 0$, there is a finite barrier, and the pore would enlarge if as a practical matter that barrier could be overcome. For $\kappa_0 < 0$, the barrier disappears completely and the pore spontaneously enlarges indefinitely. Even when an energy barrier prevents widening of the pore wall (e.g., $\kappa_0 \geq 0$), the pore length can still spontaneously increase. As the pore lengthens, the energy minimum is displaced to larger radii and the barrier against widening becomes smaller and eventually disappears (Chizmadzhev et al., 1995, 2000). A pore that forms at the site of a nipple requires less energy to bend the membranes than a pore forming between flat, parallel membranes. In effect, the energy required to bend the membranes has already been expended in creating the nipple. Thus, the existence of the nipple helps to ensure pore growth.

Freedom in pore geometry facilitates pore growth. When the pore wall is completely free to assume any geometry, the pore can more readily expand, because the wall can increase its area by assuming shapes that do not require additional energy. The pore would readily expand even in the absence of nipples. For example, if the pore connected parallel membranes and $\kappa_0 = 0$, then if the pore assumed the shape of a catenoid where $\kappa_1 = -\kappa_2$, the pore would spontaneously enlarge, because energy would not be required to bend the increased area of the pore wall.

XII. OUTLOOK

Membrane fusion is, at least for virus, a biophysical rather than a biochemical event. The essential features of this biologically important process should be understandable in true physical terms using well-established equations of physics. To reach this level of understanding, the critical work will be to determine how specific conformational changes of fusion proteins couple to the lipid rearrangements responsible for membrane fusion at each point in the process.

It has not been rigorously established that fusion proceeds sequentially through nipple formation, transitional hemifusion, and pore formation, but this sequence is currently the best model to fit most data and observations. From where we stand today, the following questions must be answered: What are the associations between fusion proteins that cause them to create nipples? Are the associations driven by physical forces alone or are they due to specific chemical interactions? What are the lipid rearrangements that initiate stalk formation and how do the proteins control them? Why does lipid not pass through the hemifusion connection and move only poorly (if at all) through the walls of the initial pore? Is this confinement of lipid critical to, and therefore somehow part of, the transition from hemifusion to fusion? How do the fusion proteins induce the formation of pores from the point of hemifusion?

Acknowledgment

This work was supported by NIH grants GM-27367 and GM-54787.

References

Almers, W., and Tse, F. W. (1990). Transmitter release from synapses: Does a preassembled fusion pore initiate exocytosis? *Neuron* **4**, 813–818.

Armstrong, R. T., Kushnir, A. S., and White, J. M. (2000). The transmembrane domain of the influenza hemagglutinin exhibits a stringent length requirement to support the hemifusion to fusion transition. *J. Cell Biol.* **151**, 425–437.

Bentz, J. (2000). Minimal aggregate size and minimal fusion unit for the first fusion pore of influenza hemagglutinin-mediated membrane fusion. *Biophys. J.* **78**, 227–245.

Bizebard, T., Gigant, B., Rigolet, P., Rasmussen, B., Diat, O., Bosecke, P., Wharton, S. A., Skehel, J. J., and Knossow, M. (1995). Structure of influenza virus haemagglutinin complexed with a neutralizing antibody. *Nature* **376**, 92–94.

Blumenthal, R., Sarkar, D. P., Durell, S., Howard, D. E., and Morris, S. J. (1996). Dilation of the influenza hemagglutinin fusion pore revealed by the kinetics of individual cell–cell fusion events. *J. Cell Biol.* **135**, 63–71.

Bullough, P. A., Hughson, F. M., Skehel, J. J., and Wiley, D. C. (1994). Structure of influenza haemagglutinin at the pH of membrane fusion. *Nature* **371**, 37–43.

Carr, C. M., and Kim, P. S. (1993). A spring-loaded mechanism for the conformational change of influenza hemagglutinin. *Cell* **73**, 823–832.

Chandler, D. E., and Heuser, J. E. (1980). Arrest of membrane fusion events in mast cells by quick-freezing. *J. Cell Biol.* **86**, 666–674.

Chanturiya, A., Leikina, E., Zimmerberg, J., and Chernomordik, L. V. (1999). Short-chain alcohols promote an early stage of membrane hemifusion. *Biophys. J.* **77,** 2035–2045.

Chen, J., Skehel, J. J., and Wiley, D. C. (1999). N- and C-terminal residues combine in the fusion-pH influenza hemagglutinin HA(2) subunit to form an N cap that terminates the triple-stranded coiled coil. *Proc. Natl. Acad. Sci. USA* **96,** 8967–8972.

Chernomordik, L., Chanturiya, A., Green, J., and Zimmerberg, J. (1995a). The hemifusion intermediate and its conversion to complete fusion: Regulation by membrane composition. *Biophys. J.* **69,** 922–929.

Chernomordik, L., Kozlov, M. M., and Zimmerberg, J. (1995b). Lipids in biological membrane fusion. *J. Membrane Biol.* **146,** 1–14.

Chernomordik, L. V., Leikina, E., Frolov, V., Bronk, P., and Zimmerberg, J. (1997). An early stage of membrane fusion mediated by the low pH conformation of influenza hemagglutinin depends upon membrane lipids. *J. Cell Biol.* **136,** 81–93.

Chernomordik, L. V., Frolov, V. A., Leikina, E., Bronk, P., and Zimmerberg, J. (1998). The pathway of membrane fusion catalyzed by influenza hemagglutinin: Restriction of lipids, hemifusion, and lipidic fusion pore formation. *J. Cell Biol.* **140,** 1369–1382.

Chizmadzhev, Y. A., Cohen, F. S., Shcherbakov, A., and Zimmerberg, J. (1995). Membrane mechanics can account for fusion pore dilation in stages. *Biophys. J.* **69,** 2489–2500.

Chizmadzhev, Y. A., Kuzmin, P. I., Kumenko, D. A., Zimmerberg, J., and Cohen, F. S. (2000). Dynamics of fusion pores connecting membranes of different tensions. *Biophys. J.* **78,** 2241–2256.

Cleverley, D. Z., and Lenard, J. (1998). The transmembrane domain in viral fusion: Essential role for a conserved glycine residue in vesicular stomatitis virus G protein. *Proc. Natl. Acad. Sci. USA* **95,** 3425–3430.

Cohen, F. S., and Melikyan, G. B. (1998). Methodologies in the study of cell–cell fusion. *Methods* **16,** 215–226.

Danieli, T., Pelletier, S. L., Henis, Y. I., and White, J. M. (1996). Membrane fusion mediated by the influenza virus hemagglutinin requires the concerted action of at least three hemagglutinin trimers. *J. Cell Biol.* **133,** 559–569.

Deuticke, B. (1968). Transformation and restoration of biconcave shape of human erythrocytes induced by amphipathic agents and changes in ionic environment. *Biochim. Biophys. Acta* **163,** 494–500.

Durell, S. R., Martin, I., Ruysschaert, J. M., Shai, Y., and Blumenthal, R. (1997). What studies of fusion peptides tell us about viral envelope glycoprotein-mediated membrane fusion. *Mol. Membrane Biol.* **14,** 97–112.

Ellens, H., Bentz, J., Mason, D., Zhang, F., and White, J. M. (1990). Fusion of influenza hemagglutinin-expressing fibroblasts with glycophorin-bearing liposomes: Role of hemagglutinin surface density. *Biochemistry* **29,** 9697–9707.

Epand, R. F., Macosko, J. C., Russell, C. J., Shin, Y. K., and Epand, R. M. (1999). The ectodomain of HA2 of influenza virus promotes rapid pH dependent membrane fusion. *J. Mol. Biol.* **286,** 489–503.

Frolov, V. A., Cho, M.-S., Bronk, P., Reese, T. S., and Zimmerberg, J. (2000). Multiple local contact sites are induced by GPI-linked influenza hemagglutinin during hemifusion and flickering pore formation. *Traffic* **1,** 622–630.

Georgiou, G. N., Morrison, I. E., and Cherry, R. J. (1989). Digital fluorescence imaging of fusion of influenza virus with erythrocytes. *FEBS Lett.* **250,** 487–492.

Gething, M. J., Doms, R. W., York, D., and White, J. (1986). Studies on the mechanism of membrane fusion: Site-specific mutagenesis of the hemagglutinin of influenza virus. *J. Cell Biol.* **102,** 11–23.

Gingell, D., and Ginsberg, L. (1978). Problems in the physical interpretation of membrane interaction and fusion. *In* "Membrane Fusion" (G. Poste and G. L. Nicholson, eds.), pp. 791–833. Elsevier/North-Holland, Amsterdam.

Hanson, P. I., Heuser, J. E., and Jahn, R. (1997). Neurotransmitter release: Four years of SNARE complexes. *Curr. Opin. Neurobiol.* **7,** 310–315.

Harter, C., James, P., Bachi, T., Semenza, G., and Brunner, J. (1989). Hydrophobic binding of the ectodomain of influenza hemagglutinin to membranes occurs through the "fusion peptide." *J. Biol. Chem.* **264,** 6459–6464.

Helfrich, W. (1973). Elastic properties of lipid bilayers: Theory and possible experiments. *Z. Naturforsch.* **28c,** 693–703.

Hernandez, L. D., Hoffman, L. R., Wolfsberg, T. G., and White, J. M. (1996). Virus–cell and cell–cell fusion. *Annu. Rev. Cell. Dev. Biol.* **12,** 627–661.

Israelachvili, J. N., Mitchell, D. J., and Ninham, B. W. (1976). Theory of self-assembly of hydrocarbon amphiphiles into micelles and bilayers. *J. Chem. Soc. Faraday Trans. 2* **72,** 1525–1568.

Kemble, G. W., Henis, Y. I., and White, J. M. (1993). GPI- and transmembrane-anchored influenza hemagglutinin differ in structure and receptor binding activity. *J. Cell Biol.* **122,** 1253–1265.

Kemble, G. W., Danieli, T., and White, J. M. (1994). Lipid-anchored influenza hemagglutinin promotes hemifusion, not complete fusion. *Cell* **76,** 383–391.

Kozerski, C., Ponimaskin, E., Schroth-Diez, B., Schmidt, M. F. G., and Herrmann, A. (2000). Modification of the cytoplasmic domain of influenza virus hemagglutinin affects enlargement of the fusion pore. *J. Virol.* **74,** 7529–7537.

Kozlov, M., and Chernomordik, L. (1998). A mechanism of protein-mediated fusion: Coupling between refolding of the influenza hemagglutinin and lipid rearrangements. *Biophys. J.* **75,** 1384–1396.

Kozlov, M. M., Leikin, S. L., Chernomordik, L. V., Markin, V. S., and Chizmadzhev, Y. A. (1989). Stalk mechanism of vesicle fusion. Intermixing of aqueous contents. *Eur. Biophys. J.* **17,** 121–129.

Lanzrein, M., Esermann, N. K., Weingart, R., and Kempf, C. (1993). Early events of Semliki Forest virus-induced cell–cell fusion. *Virology* **196,** 541–547.

Lee, J., and Lentz, B. R. (1997). Evolution of lipidic structures during model membrane fusion and the relation of this process to cell membrane fusion. *Biochemistry* **36,** 6251–6259.

Leikina, E., and Chernomordik, L. V. (2000). Reversible merger of membranes at the early stage of influenza hemagglutinin-mediated fusion. *Mol. Biol. Cell* **11,** 2359–2371.

Lindau, M., and Almers, W. (1995). Structure and function of fusion pores in exocytosis and ectoplasmic membrane fusion. *Curr. Opin. Cell Biol.* **7,** 509–517.

Lowy, R. J., Sarkar, D. P., Chen, Y., and Blumenthal, R. (1990). Observation of single influenza virus–cell fusion and measurement by fluorescence video microscopy. *Proc. Natl. Acad. Sci. USA* **87,** 1850–1854.

Lowy, R. J., Sarkar, D. P., Whitnall, M. H., and Blumenthal, R. (1995). Differences in dispersion of influenza virus lipids and proteins during fusion. *Exp. Cell Res.* **216,** 411–421.

Markin, V. S., Kozlov, M. M., and Borovjagin, V. L. (1984). On the theory of membrane fusion: The stalk mechanism. *Gen. Physiol. Biophys.* **3,** 361–377.

Markosyan, R. M., Melikyan, G. B., and Cohen, F. S. (1999). Tension of membranes expressing the hemagglutinin of influenza virus inhibits fusion. *Biophys. J.* **77,** 943–952.

Markosyan, R. M., Cohen, F. S., and Melikyan, G. B. (2000). The lipid-anchored ectodomain of influenza virus hemagglutinin (GPI-HA) is capable of inducing nonenlarging fusion pores. *Mol. Biol. Cell* **11,** 1143–1152.

Markosyan, R. M., Melikyan, G. B., and Cohen, F. S. (2001). Evolution of intermediates of influenza virus hemagglutinin-mediated fusion revealed by kinetic measurements of pore formation. *Biophys. J.* **80,** 812–821.

McNew, J. A., Weber, T., Parlati, F., Johnston, R. J., Melia, T. J., Sollner, T. H., and Rothman, J. E. (2000). Close is not enough: Snare-dependent membrane fusion requires an active mechanism that transduces force to membrane anchors. *J. Cell Biol.* **150,** 105–118.

Melikyan, G. B., Niles, W. D., Peeples, M. E., and Cohen, F. S. (1993). Influenza hemagglutinin-mediated fusion pores connecting cells to planar membranes: Flickering to final expansion. *J. Gen. Physiol.* **102,** 1131–1149.

Melikyan, G. B., Niles, W. D., and Cohen, F. S. (1995a). The fusion kinetics of influenza hemagglutinin expressing cells to planar bilayer membranes is affected by HA density and host cell surface. *J. Gen. Physiol.* **106**, 783–802.

Melikyan, G. B., Niles, W. D., Ratinov, V. A., Karhanek, M., Zimmerberg, J., and Cohen, F. S. (1995b). Comparison of transient and successful fusion pores connecting influenza hemagglutinin expressing cells to planar membranes. *J. Gen. Physiol.* **106**, 803–819.

Melikyan, G. B., White, J. M., and Cohen, F. S. (1995c). GPI-anchored influenza hemagglutinin induces hemifusion to both red blood cell and planar bilayer membranes. *J. Cell Biol.* **131**, 679–691.

Melikyan, G. B., Deriy, B. N., Ok, D. C., and Cohen, F. S. (1996). Voltage-dependent translocation of R18 and DiI across lipid bilayers leads to fluorescence changes. *Biophys. J.* **71**, 2680–2691.

Melikyan, G. B., Brener, S. A., Ok, D. C., and Cohen, F. S. (1997a). Inner but not outer membrane leaflets control the transition from glycosylphosphatidylinositol-anchored influenza hemagglutinin-induced hemifusion to full fusion. *J. Cell Biol.* **136**, 995–1005.

Melikyan, G. B., Jin, H., Lamb, R. A., and Cohen, F. S. (1997b). The role of the cytoplasmic tail region of influenza virus hemagglutinin in formation and growth of fusion pores. *Virology* **235**, 118–128.

Melikyan, G. B., Lin, S., Roth, M. G., and Cohen, F. S. (1999). Amino acid sequence requirements of the transmembrane and cytoplasmic domains of influenza virus hemagglutinin for viable membrane fusion. *Mol. Biol. Cell* **10**, 1821–1836.

Melikyan, G. B., Markosyan, R. M., Brener, S. A., Rozenberg, Y., and Cohen, F. S. (2000a). Role of the cytoplasmic tail of ecotropic Moloney murine leukemia virus Env protein in fusion pore formation. *J. Virol.* **74**, 447–455.

Melikyan, G. B., Markosyan, R. M., Roth, M. G., and Cohen, F. S. (2000b). A point mutation in the transmembrane domain of the hemagglutinin of influenza virus stablizes a hemifusion intermediate that can transit to fusion. *Mol. Biol. Cell* **11**, 3765–3775.

Melikyan, G. B., Markosyan, R. M., Hemmati, H., Delmedicao, M. K., Lambert, D. M., and Cohen, F. S. (2000c). Evidence that the transition of HIV-1 gp41 into a six-helix bundle, not the bundle configuration, induces membrane fusion. *J. Cell Biol.* **151**, 413–423.

Niles, W. D., and Cohen, F. S. (1991). Fusion of influenza virions with a planar lipid membrane detected by video fluorescence microscopy. *J. Gen. Physiol.* **97**, 1101–1119.

Nüssler, F., Clague, M. J., and Herrmann, A. (1997). Meta-stability of the hemifusion intermediate induced by glycosylphosphatidylinositol-anchored influenza hemagglutinin. *Biophys. J.* **73**, 2280–2291.

Odell, D., Wanas, E., Yan, J., and Ghosh, H. P. (1997). Influence of membrane anchoring and cytoplasmic domains on the fusogenic activity of vesicular stomatitis virus glycoprotein G. *J. Virol.* **71**, 7996–8000.

Ornberg, R. L., and Reese, T. S. (1981). Beginning of exocytosis captured by rapid-freezing of *Limulus* amebocytes. *J. Cell Biol.* **90**, 40–54.

Palade, G. (1975). Intracellular aspects of the process of protein synthesis. *Science* **189**, 347–358.

Plonsky, I., and Zimmerberg, J. (1996). The initial fusion pore induced by baculovirus GP64 is large and forms quickly. *J. Cell Biol.* **135**, 1831–1839.

Poirier, M. A., Xiao, W., Macosko, J. C., Chan, C., Shin, Y. K., and Bennett, M. K. (1998). The synaptic SNARE complex is a parallel four-stranded helical bundle. *Nat. Struct. Biol.* **5**, 765–769.

Qiao, H., Pelletier, S. L., Hoffman, L., Hacker, J., Armstrong, R. T., and White, J. M. (1998). Specific single or double proline substitutions in the "spring-loaded" coiled-coil region of the influenza hemagglutinin impair or abolish membrane fusion activity. *J. Cell Biol.* **141**, 1335–1347.

Qiao, H., Armstrong, R. T., Melikyan, G. B., Cohen, F. S., and White, J. M. (1999). A specific point mutant at position 1 of the influenza hemagglutinin fusion peptide displays a hemifusion phenotype. *Mol. Biol. Cell* **10**, 2759–2769.

Razinkov, V. I., Melikyan, G. B., Epand, R. M., Epand, R. F., and Cohen, F. S. (1998). Effects of spontaneous bilayer curvature on influenza virus-mediated fusion pores. *J. Gen. Physiol.* **112**, 409–422.

Razinkov, V. I., Melikyan, G. B., and Cohen, F. S. (1999). Hemifusion between cells expressing hemagglutinin (HA) of influenza virus and planar membranes can precede the formation of fusion pores that subsequently fully enlarge. *Biophys. J.* **77**, 3144–3151.

Razinkov, V. I., and Cohen, F. S. (2000). Sterols and sphingolipids strongly affect the growth of fusion pores induced by the hemagglutinin of influenza virus. *Biochemistry* **39**, 13462–13468.

Rey, F. A., Heinz, F. X., Mandl, C., Kunz, C., and Harrison, S. C. (1995). The envelope glycoprotein from tick-borne encephalitis virus at 2 Å resolution. *Nature* **375**, 291–298.

Safran, S. A. (1994). "Statistical Thermodynamics of Surfaces, Interfaces, and Membranes." Addison-Wesley, Reading, MA.

Schoch, C., and Blumenthal, R. (1993). Role of the fusion peptide sequence in initial stages of influenza hemagglutinin-induced cell fusion. *J. Biol. Chem.* **268**, 9267–9274.

Schroth-Diez, B., Ponimaskin, E., Reverey, H., Schmidt, M. F., and Herrmann, A. (1998). Fusion activity of transmembrane and cytoplasmic domain chimeras of the influenza virus glycoprotein hemagglutinin. *J. Virol.* **72**, 133–141.

Sheetz, M. P., and Singer, S. J. (1974). Biological membranes as bilayer couples: A molecular mechanism of drug–erythrocyte interactions. *Proc. Natl. Acad. Sci. USA* **71**, 4457–4461.

Siegel, D. P. (1993a). Modeling protein-induced fusion mechanisms: Insights from the relative stability of lipidic structures. *In* "Viral Fusion" (J. Bentz, ed.), pp. 477–512. CRC Press, Boca Raton, FL.

Siegel, D. P. (1993b). Energetics of intermediates in membrane fusion: Comparison of stalk and inverted micellar intermediate mechanisms. *Biophys. J.* **65**, 2124–2140.

Simons, K., and Ikonen, E. (1997). Functional rafts in cell membranes. *Nature* **387**, 569–572.

Simpson, D. A., and Lamb, R. A. (1992). Alterations to influenza virus hemagglutinin cytoplasmic tail modulate virus infectivity. *J. Virol.* **66**, 790–803.

Skehel, J. J., and Wiley, D. C. (1998). Coiled coils in both intracellular vesicle and viral membrane fusion. *Cell* **95**, 871–874.

Slotte, J. P., Östman, A.-L., Kumar, E. R., and Bittman, R. (1993). Cholesterol interacts with lactosyl and maltosyl cerebrosides but not with glucosyl or galactosyl cerebrosides in mixed monolayers. *Biochemistry* **32**, 7886–7892.

Spruce, A. E., Iwata, A., and Almers, W. (1991). The first milliseconds of the pore formed by a fusogenic viral envelope protein during membrane fusion. *Proc. Natl. Acad. Sci. USA* **88**, 3623–3627.

Stegmann, T., Delfino, J. M., Richards, F. M., and Helenius, A. (1991). The HA2 subunit of influenza hemagglutinin inserts into the target membrane prior to fusion. *J. Biol. Chem.* **266**, 18404–18410.

Steinhauer, D. A., Wharton, S. A., Skehel, J. J., and Wiley, D. C. (1995). Studies of the membrane fusion activities of fusion peptide mutants of influenza virus hemagglutinin. *J. Virol.* **69**, 6643–6651.

Südhof, T. C. (1995). The synaptic vesicle cycle: A cascade of protein–protein interactions. *Nature* **375**, 645–653.

Sutton, R. B., Fasshauer, D., Jahn, R., and Brunger, A. T. (1998). Crystal structure of a SNARE complex involved in synaptic exocytosis at 2.4 Å resolution. *Nature* **395**, 347–353.

Tatulian, S. A., Hinterdorfer, P., Baber, G., and Tamm, L. K. (1995). Influenza hemagglutinin assumes a tilted conformation during membrane fusion as determined by attenuated total reflection FTIR spectroscopy. *EMBO J.* **14**, 5514–5523.

Tse, F. W., Iwata, A., and Almers, W. (1993). Membrane flux through the pore formed by a fusogenic viral envelope protein during cell fusion. *J. Cell Biol.* **121**, 543–552.

Wang, H. Z., and Veenstra, R. D. (1997). Monovalent ion selectivity sequences of the rat connexin43 gap junction channel. *J. Gen. Physiol.* **109**, 491–507.

Wharton, S. A., Calder, L. J., Ruigrok, R. W. H., Skehel, J. J., Steinhauer, D. A., and Wiley, D. C. (1995). Electron microscopy of antibody complexes of influenza virus haemagglutinin in the fusion pH conformation. *EMBO J.* **14**, 240–246.

White, J. M. (1994). Fusion of influenza virus in endosomes: Role of the hemagglutinin. *In* "Cellular Receptors for Animal Viruses" (E. Wimmer, ed.), pp. 281–301, Cold Spring Harbor Laboratory Press, Cold Spring Harbor, NY.

Wilk, T., Pfeiffer, T., Bukovsky, A., Moldenhauer, G., and Bosch, V. (1996). Glycoprotein incorporation and HIV-1 infectivity despite exchange of the gp160 membrane-spanning domain. *Virology* **218,** 269–274.

Wilson, I. A., Skehel, J. J., and Wiley, D. C. (1981). Structure of the haemagglutinin membrane glycoprotein of influenza virus at 3 Å resolution. *Nature* **289,** 366–373.

Zimmerberg, J., Curran, M., and Cohen, F. S. (1991). A lipid/protein complex hypothesis for exocytotic fusion pore formation. *Ann. N. Y. Acad. Sci.* **84,** 307–317.

Zimmerberg, J., Blumenthal, R., Sarkar, D. P., Curran, M., and Morris, S. J. (1994). Restricted movement of lipid and aqueous dyes through pores formed by influenza hemagglutinin during cell fusion. *J. Cell Biol.* **127,** 1885–1894.

CHAPTER 19

Prenylation of CaaX-Type Proteins: Basic Principles through Clinical Applications

Herbert I. Hurwitz* and **Patrick J. Casey†**

Departments of *Medicine, †Pharmacology and Cancer Biology, and †Biochemistry, Duke University Medical Center, Durham, North Carolina 27710

I. Introduction
II. Enzymology of the CaaX Prenyltransferases: FTase and GGTase-I
   A. General Features of the CaaX Prenyltransferases
   B. Enzymatic Mechanism of the CaaX Prenyltransferases
   C. Structural Biology of the CaaX Prenyltransferases
III. Development of FTase Inhibitors as Anticancer Agents
   A. Rationale for Development of FTase Inhibitors as Anticancer Agents
   B. Biology of FTI Effects on Cells
IV. Clinical Results from Evaluation of FTIs as Antitumor Agents
V. Summary and Conclusions
   References

Post-translational modification by attachment of lipid moieties is critical for the biological activity of many membrane-associated proteins. This process, termed lipidation, serves to direct and anchor specific proteins to the cell membrane and can also play a role in important protein–protein interactions. A wide variety of lipids can be attached to proteins, including saturated acyl groups such as myristoyl and palmitoyl chains, glycosylphosphatidylinositol (GPI) moieties, and the 15- and 20-carbon isoprenoid groups farnesyl and geranylgeranyl, respectively. Some lipidated proteins have only a single modification, whereas many others may have multiple, in some cases variable, modifications, in which each permuation confers a unique chemical property and distinct functional characteristic to the parent protein. This chapter deals primarily with the biochemistry and biology of protein prenylation, focusing on those proteins that contain a so-called CaaX motif at their C-terminus. The C-terminal processing of CaaX proteins is initiated by addition of either a farnesyl or geranylgeranyl isoprenoid to the conserved cysteine of the CaaX motif. Farnesylation of Ras proteins in particular has attracted a great deal of attention and is the primary reason why the enzyme responsible, termed protein farnesyltransferase (FTase), has been targeted for development of inhibitors that are currently being evaluated in clinical trials as anticancer agents.

I. INTRODUCTION

The field of protein prenylation began with studies on the processing of fungal mating pheromones, which led to the realization that the presence of a specific C-terminal motif, now referred to as a CaaX motif, led to the protein being modified by an isoprenoid lipid (Glomset *et al.*, 1990; Schafer and Rine, 1992). In this motif, C is the cysteine residue to which the isoprenoid is attached, the a residues are generally aliphatic, and the X residue is the C-terminal amino acid. Prenylated CaaX proteins contain either a 15-carbon farnesyl or a 20-carbon geranylgeranyl isoprenoid attached to the cysteine residue via a thioether linkage (Glomset *et al.*, 1990; Zhang and Casey, 1996b). The majority of these proteins can be found associated with the cytoplasmic leaflet of cellular membranes for at least part of their lives, where they participate in signal transduction, protein trafficking, and other biological regulatory processes (Fu and Casey, 1999; Marshall, 1993; Sinensky, 2000). The near completion of the sequence of the human genome has led to the realization that there are probably more than 150 prenylated CaaX proteins in animal cells, including most members of the Ras superfamily of monomeric GTP-binding regulatory proteins (G proteins), the γ subunits of heterotrimeric G proteins, nuclear lamins, and a wide variety of other important regulatory proteins.

The prenylation-dependent processing of CaaX proteins is generally a three-step event (Fig. 1). Addition of the isoprenoid is accomplished by one of two distinct cytosolic enzymes termed CaaX prenyltransferases, protein farnesyltransferase (FTase) and protein geranylgeranyltransferase type I (GGTase-I) (Casey and Seabra, 1996; Schafer and Rine, 1992; Yokoyama and Gelb, 2000). Following prenylation, the modified proteins are directed to the endoplasmic reticulum (ER), where the last three amino acids (i.e., the –aaX) are removed by an enzyme termed the Rce1 CaaX protease, and the carboxyl group of the now C-terminal isoprenoid-modified cysteine is subjected to methylation by a prenylcysteine-specific methyltransferase termed Icmt (Ashby, 1998; Clarke, 1992). This trio of modifications produces a protein that exhibits a high affinity for cellular membranes and also produces a unique structure at the C-terminus, which can apparently serve as

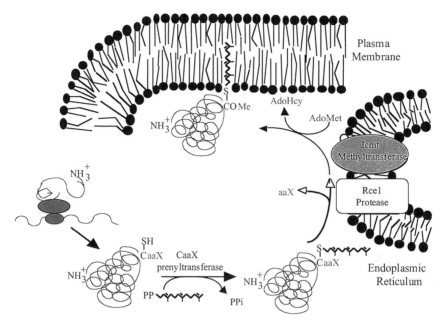

FIGURE 1   Processing pathway for CaaX prenyl proteins. CaaX proteins are synthesized as cytosolic precursors, which are then subject to modification by one of two CaaX prenyltransferases, FTase and GGTase-I, which attach either the 15-carbon farnesyl or 20-carbon geranylgeranyl, isoprenoid, respectively. After addition of the isoprenoid, the CaaX protein is directed to the endoplasmic reticulum, where the C-terminal tripeptide is removed by the Rce1 protease. The carboxyl group of the C-terminal prenylcysteine is then methylated by the Icmt methyltransferase and the mature CaaX protein is directed to its final destination, which is the plasma membrane in the case of the protein shown.

a specific recognition motif in certain protein–protein interactions (Gelb, 1997; Marshall, 1993).

The discovery of the importance of prenylation in CaaX protein function, most notably the oncogenic potential of the Ras proteins, has provoked a major effort to identify inhibitors of enzymes involved for evaluation as therapeutic agents (Gibbs, 1991; Johnston, 2001; Sebti and Hamilton, 1997). The majority of these studies have focused on FTase, because this enzyme modifies Ras proteins, and early preclinical studies indicated significant anticancer potential for FTase inhibitors (FTIs). A wide variety of FTIs have been developed, including some very promising ones, which possess antitumor activity in animal models and are now in clinical development (Crul *et al.*, 2001; Johnston, 2001). In addition, the success of FTIs in preclinical models of tumorigenesis and the increasing realization that proteins modified by GGTase-I play important roles in oncogenesis and that postprenylation processing by Rce1 and Icmt could be important in the function of Ras and other

CaaX proteins has led to the current situation in which all of the enzymes involved in CaaX protein processing are viewed as potential therapeutic targets (Bergo *et al.*, 2000; Kim *et al.*, 1999; Sebti and Hamilton, 2001).

II. ENZYMOLOGY OF THE CaaX PRENYLTRANSFERASES: FTase AND GGTase-I

A. General Features of the CaaX Prenyltransferases

FTase and GGTase-I are heterodimeric proteins containing identical 48-kDa α subunits and distinct, but closely related, β subunits of 46 and 43 kDa, respectively (Moomaw and Casey, 1992; Reiss *et al.*, 1990; Seabra *et al.*, 1991). The isoprenoid substrates for the two enzymes are farnesyl diphosphate (FPP) and geranylgeranyl diphosphate (GGPP), respectively (Table I). The two CaaX prenyltransferases are zinc metalloenzymes (Chen *et al.*, 1993; Moomaw and Casey, 1992; Reiss *et al.*, 1992). These findings initially suggested a role for the zinc ion in catalysis by these enzymes, and there is now unambiguous evidence that this is indeed the case (see below).

In terms of their protein substrates, FTase and GGTase-I recognize the canonical CaaX motif at the C-terminus of these substrates, with the C-terminal X residue of the protein substrate being the primary determinant of which of the two enzymes will modify that particular protein (Casey *et al.*, 1991; Yokoyama *et al.*, 1991) (Table I). When X is Ser, Ala, Met, or Gln, the protein is modified by FTase, whereas Leu directs modification by GGTase-I (Zhang and Casey, 1996b). Proteins containing Phe are also processed, although in this case there does not appear to be specificity for recognition, and either FTase or GGTase-I can be involved (Carboni *et al.*, 1995; Moomaw and Casey, unpublished results). Another example in this regard occurs with a Ras-related protein termed RhoB, which has been shown to be farnesylated as well as geranylgeranylated, even though its C-terminal residue is Leu (Adamson *et al.*, 1992; Armstrong *et al.*, 1995); farnesylation of this protein is apparently due to an ability to be processed by FTase (Lebowitz *et al.*, 1997a).

TABLE I

Properties of Protein Prenyltransferases

Protein prenyltransferase	Protein substrate C-terminus	Isoprenoid substrate	Metal requirements	Subunit composition (mammalian) (kDa)
FTase	-CaaX (X = M, S, Q, A)	FPP	Zn^{2+}, Mg^{2+}	48 (α)
				46 (β)
GGTase-I	-CaaX (X = L)	GGPP	Zn^{2+}	48 (α)
				43 (β)

With the exception of proteins containing Phe at the X position and the RhoB case, substrate recognition by the CaaX prenyltransferases is generally quite specific, although cross-specificity has been observed both *in vitro* and in cells. Both N-Ras and K-Ras4B can serve as substrates for both FTase and GGTase-I *in vitro*, although they are much better substrates for FTase (James *et al.*, 1995; Zhang *et al.*, 1997). The primary determinant for this type of cross-prenylation, which probably occurs only when FTase activity is reduced in cells (Rowell *et al.*, 1997; Whyte *et al.*, 1997), appears to be the existence of a Met as the C-terminal residue of the substrate protein. Although such types of cross-prenylation are probably not frequent events in cells, there is a great deal of interest in these events, because they may represent an important mechanism of resistance to FTIs (Gibbs and Oliff, 1997; Sebti and Hamilton, 2001).

B. Enzymatic Mechanism of the CaaX Prenyltransferases

FTase and GGTase-I are quite slow enzymes, with steady-state k_{cat} values of approximately 0.05 s^{-1} (Furfine *et al.*, 1995; Pompliano *et al.*, 1992; Yokoyama *et al.*, 1995; Zhang *et al.*, 1994b). The kinetic mechanism of FTase has been extensively studied and is functionally ordered, with FPP binding first to create an FTase · FPP binary complex, which then reacts quite rapidly with a CaaX substrate; kinetic and spectroscopic studies demonstrated that product formation is more than 200-fold faster than product dissociation (Furfine *et al.*, 1995; Huang *et al.*, 1997; Pompliano *et al.*, 1992). In the absence of excess substrate, the dissociation rate is so slow that FTase–product complexes can be isolated (Tschantz *et al.*, 1997). Although it is not as well studied as FTase, mammalian GGTase-I seems to have similar enzymatic mechanisms (Yokoyama *et al.*, 1997; Zhang *et al.*, 1994b), although catalysis by this enzyme is not Mg^{2+}-dependent for reasons that are as yet unclear (Zhang and Casey, 1996a).

Product release from mammalian FTase requires the binding of an additional FPP molecule (Tschantz *et al.*, 1997) (Fig. 2). This may have implications for postprenylation targeting of CaaX proteins. Although the CaaX prenyltransferases behave as cytosolic enzymes, their isoprenoid substrates are highly hydrophobic and quite possibly associated with membranes to a large degree. Hence, it has been proposed that FTase and GGTase-I may possess an additional function as a type of chaperone that holds onto its prenylated protein until the complex encounters a membrane compartment where isoprenoid binding and product dissociation would take place (Fu and Casey, 1999). This additional function of the enzymes is consistent with the presumed trafficking of prenyl proteins to the endoplasmic reticulum for completion of C-terminal processing by the Rce1 protease and Icmt methyltransferase and would require that the CaaX prenyltransferases contain information within their structure that targets them selectively to the endoplasmic reticulum.

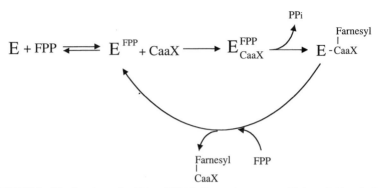

FIGURE 2   Kinetic scheme for FTase. FPP binding to free enzyme (E) is probably of minimal importance, except immediately following synthesis, because FTase cycles directly to the E–FPP complex after catalyzing the farnesylation of a protein substrate (CaaX). The direct conversion of the FTase–product complex to the FTase–FPP complex is consequence of FPP-triggered release of the product (farnesyl-CaaX).

In terms of their chemical mechanisms, it is now generally accepted that the transition state of catalysis for the CaaX prenyltransferases has both electrophilic and nucleophilic character, that is, these are combined carbocation–nucleophile reactions (Hightower and Fierke, 1998; Hightower *et al.*, 1998; Huang *et al.*, 2000) (Fig. 3). An electrophilic mechanism was suggested by the mechanisms for similar enzymes that used FPP or GGPP as substrates, such as prenyl diphosphate synthases and cyclases (Gebler *et al.*, 1992; Poulter *et al.*, 1981, 1989). In this type of mechanism, the substitution or addition of electron-withdrawing groups at the C3 position of FPP would be expected to inhibit the formation of a carbocation intermediate (Dolence and Poulter, 1995). Consistent with this model, single and multiple fluorine substitutions at C3 decreased the activity of yeast and mammalian FTase in an amount proportional to the number of added fluorines (Dolence and Poulter, 1995; Huang *et al.*, 2000). A nucleophilic mechanism for FTase is also supported by several lines of evidence. This type of mechanism is used frequently by zinc-containing enzymes that catalyze protein S-alkylation reactions, such as methanol:M methyltransferase, methionine synthases, and Ada, a DNA repair enzyme that removes methyl groups from DNA methylphosphodiesterases (LeClerc and Grahame, 1996; Matthews and Goulding, 1998; Myers *et al.*, 1993). Binding of a CaaX peptide to the FTase–isoprenoid complex lowers the pK of the cysteine thiol from 8.3 to $\sim$6.7, resulting in a dramatic increase in the amount of the nucleophilic thiolate anion at neutral pH (Hightower *et al.*, 1998). Studies with FTase in which the catalytic zinc ion was replaced with cadmium also support a nucleophilic character of the mechanism. Cadmium is a softer metal than zinc and, as a result, has greater affinity for sulfur. Substitution of Cd^{2+} for Zn^{2+} with FTase resulted in markedly enhanced metal–thiol binding, increased binding

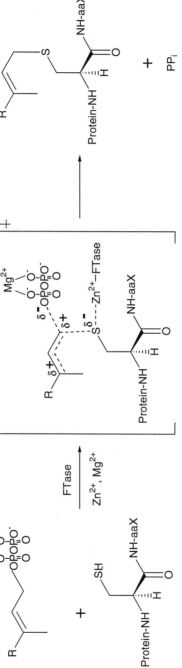

FIGURE 3 Chemical mechanism of the reaction catalyzed by FTase. The transition state for the reaction (‡) contains elements of both carbocation formation at carbon-1 of the isoprenoid and nucleophilic attack by the metal-activated cysteine thiolate. See text for further details.

of the peptide thiol to the FTase–isoprenoid binary complex, and decreased product release from the enzyme (Huang *et al.,* 2000). Overall, these data are most consistent with a carbocation–nucleophile combination reaction that emphasizes both preassociative and dissociative mechanisms (Fig. 3). Additional study will be needed to further clarify whether, and under what conditions, this reaction proceeds via a stepwise fashion with a stable intermediate or in a concerted fashion with no intermediate.

C. Structural Biology of the CaaX Prenyltransferases

The first X-ray crystal structure of FTase was of the unliganded rat enzyme and was determined at 2.2 Å resolution (Park *et al.,* 1997). Rat FTase is 93% homologous with human FTase and all residues in the active-site region are conserved, hence the mammalian enzymes are predicted to have identical active sites. The α subunit is folded into a crescent-shaped domain composed of seven successive pairs of coiled coils, so-called helical hairpins, that wrap around a significant portion of the β subunit. Pro–X–Asn–Tyr sequences were found in the turns connecting the two helices of the coiled-coil. These residues formed part of an extensive α–β subunit interface, which involved over 19.5% of the accessible surface area of the α subunit and 17% of the accessible surface area of the β subunit. The secondary structure of the β subunit of FTase was also largely helical, with the majority of the helices folded into an α–α barrel structure. One end of the barrel was blocked by a short stretch of residues near the C-terminus of the subunit, with the opposite end of the barrel open to the solvent. A deep cleft in the center of the barrel contained all of the features expected for the active site of the enzyme, including the zinc atom that participates in catalysis. The zinc was coordinated by three β-subunit residues that are conserved in all known protein prenyltransferases, Asp297, Cys299, and His362, and by a well-ordered water molecule. The central cavity of the α–α barrel in the β subunit was lined with 10 aromatic residues that are highly conserved in protein prenyltransferases, making this region highly hydrophobic. Because a similar hydrophobic cavity had been reported in the crystal structure of FPP synthase, it was suggested that this pocket plays a role in isoprenoid binding (Park and Beese, 1997).

The early finding that the two CaaX prenyltransferases contain identical α subunits suggested that the structural basis for substrate specificity of prenylation, both in terms of the isoprenoid as well as the protein substrate, must primarily be a property of the distinct β subunits of the enzymes (Seabra *et al.,* 1991; Zhang *et al.,* 1994a). In support of this hypothesis, early studies using both full-length Ras proteins and photoaffinity-labeled peptide substrates (Reiss *et al.,* 1991), as well as photoactivatable isoprenoid analogues of FPP (Pellicena *et al.,* 1996; Ying *et al.,* 1994), showed that these substrates could be crosslinked to the β subunit of FTase.

Some labeling of the α subunit was also observed in some of these studies, suggesting that the binding site of peptide substrate might be located near an interface of the α and β subunits. These predictions, as well as those noted above in terms of the hydrophobic cavity observed in the structure of the unliganded enzyme, were confirmed with the publication of crystal structures of the FTase–FPP complex as well as ternary complexes of FTase containing both an isoprenoid analogue and a CaaX peptide bound in the active site (Long *et al.,* 1998, 1999; Strickland *et al.,* 1998) (Fig. 4). These structures revealed that the isoprenoid substrate bound in an extended conformation along one side of the hydrophobic cavity formed within the α–α barrel. The diphosphate moiety of FPP was positioned in a positively charged pocket at the top of the hydrophobic cleft near the α–β subunit, whereas the CaaX motif of the peptide substrate also bound in an extended conformation in the hydrophobic cavity, with the C-terminal carboxylate near the bottom of the cavity. A large part of the binding surface of the CaaX motif was formed by the bound isoprenoid, a structure consistent with the observation noted above that substrate binding by FTase is functionally ordered, with FPP binding preceding that of the CaaX peptide.

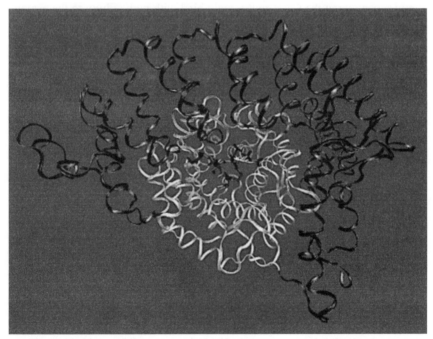

FIGURE 4   The three-dimensional structure of mammalian FTase, with the α subunit shaded dark and the β subunit shaded light. See text for details.

III. DEVELOPMENT OF FTase INHIBITORS AS ANTICANCER AGENTS

A. Rationale for Development of FTase Inhibitors as Anticancer Agents

Many growth factors signal through tyrosine kinases, the majority of which subsequently signal through Ras proteins (Boguski and McCormick, 1993). Importantly, Ras proteins are modified by the farnesyl isoprenoid and this farnesylation is an absolute requirement for their participation in signal transduction (Hancock *et al.*, 1989; Jackson *et al.*, 1990; Schafer *et al.*, 1989). Mutations of Ras that lead to production of a constitutively active protein are some of the most common genetic alterations in human cancers and occur in approximately one-third of all cancers, including 95% of pancreatic, 30–50% of colorectal, and a significant fraction of bladder, thryroid, and non-small-cell lung cancers and myelomas (Barbacid, 1987; Cox and Der, 1997). Additionally, mutation or overexpression of many tyrosine kinases that signal through Ras has also been associated with human cancers (Johnston, 2001). For these reasons, inhibition of Ras farnesylation has become an important strategy in development of potential cancer therapeutics.

The 1990 report of the identification of FTase, which also reported that short peptides containing the CaaX motif could bind with high affinity to the enzyme (Reiss *et al.*, 1990), triggered the efforts that continue to this day to develop specific FTase inhibitors, now termed FTIs, for evaluation as anticancer therapeutics. Numerous inhibitors of FTase with good selectivity for this enzyme over the related enzyme GGTase-I were quickly identified using several strategies, including design of analogues of the CaaX peptide and FPP substrates and high-throughput screening of natural product and compound libraries (Gibbs and Oliff, 1997; Graham and Williams, 1996; Sebti and Hamilton, 1997). These compounds can be generally placed into four distinct categories: mimics of CaaX tetrapeptides, mimics of FPP, bisubstrate analogues, and general organic compounds selected from the natural product and compound libraries.

Most CaaX tetrapeptides derived from natural FTase substrates bind the enzyme with affinities that are essentially identical to those of the full-length protein substrates (Kohl *et al.*, 1991; Zhang and Casey, 1996b; Zhang *et al.*, 1997). However, simple CaaX peptides suffer from lack of cell permeability and rapid degradation, which plague essentially all short peptides introduced into biological systems. The primary strategy around these problems has been to produce structurally altered forms of the peptides that exhibit greater cell permeability and stability. Analysis of structure–function relationships, as well as data from the FTase crystal structures in recent years, has allowed the development of rational peptide analogues. The prototype for the first generation of the peptidomimetic class of inhibitors was the tetrapeptide CVFM, because this CaaX peptide bound FTase with high affinity, but was not a substrate for the enzyme (Goldstein *et al.*, 1991). Crystallographic analysis of complexes of FTase containing CaaX peptides has strongly suggested that peptide substrates and inhibitors that bind in that site do so in an extended conformation (Long *et al.*, 1999; Strickland *et al.*, 1998).

FPP mimics were among the first inhibitors of FTase to be developed, although none has proceeded into clinical testing. One of the first compounds in this class was α-hydroxyfarnesylphosphonic acid, which inhibits FTase with a K_i of 5 nM and inhibits Ras processing in H-Ras-transformed NIH 3T3 cells in the 1 μM range (Gibbs *et al.*, 1993). Other compounds with squalene backbones attached to carboxylic or phosphonic acid moieties have also demonstrated potent inhibition of FTase (Graham and Williams, 1996; Patel *et al.*, 1995). However, the need for charged moieties that mimic the phosphate group, coupled with the finding that most FPP mimics also inhibit other FPP-utilizing enzymes, such as squalene synthase, has limited the development of this class of FTIs.

Several FTIs have been identified that contain features of both the CaaX peptide and FPP substrates. The parent of this series was a compound from Bristol Myers Squibb termed BMS185878, which was a low-nanomolar inhibitor of FTase (Manne *et al.*, 2001). Several additional agents in this series were synthesized that were highly selective for FTase over GGTase-I, and showed activity in Ras-transformed NIH 3T3 cells (Manne *et al.*, 2001). However, for broad antitumor activity, particularly *in vivo*, these agents did not compare well to several of the other FTIs, and have not continued into clinical development.

Several distinct classes of FTIs were identified after screening of compound libraries; one of the first reported was a tricyclic compound identified by Schering–Plough termed SCH44342 (Bishop *et al.*, 1995). Multiple modifications to improve potency and selectivity led the synthesis of SCH66336, which inhibited FTase at nanomolar ranges, but was inactive against GGTase-I at concentrations up to 50 μM (Bishop *et al.*, 2001). These compounds are nonpeptidic and do not contain sulfhydryl groups. SCH66336 was quite cell-permeable, and significant antitumor activity was seen with a variety of *in vitro* and *in vivo* tumor models, including xenografts of bladder, prostate, colon, pancreas, and lung cancers (Bishop *et al.*, 2001). This compound also exhibited favorable pharmacokinetics and good oral bioavailability and is currently in clinical trials.

The histidylbenzylglycinamides are peptidic in nature and inhibit FTase in a fashion that is competitive with respect to FPP and noncompetitive with respect to the CaaX protein substrate. The lead compound in this series, the pentapeptide PD083176, was a low-nanomolar inhibitor of FTase with good selectivity over GGTase I in cell-free systems (Sebolt-Leopold *et al.*, 2001). Multiple structure activity studies led to the compound PD169451, which was more potent and had greater cell permeability than PD083176. In both soft agar and xenograft experiments, PD169451 demonstrated activity against almost all tumor lines tested, although the mechanism appeared to be primarily cytostatic (Sebolt-Leopold *et al.*, 2001).

As with the tricyclic FTIs, the Janssen compound R115777 was identified by screening a compound library (Hudes and Schol, 2001). The lead compound was an oral quinolone analogue of a heterocyclic antifungal agent that inhibited FTase with low nanomolar affinity. A broad panel of tumor lines was sensitive to R115777, and this compound has also advanced to clinical development (Hudes and Schol, 2001).

B. Biology of FTI Effects on Cells

At least 30 cellular proteins are substrates for FTase (Cox and Der, 1997; Gibbs *et al.,* 1997). These include the four Ras isoforms and several Ras-related GTP-binding proteins that are involved in signal transduction and cytoskeletal organization, at least two γ subunits of heterotrimeric G proteins, nuclear lamins A and B, which are involved in the scaffolding of the nuclear membrane, Pxf, which is associated with peroxisome function, DNA-J, which is a chaperone protein, phosphorylase kinase, which is involved in skeletal muscle function, rhodopsin kinase, which is involved in visual signal transduction, two protein tyrosine phosphatases termed PTPCAAX-1 and -2, respectively, and the centromere-binding proteins CENP-E and -F. By comparison, an estimated 70–80 proteins are substrates for GGTase-I (Sinensky, 2000).

Despite the complexities associated with the large number of substrates for FTase, the initial evaluations of FTIs demonstrated they had potent and selective activity against Ras-transformed cells. Based on work with H-Ras-transformed cell lines, such as engineered NIH 3T3 fibroblasts and COS kidney cells, FTIs were shown to block Ras farnesylation and membrane association as well as cell proliferation and anchorage-independent growth (James *et al.,* 1993; Kohl *et al.,* 1993). Moreover, in several of these studies, the FTIs had essentially no effect on either the nontransformed parental lines or matched lines transformed with other oncogenes that bypass Ras signaling, such as Raf-1 or Mos. In addition, in many cell types, FTIs exerted their effects at least in part by inducing apoptosis (Crul *et al.,* 2001; Manne *et al.,* 2001; Prendergast, 2001). However, in tumor cell lines not transformed by H-Ras, including most human-derived tumor lines and xenografts, the response to FTIs was primarily inhibition of proliferation. FTI treatment could, however, induce apoptosis in these lines if cells were deprived of substratum attachment or serum growth factors, suggesting the overall cellular effects of FTIs may depend significantly upon cellular context (Lebowitz *et al.,* 1997b; Suzuki *et al.,* 1998).

While most early work with FTIs focused on H-Ras models, far more human solid tumors are associated with mutations in K-Ras, and human leukemias are associated almost exclusively with N-Ras (Bos, 1989; Gibbs *et al.,* 1997). The different Ras isoforms also have distinct expression patterns in normal tissues, distinct sites of cellular localization, and distinct modes of regulation (Jones and

Jackson, 1998; Koera *et al.*, 1997). Both K-Ras and N-Ras have significantly higher binding affinities to FTase than does H-Ras (James *et al.*, 1996; Zhang *et al.*, 1997), making them more difficult substrates to inhibit when using compounds that target the CaaX-binding site on the enzyme (as essentially all in clinical development do). In contrast to H-Ras, K-Ras and N-Ras can also be alternatively prenylated by GGTase-I, albeit with low efficiency, providing a potential mechanism to bypass FTase inhibition (James *et al.*, 1995; Zhang *et al.*, 1997). In fact, there is much conflicting data on the importance of oncogenic Ras in the response of a tumor cell to FTIs; in the two studies where this has been most closely examined, the presence and type of Ras mutation correlated poorly with sensitivity to FTIs (Nagasu *et al.*, 1995; Sepp-Lorenzino *et al.*, 1995). These types of studies have highlighted the potential, as well as the limitations, of FTIs as antitumor agents and have also reinforced the limitations of the models used to develop such types of agents.

These complexities of FTase biology and the preclinical studies of FTIs have suggested that inhibition of Ras might be neither necessary nor sufficient to account for many of the effects of these compounds observed in particular cell types or tissues (Cox and Der, 1997; Gibbs *et al.*, 1997). In addition to Ras proteins, other FTase substrates have been suggested to be major targets of cellular responses to FTIs. One such protein is the Ras relative RhoB. Cellular processes in which RhoB has been implicated include integrin-dependent signaling, actin stress fiber formation, trafficking of the epidermal growth factor receptor, and activation of the protein Akt (Prendergast and Oliff, 2000). As noted above, RhoB is unusual, in that it appears to be a substrate for both FTase and GGTase-I under normal conditions, and both farnesylated and geranylgeranylated forms can be found in cells (Adamson *et al.*, 1992). RhoB has a much shorter half-life than Ras, which is more compatible with the time course for FTI effects on cells. In this regard, several studies have shown that FTase inhibitors suppress Ras transformation at least in part by interfering with RhoB activity, and that this effect correlates with altering the prenylation of RhoB (Lebowitz *et al.*, 1995, 1997a). Other candidates for mediating the growth-suppressing effects of FTIs are a Ras-related protein termed Rheb (Clark *et al.*, 1997) and the centromere-binding protein CENP-E (Ashar *et al.*, 2000).

Because farnesylation is required for the function of many cellular proteins, how FTIs decrease the growth of many types of transformed cells with little effect on normal cells remains to be defined. One possibility is that when FTase is inhibited, GGTase-I might be able to modify an important subset of FTase substrates that are required for normal cell function and that geranylgeranylation restores at least some of their function. Another possibility is that the FTase substrates that are the most critical for normal growth are also the highest affinity substrates of the enzyme, allowing them to effectively outcompete the inhibitors at the doses that are sufficient to substantially reduce the activities of those substrates most important for aberrant proliferation. Clearly, the complex mechanism through which FTase inhibitors cause tumor regression needs further investigation.

IV. CLINICAL RESULTS FROM EVALUATION OF FTIs AS ANTITUMOR AGENTS

The first FTI that went into clinical testing was R115777, followed soon after by SCH66336, L778123, and BMS214662. Other compounds are still entering clinical trials. Expectations were very high for this class of compounds, which many thought would be among the first examples of rationally developed anticancer drugs. Ras and its signaling pathways had been implicated directly or indirectly in over half of human cancers, and farnesylation was seen as the Achilles' heel of Ras. In addition, the preclinical activity of these compounds was better than most anticancer drugs previously developed, and toxicity in the mouse (the primary species used in the preclinical studies) was remarkably modest at highly effective doses. However, from the limited number of human trials for which data have been reported, it seems that toxicity of FTIs is going to be an important issue (Adjei *et al.*, 2000; Hudes and Schol, 2001; Johnston, 2001). Additionally, although early hints of clinical utility have been noted, it is clear that the preclinical models used to assess these agents have been only modestly predictive of their effects in humans.

Phase I dose escalation, pharmacological, and biomarker studies have been reported for most of the FTIs in clinical development, and a limited amount of data are available on phase II studies. At least one phase III efficacy study is now underway, but no results have been reported. Despite distinct chemotypes, these compounds have had consistent toxicity profiles, which while not identical, have been similar enough to suggest they are all inhibiting FTase and that the toxicities observed are mechanism-based (i.e., due to effects on FTase). Most FTIs have been associated with a consistent pattern of toxicities, which include gastrointestinal problems such as nausea and diarrhea, myelosuppression, and neurotoxicities such as confusion and lethargy (Crul *et al.*, 2001; Johnston, 2001). In most cases, these toxicities were reversible within a few days after drug withdrawal. The toxicity of a drug depends upon dose, schedule, and route of administration and multiple pharmacokinetic properties, sensitivities of different tissues, as well as many other known and unknown factors. However, the relatively reproducible timing for the development and resolution of toxicities associated with FTI treatment suggests that prenylated proteins with moderate half-lives may be responsible for some of these toxicities. Because of concern for altering farnesylation of proteins related to vision, most studies included rigorous ophthalmologic surveillance, including electroretinograms. However, no consistent or clinically significant ophthalmologic toxicities have been reported.

Data on the efficacy of FTIs in phase II studies have been modest. Although activity has been noted in a variety of human cancers, efficacy of FTIs as single agents has been limited, with only occasional partial responses and disease stabilization that to date has been of short duration. This may be partially related to the fact that the early clinical studies targeted tumor types that are notoriously refractory to most treatments, such as pancreatic cancer. Pancreatic cancer has been targeted primarily because 90% of pancreatic adenocarcinomas harbor a K-Ras mutation, and Ras was initially thought to be the main target of FTIs. Another reason for targeting pancreatic cancer is that the limited treatment options for this cancer make it attractive for FDA registration. It may be that better clinical activity will be seen with cancers that contain oncogenic H-Ras or N-Ras, because both the Ras isoforms are easier substrates to inhibit and are less prone to cross-prenylation than is K-Ras. Approximately 40% of bladder cancers have H-Ras mutations, and an equal fraction of leukemias harbor N-Ras mutations. Because of special clinical problems associated with leukemias, information about toxicities for a new agent needs to be gathered first in patients with solid tumors. However, clinical testing in leukemia has begun, and optimism is high for activity for FTIs in this disease. In addition, the ability to repeatedly sample the tumor tissue in leukemia patients by straightforward phlebotomy or bone marrow aspirations is allowing investigation into the mechanisms of action for FTIs in human cancers. Due to the limited responses observed to date, and because of preclinical evidence of synergy with traditional anticancer agents such as taxol, FTIs are currently being testing in combination protocols, particularly in tumors where both agents have shown signs of activity (Johnston, 2001). Such phase II studies have been sufficiently encouraging to justify a few large phase III studies of FTIs in solid tumors.

The limited information available on biomarker assessments of solid tumors (which are not usually amendable to repeated biopsies) in the clinical trials have demonstrated that FTIs can inhibit farnesylation of some proteins, such as DNA-J in circulating monocytes and prelamin-A in buccal mucosa cells (Adjei *et al.,* 2000; Crul *et al.,* 2001). These analyses are not highly quantitative and clear dose–response relationships have not been seen, in part because of small numbers of patients treated at a given dose. Because of numerous technical factors, including the limited number of antibodies that can distinguish farnesylated from unfarnesylated proteins, it has also been difficult to assess farnesylation of other proteins, such as the Ras isoforms, even from monocytes. In addition, since it is not known which farnesylated proteins are most responsible for either efficacy or toxicity, these types of surrogate marker studies are only useful (at least at present) to confirm that the target (FTase) is actually being inhibited in the tumor tissue.

V. SUMMARY AND CONCLUSIONS

A remarkable amount of progress in terms of both the biochemistry and biology of the system has been achieved in the decade since the discovery of CaaX protein prenylation. However, despite the remarkable efficacy and minimal toxicity of FTIs in most *in vitro* and *in vivo* studies, the value of inhibiting farnesylation in patients is likely to be complicated, due at least in part to the greater biological and pharmacological complexity of humans and human cancers compared to the mouse models that have been the mainstay of the preclinical *in vivo* studies. The fact that multiple proteins besides Ras are farnesylated may lead to unexpected toxicities if a subset of these proteins is more important in cell physiology in humans than in mice. In addition, since many farnesylated proteins can undergo geranylgeranylation or other lipid modifications, just what level of efficacy will be achieved with FTIs alone is still unclear. Further studies, many of which are already in progress, will likely be able to address these issues more completely. The significant complexities of lipid–protein modifications, both in membrane targeting and protein–protein interactions, which are far better appreciated now than when the biomedical effort to develop FTIs started less than a decade ago, will also undoubtedly play a role in these efforts. In spite of all these concerns, there is still great optimism that FTIs will soon enter the arsenal of the oncologist.

Acknowledgments

We thank Ashley Butler for editorial assistance and Kendra Hightower, Gary McCauley, and Laura Stemmle for preparation of figures. Work from the authors' laboratory was supported by research grants from the NIH (P.J.C.) and Duke Rankin NIH General Clinical Research Unit (H.I.H.).

References

Adamson, P., Marshall, C. J., Hall, A., and Tilbrook, P. A. (1992). Post-translational modifications of p21rho proteins. *J. Biol. Chem.* **267**, 20033–20038.

Adjei, A. A., Erlichman, C., Davis, J. N., Cutler, D. L., Sloan, J. A., Marks, R. S., Hanson, L. J., Svingen, P. A., Atherton, P., and Bishop, W. R., *et al.* (2000). A phase I trial of the farnesyl transferase inhibitor SCH66336: Evidence for biological and clinical activity. *Cancer Res.* **60**, 1871–1877.

Armstrong, S. A., Hannah, V. C., Goldstein, J. L., and Brown, M. S. (1995). CaaX geranylgeranyl transferase transfers farnesyl as efficiently as geranylgeranyl to RhoB. *J. Biol. Chem.* **270**, 7864–7868.

Ashar, H. R., James, L., Gray, K., Carr, D., Black, S., Armstrong, L., Bishop, W. R., and Kirschmeier, P. (2000). Farnesyl transferase inhibitors block the farnesylation of CENP-E and CENP-F and alter the association of CENP-E with the microtubules. *J. Biol. Chem.* **275**, 30451–30457.

Ashby, M. N. (1998). CaaX converting enzymes. *Curr. Opin. Lipidol.* **9**, 99–102.

Barbacid, M. (1987). Ras genes. *Annu. Rev. Biochem.* **56**, 779–827.

Bergo, M. O., Leung, G. K., Ambroziak, P., Otto, J. C., Casey, P. J., and Young, S. G. (2000). Targeted inactivation of the isoprenylcysteine carboxyl methyltransferase gene causes mislocalization of K-Ras in mammalian cells. *J. Biol. Chem.* **275**, 17605–17610.

Bishop, W. R., Bond, R., Petrin, J., Wang, L., Patton, R., Doll, R., Njoroge, G., Catino, J., Schwartz, J., and Windsor, W., *et al.* (1995). Novel tricyclic inhibitors of farnesyl protein transferase. *J. Biol. Chem.* **270**, 30611–30618.

Bishop, W. R., K. Pai, J. J., Armstrong, L., Dalton, M., Doll, R. J., Taveras, A., Njoroge, G., Sinensky, M., Zhang, F., Liu, M., and Kirschmeier, P. (2001). Tricyclic farnesyl protein transferase inhibitors. *In* "Farnesyltransferase Inhibitors in Cancer Therapy" (S. M. Sebti and A. D. Hamilton, eds.), pp. 87–101. Humana Press, Tampa, FL.

Boguski, M. S., and McCormick, F. (1993). Proteins regulating Ras and its relatives. *Nature* **366**, 643–654.

Bos, J. L. (1989). Ras oncogenes in human cancer: A review. *Cancer Res.* **49**, 4682–4689.

Carboni, J. M., Yan, N., Cox, A. D., Bustelo, X., Graham, S. M.; Lynch, M. J., Weinmann, R., Seizinger, B. R., Der, C. J., Barbacid, M., and Manne, V. (1995). Farnesyltransferase inhibitors are inhibitors of Ras, but not R-Ras2/TC21, transformation. *Oncogene* **10**, 1905–1913.

Casey, P. J., and Seabra, M. C. (1996). Protein prenyltransferases. *J. Biol. Chem.* **271,** 5289–5292.

Casey, P. J., Thissen, J. A., and Moomaw, J. F. (1991). Enzymatic modification of proteins with a geranylgeranyl isoprenoid. *Proc. Natl. Acad. Sci. USA* **88,** 8631–8635.

Chen, W.-J., Moomaw, J. F., Overton, L., Kost, T. A., and Casey, P. J. (1993). High-level expression of mammalian protein farnesyltransferase in a baculovirus system: The purified protein contains zinc. *J. Biol. Chem.* **268,** 9675–9680.

Clark, G. J., Kinch, M. S., Rogers-Graham, K., Sebti, S. M., Hamilton, A. D., and Der, C. J. (1997). The Ras-related protein Rheb is farnesylated and antagonizes Ras signaling and transformation. *J. Biol. Chem.* **272,** 10608–10615.

Clarke, S. (1992). Protein isoprenylation and methylation at carboxyl-terminal cysteine residues. *Annu. Rev. Biochem.* **61,** 355–386.

Cox, A. D., and Der, C. J. (1997). Farnesyltransferase inhibitors and cancer treatment: Targeting simply Ras? *Biochim. Biophys. Acta* **1333,** F51–F71.

Crul, M., de Klerk, G. J., Beijnen, J. H., and Schellens, J. H. (2001). Ras biochemistry and farnesyl transferase inhibitors: A literature survey. *Anticancer Drugs* **12,** 163–184.

Dolence, J. M., and Poulter, C. D. (1995). A mechanism for posttranslational modifications of proteins by yeast protein farnesyltransferase. *Proc. Natl. Acad. Sci. USA* **92,** 5008–5011.

Fu, H. W., and Casey, P. J. (1999). Enzymology and biology of CaaX protein prenylation. *Rec. Prog. Hormone Res.* **54,** 315–343.

Furfine, E. S., Leban, J. J., Landavazo, A., Moomaw, J. F., and Casey, P. J. (1995). Protein farnesyl-transferase: Kinetics of farnesyl pyrophosphate binding and product release. *Biochemistry* **34,** 6857–6862.

Gebler, J. C., Woodside, A. B., and Poulter, C. D. (1992). Dimethylallytryptophan synthase: An enzyme-catalyzed electrophilic aromatic substitution. *J. Am. Chem. Soc.* **114,** 7354–7360.

Gelb, M. H. (1997). Protein prenylation: Signal transduction in two dimensions. *Science* **275,** 1750–1751.

Gibbs, J. B. (1991). Ras C-terminal processing enzymes: New drug targets? *Cell* **65,** 1–4.

Gibbs, J. B., and Oliff, A. (1997). The potential of farnesyltransferase inhibitors as cancer chemotherapeutics. *Annu. Rev. Pharmacol. Toxicol.* **37,** 143–166.

Gibbs, J. B., Pompliano, D. L., Mosser, S. D., Rands, E., Lingham, R. B., Singh, S. B., Scolnick, E. M., Kohl, N. E., and Oliff, A. (1993). Selective inhibition of farnesyl–protein transferase blocks ras processing *in vivo. J. Biol. Chem.* **268,** 7617–7620.

Gibbs, J. B., Graham, S. L., Hartman, G. D., Koblan, K. S., Kohl, N. E., Omer, C. A., and Oliff, A. (1997). Farnesyltransferase inhibitors versus Ras inhibitors. *Curr. Opin. Chem. Biol.* **1,** 197–203.

Glomset, J. A., Gelb, M. H., and Farnsworth, C. C. (1990). Prenyl proteins in eukaryotic cells: A new type of membrane anchor. *Trends Biochem. Sci.* **15,** 139–142.

Goldstein, J. L., Brown, M. S., Stradley, S. J., Reiss, Y., and Gierasch, L. M. (1991). Nonfarnesylated tetrapeptide inhibitors of protein farnesyltransferase. *J. Biol. Chem.* **266,** 15575–15578.

Graham, S. L., and Williams, T. M. (1996). Inhibitors of protein farnesylation. *Exp. Opin. Therapeutic Patents* **6,** 1295–1304.

Hancock, J. F., Magee, A. I., Childs, J. E., and Marshall, C. J. (1989). All Ras proteins are polyiso-prenylated but only some are palmitoylated. *Cell* **57,** 1167–1177.

Hightower, K. E., and Fierke, C. A. (1998). Zinc-catalyzed sulfur alkylation: Insights from protein farnesyltransferase. *Curr. Opin. Chem. Biol.* **3,** 176–181.

Hightower, K. E., Huang, C.-C., Casey, P. J., and Fierke, C. A. (1998). H-Ras peptide and protein substrates bind protein farnesyltransferase as an ionized thiolate. *Biochemistry* **37,** 15555–15562.

Huang, C.-C., Casey, P. J., and Fierke, C. A. (1997). Evidence for a catalytic role of zinc in protein farnesyltransferase: Spectroscopy of Co^{2+}-FTase indicates metal coordination of the substrate thiolate. *J. Biol. Chem.* **272,** 20–23.

Huang, C.-C., Hightower, K. E., and Fierke, C. A. (2000). Mechanistic studies of rat protein farnesyl-transferase indicate an associative transition state. *Biochemistry* **39**, 2593–2602.

Hudes, G. R., and Schol, J. (2001). Phase I trial of oral R115777 in patients with refractory solid tumors. *In* "Farnesyltransferase Inhibitors in Cancer Therapy" (S. M. Sebti and A. D. Hamilton, eds.), pp. 251–254. Humana Press, Tampa, FL.

Jackson, J. H., Cochrane, C. G., Bourne, J. R., Solski, P. A., Buss, J. E., and Der, C. J. (1990). Farnesol modification of Kirsten-Ras exon 4B protein is essential for transformation. *Proc. Natl. Acad. Sci. USA* **87**, 3042–3046.

James, G. L., Goldstein, J. L., Brown, M. S., Rawson, T. E., Somers, T. C., McDowell, R. S., Crowley, C. W., Lucas, B. K., Levinson, A. D., and Marsters, J. C., Jr. (1993). Benzodiazepine peptidomimetics: Potent inhibitors of Ras farnesylation in animal cells. *Science* **260**, 1937–1942.

James, G. L., Goldstein, J. L., and Brown, M. S. (1995). Polylysine and CVIM sequences of K-RasB dictate specificity of prenylation and confer resistance to benzodiazepine peptidomimetic *in vitro*. *J. Biol. Chem.* **270**, 6221–6226.

James, G., Goldstein, J. L., and Brown, M. S. (1996). Resistance of K-RasBv12 proteins to farnesyl-transferase inhibitors in Rat1 cells. *Proc. Natl. Acad. Sci. USA* **93**, 4454–4458.

Johnston, S. R. (2001). Farnesyl transferase inhibitors: A novel targeted therapy for cancer. *Lancet Oncol.* **2**, 18–26.

Jones, M. K., and Jackson, J. H. (1998). Ras-GRF activates Ha-Ras, but not N-Ras or K-Ras 4B, protein *in vivo*. *J. Biol. Chem.* **273**, 1782–1787.

Kim, E., Ambroziak, P., Otto, J. C., Taylor, B., Ashby, M., Shannon, K., Casey, P. J., and Young, S. G. (1999). Disruption of the mouse Rce1 gene results in defective ras processing and mislocalization of ras within cells. *J. Biol. Chem.* **274**, 8383–8390.

Koera, K., Nakamura, K., Nakao, K., Miyoshi, J., Toyoshima, K., Hatta, T., Otani, H., Aiba, A., and Katsuki, M. (1997). K-Ras is essential for the development of the mouse embryo. *Oncogene* **15**, 1151–1159.

Kohl, N. E., Diehl, R. E., Schaber, M. D., Rands, E., Soderman, D. D., He, B., Moores, S. L., Pompliano, D. L., Ferro-Novick, S., and Powers, S., *et al.* (1991). Structural homology among mammalian and *Saccharomyces cerevisiae* isoprenyl–protein transferases. *J. Biol. Chem.* **266**, 18884–18888.

Kohl, N. E., Mosser, S. D., deSolms, S. J., Giuliani, E. A., Pompliano, D. L., Graham, S. L., Smith, R. L., Scolnick, E. M., Oliff, A., and Gibbs, J. B. (1993). Selective inhibition of Ras-dependent transformation by a farnesyltransferase inhibitor. *Science* **260**, 1934–1937.

Lebowitz, P. F., Davide, J. P., and Prendergast, G. C. (1995). Evidence that farnesyltransferase inhibitors suppress ras transformation by interfering with Rho activity. *Mol. Cell. Biol.* **15**, 6613–6622.

Lebowitz, P. F., Casey, P. J., Prendergast, G. C., and Thissen, J. A. (1997a). Farnesyltransferase inhibitors alter the prenylation and growth-stimulating function of RhoB. *J. Biol. Chem.* **272**, 15591–15594.

Lebowitz, P. F., Sakamuro, D., and Prendergast, G. C. (1997b). Farnesyl transferase inhibitors induce apoptosis of Ras-transformed cells denied substratum attachment. *Cancer Res.* **57**, 708–713.

LeClerc, G. M., and Grahame, D. A. (1996). Methylcobamide:coenzyme M methyltransferase isozymes from *Methanosarcina barkeri J. Biol. Chem.* **271**, 18725–18731.

Long, S. B., Casey, P. J., and Beese, L. S. (1998). Co-crystal structure of protein farnesyltransferase complexed with a farnesyl diphosphate substrate. *Biochemistry* **37**, 9612–9618.

Long, S. B., Casey, P. J., and Beese, L. S. (1999). The basis for K-Ras4B binding specificity to protein farnesyltransferase revealed by 2 Å resolution ternary complex structures. *Structure* **8**, 209–222.

Manne, V., Lee, F., Yan, N., Fairchild, C., and Rose, W. C. (2001). Development of farnesyltransferase inhibitors as potential intitumor agents. *In* "Farnesyltransferase Inhibitors in Cancer Therapy" (S. M. Sebti and A. D. Hamilton, eds.), pp. 71–86. Humana Press, Tampa, FL.

Marshall, C. J. (1993). Protein prenylation: A mediator of protein–protein interactions. *Science* **259**, 1865–1866.

Matthews, R. G., and Goulding, C. W. (1998). Enzyme-catalyzed methyl transfers to thiols: The role of zinc. *Curr. Opin. Chem. Biol.* **1**, 332–339.

Moomaw, J. F., and Casey, P. J. (1992). Mammalian protein geranylgeranyltransferase: Subunit composition and metal requirements. *J. Biol. Chem.* **267,** 17438–17443.

Myers, L. C., Terranova, M. P., Ferentz, A. E., Wagner, G., and Verdine, G. L. (1993). Repair of DNA methylphosphotriesters through a metalloactivated cysteine nucleophile. *Science* **261,** 1164–1167.

Nagasu, T., Yoshimatsu, K., Rowell, C., Lewis, M. D., and Garcia, A. M. (1995). Inhibition of human tumor xenograft growth by treatment with the farnesyl transferase inhibitor B956. *Cancer Res.* **55,** 5310–5314.

Park, H.-W., and Beese, L. S. B. (1997). Protein farnesyltransferase. *Curr. Opin. Struct. Biol.* **7,** 873–880.

Park, H.-W., Boduluri, S. R., Moomaw, J. F., Casey, P. J., and Beese, L. S. (1997). Crystal structure of protein farnesyltransferase at 2.25 Å resolution. *Science* **275,** 1800–1804.

Patel, D. V., Schmidt, R. J., Biller, S. A., Gordon, E. M., Robinson, S. S., and Manne, V. (1995). Farnesyl diphosphate-based inhibitors of Ras farnesyl protein transferase. *J. Med. Chem.* **38,** 2906–2921.

Pellicena, P., Scholten, J. D., Zimmerman, K., Creswell, M., Huang, C. C., and Miller, W. T. (1996). Involvement of the alpha subunit of farnesyl–protein transferase in substrate recognition. *Biochemistry* **35,** 13494–13500.

Pompliano, D. L., Rands, E., Schaber, M. D., Mosser, S. D., Anthony, N. J., and Gibbs, J. B. (1992). Steady-state kinetic mechanism of ras farnesyl:protein transferase. *Biochemistry* **31,** 3800–3807.

Poulter, C. D., Wiggins, P. L., and Le, A. T. (1981). Farnesylpyrophosphate synthetase: A new stepwise mechanism for the 1′–4 condensation reaction. *J. Am. Chem. Soc.* **103,** 3926–3927.

Poulter, C. D., Capson, T. L., Thompson, T. D., and Bard, R. S. (1989). Squalene synthetase: Inhibition by ammonium analogues of carbocationic intermediates in the conversion of presqualene diphosphate to squalene. *J. Am. Chem. Soc.* **111,** 3734–3739.

Prendergast, G. C. (2001). Effects of farnesyltransferase inhibitors on cytoskeleton, cell transformation, and tumorigenesis. *In* "Farnesyltransferase Inhibitors in Cancer Therapy" (S. M. Sebti and A. D. Hamilton, eds.), pp. 159–169. Humana Press, Tampa, FL.

Prendergast, G. C., and Oliff, A. (2000). Farnesyltransferase inhibitors: Antineoplastic properties, mechanisms of action, and clinical prospects. *Semin. Cancer Biol.* **10,** 443–452.

Reiss, Y., Goldstein, J. L., Seabra, M. C., Casey, P. J., and Brown, M. S. (1990). Inhibition of purified p21ras farnesyl:protein transferase by Cys-aaX tetrapeptides. *Cell* **62,** 81–88.

Reiss, Y., Seabra, M. C., Armstrong, S. A., Slaughter, C. A., Goldstein, J. L., and Brown, M. S. (1991). Nonidentical subunits of p21 H-Ras farnesyltransferase: Peptide binding and farnesyl pyrophosphate carrier functions. *J. Biol. Chem.* **266,** 10672–10677.

Reiss, Y., Brown, M. S., and Goldstein, J. L. (1992). Divalent cation and prenyl pyrophosphate specificities of the protein farnesyltransferase from rat brain, a zinc metalloenzyme. *J. Biol. Chem.* **267,** 6403–6408.

Rowell, C. A., Kowalczyk, J. J., Lewis, M. D., and Garcia, A. M. (1997). Direct demonstration of geranylgeranylation and farnesylation of Ki-Ras *in vivo. J. Biol. Chem.* **272,** 14093–14097.

Schafer, W. R., and Rine, J. (1992). Protein prenylation: Genes, enzymes, targets and functions. *Annu. Rev. Genet.* **25,** 209–238.

Schafer, W. R., Kim, R., Sterne, R., Thorner, J., Kim, S.-H., and Rine, J. (1989). Genetic and pharmacological suppression of oncogenic mutations in Ras genes of yeast and humans. *Science* **245,** 379–385.

Seabra, M. C., Reiss, Y., Casey, P. J., Brown, M. S., and Goldstein, J. L. (1991). Protein farnesyltransferase and geranylgeranyltransferase share a common alpha subunit. *Cell* **65,** 429–434.

Sebolt-Leopold, J. S., Deonard, D. M., and Leopold, W. R. (2001). Histidylbenzylglycinamides: A novel class of farnesyl diphosphate-competitive peptidic farnesyltransferase inhibitors. *In* Farnesyltransferase Inhibitors in Cancer Therapy" (S. M. Sebti and A. D. Hamilton, eds.), pp. 103–114. Humana Press, Tampa, FL.

Sebti, S. M., and Hamilton, A. D. (1997). Inhibition of ras prenylation: A novel approach to cancer chemotherapy. *Pharmacol. Therapeutics* **74,** 103–114.

Sebti, S. M., and Hamilton, A. D. (2001). Farnesyltransferase and geranylgeranyltransferase I inhibitors as novel agents for cancer. *In* "Farnesyltransferase Inhibitors in Cancer Therapy" (S. M. Sebti and A. D. Hamilton, eds.), pp. 197–219. Humana Press, Tampa, FL.

Sepp-Lorenzino, L., Ma, Z., Rands, E., Kohl, N. E., Gibbs, J. B., Oliff, A., and Rosen, N. (1995). A peptidomimetic inhibitor of farnesyl:protein transferase blocks the anchorage-dependent and -independent growth of human tumor cell lines. *Cancer Res.* **55**, 5302–5309.

Sinensky, M. (2000). Recent advances in the study of prenylated proteins. *Biochim. Biophys. Acta* **1484**, 93–106.

Strickland, C. L., Windsor, W. T., Syto, R., Wang, L., Bond, R., Wu, Z., Schwartz, J., Le, H. V., Beese, L. S., and Weber, P. C. (1998). Crystal structure of farnesyl protein transferase complexed with a CaaX peptide and farnesyl diphosphate analogue. *Biochemistry* **37**, 16601–16611.

Suzuki, N., Del Villar, K., and Tamanoi, F. (1998). Farnesyltransferase inhibitors induce dramatic morphological changes of KNRK cells that are blocked by microtubule interfering agents. *Proc. Natl. Acad. Sci. USA* **95**, 10499–10504.

Tschantz, W. R., Furfine, E. S., and Casey, P. J. (1997). Substrate binding is required for release of product from mammalian protein farnesyltransferase. *J. Biol. Chem.* **272**, 9989–9993.

Whyte, D. B., Kirschmeier, P., Hockenberry, T. N., Nunez-Oliva, I., James, L., Catino, J. J., Bishop, W. R., and Pai, J.-K. (1997). K- and N-Ras are geranylgeranylated in cells treated with farnesyl protein transferase inhibitors. *J. Biol. Chem.* **272**, 14459–14464.

Ying, W., Sepp-Lorenzino, L., Cai, K., Aloise, P., and Coleman, P. S. (1994). Photoaffinity-labeling peptide substrates for farnesyl–protein transferase and the intersubunit location of the active site. *J. Biol. Chem.* **269**, 470–477.

Yokoyama, K., and Gelb, M. H. (2000). Protein geranylgeranyltransferase type I. *In* "The Enzymes" 3rd. ed., Vol. 21 (F. Tamanoi and D. S. Sigman, eds.), pp. 105–130. Academic Press, San Diego, CA.

Yokoyama, K., Goodwin, G. W., Ghomashchi, F., Glomset, J. A., and Gelb, M. H. (1991). A protein geranylgeranyltransferase from bovine brain: Implications for protein prenylation specificity. *Proc. Natl. Acad. Sci. USA* **88**, 5302–5306.

Yokoyama, K., McGeady, P., and Gelb, M. H. (1995). Mammalian protein geranylgeranyltransferase-I: Substrate specificity, kinetic mechanism, metal requirements, and affinity labeling. *Biochemistry* **34**, 1344–1354.

Yokoyama, K., Zimmerman, K., Scholten, J., and Gelb, M. H. (1997). Differential prenyl pyrophosphate binding to mammalian protein geranylgeranyltransferase-I and protein farnesyltransferase and its consequences on the specificity of protein prenylation. *J. Biol. Chem.* **272**, 3944–3952.

Zhang, F. L., and Casey, P. J. (1996a). Influence of metals on substrate binding and catalytic activity of mammalian protein geranylgeranyltransferase type-I. *Biochem. J.* **320**, 925–932.

Zhang, F. L., and Casey, P. J. (1996b). Protein prenylation: Molecular mechanisms and functional consequences. *Annu. Rev. Biochem.* **65**, 241–269.

Zhang, F. L., Diehl, R. E., Kohl, N. E., Gibbs, J. B., Giros, B., Casey, P. J., and Omer, C. A. (1994a). cDNA cloning and expression of rat and human protein geranylgeranyltransferase type I. *J. Biol. Chem.* **269**, 3175–3180.

Zhang, F. L., Moomaw, J. F., and Casey, P. J. (1994b). Properties and kinetic mechanism of recombinant mammalian protein geranylgeranyltransferase type I. *J. Biol. Chem.* **269**, 23465–23470.

Zhang, F. L., Kirschmeier, P., Carr, D., James, L., Bond, R. W., Wang, L., Patton, R., Windsor, W. T., Syto, R., Zhang, R., and Bishop, W. R. (1997). Characterization of Ha-Ras, N-Ras, Ki-Ras4A, and Ki-Ras4B as *in vitro* substrates for farnesyl protein transferase and geranylgeranyl protein transferase type I. *J. Biol. Chem.* **272**, 10232–10239.

CHAPTER 20

Peptides as Probes of Protein–Protein Interactions Involved in Neurotransmitter Release

Thomas Kuner,[*,†,‡] **Hiroshi Tokumaru,**[†,‡] **and George J. Augustine**[†,‡]

*Department of Neurobiology, Duke University Medical Center, Durham, North Carolina 27710;
†Marine Biological Laboratory, Woods Hole, Massachusetts 02543; and ‡Max-Planck-Institut für
medizinische Forschung, 69120 Heidelberg, Germany

I. Introduction
II. Peptide Design
    A. Criteria for Useful Peptides
    B. Design of Effective Peptides
III. Characterization of Peptide Activity on Protein–Protein Interactions
IV. Functional Analysis of Peptides in a Presynaptic Terminal
  V. Summary and Outlook
    References

One of the central problems in neuroscience is to define the roles of presynaptic proteins in neurotransmitter release. Peptides from the binding sites of these proteins serve as valuable probes of protein function. Important criteria for such peptides include their ability to specifically perturb a defined protein–protein interaction and to be sufficiently soluble to permit microinjection. A number of strategies have been employed to design effective peptides. Successful strategies include making educated guesses based on protein primary sequence, using binding studies to define sites of protein–protein interaction, and designing peptides based on the three-dimensional structure of known sites of interaction. Once peptides have been synthesized, it is important to characterize their actions *in vitro* before introducing them into cells. Particularly valuable in this regard are affinity chromatography, to define the binding partners of the peptides, and measurements of the actions of peptides upon binary interactions between presynaptic proteins. Finally, a synapse suited to microinjection, such as the squid giant synapse, can be used to evaluate the functional and structural actions of the peptides. This chapter describes our recent work using peptides to elucidate pre-fusion roles of *N*-ethylmaleimide-sensitive factor (NSF), α-soluble NSF-attachment protein (α-SNAP), and complexin in regulating SNAP receptor (SNARE)-dependent membrane fusion at this synapse.

I. INTRODUCTION

Communication between nerve cells in the brain arises from the release of neurotransmitter molecules, which are stored within membrane-bound organelles called synaptic vesicles (Katz, 1969). Electrical excitation causes calcium ions to enter the presynaptic terminal region of a nerve cell and the resulting intracellular calcium signal serves as a trigger for the fusion of synaptic vesicles with the plasma membrane. Such membrane fusion is only one of many reactions involving synaptic vesicles; these structures participate in a complex and precisely regulated cycle of membrane trafficking reactions within the presynaptic terminal (Fig. 1A). Although the general properties of the synaptic vesicle cycle have been known for nearly 30 years (Heuser and Reese, 1973), the molecular underpinnings of this cycle are only beginning to be understood. Biochemical and molecular biological analyses have identified many of the proteins present on synaptic vesicles (Lin and Scheller, 2000; Südhof, 1995). Further, the interactions of these proteins with a large number of other presynaptic proteins have also been established. Despite this abundance of information, the molecular basis of the vesicle trafficking reactions remains largely a matter of speculation (Augustine *et al.*, 1996).

The key to understanding the molecular basis of neurotransmitter release lies in defining the spatiotemporal dynamics of protein–protein interactions within the presynaptic terminal. One of the most powerful approaches for examining protein–protein interactions in living cells is to use synthetic peptides derived from the protein-binding domains involved in these interactions. The basic rationale behind this strategy is that a short stretch of amino acid residues of a protein can, under certain conditions, mimic structures found in the full-length proteins. This property can then be utilized to disrupt a protein–protein interaction by simple competition of the peptide and the native protein with their common binding site.

Our laboratory has used such peptides to systematically study protein–protein interactions underlying exocytosis of synaptic vesicles (Augustine *et al.*, 1999; Burns and Augustine, 1999). The purpose of this chapter is to review some of the theoretical and practical aspects of the use of peptides to study protein–protein interactions involved in membrane fusion within the presynaptic terminal. We will concentrate on one particular set of protein–protein interactions related to the SNARE proteins (Rothman, 1994). Presynaptic terminals possess at least three

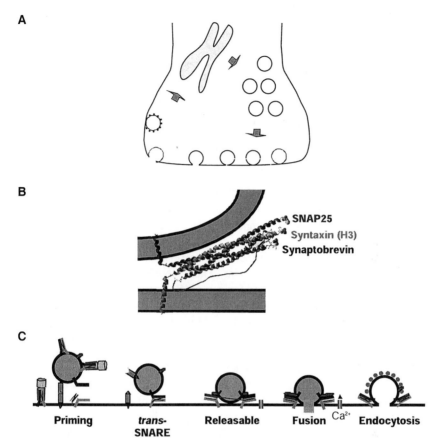

FIGURE 1 Protein–protein interactions involved in synaptic vesicle trafficking. (A) The synaptic vesicle cycle. Vesicles bud from the endosome and translocate into a reserve pool of vesicles. As releasable vesicles (*bottom*) are consumed by synaptic transmission, release sites are filled with vesicles from the reserve pool. Vesicles fused with the membrane are recycled via endocytosis (*bottom left*). (B) Molecular structure of the SNARE complex. Vesicle membrane and presynaptic plasma membrane are shown schematically. Transmembrane anchors of synaptobrevin and syntaxin are modeled as alpha helices. (C) Final stages in neurotransmitter release. Color scheme: synaptobrevin, blue; syntaxin, orange; SNAP25, violet; SNAP, brown; NSF, yellow; tethering complex, cyan; Ca^{2+} sensor, red; Ca channel, green; clathrin, pink. (See color plate.)

SNARE proteins: synaptobrevin, found on the vesicle membrane, and syntaxin and synaptosome-associated protein of 25 kDa molecular mass (SNAP-25) on the plasma membrane. Coiled-coil domains present on each of these proteins allows the three of them to form a four-helix bundle (Fig. 1B) (Sutton *et al.,* 1998). This complex is so stable that it can withstand exposure to harsh experimental conditions, such as denaturing detergents (e.g., sodium dodecylsulfate [SDS]; Hayashi *et al.*, 1994) or heating in excess of 80°C (Chen *et al.*, 1999). Within cells, dissociation of the SNARE complex is accomplished by an ATPase called NSF (Söllner *et al.*, 1993a,b). NSF is linked to the SNARE complex via another group of proteins, called SNAPs (Clary *et al.*, 1990). In addition, other proteins, such as complexin (also called synaphin; Ishizuka *et al.*, 1995; McMahon *et al.*, 1995), can bind to the SNARE complex and regulate its function. A large abundance of evidence supports the conclusion that SNARE complex proteins, as well as NSF, SNAP, complexin, and others, are essential for fusion of synaptic vesicles with the plasma membrane during neurotransmitter release (Rothman, 1994; Jahn and Südhof, 1999; Tokumaru and Augustine, 2002).

A specific model for the roles of SNARE proteins in synaptic vesicle trafficking is shown in Fig. 1C. Our current understanding is that when a vesicle docks at the presynaptic plasma membrane, the SNARE proteins on either membrane may be complexed with each other to form *cis* SNARE complexes (Otto *et al.*, 1997). SNAP then attaches NSF to such complexes during the process of priming, to yield uncomplexed SNARE proteins. These SNAREs can then bind to each other across the two membranes to form *trans* SNARE complexes. In some circumstances, *trans* SNARE complexes are sufficient to catalyze membrane fusion (Weber *et al.*, 1998); for such cases, the tight binding of the membrane-anchored SNARE proteins probably plays an active role by pulling the two membranes close together (Jahn and Hanson, 1998). However, neurotransmitter release requires the additional action of calcium ions, which presumably work by activating calcium-binding proteins, such as synaptotagmin, to trigger the final fusion of the two membranes (Augustine, 2001; Davis *et al.*, 1999; Südhof and Rizo, 1996). Recent work indicates that complexin also may play a role in conferring calcium sensitivity to transmitter release (Reim *et al.*, 2001), perhaps by oligomerizing SNARE complexes into higher order structures (Tokumaru *et al.*, 2001).

A large amount of the experimental support for this model, from our labs as well as others, derives from the use of peptides that selectively perturb the interactions among SNARE proteins. Such an experimental approach relies upon successful completion of three tasks: peptide design, biochemical characterization, and functional analysis in living cells. These tasks are considered in turn in the next three sections.

II. PEPTIDE DESIGN

A. Criteria for Useful Peptides

Specificity and affinity of a peptide for its target define its value as a probe of intracellular function. Because most peptides have much lower binding affinities than a full-length protein, the peptide needs to be present at large excess. In addition, in microinjection experiments peptide solutions must be concentrated further because of dilution upon intracellular injection. Therefore, solubility is a key issue when designing peptides. To be soluble, the peptide needs to be sufficiently hydrophilic, requiring an excess of polar residues over hydrophobic ones. In most cases, peptides from surface-exposed regions of a protein must interact with an aqueous medium and will be sufficiently hydrophilic. Hydrophobic peptides tend to form aggregates that clog microinjection pipettes. Most of the inhibitory peptides we have used successfully consist of 15–20 amino acid residues. Increasing the length of these peptides may improve affinity somewhat, but longer peptides do not always have improved solubility. It also is sometimes desirable to trace the diffusion of peptides within cells. For this purpose, peptides can be tagged with fluorescent dyes by incorporating a cysteine residue, to which a variety of dyes can be covalently linked.

B. Design of Effective Peptides

The key aspect of peptide design is to identify the regions of a protein that serve as binding sites for a given interaction. Once such regions have been identified, it is usually straightforward to develop biologically active peptides. Three general strategies to achieve this goal have been developed.

1. The "Educated Guess" Approach

The earliest efforts of our laboratory were done before information about protein-binding sites was available. In the absence of this structural information, peptides were designed based on homology mapping of proteins: Sequence alignments of related proteins were used to find structurally conserved regions. Often, conserved regions serve important functions for the protein, such as forming part of an enzymatically active site, a substrate-binding site, a site for oligomerization, or a site of interaction with other proteins. An example of a sequence alignment is illustrated in Fig. 2 for the D1 domain of seven forms of NSF. Amino acid residues that are identical across these varieties of NSF are indicated by dots in the figure. Among the several stretches of conserved residues, peptides were synthesized from the six regions shown by the outlined boxes in Fig. 2. In addition to sequence similarity, sequences can be examined with structure prediction algorithms to find known motifs (e.g., ATP-binding sites, shown in the shaded area labeled "Walker A" in Fig. 2), clusters of charged residues, surface exposure, and other potentially important sites. For peptides designed to interfere with protein–protein interactions, which are likely to occur at the solvent-exposed surfaces of these proteins, it is important to determine whether a given sequence is likely to be exposed on the surface of the protein. A predicted surface exposure pattern (line above the alignment in Fig. 2) can be useful in the absence of direct structural information.

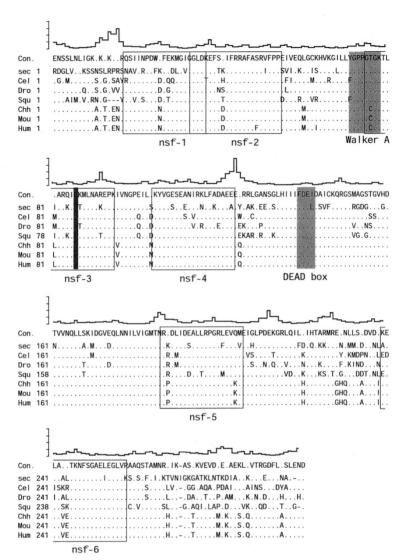

FIGURE 2 Sequence alignment of NSF D1 domains. The alignment includes the D1 domains of NSF from *Saccharomyces cerevisiae* (sec), *Caenorhabditis elegans* (Cel), *Drosophila melanogaster* (Dro), *Loligo pealei*(= squid, Squ), *Cricetulus griseus*(= Chinese hamster, Chh), *Mus musculus* (Mou), and *Homo sapiens* (Hum). Peptides derived from the sequence are marked with open boxes, and specific motifs are shaded (*light gray:* Walker, DEAD box; *dark gray: comatose* mutation). Shown above the alignment is the Emini surface exposure index (scale on the left 0–8).

This "educated guess" approach to the design of peptides has, in many cases, proven successful. For example, two of the six peptides prepared from the D1 domain of NSF have substantial effects on the ATPase activity of NSF *in vitro* and also produce effects on the magnitude and the kinetics of transmitter release *in vivo* (Schweizer *et al.*, 1998). A third peptide, NSF1, had weak physiological effects and did not affect the ATPase activity of NSF. In our hands, this one-third rate of producing active peptides is rather typical for this approach to peptide design. Mapping of biologically active peptides onto crystal structures confirms that this strategy is capable of identifying important structural regions of these proteins (see below). In some instances, mutagenesis studies can also identify important structural determinants of function. For example, a naturally occurring mutation of NSF found in *Drosophila* gives rise to a temperature-sensitive phenotype termed *comatose* (Pallanck *et al.*, 1995; Schweizer *et al.*, 1998). The mutation was traced to a Gly position in a highly conserved region of the D1 domain, located close to the ATP-binding site (Fig. 2). Such mutations can provide very useful hints regarding the design of inhibitory peptides. The main limitation of these approaches is that there is no *a priori* way to know the protein–protein interaction that is being disrupted by a biologically active peptide.

2. Biochemical Identification of Sites of Protein–Protein Interaction

A second means of peptide design avoids this limitation by using biochemical or molecular genetic assays to define the interacting partners and sites of protein–protein interaction. Powerful methods, such as two-hybrid screens and direct peptide sequencing with mass spectrometry, are available to identify binding partners. Another widely used approach uses pulldown assays performed by attaching a recombinant protein (typically fused to glutathione-S-transferase [GST]) to beads and adding other proteins to determine which associate with the beads. Once a binding partner is identified, it is necessary to define the sites that allow the binding interaction to occur. The usual approach to this problem is to use truncation mutagenesis, in conjunction with *in vitro* binding assays, to define binding sites.

For the case of complexin, a protein that binds to the plasma membrane SNARE syntaxin (Ishizuka *et al.*, 1995; McMahan *et al.*, 1995), truncation mutagenesis indicates that syntaxin binds to the highly conserved central domain of this protein (Pabst *et al.*, 2000; Tokumaru *et al.*, 2001). Three GST-fusion constructs representing the NH_2-terminal region (residues 1–51), the central region (residues 52–102), or the COOH-terminal region (residues 103–152) of complexin were incubated with syntaxin-containing detergent extracts of squid synaptosomes. Glutathione beads were then used to pull down the complexin fusion proteins, and the amount of bound syntaxin was analyzed by Western blotting. Whereas the central region of complexin bound to syntaxin, there was no detectable binding of syntaxin to

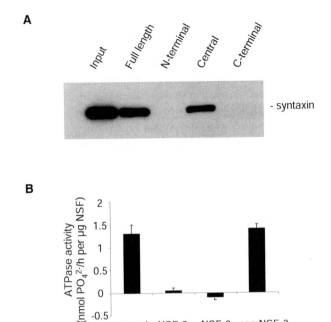

FIGURE 3 *In vitro* characterization of binding domains and effects of peptides on protein function. (A) Western blot showing interaction of complexin with syntaxin in a GST pulldown assay.
(B) SNAP-stimulated ATPase function of NSF. Two peptides, NSF-2 and NSF-3, inhibit ATPase activity. A point mutation analogous to the *comatose* mutation in the NSF-3 peptide (comNSF-3) renders
the peptide inactive.

the NH_2- or COOH-terminal constructs (Fig. 3A). Thus, the central region of
complexin is responsible for binding to syntaxin. Subsequent synthesis of smaller
peptides within this syntaxin-binding domain indicated that binding activity resides
within a small region called SBD-2 (Tokumaru *et al.*, 2001). The SBD-2 peptide
is highly active in preventing the interaction between complexin and syntaxin and
also is a potent inhibitor of neurotransmitter release (see below).

   As an alternative to measurements of protein binding, *in vitro* measurements of
the biological activity of proteins can also be used to define the sites of protein–
protein interaction. An example of such a screen comes from our studies of the
ATPase activity of NSF. In this case, an *in vitro* assay of the SNAP-stimulated
ATPase activity of NSF (Morgan and Burgoyne, 1995) was used to determine
whether the NSF peptides illustrated in Fig. 2 affected ATPase activity (Schweizer
et al., 1998). Of the six NSF peptides synthesized, only two altered the ATPase
activity of NSF (Fig. 3B). This inhibition of ATPase activity was concentration-
dependent and was well correlated with the activity of these peptides in inhibiting

transmitter release. The fact that only NSF and SNAP were present in the reaction indicates that the peptide either affected the interaction of the two proteins or prevented a conformational change occuring in either one of the proteins. Although such information does not identify the binding sites of the peptides, it clearly demonstrates that the peptide interferes with a function of the protein and can be a useful means of facilitating peptide design by defining functionally relevant regions of the protein.

In summary, this second approach to peptide design is labor-intensive, but has the advantage of generating peptides that are from known sites of binding interactions and are therefore very likely to perturb these interactions.

3. Design of Peptides Based on Protein Structure

The final, and perhaps ideal, approach to peptide design is based upon knowledge of the three-dimensional structure of proteins and interaction sites. With an increasing number of crystal structures becoming available (Brunger, 2001), it now is possible to use structural features of the protein to rationally design binding-site peptides. These structures unambiguously define which parts of the protein are exposed at the surface, and sites of protein–protein interaction can be imaged in cases where the two proteins are cocrystallized. Even for cases where the binding partner is not simultaneously imaged, it often is possible to make accurate estimates of the sites of interaction. We anticipate that the rapid current expansion in our database of structures of presynaptic proteins will yield a significant improvement in the efficient design of peptides that are specific for defined protein–protein interactions.

Recent solutions of the crystal structures of several SNARE-related proteins provide opportunities to correlate protein structure with function of peptides that were developed by previous, nonstructural approaches. Mapping the location of functional peptides can reveal important structural determinants for sites of interaction with effector proteins and provide the structural basis for the physiological effects produced by a peptide. Furthermore, knowing the structural properties of active peptides will improve future efforts for the structure-based design of new effective peptides. Here we consider two examples of retrospective structural analysis of active peptides.

As shown in Fig. 2, six peptides were synthesized from the D1 domain of NSF. Two of these showed robust effects on synaptic transmission, NSF-2 and NSF-3; one NSF-1, showed weaker effects; and the other three, NSF-4, NSF-5, and NSF-6, had no discernible biological activity. Although the D1 domain has yet not been crystallized, this domain shows high sequence identity to the D2 domain, which can therefore be used as a model to predict the likely structure of the D1 domain. Mapping these peptides onto the structure of the D2 domain reveals an interesting correlation: The three peptides with biological activity map onto a single region exposed at the surface of the protein (Fig. 4A). The two most effective peptides, NSF-2 and NSF-3, are from regions that form α helices in the mature protein,

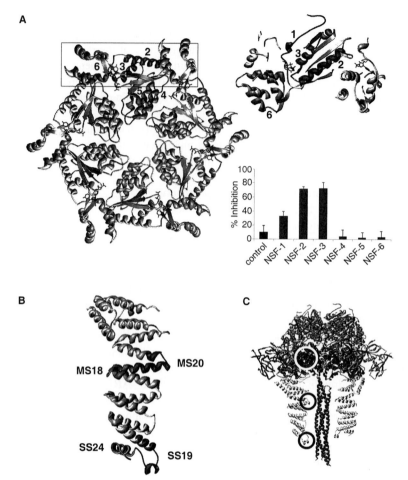

FIGURE 4 Mapping peptides on crystal structures. (A) NSF peptides mapped on the D1 domain. Based on the high sequence identity of the D1 and the D2 domains, their structures are likely to be very similar. Therefore, we used the coordinates of the D2 domain as a model for the D1 domain. Left: View from bottom. Right: Side view of 20-Å slab indicated by the box in the left panel. Active peptides are colored red, inactive peptides are green (bar diagram on the right). (B) Peptides derived from α-SNAP. Peptides shown in red affect function, whereas those shown in green do not. (C) Hypothetical model of 20S complex assembled from individual crystal structures. NSF is depicted in blue colors (N domain: violet; D1 domain: dark blue; D2 domain: light blue), α-SNAP in yellow, and the SNARE complex in red. The left α-SNAP molecule is oriented as shown in B. Circles highlight functionally important sites identified by peptide experiments. (See color plate.)

whereas the weakly effective NSF-1 peptide is derived from a region with little secondary structure. The ineffective peptides NSF-4 and NSF-5 were derived from a region located in the interior of the hexameric NSF complex, whereas the amino acid residues of NSF-6 are only partially exposed to the lateral surface of the protein.

This pattern suggests that the region formed by residues of NSF-2 and NSF-3 forms a binding site, either for another protein, one that binds to NSF, or to another region within NSF. The latter seems likely, because NSF is known to undergo significant conformational changes upon ATP hydrolysis (Hanson *et al.*, 1997). The glycine residue changed in the temperature-sensitive *comatose* mutation (G274E) (Pallanck *et al.*, 1995) is located at the buried face of the short helix that forms part of the NSF-3 peptide. We hypothesize that the glutamate residue changes the function of NSF by hindering local folding of the polypeptide, which is likely to be more strongly impaired at higher temperatures. It is also interesting to note that the NSF-3 peptide is close to the ATP-binding site and that residues preceding the N-terminal part of the peptide directly contribute to the ATP-binding site. Therefore, a local conformational change may translate into a decreased binding of ATP in the *comatose* mutated form of NSF. The effect of the NSF-3 peptide may not be connected to such a mechanism, but instead may arise from perturbing an interaction of NSF with an as-yet-unknown effector protein or preventing an interdomain interaction of the N domain with the D1 or the D2 domain. The latter seems consistent with the finding that NSF-3 abolishes ATPase activity of NSF and is consistent with the hypothesis that α-SNAP binds to the bottom face of the NSF hexamer. In conclusion, a remarkably consistent picture of the structure and function of the NSF D1 domain arises from combined data obtained from *Drosophila* genetics, crystallography, *in vitro* assays of NSF function, and peptide microinjection experiments.

Another good example of the structural basis for peptide activity comes from studies of α-SNAP. The crystal structure of this protein has recently been solved (Rice and Brunger, 1999) and four peptides from α-SNAP have been assayed for effects on membrane fusion *in vitro* and for synaptic transmission *in vivo* (DeBello *et al.*, 1995). Two of these peptides (SS19, MS20) were biologically active, whereas the other two (SS24, MS18) were not. Interestingly, the two effective peptides are located on the same face of α-SNAP (Fig. 4B). This surface has been suggested to interact with the SNARE complex (Rice and Brunger, 1999), suggesting a basis for the inhibition of SNARE-based membrane trafficking (DeBello *et al.*, 1995). In contrast, both the ineffective peptides map onto a face of α-SNAP pointing away from the central axis of the SNARE complex (Figs. 4B, and 4C). For the design of future peptides it might be interesting to note that both effective peptides consist of two helical components connected by a short loop. Such loops are often involved in forming catalytically active sites or binding sites of proteins and therefore may serve as sites for attaching α-SNAP to the SNARE complex.

III. CHARACTERIZATION OF PEPTIDE ACTIVITY ON PROTEIN–PROTEIN INTERACTIONS

Once a peptide has been designed and synthesized, it is important to document its action upon the interactions of presynaptic proteins. It is useful to begin our consideration of this topic by distinguishing two types of peptides: those that are derived from known binding sites and those that are not. For the former case, it is likely that the peptide works by binding to the same protein that the full-length protein binds. An *in vitro* characterization can test this possibility by determining whether this is the only binding partner of the peptide. In addition, biochemical analysis can define whether the peptide acts as a simple competitive inhibitor or has additional biological actions upon binding to its partner. For the case of peptides that are not derived from defined binding sites, biochemical characterization can provide these types of information and also is the only way to define the binding target of the peptide.

As mentioned in the previous section, a plethora of biochemical techniques are available to study peptide–protein interactions *in vitro,* and a detailed discussion of these methods is beyond the scope of this chapter. We will instead focus on two specific types of applications: using affinity chromatography to define the binding partners of a peptide and using GST pulldown assays to measure the effects of peptides on defined protein–protein interactions. We illustrate these two types of *in vitro* characterization with examples taken from our recent studies of the SBD-2 peptide from complexin.

Affinity chromatography is a valuable means of defining the binding partner of a peptide. For the case of the SBD-2 peptide from complexin, an affinity column was prepared by immobilizing SBD-2 on agarose beads. These beads were then treated with detergent extracts of squid optic lobe synaptosomes and SDS polyacrylamide gel electrophoresis was used to analyze proteins that bound to the beads. Staining with Coomassie blue revealed four bands, representing bound proteins (Fig. 5A, lane 1). Subsequent characterization of these bands, via Western blotting, revealed that they represent SNARE proteins: syntaxin, SNAP-25, synaptobrevin, and a proteolytic fragment of synaptobrevin (Fig. 5A, lanes 2–4). Using recombinant SNARE proteins, we found that syntaxin, but not SNAP-25 or synaptobrevin, bound directly to the SBD-2 beads (Tokumaru *et al.,* 2001). Such information leads to the conclusion that SBD-2 peptide binds directly to syntaxin and that SNAP-25 and synaptobrevin associate indirectly, by binding to syntaxin.

Once binding partners have been identified, the activity of peptides upon defined protein–protein interactions can be evaluated in binding assays. In the case of the SBD-2 peptide, which comes from the syntaxin-binding domain of complexin, we presumed that SBD-2 should act as a competitive inhibitor of the interaction of these two proteins. This hypothesis was tested by using GST pulldown assays,

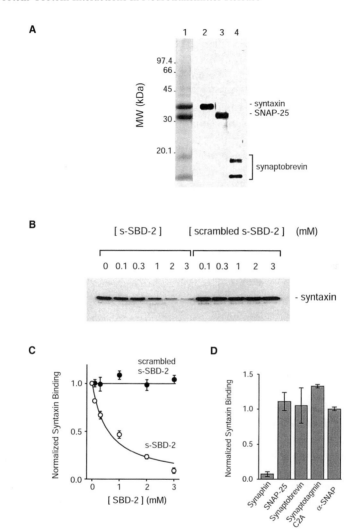

FIGURE 5 Peptide binding partners and specificity. (A) SBD-2 peptide affinity column. Lane 1 shows a Coomassie-stained SDS gel of proteins from a squid optic lobe synaptosomal extract that bind to SBD-2 peptide immobilized on a column. Lanes 2–4 show Western blots of the proteins in lane 1 with specific antibodies directed toward syntaxin, SNAP-25, and synaptobrevin respectively. (B) The amount of syntaxin bound to complexin, visualized by Western blotting, is reduced by increasing concentrations of free SBD-2 peptide. A scrambled version of this peptide has no effect. (C) Concentration dependence of the effect of SBD-2 upon the interaction of complexin with syntaxin, taken from the data shown in B. (D) Specificity of SBD-2 in blocking the interaction of complexin and syntaxin. From Tokumaru *et al.* (2001).

similar to those illustrated in Fig. 3A. Addition of SBD-2 peptide reduced the binding of squid syntaxin to GST-tagged complexin (Fig. 5B). This inhibition was concentration-dependent and half-maximal at a SBD-2 concentration of 0.6 mM (Fig. 5C). Blockade of binding of complexin to syntaxin was sequence-specific, because neither scrambled SBD-2 peptides nor SBD-1, from an adjacent region of complexin, interfered with binding of complexin to syntaxin (Figs. 5B and 5C).

Showing that a peptide prevents a given protein–protein interaction, via the sorts of experiments described above, does not eliminate the possibility that this peptide can interfere with other protein–protein interactions within a cell. *In vitro* binding experiments can be very valuable for further evaluating the specificity of action of peptides. For the case of SBD-2, further biochemical assays were performed to measure the actions of this peptide upon other interactions involving syntaxin. Such analyses demonstrated that SBD-2 did not affect the interaction of syntaxin with the other SNAREs, synaptobrevin and SNAP-25 (Fig. 5D). Further, SBD-2 did not prevent syntaxin from binding to α-SNAP or to synaptotagmin, a calcium-binding protein that is important for neurotransmitter release (Augustine, 2001). These results indicate that SBD-2 peptide acts as a selective antagonist of the binding of complexin to syntaxin.

In conclusion, defining the biochemical properties of a peptide is important when interpreting functional data and, ultimately, in defining the molecular mechanism of protein function within cells. Fortunately, a variety of techniques are available for this purpose and have proven reliable in a number of applications.

IV. FUNCTIONAL ANALYSIS OF PEPTIDES IN A PRESYNAPTIC TERMINAL

The final goal of studies such as those described in this chapter is to use peptides to study the roles of specific proteins and protein–protein interactions in cellular functions. For the case of membrane fusion, which involves intracellular proteins, these peptides must be introduced within the cell. Although a variety of procedures are available for this purpose, our lab delivers peptides by microinjection via "sharp" microelectrodes or by microdialysis through patch-clamp pipettes. The experiments described here employ the use of microinjection of peptides into the giant presynaptic terminal of the squid (*Loligo pealei*). This terminal is 1 mm long and has an average thickness of about 50 μm, a size that is sufficiently large to permit microinjection via sharp electrodes (Fig. 6A). At the same time, synaptic transmission can be monitored simultaneously by recording electrical signals from both presynaptic and postsynaptic sides, allowing a precise determination of the functional state of the synapse (Fig. 6B). Such experiments have been performed since the early 1970s (Llinas, 1999) and have proven very useful for elucidating

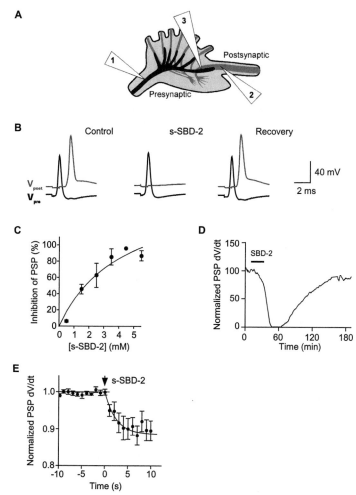

FIGURE 6 Functional analysis in squid giant synapse. (A) Anatomy of the giant synapse in the stellate ganglion of the squid, *Loligo pealei*. Action potentials are evoked in the presynaptic axon with electrode 1 and recorded in the presynaptic terminal with electrode 3, which also serves as a microinjection electrode. Postsynaptic responses are recorded with electrode 2. (B) Simultaneous recordings of presynaptic (V_{pre}) and postsynaptic (V_{post}) responses with the configuration shown in A. Injection of the SBD-2 peptide results in a complete block of transmission (*center*), which reverses when peptide injection is stopped (*right*). (C) Concentration dependence of the action of SBD-2 on synaptic transmission. (D) Time course of reversibility of SBD-2 peptide effect, measured as the slope (dV/dt) of postsynaptic potential responses (PSPs). (E) Rapid inhibition produced by a brief injection of SBD-2 peptide. An exponential function with a time constant of 2 s describes the decay of transmission after injection of SBD-2 peptide at time 0.

the function of various proteins involved in neurotransmitter release (Augustine *et al.*, 1999).

An example of such a microinjection experiment is shown in Fig. 6B. This experiment illustrates the changes in synaptic transmission that arise from injecting the SBD-2 peptide. Before injecting this peptide, an action potential evoked in the giant presynaptic terminal (V_{pre}) evoked sufficient release of the neurotransmitter (which is probably the amino acid glutamate) to evoke a suprathreshold electrical signal in the postsynaptic neuron (V_{post}). After injecting SBD-2 into the presynaptic terminal, presynaptic action potentials could still be elicited. However, these signals were much less effective in releasing neurotransmitter, so that the postsynaptic response was subthreshold. This suggests that the binding of complexin to syntaxin is important for transmitter release.

Because the giant presynaptic terminal can be damaged by overzealous microinjection, it is important to confirm that the inhibition of synaptic transmission shown in Fig. 6B is due to the SBD-2 rather than damage. A simple means of distinguishing between these two possibilities is to cease microinjection and look for reversal of the inhibitory response. Given the geometry of the giant presynaptic neuron (Fig. 6A), peptide microinjected into a single presynaptic terminal eventually diffuses away into the rest of this neuron. The resulting reduction in peptide concentration should cause synaptic transmission to recover if the inhibition is due to the peptide. Indeed, this is what is observed when injection of SBD-2 ceases (Fig. 6B). This reversal provides strong evidence that the inhibition is specifically due to the SBD-2. Such reversal is usually observed when microinjecting reagents into the squid giant presynaptic terminal. A notable exception is the microinjection of clostridial neurotroxins, which serve as proteases that irreversibly cleave SNARE proteins over the time course of a microinjection experiment (e.g., Hunt *et al.*, 1994).

Other types of experiments provide additional support for the conclusion that the inhibition of synaptic transmission is due to SBD-2 preventing the interaction of complexin with syntaxin. First, the concentrations of peptide required for inhibiting synaptic transmission *in vivo* (Fig. 6C) are very similar to those required for inhibiting the interaction between the two proteins *in vitro* (Fig. 5C). Second, peptides from the SBD-2 region of several different forms of complexin had very similar abilities to inhibit synaptic transmission. Finally, peptides that have no ability to inhibit the binding of complexin to syntaxin, such as the SBD-1 peptide, were unable to inhibit synaptic transmission, even when injected at intracellular concentrations higher than those at which SBD-2 inhibits synaptic transmission. These several parallels between the behavior of synaptic transmission and the interaction of complexin with syntaxin argue that synaptic transmission requires this interaction. Such correlations between functional and biochemical analyses are generally employed when using peptides as probes of cellular function.

Additional information about the timing of the interaction between proteins can come from analyses of the kinetics of peptide action. The time course of the inhibitory effects of the SBD-2 injection are illustrated in Fig. 6D. In this example, the strength of synaptic transmission was assessed by measuring the slope of the postsynaptic response. It can be seen that the inhibitory effect of SBD-2 developed over 30 min and required 1–2 h for recovery. The long recovery time presumably reflects the time required for diffusion of sufficient amounts of SBD-2 out of the presynaptic terminal. The 30 min required for complete inhibition reflects the period of time over which the peptide was injected. This prolonged injection time was selected for this experiment because the time required for diffusion of peptides through the 1-mm-long presynaptic terminal is on the order of tens of minutes (Bommert *et al.,* 1993). However, the time required for delivery of the SBD-2 throughout the terminal severely limits the temporal resolution of the approach.

In principle, one way to overcome this temporal limitation of microinjection is to inject a large, single bolus of peptide solution into a restricted region of the presynaptic terminal (DeBello *et al.,* 1995). The assumption behind such an experiment is that nearby targets of the peptide will be reached very rapidly, although a smaller net response will be produced, because an effective concentration of the peptide reaches only a small fraction of the transmitter-releasing sites within the presynaptic terminal. Indeed, the rate of inhibition of synaptic transmission was rapid when such an injection technique was employed for the case of SBD-2 (Fig. 6E). The time constant for inhibition was on the order of 2 s, which was orders of magnitude faster than observed with slower application protocols (Fig. 6D). The rapid speed of block of transmitter release indicates that complexin binds to syntaxin within the last 2 s before a synaptic vesicle fuses (Tokumaru *et al.,* 2001). Indeed, this value must be considered an upper estimate of the speed of this interaction, because diffusion and turnover of synaptic vesicles may play roles in determining the speed of blockade.

In summary, physiological assays of the consequences of microinjecting peptides have proven valuable in establishing the function of many presynaptic proteins. When this procedure is combined with electron microscopic examination of presynaptic terminals injected with these peptides (Hess *et al.,* 1993), many of the trafficking reactions involving synaptic vesicles can be resolved to define the roles of individual proteins in these reactions.

V. SUMMARY AND OUTLOOK

Synthetic peptides have advanced our understanding of protein–protein interactions involved in neurotransmitter release. The effectiveness of a newly designed peptide can be tested in biochemical, electrophysiological, and structural assays, with each assay contributing a different set of information upon which to base deductions regarding the function of a protein–protein interaction. Knowing the binding sites, affinity constants *in vitro* and *in vivo,* specificity, and the functional effects on synaptic transmission permits detailed molecular conclusions. However, the peptide approach still does not yield much information about the kinetics of a protein–protein interaction and the relative order by which complex interaction networks of presynaptic proteins occur. Although rapid-injection experiments (e.g., Fig. 6E) offer temporal resolution on the order of seconds, this is insufficient to develop a detailed understanding of the kinetics of the more rapid interactions that immediately precede membrane fusion (see Chapter 18 in this volume). Peptides inactivated by a light-sensitive caging group will provide a powerful new tool for studying protein–protein interactions in living cells with a temporal resolution in the range of milliseconds. Crystal structures and structure–function studies will permit a rational design of such caged peptides and open the door to an exciting new round in elucidating the molecular basis of synaptic communication.

Acknowledgments

Work described in this chapter was supported by NIH grant NS-21624, an HFSP Fellowship, and Feudor Lynen Fellowship of the Alexander von Numbold Foundation.

References

Augustine, G. J. (2001). How does calcium trigger neurotransmitter release? *Curr. Opin. Neurobiol.* **11,** 320–326.

Augustine, G. J., Burns, M. E., DeBello, W. M., Pettit, D. L., and Schweizer, F. E. (1996). Exocytosis: Proteins and perturbations. *Annu. Rev. Pharmacol. Toxicol.* **36,** 659–701.

Augustine, G. J., Burns, M. E., DeBello, W. M., Hilfiker, S., Morgan, J. R., Schweizer, F. E., Tokumaru, H., and Umayahara, K. (1999). Proteins involved in synaptic vesicle trafficking. *J. Physiol. (Lond.)* **520,** 33–41.

Bommert, K., Charlton, M. P., DeBello, W. M., Chin, G. J., Betz, H., and Augustine, G. J. (1993). Inhibition of neurotransmitter release by C2-domain peptides implicates synaptotagmin in exocytosis. *Nature* **363,** 163–165.

Brunger, A. T. (2001). Structural insights into the molecular mechanism of calcium-dependent vesicle-membrane fusion. *Curr. Opin. Struct. Biol.* **11,** 163–173.

Burns, M. E., and Augustine, G. J. (1999). Functional studies of presynaptic proteins at the squid giant synapse. *In* "Neurotransmitter Release" (H. J. Bellen, ed.), pp. 237–259. Oxford University Press, Oxford.

Chen, Y. A., Scales, S. J., Patel, S. M., Doung, Y. C., and Scheller, R. H. (1999). SNARE complex formation is triggered by Ca^{2+} and drives membrane fusion. *Cell* **97,** 165–174.

Clary, D. O., Griff, I. C., and Rothman, J. E. (1990). SNAPs, a family of NSF attachment proteins involved in intracellular membrane fusion in animals and yeast. *Cell* **61,** 709–721.

Davis, A. F., Bai, J., Fasshauer, D., Wolowick, M. J., Lewis, J. L., and Chapman, E. R. (1999). Kinetics of synaptotagmin responses to Ca^{2+} and assembly with the core SNARE complex onto membranes. *Neuron* **24,** 363–376.

DeBello, W. M., O'Connor, V., Dresbach, T., Whiteheart, S. W., Wang, S. S.-H., Schweizer, F. E., Betz, H., Rothman, J. E., and Augustine, G. J. (1995). SNAP-mediated protein–protein interactions essential for neurotransmitter release. *Nature* **373,** 626–630.

Fernandez-Chacon, R., Konigstorfer, A., Gerber, S. H., Garcia, J., Matos, M. F., Stevens, C. F., Brose, N., Rizo, J., Rosenmund, C., and Sudhof, T. C. (2001). Synaptotagmin I functions as a calcium regulator of release probability. *Nature* **410,** 41–49.

Hanson, P. I., Roth, R., Morisaki, H., Jahn, R., and Heuser, J. E. (1997). Structure and conformational changes in NSF and its membrane receptor complexes visualized by quick-freeze/deep-etch electron microscopy. *Cell* **90,** 523–535.

Hayashi, T., Yamasaki, S., Nauennburg, S., Binz, T., and Niemann, H. (1994). Disassembly of the reconstituted synaptic vesicle membrane fusion complex *in vitro. EMBO J.* **14,** 2317–2325.

Hess, S. D., Doroshenko, P. A., and Augustine, G. J. (1993). A functional role for GTP-binding proteins in synaptic vesicle cycling. *Science* **259,** 1169–1172.

Heuser, J. E., and Reese, T. S. (1973). Evidence for recycling of synaptic vesicle membrane during transmitter release at the frog neuromuscular junction. *J. Cell Biol.* **57,** 315–344.

Hunt, J. M., Bommert, K., Charlton, M. P., Kistner, A., Habermann, E., Augustine, G. J., and Betz, H. (1994). A post-docking role for synaptobrevin in synaptic vesicle fusion. *Neuron* **12,** 1269–1279.

Ishizuka, T., Saisu, H., Odani, S., and Abe, T. (1995). Synaphin: A protein associated with the docking/fusion complex in presynaptic terminals. *Biochem. Biophys. Res. Commun.* **213,** 1107–1114.

Jahn, R., and Hanson, P. I. (1998). Membrane fusion: SNAREs line up in new environment. *Nature* **393,** 14–15.

Jahn, R., and Südhof, T. C. (1999). Membrane fusion and exocytosis. *Annu. Rev. Biochem.* **68,** 863–911.

Katz, B. (1969). "The Release of Neuronal Transmitter Substances." Liverpool University Press, Liverpool, England.

Lin, R. C., and Scheller, R. H. (2000). Mechanisms of synaptic vesicle exocytosis. *Annu. Rev. Cell Dev. Biol.* **16,** 19–49.

Llinas, R. R. (1999). "The Squid Giant Synapse: A Model for Chemical Transmission." Oxford University Press, Oxford.

McMahon, H. T., Missler, M., Li, C., and Südhof, T. C. (1995). Complexins: Cytosolic proteins that regulate SNAP receptor function. *Cell* **82,** 111–119.

Morgan, A., and Burgoyne, R. D. (1995). A role for soluble NSF attachment proteins (SNAPs) in regulated exocytosis in adrenal chromaffin cells. *EMBO J.* **14,** 232–239.

Otto, H., Hanson, P. I., and Jahn, R. (1997). Assembly and disassembly of a ternary complex of synaptobrevin, syntaxin, and SNAP-25 in the membrane of synaptic vesicles. *Proc. Natl. Acad. Sci. USA* **94,** 6197–6201.

Pabst, S., Hazzard, J. W., Antonin, W., Südhof, T. C., Jahn, R., Rizo, J., and Fasshauer, D. (2000). Selective interaction of complexin with the neuronal SNARE complex. Determination of the binding regions. *J. Biol. Chem.* **275,** 19808–19818.

Pallanck, L., Ordway, R. W., and Ganetzky, B. (1995). A *Drosophila* NSF mutant. *Nature* **376,** 25.

Reim, K., Mansour, M., Varoqueaux, F., McMahon, H. T., Sudhof, T. C., Brose, N., and Rosenmund, C. (2001). Complexins regulate a late step in Ca^{2+}-dependent neurotransmitter release. *Cell* **104,** 71–81.

Rice, L. M., and Brunger, A. T. (1999). Crystal structure of the vesicular transport protein Sec17: Implications for SNAP function in SNARE complex disassembly. *Mol. Cell* **4,** 85–95.

Rothman, J. E. (1994). Mechanisms of intracellular protein transport. *Nature* **372,** 55–63.

Schweizer, F. E., Dresbach, T., DeBello, W. M., O'Connor, V., Augustine, G. J., and Betz, H. (1998). Regulation of neurotransmitter release kinetics by NSF. *Science* **279,** 1203–1206.

Shao, X., Li, C., Fernandez, I., Zhang, X., Sudhof, T. C., and Rizo, J. (1997). Synaptotagmin–syntaxin interaction: The C2 domain as a Ca^{2+} dependent electrostatic switch. *Neuron* **18,** 133–142.

Söllner, T., Whiteheart, S. W., Brunner, M., Erdjument-Bromage, H., Geromanos, S., Tempst, P., and Rothman, J. E. (1993a). SNAP receptors implicated in vesicle targeting and fusion. *Nature* **362,** 318–324.

Söllner, T., Bennett, M. K., Whiteheart, S. W., Scheller, R. H., and Rothman, J. E. (1993b). A protein assembly–disassembly pathway *in vitro* that may correspond to sequential steps of synaptic vesicle docking, activation, and fusion. *Cell* **75,** 409–418.

Südhof, T. C. (1995). The synaptic vesicle cycle: A cascade of protein–protein interactions. *Nature* **375,** 645–653.

Südhof, T. C., and Rizo, J. (1996). Synaptotagmins: C2-domain proteins that regulate membrane traffic. *Neuron* **17,** 379–388.

Sutton, R. B., Fasshauer, D., Jahn, R., and Brünger, A. T. (1998). Crystal structure of a SNARE complex involved in synaptic exocytosis at 2.4 Å resolution. *Nature* **395,** 347–353.

Tokumaru, H., and Augustine, G. J. (2001). SNARE proteins and neurotransmitter release. *Trends Neurosci.* In press.

Tokumaru, H., Umayahara, K., Pellegrini, L. L., Ishizuka, T., Saisu, H., Betz, H., Augustine, G. J., and Abe, T. (2001). SNARE complex oligomerization by synaphin/complexin is essential for synaptic vesicle exocytosis. *Cell* **104,** 421–432.

Weber, T., Zemelman, B. V., McNew, J. A., Westermann, B., Gmachl, M., Parlati, F., Söllner, T. H., and Rothman, J. E. (1998). SNAREpins: Minimal machinery for membrane fusion. *Cell* **92,** 759–772.

Index

A

AcE4K
  protonation and membrane binding, 438,
    447
  secondary structure, 442
  sequence, 442
  1-stearoyl-2-oleoyl-*sn*-glycero-3-
    phosphocholine membrane
    interactions
  amphiphilic sequence, 442
  area expansion modulus, 450
  critical membrane tension and pH effects,
    451, 453
  fluorescence quenching studies of depth
    insertion, 442, 444–446
  insertion assay, 447, 449
  membrane instability induction and pH
    effects, 453–456
  pH dependence of membrane
    partitioning, 446–447, 450
  pore-forming energetics, 456
  reversibility of binding, 455–456
  work for creating vacancy for peptide
    insertion, 449–450
ACTH, *see* Adrenocorticotropin
Actin, neutron reflection studies of lipid
  interactions, 131
Acylation
  bilayer dissociation kinetics and protein
    targeting by kinetic trapping, 383–386
  lysine, 372
  peptide synthesis
    N-acylation, 375
    S-acylation, 376
  protein partitioning in rafts, 387, 389
  S-acylation, 374–375, 383
Adrenocorticotropin (ACTH), fluorescence
  quenching studies of peptide insertion
  depth in membranes, 104
a-factor, prenylation, 382
Alamethicin

fluorescence quenching studies of peptide
  insertion depth in membranes, 103
free energy of peptide association with lipids
  configuration in bilayer, 231
  flip-flop across bilayer, 231–232
  total free energy change, 229
sequence, 228
site-directed spin labeling of membrane
  interactions, 18–19, 21
Alpha helix, *see* Amphipathic peptides;
  Transmembrane alpha helix
Amphipathic peptides
  alpha helix
    algorithms for analysis
      lipid affinity, 400, 402
      localization within database, 402
      localization within sequence, 402
    apolipoprotein A-I studies
      localization of amphipathic segments,
        405, 407
      modeling of lipid complexes, 403,
        405
      synthetic peptide studies, 407, 409
    class A peptide model studies
      biological effects, 412–414
      synthesis, 409, 411
      techniques for interaction analysis,
        411–412
    class L peptide model studies
      bilayer interactions and reciprocal
        wedge hypothesis, 414–415
      lytic mechanisms, 415
    classification and features, 398–400
    definition, 398
  beta sheet in apolipoprotein B
    algorithms for analysis
      lipid affinity, 416
      localization within sequence, 416, 419
    comparison with other proteins
      lipovitellin, 419, 421, 423–424
      porins, 419, 423–424
    distribution in protein, 415–416

Amphipathic peptides (*cont.*)
  lipoprotein affinity, 415
  model peptide studies
    lipid affinity studies, 430
    secondary structure, 424, 430
    synthesis, 424
  membrane interaction modes, 397–398
Amyloid-β peptide
  infrared reflection–absorption spectroscopy
    of monolayer interactions, 81
  isothermal titration calorimetry of
    membrane interactions, 45–46
Annexin I, X-ray reflectivity studies of lipid
    interactions, 133
Annexin V(A X V), infrared
    reflection–absorption spectroscopy of
    monolayer interactions, 81–82
Apolipoprotein A-I
  amphipathic alpha helix studies
    class A peptide model studies
      biological effects, 412–414
      synthesis, 409, 411
      techniques for interaction analysis,
        411–412
    localization of amphipathic segments,
        405, 407
    modeling of lipid complexes, 403, 405
    synthetic peptide studies, 407, 409
  crystal structure, 403
  discoidal models, 402–403, 405
  isothermal titration calorimetry of
    membrane interactions, 38, 44
Apolipoprotein A-II
  coil–helix transition induced by membrane,
    49
  isothermal titration calorimetry of
    membrane interactions, 43–44
Apolipoprotein B, beta sheet
  algorithms for analysis
    lipid affinity, 416
    localization within sequence, 416, 419
  comparison with other proteins
    lipovitellin, 419, 421, 423–424
    porins, 419, 423–424
  distribution in protein, 415–416
  lipoprotein affinity, 415
  model peptide studies
    lipid affinity studies, 430
    secondary structure, 424, 430
    synthesis, 424·

Arf1, myristoylation and membrane binding,
    380
ATR-FTIR, *see* Attenuated total reflection
    Fourier transform infrared spectroscopy
Attenuated total reflection Fourier transform
    infrared spectroscopy (ATR-FTIR),
    peptides inserted in supported bilayers
  experimental configuration, 195–196
  hemagglutinin, 197–198
  melittin, 197–198
  order parameter, 196–197
  spectral characteristics, 196

B

Bacteriorhodopsin
  bilayer effects of insertion, 345–347
  hydrophobic mismatch, 350–351
  X-ray reflectivity studies of lipid
    interactions, 134
Beta sheet, *see* Amphipathic peptides
Bioinformatics, ion channel homology,
    266–267
Brownian dynamics, ion channel simulations,
    258

C

CaaX prenyltransferases, *see*
    Farnesyltransferase;
    Geranylgeranyltransferase type I
Calcium-ATPase
  conformation and bilayer thickness,
    365
  hydrophobic mismatch, 351
  transmembrane alpha helix features,
    343–345
Cecropin AD, site-directed spin labeling of
    membrane interactions, 21
Charybdotoxin, membrane binding, 288
Cholesterol
  lipid composition effects on pore flickering
    and growth, 520
  lipid rafts and protein targeting, 387,
    389–390
  modulation of peptide–lipid energetics, 310,
    314–315, 325, 330–332
  protein modification, ester linkage, 372

transmembrane helix, cholesterol effects on bilayer incorporation, 362–363

Complexin, SBD-2 peptide
functional analysis in presynaptic terminal, 566–567
SNARE interaction characterization, 562, 564

Continuum solvent model, *see* Free energy of peptide association with lipids

CTP:phosphocholine cytidylyltransferase, fluorescence quenching studies of peptide insertion depth in membranes, 101

Cytochrome *b5*, fluorescence quenching studies of peptide insertion depth in membranes, 110

Cytochrome *c*
isothermal titration calorimetry of membrane interactions, 44
site-directed spin labeling of membrane interactions with signal sequence, 17–18

D

Diffraction of synchrotron X-ray radiation upon grazing incidence (GIXD), *see* Surface-sensitive X-ray scattering

Distorted-wave Born approximation, 122

Distribution analysis, fluorescence quenching and peptide insertion depth in membranes
depth distributions in membranes, physical principles, 93, 95
quenching profile, 96–97
theory, 95–96

E

Ebola virus, fluorescence quenching studies of fusion peptide insertion depth in membranes, 109

Electron paramagnetic resonance (EPR)
bilayer effects of peptide insertion, 347–348
bilayer studies of protein insertion effects, 347–348
doxyl-labeled lipid studies of peptide interactions, 24
site-directed spin labeling, *see* Site-directed spin labeling

Endoplasmic reticulum (ER)
protein trafficking and hydrophobic mismatch, 363–365
signal sequence
bilayer as receptor, 487–490
prospects for study, 494–496
protein-based recognition
protein translocation cycle, 490–492
signal recognition particle, 491–492, 494
signal sequence receptor, 492–493
in vitro functional assay, 488–489

δ-Endotoxin, free energy of *Bacillus thuringensis* peptide association with lipids, 242–243

Epilancin K7, fluorescence quenching studies of peptide insertion depth in membranes, 103–104

EPR, *see* Electron paramagnetic resonance

ER, *see* Endoplasmic reticulum

N-Ethylmaleimide-sensitive factor (NSF)
active peptide design, 555, 557, 559, 561
ATPase activity and peptide screening, 557–559, 561
functional overview, 554

F

Farnesyltransferase (FTase)
CaaX modification, 532, 534
catalytic mechanism, 535–536, 538
inhibitors
classes, 540
clinical trials, 544–545
rationale for cancer treatment, 533–534, 540–542
safety, 545–546
transformed cell effects, 542–543
types, 541–542
structure, 534, 538–539
substrates, 534–535, 542

fd, free energy of peptide association with lipids, 223–225

FDBP method, *see* Finite-difference Poisson–Boltzmann method

Feline leukemia virus, neutron reflection studies of fusion peptide–membrane interactions, 145

Finite-difference Poisson–Boltzmann (FDBP)
 method, electrostatic free energy
 determination, 278, 282–285, 287–289,
 302
Fluid dynamic diffusion coefficient, ion
 channel simulations, 259–260
Fluorescence quenching, peptide insertion
 depth in membranes
 accuracy, factors affecting, 99–100
 AcE4K studies, 442, 444–446
 applications
  amphipathic peptides, 101–102
  bioactive peptides, 102–104
  ion channels, 110
  model helices, 105–106
  overview, 100–101
  peptide mimics, 110–111
  signal sequence peptides, 106–109
  small peptides, 106
  viral peptides, 109
 data analysis
  comparison of methods, 97
  distribution analysis
   depth distributions in membranes,
    physical principles, 93, 95
   quenching profile, 96–97
   theory, 95–96
  empirical inspection of quenching level,
   90–91
  parallax analysis, 91–93
 fluorophore depth variation, 98
 precautions, 98–99
 principles, 90
Fluorescence recovery after photobleaching
 (FRAP), tethered bilayer systems, 150
FRAP, *see* Fluorescence recovery after
 photobleaching
Free energy of peptide association with lipids
 alamethicin studies
  configuration in bilayer, 231
  flip-flop across bilayer, 231–232
  total free energy change, 229
 bilayer structure considerations, 206–207
 classification of peptides by location in
  membrane
  hydrocarbon and polar
   headgroup-interacting peptides,
   209–210
  hydrocarbon region-interacting peptides,
   208–209

  polar headgroup-interacting peptides, 209
  rationale, 207–208
 continuum solvent model studies
  backbone hydrogen bond contributions,
   226, 244
  hydrophobicity scale derivation,
   226–228
  limitations, 232, 234
  overview, 210–211
  polar headgroup-interacting peptides,
   235–238
  polyalanine alpha helices, 225–226
  Src peptide studies, 238–239
 finite-difference Poisson–Boltzmann
  method for electrostatic component
  determination, 278, 282–285, 287–289,
  302
 hemagglutinin fusion peptide
  continuum solvent model studies,
   240–241
  Monte Carlo simulations, 241–242
 molecular dynamics simulations, overview,
  210
 Monte Carlo simulations
  δ-endotoxin from *Bacillus thuringensis*,
   242–243
  fd studies, 223–225
  hydrophobicity scale incorporation,
   223–225
  overview, 211
  Pf1 studies, 223–225
 partitioning energetics
  LamB studies, *see* LamB
  overview, 310–311
 partitioning process contributions
  bilayer effects
   curvature–elastic effects, 315–316
   dipole potential, 314–315
   fixed dipoles in membrane, 317–318
   interfacial region, 312
   isothermal compressibility modulus,
    314
   lateral pressure gradients, 315
   peptide conformational changes,
    312–313
   steric barriers, 316–317
   thickness of bilayer and hydrophobic
    mismatch, 313–314
  electrostatic and hydrophobic
   interactions, 311–312

polar headgroup-interacting peptides
neuromodulin, 235–237
short basic peptides, 234–235
rationale for study, 206, 327
total free energy change components
Coulombic component, 286–287
lipid perturbation effects, 219–220
membrane deformation, 220–221
overview, 211
peptide conformation effects, 213–214
peptide immobilization effects, 214–216
peptide–peptide interactions and lipid
demixing, 222
pK_a changes, 213, 245
polar headgroup effects, 221–222,
244–245
solvation free energy
desolvation penalty, 286
electrostatic contributions to solvation,
217–218
headgroup contributions, 218
hydrophobicity scales, 216–217
magnitude, 218
nonpolar contributions to solvation,
217–219
transfer free energies of individual
groups, 216–217
specific peptide–lipid interactions, 222
Fresnel reflectivity, calculation, 124
FTase, see Farnesyltransferase

G

Gap-junction model, membrane fusion, 508
Geranylgeranyltransferase type I (GGTase-I)
CaaX modification, 532, 534
catalytic mechanism, 535–536, 538
structure, 534, 538–539
substrates, 534–535
GGTase-I, see Geranylgeranyltransferase
type I
GIXD, see Diffraction of synchrotron X-ray
radiation upon grazing incidence
Glucitol permease
fluorescence quenching studies of signal
sequence insertion depth in
membranes, 107
signal peptide insertion into supported
monolayers, 192–193

Glycosylphosphatidylinositol (GPI)
function, 372
glycosylphosphatidylinositol-hemagglutinin
induction of pores, 514–515
Golgi apparatus, protein trafficking and
hydrophobic mismatch, 363–365
Gouy–Chapman theory, peptide–membrane
interactions, 282
gp43, site-directed spin labeling of membrane
interactions, 23–24
GPI, see Glycosylphosphatidylinositol
Gramicidin
infrared reflection–absorption spectroscopy
of monolayer interactions
gramicidin A, 80
gramicidin S, 82–83
water modeling, 262–264
X-ray reflectivity studies of membrane
interactions, 144–145

H

HA, see Hemagglutinin
HBM, see Hybrid bilayer membrane
Hedgehog, posttranslational lipidation, 372
Hemagglutinin (HA)
attenuated total reflection Fourier transform
infrared spectroscopy of inserted
peptide in supported bilayer, 197–198
fluorescence quenching studies of fusion
peptide insertion depth in membranes,
109
free energy of fusion peptide association
with lipids
continuum solvent model studies,
240–241
Monte Carlo simulations, 241–242
fusion peptide–lipid interactions, 198–201,
440–441
pore growth, energy and shape
considerations, 522–523
pore region size, 450–451
site-directed spin labeling of membrane
interactions, 23
spectroscopy of membrane interactions, 240
structure, 240, 503, 505–506
subunits, 440, 503
viral fusion overview, 502–503
viral infection role, 239

Hemifusion
  candidates of transitional hemifusion,
    515–517
  chlorpromazine studies, 516–517
  definition, 507
  diaphragm, 508, 516, 520, 522
  pore growth, energy and shape
    considerations, 522–523
  pore region size, 450–451
  site-directed spin labeling of membrane
    interactions, 23
  spectroscopy of membrane interactions,
    240
  structure, 240, 503, 505–506
  subunits, 440, 503
  viral fusion overview, 502–503
  viral infection role, 239
Hemifusion
  candidates of transitional hemifusion,
    515–517
  chlorpromazine studies, 516–517
  definition, 507
  diaphragm, 508, 516, 520, 522
  fusion intermediate evidence, 508–510
  fusion protein role, 517–518
  glycosylphosphatidylinositol-hemagglutinin
    induction, 514–515
  lipid changes in fusion, 506–507
  prospects for study, 524
HIV, *see* Human immunodeficiency virus
Hodgkin–Huxley equations, ion channel
  simulations, 256
Human immunodeficiency virus (HIV)
  Env structure and fusion mechanism, 503,
    505, 518
  fluorescence quenching studies of envelope
    protein insertion depth in membranes,
    109
  Gag
    electrostatic and hydrophobic interactions
      in membrane binding, 280
    signaling, 277
  gp43
    site-directed spin labeling of membrane
      interactions, 23–24
    structure, 506
  Vpu, X-ray reflectivity studies of lipid
    interactions, 136, 138
Hybrid bilayer membrane (HBM)
  preparation, 147

reflection studies of peptide interactions,
  147–148
Hydrophobicity scale, *see* Free energy of
  peptide association with lipids
Hydrophobicity threshold, peptide insertion in
  membranes
  biological correlations, 472
  comparison with other hydropathy scales,
    472, 474–475
  database construction, 471–472
  high-performance liquid
    chromatography-determined
    hydrophobicity scale, 467–469,
    473–474
  intrinsic competence of transmembrane
    segments, 475
  overview of analysis, 465–466
  quantification, 469–471
  rationale for determination, 466, 476–477
  sequence relationship with transmembrane
    segment partitioning, 475–476
  statistical analysis of transmembrane
    segments, 471, 476
  transmembrane peptide synthesis and
    insertion in micelles, 467–469
Hydrophobic mismatch
  bacteriorhodopsin studies, 350–351
  biological consequences, 363–365
  calculation of effects, 349–350
  mattress model, 350
  model peptides, 356–359
  partitioning process contribution, 313–314
  positive versus negative mismatch, 348–349

I

Infrared reflection–absorption spectroscopy
  (IRRAS)
  applications, 57–61
  instrumentation, 61–63
  Langmuir film studies, 65–66
  monolayer film studies of peptide and
    protein conformations
    amyloid-β peptide, 81
    annexin V(A X V), 81–82
    gramicidin A, 80
    gramicidin S, 82–83
    photosystem II core complex, 81
    prospects, 83–84

principles, 61
spectral analysis
  angle of incidence, 68, 71
  frequency, 66–67
  intensity, 67–68, 71
  methylene scissoring modes, 74
  phosphate stretching vibrations, 75
  reflection coefficients, 67–68
  spectra–structure correlations, 60, 66,
    72–73
  tilt angle extraction, 71, 84
structural information available, 59–60
surfactant studies
  SP-B peptide secondary structure,
    76–78
  SP-C orientation in lipid environments,
    78, 80
water signal reduction, 63–65
Ion channel, computational analysis
effective diffusion coefficient, 260
fluid dynamic diffusion coefficient,
  259–260
gating analysis, 266
gramicidin channel water modeling,
  262–264
hierarchical strategy
  atomically detailed pseudo-classical
    description, 257–258
  Brownian dynamics simulations, 258
  diffusive level of description, 258
  overview, 255–256
  quantum mechanical description, 257
Hodgkin–Huxley equations, 256
importance sampling, 261
KcsA channel, potassium modeling,
  264–266
molecular dynamics calculation,
  259–261
sequence homology studies
  alpha-type channels, 268–269
  bioinformatics, 266–267
  overview, 257
  transport classification system, 267,
    269
  voltage-gated channels
    genome analysis, 271
    membrane topology, 269–270
    phylogenetic analysis, 269–270
    prospects for study, 271
structural data, 257

UHBD program, 261–262
weighted histogram analysis method, 261
IRRAS, *see* Infrared reflection–absorption
  spectroscopy
Isothermal compressibility modulus,
  partitioning free energy, 314, 324–326,
  331–332
Isothermal titration calorimetry (ITC)
applications, 32–33
instrumentation, 32
peptide–membrane interactions
  binding isotherm determination, 33,
    35–36
  binding models
    Langmuir adsorption equilibrium,
      37–38
    partition equilibrium, 37
    surface partition equilibrium with
      electrostatic correction, 38–40
  binding reaction thermodynamics, 33
  coil–helix transition induced by
    membrane, 48–50
  complex membrane-binding/membrane-
    perturbation equilibria
    membrane-induced peptide
      aggregation, 45–46
    membrane permeabilization,
      41, 43
    peptide/protein-induced membrane
      phase transitions, 43–45
    protonation reaction at membrane
      surface, 46–48, 54
  curved and planar membrane binding
    thermodynamics, 52–54
  reaction enthalpy determination, 33
  thermodynamic parameters, 31, 54
ITC, *see* Isothermal titration calorimetry

K

KcsA channel
potassium modeling, 264–266
transmembrane alpha helices, 342–343
K-ras4B
electrostatic and hydrophobic interactions in
  membrane binding, 279, 281
membrane targeting, 301
prenylation, 382
signaling, 277

L

β-Lactoglobulin, isothermal titration calorimetry of membrane interactions, 44–45

LamB
conformation in bilayers, 309, 327–329
fluorescence quenching studies of signal sequence insertion depth in membranes, 107–109
partitioning free energy measurement
area per lipid molecule, 324–326, 331–332
electrostatics and classical hydrophobic effect, 319–321, 327–329
interfacial dipole effects, 321–324, 329–331
isothermal compressibility modulus, 324–326, 331–332
spectroscopic techniques, 318–320
signal sequence
characteristics, 318, 321
location in bilayer, 326–327
physiochemical features in function, 487–488

Lipidation, *see* Acylation; Cholesterol; Glycosylphosphatidylinositol; Myristoylation; Palmitoylation; Prenylation

Liposome, pH sensitivity of drug delivery, 437–438, 460

Lipovitellin, amphipathic beta sheets, 419, 421, 423–424

M

Magainin 2
amide
coil–helix transition induced by membrane, 49–50
isothermal titration calorimetry of membrane interactions, 39–41, 43, 49–50
class L peptide model studies, 414–415
fluorescence quenching studies of peptide insertion depth in membranes, 104
X-ray reflectivity studies of membrane interactions, 145

Magic angle spinning nuclear magnetic resonance, lipids

dipolar recoupling on-axis with scaling and shape preservation, 179
nuclear Overhauser effect enhancement spectroscopy, 179–181
resolution of lipid resonances, 178–179

MARCKS, *see* Myristoylated alanine rich C-kinase substrate

Mastoparan
class L peptide model studies, 414–415
fluorescence quenching studies of mastoparan X insertion depth in membranes, 104

Melittin
attenuated total reflection Fourier transform infrared spectroscopy of inserted peptide in supported bilayer, 197–198
fluorescence quenching studies of peptide insertion depth in membranes, 104
site-directed spin labeling of membrane interactions, 22

Membrane protein folding, thermodynamics, 141, 143

Monolayer films, physical characterization techniques, 58–59

Monte Carlo simulation, *see* Monte Carlo simulations

Myristoylated alanine rich C-kinase substrate (MARCKS)
aromatic component of free energy change of membrane binding, 296
basic effector region, 293, 296–297
calcium–calmodulin binding, 297
electrostatic and hydrophobic interactions in membrane binding, 279–281, 293, 296–298
functions, 300
membrane targeting, 301–302
myristate in membrane binding, 278, 293
phosphatidyl inositol interactions, 297–298, 300, 300, 302
phosphorylative regulation, 280–281, 297
site-directed spin labeling of membrane interactions, 12–13, 16–17

Myristoylation
amino terminal modification, 374
Arf1, 380
bilayer dissociation kinetics and protein targeting by kinetic trapping, 385
free energy of membrane partitioning, 378–380

membrane binding enhancement
mechanisms, 380–381
protein kinase A myristoylation and
membrane binding, 377–378
Src myristoylation and membrane binding,
289–291, 380

N

Neurokinin A, neutron reflection studies of
membrane interactions, 145
Neuromodulin, free energy of peptide
association with lipids, 235–237
Neurotransmitter release
overview of presynaptic proteins,
552–554
peptides from presynaptic proteins
design
criteria for useful peptides, 554–555
educated guess approach, 555, 557
protein–protein interaction site
identification, 557–559
rationale, 551–552, 567–568
structure-based design, 559, 561
functional analysis in presynaptic
terminal, 564, 566–567
prospects for study, 567–568
protein–protein interaction effects,
characterization, 562, 564
Neutron scattering
applications in lipid/peptide interactions,
117–118
data inversion and modeling, 122–126
planar lipid model system characterization
lipid bilayers and multi-bilayer stacks
hybrid bilayer membranes, 147–148
peptide/lipid bilayer systems, 141,
143–146
pure lipid systems, 138–139, 141
SAMs, 146–147
tethered functional lipid bilayers,
148–150
surface monolayer systems
composite monolayer systems,
130–134, 136, 138
pure lipid monolayers, 127–130
types of systems, 118–119
planar lipid model systems, 118–119
theory, 119, 121–122

Nipple
fusion dependence on spontaneous
curvature of monolayers, 510–513
pore growth effects, 523
Nisin, fluorescence quenching studies of
peptide insertion depth in membranes,
103
NMR, *see* Nuclear magnetic resonance
NSF, *see* N-Ethylmaleimide-sensitive factor
Nuclear magnetic resonance (NMR),
peptide–membrane interaction studies
advantages, 164–165
high-resolution spectroscopy of peptide
structure
circular dichroism validation, 166
conformational averaging, 169
experimental conditions, 165–166
geometric constraints, 168
micelle size, 167
molecular weight limit for protein
structure resolution, 164
two-dimensional spectroscopy, 165,
167–169
lipid matrix structure and dynamics
degrees of freedom and anisotropy,
176–177
deuterium quadrupolar splitting,
177–178
magic angle spinning spectroscopy
dipolar recoupling on-axis with scaling
and shape preservation, 179
nuclear Overhauser effect enhancement
spectroscopy, 179–181
resolution of lipid resonances,
178–179
peptide binding effects on structure,
182–183
phosphorous-31 anisotropy of chemical
shift, 177
prospects, 183–184
solid-state spectroscopy
angular dependence of interactions
chemical shift interaction, 172–173
deuterium quadrupolar interactions,
173–174
dipolar interactions, 174
anisotropy correlation of chemical shift
and dipolar interactions, 175
rationale, 171
resolution, 172

Nuclear magnetic resonance (NMR) (*cont.*)
 rotational-echo double resonance,
  174–175
 rotational resonance, 174
 torsion angle measurement, 175–176
 techniques, 163–164
 weakly-interacting peptides and transferred
  NOE measurements, 170–171

P

Palmitoylation
 amino terminal in bacteria, 372
 conformational effects in proteins, 383
 protein partitioning in rafts, 389
Parallax analysis, fluorescence quenching and
  peptide insertion depth in membranes,
  91–93
PEAA, *see* Poly(2-ethylacrylic acid)
Peptide–membrane interactions, *see*
  Amphipathic peptides; Fluorescence
  quenching; Free energy of peptide
  association with lipids; Hydrophobicity
  threshold; Infrared reflection–absorption
  spectroscopy; Isothermal titration
  calorimetry; Neutron scattering; Nuclear
  magnetic resonance; Site-directed spin
  labeling; Supported bilayer;
  Surface-sensitive X-ray scattering;
  Transmembrane alpha helix
Pf1, free energy of peptide association with
  lipids, 223–225
PGLa, isothermal titration calorimetry of
  membrane interactions, 41, 43
pH-sensitive peptides and polymers, *see*
  AcE4K; Poly(2-ethylacrylic acid)
PhoE, fluorescence quenching studies of signal
  sequence insertion depth in membranes,
  107
Phosphatidic acid, transmembrane helix
  interactions, 359–361
Phosphatidylserine, transmembrane helix
  interactions, 360–361
Phospholipase A_2 (PLA$_2$), X-ray reflectivity
  studies of lipid interactions, 133
Phospholipase C (PLC)
 pleckstrin homology domain, 278
 substrate interactions with myristoylated
  alanine rich C-kinase substrate,
  297–298, 300, 300, 302

Photosystem II core complex, infrared
  reflection–absorption spectroscopy of
  monolayer interactions, 81
PKA, *see* Protein kinase A
PKC, *see* Protein kinase C
PLA$_2$, *see* Phospholipase A_2
PLC, *see* Phospholipase C
Poly(2-ethylacrylic acid) (PEAA)
 liposome incorporation and pH response,
  438–439
 liposomes and pH sensitivity of drug
  delivery, 437–438, 460
 membrane instability induction and pH
  effects
  area expansion modulus, 459–460
  cholesterol addition effects, 457,
   459–460
  energetics of membrane binding,
   457, 459
  work for vacancy formation, 459–460
 protonation and membrane binding, 438
 structure and pH dependence, 456–457
Porins, amphipathic beta sheets, 419,
  423–424
Posttranslational lipidation, *see* Acylation;
  Cholesterol;
  Glycosylphosphatidylinositol;
  Myristoylation; Palmitoylation;
  Prenylation
Prenylation
 bilayer dissociation kinetics and protein
  targeting by kinetic trapping,
  383–385
 catalysis, *see* Farnesyltransferase;
  Geranylgeranyltransferase type I
 free energy of membrane partitioning, 378
 membrane binding enhancement
  mechanisms, 381–382
 peptide synthesis, 375–376
 processing of prenylated proteins,
  532–533
 protein distribution, 374
 protein partitioning in rafts, 389
 receptors, 382–383
Protein kinase A (PKA), myristoylation and
  membrane binding, 377–378
Protein kinase C (PKC)
 myristoylated alanine rich C-kinase
  substrate regulation, 280–281, 297
 translocation, 280

R

Rac-1, electrostatic and hydrophobic interactions in membrane binding, 279

Rafts, lipidated protein targeting, 387, 389–390

Rap1, electrostatic and hydrophobic interactions in membrane binding, 279

Ras
  farnesyltransferase inhibitor suppression of transformation, 542–543
  mutations in cancer, 545

REDOR, *see* Rotational-echo double resonance

Rho-A, electrostatic and hydrophobic interactions in membrane binding, 279

ROMK 1, fluorescence quenching studies of peptide insertion depth in membranes, 110

Rotational resonance (RR), peptide–membrane interaction studies, 174

Rotational-echo double resonance (REDOR), peptide–membrane interaction studies, 174–175

RR, *see* Rotational resonance

S

SAM, protein immobilization for reflection studies, 146–147

Scattering length density (SLD), calculation, 123, 128, 147

Scherrer equation, 121

SDSL, *see* Site-directed spin labeling

Signal sequence
  bilayer as receptor, 487–490
  Blobel hypothesis, 484
  canonical structure, 484–485
  cytochrome *c*, site-directed spin labeling of membrane interactions with signal sequence, 17–18
  discovery, 483
  domains, 483
  endoplasmic reticulum *in vitro* functional assay, 488–489
  LamB, *see* LamB
  leucine abundance, 485, 493–494
  prospects for study, 494–496
  protein-based recognition

protein translocation cycle, 490–492
  signal recognition particle, 491–492, 494
  signal sequence receptor, 492–493

Signal sequence
  fluorescence quenching of peptide insertion depth in membranes, 106–109
  glucitol permease
    fluorescence quenching studies of signal sequence insertion depth in membranes, 107
    insertion into supported monolayers, 192–193

Site-directed spin labeling (SDSL)
  characteristics, 194–195
  electron paramagnetic resonance spectra analysis
    factors affecting spectra, 7
    hyperfine coupling constant, 8–9
    membrane–aqueous partitioning, 9
    power saturation, nitroxide accessibility and bilayer depth information, 10–12
    secondary structure and tertiary contact, 7–9
    hemagglutinin, 197–198
    melittin, 197–198
    order parameter, 196–197
    spectral characteristics, 196
  fusion peptide–lipid interactions, 198–201
  overview, 3–5
  peptide insertion into supported monolayers
    glucitol permease signal peptide insertion, 192–193
    partition constants, 193–194
    rationale, 191–192
    surface pressure, 192–193
  peptide synthesis with spin labels, 5–7
  preparation, 194–19h; 24
    hemagglutinin, 23
    melittin, 22
    trichogin GA IV, 22
  spin labels, 5–6

S-layer, reflection studies of lipid interactions, 134, 136, 150

SLD, *see* Scattering length density

SNAP
  active peptide design, 561
  functional overview, 554
  receptor, *see* SNARE

SNAREs
  complex affinity and dissociation,
    553–554
  complexin SBD-2 peptide
    functional analysis in presynaptic
      terminal, 566–567
    interaction assays, 562, 564
  functional overview, 554
  fusion mediation, 506
  syntaxin-binding peptides, 557–558, 562,
    564
  types, 553
Sodium channel, fluorescence quenching
  studies of peptide insertion depth in
  membranes, 110
Somatostatin, isothermal titration calorimetry
  of membrane interactions with SDZ
  analog, 47–48
Sphingomyelin, lipid composition effects on
  pore flickering and growth, 520
Src
  electrostatic and hydrophobic interactions
    in membrane binding, 279–280,
    289–291
  free energy of peptide association with
    lipids, 238–239
  membrane targeting, 301
  myristoylation and membrane binding,
    289–291
  signaling, 277
  site-directed spin labeling of membrane
    interactions, 12–13, 16–17
Stalk, fusion dependence on spontaneous
  curvature of monolayers, 511–512
Supported bilayer
  attenuated total reflection Fourier transform
    infrared spectroscopy of inserted
    peptides
    experimental configuration, 195–196
    hemagglutinin, 197–198
    melittin, 197–198
    order parameter, 196–197
    spectral characteristics, 196
  characteristics, 194–195
  fusion peptide–lipid interactions, 198–
    201
  peptide insertion into supported monolayers
    glucitol permease signal peptide
      insertion, 192–193
    partition constants, 193–194

    rationale, 191–192
    surface pressure, 192–193
  preparation, 194–195
Surface-sensitive X-ray scattering
  applications in lipid/peptide interactions,
    117–118
  data inversion and modeling, 122–126
  diffraction of synchrotron X-ray radiation
    upon grazing incidence, 119, 121, 127,
    134, 144
  planar lipid model system characterization
    lipid bilayers and multi-bilayer stacks
      hybrid bilayer membranes, 147–148
      peptide/lipid bilayer systems, 141,
        143–146
      pure lipid systems, 138–139, 141
      SAMs, 146–147
      tethered functional lipid bilayers,
        148–150
    surface monolayer systems
      composite monolayer systems,
        130–134, 136, 138
      pure lipid monolayers, 127–130
    types of systems, 118–119
  theory, 119, 121–122
Surfactant
  composition, 75–76
  infrared reflection–absorption spectroscopy
    studies
    SP-B peptide secondary structure,
      76–78
    SP-C orientation in lipid environments,
      78, 80
  pathology, 76
  X-ray reflectometry of SP-B interactions
    with lipids, 131

T

Threshold hydrophobicity, *see* Hydrophobicity
  threshold
Transmembrane alpha helix
  amino acid composition and localization,
    340–342
  anionic phospholipid interactions
    energy of ion interactions, 360
    phosphatidic acid, 359–361
    phosphatidylserine, 360–361
  bilayer effects of insertion

boundary lipids versus bulk lipids, 347–348
dynamics of lipid, 345–347
electron spin resonance studies, 347–348
lipid perturbation energy, 349
calcium-ATPase features, 343–345
cholesterol effects on bilayer incorporation, 362–363
hydrogen bond energetics, 339
hydrophobicity threshold for membrane insertion, *see* Hydrophobicity threshold
hydrophobic mismatch
    bacteriorhodopsin studies, 350–351
    biological consequences, 363–365
    calculation of effects, 349–350
    mattress model, 350
    model peptides, 356–359
    positive versus negative mismatch, 348–349
KcsA features, 342–343
model peptide studies
    L_{16}, 355
    L_{24}, 352
    P_{16}, 352–353
    P_{24}, 353
    polyalanine modeling, 354–355
phospholipid phase effects, 361–362
structure, 339–340
Transport classification system, ion channels, 267, 269
Trichogin GA IV, site-directed spin labeling of membrane interactions, 22

U

UHBD program, ion channel simulations, 261–262

V

VACF, *see* Velocity autocorrelation function
Velocity autocorrelation function (VACF), statistical mechanics, 259
Viral fusion, *see* Hemagglutinin; Hemifusion; Human immunodeficiency virus
Volume-restricted distribution function (VRDF), surface-sensitive X-ray scattering analysis, 125–126, 128
Vpu, X-ray reflectivity studies of lipid interactions, 136, 138
VRDF, *see* Volume-restricted distribution function

W

Weighted histogram analysis method (WHAM), ion channel simulations, 261
WHAM, *see* Weighted histogram analysis method

X

X-ray scattering, *see* Surface-sensitive X-ray scattering

ISBN 0-12-153352-2

9 780121 533526

90018

CHAPTER 5, FIGURE 4b   (b) Projected electron density map using the intensities shown in
(a) and phases derived from electron microscopy (Henderson *et al.*, 1986). The annular structure of
the trimeric seven helix bundles within the unit cell is clearly identified.

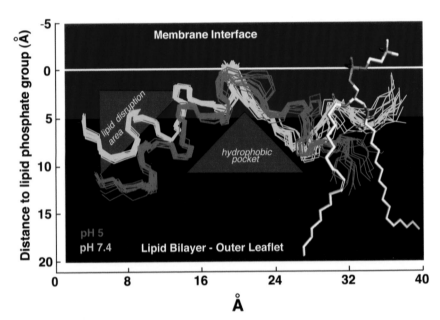

CHAPTER 7, FIGURE 6 Structure and conformational change of the fusion peptide of influenza hemagglutinin in lipid bilayers at pH 7.4 *(yellow)* and pH 5 *(red)*. The structures of the peptides in dodecylphosphocholine micelles were determined by ¹H-NMR spectroscopy. Site-directed spinlabeling at 18 positions was then used to determine the structures in lipid bilayers by electron paramagnetic resonance spectroscopy. Both structures are angled, but the fusion-triggering pH 5 structure has a sharper bend and inserts more deeply into the hydrophobic core of the bilayer than the inactive pH 7.4 structure. A phospholipid molecule in a fluid conformation is shown for reference. Adapted from Han *et al.* (2001).

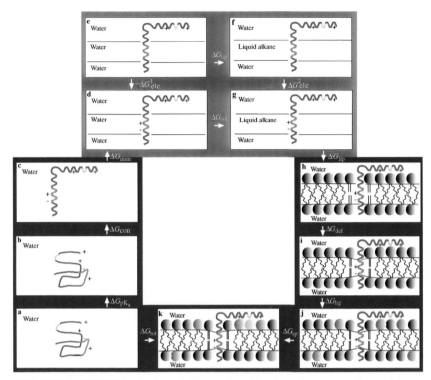

CHAPTER 8, FIGURE 2   A schematic diagram describing the various contributions to the total free energy of association of a peptide with the lipid bilayer. The association process is depicted as a sequence of different thermodynamic substates, designated a–k. The various contributions to the total free energy of peptide–membrane association characterize the passages between the different substates. The peptide is schematically depicted as a green curve (in its unfolded states a and b) or as two joined green helices (in its folded states c–k), and ΔG_{con} is the free energy due to conformational change. The black (in state a) and pink (in b–k) patches on the peptide represent two different protonation states of a titratable residue, and ΔG_{pK_a} is the free energy associated with the change of protonation state. The central hydrophobic region of the peptide is colored blue, and the two horizontal lines represent the boundaries of the hydrocarbon region of the membrane. The free energy penalty of peptide immobilization (c–d) is denoted ΔG_{imm}. The solvation free energy (d–g) is denoted ΔG_{sol}. Here d–g, surrounded by light blue shading, describe a thermodynamic cycle based on the separation of ΔG_{sol} into electrostatic and nonpolar contributions. The hydrophobic core of the peptide is discharged in water (d–e; $-\Delta G_{elc}^{1}$). It is then transferred into a liquid alkane (e–f; ΔG_{np}) and recharged again in the liquid alkane (f–g; ΔG_{elc}^{2}). The unperturbed lipid chains are represented by the black wavy lines. Peptide-induced lipid perturbation effects (g–h) are denoted ΔG_{lip}; state h shows lipid chains in the vicinity of the peptide, which become more rigid due to their interactions with the peptide core; the ΔG_{lip} contributions reflect the associated entropic penalty. Membrane deformation effects (h–i) are denoted ΔG_{def}. The neutral and acidic headgroups are represented by gray and red spheres, respectively. Effects resulting from the interactions between the peptide and lipid headgroups, for example, positively charged residues of the peptide with negatively charged lipids (i–j), are denoted ΔG_{hg}. The yellow sphere represents a polar headgroup that interacts specifically with the peptide (the peptide segment that is involved in this interaction is colored yellow). Effects resulting from specific peptide–lipid interactions (j–k) are denoted ΔG_{sp}.

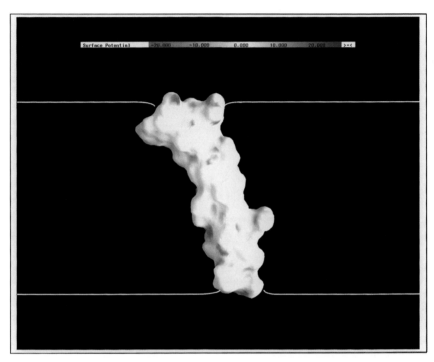

CHAPTER 8, FIGURE 5 Schematic representation of the orientation of alamethicin in the 2-Å deformed bilayer. Alamethicin is depicted with the electrostatic potential ϕ color-coded on its molecular surface. This was calculated using DelPhi (Nicholls and Honig, 1991) and is displayed on the molecular surface using GRASP (Nicholls *et al.,* 1991). Negative potentials ($0 > \phi > -20\ kT/e$) are red, positive potentials ($0 < \phi < 20\ kT/e$) are blue, and neutral potentials are white. Calculations have been carried out for the conformation found in the crystal structure (Protein Data Bank, accession number 1AMT). The peptide is shown with its N-terminus pointing down and its more polar C-terminus pointing up. The two white lines represent the boundaries of the hydrocarbon region of the lipid bilayer. The polar termini of alamethicin are outside the hydrocarbon region of the deformed lipid bilayer, and the hydrophobic core of the peptide is immersed inside this region.

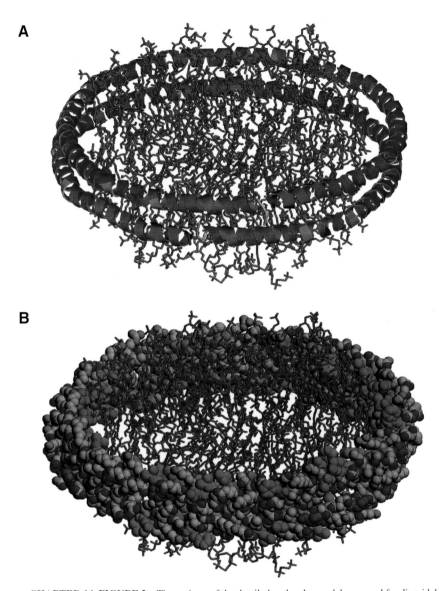

CHAPTER 14, FIGURE 2 Three views of the detailed molecular model proposed for discoidal human apo A-I:phospholipid complexes. The molecular models (Segrest *et al.*, 1999b) were generated with Rasmol (Sayle and Milner-White, 1995). (A) Phospholipid bilayer disc displayed in molecular stick mode and protein displayed in cartoon mode. Color code: phospholipid, gold; helixes 1–10 (residues 44–243) of apo A-I, red; residues 33–43 (one of three G* amphipathic α-helical repeats at the N-terminal end), blue; prolines, green. (B) Lipid displayed in molecular stick mode and protein displayed in space-filling mode. Color code: phospholipid, gold; hydrophobic residues (L, F, M, V, W, Y, I), green; prolines, magenta; other residues, CPK. (C) Lipid and protein displayed in space-filling modes. Color code: phospholipid, gold; hydrophobic residues (L, F, M, V, W, Y, I), green; prolines, magenta; other residues, CPK.

C

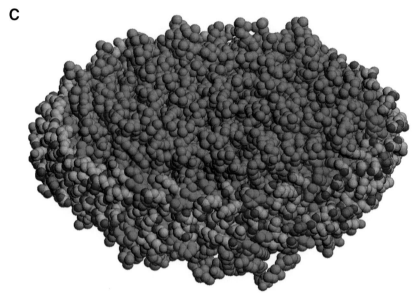

CHAPTER 14, FIGURE 2 *(continued)*

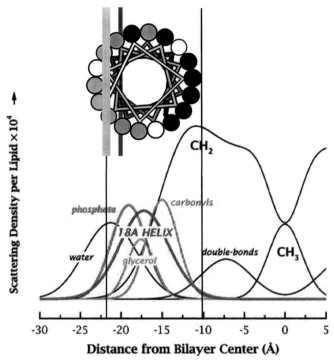

CHAPTER 14, FIGURE 6 Transbilayer position and orientation of the peptide Ac-18A-NH$_2$ as determined by a novel X-ray diffraction method in oriented fluid-state dioleoylphosphatidylcholine bilayers. Summary of the results of the structural refinement of Ac-18A-NH$_2$ in DOPC bilayers showing the transbilayer distribution of the amphipathic α helix in the context of the structure of the fluid DOPC bilayer. The inset shows a helical wheel plot of Ac-18A-NH$_2$ produced by the $\Lambda\alpha$ algorithm, which is aligned parallel to the bilayer plane with the nonpolar surface facing the hydrocarbon core of the bilayer (to scale). Color code for the amino acid residues: hydrophobic (L, F, W, Y), black; basic, blue; acidic, red; other, white. The Gaussian distribution of the helix indicates that the thermally disordered surface of the helix penetrates the hydrocarbon core to the level of the double bonds. The helix axis, located 17.1 Å from the bilayer center, coincides closely with the mean position of the DOPC glycerol groups, located about 17.6 Å from the center. The thin vertical lines indicate the calculated position of the water/lipid interface and the most deeply buried residue. Figure modified from Hristova *et al.* (1999) with permission.

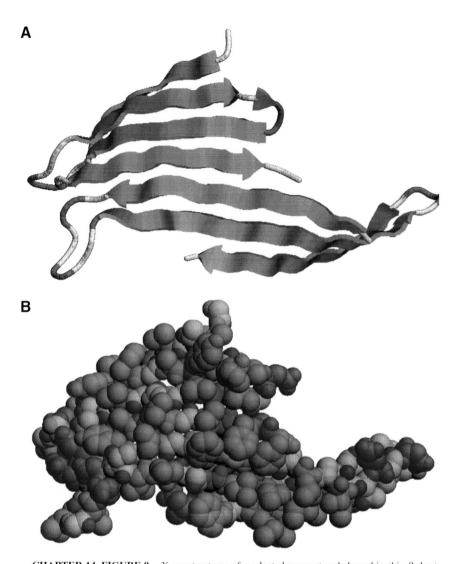

CHAPTER 14, FIGURE 9 X-ray structures of a selected seven-stranded amphipathic β sheet from porin (*Rhodopseudomonas blastica*) and from lipovitellin (lamprey). The molecular models were generated with Rasmol (Sayle and Milner-White, 1995). Color code for the space-filling models: hydrophobic residues (L, F, M, V, W, Y, A, I), gold; basic residues (L, R, H), blue; acidic residues (D, E), red; other, CPK. The X-ray structure for porin is from Kreusch *et al.* (1994) and that for lipovitellin is from Anderson *et al.* (1998). (A) A 113-residue-long, seven-stranded amphipathic β sheet, residues 2–57 and 233–289, from porin displayed in cartoon mode and viewed from the hydrophobic (convex) face. (B) A seven-stranded amphipathic β sheet from porin displayed in space-filling mode and viewed from the hydrophobic (convex) face. (C) A seven-stranded amphipathic β sheet from porin displayed in space-filling mode and viewed from the hydrophilic (concave) face. (D) A 96-residue-long, seven-stranded amphipathic β sheet, residues 1374–1458 and 1493–1503, from lipovitellin displayed in cartoon mode and viewed from the hydrophobic (concave) face. (E) A seven-stranded amphipathic β sheet from lipovitellin displayed in space-filling mode and viewed from the hydrophobic (concave) face. (F) A seven-stranded amphipathic β sheet from lipovitellin displayed in space-filling mode and viewed from the hydrophilic (convex) face.

C

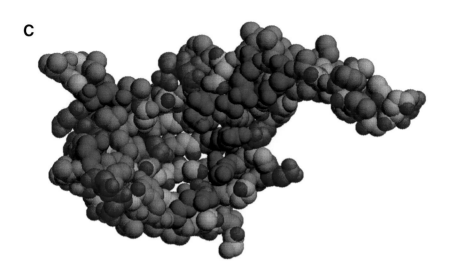

D

CHAPTER 14, FIGURE 9 *(continues)*

E

F

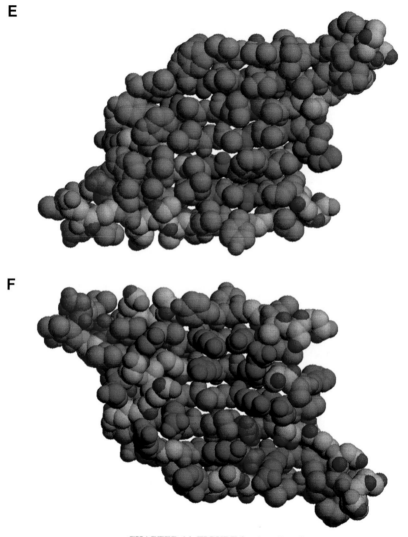

CHAPTER 14, FIGURE 9 *(continued)*

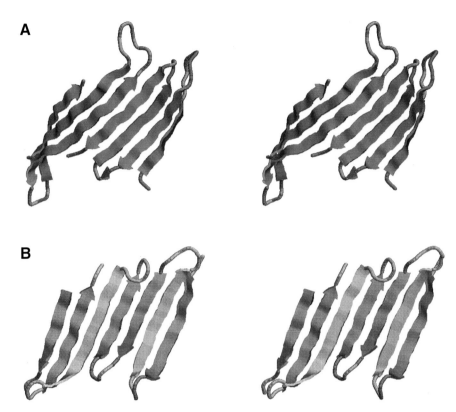

CHAPTER 14, FIGURE 10   Stereo models of the seven-stranded amphipathic β sheets from porin and lipovitellin showing the location of amphipathic β strands predicted by the current version of βLOCATE. The two cross-eyed stereo models were created by Rasmol (Sayle and Milner-White, 1995) and are displayed in cartoon mode. Amphipathic β strands selected by the current version of βLOCATE for a $\Lambda\beta \geq 5$ are highlighted in gold. The X-ray structure for porin is from Kreusch *et al.* (1994) and that for lipovitellin is from Anderson *et al.* (1998). (A) A 113-residue-long, seven-stranded amphipathic β sheet, residues 2–57 and 233–289, from porin (*Rhodopseudomonas blastica*) displayed in cartoon mode and viewed from the hydrophobic (convex) face. (B) A 96-residue-long, seven-stranded amphipathic β sheet, residues 1374–1458 and 1493–1503, from lipovitellin (lamprey) displayed in cartoon mode and viewed from the hydrophilic (convex) face. Two amphipathic β strands selected by the current version of βLOCATE for $\Lambda\beta < 5.0$ but ≥ 2.5 are highlighted in yellow.

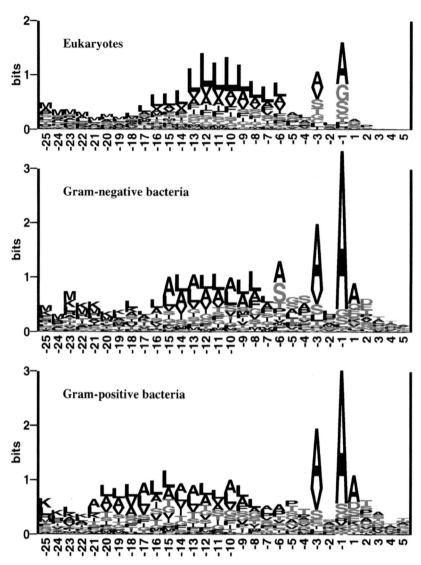

CHAPTER 17, FIGURE 2 Sequence logos of signal peptides aligned by signal peptide cleavage site. The total height of the letter stack at each position is representative of the information content, whereas the relative amino acid abundance at that residue is displayed by the letter height. The information is defined as the difference between the maximal and the actual entropy. Charged residues are shown in blue and red, uncharged in green, and hydrophobic in black. Reprinted with permission from Nielsen *et al.* (1997). Copyright © 1997 Oxford University Press.

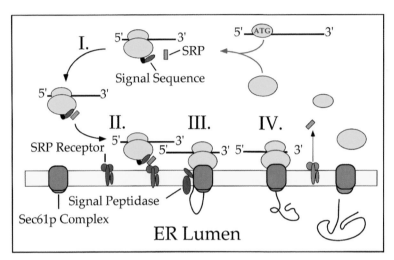

CHAPTER 17, FIGURE 3 Schematic illustration of the protein translocation cycle in the mammalian endoplasmic reticulum. In current views, the process of protein translocation begins in the cytosol. After the initiation of protein synthesis, the signal sequence is recognized in the context of the ribosome by the signal recognition particle (SRP), to yield the formation of a ribosome/nascent chain/SRP complex (stage I). This complex is targeted to the ER membrane through interaction with the SRP receptor (stage II). Subsequently, the ribosome/nascent chain complex binds to the translocon, the protein complex responsible for protein translocation, and translocation across the ER membrane ensues (stage III). The termination of protein synthesis is thought to result in the release of the ribosomal subunits back to the cytosol and the inactivation of the translocon (stage IV).

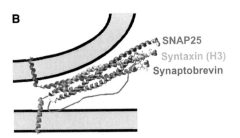

B

SNAP25
Syntaxin (H3)
Synaptobrevin

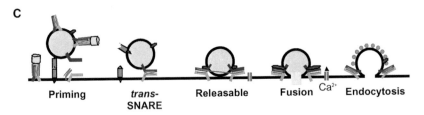

C

Priming   *trans*-SNARE   Releasable   Fusion Ca²⁺   Endocytosis

CHAPTER 20, FIGURE 1 (B) Molecular structure of the SNARE complex. Vesicle membrane and presynaptic plasma membrane as well as transmembrane anchors of synaptobrevin and syntaxin are shown schematically. (C) Final stages in neurotransmitter release. Color scheme: synaptobrevin, blue; syntaxin, orange; SNAP25, violet; SNAP, brown; NSF, yellow; tethering complex, cyan; Ca^{2+} sensor, red; Ca channel, green; clathrin, pink.

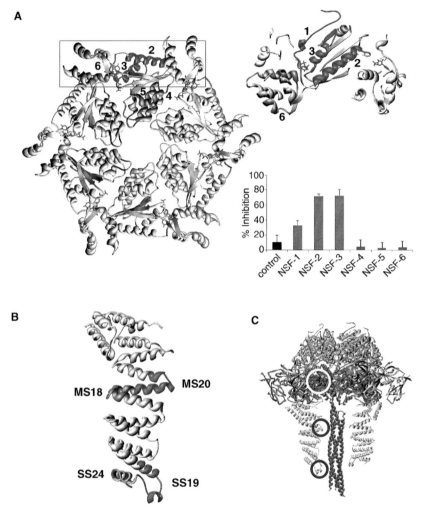

CHAPTER 20, FIGURE 4   Mapping peptides on crystal structures. (A) NSF peptides mapped on the D1 domain. Based on the high sequence identity of the D1 and the D2 domains, their structures are likely to be very similar. Therefore, we used the coordinates of the D2 domain as a model for the D1 domain. **Left:** View from bottom. **Right:** Side view of 20-Å slab indicated by the box in the left panel. Active peptides are colored red, inactive peptides are green (bar diagram on the right). (B) Peptides derived from α-SNAP. Peptides shown in red affect function, whereas those shown in green do not. (C) Hypothetical model of 20S complex assembled from individual crystal structures. NSF is depicted in blue colors (N domain: violet; D1 domain: dark blue; D2 domain: light blue), α-SNAP in yellow, and the α-SNARE complex in red. The left α-SNAP molecule is oriented as shown in B. Circles highlight functionally important sites identified by peptide experiments.